D0643648

The Mammalian Oviduct

The Mammalian Oviduct

Comparative Biology and Methodology

EDITED BY

E. S. E. Hafez and R. J. Blandau

The University of Chicago Press

Chicago and London

Library of Congress Catalog Card Number: 68-29936

THE UNIVERSITY OF CHICAGO PRESS, CHICAGO 60637
The University of Chicago Press, Ltd., London W.C.1

Printed in the United States of America

Preface

There have been numerous conferences on the ovary, uterus, vagina, placenta, and reproductive tract as a whole, but the biology of the oviduct has not received similar emphasis. Now that enough interest and research have shown its intriguing structure and its fascinating functions concerned with the survival and final maturation of the gametes, as well as the early development of zygote and embryo, it was decided that a symposium should be devoted to it.

In the last decade, substantial progress has been made in understanding the biology of the oviduct. This progress resulted from modern techniques and instrumentation in microanatomy, neurophysiology, genetics, endocrinology, biochemistry, and biophysics. These studies are scattered in such a wide spectrum of journals that the scientist can hardly keep abreast of the advances. The purpose of this symposium is to familiarize investigators in biological, medical, and veterinary sciences with the application of techniques of these disciplines, and with the results to date.

A serious gap exists in communication among the disciplines. For example, recent advances in human medicine are not always appreciated by veterinary colleagues; likewise, the knowledge of animal reproduction is not fully exploited for a better understanding of possible applications for human reproduction.

This monograph presents the results of an international symposium, "The Mammalian Oviduct," held at Washington State University, Pullman, Washington, July 31–August 4, 1967. The symposium was sponsored by the Department of Animal Sciences and Department of Clinical Medicine and Surgery, Washington State University, Pullman; and the Department of Biological Structure, University of Washington, Seattle. Each paper, presented in thirty minutes, either reviews a larger field of endeavor, or reports restricted investigations pertinent to the general subject. The papers include a brief description of the fundamentals and modern concepts, general principles of research methodology, and gaps in our knowledge.

The symposium was divided into eight sessions which covered the structure, embryology, development, physiology, endocrinology, pharmacology, biochemistry, immunology, and pathology of the oviduct tissues, fluids, and eggs. The contributors were from various disciplines: anatomy, animal sciences, clinical medicine, veterinary medicine, and pathology. The contributors were from England, the Netherlands, Puerto Rico, Scotland, Sweden, and the United States. Among the 210 who attended the symposium were persons from Argentina, Canada, Ceylon, France, Japan, New Zealand, and the United States.

Each invited speaker was asked to submit an outline of his paper before the symposium. These outlines were edited carefully to avoid unnecessary duplication and to make this volume as comprehensive as possible. For uniformity, a detailed guide was prepared for the manuscript and illustrations. The papers submitted for editing before the symposium were extensively reviewed by at least three scientists.

This volume has a twofold purpose: to stress the need to study the entire oviduct in mammalian species at all levels of evolution in order to gain a full understanding of their function, and to explore how far such studies in laboratory animals can help clarify similar problems in domestic animals and in man.

E. S. E. Hafez
R. J. Blandau

July 1968

Acknowledgments

The symposium was generously supported by a contract (PH-43-67-672) from the National Institute of Child Health and Human Development, United States Public Health Service, and by a grant from the Population Council (M 67.011).

Donations from the following pharmaceutical companies and foundations are acknowledged: Ayerst Laboratories, New York, New York; Bay Histology Service, San Rafael, California; Ciba Pharmaceutical Company, Summit, New Jersey; Diamond Laboratories, Des Moines, Iowa; Hoyt Foundation, Chicago, Illinois; Marion Laboratories, Inc., Kansas City, Missouri; Microbiological Associates, Inc., Bethesda, Maryland; Ortho Research Foundation, Raritan, New Jersey; Schering Corporation, East Orange, New Jersey; Searle & Company, Chicago, Illinois; Squibb & Sons, New Brunswick, New Jersey; Syntex Research, Palo Alto, California; and Upjohn Company, Kalamazoo, Michigan.

The sessions at the symposium were chaired by Drs. J. D. Biggers, R. J. Blandau, R. I. Dorfman, A. C. Enders, D. L. Moyer, D. L. Odor, C. Polge, and W. A. Wimsatt.

The organizers at Washington State University, Pullman, were Dr. L. C. Strait, coordinator for local arrangements; Dr. and Mrs. T. H. Blosser, general organizers; Drs. M. H. Ehlers, P. A. Klavano, and D. R. Lingard, in charge of demonstrations; Drs. L. C. Strait and J. A. Davis, organizers of film sessions; and Drs. V. L. Estergreen, H. Eastlick, and L. P. Beck, in charge of exhibits.

The editors wish to thank the contributors, who prepared their chapters meticulously. Thanks are due to Dr. Penelope Gaddum, Lynn Goldner, Nancy Hansen, Annie Kennedy, Mike Kern, Laila Moustafa, Elizabeth Phinney, Grace Urdal, and Gina Whest, who painstakingly helped with the galley proofs and assisted with the author and subject indexes.

Contents

Part IV Experimental Manipulation of Oviductal Eggs and Oviduct

Part V Clinical Aspects

Introduction

History of the Mammalian Oviduct

C. W. Bodemer

Department of Biomedical History
School of Medicine
University of Washington

Contemporary knowledge of the mammalian oviduct derives in great part from twentieth-century advances in technology and the application of endocrinological concepts and techniques. Yet long before our time the mammalian oviduct was identified, its structure described, and its role in generation defined. A historical review quickly reveals that interpretations of the oviduct and its function relate most immediately to existing concepts of generation and development of the embryo. Indeed, the history of the knowledge of the oviduct reflects directly and vividly the various and diverse theories of development engendered within Western biological thought, and an understanding of past attitudes toward this structure is necessarily incomplete outside the context of contemporary concepts of generation and embryogeny. This essay therefore describes the growth of knowledge regarding the structure and function of the mammalian oviduct and endeavors to elucidate the intimately interdependent relationship of ideas of oviductal function and physiological theory from the period of classical antiquity to the mid-nineteenth century.

I. Knowledge of the Mammalian Oviduct in Antiquity

The science of antiquity and the medieval period was stamped indelibly by the works of Aristotle, and his profound influence upon biology is manifest in Greco-Roman interpretations of the female reproductive organs. Aristotle's description of the uterus in *Historia animalium* reveals his wide comparative anatomical experience and his belief that the uterus is always double:

> The uterus is not of identical formation in those animals which possess one, nor is it similar in all; differences are found both among the Vivipara and Ovipara. In all animals which have the uterus close to the generative organs the uterus is forked . . . its starting point, however, is single, and so is its opening, as it were a tube. . . .

There is little question that Aristotle includes the oviducts within the term "uterus." He remarks that "in most animals, at the extremities of the so-called *keratia* (horns), the uterus has a convolution," but he clearly does not describe the oviduct as an entity distinct from the uterus.

The Aristotelian concept of embryogenesis was an enduring influence in biological thought. According to this concept, in animals which are generated by sexual reproduction the male contributes seminal fluid, which is formed in the seminal ducts, and the maternal contribution to the embryo is furnished by the oviduct in ovipara and by the uterus in mammals. The female semen is the *catamenia*, which correspond to the menstrual blood, and the process of reproduction consists in the admixture of the male and female semina. This general doctrine appears also in the Hippocratic treatise on generation, but Aristotle introduces a new idea in asserting that the male semen imparts only the formative impulse which initiates growth and determines the form of the embryo, whereas the undischarged residue of the *catamenia* in the female contributes the substance of the embryo.

The *catamenia*, which in oviparous animals form the egg, are in all animals secreted by the uterus. The *catamenia* are analogous to the male semen, and thus the uterus is analogous to the testes. This relationship is revealed in the bicornuate character of the uterus. "The uterus is always double," writes Aristotle in *De generatione animalium*, "just as the testes are always two in the male." Aristotle's concept of reproduction requires only a secretory uterus. It is the female gonad and it secretes the *catamenia* directly at the site of generation; hence, no ducts are required for transport of the female contribution to the embryo.

The human ovaries were first described during the third century B.C. by the Alexandrian anatomist, Herophilus of Chalcedon, whose treatise on midwifery was widely known during antiquity. The anatomical investigations of Herophilus had particular impact upon Galen, who studied at Alexandria around 152–57 A.D. This was an especially felicific circumstance, because it is primarily through Galen's extensive quotation of Herophilus that the latter's views are known today. According to Galen's account in *De semine*, Herophilus does not consider the uterus analogous to the testes, ascribing production of female semen to the *testiculi foeminis*, or ovaries. Herophilus also describes a duct associated with the "female testes":

> A very small seminal duct, difficult to see, occurs on each side, arising from the uterus. The first part of this duct is much folded, and, as in males, it runs from the testicle to the fleshy part of the neck of the bladder.

Herophilus may have been describing an anomaly, e.g., a persistent Wolffian duct, or, conceivably, the oviduct; whatever the case, the seminal duct as he defined it possessed surprising longevity.

The anatomical observations of Herophilus were basal to works of Rufus of Ephesus, Greek anatomist and physician who flourished under Emperor Trajan. Rufus first described the sheep oviduct, and a short treatise attributed to him includes an account of the human oviducts as "antennae or octopus-like arms extending as prolongations from each side of the uterus" (Daremberg 1879).

Herophilus' description of the female generative organs is evident in the works of Soranus of Ephesus, the leading authority on gynecology, obstetrics, and pediatrics in antiquity. The most important surviving work by Soranus is *Gynecology*, a treatise of such merit that it served as the original for such famous works as Rösslin's *Der swangern frawen und hebammen roszgarten* (1513) and Raynalde's *Byrth of Mankynde* (1540). Soranus was much esteemed during the early Christian era: Tertullian regarded him quite highly, Saint Augustine called him "*medicinae auctor nobilissimus*," and he is mentioned by John of Salisbury on a par with Socrates, Plato, Aristotle, and Seneca. Such widespread approbation and the outstanding quality of *Gynecology* account for the survival of many of Soranus' views and practices into the early modern period. Soranus' description of the female organs of generation therefore represents an important Roman legacy to the Latin West:

> . . . the didymi (ovaries) are attached to the outside of the uterus, near its isthmus, one on each side. . . . The seminal duct runs from the uterus through each didymus and extending along the sides of the uterus as far as the bladder, is implanted in its neck. Therefore, the female seed seems not to be drawn upon in generation since it is excreted externally. . . .

Galen, born the year of Soranus' death, was the last really important thinker and writer on generation in antiquity, and his ideas dominated biological thought for centuries. Galen's views on generation are developed primarily in *De semine*, *De foetuum formatione*, and, to a lesser extent, *De usu partium*. Galen believes the male semen to be derived from the blood by an action of the spermatic vessels and reproductive glands. He opposes Aristotle in insisting that the female produces semen, a special substance which resembles male semen and is produced in comparable manner within the "ovaries." The ovary helps to produce the female semen, which is conveyed through the oviduct to the uterus, where the male and female fluids mix, coagulate, and evolve the embryo. Galen indicates clearly in *De usu partium* that the oviduct is the efferent vessel for the female semen, and he corrects Herophilus in *De semine:*

> . . . when he says it is very small he errs . . . and he errs even more when he says that it is inserted into the neck of the bladder in the same ways as in males, for the seminal duct is not in any female animal inserted into the neck of the bladder nor into the neck of the uterus, although it is much nearer to the uterus than to the bladder . . . it actually runs to the apex or cornu of the uterus on each side, and there deposits the semen.

Aristotle, believing the female material contribution to the embryo to be secreted in the uterus, does not describe any form of seminal passageway in the female. Soranus believes that the seminal duct passes from the ovary to the bladder, and he therefore infers that the female semen bears no relation to generation. Galen posits the *in utero* union of male and female semina in formation of the embryo and regards the oviduct as the passageway transporting the female semen to the uterus. All these ideas were abroad at the end of the Greco-Roman period and continued into medieval biomedical thought.

II. Medieval Interpretations of the Mammalian Oviduct

The anatomy of the female reproductive organs presented in medieval works is generally an elaboration of views of the writers of antiquity. The oldest and most numerous accounts of the female organs of generation available during the medieval period represent the uterus as a more or less paired organ. The recurrent descriptions of this bent rested upon the authority of Aristotle and Galen, and the descriptions of the human uterus as a simplex uterus by Rufus, Soranus, and, occasionally and inconsistently, Galen, had little apparent impact. Antiquity also bequeathed to the Latin West the idea that the uterus of multiparous animals is divided into several cavities. This notion was granted wide, if not universal, application, and medieval anatomical literature abounds with representations of multichambered uteri. Copho, Ricardus Salernitanus, Wilhelm von Conches, Michael Scotus, and

Mundinus, for example, declare the human uterus a seven-chambered organ, and Bartholomus Anglicus and Lanfrancus describe three manifest uterine cavities.

The Greco-Roman influence is everywhere apparent, and the seminal duct is a common feature of medieval descriptions of the human generative organs. In a typical graphic representation of the human generative organs (Fig. 1) the uterus is depicted as bicornuate, and the ovaries (*testes muliebres*), located medial to the uterine cornua, join the body of the uterus by spermatic vessels (*vasa spermatica*). The oviducts are presumably included within the uterine horns. This anatomical rendering thus combines elements of Aristotle, Soranus, and Galen. Another medieval representation of the female reproductive organs (Fig. 2) reveals the additional influence of Gilbertus Anglicus and Constantinus Africanus. The vagina and uterus are depicted in longitudinal section, and a fetus resides within a chamber at the top of the uterine cavity. The oviduct passes from the upper poles of the uterus to the ovary, which is marked "Here is the testicle, from which the semen flows." On the right the seminal portion of the ovary connects with a part designated "stationary blood," and the notation indicates that the broad connecting duct is the "path of the menstrual blood."

The *vasa spermatica* of medieval writers generally correspond to the seminal ducts de-

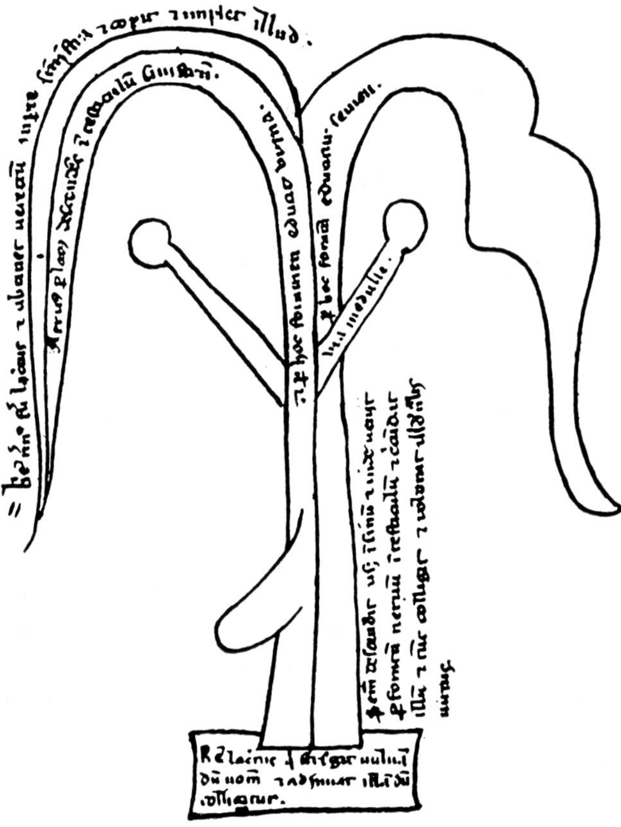

FIG. 1. A medieval representation of the human generative organs. The uterus is depicted as bicornuate, and the oviduct is presumably included within the uterine cornua. The ovaries are shown situated medial to the uterine horns and connected to the uterus by *vasa spermatica*. (From Ferckel 1917. Courtesy of Franz Steiner Verlag.)

fined by Soranus and Galen, but they do not necessarily represent the oviduct. For example, a twelfth-century Salernitian text (de Renzi 1852–57) describes the ovaries that "lie under the trumpet-like extremities of the uterus" and from which "there goes a stem-like cord, through which the testicle ejects semen into the spermatic vessel." A thirteenth-century text (von Töply 1902) compares spermatic vessels in the male and female, and notes that because the female semen is coarser and moister, "passage of the female semen from the testicles to the uterus is short, and the semen enters the uterus immediately at the lower part of the uterus on either side." It is apparent that medieval anatomists did little but formalize the observations of Greco-Roman writers and inject the conceit of the multichambered uterus. The designation of a spermatic vessel or seminal duct as something distinct from the oviduct, however, tended to crystallize during the medieval period. This obscured the true identity and function of the oviduct for centuries.

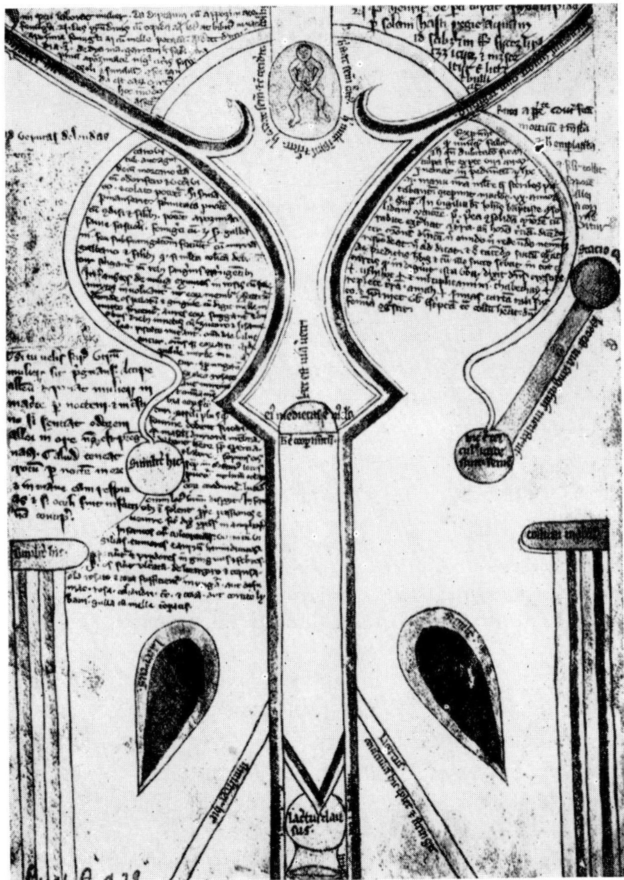

FIG. 2. A medieval representation of the human generative organs. The vagina and uterus are shown in longitudinal section, and a fetus resides in a chamber at the top of the uterine cavity. On each side the oviduct passes from the upper poles of the uterus to the ovary, which is marked, "Here is the testicle, from which the semen flows." On the right side of the illustration the seminal portion of the ovary connects with a part designated "stationary blood," and the notation indicates that the broad connecting duct is the "path of the menstrual blood." (From Ferckel 1917. Courtesy of Franz Steiner Verlag.)

III. Knowledge of the Mammalian Oviduct during the Early Modern Period

A. *The Sixteenth Century*

The transition from medieval to modern biology is apparent in Berengario da Carpi's *Isagogae breves* (1522). The transitional character of Berengario's treatise is conveyed in his description of the human uterus (Fig. 3). Berengario concurs with Rufus and Soranus in describing a *simplex* uterus, but a cautious qualification maintains compatibility with the medieval concepts of a bicornuate or chambered uterus. "Perhaps he does not err," writes Berengario, "who says there are two uteruses, because there are two cavities like two hollow hands touching each other and covered by one panniculus terminated at one canal."

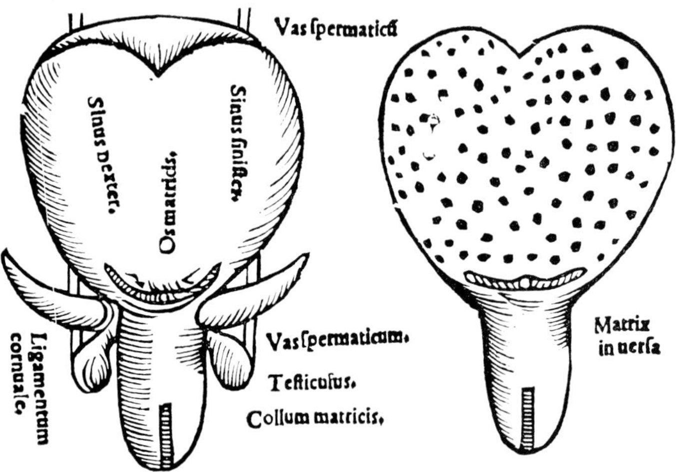

FIG. 3. Illustration of the human uterus and ducts in Berengario's *Isagogae breves* (1522). On the left, the uterus is depicted as divisible into left and right halves and possessing an *os matricis* "through which pass the menses and the foetus and through which the male semen enters." A stout *ligamentum cornuale* is present on each side, and the *vasa spermatica* are shown entering the ovaries (*Testiculus*). In this scheme these latter vessels are considered as vessels of preparation, descending from the aorta and vena cava; from the ovaries deferent vessels pass to the cavity of the uterus near its mouth. The right side of the figure illustrates the interior of the uterus and the fossulae, or cotyledons, through which the menses flow into the uterine cavity.

Berengario's analysis of the female reproductive organs reflects the prevailing concept of generation. According to this view, certain spermatic vessels are responsible for the formation of female semen from the blood. These *vasa praeparantia* carry the seminal material to the ovary, which completes formation of the semen and endows it with a "generative force." The completed semen is transported thence by other spermatic vessels, *vasa deferentia*, into the uterus near its mouth.

The account of the anatomy of the generative organs by Vesalius in *De humani corporis fabrica* (1543) emphasizes the then current physiological theory. In this theory the precursor of the semen and nourishment is thought to be carried to the ovary by the spermatic vessels. The ovary is considered analogous to the testis, and it is therefore depicted surrounded by an epididymis. The true semen is filtered off in the epididymis and transported to the uterus through a deferent duct. This deferent duct appears to be the *ligamentum ovarii proprium;* what is probably the oviduct and its vessels Vesalius describes as a portion of the seminal artery and vein. Vesalius had few opportunities to dissect the female body, and his analysis of the female generative organs is understandably weak; it is noteworthy that he attempts to show that the traditional concept of the bicornuate uterus was derived from animal anatomy.

It remained for Gabriel Fallopius to describe the organs which now bear his name. Fallopius published only one book, *Observationes anatomicae* (1561), an unillustrated commentary or series of observations on Vesalius' *Fabrica*, in which Fallopius seeks to correct errors or to present new material overlooked by his predecessor at Padua. It includes an account of the oviduct, which Fallopius regards as the passageway for the female semen:

> The slender and narrow seminal passage arises very white and sinewy from the uterine horn, but after it has passed outward a little way it becomes gradually broader and curls like the tendrils of a vine until near the end when the tendril-like curls spread out, and it terminates in a very broad ending which, because of its reddish color, appears membranous and fleshy. This ending is quite shredded and worn, as if it were the fringe of a worn piece of cloth, and it has a broad opening which always lies closed because of the apposition of those fringed ends. However, if they be carefully opened and spread apart they form, as it were, the bell-like mouth of a bronze trumpet. Consequently since, where the tendril-like curls may be removed or even added to this classical instrument the seminal passage will extend from its head even to its uttermost ending, and so it has been designated by me the trumpet of the uterus. They are arranged this way in all animals, not only in man, but also in fowls and cattle, and in the bodies of all other animals I have studied.

Fallopius established finally the existence and anatomical relationships of the oviduct, but his work comprised no greater departure from contemporary physiological dogma than that of Vesalius. The limits to the advance in understanding oviductal function were imposed by traditional concepts of generation and embryogenesis.

B. *The Seventeenth Century*

Comprehension of the full significance of Fallopius' observations was hindered by the force of tradition and, to a lesser extent, by the plethora of theories of generation abroad during the sixteenth and seventeenth centuries. Some (Scaliger 1592) might disagree, but most seventeenth-century investigators concurred with Hippocrates, Herophilus, and Galen in supposing the production of semen by both sexes (Benedetti 1502; Fernel 1554; Colombo 1559; Aranzio 1564; Eustachio 1564; Riolan the Elder 1578; Plater 1586; Piccolomini 1586; Bonacciuoli 1586; Du Laurens 1593; Capivaccio 1603; Riolan the Younger 1650; Vesling 1659; Cornelio 1663). The female semen was variously described as a vague material, a kind of liquid, or a favorable concatenation of particles. Parisano (1621) and Everard (1661), for example, refer to a liquid containing a seminal virtue or prolific spirit, whereas other students of generation (Digby 1644; Browne 1646; Highmore 1651; Gassendi 1658; Descartes 1664) favor a mechanistic rationalism in which atoms or corpuscles form the basis of generation. In some theories the semen was believed to arise independently of the *testes muliebres*, in others the ovary was little more than a relay or assembly station. Within this milieu provocation to study of the oviduct was slight.

The idea of *vasa spermatica* was garbed in the robes of tradition and enjoyed widespread support in the form suggested by Fernel (1554), Varolio (1591), and particularly Du Laurens (1593). The latter describes two *vasa spermatica deferentia* devoted to transmitting the female semen (Fig. 4). From a common trunk (*Q*) a short duct (*R*) leads to the oviduct and a longer duct (*S*) passes through the sides of the uterus, ending at the cervix and pudendum. It is most likely that these ducts are the *ligamentum ovarii proprium* and the *ligamentum teres uteri*, respectively. According to Du Laurens semen normally passes into the uterus through the upper duct, but during pregnancy it is evacuated through the other duct. Most seventeenth-century investigators agreed with the general features of Du Laurens' system (Piazzoni 1644; Riolan 1650; Highmore 1651; Wharton 1656; Everard 1661; Barbato 1676).

François Mauriceau was perhaps the leading representative of obstetrical knowledge during the seventeenth century, and his obstetrical treatise (1668), which passed through seven editions and various translations, was a kind of canon of the art of his time (Fig. 5). Mauriceau's description of the female generative organs and their function is thus of special interest. Following Du Laurens, Mauriceau describes two deferent ducts (Fig. 6). He illustrates the oviduct, but, disturbed by its lack of continuity with the ovary, he declares for the *ligamentum ovarii proprium* as the passageway for the female semen. "The oviducts," he writes:

> serve (according to common opinion) as pores for the semen and to discharge it into the uterus. Yet, their origin makes me doubt this use, because they do not derive it from the testicle, not touching it in the least. I much prefer to believe that women usually discharge it by another vessel, coming directly from the testicle to the side of the uterus near its horn [oviduct], which, in truth, does not seem manifestly hollow, although it is big enough. But it is not necessary that it be so, because the semen, being very spiritous, may easily pass through its porous substance.

Thus, Mauriceau, like so many of his contemporaries, accommodates his concept of generation by incorporating the oviduct into the traditional system of deferent ducts.

Initial insight into the significant oviductal secretory function was achieved by Fabricius (1621), who gives the first essentially accurate account of the role of the oviducts in the formation of the complete egg of the hen. Coiter (1572) had indicated that the egg acquires an albumen coat in the uterus, but did not suspect that the albumen was produced by the oviduct. Fabricius is the first clearly to indicate that the albumen, chalazae, shell membranes, and shell are secreted by the oviduct. The term *oviductus* appears first in Stensen's *De anatome rajae epistola* (in *Opera philosophica*, 1910); hence Fabricius speaks of an *inferior uterus* in which the upper part secretes the chalazae and albumen.

Fabricius' interpretation of the oviduct as an organ of secretion did not receive universal acceptance. Even his student, Harvey (1651), seems confused about oviductal function relative to formation of the hen's egg. Harvey believes it improbable that, "the yolk is first made, and then the white adjoyned to it," postulating instead that, "the yolk findeth out and concocteth its white." And further, he concludes, "the egge is neither the workmanship of the uterus, nor controuled or governed by it."

The role of the oviduct in formation of the avian egg received belated acceptance, but oviductal function was understood better in birds than in mammals. This stemmed in part from an inadequate appraisal of the mammalian ovaries and their primary function. For example, Harvey, associated forever with the slogan, *Omne vivum ex ovo*, considers them to be "mere little glands" with an insignificant role in generation. It is indicative that Fabricius, who introduced the term *ovarium*, labels the human ovary *testis* and the ovaries of the sow as *glandulae*, for no one related the *testes muliebres* of mammals with the ovaries of ovipara until Stensen (1667) proposed their equivalence.

Harvey's (1651) observations on reproduction in the deer led him to deny the existence of *vasa praeparantia* and *vasa deferentia seu ejaculantia* in any animal. He also comments upon the motility of the oviduct and its altered appearance at estrus. However, Harvey's investigation also contributed to the confusion of subsequent investigators. He failed to observe semen in the uterus of various mammals following coitus and he was also unable to find, "even a trace of the conception . . . after a fruitful union of the sexes" within the uterus for some days. Harvey therefore concludes that the female semen is a fiction and deems unnecessary any prepared maternal matter in the uterus. Harvey proposes that animals arise from an inherent primordium which is the ovum, "some corporeal substance having life in it in potentia, or something subsisting of it selfe, which is apt to be transformed into a vegetative form, by some internal principle acting in it." Harvey's vague generalization scarcely clarifies generation; but it affected interpretations of the oviducts less than his failure to observe semen or a conceptus in the uterus following coitus, for the mammalian ovary was soon shown to produce "ova."

The first reference to ova contained in mammalian ovaries appears in a letter written by Johannis van Horne (1668): "That which is the ovarium of ovipara are the *testes muliebres*, inasmuch as they contain within themselves perfect ova, surrounded by fluid and enclosed by individual sheaths of such a kind that they are maintained in their place in a swollen condition." Noting that the uterine tube is often greatly dilated and filled with

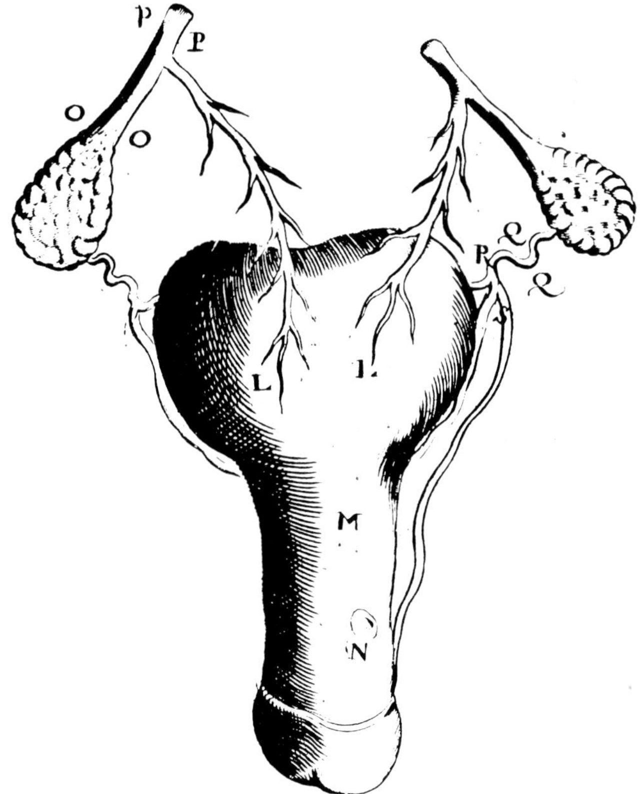

FIG. 4. Illustration of human reproductive organs according to Du Laurens, showing the relationships of various ducts to the uterus. Preparatory vessels (*P*) enter the ovary (*O*), and *vasa spermatica deferentia* are depicted leaving the inferior pole of the ovary by a common trunk (*Q*), then dividing into a superior branch (*R*) and an inferior branch (*S*). The superior branch enters the oviduct and the longer inferior branch passes through the sides of the uterus, ending at the cervix and pudendum. Female semen is considered to pass normally through the upper duct, but during pregnancy to pass through the lower one. (From Du Laurens 1778. From a copy in the Wellcome Historical Medical Museum and Library. By courtesy of The Wellcome Trustees.)

fluid toward its fimbriated end, van Horne suggests that, since the oviduct is hollow and communicates with the uterine cavity, it may carry the ova to the uterus.

Death prevented van Horne from bringing his work to perfection, but it incited others to pursue the subject further (Kerckring 1671, 1672). Definition of the ovarian follicle of mammals was soon forthcoming (de Graaf 1671, 1672; Swammerdam 1672) (Figs. 7, 8),

FIG. 5. Frontispiece from Mauriceau's obstetrical classic. (From Mauriceau 1668. From a copy in the Wellcome Historical Medical Museum and Library. By courtesy of The Wellcome Trustees.)

and confirmation of the Graafian follicle (Langly 1674; Bartholin 1678), combined with reports on lower forms (Lorenzini 1678; Tauvry 1690) clarified ovarian function in mammals and established securely the homology of the ovary in vivipara and ovipara.

The discovery of the ovarian follicle did not lead quickly to appreciation of the transport function of the mammalian oviduct. In part this was a consequence of the prevailing theories of fertilization, and in part it stemmed from the difficulty of ascertaining the presence of semen or ova in the female reproductive tract following coitus. For some years the prevailing concept of fertilization was that accepted by Harvey following his failure to observe male semen in the female ducts. Harvey concurs with Fabricius that the male semen never reaches the seat of generation and attributes fecundation to a seminal aura or effluvium which, absorbed by the blood and conveyed by the circulation to the ovary, makes the ova fertile and able to engender an embryo. Through a kind of contagion, then, the ova are rendered

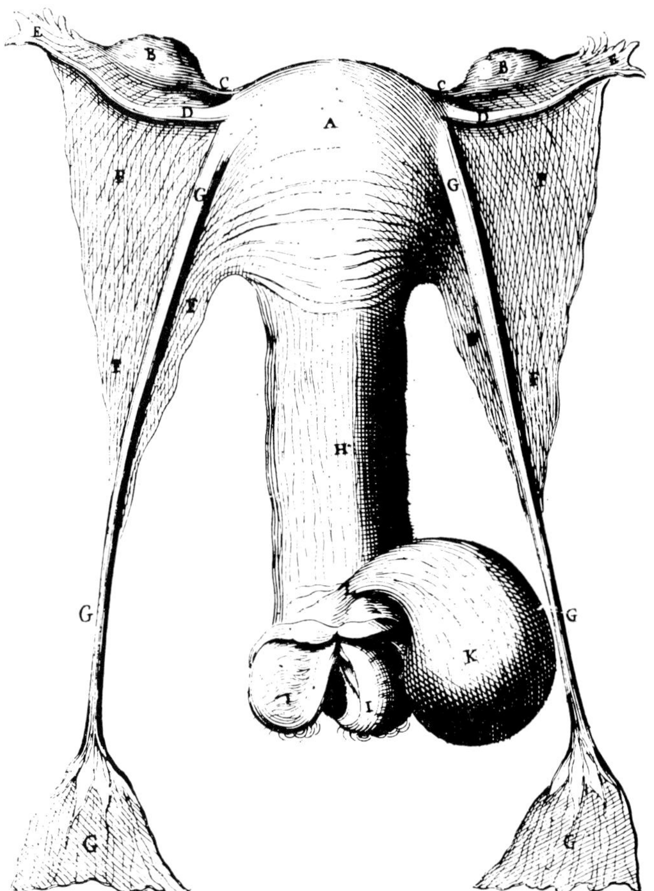

FIG. 6. Mauriceau's representation of the human generative organs, illustrating a modification of the duct system outlined by Du Laurens. The oviduct is illustrated (*D*), but Mauriceau considers the structure marked *C* (*ligamentum ovarii proprium*) to be the passageway for the female semen. (From Mauriceau 1668. From a copy in the Wellcome Historical Medical Museum and Library. By courtesy of The Wellcome Trustees.)

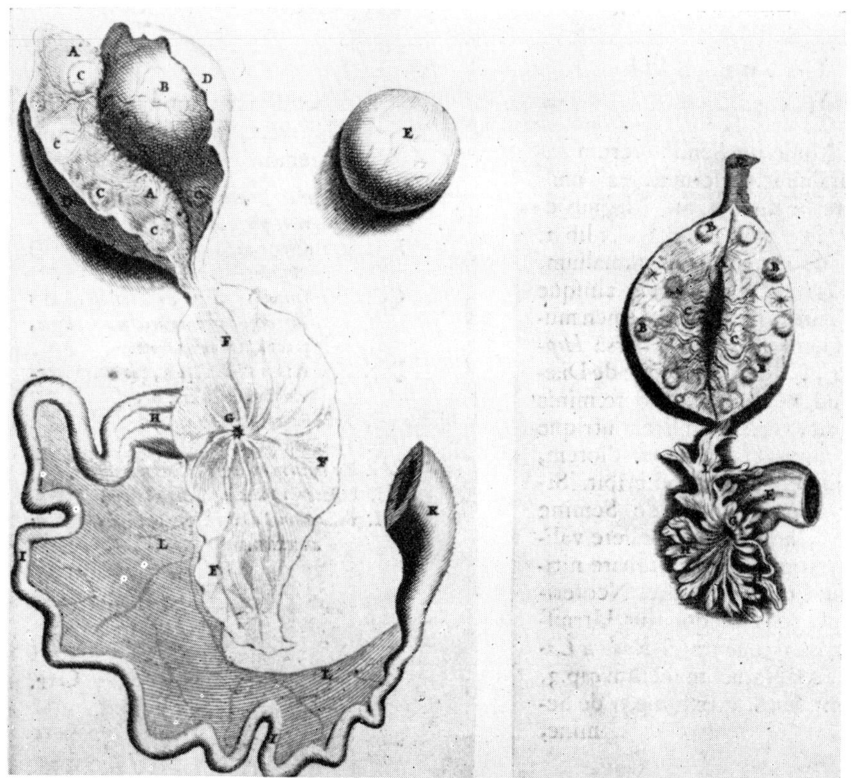

FIG. 7. Illustrations of reproductive organs in de Graaf's treatise demonstrating the ovarian follicles in mammals. The opened ovary of a cow is shown on the left; the human ovary and oviduct are depicted on the right. The ovarian follicles and the relationship of the oviduct to the ovary are shown. (From de Graaf 1672.)

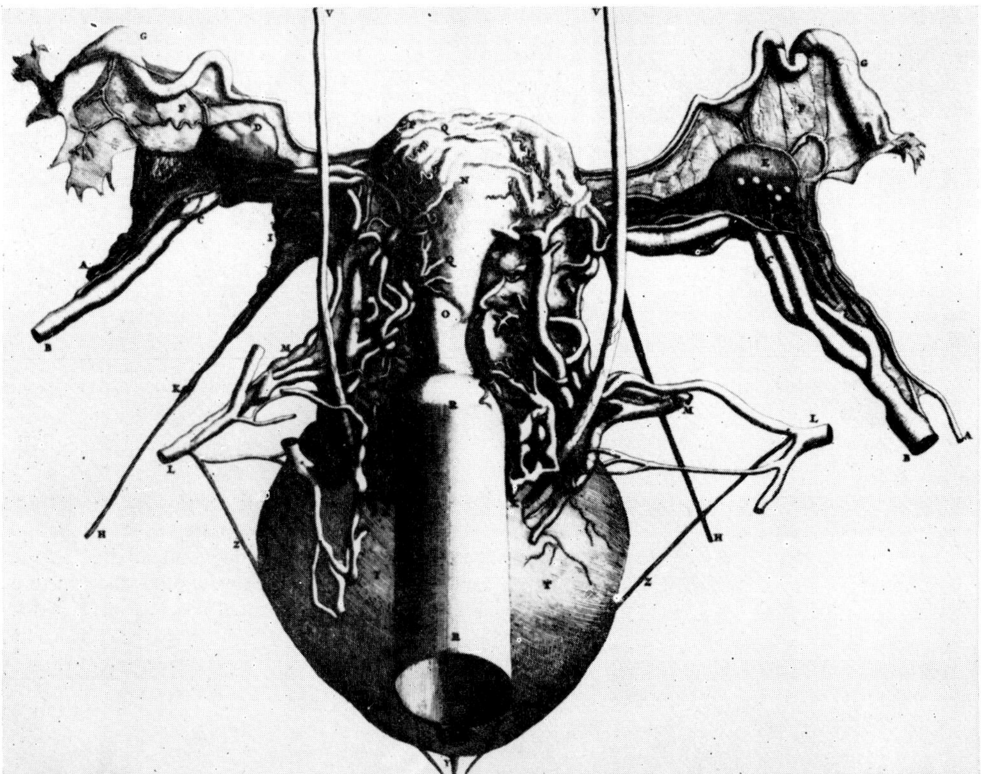

FIG. 8. Swammerdam's illustration of human organs of generation depicting the relation of ovaries, oviducts and uterus. (From Swammerdam 1717. From a copy in the Wellcome Historical Medical Museum and Library. By courtesy of The Wellcome Trustees.)

fecund before they leave the ovary. This view, which received considerable support (Bartholin 1679; Swedenborg 1740), had significant effect upon interpretations of oviductal function.

The difficulty of observing semen, male or female, within the female reproductive organs was a salient cause of confusion about the nature of fertilization and enhanced continued belief in the traditional system of ducts. Semen was reported in the female ducts by Galen and others (Fallopius 1561; Riolan 1618) prior to Harvey's observations on the deer; but thenceforth most investigators failed to locate either male or female semen in the female generative organs. Thus, Everard (1661) dissected Dutch rabbits at diestrus and various intervals from 3 to 25 days *post coitum*, and finding nothing in the uterus, excludes female semen from generation, "not as useless, but as not prolific, and as contributing nothing to it." Everard, apparently the first to employ rabbits in a systematic study of reproduction, describes distinctive changes in the oviduct at diestrus, at estrus, and during pregnancy, observing that in the doe not in heat, "the uterine horns appear pale and rather compact and contracted," whereas on the third day *post coitum*, "they are more expanded and reddish, and in some places somewhat raised to form an elongate swelling . . . where . . . the inner membrane is more reddish and uneven, as if filled with small glands."

The death knell of the traditional duct system sounded later, when de Graaf (1672) confirmed earlier suggestions (Riolan 1650; Wharton 1656; Kerckring 1671) by following the course of the "ova" through the oviduct to the uterus. De Graaf points out the errors of his predecessors regarding the deferent ducts:

> We declare that those processes . . . are nothing but the ligaments of the testes, and that they can by no means be regarded as vasa deferentia . . . you will not find them containing semen or anything analogous to it.

The end of the seventeenth century witnessed publication of the noteworthy experimental study of Nuck (1691, Fig. 9). This investigator applied a ligature approximately halfway along the length of the left uterine horn of a dog 3 days after coitus, and 20 days later he found two fetuses in the horn above the ligature; below the ligature the horn was empty. He concludes that these two fetuses originated from two "fecundated ova" he observed in the left ovary at the time of operation. Nuck's study is the first experimental evidence in support of the role of the mammalian egg in embryogenesis.

The idea that the mammalian egg was fecundated prior to leaving the ovary was supported in one way or another by many eighteenth-century investigators (Andry 1700; Geoffroy and Du Cerf 1704; Bellefontaine 1712; Maître-Jan 1722; Vallisnieri 1721; Haller 1766; Cruikshank 1797). Attendant upon this concept the oviduct was believed to provide passage for the zygote, and it was excluded as the fertilization site. But the formidable animalculistic concept of development involved a different view of oviductal function. The first published account of spermatozoa appeared in 1678, and within five years these spermatic *animalcula* had been demonstrated in the semen of animals of every class (Bartholin 1680; Hartsoeker 1678; Huygens 1678; Leeuwenhoek 1679, 1683; Schrader 1681). A system of generation from *animalcula*, in contradistinction to generation from ova, soon developed. This system impressed some (James 1745) as "utterly romantic, and inconsistent with the conduct of Providence," but none could contest the sentiment (Folkes 1724) that the discovery of seminal *animalcula* had "given a perfectly new turn to the theory of generation."

It gave a perfectly new turn to the study of the mammalian oviduct also. Leeuwenhoek (1683) set the tone in an argument against generation from eggs:

> . . . I need not believe that these round Bodies should be drawn down from the imagined Egg-branch, thru the long and very narrow passage of the Tuba Fallopiana, Because some of the Bodies are as big as a Pease (nay, as the whole Egg-branch); and of a very firme and compacte Substance. But the way thru which they should pass is no wider than the Compass of a small Pin.

This proposition had its supporters (Sbaragli 1701; La Motte 1718), although Leeuwenhoek's later (1722, 1724) reference to an imaginary opening in the oviduct received little attention.

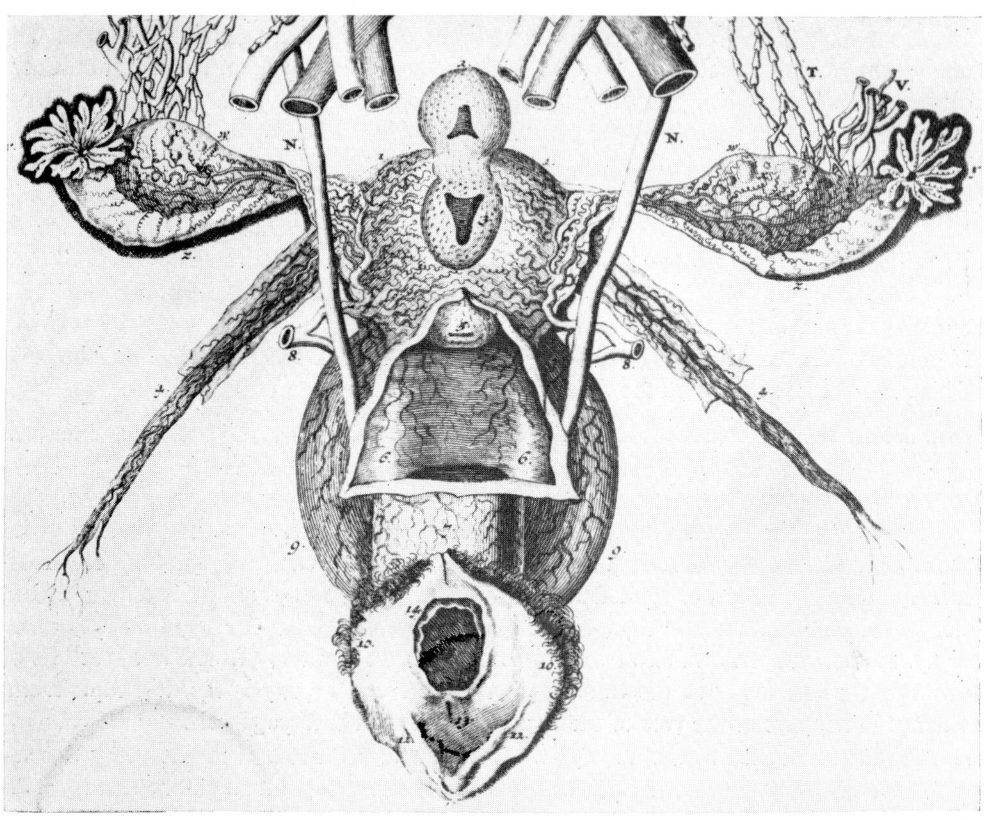

FIG. 9. Illustration of the human reproductive organs and their lymphatic vessels. Note the rich supply of vessels to the ovary and oviduct. (From Nuck 1722. From a copy in the Wellcome Historical Medical Museum and Library. By courtesy of The Wellcome Trustees.)

IV. The Eighteenth and Nineteenth Centuries: Foundation of Modern Interpretations of the Mammalian Oviduct

A. *The Eighteenth Century*

Various concepts of generation, each with a different view of the oviduct, were abroad in the eighteenth century. Some investigators (Vallisnieri 1721) continue to posit fertilization of the ovum *in situ* through a seminal aura; in this case the oviduct is merely a channel permitting the effluvium access to the ovum. For others (Geoffroy and Du Cerf 1704) the oviduct is an avenue leading the spermatozoa to the ovary. On the other hand, many biologists support the idea of conception in the uterus (Andry 1700; Astruc 1765; Boerhaave 1744; Buffon 1785; Drelincourt 1684, 1685; Haller 1766; Le Grand 1672), and consider the oviduct as a passageway for some sort of seed. Other theories of generation, however, do not require the oviduct (d'Agoty 1750) or deny its function in transport of the ovum (Maupertuis 1744).

Most eighteenth-century theories of development include the tenet that the "ovum" or its precursor passes to the uterus via the oviduct. Accordingly, explanations of the mechanism of egg transfer were attempted. Méry (1704), perhaps the only opponent of ovism in the Académie des Sciences, maintains that the eggs cannot be detached from the mammalian ovary, and even Malpighi (1686) appears to have had some doubts. But as the ovist position became ascendant, various mechanisms were proposed to account for ovulation and egg transport. Ovulation was ascribed to internal processes within the ovary (Andry 1700; Geoffroy and Du Cerf 1704), to stimulation by seminal aura (Vallisnieri 1721; Cooke 1762; Haighton 1797; Hildebrandt 1828), or to coitus (Muralt 1677; Saumarez 1798). The oviduct was frequently assigned a passive role, merely catching the ova; but the oviduct was sometimes considered the prime factor in ovulation. Thus according to one view (Drelincourt 1685; Blundell 1834) at certain times the oviducts embrace the ovaries and by "suction" actively extract eggs from the ovary.

Oviductal transport of the ovum was often disclaimed on the basis of the structural discontinuity of the ovaries and oviducts. Opponents of ovism fancied necessary the possession by the oviducts of some instinct to ensure their union with the ovary at the appropriate time. Dionis (1724) counters this argument with a thoroughly mechanical scheme. He suggests that at coition the round ligaments contract to pull the uterus toward the vagina, thus pulling the extremities of the oviduct toward the ovary. Concomitantly, the broad ligaments pull the ovary toward the oviduct, and, "the extremities of the tubae mounting up and the ovaries descending by one and the same motion, they are brought together." Dionis emphasizes the importance of oviductal motility, stating that having received the egg, the oviducts "embrace it, and by a vermicular motion, like that of the intestine, press it gently and carry it to the uterus."

In Dionis' time the muscularity of the uterus was subject to close attention (Morgagni 1719; Haller 1766), and longitudinal and circular "fibers" had been described in the oviduct (Verduc 1704). Motility of the oviduct received increasing emphasis during the eighteenth century, and peristalsis became part of most interpretations of oviductal function (Saumarez 1798; Sauvages 1755; Blumenbach 1795; Haighton 1797; Cruikshank 1797; Burdin 1803; Blundell 1834; Coste 1834; Burdach 1835; Dugés 1839).

Progress in analysis of oviductal function in generation diminished consequent to Haller's (1752, 1766) study of reproduction in sheep. Systematically examining ewes sacrificed serially after mating, Haller found nothing in the oviducts, nor could he find a recognizable embryo in the uterus before the 13th day. He therefore concludes that something corporeal, perhaps in a fluid state, passes from the ovary to the uterus, and there coagulates into an embryo. The concept, proposed by the outstanding biologist of the century, appears to have inhibited comparable investigations for several decades.

In 1788, William Cruikshank, assistant to John Hunter at the Great Windmill Street School of Anatomy, began a study of great import. Cruikshank observed blastocysts in the rabbit oviduct on the 3d day *post coitum* and in the uterine ducts on subsequent days, in numbers corresponding with the number of corpora lutea in the ovaries (Fig. 10). Hence, he concludes that, "the ovum is formed in, and comes out of the ovarium after conception

. . . passes down the fallopian tube, and is some days in coming through it . . . [and] . . . comes into the uterus on the fourth day." Cruikshank's study clearly confirmed de Graaf's investigation 106 years earlier; and the relatively slow passage of the ovum through the oviduct, now revealed, explained Haller's observation of the delay between the escape of the ovarian substance and the appearance of an embryo in the uterus.

Coincidentally, Cruikshank and Haighton published their results in 1797. The latter's observations are quite consistent with those of Haller, and this may have dampened the impact of Cruikshank's findings. Haighton's study is, however, of special interest as the first to utilize surgical methods in the study of mammalian embryology.

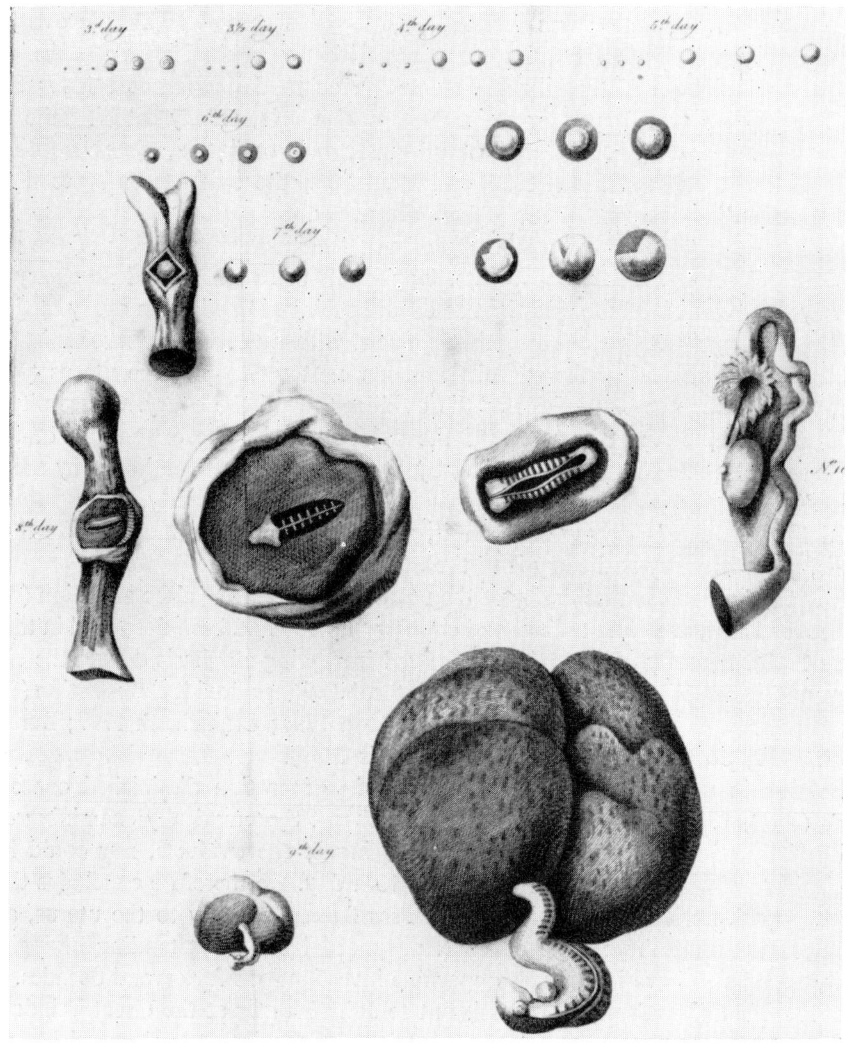

FIG. 10. Illustrations accompanying Cruikshank's observations on tubal ova in the rabbit. The oviduct and its relation to the ovary are illustrated on the right middle aspect of the figure. (From Cruikshank 1797.)

B. *The Nineteenth Century*

The mystery surrounding the events occurring during the 3 days between release of the "ovum" and its first discovery in the oviducts contributed to the longevity of the Harvey-Haller concept of an ovum formed *in utero*. However, during the year that the Göttingen Academy of Sciences awarded its prize to Hausmann for a paper affirming this concept, an article appeared which presaged the doctrine's demise (Prévost and Dumas 1824). These investigators propose, and support with experimental data, that fertilization involves contact of ovum and seminal fluid. They also support earlier suggestions (Cooke 1762) that fertilization occurs in the oviduct (Dumas 1827).

The belated discovery of the true mammalian ovum (von Baer 1827) dispelled ideas of "uterine ova" and clarified various obscurities resulting from the identification of ovarian follicles as ova. The emphasis in the study of oviductal functions shifted accordingly. Movement of the conceptus through the oviduct had long been explained in terms of oviductal activity; the lymphatics and rich capillary network of the oviduct wall were appreciated and the gross changes in the oviduct at different stages of the estrus cycle and during pregnancy were well known (Peyer 1685; de Graaf 1672; Burdach 1831; Carus 1835). Now another aspect of oviductal structure and function was introduced with the demonstration of cilia on the inner surface of the oviduct in amphibia, birds, and mammals (Purkinje and Valentin 1834). The discovery was made during a search for tubal ova in the rabbit. "One of us," they wrote:

> noticed under the microscope that small particles on the mucus membrane of the oviduct moved briskly and rotated around their axes. The other confirmed this phenomenon and instantly recognized it as ciliary movement. The entire uterus, as well as the inner parts of the genitals were then examined carefully, and it was found that these movements were nowhere lacking, although they were of different intensity at different sites. They were most lively in the tubes, less strong in the horns, and even less strong in various regions of the uterus.

Within several years the presence of ciliated epithelium in the human oviduct was described, and movement of the mammalian blastocyst consequent to activity of oviductal cilia had been observed (Barry 1839; Bischoff 1842; Henle 1838). For Bischoff, as for later investigators (Donné 1845; Müller 1840; Pouchet 1847):

> The forces which lead the egg out of the ovary into and then through the oviduct are in part the vibratory movement of the cilia of the epithelium of the infundibulum and the mucus membrane of the oviduct, and in part the independent motion of the latter.

Some questioned the mechanism through which the ovum enters the oviduct, particularly in lower forms where the ovary and oviduct are widely separated. However, the demonstration of cilia on the surface of the coelomic peritoneum in sharks, teleosts, and frogs by Valentin, Vogt, and Mayer (Müller 1840) revealed that the transport of the ovum to the oviduct could be explained on the basis of ciliary activity within the oviduct and the nearby peritoneum. This information derived from lower vertebrates supports Müller's (1840) proposal that ciliary motion existing on the fimbria and inner surface of the mammalian oviduct is a "great influence in aiding the entrance of the ovum into the tube." By the mid-nineteenth century the role of cilia in oviductal function was sufficiently established to prompt the following analysis (Todd and Bowman, 1845–56):

> The mucous membrane [of the oviduct] is disposed in longitudinal folds, upon which lies a single layer of columnar epithelium. These cells are ciliated, and by the vibration of the cilia, a current

is produced, the direction of which is from the ovaries toward the uterus; so that the transmission of the ovum into the uterus would be favoured, while the passage of the spermatozoa along the tube would be retarded.

Oviductal secretions also received attention at this time. Despite knowledge of the secretory activity of the hen's oviduct in relation to formation of the egg, there seems to have been little thought that the oviduct might perform a similar office in mammals until fairly recent times. Verduc (1704), believing that the albumen secreted by the oviduct contributes toward perfection of the chick egg, suggests that the mammalian oviductal fluid may "furnish white for the yellow, as it happens in birds." Pursuing the analogy, Verduc proposes that this oviductal secretion serves to nourish and protect the egg from mechanical injury. Smellie (1766) later applies a comparable idea to the human, stating that small vessels opening on the surface of the ovum "absorb the surrounding fluid which is secreted by the glands in the cavity of the tube and Uterus, or forced into them by motion, heat, and rarefaction, and carried along the umbilical vein, for the nourishment and increase of the impregnated mass."

During the first half of the nineteenth century the oviductal fluid was considered less a source of embryonic nutrition than a vitalizing influence upon the egg (Prévost and Dumas 1824; Coste 1837) or a barrier to spermatozoa (Pouchet 1847). There was, however, no demurral about the addition of oviductal secretions to the ova of lower animals, including egg-laying mammals (Wagner 1834). Thus, Home (1814) suggests the equivalence of oviductal function in birds and monotremes, indicating that the ovum of *Ornithorhyncus* receives its albumen in the oviduct. And Blundell (1834) amplifies significantly Verduc's suggestion of secretion by the mammalian oviduct:

> I have sometimes thought that, as in birds, the oviducts are superadding to the yolks certain parts derived from the ovaries, which render them more perfect for generation. . . . It may not be impossible that the fallopian tubes may add something too; and this is more probable, first, because we find the inner sides of the fallopian tubes vascular in a high degree, and secondly, because their inner membrane is folded longitudinally, as if nature intended to spread them out for the purposes of secretion.

As knowledge of early development increased, it was deemed possible that the ovum acquires an additional coat during its passage through the oviduct. Bischoff (1845), for example, refers to tubal ova in the rabbit in which the "discus proligerus has disappeared, and in its place a slight layer of albumen has formed around the zona." From a similar viewpoint Barry (1839, 1840) believes that oviductal secretions operate in establishment of the two membranes ("chorion and blastoderma") of tubal ova. He suggests that "the space between the two membranes is filled with a thin layer of albumen, which the ovulum has acquired in the oviduct and uterus, and which swells by imbibition of water, producing a greater separation of the two membranes."

During the first years of the second half of the nineteenth century it was established that the spermatozoon enters into the ovum at fertilization (Bischoff 1854), and thenceforth investigation centered upon fertilization and early cleavage, rather than upon the significance of oviductal secretions. Various studies established the maturation of the ovum independently of the male semen, and evidence favored the oviduct as the site of fertilization and early development (Bischoff 1845; Pouchet 1847).

At mid-century knowledge of the mammalian oviduct was relatively complete. The general structure of the organ was known; satisfactory mechanisms were available to account for egg transport; the oviduct was recognized as the site of fertilization and early embryonic development; and its role as a secretory organ was suspected. Uniformity of interpretation was incubating; and unfounded, speculative theories were in decline, partly because of more rigorous and extensive scientific methods and techniques, and partly because of increased factual knowledge. A contemporary (Thomson 1839) remarked, "As the knowledge of minute anatomy and physiology has increased, and the accurate observation of the process of development has been more extended, the number of . . . hypotheses has gradually diminished." This could have none but a salutary effect upon continued growth of knowledge of the oviduct and its functions; for the interpretations of the oviduct have been inextricably linked with concepts of generation and embryogenesis since the period of classical antiquity. But that is part of its charm, for as Pulley (1801) said,

> The mystery of nature in the wonderful operation of animal reproduction has from the earliest period courted the attention of physiologists, and though experiment and imagination have toiled and fancied through all the ages, and opinions have been various and ingenious, the present day still finds the subject unconcluded.

Acknowledgment

This research was supported by a grant from the National Science Foundation. For the privilege of using the resources of their institutions and for the many courtesies extended to him, the author wishes to express his gratitude to the Director of the Wellcome Library, the Directors of the British Museum, and the Secrétaires perpétuels of the Académie des Sciences, Paris.

References

1. Andry, N. 1700. *De la génération des vers dans le corps de l'homme.* Paris.
2. Aranzio, G. C. 1564. *De humano foetu libellus.* Bologna.
3. Aristotle. 1912. *De generatione animalium*, trans. and ed. A. Platt. Oxford.
4. ———. 1965. *Historia animalium*, trans. A. L. Peck, Cambridge.
5. Astruc, J. 1765. *Traité des maladies des femmes.* Paris.
6. Baer, C. E. von. 1827. *De ovi mammalium et hominis genesi.* Leipzig.
7. Barbato, G. 1676. *De formatrice, conceptu organizatione, & nutritione foetus in utero.* Padua.
8. Barry, M. 1839. Researches in embryology. Second series. *Phil. Trans. Roy. Soc. London* Part 2: 307.
9. ———. 1840. Researches in embryology. Third series: A Contribution to the Physiology of Cells. *Phil. Trans. Roy. Soc. London* Part 2: 529.
10. Bartholin, C., the Younger. 1678. *De ovariis mulierum et generationis historia, epistola anatomica.* Amsterdam.
11. ———. 1680. De ovariis mulierum. In *Acta medica et philosophica Hafniensia*, ed. T. Bartholin, The Elder. Copenhagen.
12. Bartholin, T. the Elder. 1679. *Acta medica et philosophica Hafniensia.* Copenhagen.
13. Bellefontaine, L. 1712. *La médecine dogmatique, méchanique, et pharmacopée rationelle.* Amsterdam.
14. Benedetti, A. 1502. *Historia corporis humani sive anatomice.* Venice.
15. Berengario da Carpi, J. 1522. *Isagogae breves perlucide ac uberime in anatomiam humani corporis a communi medicorum academia usitatum.* Bologna.

16. Bischoff, T. L. W. von. 1842. *Entwicklungsgeschichte des Kaninchen-Eies.* Brunswick.
17. ———. 1845. *The periodical maturation and extrusion of ova, independently of coitus, in mammalia and man, proved to be the primary condition to their propagation*, trans. H. Smith. London.
18. ———. 1854. *Bestätigung des von Dr. Newport bei den Batrachiern und Dr. Barry bei den Kaninchen behaupteten Eindringens der Spermatozoiden in das Ei.* Giessen.
19. Blumenbach, J. F. 1795. *De vi vitali sanguini denaganda.* Göttingen.
20. Blundell, J. 1834. *The principles and practice of obstetricy.* London.
21. Boerhaave, H. 1744. *Praelectiones academicae in proprias institutiones rei medicae edidite, et notas addidit Albertus Haller.* Göttingen.
22. Bonacciuoli, L. 1586. Enneas muliebris; qua multa variaque de conceptione, uteri gestatione, abortu, partu, obstetricatu, puerperie, nutricum, & infantium cura, aliaque huiusmodi copiose & erudite disseruntur, "in *Gynaeciorum, sive de mulierum affectibus commentarii*, ed. C. Wolf. Basel.
23. Browne, T. 1646. *Pseudodoxia Epidemica, or enquiries into very many received tenets and commonly presumed truths.* London.
24. Buffon, Comte G. L. L. de. 1785. *Histoire naturelle.* Paris.
25. Burdach, K. F. 1829–40. *Die Physiologie als Erfahrungswissenschaft.* Leipzig.
26. Burdin, J. 1803. *A course of medical studies: Containing a comparative view of the anatomical structure of man and of animals; a history of diseases; and an account of the knowledge hitherto acquired with regard to the regular action of the different organs.* London.
27. Capivaccio, G. 1603. *Opera omnia quinque section.* Frankfurt.
28. Carus, C. G. 1835. *Traité d'anatomie comparée.* Paris.
29. Coiter, V. 1572. *Externarum et internarum principalium humani partium tabulae.* Nuremberg.
30. Colombo, R. 1559. *De re anatomica libri XV.* Venice.
31. Cooke, J. 1762. *The new theory of generation, according to the best and latest discoveries in anatomy, farther improv'd and fully display'd.* London.
32. Cornelio, T. 1663. *Progymnasmata physica.* Venice.
33. Coste, J. J. M. C. V. 1834. *Recherches sur la génération des Mammifères.* Paris.
34. ———. 1837. *Embryogénie Comparée.* Paris.
35. Cruikshank, W. C. 1797. Experiments in which, on the third day after impregnation, the ova of rabbits were found in the Fallopian tubes; and on the fourth day after impregnation in the uterus itself; with the first appearances of the foetus. *Phil. Trans. Roy. Soc. London* 87:197.
36. Descartes, R. 1664. *L'homme, et un traité de la formation du foetus.* Paris.
37. Digby, K. 1644. *Two treatises, in the one of which, the nature of bodies; in the other, the nature of mans soule; is looked into.* Paris.
38. Dionis, P. 1724. *Traité général des Accouchemens, qui instruit de tout ce qu'il faut faire pour être habile Accoucheur.* Paris.
39. Donné, A. 1845. *Cours de microscopie complémentaire des études médicales.* Paris.
40. Drelincourt, C. 1684. *De foeminarum ovis, tam intra testiculos et uterum, quam extra, ab anno 1666 ad retro secula.* Leiden.
41. ———. 1685. *De conceptione adversaria.* Leiden.
42. Dugés, A. 1839. *Traité de physiologie comparée de l'homme et des animaux.* Montpellier and Paris.

43. Du Laurens, A. 1593. *Opera anatomica.* Lyons.
44. ———. 1599. *Historia anatomica humani corporis et singularem eius partium multis controversys & observationibus novis illustrata.* Frankfurt.
45. Dumas, J. B. A. 1827. Note sur la théorie de la génération. *Ann. Sci. Nat.* 12: 443.
46. Eustachio, B. 1564. *Opuscula anatomica.* Venice.
47. Everard, A. 1661. *Novus et genuinus hominis brutique animalis exortus.* Middleburg.
48. Fabricius ab Aquapendente, H. 1621. *De formatione ovi et pulli tractatus.* Padua.
49. Fallopius, G. 1561. *Observationes anatomicae.* Venice.
50. Fernel, J. 1554. *Medicina.* Paris.
51. Folkes, M. 1724. Some account of Mr. Leeuwenhoek's curious microscopes lately presented to the Royal Society. *Phil. Trans. Roy. Soc. London* 32:446.
52. Galen. 1821–33. *Opera omnia*, ed. C. G. Kühn. Leipzig.
53. Gassendi, P. 1658. *Opera omnia.* Lyons.
54. Gautier d'Agoty, J. 1750. *La zoogénie, ou génération des animaux.* Paris.
55. Geoffroy, E. F., and du Cerf, C. 1704. *Quaestio: Ergo hominis primordia vermis.* Paris.
56. Graaf, R. de. 1671. *Epistola ad virum clarissimum D. Lucam Schacht . . . de partibus genitalibus mulierum.* Leiden.
57. ———. 1672. *De mulierum organis generationi inservientibus tractatus novus, demonstrans tam homines et animalia, caetera omnia quae vivipara dicuntur, haud minus quam ovipara, ab ovo originem ducere.* Leiden.
58. Haighton, J. 1797. An experimental inquiry concerning animal impregnation. *Phil. Trans. Roy. Soc. London* Part 1:159.
59. Haller, A. von. 1746–52. *Disputationum anatomicarum selectarum.*
60. ———. 1757–66. *Elementa physiologiae corporis humani.* Lausanne.
61. Hartsoeker, N. 1678. Extrait d'une lettre de M. Nicolas Hartsoker écrit à l'auteur de Journal touchant la manière de faire les nouveaux microscopes, dont il a esté parlé dans le Journal il y a quelques jours. *J. Ŝcavans.* 29 August.
62. Harvey, W. 1651. *Exercitationes de generatione animalium.* London.
63. Henle, F. G. J. 1838. Ueber die Ausbreitung des Epithelium im menschlichen Körper. *Arch. Anat. Phys. Wiss. Med.* 2:103.
64. Highmore, N. 1651. *The history of generation, examining the severall opinions of divers authors, especially that of Sir Kenelm Digby in his discourse of bodies.* London.
65. Hildebrandt, F. 1828. *Lehrbuch der Physiologie.* Erlangen.
66. Hoffmann, C. 1645. *Institutionum medicarum libri sex.* Lyons.
67. Home, E. 1814–28. *Lectures on comparative anatomy; in which are explained the propagation in the Hunterian collection. Supplement to the foregoing lectures on comparative anatomy.* London.
68. Horne, J. van. 1668. *Suarum circa partes generationis in utroque sexu observationum prodromus.* Leiden.
69. ———. 1675. *Microcosmos, seu brevis manuductio ad historiam corporis humani . . . Accessit huic editioni Epistola ad virum celeberrimum D. D. Guernerum Rolfincium . . . observationum in sexus utriusque partibus genitalibus specimen exhibens.* Lyons.
70. Huygens, C. 1678. Extrait d'une lettre de M. Huyghens de l'Acad. R. des Sciences à l'auteur du Journal, touchant une nouvelle manière de microscope qu'il a apporté de Hollande. *J. Ŝcavans.* 15 August.
71. James, R. 1745. *A medical dictionary.* London.

72. Kerckring, T. 1671. *Anthropogeniae ichnographia.* Amsterdam.
73. ———. 1672. An account of what hath been of late observed by Dr. Kerkringius concerning eggs to be found in all sorts of females. *Phil. Trans. Roy. Soc. London.* No. 81, p. 4018.
74. Langly, W. 1674. *Observationes et historiae omnes & singulae e Guiljelmi Harvei libello De Generatione Animalium excerptae et in accuratissimum ordinem redactae: item Wilhelmi Langly De Generatione Animalium observationes quaedam. Accedunt ovi faecundi singulis ab incubationis diebus factae inspectiones; ut et observationum anatomico-med. decades quatuor; denique cadavera balsamo coniendi modus, studio Justi Schraderi.* Amsterdam.
75. Leeuwenhoek, A. van. 1679. Observationes de natis è semine genitali animalculis. *Phil. Trans. Roy. Soc. London* 12:1040.
76. ———. 1683. No title. *Phil. Trans. Roy. Soc. London* 13:74.
77. ———. 1722. Observations upon a foetus, and the parts of generation of a sheep· *Phil. Trans. Roy. Soc. London* 32:151.
78. ———. 1724. Of the generation of animals. *Phil. Trans. Roy. Soc. London* 32:438.
79. Le Grand, A. 1672. *Institutio philosophiae, secundum principia Renati Descartes.* London.
80. Lorenzini, S. 1678. *Osservazioni intorno alle torpedini.* Florence.
81. Maître-Jan, A. 1722. *Observations sur la formation du poulet où les divers changemens qui arrivent à l'oeuf à mesure qu'il est couvé sont exactement expliqués et representés en figures.* Paris.
82. Malpighi, M. 1686. *Opera omnia.* London.
83. Maupertuis, P. L. M. de. 1744. *Venus physique.* Leiden.
84. Mauriceau, F. 1668. *Des maladies des femmes grosses et accouchées.* Paris.
85. Méry, J. de. 1704. Sur la génération de l'homme par des oeufs. *Hist. Acad. Roy. Sci. Ann.* 1701. Paris.
86. Morgagni, J. B. 1719. *Adversaria anatomica omnia.* Padua.
87. Motte, G. M. de la. 1718. *Dissertations sur la génération.* Paris.
88. Müller, J. 1834–40. *Handbuch der Physiologie des Menschen für Vorlesungen.* Koblenz.
89. Muralt, J. 1677. *Vade mecum anatomicum sive clavis medicinae, pandens experimenta de humoribus, partibus, et spiritibus.* Basel.
90. Nuck, A. 1691. *Adenographia curiosa et uteri foeminei anatome nova cum epistola ad amicum de inventis novis.* Leiden.
91. Parisano, E. 1621. *Nobilium exercitationum libri duodecim. De Subtilitate.* Venice.
92. Peyer, J. C. 1685. *Merycologia.* Basel.
93. Piazzoni, F. 1644. *De partibus generationi inservientibus libri duo . . . Item Arantii de humano foetu libellus. Item Gregorii Nymmani de vita foetus in utero dissertatio.* Leiden.
94. Piccolomini, A. 1586. *Anatomicae preaelectiones.* Rome.
95. Plater, F. 1586. De mulierum partibus generationi dicatis tabulae iconibus illustratae structuram usumque explicantes. In *Gynaeciorum, sive de mulierum affectibus commentarii.* Basel.
96. Pouchet, F. A. 1847. *Théorie positive de l'ovulation spontanée et de la fécondation des mammifères et de l'espèce humaine basée sur l'observation de toute la série animale.* Paris.

97. Prévost, J. L., and Dumas, J. B. A. 1824. De la génération dans les mammifères, et des premiers indices du développement de l'embryon. *Ann. Sci. Nat.* 3:113.
98. Pulley, J. 1801. *An essay on the proximate cause of animal impregnation; being the substance of a paper read and discussed in the medical society at Guy's Hospital, in October 1799.* London.
99. Purkinje, J. E., and Valentin, G. G. 1834. Entdeckung continuirlicher durch Wimperhaare erzeugter Flimmerbewegungen, als eines allgemeinen Phänomens in den Klassen der Amphibien, Vögel und Säugethiere. *Arch. Anat. Phys. Wiss. Med.* p. 391.
100. Renzi, S. de. 1852–57. *Collectio Salernitana. Ossia documenti inediti e trattati di medicina appartenenti alla scuola medica Salernitana.* Naples.
101. Riolan, J. the Elder. 1578. *Ad librum Fernelii de procreatione hominis commentarius.* Paris.
102. Riolan, J. the Younger. 1618. *Anthropographia.* Paris.
103. ———. 1650. *Opera anatomica.* Paris.
104. Rösslin, E. 1513. *Der swangern frawen und hebammen roszgarten.* Hagenau.
105. Rufus. 1879. *Oeuvres, texte collationné sur les MSS, traduit pour la première fois en français avec une introduction. Publication commencée par Ch. Daremberg, continuée et terminée par Ch. É. Ruelle.* Paris.
106. Saumarez, R. 1798. *A new system of physiology, comprehending the laws by which animated beings in general, and the human species in particular, are governed, in their several states of health and disease.* London.
107. Sauvages, F. 1755. *Physiologiae elementa.* Amsterdam.
108. Sbaragli, G. G. 1701. *Exercitationes physico-anatomicae . . . quibus . . . accesserunt ad epistolares de recentiorum medicorum studio dissertationes appendix, et de vivipara generatione altera scepsis.* Bologna.
109. Scaliger, J. C. 1592. *Exotericarum exercitationum liber XV. de subtilitate, ad Hieronymum Cardanum.* Frankfurt.
110. Schrader, J. 1674. *Observationes et historiae omnes et singulae e Guiljelmi Harvei libello De Generatione Animalium excerptae et in accuratissimum ordinem redactae: item Wilhelmi Langly De Generatione Animalium observationes quaedam. Accedunt ovi faecundi singulis ab incubatione diebus factae inspectiones; ut et observationum anatomico-medicarum decades quatuor denique cadavera balsamo condiendi modus studio Justi Schraderi.* Amsterdam.
111. Schrader, F. 1681. Dissertatio epistolica da microscopiorum usu in naturli scientia et anatome. Göttingen.
112. Smellie, W. 1766. *A treatise on the theory and practice of midwifery.* London.
113. Soranus. 1894. *Die Gynäkologie, Geburtshilfe, Frauen- und Kinder-Krankheiten, Diätetik der Neugeborenen,* trans. H. Lüneberg. Munich.
114. Steno, N. 1667. *Elementorum myologiae specimen . . . Cui accedunt canis carchariae dissectum caput, et dissectus piscis ex canum genere.* Florence.
115. ———. 1910. De anatome rajae epistola. In *Opera philosophica,* 5, ed. Maar. Copenhagen.
116. Swammerdam, J. 1672. *Miraculum naturae, sive uteri muliebris fabrica.* Leiden.
117. Swedenborg, E. 1740. *Oeconomia regni animalis . . . Accedit introductio ad psychologiam rationalem.* London, Amsterdam.

118. Tauvry, D. 1690. *Nouvelle Anatomie raisonnée*. Paris.
119. Thomson, A. 1839. Generation. In *Todd's Cyclopedia of Anatomy and Physiology*. London.
120. Todd, R. B., and Bowman, W. 1845–56. *The physiological anatomy and physiology of man*. London.
121. Töply, R. von. 1902. *Anatomia Ricardi Anglici (ca. 1242–1252) ad fidem codicis MS. n. 1634 in Bibliotheca Palatina Vindobonensi. . . .* Vienna.
122. Vallisnieri, A. 1721. *Isotoria della generazione dell'uomo e degli animali se sia da' vermicelli spermatici, o dalle uova*. Venice.
123. Varolio, C. 1591. *Anatomiae, sive de resolutione corporis humani . . . libri IIII . . . Eiusdem Varolii & Hier. Mercurialis de nervis opticis, nonnullisque aliis, . . . epistolae*. Frankfurt.
124. Verduc, J. B. 1704. *A treatise of the parts of a humane body: With their respective uses; in which their particular functions are clearly and fully explain'd*, trans. J. Davis. London.
125. Vesalius, A. 1543. *De humani corporis fabrica libri septem*. Basel.
126. Vesling, J. 1659. *Syntagma anatomicum, commentariis illustratum A Gerardo Leonardi Blasio*. Amsterdam.
127. Wagner, R. 1834. *Lehrbuch der vergleichenden Anatomie*. Leipzig.
128. Wharton, T. 1656. *Adenographia: sive, glandularum totius corporis descriptio*. London.

Part I Development and Structure

1

Prenatal Development of the Oviduct *in Vivo* and *in Vitro*

D. Price/J. J. P. Zaaijer/E. Ortiz

Department of Zoology
University of Chicago
Laboratory for Cell Biology and Histology
University of Leiden, The Netherlands
Biology Department
University of Puerto Rico, Río Piedras

I. Survey of Embryological Studies

A. *Historical Background*

The oviducts (uterine tubes or Fallopian tubes) of adult mammals are highly specialized structures which derive embryonically from the cranial region of the primitive Müllerian ducts. Postnatally they assume, during the reproductive period in females, one of the most basic roles in the reproductive process—the assurance that sperm and eggs meet under optimal conditions for fertilization and that, subsequently, there is safe transport of the zygote.

The Müllerian (paramesonephric) ducts of male and female fetuses have been of intense interest to the embryologists ever since they were described by Johannes Müller in his classic publication in 1830. Their development in both sexes and their subsequent retention in females and degeneration in males have been studied in many species of mammals and in nonmammalian vertebrates. Studies of their counterparts, the Wolffian (mesonephric) ducts, have established that an ambisexual stage of the gonaducts exists in all vertebrate fetuses. Purely descriptive work was done for many years, but after Lillie (1917, 1923) proposed a hormonal theory for sex differentiation, a spate of experimental studies followed in which a variety of techniques was employed in many species of vertebrates. In mammals, critical studies on sex differentiation have demonstrated the importance of fetal testicular androgen in maintenance of Wolffian ducts in males, and have indicated that retrogression of Müllerian ducts in this sex may be caused by hormone from fetal testes. There is convincing evidence that the retention of Müllerian ducts in females beyond the ambisexual stage is hormonally independent. The mass of literature that has now accumulated has been the subject of frequent reviews, among them those by Willier (1939), Burns (1955, 1961), Jost (1953, 1965), Raynaud (1962), Wolff (1962), Wells (1962, 1965) and Price and Ortiz (1965).

In all these studies, prenatal development of the oviductal segment of the primitive Müllerian ducts has received far less attention than that of the uterus, and still less than the question of the contribution of the Müllerian ducts to the formation of the vagina (Zuckerman 1940; Raynaud 1962; Forsberg 1963). Descriptive studies of oviductal development are available for a few species, but little experimental work has been directed specifically to the question of hormonal influence. The purpose of the present paper is to review briefly some general aspects of early formation of Müllerian ducts, to outline the stages in oviductal development and compare the patterns in several mammals, and to report our studies on the development of the oviduct *in vivo* and *in vitro* in the guinea pig fetus.

B. *General Pattern of Müllerian Duct Development in Mammals*

1. *Formation of the Ostium and the Early Growth of the Ducts in Both Sexes.* The early stages in the development of the mammalian Müllerian ducts suggest that a close relationship exists between the Müllerian ducts and the nephric (Wolffian) system, although the evidence is not so direct as in lower vertebrates (Burns 1955; Witschi 1959). The ostium in mammals is considered to form as an invagination of the coelomic epithelium in a special area—the funnel field on the primitive urogenital ridge. One ostium usually forms on each side, but supernumerary ones may form and may remain as hydatid cysts or as appendages associated with the oviduct in females or with the testis or epididymis in males.

After the formation of an ostium, the blind tip of the forming Müllerian duct becomes very closely applied to the mesonephric (Wolffian) duct and then grows caudally to the urogenital sinus in close association with the preexisting nephric duct. So close is this association that it suggests dependence of the Müllerian duct on the Wolffian duct for directional guidance or an even more intimate relation, such as actual contribution of cells from the Wolffian duct to the growing Müllerian duct (Gruenwald 1941; Burns 1955).

In mammals, formation of the ostium, growth of the Müllerian duct, and the formation of a uterovaginal canal (or prostatic utricle) are usually similar in the two sexes, and progress at essentially the same speed. However, in some species the cranial region of the Müllerian duct in males is not so well developed as the corresponding region in females (Price and Ortiz 1965).

2. *Fate of the Oviductal Segment of the Müllerian Duct.* The end of the ambisexual stage of the gonaducts in the male fetus is marked by degeneration of the Müllerian ducts. This process usually begins at the cranial end, although, typically, a small piece of the duct remains as an embryonic rest, the so-called testicular appendage. In females, the cranial portion, the prospective oviductal region, soon begins to undergo a series of morphogenetic and histogenetic changes. It becomes distinguishable from the uterus by the development of coils and by an abrupt difference in size of the two segments as the uterus enlarges in diameter. Differentiation of epithelium and muscles then progresses, mucosal folds are elaborated, fimbriae develop, the segments—infundibulum, ampulla, isthmus—are identifiable and the uterotubal junction is marked by folds. The degree to which this differentiation progresses prenatally differs rather widely among mammalian species.

In adult females, there are well-recognized differences between species with regard to structural details of the oviductal segments, the complexity of the uterotubal junction, and the degree of development of the ovarian bursa (Strauss 1964, 1966). In this paper, the embryonic development of the oviduct in five species—hamster, mouse, rat, guinea pig, and man—will be discussed and compared. These species differ, more or less, in anatomical structure of the oviduct and the ovarian bursa. They differ significantly in the length of the gestation period and consequently in the degree of maturity at birth. As a model, oviductal development in the fetal guinea pig will be described in detail from our published and unpublished observations and compared with developmental stages that have been described for the other four species.

II. Development of the Oviduct *in Vivo*

A. *Morphogenesis and Histogenesis in the Guinea Pig*

1. *Background.* The early development of the Müllerian ducts in fetal guinea pigs had been only slightly touched upon (Scott 1937) until our recent studies of sex differentiation of the Müllerian and Wolffian duct systems. A timetable of progressive development of Müllerian ducts in females (and retrogression in males) was presented for the span from 22 days of fetal life, the beginning of ostium formation, to 38 days, the end of the critical stages in sex differentiation (Price and Ortiz 1965; Price, Ortiz, and Zaaijer 1967). Our detailed study of the morphogenesis and histogenesis of the oviduct, from the time when it is first identifiable as a specific segment of the Müllerian duct to the stage just after birth, will be presented here.

2. *Prenatal and Neonatal Development.* The first indication of morphogenesis of the oviductal segment occurs by 31 days of age with the inception of coiling—a process which advances rapidly as the oviduct increases in length (Table 1). The stages in coiling can be followed clearly in Figures 1*a*–*e*. The simple Müllerian duct at 29 days (Fig. 1*a*) develops slight undulations in the oviductal region by 31 days (Fig. 1*b*) which by 34½ days (Fig. 1*c*) constitute a series of wavy folds. At 36 days, there are deep coils (Fig. 1*d*) which enlarge and become loose loops within the mesosalpinx at 54 days (Fig. 1*e*) and appear as though suspended in a fluid-filled bursa. Subsequent increase in length and diameter of the oviduct results, by 64 days, in tightly packed coils which partially overlie the ovary.

In the guinea pig, the development of the ovarian bursa follows rather closely that of the rat (Kellogg 1941), although the bursa of the guinea pig is incomplete and remains an open sac (Sobotta 1917). As in the rat, the bursa seems to be formed mainly by mesosalpinx (mesotubarium). There is no indication of bursal formation at 30–31 days (Fig. 2*a*), but the primitive mesosalpinx is forming. This double-walled mesentery which carries the oviduct presumably receives contributions from the peritoneum of the degenerating Wolffian body and from the tubal ridge, as described for the rat. By 34½ days (Fig. 2*b*) the mesosalpinx is being drawn mesially and ventrally across the surface of the ovary as the ovary changes position, so that by 54 days (Fig. 2*c*) the bursal sac is well formed and the infundibular funnel lies against the mesial side of the ovary, to which the fimbriae become partially attached (Fig. 2*f*). The anatomical relationships that obtain at 64 days and at 2 hr after birth are similar to those described for the adult (Sobotta 1917).

Soon after the oviduct starts to coil and is identifiable as a discrete region, its caudal limits are demarcated by an enlargement of the uterine segment which results in a rather abrupt increase in diameter at the uterotubal junction (Figs. 1*c*, *d*). At the same stage, morphogenesis of the infundibular end of the oviduct begins with the formation of primordia of fimbriae.

Histogenesis, in the meantime, has been progressing hand-in-hand with the obvious morphogenetic changes (Table 1). The cells of the simple primitive epithelium of the duct (Fig. 2*d*) undergo rapid mitotic activity. As the oviduct grows in length and increases in diameter, the epithelium becomes higher (Fig. 2*e*) and small crypts appear (Fig. 2*g*) which presage the development of longitudinal folds. By 38–40 days, the infundibular and ampullary regions are differentiating, and the epithelium is pseudostratified and contains peg

TABLE 1

DEVELOPMENTAL STAGES IN THE OVIDUCT OF THE FETAL GUINEA PIG

FETAL AGE IN DAYS	
	Müllerian Duct
22–23	Beginning to develop; ostium and anterior segment present
26–27	Anteroposterior growth complete; caudal ends reach wall of the urogenital sinus; uterovaginal canal almost formed
29–30	Uterovaginal canal completely formed
30–31	Slight undulations in oviductal segment (Fig. 1*b*)
	Oviduct
34–36	Demarcated from uterine segment (Figs. 1*c*, *d*) Coiling more advanced, deep undulations; ovarian bursa beginning to develop (Figs. 1*c*, *d*) Infundibular fimbriae beginning to develop Slight crypts in epithelium mark beginning of fold formation (Fig. 2*g*)
38–40	Coils and bursa much further developed (Fig. 5*c*) Infundibular fimbriae more complex; epithelium mainly pseudostratified, some peg cells Ampulla with wider lumen, epithelium mainly pseudostratified, some peg cells Epithelium throughout length of duct has 4 deep crypts marking off longitudinal folds; crypts deeper near junction with uterus
53–54	Extensive growth and coiling of duct have occurred within a well-developed bursa (Fig. 1*e*) Infundibular fimbriae with further branching; many peg cells, many ciliated cells Ampulla with complex epithelial folding; epithelium with many peg cells, a few ciliated cells Isthmus a distinct region with star-shaped lumen resulting from 5 deep longitudinal folds of mucosa; epithelium simple columnar, no ciliated cells; longitudinal and circular muscle layers differentiated
63–64	Ampulla with more complex mucosal folding; evidence of secretion Isthmus with higher epithelium, no cilia Further development of folds at the uterotubal junction
2 hr after birth	Ampullary epithelium higher and becoming simple columnar (Fig. 2*h*)

cells. The isthmus is less developed, but marked crypts in the caudal region represent the beginning of the deep folds characteristic of the uterotubal junction. Further advances in all aspects of differentiation are rapid (Table 1). The epithelium acquires ciliated cells in the infundibulum and ampulla by 53–54 days, and these increase in number during fetal life, although no cilia appear in the isthmus. Secretion is evident in the lumen by 63–64 days. At 2 hr after birth, the fimbriae are well developed and the ampullary region is characterized by complicated mucosal folds (Figs. 2*f*, *h*) lined by a high columnar epithelium containing many ciliated and nonciliated cells and conspicuous peg cells. Secretory activity is indicated, and the oviduct appears very well developed.

The important events in the early development of the Müllerian ducts and the later transformation of their cranial region into oviducts can be arranged chronologically as

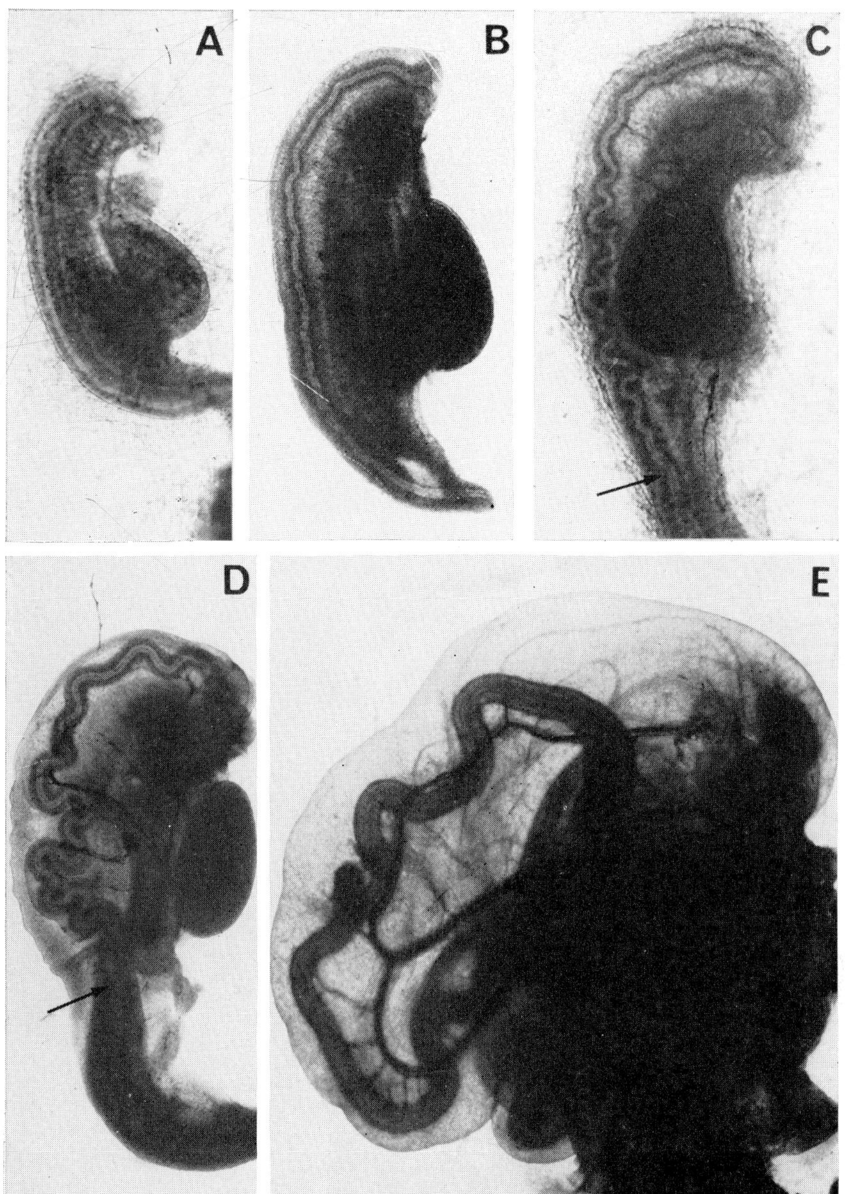

FIG. 1. Developmental series of female reproductive tracts from guinea pig fetuses of different ages. Tracts were excised, stretched on lens paper on the culture medium and photographed. Figures show the right side of each tract. Magnifications: *a*, *b*, and *c*, ×16; *d* and *e*, ×13. (*a*) From fetus 29 days old. Note the conspicuous Müllerian duct and the slender Wolffian duct, which is beginning to degenerate. (*b*) From fetus 30–31 days old. Note the slight undulations in the oviductal segment of the Müllerian duct. (*c*) From fetus 34½ days old. Coiling is more advanced, and the oviduct appears demarcated from the uterine segment, as indicated by the arrow. The ovarian bursa is beginning to develop. (*d*) From fetus 36 days old. Note increase in diameter of oviduct, complex coiling, conspicuous bursa, and further enlargement of the uterus. (*e*) From fetus 54 days old, showing advanced development of the oviduct and the bursa.

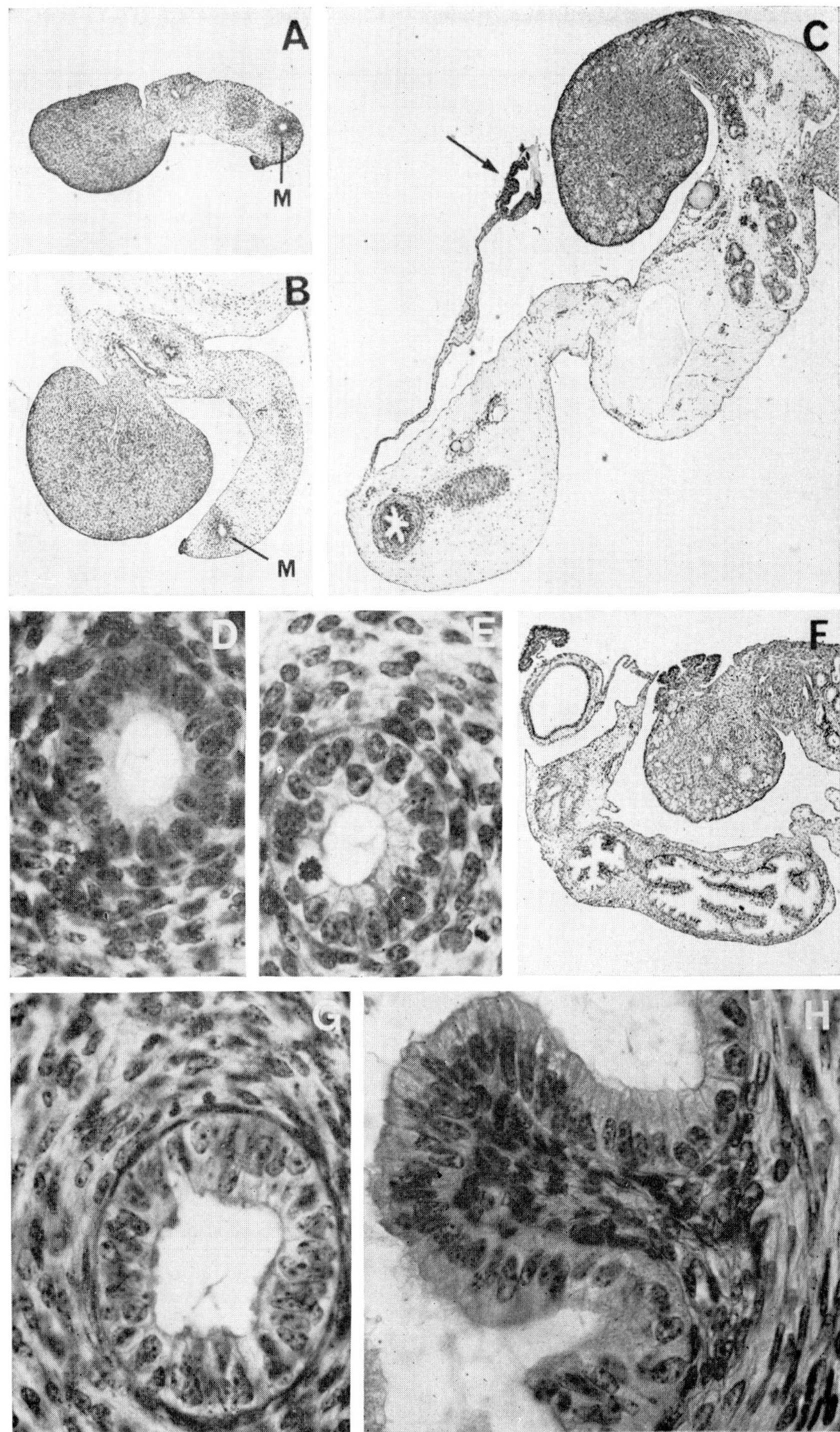

FIG. 2. Photomicrographs of cross sections through various regions of the reproductive tract of female guinea pig fetuses at different ages. Magnifications: *a*, *b*, *c*, and *f*, ×35; *d*, *e*, *g*, and *h*, ×500. (*a*) Tract from fetus 30–31 days old, through the middle of the ovary. Note the small diameter of the Müllerian duct (*M*). (*b*) Tract from fetus 34½ days old, through the middle of the ovary to show the developing bursa, as represented by the curvature in the growing mesosalpinx. (*c*) Tract of fetus 54 days old, through the ovary, showing extensive growth and development of the oviduct and the ovarian bursa. The arrow points to the infundibulum and ostium; at the lower left is a cross section of the isthmus. Note the conspicuous mesonephric tubules. (*d*) Müllerian duct from fetus 29 days old, showing undifferentiated epithelium and surrounding mesenchyme. (*e*) High-power view of Müllerian duct shown in *b*, at 34½ days. (*f*) Tract of newborn 2 hr old, through the ampullary region of the oviduct to show complex folds and fimbriae (upper left), some of which are attached to the ovary. (*g*) Oviduct from fetus 36 days old to show the beginning of fold formation as indicated by 4 slight epithelial crypts. (*h*) High-power view of one mucosal fold in ampullary region of the oviduct shown in *f*, at 2 hr after birth. Note the high, simple columnar epithelium.

a sequential pattern including certain other landmarks (Fig. 3). It is clear then that the oviducts are established as such and have begun their typical development—morphogenesis and histogenesis—by the end of the first half of the gestation period.

B. *Comparative Stages of Oviductal Development in Other Mammals*

1. *Hamster, Mouse, and Rat.* The hamster is characterized by an unusually short gestation period of 16 days. During the prenatal period, no morphogenesis of the oviductal region of the Müllerian duct takes place (Ortiz 1945), and the oviduct is not demarcated from the uterus until after birth (LaVelle 1951).

In the mouse, a number of developmental events occur in the oviduct just before birth (Agduhr 1927). These include small folds (fimbriae) in the infundibular region, primordia of folds in the enlarged ampullary region, and an enlargement indicating where demarcation of uterus and oviduct will be (the sharp boundary is not formed until after birth). The epithelium shows some differentiation—increase in height, pseudostratification, presence of peg cells, and evidence of secretory activity. Cilia and coiling of the duct are not found until after birth.

The rat has a limited degree of oviductal differentiation prenatally. Postnatal study of the oviducts (Kellogg 1945) revealed that anlagen of the folds can be found at birth and that they are more prominent at the opposite ends of the duct. This suggests the first evidence of formation of fimbriae in the infundibular region and the beginning of the folds of the uterotubal junction at the caudal end. The epithelium contains columnar cells of varying height and is pseudostratified or stratified toward the upper end of the oviduct. However the epithelium is, in general, rather undifferentiated. Coiling begins as slight undulations at 3 days after birth; the uterotubal junction is sharply defined early in the first week, and cilia develop by 6 days.

In all three of these short-term rodents it should be noted (Fig. 3) that the developmental stages in morphogenesis and histogenesis fall into essentially the same sequential pattern as that described for the guinea pig; the marked difference is in the degree of differentiation of the oviduct that is achieved prenatally in the long-term rodent.

2. *Man.* It is difficult to compare the stages of development of the human oviduct with the findings in laboratory animals because data are sparse and the estimation of fetal age is inexact. From various sources (Hunter 1930; Glenister and Hamilton 1963; Horstmann and Stegner 1966; Zaaijer, unpublished) comes evidence that the major steps in morphogenesis and histogenesis of the oviduct—coiling, formation of folds and fimbriae—all occur during the first half of the gestation period. Ciliated cells, already present in the epithelium of the ampullary region at 4 lunar months, are numerous by the 7th month. No evidence of secretory activity is detectable at that time in the nonciliated cells. In the newborn, ciliated cells and secretory cells are well developed, and secretory activity is unmistakable (Stegner 1961, 1962). Thus in man, as in the guinea pig, most of the steps in oviductal differentiation are reached relatively early in the gestation period. By birth, the oviduct is very well developed.

The order in which the various developmental stages occur in man resembles that found in the four rodents (Fig. 3). It illustrates quite diagrammatically an orderly pattern in the organogenesis of the mammalian oviduct. An equally orderly pattern of events in sex differentiation in these four rodents and in the rabbit has been described (Price and Ortiz 1965).

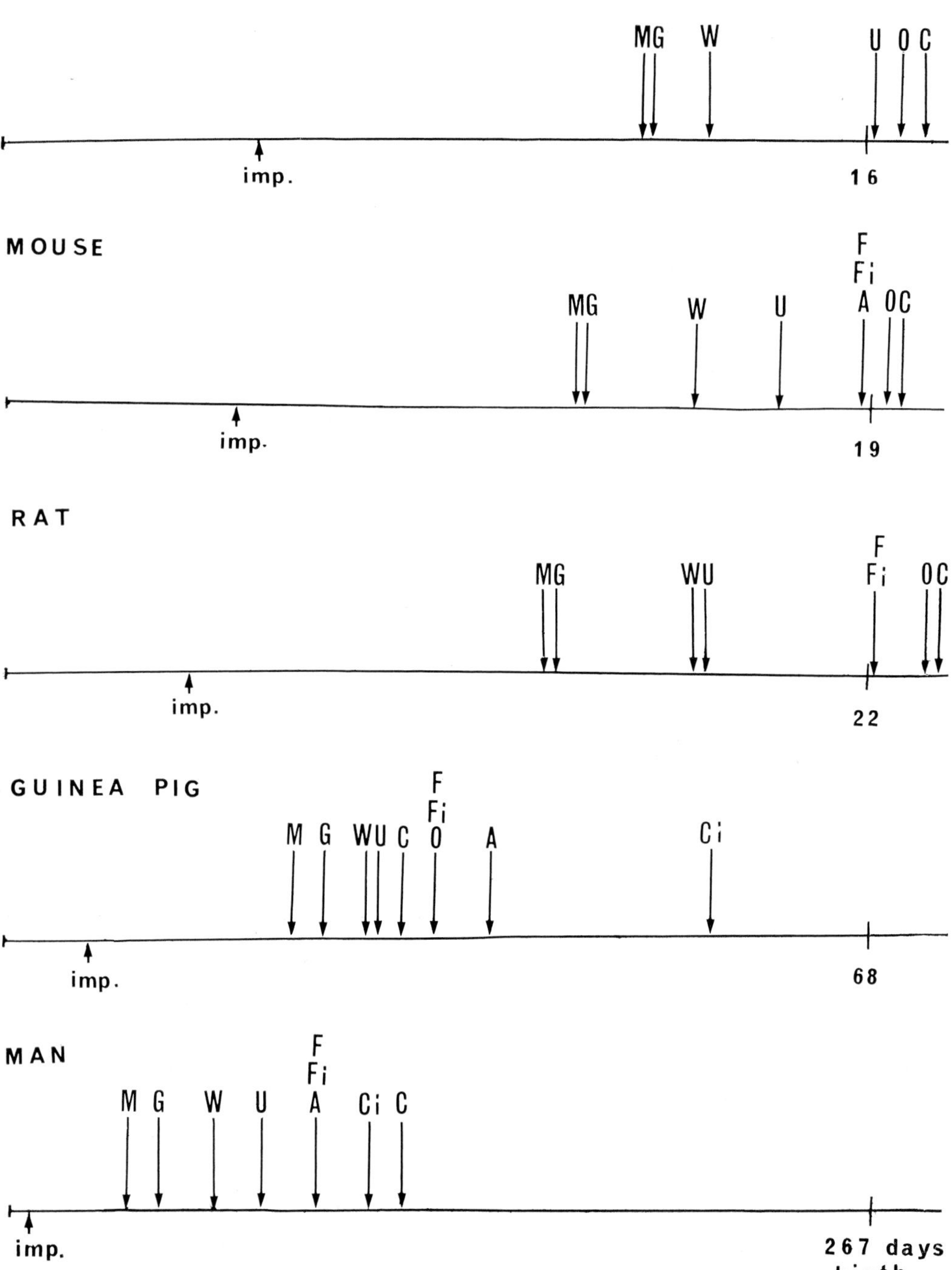

FIG. 3. Comparative sequence of developmental stages in the female reproductive tract of different mammals. Arrows mark the beginning of certain important events, such as implantation (*imp.*), Müllerian duct development (*M*), gonadal sex differentiation (*G*), uterovaginal canal (*U*), and Wolffian duct degeneration (*W*), as well as the beginning of the following events in oviductal differentiation: *O*, demarcation from uterus; *C*, coiling; *F*, epithelial folds; *Fi*, fimbriae; *A*, ampulla differentiation; *Ci*, ciliated cells.

The advanced degree of differentiation of the oviduct that is found at birth in guinea pig and man suggests that hormonal stimulation might have played a part, at least just before birth.

III. The Question of Influence of Hormones on Prenatal Development

A. *Evidence for Stimulating Effects of Hormones in Man and Guinea Pig*

Early studies, reviewed by Courrier (1945) and Price (1947), reported evidence that the reproductive tract of the human fetus is stimulated prenatally by hormones. The fetal vagina, uterus, and mammary glands show the effect of hormonal stimulation by the 7th lunar month. In the newborn, the stimulation is very marked, but it disappears later. It was suggested that estrogen, progestin, gonadotropin, and mammotropin might be implicated, and that maternal, placental, and fetal sources might contribute the hormones. Evidence obtained with chemical methods now attests to the ability of the human placenta, fetal testes, and fetal adrenals to synthesize a variety of steroids (Bloch 1964; Solomon 1966).

Whatever their nature and their source, the hormones affecting the female fetus are present by the 7th month, and the oviduct may be responding. The functional differentiation present in the oviduct at birth may reflect hormonal stimulation, although it need not be assumed that hormones play an essential role in oviductal development. The most important stages in morphogenesis and histogenesis are completed by the end of the 5th month, and it appears probable that early development is hormone-independent.

The guinea pig, also, shows clear evidence at birth of prenatal stimulation of the reproductive tract (Courrier 1945). The newborn female is in a state which has been described as a temporary puberty and is characterized by uterine hypertrophy and marked vaginal changes similar to those observed in the human female at birth. The stimulation of the vagina and the uterus was ascribed mainly to female hormone from the mother's ovaries.

Additional evidence of effects of hormones prenatally comes from the observation of a special type of behavior—the heat response—in newborn female and male guinea pigs when they are stroked (Boling *et al.* 1939). Responses can be elicited almost immediately after birth, but for only a period of 2 or 3 hr. Significantly, ovariectomy of the mother does not abolish the heat response in the newborn, and this rules out the maternal ovary as a source of the hormones that stimulate the behavioral response.

Hormones from some source are certainly present in the female prenatally, but the important question is whether they are there early in gestation and might be responsible for any important aspect of oviductal development. In our studies, we noted that the uterus of fetuses about 60 days old appears large and hyperemic, but no hyperemia was observed at younger ages. This indicates that the uterus is responding to hormones in late gestation and suggests that the oviduct may also be responding. The absence of uterine hyperemia in younger fetuses may mean that hormones were not present then in effective amounts and provides some evidence against the idea that hormonal intervention is important in the early organogenesis of the oviduct.

The source of the hormones that stimulate the vagina, uterus, and, probably, oviduct in late fetal life in the guinea pig is an open question. Obvious sources, other than maternal, would be the placenta, fetal ovary, or fetal adrenal gland. The guinea pig placenta was found incapable of synthesizing estrogen from ^{19}C substrates (Ainsworth and Ryan 1966) or converting androstenedione to dehydroepiandrosterone (Bloch and Newman 1966).

Ovaries and adrenals of fetal guinea pigs have apparently been little studied for their capacity for steroid metabolism. However, the presence of 3β-hydroxysteroid dehydrogenase activity was demonstrated in the fetal testis and adrenal but not in the fetal ovary (Price, Ortiz, and Deane 1964).

B. *Evidence for Hormone Secretion in Organs of Fetal Guinea Pigs*

1. In Vitro *Method for Detecting Androgenic Secretion.* A test was devised for detecting the presence of androgenic hormone in gonads and adrenals of fetal guinea pigs in order to explore the problem of secretory activity in the developing organs (Zaaijer, Price, and Ortiz 1966). For this test, the organ is explanted and cultured in close contact with an explant of ventral prostate tissue from a young (postnatal) rat. The prostate, a highly androgen-dependent organ, retrogresses rapidly in the absence of male hormone *in vitro* as well as *in vivo.* Thus, the maintenance or retrogression of normal histological structure provides dependable criteria for judging the results of the test.

TABLE 2

SUMMARY OF RESULTS OF ANDROGENIC TEST ON ORGANS OF FETAL GUINEA PIGS AT DIFFERENT AGES

ORGANS	NUMBER OF EXPLANTS AND RESULTS AT:		
	22 to 36 Days	41 to 46 Days	60 to 62 Days
Adrenal of female	63++	3++	16++
Ovary	62−	1−; 14+	22++
Adrenal of male	50++	3++	12++
Testis	63++	8++	14++
Mesonephros	10−		
Metanephros	24−		
Anterior pituitary			16−

SOURCE: Zaaijer, Price, and Ortiz 1966; Ortiz, Price, and Zaaijer 1966; Ortiz, Zaaijer, and Price 1967.
NOTE: ++, markedly androgenic; +, mildly androgenic; −, not androgenic.

Our method of organ culture is a modified watch-glass technique using disposable dishes (Falcon Plastics) and a medium containing 3 parts chick embryo extract, 1 part cock plasma, and 4 parts Hanks' saline solution. Organs are dissected from the fetus and put on a lens paper raft on top of the medium in the culture dish. Next to the organ is placed a piece of prostate approximately 1 mm^3 in size, from a rat 21 to 23 days old. Culturing is done at a temperature of 35°C, and after 3 days the explants are washed in Hanks' solution and transferred, on their lens paper raft, to fresh medium. At the end of the culture period of 5 days, the explants are fixed, sectioned, and studied. Androgenicity of the organs is judged by the histological structure of the prostatic tissue.

2. *Results of Androgenic Test.* The results of these tests are presented in Table 2. They demonstrate that hormone with androgenic activity is present in testes and in adrenals of fetuses ranging in age from 22 to 62 days. On the other hand, control organs, mesonephros, metanephros, and anterior pituitary have no effect on the prostate, which retrogresses to the same degree as when it is cultured alone. The ovary presents particularly interesting results—no androgenic activity between 22 and 36 days of age, mild androgenicity by 41–46 days and marked androgenic effects at 60–62 days. We conclude from this

that the ovary of the guinea pig fetus is metabolically active in the last third of the gestation period (Ortiz, Zaaijer, and Price 1967). What hormones are being secreted is not known, but the androgenic steroids produced as steps in the biosynthetic pathways to estrogen seem probable. There is no direct evidence for estrogen secretion by the ovary, but the hyperemia of the uterus at 60 days and the degree of differentiation of the oviduct toward birth suggest an estrogen source.

These results show that, although testes and adrenals of fetal guinea pigs have the ability to secrete hormones from very early stages, the ovary apparently does not possess this capacity until late in fetal life.

IV. Development of the Guinea Pig Oviduct *in Vitro*

A. *Technique of Culturing*

Organ culture of fetal reproductive tracts has proved a useful method for studying sex differentiation in the fetal rat (Pannabecker 1957; Price and Pannabecker 1959), the mouse (Brewer 1962) and the guinea pig (Price, Ortiz, and Zaaijer 1967; Ortiz, Zaaijer, and Price 1967). Isolating the whole reproductive tract which is developing in a complex fetal environment and culturing it in an environment free from placental and maternal hormones have obvious advantages. The development and differentiation of the gonaducts can be observed and studied under a variety of hormonal conditions. This method of organ culture provides a means for analyzing the factors that are operating in sex differentiation *in vivo* and for observing the responsiveness of the ducts to fetal hormones.

In the experiments to be discussed here, the details of technique are identical to those described for the androgenic test with regard to culture dishes, medium, and oven temperature. The reproductive tracts of female fetuses were dissected, trimmed, and laid out (either whole or as half tracts) on lens paper on the medium. The tracts were cultured with the ovaries removed or left *in situ* or replaced by adrenals or testes. When the tracts were to be cultured without ovaries, these were cut away with cataract knives before explantation. If testes or adrenals were to be substituted for ovaries, they were explanted next to the tract in the position of the excised ovary. Transfers were made every 3 days, and the culture period terminated at 9 to 11 days, when explants were fixed for microscopic study.

On the day of explantation, the image of the tracts was traced in detail by projecting it on transparent paper. The tracts were also photographed. Tracing and photography were repeated every 3 days and at the termination of the experiment, just before the explants were fixed. In this way, the sequential changes in size (length and width) in the gonaducts and changes in configuration could be followed.

The experiments were done in two series: in one, the reproductive tracts were from fetuses 26–27 days of age (the ambisexual stage in the gonaducts); in the second, they were from fetuses 29–30 days old.

B. *Results on the Oviduct*

Studies on sex differentiation of the Wolffian and Müllerian ducts were made on these cultured tracts, and the results were reported by Price, Ortiz, and Zaaijer (1967) and Ortiz, Zaaijer, and Price (1967). We now present a more detailed analysis of the morphogenesis and histogenesis of the oviduct in the female tracts. These results are summarized in Table 3.

The Müllerian duct as a whole was equally well maintained in all explants at the two ages under all endocrine conditions. However, morphogenesis of the oviductal segment, as shown by coiling, was limited to oviducts in the older series of 29–30 days. In this series, the coiling of the oviductal region during the culture period was essentially similar in tracts cultured in the absence of the ovary or with the ovary left *in situ* (Fig. 4*b*) or with testes (Fig. 4*e*) or adrenals (Figs. 5*b*, *e*). The development of oviductal coiling in culture can be followed by comparing the simple Müllerian duct at 29 days (Fig. 1*a*), the age at explantation, with the undulating configuration that has developed in the oviductal segment after 3 days of culture (Fig. 5*a*). These undulations are far more marked than in the uncultured tract at 30–31 days (Fig. 1*b*). With another 6 days of culture (Fig. 5*b*), deep coils are evident, and the diameter of the oviduct is similar to that of the uncultured tract at 38 days of age (Fig. 5*c*). The diameter increase in culture is more apparent in cross section, as can be observed in the figures.

TABLE 3

MORPHOGENESIS AND HISTOGENESIS OF OVIDUCTS IN EXPLANTED REPRODUCTIVE TRACTS OF FETAL GUINEA PIGS (TRACTS CULTURED WITHOUT GONADS OR WITH OVARIES, TESTES, OR ADRENALS)

Number of Tracts, Age, and Culture Period	Results
9 Explanted at 26–27 days, cultured until 35–38 days	Similar under all endocrine conditions: 2× diameter increase No increase in length, no coiling Beginning formation of tiny fimbriae Very slight epithelial crypts No differentiation into regions
11 Explanted at 29–30 days, cultured until 38–40 days	Similar under all endocrine conditions: More than 2× diameter increase Approximately 2× increase in length, marked coiling Marked development of fimbriae Ampulla enlarged, epithelium pseudostratified with peg cells, marked crypts Isthmus with slight epithelial crypts

The pronounced mitotic activity that is so evident in the epithelium results not only in diameter increase but also in marked increase in length. In some explants, for example the one in Figure 5*e*, both right and left oviducts doubled their length after explantation. In contrast, the uterine segment grows very little in culture, and the abrupt increase in diameter (Figs. 1*c*, *d*) that marks the cranial boundary of the uterine horn never appears.

Differentiation progresses in culture, but it is more limited in tracts explanted at the younger age (Table 3). In the older series, the stage of development reached by oviducts in culture by 38–40 days is similar in kind, if not in degree, to the differentiation at the same age *in vivo* (Table 1). The ampullary region undergoes more differentiation in culture than the isthmus.

A striking point is that coiling of the oviduct progresses in a symmetrical pattern so similar to the normal, in spite of the fact that the primitive mesosalpinx does not develop further and no bursa is formed in culture.

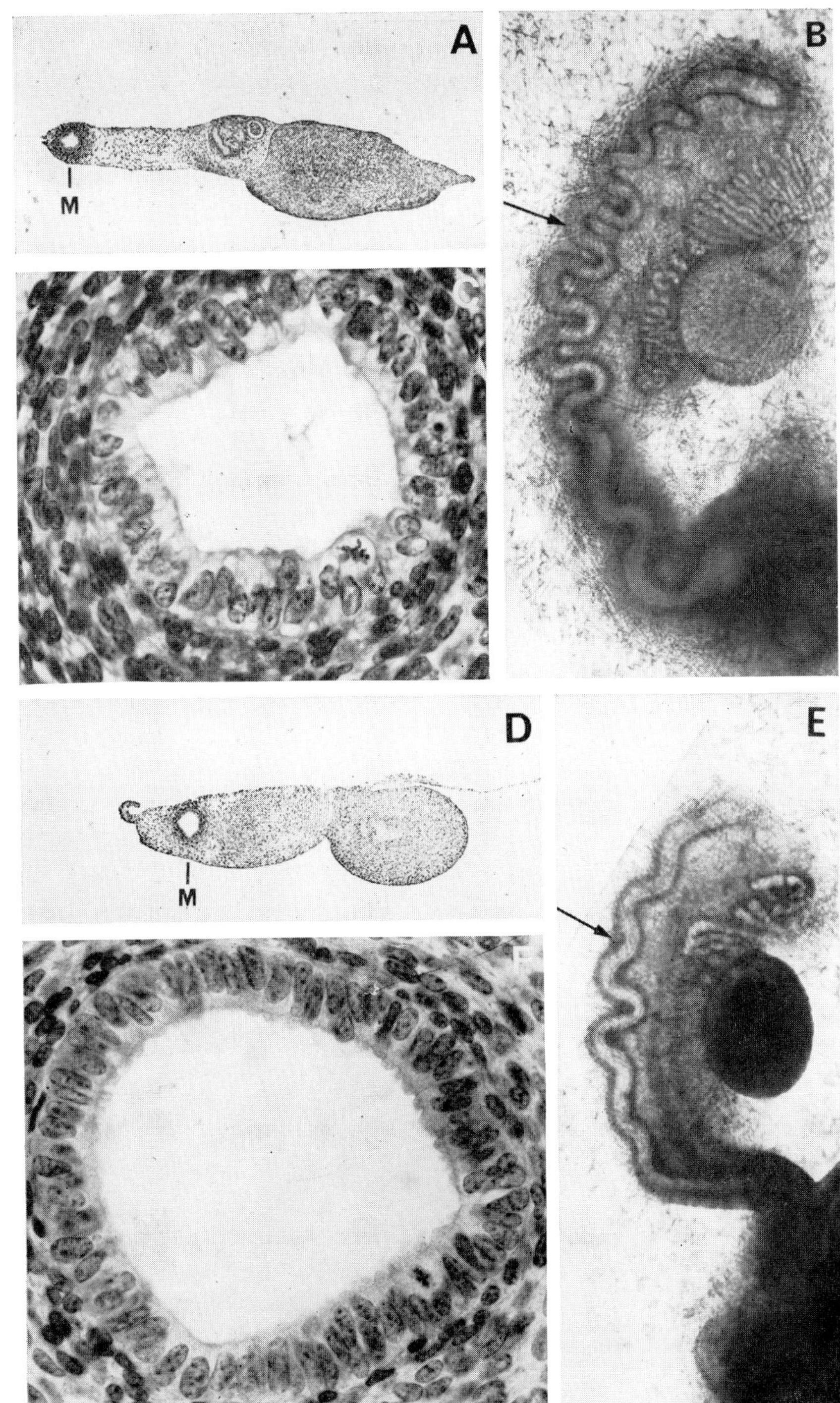

FIG. 4. Development of the guinea pig oviduct in culture. Explanted reproductive tracts from fetuses 29 days of age, cultured 9 days; age, 38 days. Magnifications: *a* and *d*, ×35; *b* and *e*, ×16; *c* and *f*, ×500. (*a*) Cross section of tract in *b*, at the level of the arrow, showing Müllerian duct (*M*), mesonephric tubules and ovary. (*b*) Half tract cultured with its own ovary *in situ*. There is marked increase in length and diameter of oviduct in culture over the initial stage of 29 days, shown in Fig. 1*a*. Note development of coils. (*c*) High-power view of cross section of oviduct in *a* to show diameter increase over the stage at 29 days shown in Fig. 2*d*. Crypts are similar to those *in vivo* at 36 days (Fig. 2*g*). (*d*) Cross section of reproductive tract in *e* at the level of the arrow, showing Müllerian duct (*M*) and testis. The Wolffian duct was not maintained. (*e*) Right side of whole tract cultured with testes of littermate. (*f*) High-power view of cross section of oviduct in *d*. Compare with *c* and Figs. 2*d*, *g*.

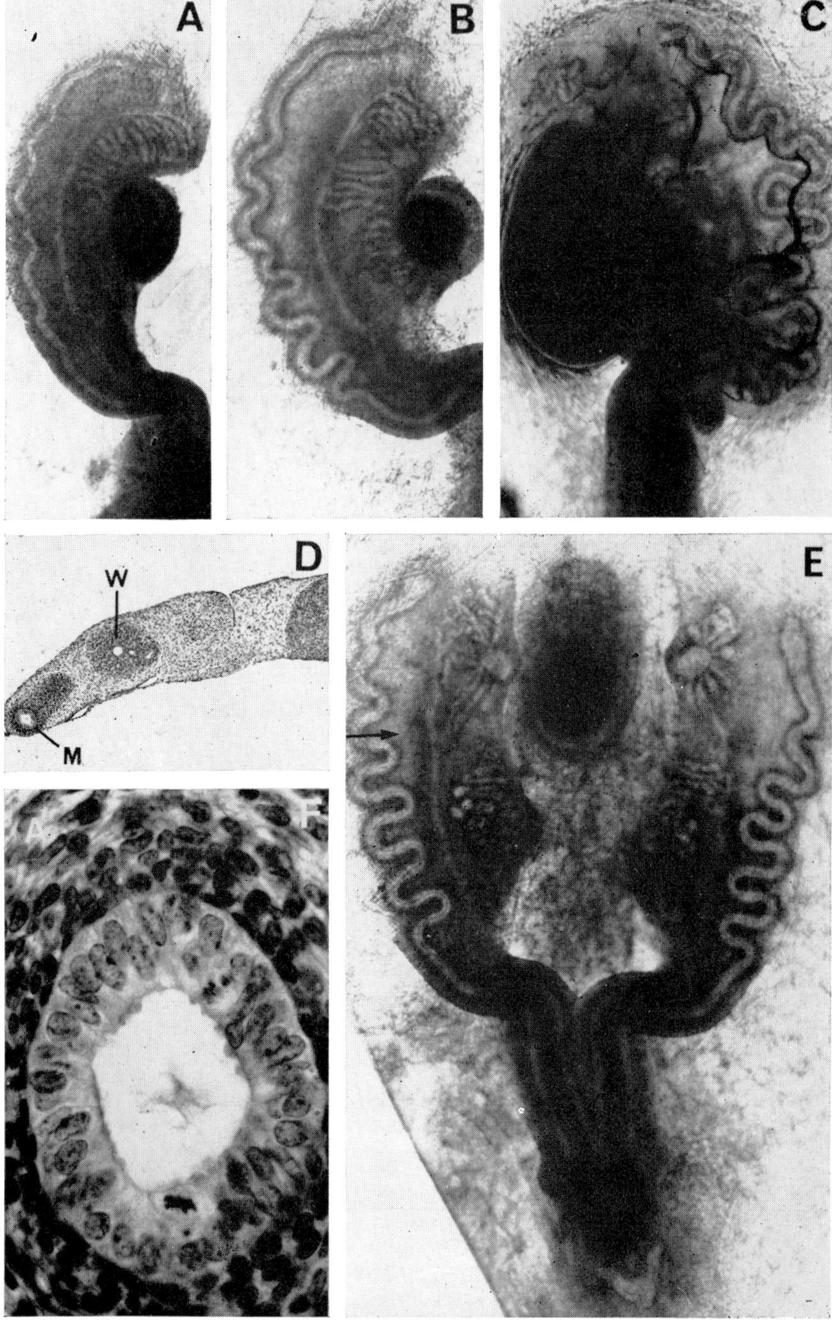

FIG. 5. Growth and coiling of the guinea pig oviduct in culture as compared with development *in vivo*. Magnifications: *a*, *b*, *c*, and *e*, ×16; *d*, ×35; *f*, ×500. (*a*) Right side of tract explanted at 29–30 days, after the first 3 days of culture with 2 pieces of adrenals from the same fetus. Compare degree of oviductal coiling with that in Figs. 1*b*, *c*. (*b*) The same explant as in *a*, after 9 days of culture, age, 38 days. Note maintenance of the Wolffian duct. Compare oviductal diameter and coiling with those in the uncultured tract at 38 days (*c*). (*c*) Portion of uncultured reproductive tract from female fetus to show degree of oviductal growth and coiling *in vivo* at 38 days of age. (*d*) Cross section of cultured tract in *e*, at the level of the arrow, showing Müllerian duct (*M*) and partial maintenance of Wolffian duct (*W*). (*e*) Whole female tract explanted at 29 days and cultured for 11 days with one piece of adrenal from the same fetus. Age, 40 days. Compare degree of growth and coiling of the oviduct with that *in vivo*, shown in *c*. (*f*) High-power view of cross section of oviduct in *d*, showing approximately 2× diameter increase over the initial stage at 29 days (Fig. 2*d*) and slight crypts in the epithelium.

C. *Discussion*

Our studies on the primitive Müllerian duct of the guinea pig have shown that early stages in the maintenance and development of this duct are independent of hormones from maternal, placental, and fetal sources (Price, Ortiz, and Zaaijer 1967; Ortiz, Zaaijer, and Price 1967). The female Müllerian duct is consistently maintained under all experimental conditions in reproductive tracts explanted either at the ambisexual stage in sex differentiation of the gonaducts or later. This finding for the guinea pig fetus corroborates earlier observations in cultured reproductive tracts of the rat and mouse (Price and Ortiz 1965; Brewer 1962).

The marked ability of the oviductal segment of the Müllerian duct to undergo growth and differentiation when isolated from the fetal environment suggests that the early stages of development *in vivo* are hormone-independent. This, of course, does not preclude the possibility that hormones in the fetus may have some influence on growth and differentiation at early or later stages, but it indicates that hormone influence is not essential for the early organogenesis of the oviduct.

When the tracts were explanted with testes or adrenals, the androgenic hormone that stimulated the Wolffian ducts in some explants (depending on the age), as in the one shown in Figures 5*b*, *e*, had apparently no marked stimulating or inhibiting effects on the oviducts.

This might be interpreted as inability of the fetal oviduct to respond to hormones, but this is unlikely. There is ample evidence that the fetal Müllerian ducts of many species respond to exogenous androgen and estrogen (Burns 1961). A pertinent observation is the precocious growth and coiling of fetal oviducts in explants of rat tracts when testosterone or estradiol was added to the culture medium (Pannabecker 1957; Price and Pannabecker 1959). Coiling was observed in these tracts at 20½ days of fetal age, whereas it is not present *in vivo* until 3 days after birth. The apparent lack of response of the fetal oviducts in the guinea pig to testicular and adrenal androgens requires further analysis. In any case, the early morphogenesis does not depend on specific hormonal stimulation.

V. Conclusions

The similar sequence of stages in the prenatal development of the oviduct in hamster, mouse, rat, guinea pig, and man demonstrates a basic orderly pattern in oviductal organogenesis. The Müllerian duct of the fetal female guinea pig is not dependent on maternal, placental, or fetal hormones for maintenance and early development. The oviductal segment of the primitive Müllerian duct is capable of a considerable degree of hormone-independent growth, coiling, and epithelial differentiation.

The stage of development of the oviduct of guinea pig and man at birth suggests prenatal hormone influence in the latter part of the gestation period. In the guinea pig, the ovary, which becomes metabolically active late in fetal life, and the adrenal gland are possible sources of hormones that might stimulate the oviduct.

Acknowledgment

Our research on the guinea pig has been aided by research grants GM-05335 and AM-03628 from the National Institutes of Health, United States Public Health Service, and by the Nederlandse Organisatie voor Zuiver-Wetenschappelijk Onderzoek. We are greatly indebted to Miss J. D. Young, Miss M. J. Verhoog and Miss A. M. Wassenaar for invaluable assistance in several aspects of the work.

References

1. Agduhr, E. 1927. Studies on the structure and development of the bursa ovarica and the tuba uterina in the mouse. *Acta Zool.* 8:1.
2. Ainsworth, L., and Ryan, K. J. 1966. Steroid hormone transformations by endocrine organs from pregnant mammals. I. Estrogen biosynthesis by mammalian placental preparations *in vitro*. *Endocrinology* 79:875.
3. Bloch, E. 1964. Hormone production by the foetal adrenocortical gland. *Proc. Second Int. Cong. Endocrin.* (*Lond.*) 785.
4. Bloch, E., and Newman, E. 1966. Comparative placental steroid synthesis. I. Conversion of (7-^{3}H)-dehydroepiandrosterone to (^{3}H)-androst-4-ene-3, 17-dione. *Endocrinology* 79:524.
5. Boling, J. L.; Blandau, R. J.; Wilson, J. G.; and Young, W. C. 1939. Post-parturitional heat responses of newborn and adult guinea pigs: Data on parturition. *Proc. Soc. Exper. Biol. Med.* 42:128.
6. Brewer, N. L. 1962. Sex differentiation of the fetal mouse *in vitro*. Ph.D. diss., University of Chicago.
7. Burns, R. K. 1955. Urogenital system. In *Analysis of development*, ed. B. H. Willier, P. A. Weiss, and V. Hamburger, chap. 6. Philadelphia: W. B. Saunders.
8. ———. 1961. Role of hormones in the differentiation of sex. In *Sex and internal secretions*, 3d ed., ed. W. C. Young, vol. 1, chap. 2. Baltimore: Williams & Wilkins Co.
9. Courrier, R. 1945. *Endocrinologie de la Gestation*. Paris: Masson & Cie.
10. Forsberg, J.-G. 1963. *Derivation and differentiation of the vaginal epithelium*. Håkan Ohlossons Boktryckeri, Lund.
11. Glenister, T. W., and Hamilton, W. J. 1963. The embryology of sexual differentiation in relation to the possible effects of administering steroid hormones during pregnancy. *Amer. J. Obstet. Gynec.* 70:13.
12. Gruenwald, P. 1941. The relation of the growing Müllerian duct to the Wolffian duct and its importance for the genesis of malformations. *Anat. Rec.* 81:1.
13. Horstmann, E., and Stegner, H.-E. 1966. Tube, Vagina und äussere weibliche Genitalorgane. In *Handbuch der Mikroskopischen Anatomie des Menschen*, ed. W. von Möllendorff and W. Bargmann, vol. 7, part 4. New York: Springer-Verlag.
14. Hunter, R. H. 1930. Observations on the development of the human female genital tract. *Contrib. to Emb. Carnegie Inst.* 129:91.
15. Jost, A. 1953. Problems of fetal endocrinology: The gonadal and hypophyseal hormones. In *Recent progress in hormone research*, 8:379. New York: Academic Press.
16. ———. 1965. Gonadal hormones in the sex differentiation of the mammalian fetus. In *Organogenesis*, ed. R. L. DeHaan and H. Ursprung, chap. 24. New York: Holt, Rinehart & Winston.
17. Kellogg, M. 1941. The development of the periovarial sac in the white rat. *Anat. Rec.* 79:465.
18. ———. 1945. The postnatal development of the oviduct of the rat. *Anat. Rec.* 93:377.
19. LaVelle, F. W. 1951. A study of hormonal factors in the early sex development of the golden hamster. *Contrib. to Emb. Carnegie Inst.* 34:19.
20. Lillie, F. R. 1917. The free-martin: A study of the action of sex hormones in the foetal life of cattle. *J. Exp. Zool.* 23:371.
21. ———. 1923. Supplementary notes on twins in cattle. *Biol. Bull.* 44:47.

22. Ortiz, E. 1945. The embryological development of the Wolffian and Müllerian ducts and the accessory reproductive organs of the golden hamster (*Cricetus auratus*). *Anat. Rec.* 92:371.
23. Ortiz, E.; Price, D.; and Zaaijer, J. J. P. 1966. Organ culture studies of hormone secretion in endocrine glands of fetal guinea pigs. II. Secretion of androgenic hormone in adrenals and testes during early stages of development. *Proc. Kon. Ned. Akad. Wet. C* 69:400.
24. Ortiz, E.; Zaaijer, J. J. P.; and Price, D. 1967. Organ culture studies of hormone secretion in endocrine glands of fetal guinea pigs. IV. Androgens from fetal adrenals and ovaries and their influence on sex differentiation. *Proc. Kon. Ned. Akad. Wet. C* 70:475.
25. Pannabecker, R. 1957. An analysis of sex differentiation in the fetal rat by means of organ culture studies. Ph.D. diss., University of Chicago.
26. Price, D. 1947. An analysis of the factors influencing growth and development of the mammalian reproductive tract. *Physiol. Zool.* 20:213.
27. Price, D., and Ortiz, E. 1965. The role of fetal androgen in sex differentiation in mammals. In *Organogenesis*, ed. R. L. DeHaan and H. Ursprung, chap. 25. New York: Holt, Rinehart & Winston.
28. Price, D.; Ortiz, E.; and Deane, H. W. 1964. The presence of Δ^5-3β-hydroxysteroid dehydrogenase in fetal guinea pig testes and adrenal glands. *Amer. Zool.* 4:327.
29. Price, D., and Pannabecker, R. 1959. Comparative responsiveness of homologous sex ducts and accessory glands of fetal rats in culture. *Arch. Anat. Micr. Morph. Exp.* 48:223.
30. Price, D.; Ortiz, E.; and Zaaijer, J. J. P. 1967. Organ culture studies of hormone secretion in endocrine glands of fetal guinea pigs. III. The relation of testicular hormone to sex differentiation of the reproductive ducts. *Anat. Rec.* 157:27.
31. Raynaud, A. 1962. The histogenesis of urogenital and mammary tissues sensitive to oestrogens. In *The ovary*, ed. S. Zuckerman, vol. 2, chap. 15. New York: Academic Press.
32. Scott, J. P. 1937. The embryology of the guinea pig. I. A table of normal development. *Amer. J. Anat.* 60:397.
33. Sobotta, J. 1917. Über den Mechanismus der Aufnahme der Eier der Säugetiere in dem Eileiter und des Transportes durch diesen in den Uterus. Nach Untersuchungen bei Nagetieren (Maus, Ratte, Kaninchen, Meerschweinchen). *Anat. Hefte* 54:359.
34. Solomon, S. 1966. Formation and metabolism of neutral steroids in the human placenta and fetus. *J. Clin. Endocrin. Metab.* 26:762.
35. Stegner, H.-E. 1961. Das Epithel der Tuba Uterina des Neugeborenen. Elektronenmikroskopische Befunde. *Z. Zellforsch.* 55:247.
36. ———. 1962. Elektronenmikroskopische Untersuchungen über die Sekretionsmorphologie des menschlichen Tubenepithels. *Arch. Gynaek.* 197:351.
37. Strauss, F. 1964. Weibliche Geschlechtsorgane (I). In *Handbuch der Zoologie*, ed. J.-G. Helmcke, H. v. Lengerken, D. Starck, and H. Wermuth, 8:9(3). Berlin: Walter de Gruyter.
38. ———. 1966. Weibliche Geschlechtsorgane (II). In *Handbuch der Zoologie*, ed. J.-G. Helmcke, H. v. Lengerken, D. Starck, and H. Wermuth, 8:9(3). Berlin: Walter de Gruyter.
39. Wells, L. J. 1962. Experimental studies of the role of the developing gonads in mam-

malian sex differentiation. In *The ovary*, ed. S. Zuckerman, vol. 2, chap. 14*A*. New York: Academic Press.

40. Wells, L. J. 1965. Fetal hormones and their role in organogenesis. In *Organogenesis*, ed. R. L. DeHaan and H. Ursprung, chap. 27. New York: Holt, Rinehart & Winston.
41. Willier, B. H. 1939. The embryonic development of sex. In *Sex and internal secretions*, 2d ed., ed. E. Allen, C. H. Danforth, and E. A. Doisy, chap. 3. Baltimore: Williams & Wilkins Co.
42. Witschi, E. 1959. Embryology of the uterus: Normal and experimental. *Ann. N.Y. Acad. Sci.* 75:412.
43. Wolff, E. 1962. The effect of ovarian hormones on the development of the urogenital tract and mammary primordia. In *The ovary*, ed. S. Zuckerman, vol. 2, chap. 14*B*. New York: Academic Press.
44. Zaaijer, J. J. P.; Price, D.; and Ortiz, E. 1966. Organ culture studies of hormone secretion in endocrine glands of fetal guinea pigs. I. Androgenic secretion as demonstrated by a bioindicator method. *Proc. Kon. Ned. Akad. Wet. C* 69:389.
45. Zuckerman, S. 1940. The histogenesis of tissues sensitive to oestrogens. *Biol. Rev.* 15:231.

2

Comparative Morphology and Anatomy of the Oviduct

A. V. Nalbandov
Department of Animal Science
University of Illinois, Urbana

Andreas Vesalius published the first, albeit incorrect, description of the oviduct. He admits his error to Fallopius in the free and easy style of the times. He wrote: "I willingly confess that in dissections of healthy women I have observed nothing corresponding to what you have found outside the peritoneum, although I have dissected the abdominal muscles in several women, and especially in the one which first fell to my lot for anatomy in Paris. She was hanged by a noose and had a most attractive figure. Her body exhibited . . . a notably fleshy likeness of the vessels carrying semen from the (female) testes into the uterus; consequently, as I still was a tyro in dissection and completely lacked guidance . . . , I wrote that I believed those vessels were the cornua of the uterus." (Cushing 1962.)

Fallopius, who was born in 1523 and died in 1562 of tuberculosis, had made most notable contributions to anatomy while he was professor of anatomy, surgery, and botany at the University of Padua, which was then known as "the nurse of genius." His contributions which concern us most are the correct and accurate description of the female reproductive system including the oviducts, which today are known as the "Fallopian tubes." He is also remembered today because he corrected Vesalius' incorrect account of the vessels of the penis, and because he even then recognized the clitoris as the homologue of the penis. He also named the ligamentum teres of the uterus and described the hymen, noting that it is present in virgins and absent in married women.

Not until 1873 do we find the next serious study of the Fallopian tubes, this one by Henle (of the loop of Henle fame), who gives a little more detail than did Fallopius, to be followed in 1891 by a still more detailed discussion of the intricacies of the oviduct by Williams. Gradually the tube of Fallopius acquired three more names, the term "oviduct" being more favored by American writers, while the British prefer to call them uterine tubes. The term salpinx has apparently fallen into disuse except in connection with various abnormalities of the oviduct.

In the present discussion I shall begin by giving a generalized description of the anatomy of the mammalian oviduct which could be used as the prototype for all mammals, mentioning the work on the lymphatic system, proceeding to a catalog of the various modifications adopted by different species, and conclude by discussing anatomical abnormalities of the oviduct. In the final portion I shall mention the avian oviduct. Since both the histology and the chemical function of the oviduct will be discussed by others, I shall mention these subjects only in passing.

I. General Anatomy of the Oviduct

The paired connecting oviducts between the ovaries and the uterus are long convoluted derivatives of the Müllerian ducts. The degree of convolution varies between species. In some of them (rabbits), the duct is almost completely straight while in others (mares) it shows extreme convolution (Lee 1928). The oviduct is suspended by the mesosalpinx, a derivative of the broad ligament. The fimbriae envelop the ovary more or less intimately. In pigs they cover the ovary completely, while in mares they embrace only the ovulation fossa. At the time of heat and ovulation the fimbriated end of the oviduct shows great motility, which led older anatomists to suspect that this "massaging action," as it was called, facilitated or even caused ovulation (see Grosser 1919). Since ovulation proceeds normally in animals with amputated fimbriae, this theory can be discounted, but the action of the fimbriae without doubt does aid the eggs in finding their way into the oviduct. In fact, the classic work of Spallanzani has shown that unilaterally castrated females in which the contralateral oviduct was removed showed some fertility, indicating that the eggs dropped into the peritoneal cavity could be swept up by the fimbriae of the intact oviduct.

Next to the fimbriae is the ampulla, where fertilization takes place, which, in most animals, is also the part with the largest diameter and the most intricate folding of the mucous membrane. The ampulla tapers down to the isthmus, which in turn enters the uterus in some animals at the top (for instance, sheep and cow); in others, it enters at the mesenteric side of the horn.

The musculature of the oviduct consists of the inner circular and the outer longitudinal layers. Both peristalsis and antiperistalsis take place, the latter occurring infrequently, thus suggesting that it may be an artifact. The mucosal lining of the oviduct consists of an intricately folded epithelium which has a simple columnar epithelium. Interspersed between the columnar cells are modified cells of the goblet cell variety which presumably secrete the polysaccharide responsible for the coating of rabbit eggs. The columnar cells are ciliated, their beat being abovarian. Both the height of the cells and the number of cilia vary with the reproductive state of the female. In castrates the cells become cuboidal and the cilia disappear. There are no true glands in the oviduct except at the uterotubal junction.

In the serous layer there are large nerve bundles which extend into the peripheral layers of the longitudinal muscle layer. The circular layer contains a dense plexus of thin nerve bundles extending into the mucous membrane and into the muscle fibers.

II. Adaptations of the Oviduct

A. *Fimbriae*

In most species the fimbriae form a fringed open funnel which envelops the ovary more or less completely. In rats and mice it forms a sac around the ovary and is then called the bursa ovarii. In these species the bursa is perforated by a small hole. The dog, fox, mink, and raccoon also have a bursa, but in them the bursa has a longitudinal slit through which the whole ovary can be extruded. The significance of either the bursa or of its perforations is obscure since other litter-bearing animals, such as pigs or rabbits which have open fimbriae, appear to be able to reproduce just as efficiently as do those with a bursa.

III. The Uterotubal Junction

In 1920, Rubin introduced the oviductal insufflation test in women to establish whether the oviducts were patent. He found that in women, fluid could be forced from the uterus into the oviduct. Subsequent work showed (Lee 1925, 1928) that in some animals, notably the mouse (Allen 1922) and the guinea pig (Kelly 1927), tubal insufflation is impossible since the uterus will rupture before fluid will enter the oviduct. Careful examination of the uterotubal junction showed that in some animals this area contains either a sphincter muscle or folds and polyp-like projections which are arranged in such a way as to prevent the passage of fluid and uterine contents, in general, from the uterus into the oviduct. According to Lee (1925, 1928), in the cat, dog, rabbit, rat, guinea pig, and sow (note that they are all litter-bearing animals), the opening of the oviduct is always guarded by special folds. He also pointed out that in these species fluid could be forced from the uterus into the oviduct when the ovaries contained no large follicles. Subsequently, Anderson (1927) conducted a study on the ease with which oviductal insufflation could be accomplished throughout the estrous cycle using excised uteri from slaughtered sows. She showed that the pressure required to force liquid into the oviducts was many times greater when eggs were present in the oviducts. To my knowledge such experiments have not been performed *in vivo*.

Anderson has also made the following classification of oviducts, according to the type of uterus present:

1) In marsupials the isthmus is tortuous, and has a wide lumen which narrows down somewhat at the point of entry into the uterus. There is neither a sphincter nor folds.

2) In all animals with bicornuate uteri, the isthmus is either straight or tortuous and always has a thick muscular coat and a narrow lumen. In sheep and cows, the oviduct enters the tip of the uterus and the oviductal opening is not protected by either folds or sphincter. In these species fluid can freely enter the oviduct from the uterine lumen. In all other bicornuate species (those named by Lee), the oviduct enters the uterus on the mesenteric side and is always protected by either a sphincter, villi, or mucosal folds.

3) In animals with a simplex uterus, the oviductal opening has no protected villi or folds, although Williams (1891) states that in women the uterine ostium has a sphincter which, however, is not sufficiently strong to prevent passage of either gases or liquids from uterus to oviducts. Neither does it have the polyp-like projections (see also Snyder 1923, 1924).

In view of these findings it is difficult to generalize concerning the possible function of the barrier between oviducts and uterus, which seems to be generally present in polytocous animals and absent in monotocous species. I would like to emphasize that, contrary to statements in many textbooks, a uterotubal barrier depends on the reproductive stage of the female, being greatest when eggs are present in the oviducts and totally nonfunctional during the rest of the cycle. Furthermore, it appears that even if the eggs have been fertilized, making pregnancy highly probable, the barrier relaxes as soon as the eggs have passed out of the oviducts. Also unsupported is the guess of some workers (Sobrero 1963) that the uterotubal junction may be concerned with the control of passage of sperm from the uterus into the oviduct. Since, as we have seen, such a barrier is absent in many species, it could play a role of a gatekeeper only in the polytocous species named above. Final judgment concerning the role of this oviductal adaptation must apparently await further research work.

IV. The Lymphatics

The available studies which have concerned themselves with the blood vascular supply of the oviduct show that the whole length of the oviduct is uniformly supplied with arteries and veins. In contrast, the only detailed study of the lymphatic system shows striking differences between the ovarian and uterine halves of the oviduct. This is, of course, the classic work of Andersen (1927). Its only drawback lies in the fact that it is restricted to the sow and does not permit us to judge how typical this system is of monotocous or other polytocous species. Figure 1 reproduces one of Andersen's plates and clearly shows that the isthmus, the uterotubal junction, and that portion of the uterine horn immediately adjacent to the oviduct have a distinctly more elaborate lymphatic system than does the ovarian half of the oviduct. On the basis of the work of her predecessors, which is far less elaborate, Andersen is inclined to think that this difference in the degree of development of the lymphatics is typical of other species, especially of women.

Of great interest is her observation that the lymphatic system is not static but changes in the different stages of the estrous cycle. First, she noted that the lymphatic vessels of the isthmus and the ampulla were more easily injected during estrus, as long as the corpora lutea were small (less than 6 mm in diameter). In contrast, the lymphatics of the uterotubal junction and of the adjacent piece of the uterine horn were more easily injected when the corpora lutea were mature or on days 8 to 14 of the cycle. She followed up these initial observations by actual measurements of the degree of development of the lymphatics during the various stages of the cycle using the width of the vessels as an index of the degree of development. She was working with slaughterhouse material and could not have known whether the donors of the specimens she collected were actually in "estrus" or "mid-estrus"—the classification she uses in her description. Thus, her findings should probably be accepted with some reservations and should be interpreted to mean that she was dealing with specimens whose ovaries contained predominantly follicles or corpora lutea. Furthermore, she based her conclusions on only 8 sets of oviducts. These disclaimers should be kept in mind in evaluating the description of her data. She found that the vessels in the ampulla and the isthmus were twice as wide in estrus (read: follicular phase), as they were in mid-estrus (luteal phase). In contrast, the lymphatics of the uterotubal junction and of the proximal part of the uterus showed the greatest development during the luteal phase.

These observations, if confirmed, imply that the lymphatics of the uterus and of the uterotubal junction, respond to trophic hormones in a different manner, e.g., one part of the lymphatics of the oviductal system enlarges in response to estrogen while another responds to progestin. At first glance one finds this difficult to understand, and this whole matter, which has received remarkably little attention, should be investigated on castrated females treated with estrogen or progesterone. If this claim can be substantiated, the possibility arises that cyclic contraction and expansion of the lymphatic system may play a role in the ability of the oviduct to permit passage of sperm and eggs and may explain such

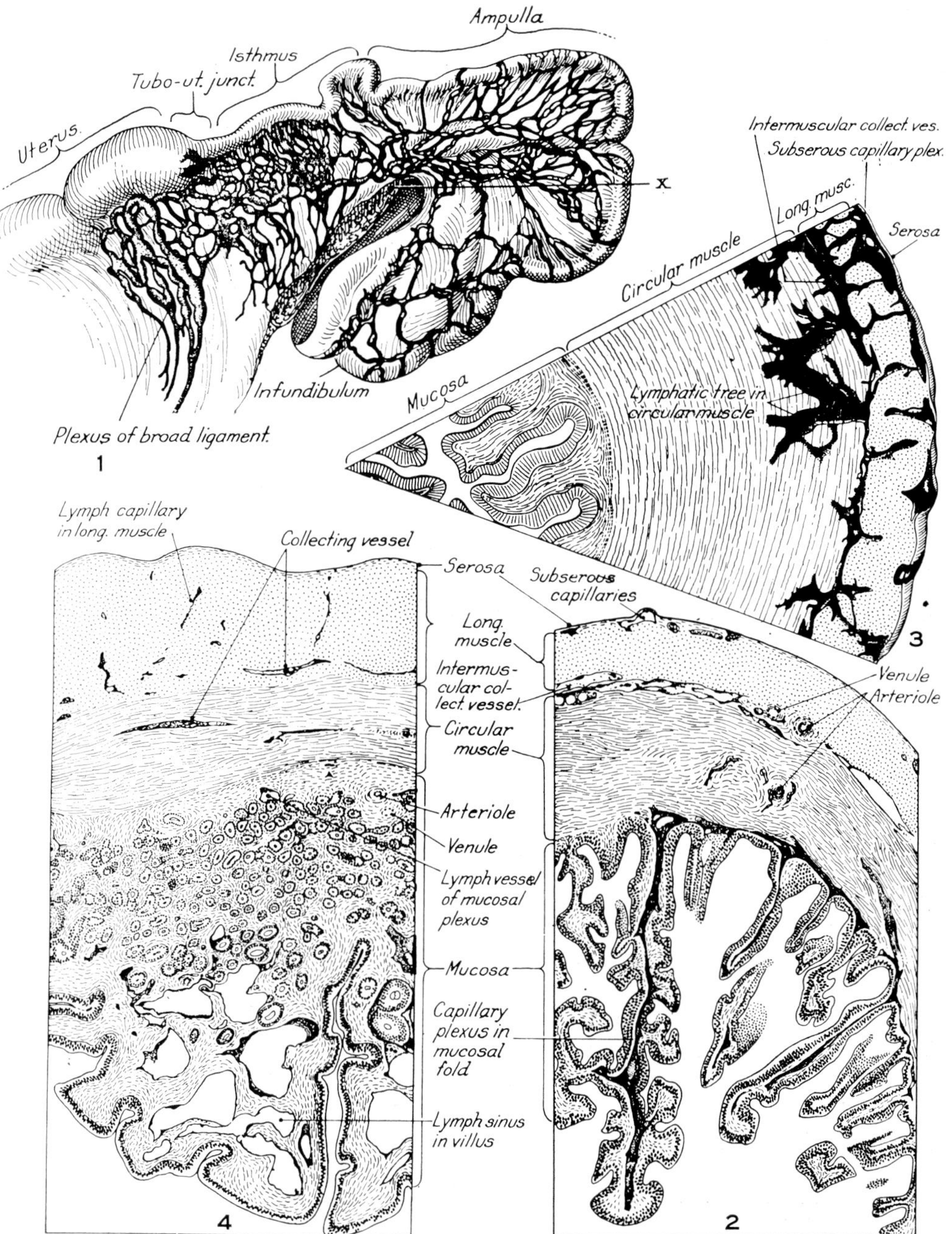

FIG. 1. The lymphatic system of the oviduct of the sow. (From Andersen 1928.)

phenomena as "tube-locking" of eggs, or their extremely rapid passage through the oviduct under certain other hormonal conditions. The elaborateness of the lymphatic system of the oviduct, the fact that it is better developed in its uterine than in its ovarian end, and the possibility that it may respond to hormonal influences, suggest that there is room for much fruitful research in this area which, to my knowledge, has remained undone to the present time.

V. Tubal Abnormalities

For reasons which are completely unclear, the pig is very prone to develop obstructions of the oviducts, usually bilateral, which cause pyo- or hydrosalpinx. In fact, these oviductal obstructions are the second most important cause of sterility in swine in North America. In one sample (Nalbandov 1952) of sterile swine, 35% of the females showed oviductal obstructions, of which 31% were found in gilts and only 1 case in a sow. In contrast, a similar study conducted in Europe revealed not a single case of oviductal obstructions in 1,000 sterile swine (Goethals 1951); and only 3 cases in 83 sterile females were reported in another study (Perry and Pomeroy 1956). This significant difference between the European and American breeds suggests that the large incidence of oviductal nonpatency in the American study (Nalbandov 1952) may be ascribed to genetic difference. Both pyo- and hydrosalpinges are fairly common in women and are usually thought to be an aftermath of gonorrheal infection of the reproductive tract. Attempts to isolate infectious organisms from either hydro- or pyosalpinges of swine were all negative and, in fact, these enlarged oviducts were found to be bacteriologically completely sterile. Conversely, attempts to cause salpingitis by scarifying the oviductal lumen and instilling a variety of infectious organisms into it also failed, in spite of an adequate number of tries. The fact that in pigs this condition is almost totally restricted to gilts also speaks against infections as a causative agent, since one could reasonably assume that sows would be exposed to infections as readily as are gilts. Thus, while we can probably rule out infection as a cause of oviductal obstruction in pigs, the real explanation of its occurrence remains unknown.

Hydro- and pyosalpinges are rare in other species. Furthermore, when they do occur in sheep or cattle, they are quite frequently unilateral, while in swine and women they are almost invariably bilateral. In spite of special vigilance over a period of years, no enlarged oviducts have been seen among many thousands of rabbits, guinea pigs, or rats autopsied in this laboratory.

In summary, then, it appears that the formation of oviductal abnormalities in women, but perhaps not in swine, is frequently caused by infection. It seems improbable that the uterotubal junction plays a role in this abnormality because, as was pointed out, several other polytocous animals have a uterotubal barrier as pronounced as that of pigs and yet very rarely develop this abnormality.

VI. The Avian Oviduct

In birds, the usually unilateral oviduct is also of Müllerian origin, but it performs a totally different function than does the oviduct of mammals. All of the portions of the avian oviduct have secretory glands which differ from one portion to the next. A classic, and still unsurpassed, description of the chicken oviduct is that of Richardson (1935).

Occasionally the infundibulum has been divided into two parts. The fimbriae are ciliated, embrace the ovary, and show great activity shortly before and during ovulation.

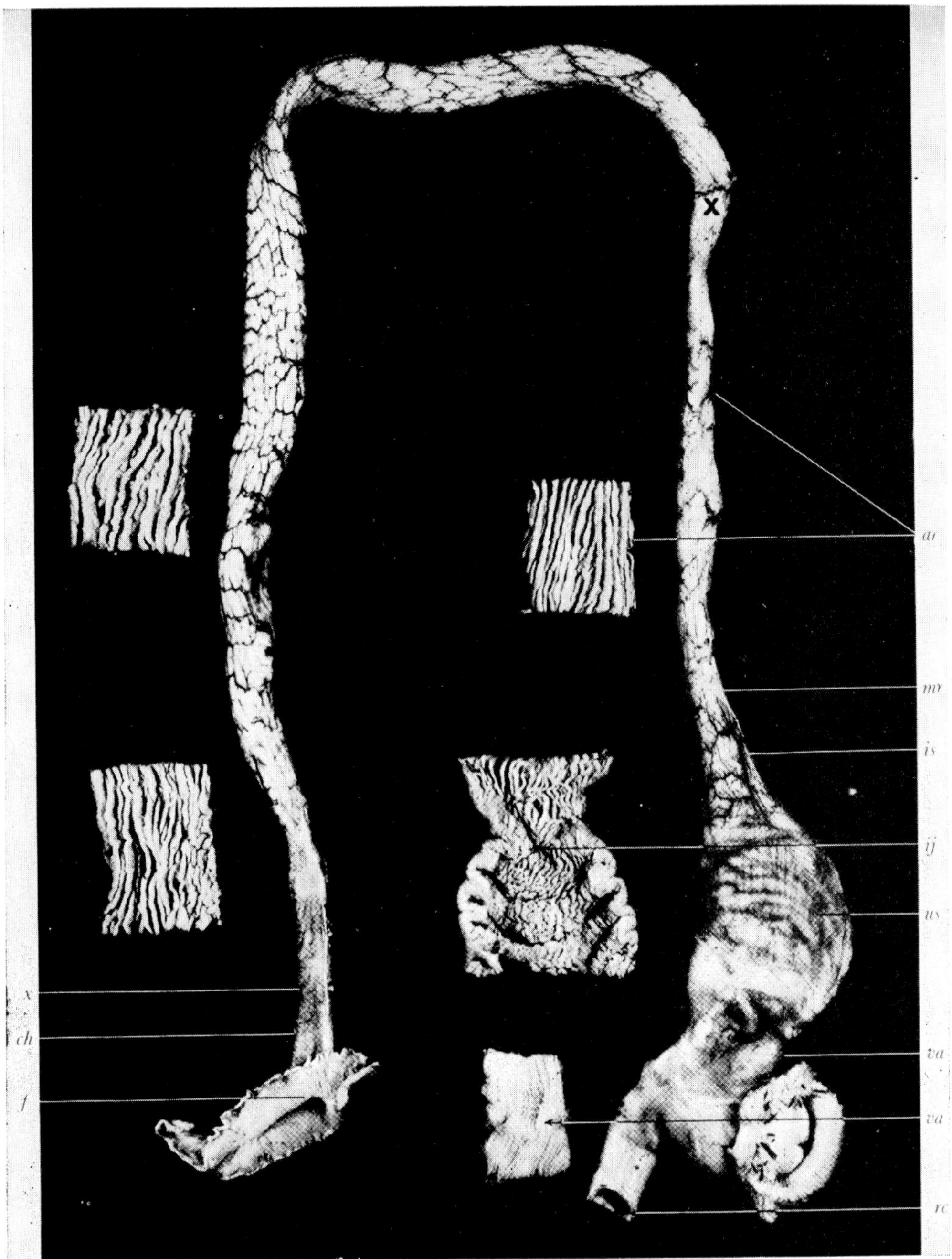

FIG. 2. The female duct system of the chicken: *f*, fimbria; *ch*, chalazal region; *x—x*, magnum (note the line separating the magnum from the isthmus); *x—ij*, isthmus; *us*, uterus or shell gland; *va*, vagina. (From Richardson 1935.)

They engulf the ovulated egg as it emerges from the follicle, and if the egg escapes into the body cavity the fimbriae can retrieve it from there because they are capable of vigorous motility. The remainder of the infundibulum is ciliated and contains mucus-secreting glands from which the spiral-like chalaza are formed around the egg. For this reason this portion is called the chalaziferous region. Next, the egg passes into the magnum portion of the oviduct where ovalbumen, the glucoprotein of albumen and mucus, is secreted around the egg. This part is also ciliated and extremely glandular. The glands are built by estrogen alone, and they become capable of secretory activity when either androgen or progesterone acts on them (Nalbandov 1959).

The next portion, the isthmus, is histologically distinctly different from the magnum and is anatomically separated from the latter by the magnum-isthmus line. In the isthmus, the epithelial cells are ciliated and the tubular glands are responsible for the secretion of the shell membranes. Not much is known about the endocrine mechanisms controlling this portion of the oviduct. It is remarkable that a total of 33 gm of material is secreted by the infundibulum and the magnum in the amazingly short time of 4½ hr. The calcareous shell is deposited in the next segment, the shell gland. Some of the epithelial cells of the shell gland are ciliated and the gland cells, prior to secreting the calcium-containing shell around the egg, also add thin fluid (albumen) which causes the finished egg to "plump." The secretion of the shell requires about 20 hr. The final portion of the oviduct is the vagina, which has ciliated epithelial cells and contains glands which secrete mucus, which facilitates the laying of the finished egg and seals the pores of the shell, protecting its contents against bacterial invasion and evaporation.

VII. Conclusions

In reading for this assignment, I was struck by the wealth of folklore which surrounds the tube of Fallopius and which stretches the credibility of the reader. Thus, in some of the older literature one finds statements, unsupported by descriptions of how this evidence was obtained, that the oviduct has the ability to sense which ovary contains the ovulatory follicle (Grosser 1919). It is said that the corresponding oviduct becomes erect and shows enormous activity while the contralateral duct is relaxed and inactive. In a 1962 German paper (Schilling) the ostium of the oviduct is credited with the ability to line itself up exactly with the location of the ovulatory follicle in the ovary in sheep and in cattle. Into the most recent times the story is repeated, especially in texts on obstetrics and gynecology, that the fimbriae cause ovulation by massaging the ovary. Out of the mass of seemingly unfounded and, in some instances, totally incredible assertions, I was fascinated by one of the facts which, in my opinion, deserves further study—the lymphatic system of the oviduct and its possible responsiveness to steroid hormones. Although I have not made attempts to inform myself completely on the subject of the lymphatics of the uterus, it is my impression that it is much less well endowed than the oviduct. Such investigations seem especially pertinent in view of the increasing awareness, amply documented by recent studies, that the uterus, and perhaps even the oviduct, play(s) an important role in the control of ovarian function. Both luteolytic and luteotrophic effects of uteri and of uterine contents (IUD's or embryos) have been recorded. These effects are frequently local and sometimes they do not seem to be mediated systemically. The question is whether the conundrum of the local effect may be resolved by more thorough studies of the oviductouterine, vascular, lymphatic, and nervous systems.

References

1. Allen, E. 1922. The estrous cycle in the mouse. *Amer. J. Anat.* 30:297.
2. Andersen, D. H. 1927. Lymphatics of the Fallopian tube of the sow. *Contrib. to Embryol. Carnegie Inst.* 19:135.
3. ———. 1928. Comparative anatomy of the tubo-uterine junction: Histology and physiology in the sow. *Amer. J. Anat.* 42:255.
4. Cushing, H. 1962. *A bio-bibliography of Andreas Vesalius*, 2d ed. Hamden, Conn.: Archon Books.
5. Goethals, P. 1951. Sterilität. *Vlaam. diergeneesk. Tijdschr.* 20:155.
6. Grosser, O. 1919. Ovulation und Implantation und die Funktion der Tube beim Menschen. *Arch. Gynaek.* 110:297.
7. Henle, J. 1873. Eingeweidelehre. *Handbuch der Anatomie des Menschen* (*Zweite Aufl.*) 2:485.
8. Kelly, G. L. 1927. The uterotubal junction in the guinea pig. *Amer. J. Anat.* 40:373.
9. Lee, F. C. 1925. A brief note on the anatomy of the uterine opening of the Fallopian tube. *Proc. Soc. Exp. Biol. Med.* 22:470.
10. ———. 1928. The tubo-uterine junction in various animals. *Bull. Johns Hopkins Hosp.* 42:335.
11. Nalbandov, A. V. 1952. Anatomic and endocrine causes of sterility in female swine. *Fertil. Steril.* 3:100.
12. ———. 1959. Role of sex hormones in the secretory function of the oviduct. In *Comparative endocrinology*, ed. A. Gorbman, chap. 32. New York: John Wiley & Sons.
13. Perry, J. S., and Pomeroy, R. W. 1956. Abnormalities of the reproductive tract of the sow. *J. Agric. Sci.* 47:238.
14. Richardson, K. C. 1935. The secretory phenomena in the oviduct of the fowl, including the process of shell formation examined by the microincineration technique. *Phil. Trans. Roy. Soc.* (*Lond.*) *Ser. B* 225:149.
15. Rubin, I. C. 1920. Nonoperative determination of patency of Fallopian tubes by means of intra-uterine inflation with oxygen and production of an artificial pneumoperitoneum. *J.A.M.A.* 75:661.
16. Schilling, E. 1962. Untersuchung über den Bau und die Arbeitsweise des Eileiters vom Schaf und Rind. *Zbl. Veterinärmed.* 9:805.
17. Snyder, F. F. 1923. Changes in the Fallopian tube during the ovulation cycle and early pregnancy. *Bull. Johns Hopkins Hosp.* 34:121.
18. ———. 1924. Changes in the human oviduct during the menstrual cycle and pregnancy. *Bull. Johns Hopkins Hosp.* 35:141.
19. Sobrero, A. J. 1963. Sperm migration in the female genital tract. In *Mechanisms concerned with conception*, ed. C. G. Hartman, chap. 4. New York: Macmillan Co.
20. Williams, J. W. 1891. Contributions to the normal and pathological histology of the Fallopian tube. *Amer. J. Med. Sci.* 102:377.

3

Light and Electron Microscopic Structure of the Oviduct

O. Nilsson/S. Reinius

Department of Human Anatomy
University of Uppsala, Sweden

Several reviews of the physiology and light microscopic morphology of the reproductive system have been published (Austin 1963; Blandau 1961; Eckstein and Zuckerman 1956). The few published papers on electron microscopy of the oviduct have been compiled by Horstmann and Stegner (1966). Electron microscopy of the epithelial lining has revealed that there are greater differences than expected in the oviducts of various species, among varied segments of a single oviduct, and in the same segment of an oviduct under various hormonal influences. Therefore, the present review will be confined mostly to oviductal studies by electron microscopy. The species to be described are mouse, rat, guinea pig, ruminants, rabbit, and man.

The technical improvements that have made present-day studies possible will be described first. We will next consider the interrelationship of the epithelial lining of the oviduct with the gametes or eggs, in an attempt to correlate the structure of the oviductal epithelium to the reproductive stages. Finally, the structural background to some physiological processes in the oviduct will be outlined.

I. Methods

In our experience, the use of light and electron microscopy in reproductive research sets certain requirements with reference to the mode of fixation, embedding, and sectioning (Mayer, Nilsson, and Reinius 1967; Nilsson 1966*a*, 1967; Reinius 1965, 1967).

The *fixation* should be by the perfusion method, using a rapidly penetrating fixative in order to disturb the gross anatomical features of the tissue as little as possible. *In situ* preservation is necessary when one wishes to establish the width of the oviductal lumen or evaluate the interrelationship between the epithelium of reproductive tract and eggs. However, the requirement of perfusion for fixation cannot always be fulfilled, and other types of fixation must be used. This increases the risk of artifacts and makes it difficult to compare specimens obtained by various techniques of fixation. Fixation by perfusion is preferred in our laboratory for small animals (mouse, rat, guinea pig, and rabbit); for larger animals the oviductal tissue, rapidly cut into small pieces with a razor blade, is fixed by immersion. The fixative is a 2.5% solution of glutaraldehyde in Sørensen's phosphate buffer, pH 7.4 (Sabatini, Bensch, and Barrnett 1963).

The procedure for fixation by perfusion is as follows:

1) Anesthetize the animal by an i.p. injection of Nembutal (Abbott).

2) Cut through the ventral abdominal wall and move the intestines carefully aside and avoid muscular contractions of the reproductive tract by mechanical irritation.

3) Expose the abdominal aorta and the inferior vena cava in the retroperitoneal space by splitting and retracting the peritoneum covering them.

4) Clamp the aorta distal to the renal vessels, then insert a cannula of appropriate size, connected with an apparatus for injection under pressure, into the abdominal aorta just distal to the clamp, aided by a magnifying glass if necessary.

5) The vena cava is cut and the fixative is injected into the aorta under pressure of 100 to 200 mm Hg for a period of 5 min. With this setup the fixative will perfuse through the vessels of the lower half of the body only.

A successful fixation is indicated by an immediate disappearance of blood from small vessels and a stiffening and yellowing of the perfused tissues (Hopwood 1967). Diffusion of fixative from the capillaries quickly fixes the epithelial lining and the gametes or eggs, if present. When the perfusion is completed, the oviduct is removed carefully, avoiding any pinching or tearing. Immediately, or after storing in the glutaraldehyde fixative, the oviduct is cut into appropriate sizes, rinsed in Sørensen's phosphate buffer, and postfixed for 3 to 4 hr in a 1% solution of osmium tetroxide in Sørensen's phosphate buffer. A similar postfixation is used also after the immersion fixation. The purpose of the postfixation is to facilitate later orientation and trimming of the specimen embedded in plastic, and to contribute to the contrast of the section when studied with the electron microscope by adding heavy metal to the tissue.

The *embedding* in some of the newer plastic materials is preferred. Epon (Luft 1961) after dehydration in ethanol gives minimal distortion of tissues, and permits the cutting of larger blocks of tissue. Further, the same embeddings may be used for both light and electron microscopy. This is important when searching for eggs in the oviduct. The use of sections of about 1 μ thickness for light microscopy also gives a better result than can be obtained with paraffin embeddings.

The *sectioning* for light microscopy is done with a dry glass knife on a Porter-Blum ultramicrotome. The sections are then carefully transferred with forceps to a slide. This method makes possible plastic sections of a maximum size of 4 $\times$ 7 mm, with a thickness of 1 to 10 μ. A fairly good view of the tissue in the sections is obtained by observing them through the stereoscope of the ultramicrotome. This is a great advantage, making possible a rapid search for eggs without mounting, staining, and then observing each section under the microscope. The sections intended for further study are stained by heating on the slide in a few drops of solution—1% toluidine blue with 1% borax or sodium bicarbonate—for about 0.5 min. (Reinius 1966*a*).

For electron microscopy, the blocks are sectioned with an ordinary through-mounted glass knife on an Ultrotome ultramicrotome; the sections are mostly collected on one-hole grids (Galey and Nilsson 1966). For staining, a solution of uranyl acetate (Watson 1958) followed by lead citrate (Reynolds 1963) is used.

II. Oviductal Segments

The classification of the parts of the oviduct used in the present treatise is based mainly upon the microscopic appearance of the oviduct. In general, this mode of classification is consistent with the common macroanatomical mode, but some discrepancies exist (Figs. 1, 2). We suggest this new classification since, in our opinion, it forms a more appropriate basis for correlations between morphology on the one hand and biochemistry, physiology, pharmacology, etc., on the other.

According to the structure of the epithelium, 4 segments of the oviduct may be distinguished: the preampulla, the ampulla, the isthmus, and the junctura. The preampulla includes both the fimbriae and the infundibulum which merge gradually into the ampulla. It appears that no great differences in the type of cells exist between the preampulla and the ampulla; the differences seem to involve the number of ciliated cells and the number of granules in the nonciliated cells. The nonciliated cells of the ampulla might differ from those of the isthmus, although the cells in the border region between the two segments are less characteristically outlined. The last segment, the junctura, includes not only the intramural part of the oviduct, but also a portion of its extramural part. The junctura is not the same anatomical segment as the uterotubal junction.

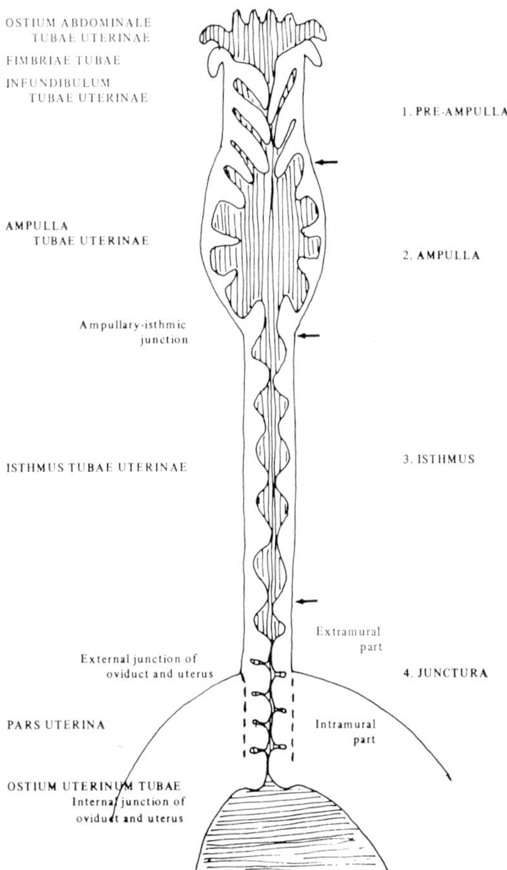

FIG. 1. Schematic drawing of the oviduct

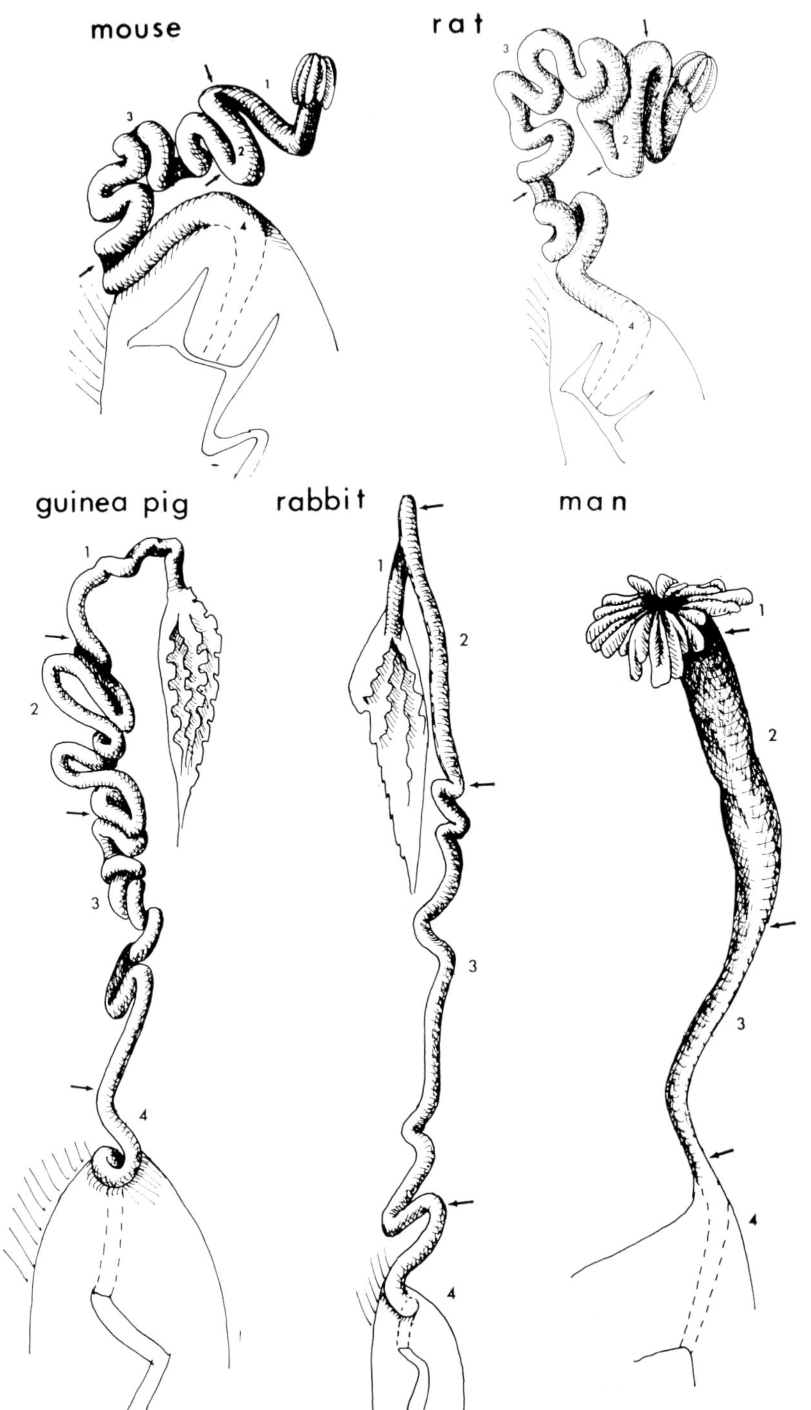

FIG. 2. Drawings of the oviducts of the mouse, rat, guinea pig, rabbit, and man; 1, preampulla; 2, ampulla; 3, isthmus; 4, junctura.

The segments of the oviduct have the following general characteristics:

1) *preampulla*—thin wall, epithelium almost entirely ciliated cells; mainly concerned with the transport of eggs from the ovary into ampulla;

2) *ampulla*—slightly thicker wall, fewer ciliated cells than the preampulla; fertilization occurs here; forms nutrients;

3) *isthmus*—pronounced muscular coat, more nonciliated cells, ciliated cells may even be rare; affects the nutrition and transport of sperm and eggs;

4) *junctura*—well-developed muscle layers, a more varying proportion between ciliated and nonciliated cells than the isthmus, narrowest lumen of the whole oviduct.

III. Oviductal Wall

Tunica serosa is the outermost layer of the oviduct, composed of mesothelium continuous with that of peritoneum, and connective tissue. The serosal layer is well vascularized. Among the components of the connective tissue, mast cells might occur in a relatively high number. Smooth muscle fibers are present both subperitoneally and around the vessels.

Tunica muscularis lies inside the serosa. The thickness of this layer varies in the different segments of the oviduct, generally being thickest in the isthmus. In the mouse and rat, the isthmus has a thin outer longitudinal layer and a thicker inner circular layer; in the guinea pig, an outer circular and an inner longitudinal; and in man, a circular layer between an outer and an inner longitudinal layer. At the junction of the oviduct and uterus, the oviductal muscles merge partly with the uterine muscles.

The adrenergic nerves of the rabbit oviduct, as demonstrated by a histochemical fluorescence method, supply the vessels and the muscles. The amount of muscular innervation in the ampulla is moderate and rather constant; at the ampullary-isthmic junction there is a marked and local increase in density, while the remaining part of the isthmus has a moderate innervation. No increased amount of innervation is present in the external junction of the oviduct and uterus (Brundin 1965; Owman and Sjöberg 1966). Peristalsis or irregular contractions of the tunica muscularis influence the width of the oviduct lumen. This factor also contributes to egg transport.

Tunica mucosa classically consists of a lamina propria and a lamina epithelialis. Mucosal muscle fibers have hardly been mentioned, but we believe that future electron microscopy will reveal that they are present, intermingled with the connective tissue elements. If so, the activity of the mucosal muscles will have to be considered when discussing oviductal transport mechanisms.

The lamina propria forms the framework of the mucosal folds in the oviduct. The folds generally occur as branching longitudinal ridges; however, in the isthmus of the mouse they seem to be circular. The degree of folding varies among the species and among the segments of the oviduct, being most pronounced in the preampulla. The volume of the folds, determined by such factors as the width of the mucosal vessels and the amount of interstitial fluid, influences the degree of free space in the lumen.

The lamina epithelialis is dominated by two cell types: ciliated and nonciliated. These cells are the most characteristic feature of the oviduct, since they show regular and often specific changes in structure and relative numbers among the species. They probably exert an important influence on the composition of the oviductal secretion and, thereby, on the activities occurring within the lumen. The ultrastructure of these two cell types will therefore be treated in detail.

The ciliated cells have motile cilia (kinocilia) that emerge through the luminal membrane. Generally, the number of ciliated cells is greater in the preampulla and ampulla, and lower in the isthmus and the junctura. The ciliated cells in the various species examined have a rather homogeneous ultrastructure; however, there is at least one exception: the ciliated cells of the human oviduct. Other cells show large vesicles containing cilia; these structures have been regarded as early stages in the development of the kinocilia (Flerkó 1954; Overbeck 1967).

The task of the ciliated cells is to contribute to the transport of eggs and secretions, and probably also to secure a free passage through the oviduct. The frequency of the beat of the cilia in the rabbit oviduct is increased about 25% for a period after ovulation (Borell, Nilsson, and Westman 1957).

The nonciliated cells appear in many different forms, often rather specific for a certain species. The task of the nonciliated cells, broadly speaking, is to produce appropriate conditions of the oviduct during the migration of sperm, fertilization, and the transport of eggs. How this is achieved is still unknown. We cannot even be certain that all nonciliated cells really secrete; for instance, in the junctura a resorptive activity does not seem unreasonable.

Electron-dense granules occur rather often in the nonciliated cells. The granules are noticed in all parts of the cell, although they are most frequent apically. Their number varies from a few granules beneath the luminal surface (e.g., in the preampulla of mouse and rat) to many granules crowded in the cell (e.g., in the isthmus of the rabbit). The granules have varying diameters in the different animals. A border membrane, some internal structure, or varying density of the granules may be noticed.

Transitional forms between the ciliated and nonciliated cells have been reported (Flerkó 1954). However, in our ultrastructural studies we have not found any transitional cell forms. At most, solitary, aberrantly situated kinocilia have appeared on one or two nonciliated cells. Microvillous vesicles (Fredricsson and Björkman 1962) are found in nonciliated cells and may correspond to the vesicles containing cilia in the ciliated cells. Of the nonciliated cells, two types have been given special names, the "peg cells" and the "basal cells." Peg cells seem to refer to nonciliated cells when compressed in the epithelial membrane. Basal cells have also been named on a purely morphological basis. The function of these cells is not known.

IV. Changes of the Oviductal Wall

A. *The Mouse*

The following description is based mainly upon results obtained in our laboratory (Reinius 1966*b;* 1968). Additional references are given in the text or at the end of the section. The mouse oviduct is connected with the ovary by an ovarian bursa and is coiled, consisting of about 10 loops (Fig. 2). The last portion of the oviduct runs intramurally in the uterus, ending with the colliculus tubarius (Fischel 1914; Powierza 1912). The structure of the oviduct will be described chiefly with reference to the epithelial cells and the position of the eggs at that particular moment.

Virgin female mice (aged 12 to 25 weeks) of the CBA-strain have been used. The animals have been conditioned to artificial light from 5 A.M. to 7 P.M. Day 1 of pregnancy is the morning on which a vaginal plug is found.

The animals have been perfused with fixative for study of (1) sperm transport: a few hr after finding the vaginal plug (from 10 P.M. to 5 A.M.); and (2) egg transport: at days 1, 2, and 3. The junctura also has been studied at days 4 to 8.

1) The *preampulla* comprises the fimbriae and loops 1 and 2. The muscular tissue consists of only 2 or 3 layers of cells, arranged mostly longitudinally. The propria is scanty and extends into high longitudinal folds. The folds meet in the middle, appreciably restricting the free space in the lumen. The epithelium contains ciliated cells with a few nonciliated cells dispersed among them. The ciliated cells (Fig. 3*a*), 10 to 15 μ high, have interposed microvilli among the kinocilia. The endoplasmic reticulum is poorly developed, consisting of a few scattered, short, rough membranes. Some fat granules and vacuoles occur, increasing in number toward the fimbriated end. The vacuoles sometimes occupy the whole width of the cells. They can be found either apically or basally.

The nonciliated cells are 15 to 17 μ in height. The apical part of the cell is broader than the basal part and bulges more or less above the surface of the ciliated cells. Since the ciliated cells do not change form, they seem to have a more rigid construction than the nonciliated cells. The surface membranes of the nonciliated cells have few short microvilli, irregularly distributed. The cytoplasm stains darker than that of the ciliated cell. The Golgi apparatus is large, extending over the luminal side of the nucleus. The endoplasmic reticulum is moderately developed, with long rough membranes in parallel arrangement and only partly distended. A few dense granules, 0.4 μ in diameter, are situated apically. The nucleus is high, ranging into the bulging part of the nonciliated cell, and has only a few indentations.

Morphologic changes are not observed during the time of sperm and egg transport. Unfertilized eggs rarely have been found in the preampulla.

2) The *ampulla* constitutes loop 3 and is easily distinguished on day 1. It is distended by fluid and eggs. In other stages, however, no significant demarcations from the preampulla or isthmus are seen. The outer diameter of the distended ampulla is about 500 μ; the tunica muscularis contains 1 or 2 cell layers, and the tunica mucosa forms small longitudinal folds into the wide lumen. The ampulla, not distended, measures 300 μ in diameter, and many mucosal folds are noticed in a narrow lumen. The epithelium, although dominated by nonciliated cells, contains numerous ciliated cells. The latter have the structure previously described but have fewer lipid granules and vacuoles.

The nonciliated cells vary in form according to the cyclic state. On day 1, when the ampulla is distended, the cells are low (10 to 12 μ), cuboidal, and with a slightly bulging apical part. When the ampulla is not distended, the cells are cylindrical (14 to 20 μ) with a more bulging apical part which, however, is not wider than the basal part. The surface membrane of the nonciliated cells has relatively few microvilli, 0.5–0.7 μ long (Fig. 3*b*). The Golgi apparatus is large. The endoplasmic reticulum is extensive and forms irregularly distended cisternae and vesicles arranged like strings of pearls (Fig. 4*a*). The membranes have some ribosomes attached to their distended parts. The vesicles and cisternae are filled with a substance of light electron-density. Granules are seen to a much greater extent in the ampulla than in the preampulla.

During the time of sperm and egg transport, the amount of endoplasmic reticulum is rather constant; but changes in number of granules and in size of the cisternae are observed. When the ampulla is not distended, i.e., at early sperm transport and at late stages of egg transport, the granules are more numerous and the cisternae of the endoplasmic reticulum are more dilated than what is observed when the ampulla is distended by fluid, i.e., when the eggs lie in the ampulla. If the extensive endoplasmic reticulum indicates a high rate of protein synthesis, then a presence of dilated cisternae might point also to an increased

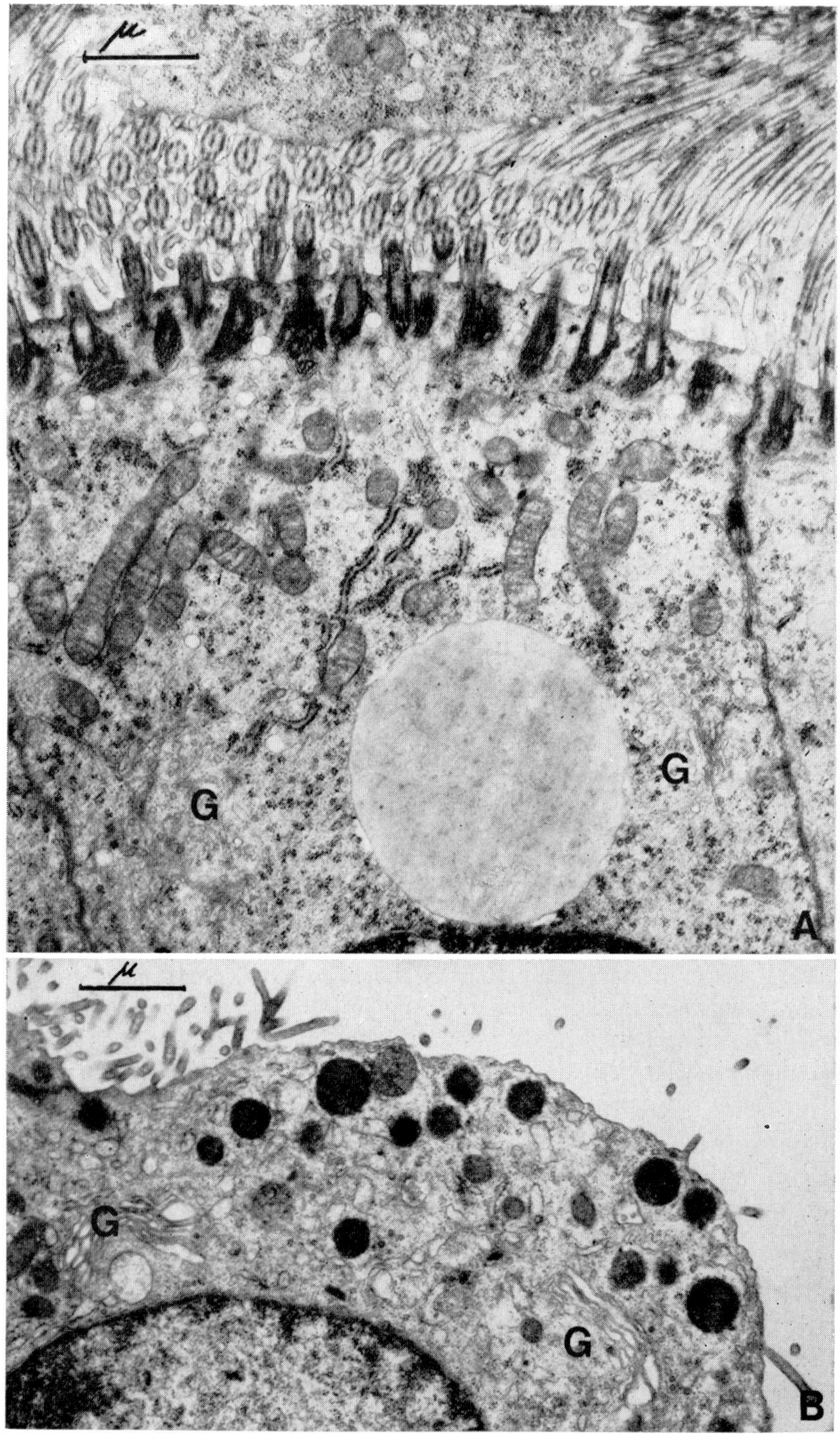

FIG. 3. (*a*) Mouse, preampulla (day 3, 4 P.M.). Apical part of a ciliated cell, showing a row of basal corpuscles with kinocilia projecting into the lumen. Golgi apparatus, several mitochondria, granular endoplasmic reticulum, and free ribosomes are seen in the cytoplasm. A large lipid inclusion is observed in the lower part of the figure. (×16,000).

(*b*) Mouse, ampulla (day 3, 5 P.M.). Apical part of a nonciliated cell. Several dense granules are visible in the cytoplasm. The Golgi apparatus is situated above the nucleus. Few microvilli are noticed. (×14,000).

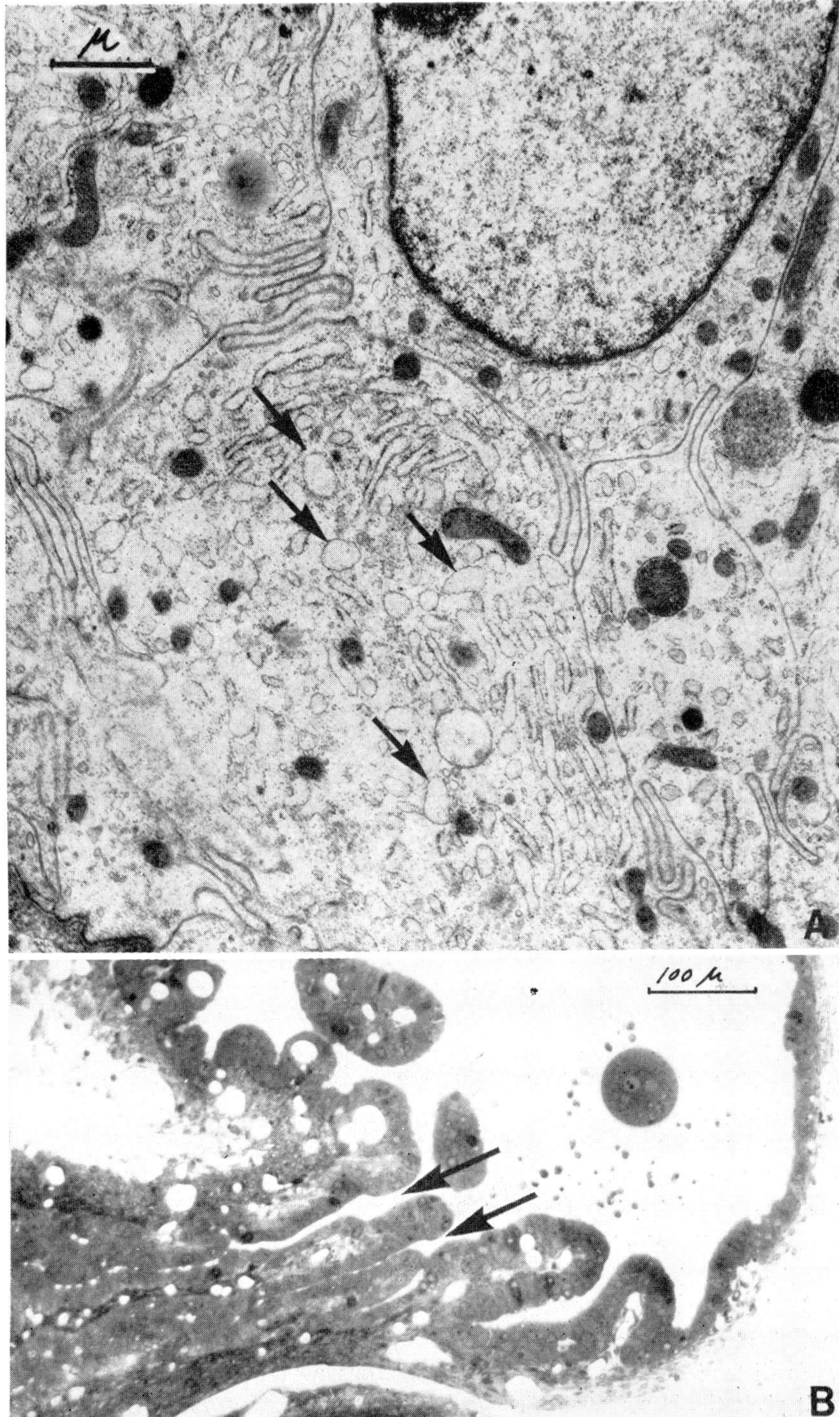

FIG. 4. (*a*) Mouse, ampulla (day 2, 3 P.M.). Basal part of nonciliated cell, demonstrating granular endoplasmic reticulum with less granular cisternae (→). Part of the nucleus in the upper, right corner of the picture. (×15,000).

(*b*) Mouse, ampullary-isthmic junction (day 1, 5 P.M.). The wide lumen of the ampulla to the right, the narrow lumen of the isthmus to the left (→). Note the thin muscular layer of the ampulla compared to the thicker one of the isthmus. A 1-cell egg in pronuclear stage surrounded by dispersed granulosa cells lies in the ampulla lumen. Light microscopy. (×150).

storage of secretory products. During the early egg transport, however, the ultrastructure is consistent with a high degree of depletion of the secretion from the cell. Thus, the fluid of the ampulla is partly a product of the mucosal membrane.

The *ampullary-isthmic junction* is formed by a narrow curve of the loop next to the ampulla (Fig. 4*b*). At the distended stage of the latter, the outer curve shows only a gradual thickening of the muscle wall, but the inner curve forms a massive thickening of most circular musculature. The opening into the isthmus has rather long mucosal folds filling its lumen and protruding into the ampullary lumen. The epithelium of these folds, like that in the rest of the isthmus, consists mostly of nonciliated cells. Ciliated cells occur sparsely, being situated in the epithelial crypts.

3) The *isthmus* forms a long and coiled part of the oviduct, consisting of the 4th to the 9th coil. The muscular tissue is mainly circular and measures 20 to 30 μ in thickness. Longitudinal epithelial folds occur only in the loop next to the ampulla. In the rest of the isthmus longitudinal folds are lacking, but low circular folds have appeared. The width of the lumen varies among the loops, but a local widening of the lumen is the rule where eggs are present. These lie no more than 2 loops apart and are often grouped 2 to 7 together (Fig. 5*a*). Solitary sperm are noticed in both the wide and narrow loops. The nonciliated cells are of 2 kinds. One is similar to that in the ampulla and is most abundant in the 1st and 2nd loop of the isthmus. The other cell type is the most common one and can be regarded as the characteristic isthmus cell (Fig. 5*b*). Its height is 10 to 30 μ, and has high (0.8 to 1.2 μ), relatively regular microvilli. The endoplasmic reticulum is moderately developed and is often localized below the nucleus, comprising groups of many granular (rough) membranes without cisternae. Some agranular (smooth) membranes are also noticed. The Golgi apparatus is well developed. Dense granules are seen rarely. Glycogen granules occur frequently at day 2 when the eggs pass through the isthmus. The nucleus is centrally placed and ovoid; the nuclear membrane has few indentations.

The only change observed in the isthmus is the presence of glycogen during day 2. The isthmic cells have the general appearance of a moderately active cell, perhaps secreting a less proteinaceous, but a more carbohydrate-rich, fluid than does the ampulla cell.

4) The *junctura* constitutes the last loop of the extramural and intramural portions of the oviduct. The external loop rests on the surface of the uterus and bends sharply toward the antimesometrial side of the uterine wall. The intramural portion will then end as the colliculus tubarius. The muscular tissue forms an even thicker coat than the isthmus and consists mostly of circularly arranged cells. The intramural part of the muscle layer is thinned toward the top of the colliculus. Where the oviduct joins the uterus, longitudinally arranged muscle cells form a direct continuity with the uterine muscle coat; thus no longitudinal muscle cells line the intramural part.

The lamina propria is a dense, fibrous tissue and forms a framework of rather high longitudinal folds along the junctura (Fig. 6*c*). In the last portion of the colliculus, the folds are lacking. The colliculus constitutes a rather rigid structure. Thus, when the lumen of the uterus is distended during estrus, the colliculus protrudes into the uterine lumen; otherwise, it gives the impression of a uterine crypt. Immersion fixation may give a false appearance of the colliculus and may give the junctura and the uterus a wide lumen. After an appropriate preservation by perfusion, the lumen of the junctura is narrow (Fig. 6*a*), resulting in a close contact between the epithelial cells and the sperm or the almost spherical eggs (Fig. 6*c*).

Ciliated cells are absent in the junctura. The nonciliated cells have a characteristic

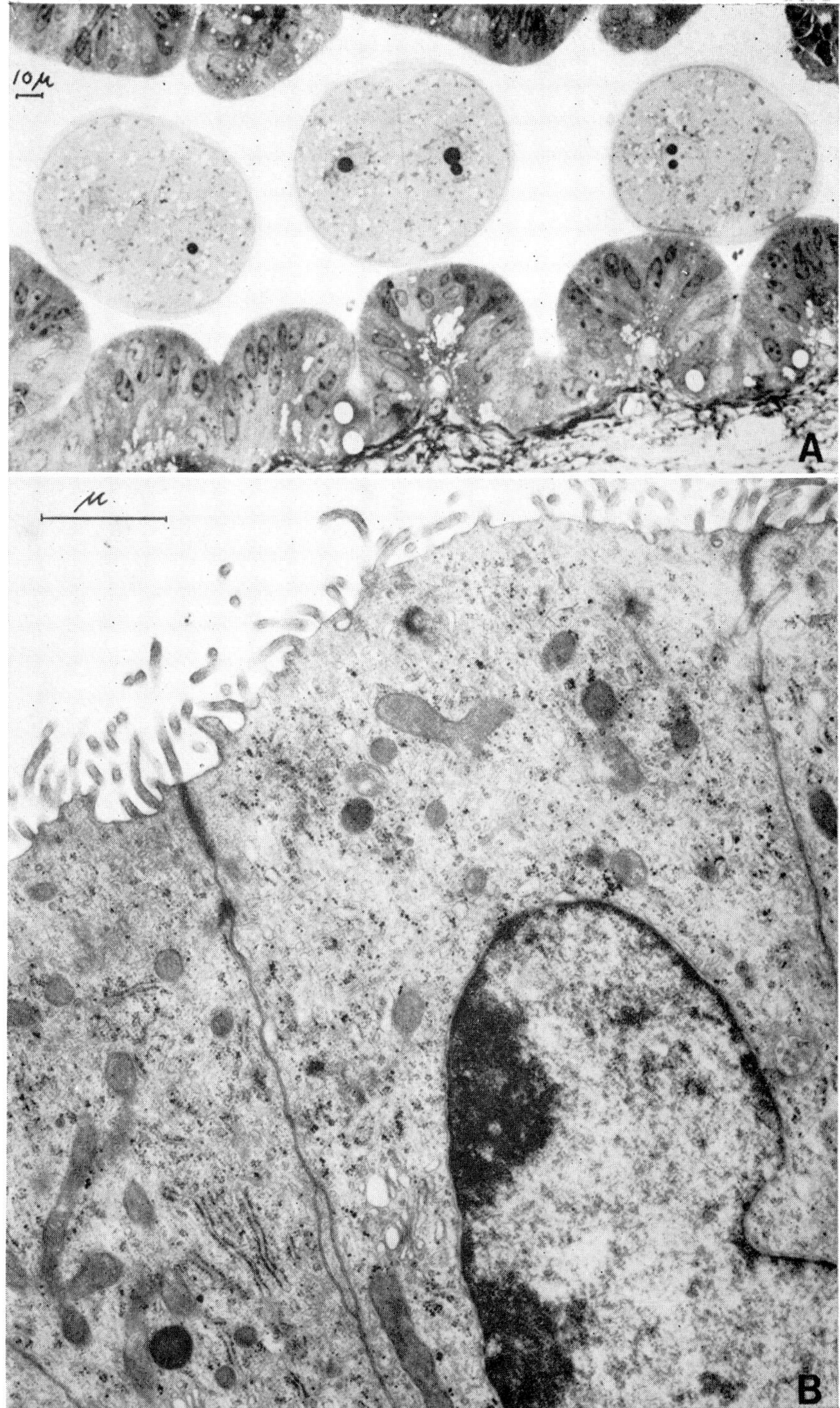

FIG. 5. (*a*) Mouse, isthmus (day 2, 4 P.M.). Longitudinal section with three 2-cell eggs in the lumen. The mucosa forms circular ridges into the lumen. Light microscopy. (×370).

(*b*) Mouse, isthmus (day 2, 5 P.M.). Apical part of nonciliated cell, demonstrating more microvilli and less developed endoplasmic reticulum compared to the nonciliated cells of the ampulla (Fig. 4*a*). Further, the isthmic cell lacks the dense granules. (×17,000).

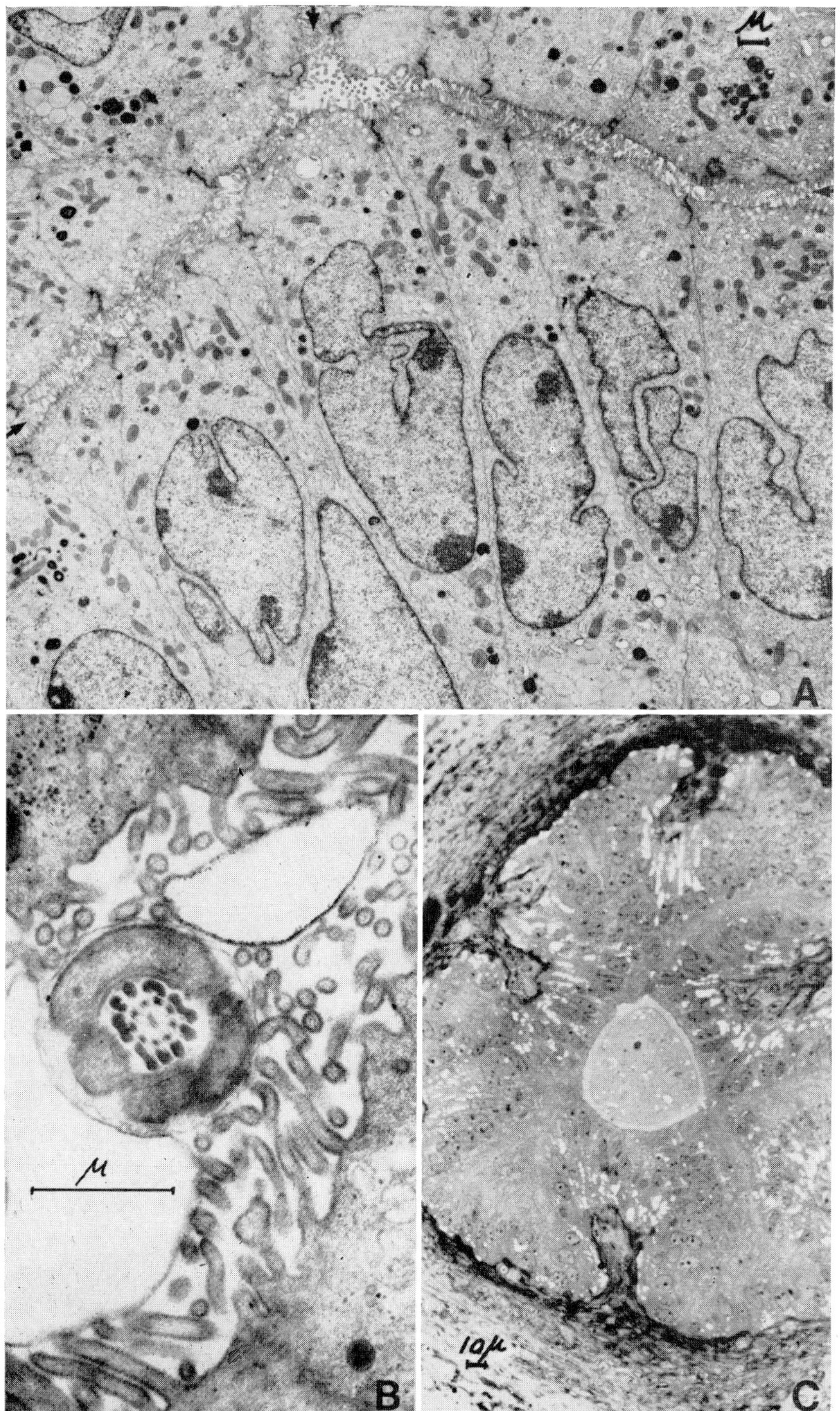

FIG. 6. (*a*) Mouse, junctura (day 5, 5 P.M.). Rows of epithelial cells, outlining a very narrow lumen (→). The epithelial cells are characterized by small apical vesicles and an indented nucleus. (×4,000).

(*b*) Mouse, junctura (day 0, 10 P.M., 45 min after copulation). A middle-piece of sperm is observed among the microvilli in the narrow lumen. Luminal blebs lie on each side of the sperm (×20,000).

(*c*) Mouse, junctura (day 3, 5 P.M.). Cross section showing an egg in very close contact with the luminal epithelium. The darkly stained material in the periphery is connective tissue of the propria. Light microscopy. (×100).

appearance. Their height is about 12 to 35 μ. Their surface membrane shows regular microvilli of 0.5 μ in length. The Golgi apparatus is moderately developed, and the endoplasmic reticulum, consisting of short, granular membranes, is scattered in the cytoplasm. The cytoplasm is rather dense. At the basal end of the nucleus, the cell is almost filled with lipid granules. Glycogen granules are apparent at the time of sperm migration. Most mitochondria are situated apically, while they are scattered in the nonciliated cells previously described. Apical vesicles, 0.1 to 0.2 μ in diameter, are present in the cytoplasm (Fig. 6*a*). They are devoid of dense material and surrounded by a single membrane. During day 3—the day of egg passage—these apical vesicles are most numerous.

The nucleus, situated in the middle third of the cell, is long and has several deep, irregular indentations of the nuclear membrane.

It is interesting to note that the nonciliated cells of the junctura are similar to the epithelial cells of the uterus during its progestational or preattachment stage (days 3 and 4). In the uterine horns, the epithelial cells later attain another structure. When the uterus turns into its gestational or attachment stage (day 5 and later), the dominating feature will be a close attachment (a 250 Å distance) between apposing cell surfaces of the luminal epithelium (Nilsson 1966*b;* Reinius 1967). We first thought that the epithelium of the junctura also had this capacity, which would have been a nice explanation of "tube locking," but we found that it neither changed in structure nor formed as closely apposed luminal membranes as in the uterus. Our hypothesis is that the epithelial lining of the junctura, like that of the uterus in progestation, inhibits implantation of the blastocysts, and that the epithelial cells of the junctura maintain this activity during the gestational stages of the endometrium. Probably the apical vesicles are important in this process by controlling, either through secretion or through resorption, the composition of the luminal fluid in the junctura and uterus (Agduhr 1927; Fekete 1941; Sobotta 1895; Wimsatt and Waldo 1945).

B. *The Rat*

The following description is based mainly upon results obtained in our laboratory. Additional references are given at the end of the section. The rat oviduct has the same anatomical parts as and a similar organization to that described for the mouse (Fig. 2). Some differences exist, however, in the ultrastructure of the nonciliated cells of the epithelial lining.

The nonciliated cells of the preampulla have essentially the same structure as those in the mouse, showing sparsely distributed microvilli, apically dense granules, and a moderately developed endoplasmic reticulum. In the ampulla, the nonciliated cells have rather regular microvilli, 2 μ long. There are few apical granules. The granular endoplasmic reticulum is moderately to extensively developed and has some dilated cisternae among the parallel membranes.

The isthmus demonstrates very long and regular stereocilia up to 5 μ in length on the nonciliated cells (Fig. 7*a*). Granules of varying density occur in great number in the slightly bulging apical cytoplasm. In light microscopy, the tufts of stereocilia and apical granules can look like kinocilia with basal corpuscles, which has caused errors in estimating the proportion of ciliated and nonciliated cells in the isthmus of the rat. The endoplasmic reticulum is rather extensively developed without cisternae and often contains some groups of parallel membranes at the basal end of the cell. The Golgi apparatus is well developed.

In the junctura (Fig. 7*b*), the microvilli are about 1 to 2 μ long. Only a few granules are noticed. Apical vesicles are numerous at day 3 of pregnancy. In contrast to the findings in

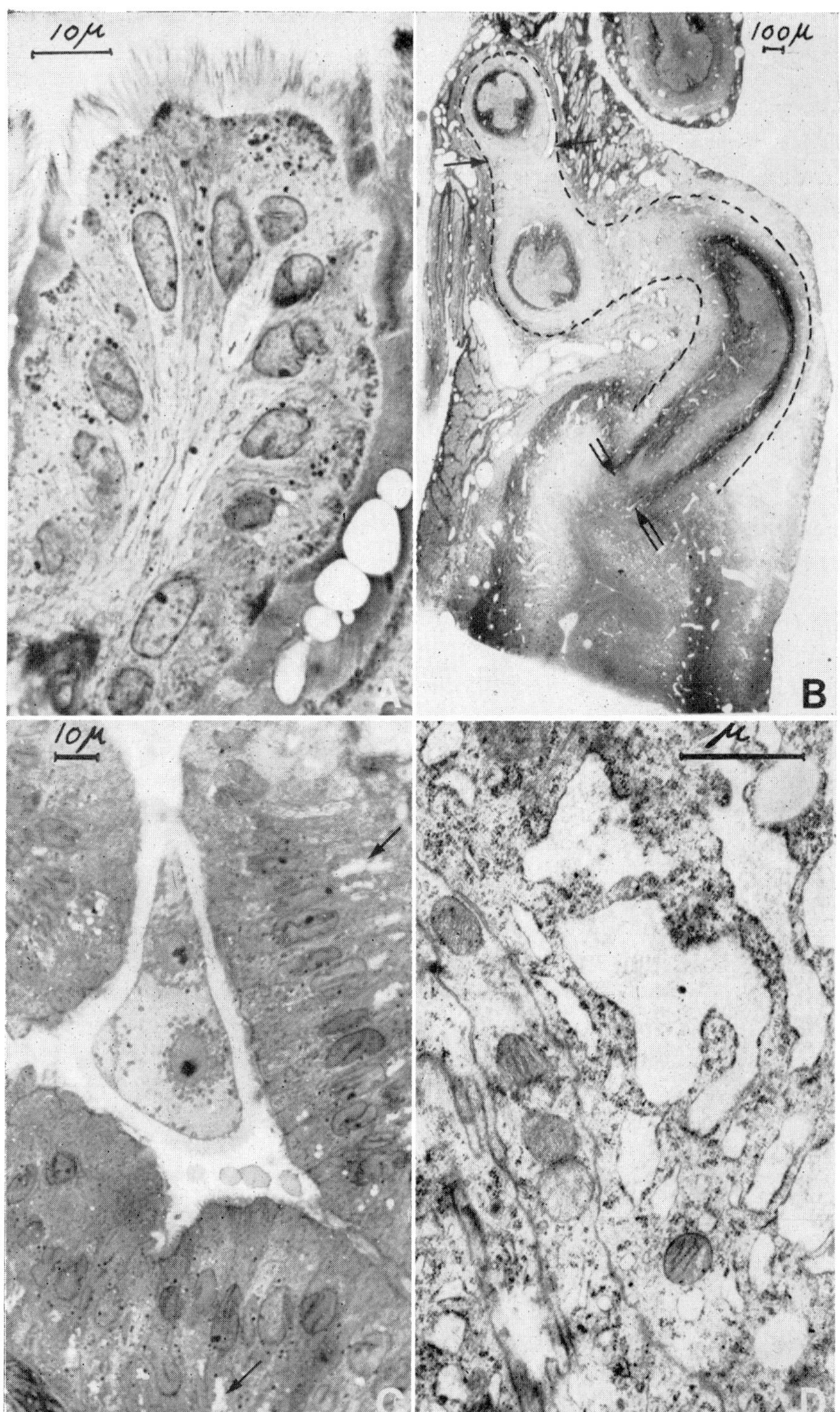

FIG. 7. (*a*) Rat, isthmus (day 4, 9 A.M.). An epithelial fold consisting of nonciliated cells characterized by long stereocilia and dark granules apically. Light microscopy. (×1,100).

(*b*) Rat, junctura (day 4, 9 A.M.). Survey picture demonstrating the extramural and intramural parts of the junctura and the opening of the junctura into the uterus: outline of the junctura (===), external junction of oviduct and uterus (→), internal junction of oviduct and uterus (⇒). The connective tissue is dark-stained. An egg lies in the uterine lumen. Light microscopy. (×27).

(*c*) Rat, junctura (day 4, 9 A.M.). An egg surrounded by zona pellucida is squeezed between apposing mucosal folds. The nonciliated cells contain dilated cisternae of the endoplasmic reticulum in their basal part (→). Light microscopy. (×600).

(*d*) Rat, junctura (day 3, 9 P.M.). Basal part of a nonciliated cell, demonstrating dilated cisternae of the endoplasmic reticulum. (×17,000).

the mouse, the endoplasmic reticulum forms extremely distended cisternae (Fig. 7*d*) localized below the nucleus. The lipid granules, observed in the corresponding segment in the mouse, are almost lacking. Also in the rat, the junctura forms a tight contact with the egg, but the rat eggs take different peculiar forms when squeezed between the epithelial membranes (Fig. 7*c*).

Thus, one outstanding difference between the two animals is the presence of long stereocilia and granules in the isthmus of the rat. Another difference is the dilated endoplasmic reticulum in the junctura of the rat. It seems that the eggs of the rat are more flexible than those of the mouse, since the eggs of the rat adapt themselves to the form of the junctural lumen, while the eggs of the mouse maintain their spherical form (Alden 1942*a*, *b*; Andersen 1928; Borell *et al.* 1959; Bronzetti *et al.* 1963; Huber 1915; Kellogg 1945; Nilsson 1957; Wolf 1960).

C. *The Guinea Pig*

The following description is based mainly upon results obtained in our laboratory. Additional references are given at the end of the section.

The oviducts have been obtained from animals in metestrus with unfertilized eggs in the ampulla. The oviduct is longer than that of the rat and is connected to an ovarian bursa by a lansiform-shaped fimbrial apparatus (Fig. 2). No outer, evident anatomical demarcations indicating the oviductal parts are seen. The tunica muscularis is thin in the ampulla but thick in the isthmus, where an outer circular and an inner longitudinal layer, both of about equal thickness, are present (Fig. 8*c*).

The preampulla, especially the fimbriae, contains a high proportion of ciliated cells, among which are solitary nonciliated cells. The nonciliated cells have a narrow basal part and a bulging apex covered with irregularly distributed, short microvilli. Granules with a maximum diameter of 0.6 μ and a varying density are more or less numerous in the apical cytoplasm. The endoplasmic reticulum consists of rough membranes with a few cisternae. The ampulla contains more nonciliated than ciliated cells (Fig. 8*a*, *b*). The character of these nonciliated cells is similar to that of the nonciliated cells of the preampulla, although they bulge less.

The isthmus (Fig. 8*c*) has a few ciliated cells, often located on the ridges of the longitudinal folds. The nonciliated cells are similar to those in the preampulla and the ampulla, though the luminal surface does not bulge. Toward the junctura, some nonciliated cells display at the luminal side of the nucleus, a large, about 2 μ, solitary vesicle, which sometimes contains membranous formations. The junctura also shows some ciliated cells. The nonciliated cells differ from the proceeding ones only in having more apical granules. The large vesicle is often seen. Thus, the oviductal epithelium of the guinea pig also has its own peculiar appearance. It differs from that of the mouse and rat by having, for instance, more granules less characteristically distributed in the different parts of the oviduct. The physiological implication of this is not yet known (Groodt *et al.* 1960; Kelly 1927; Lucas 1930).

D. *Ruminants*

1. *Sheep*. The upper half of the sheep oviduct includes the preampulla and the ampulla, the lower half of the isthmus, and the junctura. The ampullary segment is wide, and has a thin wall and many mucosal ridges in the lumen. The isthmic segment is narrow, but the wall is relatively thick and muscular, and the mucosa forms lower ridges. The epithelium,

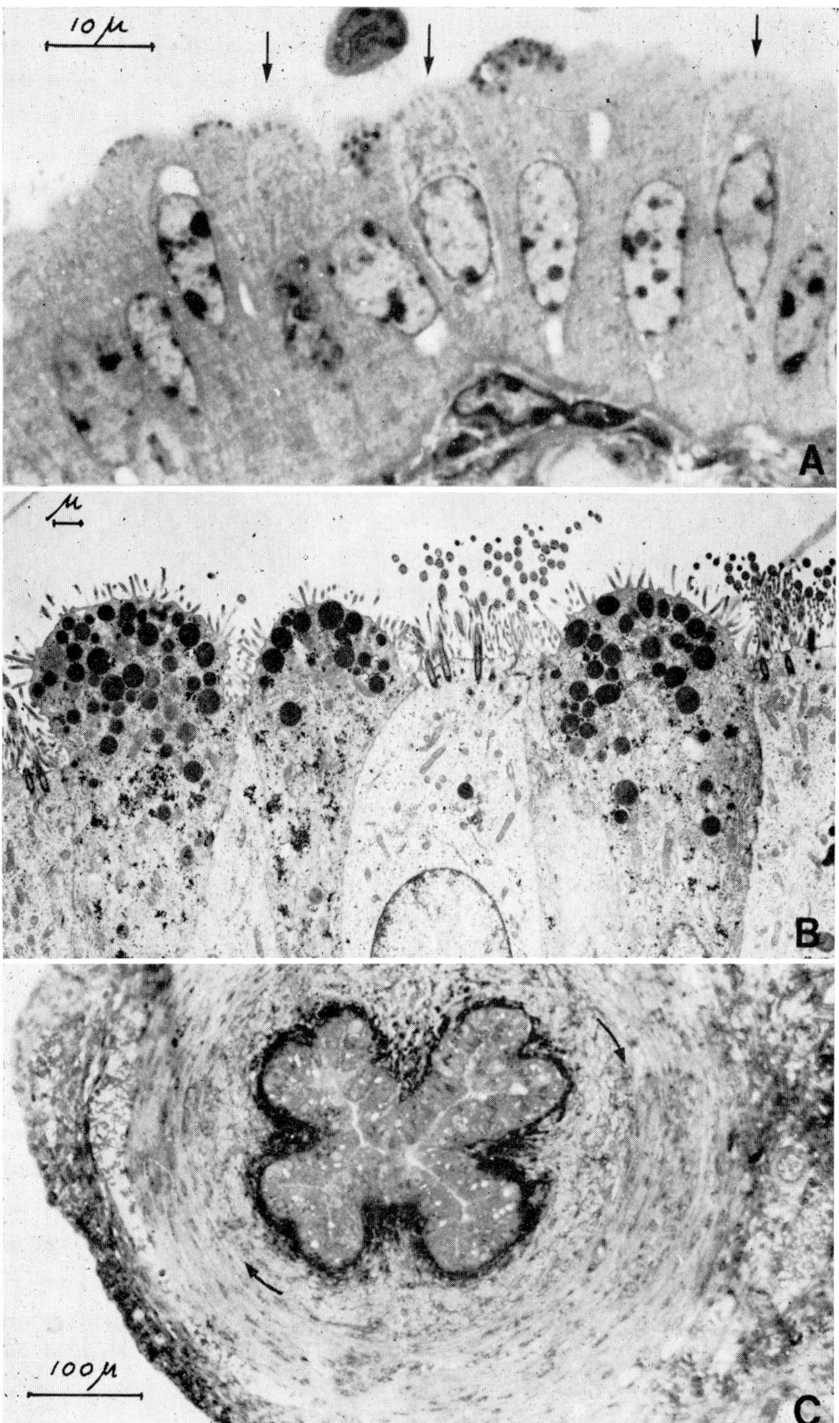

FIG. 8. (*a*) Guinea pig, ampulla (metestrus). Ciliated (→) and nonciliated cells. In some of the latter cells granules are observed. A corona radiata cell is noticed above the epithelial membrane. Light microscopy. (×1,500).

(*b*) Guinea pig, ampulla (metestrus). Ciliated and nonciliated cells with granules are observed. Electron microscopy of an area, similar to that demonstrated by light microscopy in the preceding figure. (×4,000).

(*c*) Guinea pig, isthmus (metestrus). Cross section, demonstrating the narrow, star-shaped lumen; the connective tissue of the tunica mucosa (dark-stained); the inner longitudinal and the outer circular muscle layer; and the tunica serosa. The arrows point to the region between the two muscle layers. Light microscopy. (×160).

as observed with the light microscope, consists of ciliated and nonciliated cells. Both of these are taller in the ampulla than in the isthmus. They undergo a gradual increase in height throughout the oviduct during the ovarian cycle, reaching their maximum height (30 to 35 μ) in estrus. Morphological changes among the segments of the oviduct are described only sparsely as yet; the electron microscope will probably reveal much more.

The preampulla is the part that—contrary to what holds for the other species of this review—contains the highest proportion of nonciliated cells. These cells are not as numerous in the ampulla as in the preampulla. The nonciliated cells in the preampulla and ampulla contain many granules, most abundant during estrus. Cytoplasmic projections are an additional feature of the nonciliated cells in the ampulla.

The isthmus contains proportionally fewer nonciliated cells, which differ from the nonciliated cells of the ampulla by the absence of the cytoplasmic projections. Two additional, sparsely occurring, cell types have been reported, rod cells and round cells. The rod cells appear during metestrus. They are as tall as the ciliated and nonciliated cells. They may be derived from the nonciliated cells by the extrusion of their nuclei and may represent a degenerated form of the nonciliated cells. The small round cells (probably similar to the basal cells) are observed during diestrus and proestrus periods. They lie between the bases of the other cells (Andersen 1928; Hadek 1955; Restall 1966).

2. *Cow.* The cow has two waves of follicular growth during an estrous cycle, which makes it difficult to interpret the hormonal influence on the oviductal epithelium. Moreover, different functional states seem to be represented within a single histological section. Ciliated and nonciliated cells are present. To date, only the ultrastructural characteristics of the ampulla epithelium have been published. The ciliated cells have the usual appearance and do not change during the estrous cycle. The nonciliated, granule-containing cells may have the same height as the ciliated cells or bulge over them. The endoplasmic reticulum contains cisternae. The granules are most numerous during the follicular phase. They appear as small dense granules but grow larger and less dense, attaining a membranous structure. Also in the epithelium of the cow, rounded basal cells have been noticed (Andersen 1928; Björkman and Fredricsson 1960, 1961).

E. *The Rabbit*

The ampullar region amounts to about half the length of the oviduct (Fig. 2). The border zone is visible, since the ampullar and isthmus regions differ in diameter. In the ampulla, the circular muscle layer is thin, whereas it is thicker in the isthmus. In the latter segment, inner longitudinal and some bands of outer longitudinal muscle cells are also found. The mucosal membrane of the ampulla forms high branched folds, while the mucosal folds of the isthmus are shorter and thicker. In the epithelial lining, both ciliated and nonciliated cells are present. The nonciliated cells occur in two types: one confined to the preampulla, the other distributed both in the ampulla and the isthmus.

The preampulla nonciliated cell has an upper surface that bulges into the lumen and possesses small microvilli. No granules are present in the cytoplasm; the endoplasmic reticulum has dilated cisternae. The ultrastructure of this cell appears similar both before and after ovulation.

The nonciliated cells of ampulla and isthmus also have a bulging luminal surface and a few short microvilli, but their main feature is the many granules (Fig. 9*a*). In estrus, most of the granules are dense with a maximum diameter of about 1 μ; but after ovulation, they become even larger and much lighter, containing some dark spots. The preovulatory gran-

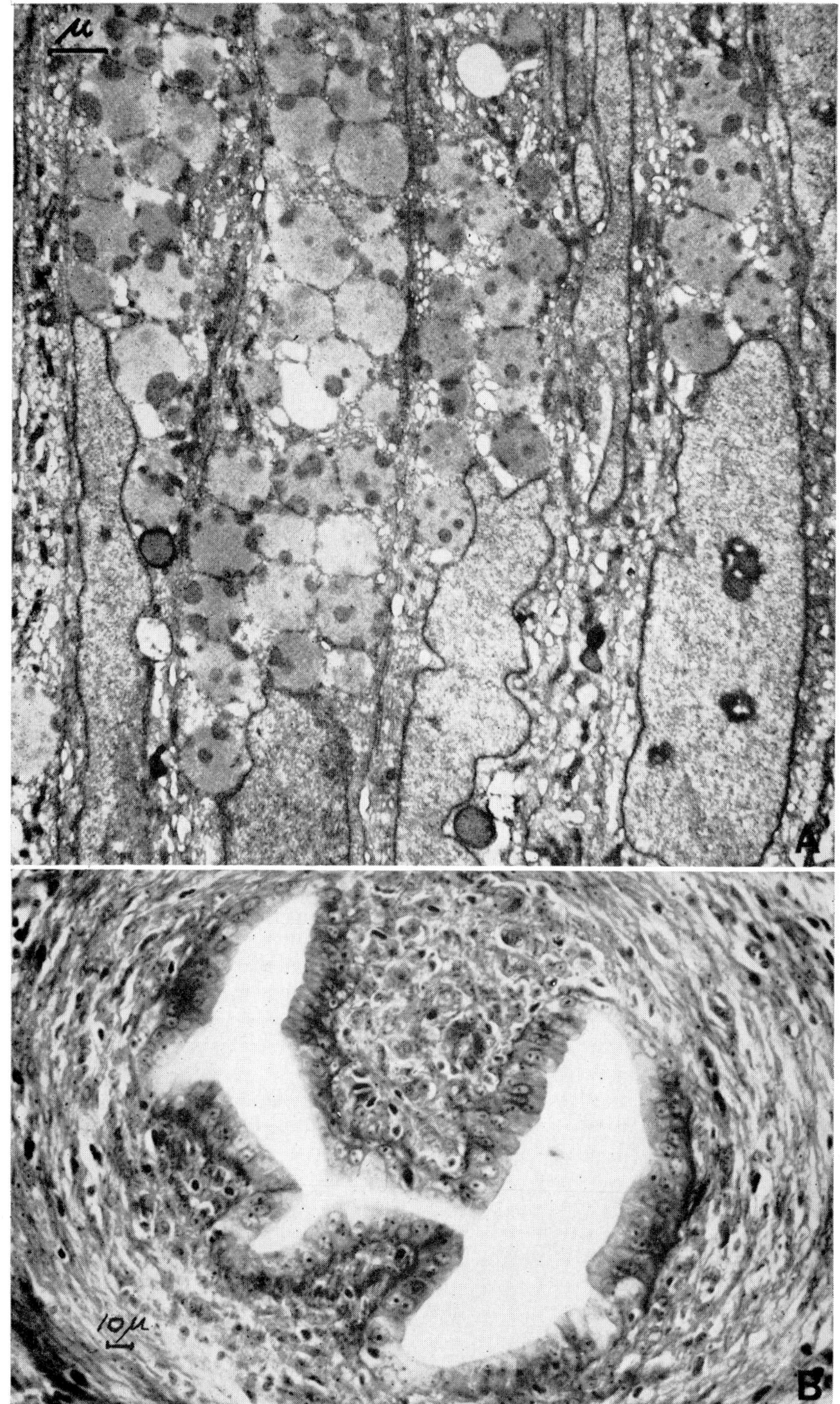

FIG. 9. (*a*) Rabbit, isthmus (day 3). Several granules of a low density and with small dark spots are present in the apical parts of the nonciliated cells. (×8,000).

(*b*) Rabbit, junctura (day 3). An ordinary immersion fixation and paraffin embedding gives a falsely wide lumen. No ciliated cells are present. Light microscopy. (×340).

ules remind one of protein-rich granules, such as the zymogen granules of the pancreas, while the postovulatory granules are of the mucinous type. The abundant occurrence of mucin granules in the oviductal epithelium during egg transport corresponds to the time that the rabbit egg is coated with mucin.

In the junctura the mucosal folds terminate in long processes, which project into the uterine lumen (Fig. 9*b*). Ciliated cells occur but are fewer than the nonciliated cells. The lumen is narrow in appropriately fixed materials (Reinius, preliminary results), but otherwise wide (Fig. 9*b*) (Andersen 1928; Borell *et al.* 1956; Fredricsson 1959; Greenwald 1961; Harper 1961*a*, *b;* Hashimoto *et al.* 1959*a*, *b;* Nilsson 1958; Nilsson and Rutberg 1960).

F. *Man*

The oviduct bends slightly, and only a gradual macroanatomical difference can be observed between the ampulla and the isthmus (Fig. 2). The tunica muscularis comprises 3 layers: an inner longitudinal, an intermediary circular, and an outer longitudinal layer. The circular layer is most extensive in the intramural part; the longitudinal layers increase toward the uterus. Ciliated and nonciliated cells occur in about the same number in the preampulla and the ampulla (Fig. 10*a*).

Ciliated cells of the human oviduct differ from those of the oviducts in preceding species mentioned in that they show cyclical changes and have a characteristic ultrastructure of the cilia. The cyclical change most apparent is in the number of apical vesicles; these are numerous during the follicular phase and are regarded as pinocytotic vesicles. The cilia have cross-striated rootlets (Fig. 10*c*); the distance between the striations is about 450 Å, suggesting a contractile protein. An important question is whether the cross-striated rootlets endow the cilia with some functional specificity, such as concerns the beating frequency.

The nonciliated cells of the preampulla have small vesicles in the cytoplasm, but no granules. The cells do not show cyclical changes. In the ampulla, similar nonciliated cells are present, but closer to the isthmus, cells containing granules appear.

The isthmus contains a nonciliated cell which, during the follicular phase, contains granules varying from small and dense to large and light. During the luteal phase the large, light type of granule is dominant. Simultaneously, the endoplasmic reticulum shows cisternae dilated by some substance. The outer surface of these nonciliated cells has a thin coat which is most conspicuous during the follicular phase, while a mucin-like material appears during the luteal phase. Signs of increased activity of apocrine secretion are evident.

The junctura forms the most narrow part of the oviduct, having only sparsely developed mucosal folds and meeting epithelial surfaces (Fig. 10*b*). Both ciliated and nonciliated cells occur, the latter ones being more numerous (Fig. 10*c*) (Björkman and Fredricsson 1962; Clyman 1966; Fredricsson and Björkman 1962; Hashimoto *et al.* 1962, 1964; Horstmann and Stegner 1966; Lisa, Gioia, and Rubin 1954).

V. Luminal Changes

The width of the oviductal lumen is determined by muscular contractions, amount of interstitial fluid and luminal secretion. Since the width of the lumen ought to have great importance for the transport of sperm and eggs, physiological and morphological methods have been used for estimations of the luminal passage. As mentioned earlier, the methods of preparation commonly used do not preserve the lumen in its natural shape. Only freeze-

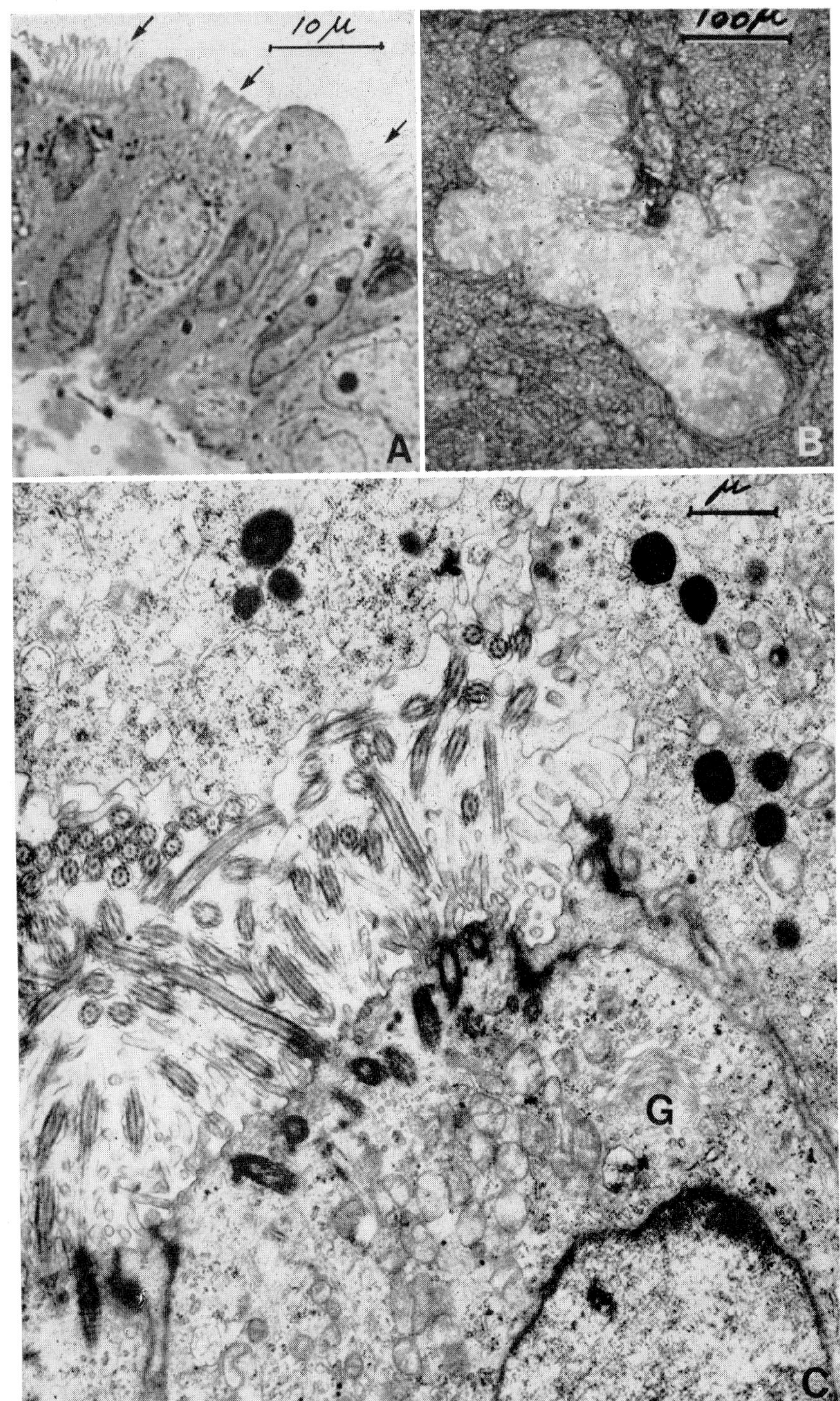

FIG. 10. (*a*) Man, ampulla (day 19 of the menstrual cycle). Epithelial membrane with ciliated (→) and nonciliated cells. Apical parts of nonciliated cells are bulging into the oviduct lumen. Light microscopy. (×1,500).

(*b*) Man, junctura (day 4 of the menstrual cycle). Cross section of the narrow lumen in the intramural part of the junctura. The inner longitudinal muscle cells, surrounded by dark-stained connective tissue, are observed. Light microscopy. (×150).

(*c*) Man, junctura (day 4 of the menstrual cycle). The narrow lumen of the intramural part is filled with kinocilia and microvilli. Ciliated cells are observed in the lower part of the figure and nonciliated cells in the upper part. The apical part of the ciliated cell contains many mitochondria and a Golgi apparatus. The striated rootlets of the basal corpuscles are seen. The nonciliated cells show some dark granules. (×12,000).

drying, cryostat sectioning, or our current technique with perfusion and plastic embedding seem to give reliable information.

In the preampulla of the mouse and rat, the lumen is always filled by epithelial folds, so that the cells on the opposite folds meet one another. In the ampulla of day 1, the ova with surrounding corona radiata cells lie free in a wide lumen. At the time of sperm migration and late egg transport only narrow luminal spaces can be observed. As yet, our only knowledge of the guinea pig is that the preampulla, ampulla, and adjoining part of the isthmus show rather wide luminal spaces in metestrus.

In the isthmus of the mouse and rat, the loops containing the eggs are distended by fluid (Fig. 5*a*), while the other loops have closed lumina.

The junctura lumen of the mouse, rat, guinea pig, and rabbit is constantly closed in perfusion-fixed material but varying in immersion-fixed material. The microvilli of the apposing nonciliated cells (and ciliated cells in guinea pig and man) meet in the middle (Fig. 6*a*). The appearance is similar when the sperm are within the junctura (Fig. 6*b*), and even when the eggs pass the segment (Figs. 6*c*, 7*c*). Both mouse and rat eggs are tightly surrounded by the mucosa, the mouse egg being spherical, the rat egg conelike deformed, always having the top of the cone directed toward the uterus.

VI. Comments

A. *Oviductal Secretion and Resorption*

The oviductal secretions no doubt serve several functions. It is apparent also that the properties of the secretions differ under various hormonal conditions, as well as in different segments of the oviduct. The mucosa is responsible for control of the secretion. Since certain functional states may be associated with definitive structural characteristics, an understanding of the morphology of the oviductal epithelium may help elucidate its physiology. It is quite probable that, when estimating a functional stage, the degree of ultrastructural change may be a more convenient parameter than, for instance, some biochemical analysis. It is, at present, difficult to relate ultrastructure to function, but some hints of the oviductal secretion and resorption have been obtained.

Secretions into the lumen of the oviduct are controlled by the surface membrane of the epithelial cells and possibly the region of the terminal bars. All substances, whether moving in or out of these cells, must pass through one of these structures. The larger the epithelial surface and the looser the underlying connective tissue, the better the possibilities for a transport of fluid out into the lumen.

Transport mechanisms available to the cell membrane are free diffusion, carrier mediated transport, and vesiculation (Csáky 1965). It might be assumed that increased activity of the first two processes would be enhanced by an increase in the cell surface area in the form of microvilli, and that the process of vesiculation could be associated with the apical vesicles. Thus, long microvilli and apical vesicles (whether secretory or pinocytotic) may indicate increased activity of these transport mechanisms. The problem that confronts us is that we cannot tell from ultrastructure alone whether what we see is a secretion, a resorption, or, most probably, a combination of the two.

The intercellular spaces are equipped with structural specializations in the form of tight junctions (Farquhar and Palade 1963). The tight junctions and similar structures are probably involved in controlling the passage of fluid in the intercellular spaces. The luminal membrane sometimes forms projections (e.g., in man) with a homogeneous granular interior, which might imply a proteinaceous material. The projections are probably signs of apocrine secretion. Circular membranous structures, "luminal blebs," are found frequently in the oviductal lumina of the mouse and rat. The apical protrusions and luminal blebs seem to increase in number in poorly fixed material and cannot conclusively be interpreted as a secretion.

The intracellular structures reported here, which can be ascribed to the secretory or resorptive functions, are (1) the Golgi apparatus and apical vesicles, (2) the granules, and (3) the endoplasmic reticulum.

1) The Golgi apparatus schematically consists of groups of parallel membranes, more or less dilated, and of small vesicles which are lying in clusters around the membranes (Figs. 3*a*, *b*). Many Golgi membranes and many Golgi vesicles probably indicate a budding off of vesicles from the membranes (for references, see Fawcett 1966). The Golgi vesicles have the capacity to absorb proteinaceous substances and transform them into granules.

2) The granules differ in structure, numbers, and location among various nonciliated cells and differ often in the same cell type during different functional states (Figs. 3*b;* 8*a*, *b;* 9*a;* 10*c*). These granules constitute a specific group of granules which differ from the lysosomes, the lipid granules, and similar well-defined groups of granules. It should be noted, however, that regular changes in the number of glycogen granules occur: the granules are most numerous in the isthmus, when the eggs pass, and in the junctura at the time of sperm migration. This points to an increased, segmentally localized, concentration of carbohydrates in the oviductal secretion during these stages, which is consistent with the energy requirements of the early mouse eggs, as studied by *in vitro* experiments (Biggers, Whittingham, and Donahue 1967, Brinster 1965, and Whitten 1957), and probably also experiments of migrating sperm cells (Mann 1964). The granules may be either a product of secretion or a concentration of absorbed substances. Therefore, the designation "secretion" granules should be used with caution. To avoid premature interpretation, we have decided to call them merely "granules."

The ultrastructure of the granules gives some clue to their composition: dense granules are proteinaceous, perhaps enzyme-rich (Fig. 3*b*), while less dense granules are mucinous (Fig. 9*a*). Dense granules can be transformed into less dense granules (e.g., in the rabbit and man).

3) The endoplasmic reticulum and the free ribosomes produce the proteins of the cell. Generally, a well-developed endoplasmic reticulum and many free ribosomes indicate more a secretory than a resorptive function.

The produced proteins might either be delivered outside the endoplasmic membranes into the cytoplasm or remain inside the endoplasmic membranes. Delivered into the cytoplasm, the proteins may form (secretory) granules; inside the endoplasmic membranes, the proteins may form granules or cisternae and be transported outside the cell, since the endoplasmic reticulum is said to open at the cell surface. There is an intricate system of free ribosomes, endoplasmic reticulum, rough or smooth, and with or without cisternae in the cytoplasm. Consideration of the interrelationship among these components will probably tell us a great deal in the future. Whatever it means, we now know that some nonciliated cells in the oviduct have (*a*) a moderately developed endoplasmic reticulum (e.g., in the preampulla of the mouse and rat, and the isthmus and junctura of the mouse [Figs. 5*b*, 6*a*]), (*b*) an extensive endoplasmic reticulum without cisternae (e.g., the ampulla and isthmus of the rat), and (*c*) an extensive endoplasmic reticulum with many cisternae (e.g., the ampulla of the mouse [Fig. 4*a*] and cow, and the junctura of the rat [Fig. 7]). The ciliated cells, on the contrary, have a poorly developed endoplasmic reticulum (Fig. 3*a*).

To conclude, the ultrastructure of the epithelium in an oviduct from a single species differs sufficiently among its segments to indicate variations in, for instance, functional processes or biochemical composition of the luminal secretions. Therefore, the ultrastructure of the oviductal epithelium will probably prove to be a valuable parameter in experiments on the oviductal function.

B. *Transport of Sperm and Eggs*

In the mouse reproductive system, when prepared for light microscopy, many sperm heads are observed at the internal junction of the oviduct and uterus, only a few in the narrow lumen of the junctura, and hardly any in the isthmus. In the electron microscope, however, tails, middle-pieces, and heads are easily distinguished in the junctura (Fig. 6*b*). A few sperm cells can also be seen in the isthmus, either free in a wide lumen or in narrow passages between the folds. The mechanisms of the sperm migration have been reviewed by Bishop (1961) and Sobrero (1963). Our observations in the mouse have added some new data on the entering and migration of sperm in the junctura.

At copulation, the oviduct protrudes into the uterine lumen as a colliculus. Sperm cells migrate randomly in the uterine lumen; many are trapped in crypts of the uterus, but some sperm cells penetrate into the narrow opening of the oviduct on the colliculus. Then they almost climb on the surrounding microvilli of the tightly apposed luminal surfaces and follow the longitudinal ridges of the junctura.

A narrow junctural lumen seems to be necessary for successful fertilization. It has to be narrow in order to take the sperm to the ovarian end of the oviduct by forcing them to keep to the initial direction of the migration. Consequently, we think that a narrow lumen of the junctura will be found to be a rather common feature among the species, provided that the fixation is of an adequate quality.

The mechanisms capable of influencing the transport of eggs are (1) muscular contractions of the tunica muscularis and of the still hypothetical muscularis mucosae; (2) volume changes of the lamina propria; (3) the beating of the kinocilia; and (4) the secretory activity of the lamina epithelialis (Austin 1963; Blandau 1961). We will examine to what extent these mechanisms are operating when considering the morphology of the different parts of the oviduct in the animals previously described.

The eggs have to be transported from the ovary through the preampulla into the ampulla, where fertilization occurs. The preampulla has a very thin wall, an extensive folding of the mucosal membrane, and mostly ciliated cells (except in the sheep). Thus, the transport of eggs in the preampulla may rely less on contractions of the tunica muscularis and more on activities of the tunica mucosa.

In the ampulla, the tunica muscularis has increased only slightly in thickness, the mucosal folding is less extensive, and the proportion of ciliated cells is lower. In the mouse and rat, the ampulla is distended by fluid at the time of fertilization. The eggs of the mouse stay up to 24 hr in the ampulla, near the junction to the isthmus (Fig. 4*b*). The activity of the secretory processes is hard to tell, but the nonciliated cells have intracellular signs of secretion. The transport mechanism in this part could be a mere flow of the eggs in the oviduct secretion with superimposed contractions of the tunica muscularis and ciliary activity.

The ampullary-isthmic junction seems to block a further transport of eggs as long as these are surrounded by granulosa cells. Since the opening of the isthmus is not wide enough even for single eggs (Fig. 4*b*), the ampullary-isthmic junction has to widen either by a relaxation of the muscular coat or a diminishing of the mucosal folds.

The isthmus has a thick muscular layer. Epithelial foldings are scanty in the mouse and rat but more extensive in the rabbit and man. The number of ciliated cells also varies; in the mouse and rat the cells are rare, in the guinea pig a few are situated on the epithelial ridges, and in the rabbit and man they are relatively numerous. Thus, muscle contractions, but not ciliary beating, should influence the transport of eggs, at least in the mouse and rat. In these two species, various parts of the isthmus have a wide lumen, implying both an active secretion and oviductal constrictions due to the muscular activity. The time spent by the eggs of the mouse in the isthmus is about 1½ days.

The junctura of the mouse and rat demonstrates higher mucosal folds than does the isthmic part, while in the guinea pig, rabbit, and man the folds are lower in the junctura than in the isthmus. Ciliated cells are absent in the mouse and rat junctura, and present in the guinea pig, rabbit, and man. Mouse eggs stay at least 6 hr, rat eggs up to 12 hr in the junctura. An evaluation of the transport of eggs in the mouse and rat junctura has to consider the following morphological features: a thick muscular coat, a thick mucosa, a very narrow lumen, and eggs tightly surrounded by luminal epithelium (Figs. 6*c*, 7*c*). The physiological significance of these various features is unknown. Further studies on the morphological relations between oviductal epithelium and sperm or eggs in animals, fixed by perfusion and embedded in plastic, should be of great value in understanding the physiological mechanisms of oviductal transport.

Acknowledgment

This work has been supported by the Swedish Medical Research Council (Project 12X-70-03) and "Sällskapet för Medicinisk Forskning." Mrs. Brita Sydh and Mrs. Ingegerd Mjöbäck assisted in compiling the manuscript. Mrs. Gunnel Lindell provided much able assistance with electron microscopy and Miss Anne-Marie Ström with the preparation of the specimens. Mr. A. Englesson skillfully did the photographic work.

References

1. Agduhr, E. 1927. Studies on the structure and development of the bursa ovarica and the tuba uterina in the mouse. *Acta Zoologica* 8:1.
2. Alden, R. H. 1942*a*. The periovarial sac in the albino rat. *Anat. Rec.* 83:421.
3. ———. 1942*b*. The oviduct and egg transport in the albino rat. *Anat. Rec.* 84:137.
4. Andersen, D. H. 1928. Comparative anatomy of the tubo-uterine junction: Histology and physiology in the sow. *Am. J. Anat.* 42:255.
5. Austin, C. R. 1963. Fertilization and transport of the ovum. In *Mechanisms concerned with conception*, ed. C. G. Hartman, p. 285. Oxford: Pergamon Press.
6. Biggers, J. D., Whittingham, D. G., and Donahue, R. P. 1967. The pattern of energy metabolism in the mouse oöcyte and zygote. *Proc. Nat. Acad. Sci. U.S.A.* 58:560.
7. Bishop, D. W. 1961. Biology of spermatozoa. In *Sex and internal secretions*, ed. W. C. Young, vol. 2, p. 707. Baltimore: Williams and Wilkins Co.
8. Björkman, N., and Fredricsson, B. 1960. The ultrastructural organization and the alkaline phosphatase activity of the bovine Fallopian tube. *Z. Zellforsch.* 51:589.
9. ———. 1961. The bovine oviduct epithelium and its secretory process as studied with electron microscope and histochemical tests. *Z. Zellforsch.* 55:500.

10. ———. 1962. Ultrastructural features of the human oviduct epithelium. *Int. J. Fertil.* 7:259.
11. Blandau, R. J. 1961. Biology of eggs and implantation. In *Sex and internal secretions*, ed. W. C. Young, vol. 2, p. 797. Baltimore: Williams and Wilkins Co.
12. Borell, U., Gustavsson, K.-H., Nilsson, O., and Westman, A. 1959. The structure of the epithelium lining the Fallopian tube of the rat in oestrus. *Acta Obstet. Gynec. Scand.* 38:203.
13. Borell, U., Nilsson, O., Wersäll, J., and Westman, A. 1956. Electron-microscope studies of the epithelium of the rabbit Fallopian tube under different hormonal influences. *Acta Obstet. Gynec. Scand.* 35:35.
14. Borell, U., Nilsson, O., and Westman, A. 1957. Ciliary activity in the rabbit Fallopian tube during oestrus and after copulation. *Acta Obstet. Gynec. Scand.* 36:22.
15. Brinster, R. L. 1965. Studies on the development of mouse embryos in vitro. II. The effect of energy source. *J. Exp. Zool.* 158:59.
16. Bronzetti, P., Mazza, E., Milio, G., and Motta, P. 1963. Su alcuni aspetti dell'attività fosfatasica alcalina e della P.A.S.-reattività dell'epitelio della tuba uterina di Mus rattus albinus. *Biol. Lat.* 16:385.
17. Brundin, J. 1965. Distribution and function of adrenergic nerves in the rabbit Fallopian tube. *Acta Physiol. Scand.* 66 (suppl. no. 259):1.
18. Clyman, M. J. 1966. Electron microscopy of the human Fallopian tube. *Fertil. Steril.* 17:281.
19. Csáky, T. Z. 1965. Transport through biological membranes. *Ann. Rev. Physiol.* 27:415.
20. Eckstein, P., and Zuckerman S. 1956. Morphology of the reproductive tract. In *Marshall's physiology of reproduction*, ed. A. S. Parkes, vol. 1, part 1, p. 43. London: Longmans.
21. Farquhar, M. G., and Palade, G. E. 1963. Junctional complexes in various epithelia. *J. Cell Biol.* 17:375, 1963.
22. Fawcett, D. W. 1966. *The cell. Its organelles and inclusions.* Philadelphia: W. B. Saunders Co.
23. Fekete, E. 1941. Histology. In *Biology of the laboratory mouse*, ed. G. D. Snell, p. 81. New York: Blakiston Co.
24. Fischel, A. 1914. Zur normalen Anatomie und Physiologie der weiblichen Geschlechtsorgane von Mus decumanus sowie über die experimentelle Erzeugung von Hydro- und Pyosalpinx. *Arch. Entwicklungsmech. Organ.* 39:578.
25. Flerkó, B. 1954. Die Epithelien des Eileiters und ihre hormonalen Reaktion. *Z. Mikroskopischanat. Forsch.* 61:99.
26. Fredricsson, B. 1959. Proliferation of rabbit oviduct epithelium after estrogenic stimulation, with reference to the relationship between ciliated and secretory cells. *Acta Morphol. Néer.-Scand.* 2:193.
27. Fredricsson, B., and Björkman, N. 1962. Studies on the ultrastructure of the human oviduct epithelium in different functional states. *Z. Zellforsch.* 58:387.
28. Galey, F. R., and Nilsson, S. E. G. 1966. A new method for transferring sections from the liquid surface of the trough through staining solutions to the supporting film of a grid. *J. Ultrastruct. Res.* 14:405.
29. Greenwald, G. S. 1961. A study of the transport of ova through the rabbit oviduct. *Fertil. Steril.* 12:80.

30. Groodt, M. de, de Rom, F., Lagasse, A., Sebruyns, M., and Thièry, M. 1960. Détails de l'ultrastructure de l'épithélium cilié de la trompe. *Bull. Soc. Roy. Belg. Gynec. Obstet.* 30:347.
31. Hadek, R. 1955. The secretory process in the sheep's oviduct. *Anat. Rec.* 121:187.
32. Harper, M. J. K. 1961*a*. Egg movement through the ampullar region of the Fallopian tube. *Proc. Fourth Int. Congr. Anim. Reprod. The Hague* 2:375.
33. ———. 1961*b*. The mechanisms involved in the movement of newly ovulated eggs through the ampulla of the rabbit Fallopian tube. *J. Reprod. Fertil.* 2:522.
34. Hashimoto, M., Shimoyama, T., Mori, Y., Komori, A., Tomita, H., and Akashi, K. 1959*a*. Electron microscopic observations on the secretory process in the Fallopian tube of the rabbit. Report I. *J. Jap. Obst. Gyn. Soc.* 6:235.
35. Hashimoto, M., Shimoyama, T., Mori, Y., Komori, A., Kōsaka, M., and Akashi, K. 1959*b*. Electron microscopic observations on the secretory process in the Fallopian tube of the rabbit. Report II. *J. Jap. Obstet. Gyn. Soc.* 6:384.
36. Hashimoto, M., Shimoyama, T., Kōsaka, M., Komori, A., Hirasawa, T., Yokoyama, Y., and Akashi, K. 1962. Electron microscopic studies on the epithelial cells of the human Fallopian tube. Report I. *J. Jap. Obst. Gyn. Soc.* 9:200.
37. Hashimoto, M., Shimoyama, T., Kōsaka, M., Komori, A., Hirasawa, T., Yokoyama, Y., Kawase, N., and Nakamura, T. 1964. Electron microscopic studies on the epithelial cells of the human Fallopian tube. Report II. *J. Jap. Obst. Gyn. Soc.* 11:92.
38. Hopwood, D., 1967. Some aspects of fixation with glutaraldehyde. *J. Anat.* 101:83.
39. Horstmann, E., and Stegner, H.-E. 1966. Der Eileiter. In *Anatomie des Menschen*, ed. W. v. Möllendorff and W. Bargmann, vol. 7:4, p. 35, Harn- u. Geschlechtsapparat. Berlin: Springer-Verlag.
40. Huber, G. C. 1915. The development of the albino rat, *Mus norvegicus albinus. J. Morph.* 26:247.
41. Kellogg, M. 1945. The postnatal development of the oviduct of the rat. *Anat. Rec.* 93:377.
42. Kelly, G. L. 1927. The uterotubal junction in the guinea-pig. *Am. J. Anat.* 40:373.
43. Lisa, J. R., Gioia, J. O., and Rubin, I. C. 1954. Observations on the interstitial portions of the Fallopian tube. *Surg. Gynec. Obstet.* 99:159.
44. Lucas, A. M. 1930. The structure and activity of the ciliated epithelium lining the vertebrate Fallopian tube. *Anat. Rec.* Abstr. 45:230.
45. Luft, J. H. 1961. Improvements in epoxy resin embedding methods. *J. Biophys. Biochem. Cytol.* 9:409.
46. Mann, T. 1964. *The biochemistry of semen and of the male reproductive tract.* p. 283. London: Methuen.
47. Mayer, G., Nilsson, O., and Reinius, S. 1967. Cell membrane changes of uterine epithelium and trophoblasts during blastocyst attachment in rat. *Z. Anat. Entwicklungsgesch.* 126:43.
48. Nilsson, O. 1957. Observations on a type of cilia in the rat oviduct. *J. Ultrastruct. Res.* 1:170.
49. ———. 1958. Electron microscopy of the Fallopian tube epithelium of rabbits in oestrus. *Exp. Cell Res.* 14:341.
50. ———. 1966*a*. Structural differentiation of luminal membrane in rat uterus during normal and experimental implantations. *Z. Anat. Entwicklungsgesch.* 125:152.

51. ———. 1966*b*. Estrogen-induced increase of adhesiveness in uterine epithelium of mouse and rat. *Exp. Cell Res.* 43:239.
52. ———. 1967. Attachment of rat and mouse blastocysts onto uterine epithelium. *Int. J. Fertil.* 12:5.
53. Nilsson, O., and Rutberg, U. 1960. Ultrastructure of secretory granules in postovulatory rabbit oviduct. *Exp. Cell Res.* 21:622.
54. Overbeck, L. 1967. Entwicklung der Kinocilien im Tubenepithel des Menschen. *Naturwissenschaften* 9:229.
55. Owman, Ch., and Sjöberg, N.-O. 1966. Adrenergic nerves in the female genital tract of the rabbit, with remarks on cholinesterase-containing structures. *Z. Zellforsch.* 74:182.
56. Powierza, S. 1912. Über Änderungen in Bau der Ausführwege des weiblichen Geschlechtsapparates der Maus während ihres postembryonalen Lebens. *Bull. Int. Acad. Sci., Krakau, B. Scienc. Nat.*, p. 349.
57. Reinius, S. 1965. Morphology of the mouse embryo from the time of implantation to mesoderm formation. *Z. Zellforsch.* 68:711.
58. ———. 1966*a*. Sectioning tissue for light microscopy with the Ultrotome ultramicrotome. *Sci. Tools* 13:10.
59. ———. 1966*b*. Ultrastructure of epithelium in mouse oviduct during egg transport. *Proc. Fifth World Congr. Fertil. Steril., Stockholm. Excerpta Med.* Internat. Congr. Ser. No. 133:199.
60. ———. 1967. Ultrastructure of blastocyst attachment in the mouse. *Z. Zellforsch.* 77:257.
61. ———. 1968. Oviductal function during gamete and zygote passage in the mouse studied by light and electron microscopy. I: Secretory activity of the oviduct epithelium. In preparation.
62. Restall, B. J. 1966. Histological observations on the reproductive tract of the ewe. *Aust. J. Biol. Sci.* 19:673.
63. Reynolds, E. S. 1963. The use of lead citrate at high pH as an electron opaque stain in electron microscopy. *J. Cell Biol.* 17:208.
64. Sabatini, D. D., Bensch, K., and Barrnett, R. J. 1963. Cytochemistry and electron microscopy. *J. Cell Biol.* 17:19.
65. Sobotta, J. 1895. Die Befurchtung und Furchung des Eies der Maus. *Arch. Mikroskopischanat. Forsch.* 45:15.
66. Sobrero, A. J. 1963. Sperm migration in the female genital tract. In *Mechanisms concerned with conception*, ed. C. G. Hartman, p. 173. Oxford: Pergamon Press.
67. Watson, M. L. 1958. Staining of tissue sections for electron microscopy with heavy metals. I. *J. Biophys. Biochem. Cytol.* 4:475.
68. Whitten, W. K. 1957. Culture of tubal ova. *Nature* (*Lond.*) 179:1081.
69. Wimsatt, W. A., and Waldo, C. M. 1945. The normal occurrence of a peritoneal opening in the bursa ovarii of the mouse. *Anat. Rec.* 93:47.
70. Wolf, J. 1960. Resorpění činnost epithelu vejcovodu z elektronově mikroskopického hldiska. *Česk. Morf.* 7:164.

4

The Mammalian Uterotubal Junction

E. S. E. Hafez/D. L. Black

Reproduction Laboratory, Department of Animal Sciences
Washington State University, Pullman

Department of Veterinary and Animal Science
University of Massachusetts, Amherst

I. Anatomy of the Uterotubal Junction

Anatomically, the term uterotubal junction as used here refers to the region where the oviduct enters the uterus. This involves the caudal end of the oviduct in transition to the rostral end of the uterine cornu. An extramural and intramural portion of the uterotubal junction may be recognized in certain species. Conventional histological techniques have been employed to study the structure of the uterotubal junction, but little attention has been paid to three-dimensional morphological studies, nor have histochemical or cytochemical techniques been applied. Rigby (1966) studied the structure of the uterotubal junction of the sow in relation to the persistence of sperm in the apex of the uterine horn, or the caudal part of the oviduct. His material was fixed in 4% formol saline and embedded in polyester wax. Van Gieson or H and E stained sections were then projected and traced on paper. The surface area of the lumen was measured planimetrically.

The following discussion is concerned with the anatomical and histological characteristics of the major uterotubal junction components in a variety of mammals.

A. *The Mucosa*

The mucosa of the extramural portion of the uterotubal junction is thrown into longitudinal folds, which may be sparse or numerous, high or low, broad or narrow, and may also show varied branching according to the species and the stage of the reproductive cycle. The mucosal folds are covered with a single layer of columnar epithelium which rests on a basement membrane (Fig. 1). Many of the epithelial cells may be ciliated. The cilia beat toward the uterus, though some have suggested that the cilia may beat toward the infundibulum, during certain stages of the sexual cycle. The underlying connective tissue has vascular channels.

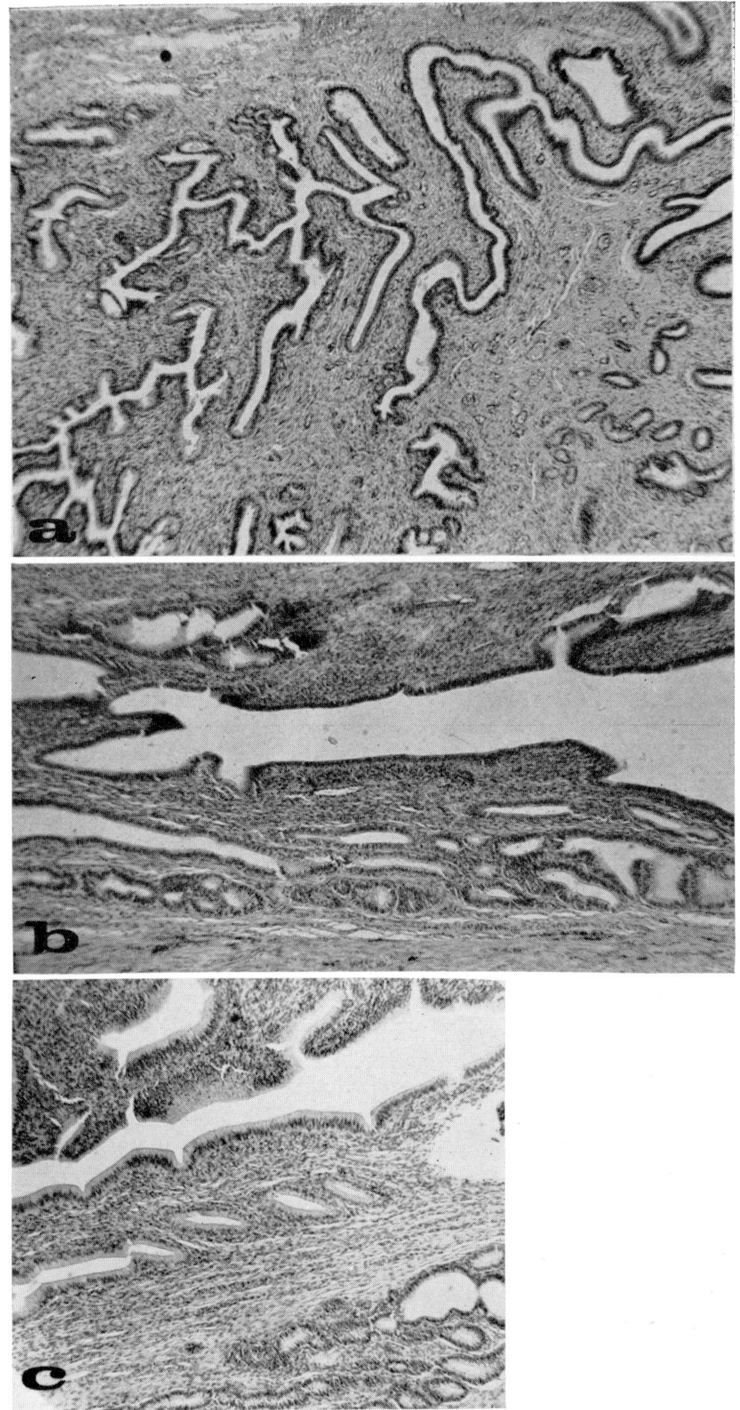

FIG. 1. Comparative histological characteristics (longitudinal sections) of the uterotubal junction in ungulates (H and E, ×175 approx.): (*a*) pig, (*b*) sheep, and (*c*) cow. Note the complexity of the lumen in the uterotubal junction in the pig. (Hafez, unpublished data.)

Normal endometrium may be present in the interstitial portion of the uterotubal junction in man and may respond to hormones in a manner similar to the endometrium. The presence of endometrium appears to be a normal developmental phenomenon and is not considered to be a pathological condition, having no deleterious effect on the occurrence of pregnancy. It has no causal relationship to extrauterine endometriosis or salpingitis isthmica nodosa (Lisa, Gioia, and Rubin 1954).

The numerous longitudinal folds of the uterotubal junction, the *plica tubarica*, rarely show the complexity of the mucosal folds of the ampulla. In a transverse section the mucosal folds have branched processes projecting into, and more or less filling, the lumen. The folds vary in size and number in different species. Another variable is the distance which they protrude into the uterine lumen. They are prominent in the rabbit (Fig. 2), pig, and guinea pig, absent in sheep and cattle, and intermediate in size in carnivores.

The mucosal folds of the uterotubal junction at different stages of the reproductive cycle have variable histological characteristics, distinguishable by the degree of edema, the dilation of the blood and lymph vessels, the abundance of ciliated and other specific cells, and the height and secretory activity of the epithelium itself. For example, in the pregnant rabbit the mucosal folds are extremely edematous and do not regress for a few days post-partum.

The administration of estrogen causes edema of the endometrium in the cow (Asdell 1955) and of the subserous and muscular layers of the uterotubal junction in the ewe (Edgar and Asdell 1960). In the cow and ewe there is a distinct flexure at the junction and it is possible that the degree of flexure, which varies with the stage of the estrous cycle, is brought about by variation in the degree of edema in this area.

B. *Musculature*

The tunica muscularis of the uterotubal junction is composed of a circular layer and a longitudinal layer of smooth muscle fibers. The thickness and the arrangement of these layers vary with the species (Fig. 3). The muscular layer in the isthmic and interstitial sections is always thicker than in the ampullary portion, where the longitudinal muscle bundles are rather sparse and more widely separated by loose, highly vascular connective tissues. With the exception of the excellent investigation of Schilling (1962) on the musculature in ungulates (Fig. 4), most of the detailed investigations in the literature are limited to the interstitial portion in man.

The autochthonous musculature of the uterus in man, which forms the inner layer of the uterine wall, consists of 4 bundles (Fig. 5). This arrangement enables a constrictor effect to be placed upon the larger part of the intramural tube. It may also influence the transport of the egg in the direction of the uterine cavity and that of sperm in the opposite direction (Vasen 1959). The inner, autochthonous muscular layer contains spiral fibers which originate from two different directions, and there are intersections at regular intervals. Each spiral traverses the inner layer from the outside inward as its convolutions gradually assume a smaller diameter. There is a gradual increase of adrenergic nerve terminals in the isthmus. These appear to follow the smooth muscle cells of the circular layer. There is an abundance of nerve terminals in the thick circular musculature of the uterotubal junction. This appears almost as a sphincter near the uterus. The myometrium, on the other hand, seems to be devoid of nerve terminals except for those seen around the blood vessels.

C. *Sphincter Nature of the Uterotubal Junction*

Frequent attempts have been made to discover a sphincter which may control the passage of gametes or even impede or prevent the passage of noxious material from the uterus into the abdominal cavity. Anatomical evidence indicates that the uterotubal junction may be closed periodically. Andersen as early as 1928 demonstrated that in several species there are arrangements of the mucosa at this site which, under certain circumstances, are capable of completely closing the uterotubal junction.

Various species have different mechanisms whereby the uterotubal junction may be closed. In the pig, the size of the lumen decreases greatly within 1 cm of the uterine ostium.

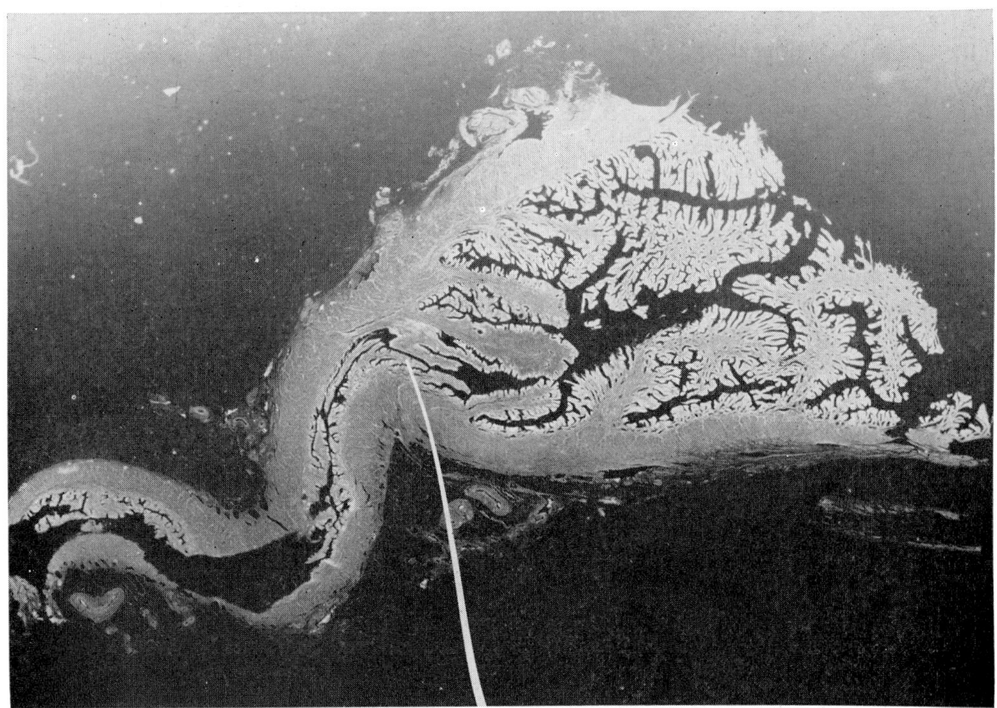

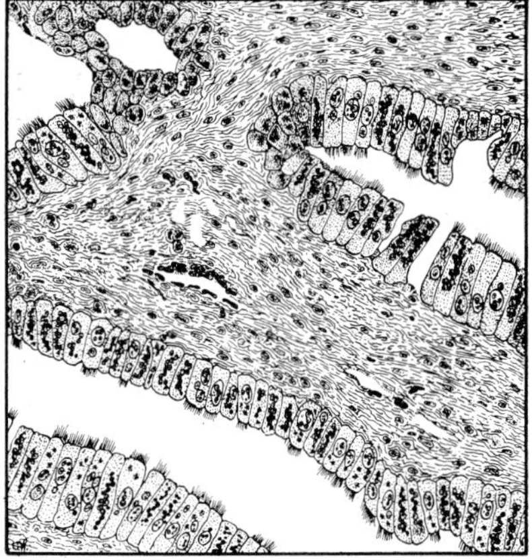

FIG. 2. Longitudinal section in the uterotubal junction of the rabbits at 9 days *post coitum*. Note the histological characteristics of the projections and the uterine folds. (Hafez, unpublished data.)

The oviduct then remains narrow for about 6 cm before widening out into the ampullary region. Within the uterotubal junction itself the rostrally directed, characteristic diverticulae appear as branches of the uterine lumen, and may have a relevant functional significance (Rigby 1966).

Black and Asdell (1959) state that a structural or physiological mechanism in the lower end of the rabbit oviduct prevents the eggs from entering the uterus until the reproductive organs come under the influence of progesterone. Edgar and Asdell (1960) suggested

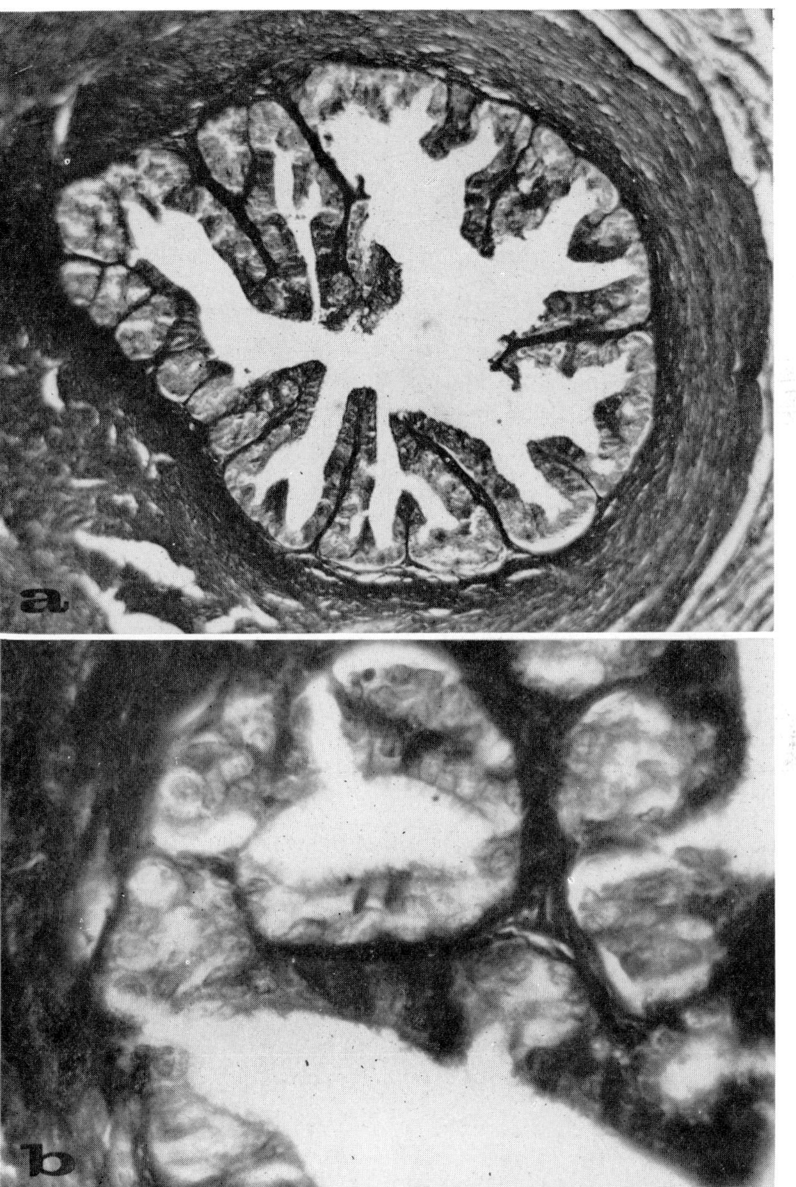

FIG. 3. Cross section in the intramural section of the oviduct of the rhesus monkey: (*a*) Note the thick muscular layer (×350 approx.). (*b*) Note the arrangement of the muscle fibers within the folds and the distribution of ciliated cells (×1,550 approx.) (Masson stain). (Hafez, unpublished data.)

that an estrogen-induced edema and flexure of the wall of the uterotubal junction in sheep cause a valvelike action which prevents movement of fluid and eggs from the oviduct into the uterus.

There are no special mucosal formations at the uterotubal junction in man. The strikingly thick muscular wall surrounding the relatively long and narrow intramural part of the oviduct (Fig. 6) may be suggestive of a sphincter. A contraction of the longitudinal musculature, uniquely located innermost along the length of the interstitial segment of the oviduct, may have a sphincteric action. The oviduct, however, is surrounded by its own circular musculature.

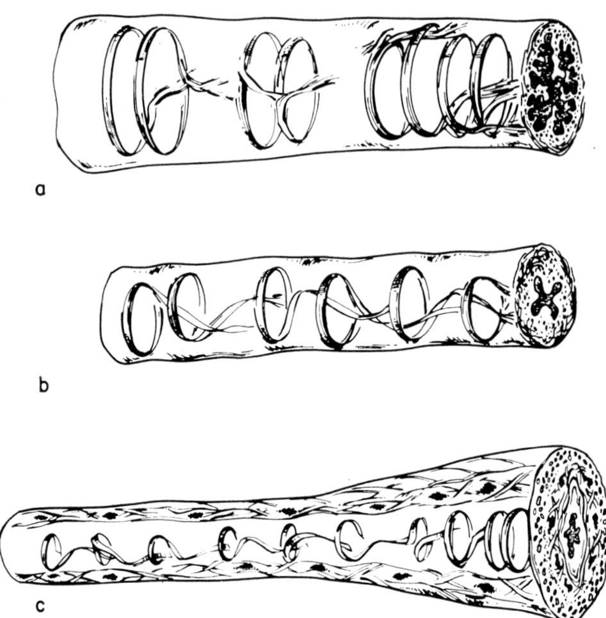

Fig. 4. Diagrammatic illustration of the musculature in ungulate oviducts: (*a*) ampulla: the musculature consists of spiral fibers arranged almost circularly; (*b*) isthmus: note differences in morphology of muscle fibers; (*c*) uterotubal junction: note the longitudinal muscle coat of uterine origin, as well as peritoneal fibers. (Schilling 1962.)

In man, the intramural (interstitial) portion traverses the uterine wall and has an ampulla-like dilation just before it connects with the uterine cavity. The lumen is either connected, as determined by X-ray analysis, to the uterine cavity by a threadlike communication or is separated from it by a narrow opaque zone. The constriction at this oviductal shadow, which is usually designated as the sphincter, is caused by an annular fold of the uterine mucosa (Novak and Rubin 1952).

D. *Vascular, Neural, and Lymphatic Supply*

The abundant blood supply of the uterotubal junction is derived from the ovarian and uterine blood vessels which anastomose within the mesosalpinx, forming a prominent utero-ovarian vascular arch.

At the external uterotubal junction in man, there is a short segment surrounded by entering and departing uterine vessels. One or two vascular rings in the vasomuscular or middle layer enclose the intramural part of the oviduct. The vessels which constitute these

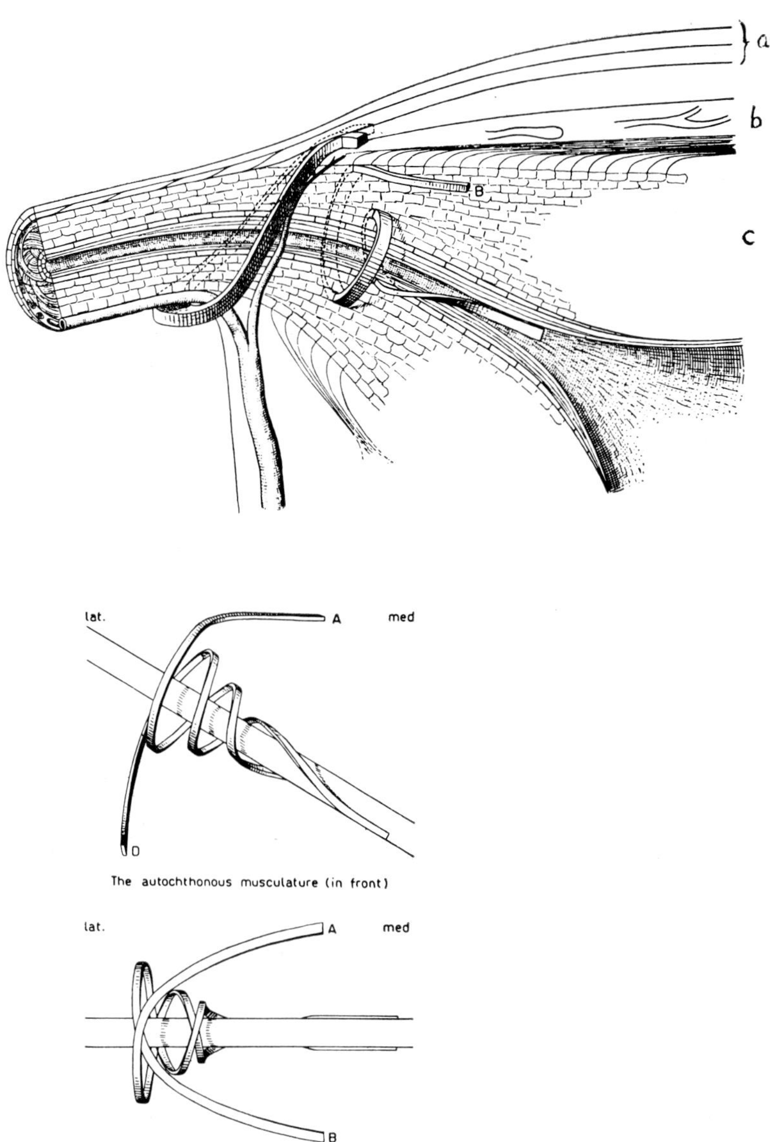

FIG. 5. *Top:* Diagrammatic illustration of the three layers of muscle wall around the intramural tubal lumen in man; (*a*) subperitoneal musculature; (*b*) vasomuscular layer; (*c*) autochthonous musculature.

Bottom: The autochthonous musculature of the uterus around the intramural portion of the tubal lumen consists of four muscle bundle systems. Three of these systems are represented by the spiral *A*, its image *B*, and spiral *D*. The bundles of the fourth system are the images of the bundles of system *D*. (Vasen 1959, by permission of *International Journal of Fertility*.)

rings arise from the ascending branches of the arteria and vena uterina and are surrounded by muscular-enveloping structures, the fibers of which are virtually parallel to the longitudinal axis of the vessels.

The autonomic nerves of the uterotubal junction are derived principally from the ovarian plexus and, to a lesser degree, from the hypogastric plexus.

The lymphatic system of the pig uterotubal junction consists of one or more long club-shaped sinuses lying in the center of each mucosal villus. These drain into a lymphatic

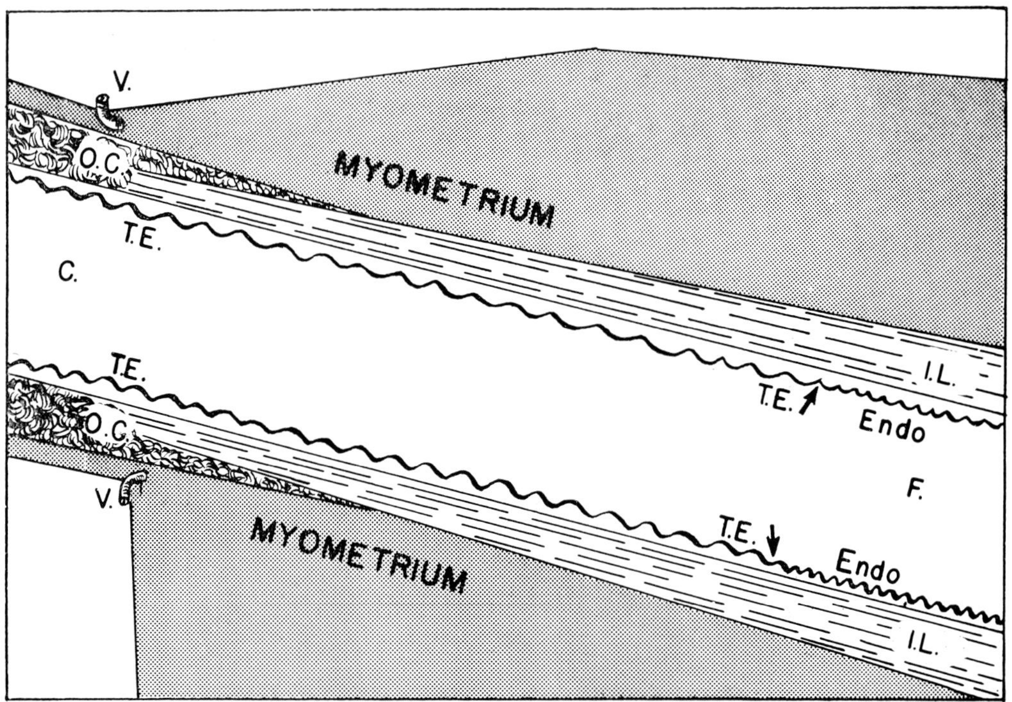

FIG. 6. Diagram of the architectural pattern of the interstitial portion of the human oviduct. *O.C.*, outer circular muscle; *I.L.*, inner longitudinal muscle; *C.*, cornual end of tube; *F.*, endometrial funnel; *T.E.*, tubal epithelium; internal ostium of tube (arrows); *Endo.*, endometrium; *V.*, vessels at uterotubal junction. (From Lisa, Gioia, and Rubin 1954, by permission of *Surgery, Gynecology and Obstetrics.*)

plexus located in the mucosa at the base of the villus and then empty via the intermuscular connecting vessels into the broad ligament lymph vessels (Andersen 1927*a*).

E. *Supporting Ligaments*

Considerable variations are observed in the anatomical and histological characteristics of the supporting uterotubal junction ligaments. In certain species, these ligaments are well developed and seem to have a major physiological significance in their mobility, flexure, and closure of the uterotubal junction (Fig. 7). A description of these ligaments in the rabbit (Westman 1926) follows: there are two different parts of the peritoneal fold or mesotubarium—a fatty part, dorsal to the oviduct, the *mesotubarium inferius*, and a ventral part containing little or no fat, the *mesotubarium superius*.

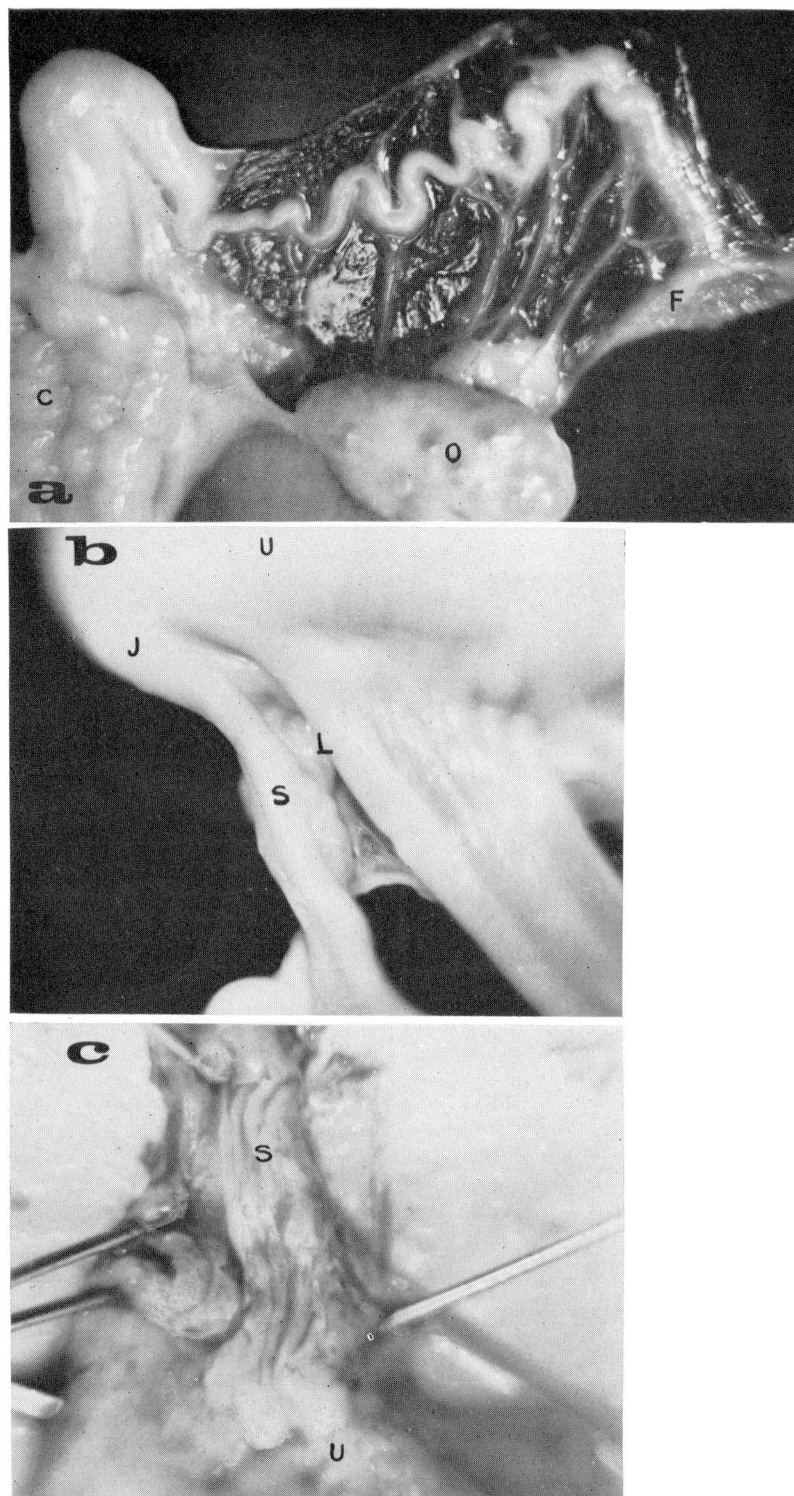

FIG. 7. The oviduct and uterotubal junction in the domestic cow (*Bos taurus*) and pig-tail monkey (*Macaca nemestrina*).

(*a*) Vascular supply and ligaments of the bovine oviduct. Note the fimbriae (*F*), the ovary (*O*), and the exposed uterus showing the caruncles (*C*).

(*b*) The uterotubal junction in the pig-tail monkey. The ligaments (*L*) attached to the isthmus (*S*), to the uterotubal junction (*J*), and to the uterus (*U*) cause different degrees of flexure according to the reproductive cycle. The significance of these ligaments in the physiology of the uterotubal junction is unknown.

(*c*) An open dissection showing the isthmic folds (*S*) tapering down in the apex of the uterine horn (*U*) in the cow. (Hafez, unpublished data.)

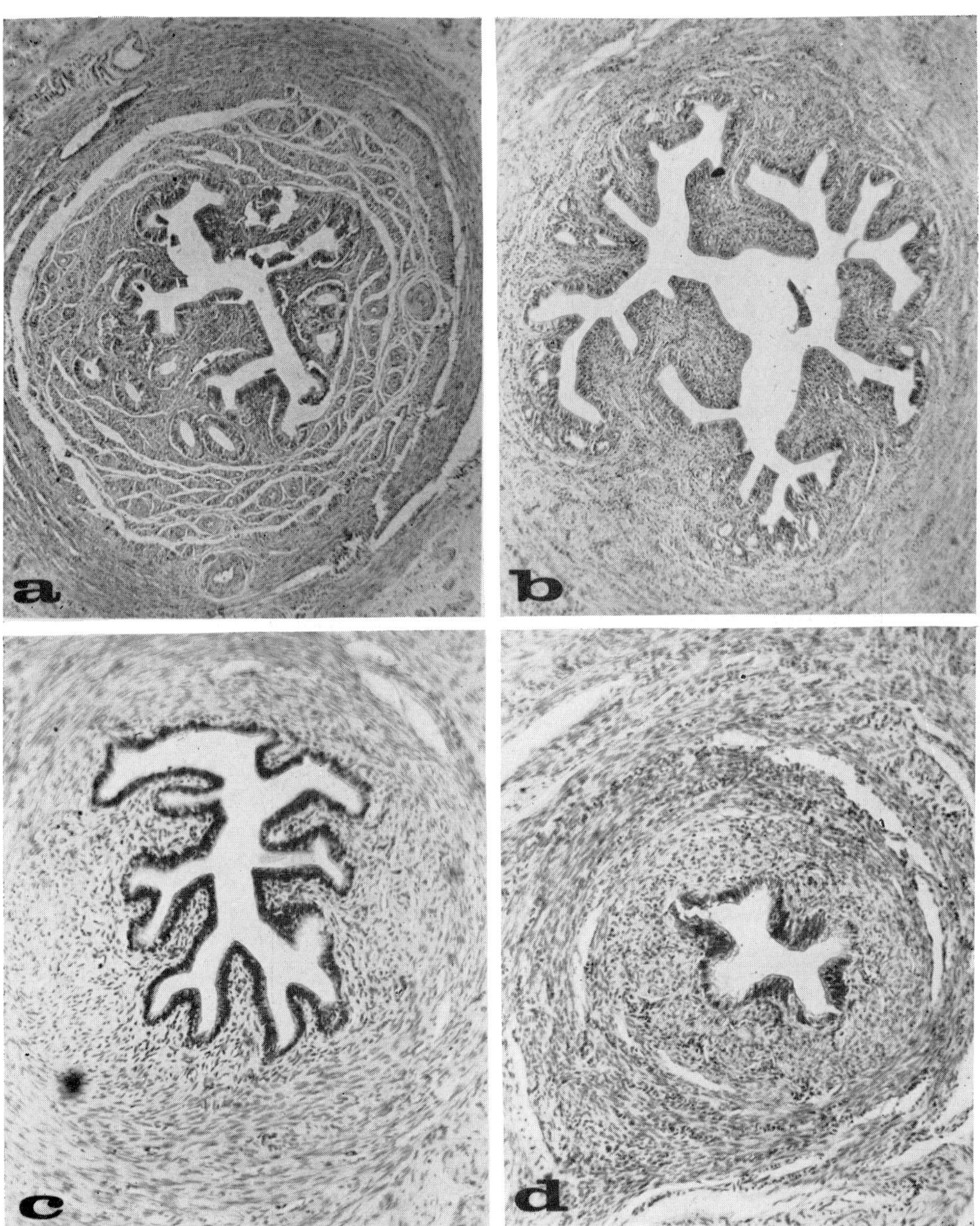

FIG. 8. Comparative histological characteristics (cross section) of the uterotubal junction in several mammals. Note differences in the degree of branching and the relative thickness of the circular and longitudinal muscles. H and E: (*a*) sheep (×175 approx.), (*b*) rabbit (×175 approx.), (*c*) rhesus monkey (×350 approx.), (*d*) pig-tail monkey (×350 approx.). (Hafez, unpublished data.)

F. *Species Anatomical Differences*

Considerable differences in the anatomical and histological characteristics of the uterotubal junction exist among different species. Differences have been noted in (1) the degree of flexure and angle of entry of the oviduct into the uterine cornu; (2) the degree of narrowing in the caudal portion of the isthmus; (3) the number and prominence of mucosal folds; (4) the vascular and lymph supplies; (5) the number and morphology of ciliated cells; (6) the cytological characteristics of the mucosal epithelium secretory cells and uterine glands; (7) the rate of ciliary movement during different stages of the reproductive cycle; (8) the relative thickness of longitudinal and circular muscular layers; and (9) the degree of continuity of the oviductal musculature with the myometrium.

Among the few mammalian species studied, 5 different types of uterotubal junctions are recognized. This arbitrary classification is based on the structure of the mucosal folds which guard the junction and the degree of flexure where it connects the apex of the uterine horn. These 5 types are summarized in Table 1 and illustrated in Figures 8, 9, and 10.

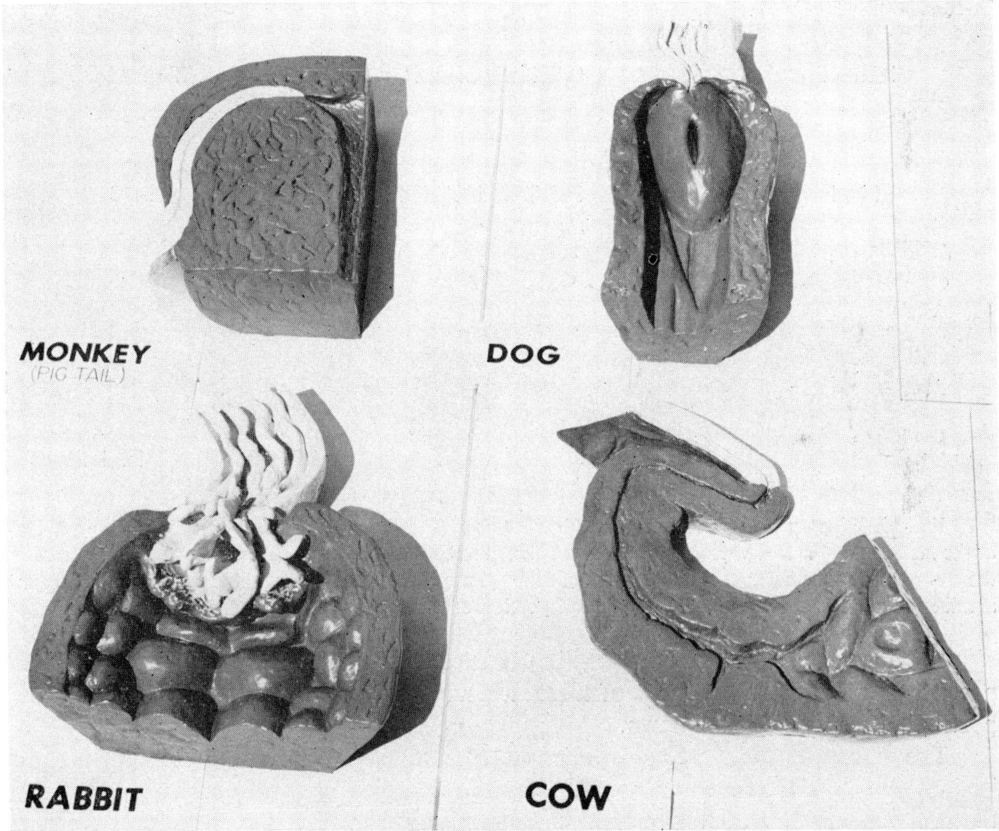

FIG. 9. Sculpture of the uterotubal junction of four mammals constructed from formalin-preserved specimens. Note the rosette structure in the domestic rabbit (*Oryctolagus cuniculus*), the marked flexure and lack of projections in the domestic cow (*Bos taurus*), the moundlike structure in the domestic dog (*Canis familiaris*), and the intramural portion in the pig-tail monkey (*Macaca nemestrina*). (Hafez and Hook, unpublished data.)

Type I: Rabbit and Pig. In the domestic rabbit (*Oryctolagus cuniculus*) the isthmus has 4 major longitudinal, parallel, mucosal folds which become taller and thinner as they approach the uterotubal junction. Between any two of these major folds, lower folds of mucosa running more or less with the major folds are present, as well as some still lower anastomosing folds. The 4 major folds become even taller and thinner about 4 to 6 mm from the end of the isthmus. Another large, short fold almost equal in size to the major folds is present close to the ostium of the isthmus between each of the 2 major folds. These folds, of short length, run parallel with the major folds and end in pendulous projections. Hence, when the entrance of the isthmus into the uterine lumen is viewed from the side, a "rosette-like" structure is observed with 6 to 8 projections (Fig. 11, *left*).

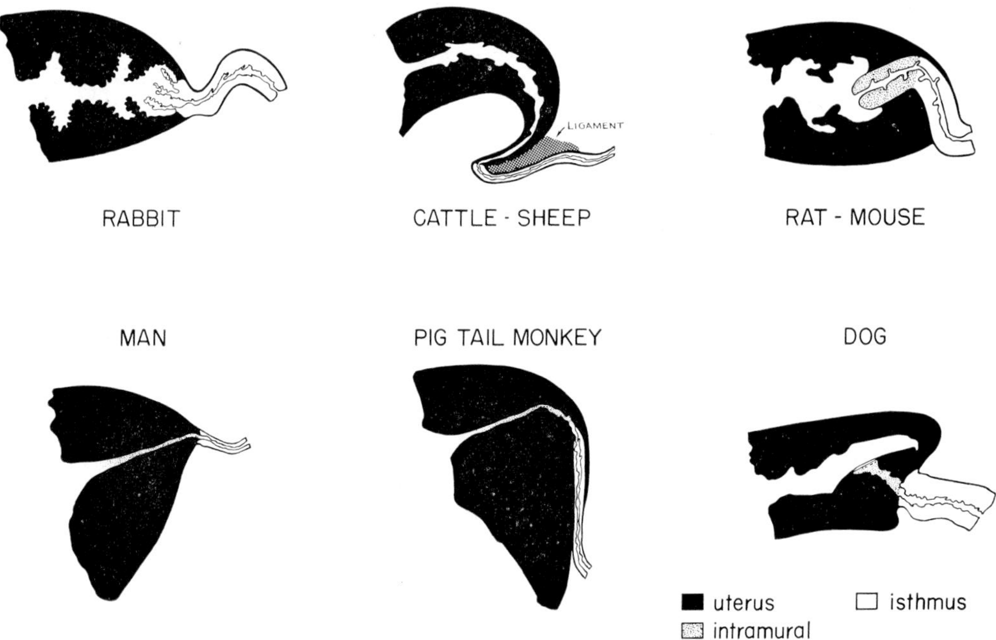

FIG. 10. Diagrammatic illustration (not drawn to scale) to show species differences in the morphology of the uterotubal junction. Varied anatomical relationships between the isthmus and the uterus are observed: the conspicuous folds in the rabbit, the flexure of the isthmus in cattle, the moundlike papilla in the dog, the archlike junction in the rat, and the intramural portion of oviduct in primates. (Hafez, unpublished data.)

Within the folds there are pocket-like diverticulae of different sizes (Fig. 11, *right*). The possible physiological significance of these structures in sperm capacitation or sperm transport into the ampulla is unknown.

The mucosa of the uterus gives rise to a transverse fold close to the base of the projections (Fig. 12). Hence, in an open dissection the projections (villi) more or less overlie the transverse fold. This fold may be discontinuous in one or two places, but seems to encircle the lumen of the uterotubal junction.

The uterotubal junction in the pig is similar to that in the rabbit. The finger-like processes pointing toward the uterus are flatter and more lobulated than in the rabbit. Pocket-like diverticulae of different sizes are found within these folds (Fig. 13). The uterotubal junction is surrounded by numerous endometrial glands (Fig. 14), a common histological characteristic in the uterotubal junction of all ungulates.

TABLE 1

ANATOMICAL AND HISTOLOGICAL CHARACTERISTICS OF THE UTEROTUBAL JUNCTION (UTJ) IN SEVERAL MAMMALS

SPECIES	ANATOMICAL CHARACTERISTICS	HISTOLOGICAL CHARACTERISTICS	
		Mucosa	Musculature
Cat (*Felis catus*)	Oviduct enters uterus obliquely at a 120° angle and is elevated within the uterine cavity Oviduct opens into the uterus through a low papilla, formed of mucosa, which protrudes out into the uterine cavity		Circular muscle of oviduct and uterus are continuous and at their junction there is a special thickening; a collar of circular musculature extends halfway up into the papilla Longitudinal musculature of the oviduct lies just outside of the circular muscle; just before uterus is reached, the two layers become separated from each other by a layer of connective tissue containing large blood vessels; this separation of muscle layers continues throughout the uterus
Cattle (*Bos taurus*)	Uterus tapers off gradually toward isthmus Oviductal folds of isthmus disappear 1–3 mm from uterus; oviductal opening smooth and funnel-shaped; tip of uterine cornu (during estrus) curves more than 180°, narrowing as it curves to isthmus; a marked flexure at the UTJ where the isthmus curves back running parallel with tip of cornu; there are no special projections of mucosa into or in the uterine lumen The ligaments passing over the UTJ are composed of connective tissue with varying amounts of smooth muscle and are covered by mesothelium which connect the uterine horn with the isthmus; one short ligament, attached on either antimesometrial side of the UTJ, may be of functional significance	Folds of oviductal mucosa blend into mucosa of uterine horn without special projections; mucosa has a smooth surface without caruncles for the 3–4 cm Isthmus covered with either nonciliated or a few ciliated cells; at estrus projections from the nonciliated cells may resemble cilia or be longer and more irregular; such projections may represent secretions from the surface of cells; cell height 36–45 μ; leukocytes and mitotic figures present Endometrial glands are present at the UTJ; the length and arrangement visible in tunica propria indicate that they empty onto the uterine surface	Two thin layers of longitudinal muscle are present on either side of the much thicker muscular layer Thickening of a circular muscle at the region of greatest flexure may be palpated
Dog (*Canis familiaris*)	At apex of uterine horn, there is a rather broadly based mound of uterine endometrium which arises from the mesometrial side of the uterus; uterine lumen contacts the mound about $\frac{3}{4}$ of circumference to uterus; viewed from antimesometrial side it looks like a papilla about 3 mm long of roughly oval appearance with the caudal end bluntly tapered; it is about 2 mm at its widest part; it has a longitudinal slit about 1 mm long which is parallel with	Ciliated cells absent; nonciliated cells oval and slender, a few leukocytes lie above or below a marked basement membrane Tunica propria dense in folds; very thin layer outside folds	Circular layer thick at the UTJ and just above it; thin layer of longitudinal muscle; blood vessels especially large on mesometrial side

SOURCE: Adapted from Alden 1942–43; Allen 1922; Andersen 1928; Edgar and Asdell 1960; Hafez, unpublished data; Hook and Hafez, unpublished data; Kellogg 1945; Kelly 1927; Kuhlmann 1964–65; Lee 1928; Lisa, Gioia, and Rubin 1954; Mastroianni 1962; Novak and Rubin 1952; Rigby and Glover 1965; Rocker 1964; Sweeney 1962; Vasen 1959; Westman 1926.

TABLE 1—*Continued*

Species	Anatomical Characteristics	Histological Characteristics	
		Mucosa	Musculature
Dog (*Canis familiaris*)—*Continued*	walls of uterine horn; uterine endometrium forms margins of slit Isthmus with 3 or 4 medium flexures has 4 major mucosal folds which terminate in the depths of the uterine mound, short intramural portion of duct overlapping the circular isthmus muscle layer at very end by a circular uteral musculature; some longitudinal fibers on one side		
Guinea pig (*Cavia porcellus*)	UTJ is very complex; isthmus is tortuous, with moderately firm muscular walls Oviduct connected to the uterus at a right angle UTJ is guarded by mucosal folds protruding into the apex of the uteral lumen; folds vary in number and size but some are always present Uterine horn is trumpet-like	Immediately behind the opening, the oviduct widens; from the ovarian side a broad rounded mucosal papilla extends, this papilla consists of a core of smooth muscle and points toward the oviductal opening	Circular muscle is especially thick opposite the oviductal opening; it extends into the lips of the mucosa and contains uterine glands
Hedgehog, African (*Erinaceus albiventris*)	Oviduct opens through the tip of the papilla by a narrow slit		Longitudinal and, to a lesser extent, circular musculature of the oviduct continue almost to the UTJ opening, forming a large part of the papilla bulk
Man (*Homo sapiens*)	Oviduct approaches the surface of the uterus slightly anterior; it then goes through the uterine muscle a short distance from the anterior, superior surface of the fundus and leaves the uterus at the junction of its superior and lateral surfaces Diameter of intramural portion is 2–4 mm; as oviduct approaches the endometrial cavity it becomes wider and appears to project slightly into the cavity Main course of the UTJ is related to the inner muscle matrix and follows a straight course until it reaches the vascular layer; at this point a constant acute change occurs; the lumen may be only 100 μ No sphincter muscle; UTJ not guarded by mucosal folds	Smooth mucosal surface of uterus merges gradually into oviductal mucosa Mucosa is thinnest at the intramural portion where it is thrown into primitive folds; these folds vary in cross section from a cloverleaf shape to a star or H-shape; it is thickest at the fimbriated end, where it composes a major part of the oviductal wall Interstitial portion similar to extramural portion of both the oviduct and the endometrium	Subperitoneal musculature of the uterus is arranged in a loop around the intramural portion of the oviduct 1 or 2 vascular rings within the intramural portion are accompanied by muscle bundles; inner layer of the musculature is arranged in spirals which encircle the intramural portion Stout inner longitudinal muscular coat; frequently there is a definite smooth-muscle sphincter about the end of the oviduct as well as a special mucosal structure, either in the form of an oviductal papilla at the tip or in the form of villi surrounding the mouth of the oviduct

TABLE 1—*Continued*

Species	Anatomical Characteristics	Histological Characteristics	
		Mucosa	Musculature
Monkey, pig-tail (*Macaca nemestrina*)	Isthmus has an intramural portion which runs alongside the uterus, penetrating the uterine wall just above the origin of round ligament in the anterior upper half of the uterus; at the top of the ligament it curves inward terminating in the funnel-shaped lateral extension of the uterine lumen	The plicae are low and rather broad and few in number (2–4); ciliated cells are outnumbered by nonciliated; both types are 22 μ in diameter Tunica propria, relatively thick and of moderate density	Circular muscle layer only in intramural portion—blending with uterine musculature at periphery of duct
Mouse (*Mus musculus*)	Oviduct enters uterus at a right angle and not at the apex; entrance is marked by a very definite moundlike protrusion; oviduct enters through the center of the mound *Colliculus tubarius*, with valvelike action, is an oviductal projection into the uterus which is surrounded by a furrow and a ring of uterine mucosa; anatomy varies with the strain		Slight increase in circular muscle at ostium; a better blood supply than the rest of the oviduct
Opossum (*Didelphis virginiana*)	Oviduct enters uterus from the antimesometrial side UTJ is not guarded by any special mucosal folds; small longitudinal folds of oviductal mucosa continue on into the uterine folds		
Pig (*Sus domesticus*)	Isthmus almost straight; slight flexure in the UTJ; enters more to mesenteric side 6–12 primary longitudinal folds bearing secondary folds and ending with finger-like processes (villi) pointing toward uterus A second row of larger, flat, branched processes; farther along the uterine wall are other villi, less regularly placed, narrower and longer, the longest being about 3 mm Each villus is supplied by one or more arterioles and several veins, both of which are located within the central portion of the villus No sphincter at the UTJ	Mucosa of oviduct and uterus are connected by a band of transitional mucosa bearing villi The presence of cilia distinguishes the uteral epithelium from the oviductal epithelium; the cilia beat from the base toward the tip of the villus	Musculature of the isthmus is relatively thick and firm Circular muscle of the oviduct and uterus are continuous; no thickening at the UTJ Blood vessels within circular layer are prominent and numerous (in late luteal phase) Longitudinal layer at the UTJ averages 211 μ thick except at mesenteric attachment; circular layer 2× as thick as long layer; basal glands in very close contact if not actually among inner circular muscle fibers Few strands of long fibers among circular fibers near isthmus
Rabbit (*Oryctolagus cuniculus*)	Oviduct joins uterus obliquely in mesometrial side Isthmic mucosa bears 4 large primary folds and a varying number of lesser folds; ends	Isthmic folds and projections are composed of a single layer of columnar epithelium with ciliated and nonciliated cells and tunica propria which varies in cell density	A thickening of the circular muscle layer in the junction area which suggests possible sphincter action

TABLE 1—*Continued*

Species	Anatomical Characteristics	Histological Characteristics	
		Mucosa	Musculature
Rabbit (*Oryctolagus cuniculas*)—*Continued*	of primary folds and 2–4 additional short folds between these long processes, which project into uterus Primary and secondary isthmic folds terminate at the UTJ in bulbous or pendulous projections (villi) which extend 6 mm into the uterine lumen, giving a "rosette" appearance when viewed from the uterine side; surrounding these projections is a discontinuous circular fold, probably of endometrial origin; these discontinuous folds may overlap at their ends; their free margins are irregular, suggesting lobes of varying size depending upon the depth of the indentations at margins Pocket-like diverticulae of different sizes and shapes (cone, semilunar) are attached to the side of folds projecting within uterus or among the attachments of the projections to the circular uterine fold	according to its location in folds of varying thickness or as a layer below the folds Uterine fold has more infolding of epithelium surface but no definite glands	
Rat (*Rattus norvegicus*)	Part of the last coil of isthmus forms an arch over tip of uterine horn with distal end of arch surrounded by uterine muscle as it terminates in a papilla projecting into uterine lumen, uterine lumen surrounds papilla (*Colliculus tubaricus*) as a fornix which varies in depth Oviduct enters through the center of papilla, on antimesometrial side of uterus, in contrast to mesenteric side as in most mammals, caudal opening of oviduct lies somewhat below apex of uterine cornu	Mucosa around the margin of the oviductal opening is slightly raised, though not enough to form villi or a papilla, papilla contains tissue from oviduct and uterus Inner mucosa is oviductal in appearance while the tunica propria of the outer mucosa resembles uterus without	Strong band of circular muscle around ostium, musculature extends a short distance into the mucosa Circular and longitudinal musculature are continuous over the uterus and oviduct and show no separation in the uterus Smooth muscle fibers extend into base of papilla from oviduct and from uterine muscularis layers
Sheep (*Ovis aries*)	Isthmic folds end at ostium of the oviduct without any special mucosal structure, 5 primary folds give the lumen a star-shaped appearance in cross sections No mucosal folds protrude into the uterine lumen	Uterine glands do not end abruptly at the mouth of the oviduct, but gradually decline in the first few millimeters of the oviduct	Abundant musculature forms a layer of similar thickness in the adjacent uterus, in proportion to the lumen it is thicker; inner thin longitudinal muscle between tunica and thicker circular layer; No muscular sphincter

Type II: Rat, Mouse, and Hamster. In the rat and mouse, the structure responsible for the valvelike action of the uterotubal junction (Fig. 15) is the *colliculus tubarius*. The *colliculus*, which is a projection of the oviduct into the uterus, may vary anatomically in different strains. In BALB/c females the projection is hemispherical and smooth and has a normal length of about 0.3 mm. In strain 129, it is almost cylindrical or conical with two or more encircling folds and is about 0.4 mm in length (Kuhlmann 1964–65). In the hamster, the uterotubal junction resembles a corkscrew in shape (Fig. 16).

Type III: Dog and Cat. In the dog and cat, the isthmus opens into the uterus through a broadly based mound formed of mucosa protruding into the uterine lumen. In the dog, the mound is ¾ of the circumference of the uterus, and the ostium forms a longitudinal slit (Figs. 17 and 18).

Type IV: Cattle and Sheep. In cattle and sheep, there is a distinct flexure where the isthmus connects the apex of the uterine horn. The extent of this flexure seems to vary with the stage of the estrous cycle. The isthmus and the apex of the uterine horn are attached to ligaments which may have some physiological significance in the control of opening and closing the uterotubal junction.

Type V: Man and Monkeys. In man and infrahuman primates which possess a uterus of the simplex type, the isthmus joins the uterine cavity near the fundus after a rather long interstitial course. Its opening is not protected by mucosal folds or villi.

In the interstitial portion in man there is an additional longitudinal muscular layer, which lies next to the mucosa (Novak and Rubin 1952). This layer is a continuation of the subendometrial muscle at the funnel of the endometrial cavity and becomes continuous with the oviductal musculature at its cornual end. The outer layer is a continuation of the extramural muscle portion of the oviduct and soon becomes part of the myometrium. The muscle bundles, lying deeper in the subperitoneal musculature, form a loop of tissue around the distal end of the intramural oviductal lumen. Contraction of this tissue may cause local constriction or even obstruct the oviductal lumen.

Individual differences, even within the same species, are evident. Sweeney (1962) classified the interstitial portion of the human oviduct into three categories: (1) tortuous course; (2) straight course; and (3) curved course with the convexity directed gently upward. The majority of the 100 specimens examined had a tortuous course and either (*a*) turned 90° halfway through the myometrium; (*b*) possessed 2 to 4 gentle convolutions; or (*c*) diagonally bent downward through the myometrium with 2–4 convolutions and kinks emerging from the uterine wall at a much lower level.

G. *Structural Abnormalities*

Different structural abnormalities in the uterotubal junction may result in different degrees of reproductive failure in man; e.g., endometrium may develop in the mucosa of the interstitial segment, or atresia from an inflammatory disease, papilloma, and carcinoma or endometrial polyps may arise near the internal ostium of the oviduct and result in stenosis by projection up into the lumen. Endometrial carcinoma of the uterine body may affect the interstitial oviduct by projection of a polypoid mass into its lumen, by spreading along the surface, or by permeation into its vascular channels (Lisa, Gioia, and Rubin 1954).

When the oviduct is distended with fluid as a result of ligation or salpingitis, the uterotubal junction undergoes flexure and edema of the wall, particularly of the subserous layer and myometrium (cf. Edgar and Asdell 1960).

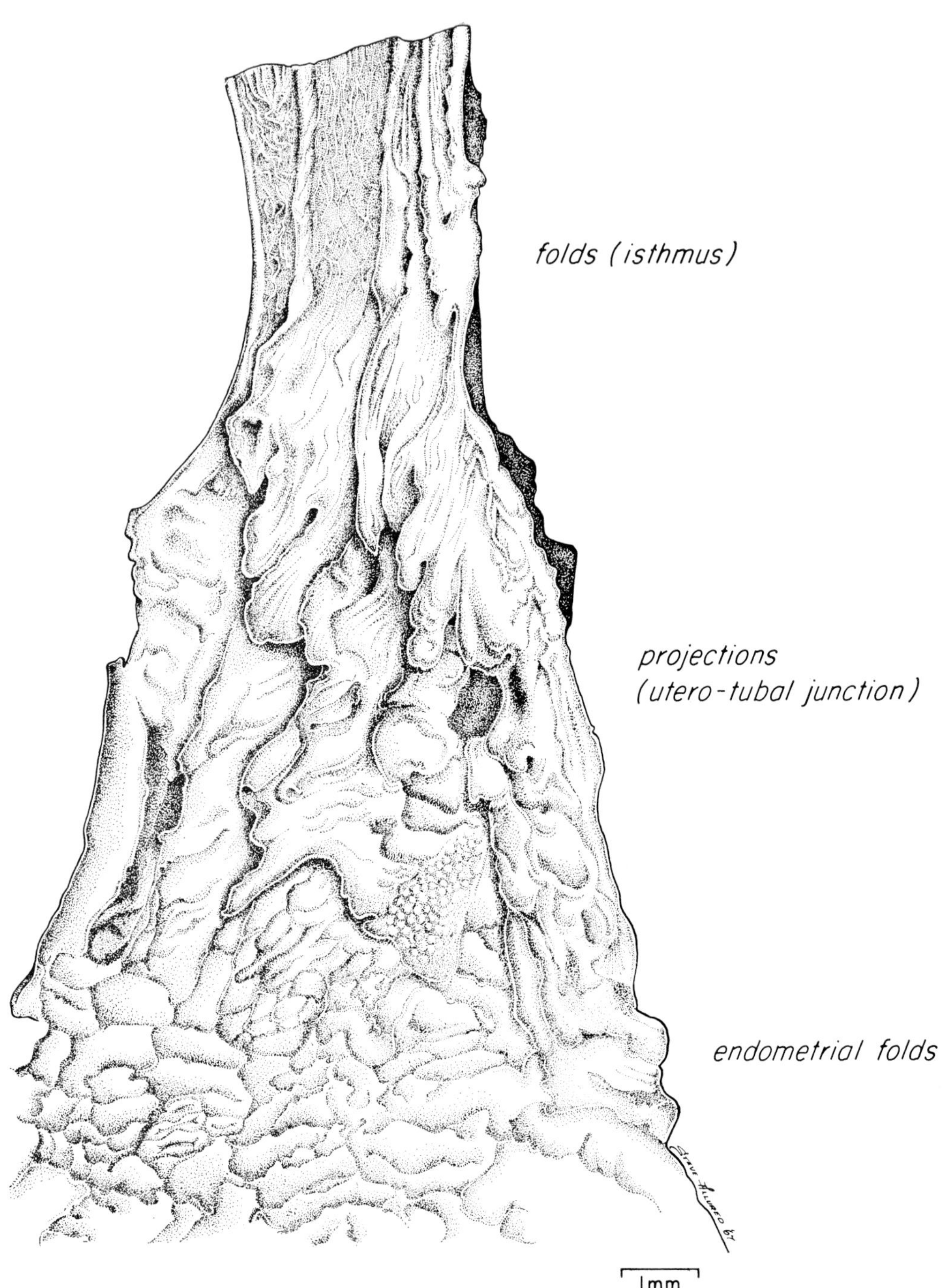

FIG. 11.

Three-dimensional drawing of the uterotubal junction in the domestic rabbit (*Oryctolagus cuniculus*) drawn from a formalin-preserved open dissection. The primary and secondary folds of the isthmus terminate in long processes which project into the lumen of the uterus giving a "rosette" appearance.

Opposite page: Pocket-like diverticulae of different sizes are found within the folds protruding into the lumen. The physiological significance of these structures is not ascertained. (Hafez and Hook, unpublished data.)

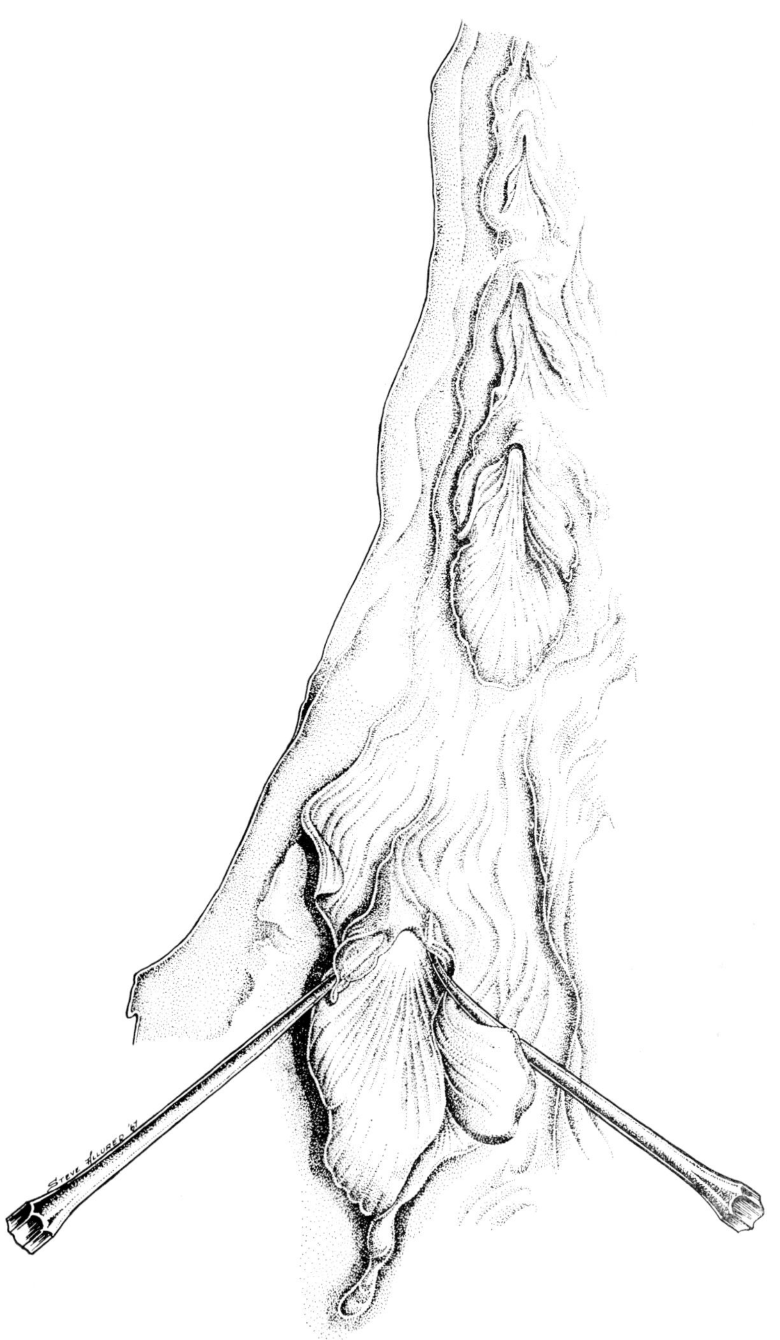
STEVE ALLURED

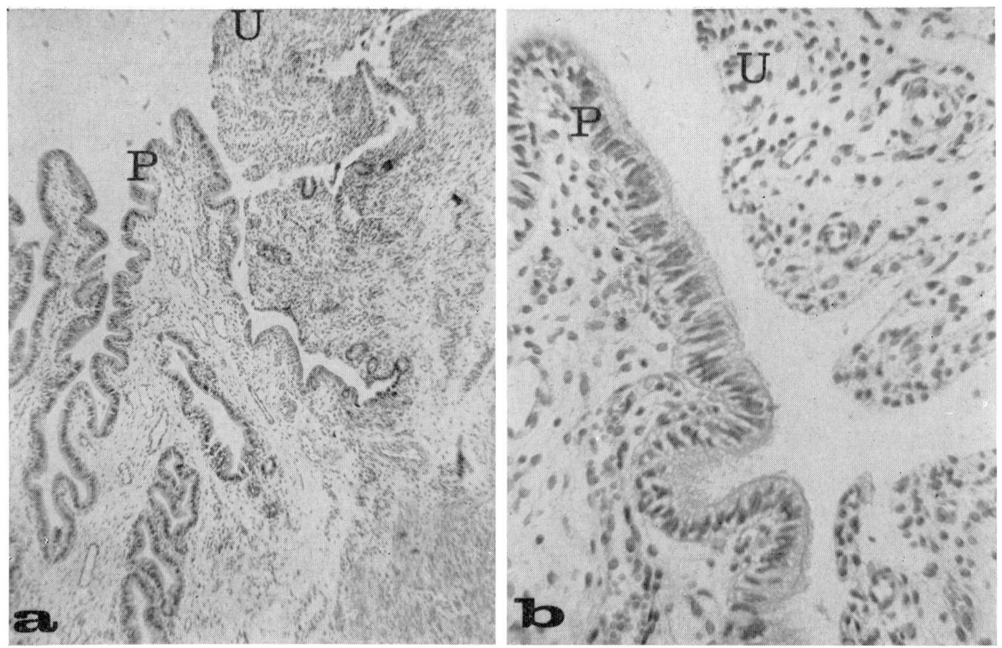

FIG. 12. Longitudinal section in the domestic rabbit. Note the projections (*P*) overlying a circular fold of the uterus (*U*); *a* (×160 approx.); *b* (×700 approx.) (H and E). (Hafez, unpublished data.)

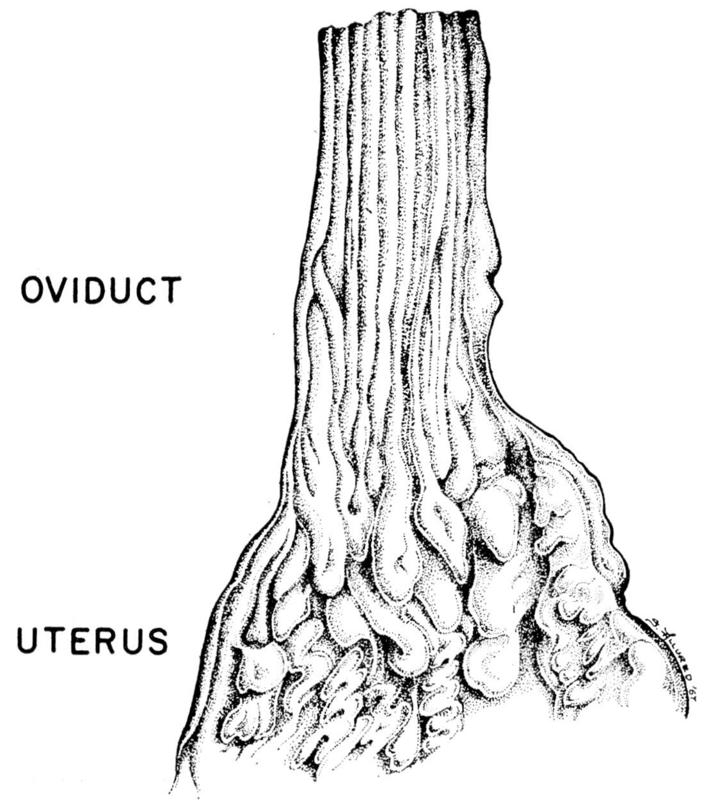

FIG. 13. Three-dimensional drawing of the uterotubal junction in the domestic pig (*Sus domesticus*) drawn from formalin-preserved open dissections. The primary and secondary folds of the isthmus terminate in finger-like processes which project into the lumen of the uterus, giving a "rosette" appearance. Pocket-like diverticulae of different sizes are found within the folds protruding into the lumen. The physiological significance of these structures is not ascertained. Note the similarity of the uterotubal junction of the pig and rabbit. (Hafez, unpublished data.)

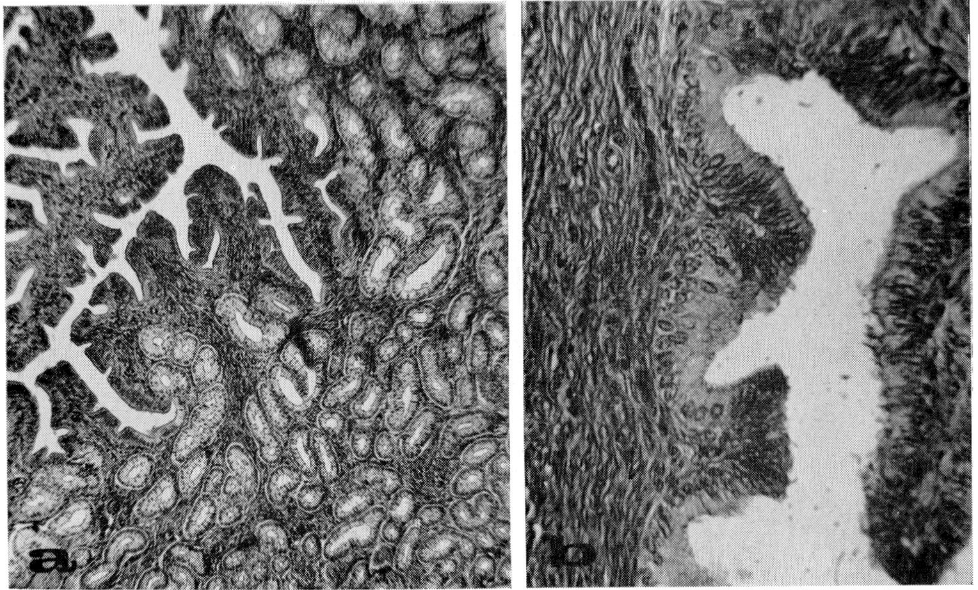

FIG. 14. Longitudinal section in the uterotubal junction of the domestic pig: (*a*) Note the abundance of the endometrial glands around the lumen (×175 approx.). (*b*) Note the arrangement of muscle fibers (×770 approx.) (Masson stain). (Hafez, unpublished data.)

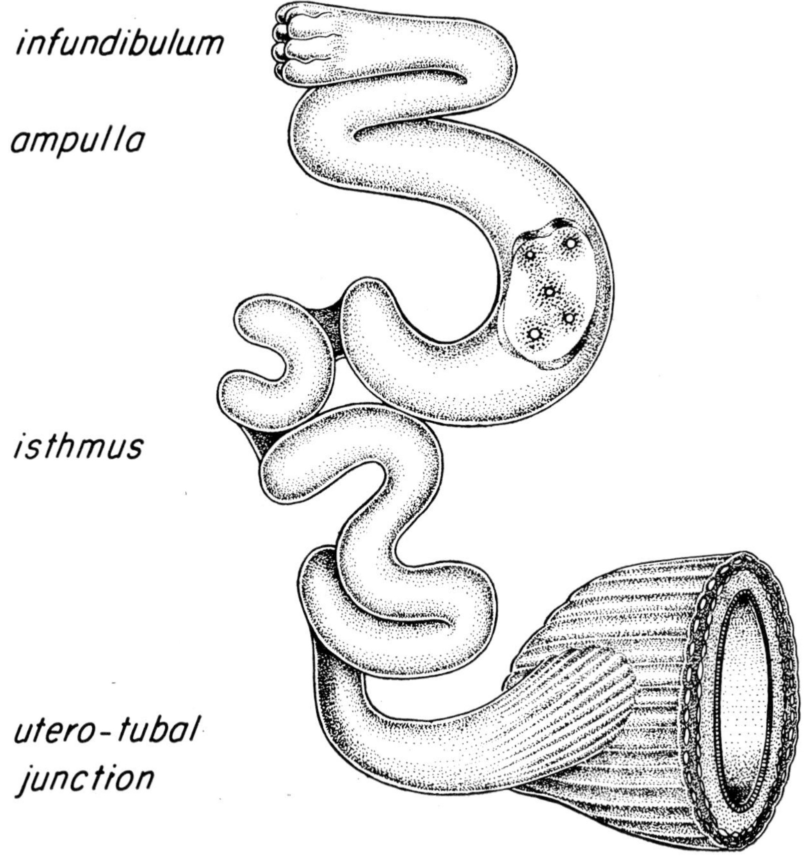

FIG. 15. Diagram of the mouse oviduct. The eggs are shown in the diagram in the ampulla on the first day of pregnancy. Note the anatomical relationship of the uterotubal junction. (Redrawn from Reinius 1966.)

II. Physiology of the Uterotubal Junction

The physiology of the uterotubal junction has lagged behind investigations concerned with other parts of the female tract. Although it has been recognized for many years that the anatomy of this junction is highly complex, few reports have appeared relating its morphology and physiology. One of the greatest difficulties encountered in studying this organ is the fact that it acts in concert with the uterus and isthmic portion of the oviduct. Thus whatever physiological functions it has must be considered in light of what is happening in the adjacent uterus and oviduct. It is therefore a shortcoming that integrative studies, which are the most meaningful, are also the least available.

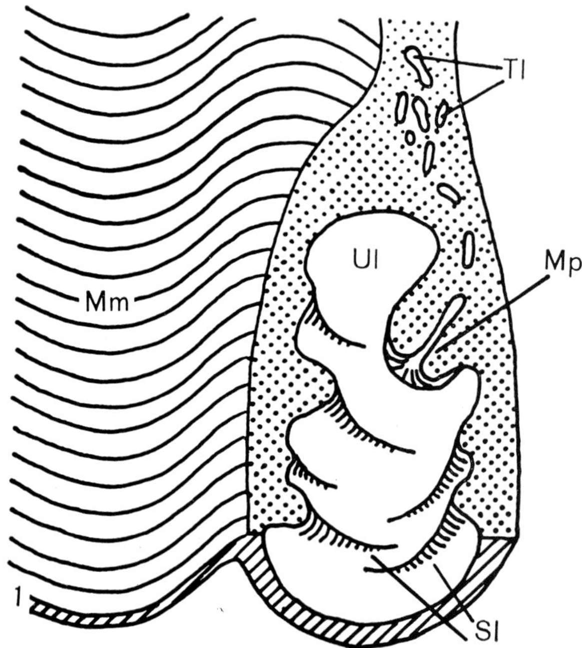

FIG. 16. Anatomical relationship of the uterotubal junction in the golden hamster (*Mesocricetus auratus*); *Mm*, mesometrium; *Mp*, orifice; *Sl*, uterine folds; *Tl*, oviductal lumen; *Ul*, uterine lumen. (From Bogli 1959.)

From the physiological standpoint, the term uterotubal junction is not meaningful, since a junction implies a point of union. At the oviduct-uterus connecting zone there is no discrete junction in the strict sense of the word; rather, there is an area of transition. In the present discussion, which is an attempt to bring the available physiological and anatomical facts together, the term uterotubal junction will refer to the proximate area of the oviduct-uterus connection and may include the extreme caudal portion of the isthmus as well as those muscular components of the uterus which surround the oviduct.

Several functions have been ascribed to the uterotubal junction, few with any supportive experimental evidence. Among these are sperm passage up the oviduct, control of egg passage into the uterus, and prevention of retrograde menstruation.

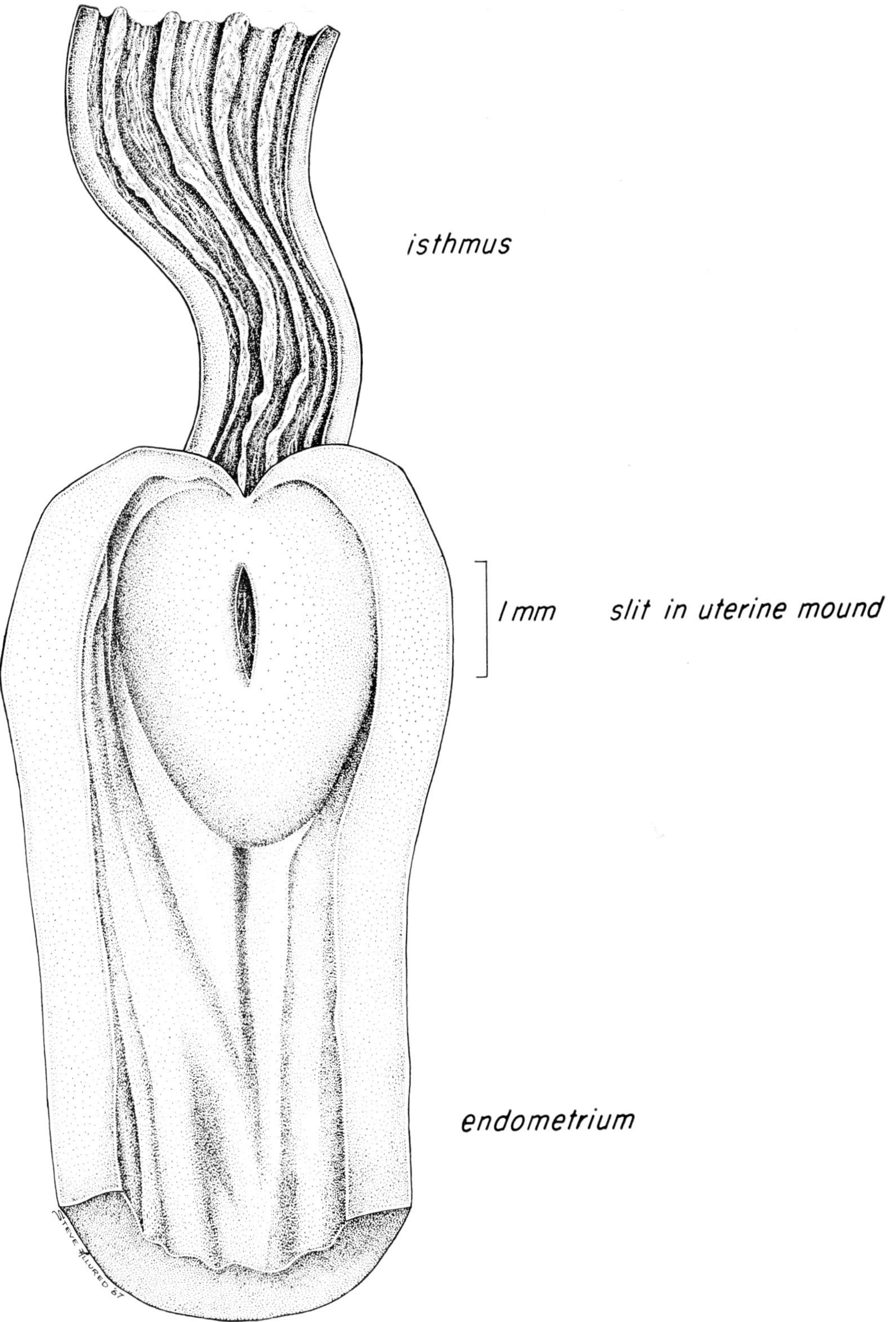

FIG. 17. Three-dimensional drawing of the uterotubal junction of the domestic dog (*Canis familiaris*). Note the broadly based mound of uteral endometrium which rises from the mesometrial; the mound has a longitudinal slit about 1 mm long which is parallel to the walls of uterine horn. The isthmic folds terminate in the uterine mound. (Hafez and Hook, unpublished data.)

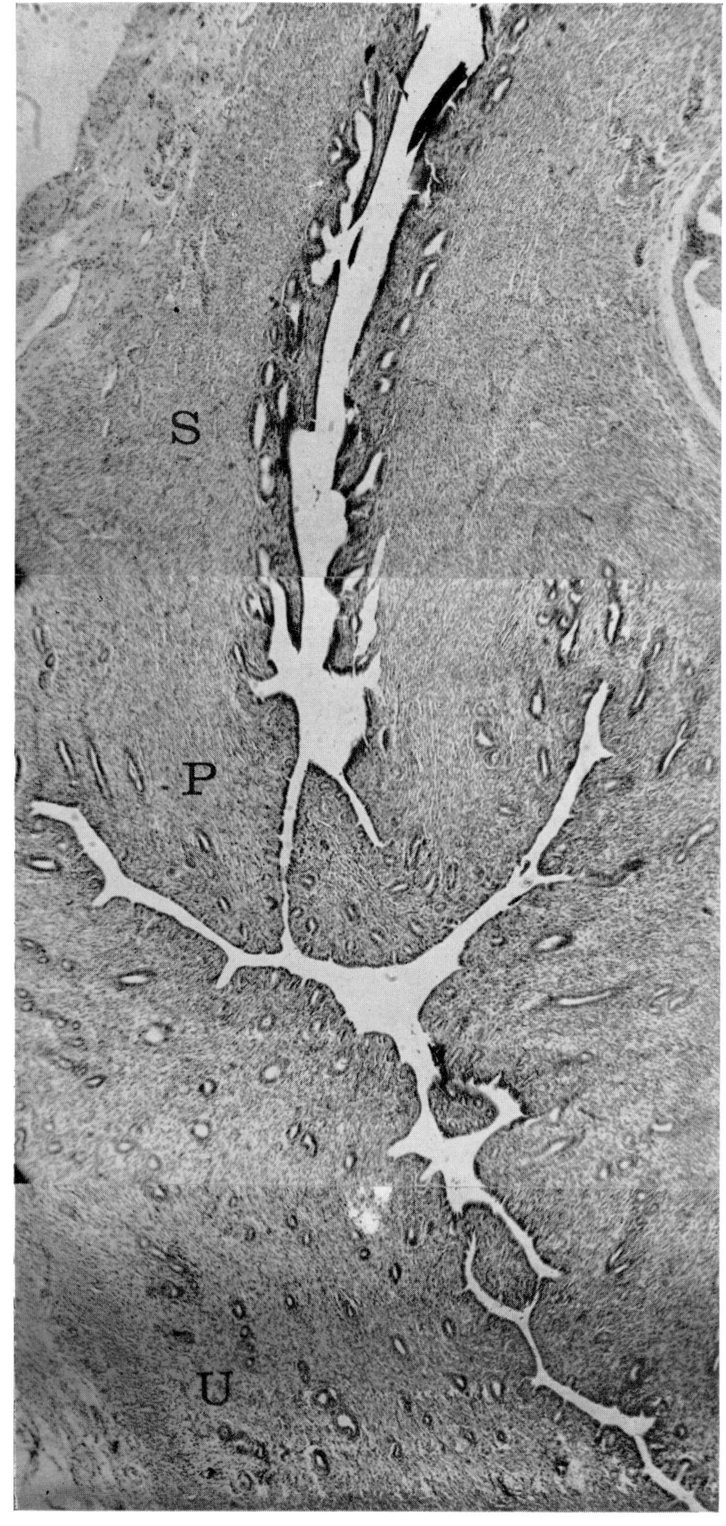

FIG. 18. Longitudinal section in the uterotubal junction in the dog. Note the papilla (*P*) connecting the isthmus (*S*) with the uterus (*U*) (H and E, ×175 approx.). (Hafez, unpublished data.)

A. *Methods of Physiological Studies*

Few techniques are available for uterotubal junction physiological investigations. The common histological and histochemical methods have been employed to relate gross anatomical changes to physiological phenomena; e.g., the presence of large folds at the oviductal orifice in some animals may account for the high pressure required in uterotubal insufflation experiments. The bases for more subtle changes, however, such as increased muscular activity or increased muscular tonus, are not readily apparent in materials examined in this manner. The chief objection to histological observations in the study of physiological phenomena is the fact that they must be made on dead tissue. Many tissues, including the oviduct and presumably the uterotubal junction, may also undergo drastic postmortem changes. At this time the oviduct thickens and contracts markedly. In addition to these changes, fixation *per se* is known to cause shrinkage, and in so doing may alter their histological picture considerably.

Recently, improved histological techniques have been used. The structure, shape, and direction of the human uterotubal junction have been determined using very thick sectional preparations (Gough 1960) and cinematography (Marshall and May 1960). A highly specific fluorescence method for the histochemical demonstration of monoamines, including norepinephrine in adrenergic nerve terminals (Falck 1962; Falck *et al.* 1962), has been applied to the oviduct. At present, however, little information is available at the submicroscopic level.

The distribution of radio-opaque substances, following the injection of such substances into the uterine and oviductal lumens, as shown by serial X-ray or fluoroscopy methods, has been fruitful. An advantage of this method is that it can be used repeatedly in living animals. However, in the past, many of these radio-opaque substances were tissue irritants, and as a result, stimulated oviductal motility. A further disadvantage with this technique is that it is often difficult to judge the plane at which a particular portion of the oviduct lies. A break in the continuity of the radio-opaque column, which may be interpreted as muscular contraction closing the lumen, may in reality be an abrupt flexure of the oviduct (Schneider 1942).

The motility patterns of the oviduct are most commonly studied by insufflation techniques. As with the radio-opaque injection method, insufflation can be used repeatedly *in vivo*. The insufflation technique has as its theoretical basis the principles of fluid flow through tubes. In rigid tubes, provided the pressure applied and the viscosity of the fluid remain constant, a decrease to one-half the radius of the tube will result in the flow being one-sixteenth the original. A 16% decrease in radius will halve the flow. It is quite apparent that the oviduct is not a rigid tube and certain corrections must be made to account for wall elasticity. However, the same general principles apply; the pressure rises inversely proportional to the luminal diameter.

Fluctuation in pressure required to maintain fluid flow (either gas or liquid) through the uterotubal junction has therefore been used as an index of luminal diameter changes caused by muscular activity. It should be emphasized that to compare results obtained with this technique, careful attention must be given to the procedural details; in particular the flow rate must remain constant.

Techniques employed for investigating the role of the uterotubal junction in egg or sperm movement revolve around an estimation of their position in the oviduct. Such determinations are made by dividing the oviduct into several segments and examining the flushing of each segment for eggs or sperm. If quantitation of sperm is desired, counts are made with a hemocytometer. For more precise location of eggs, chemical clearing of the oviducts is often a useful technique; the eggs can then be observed directly through the oviduct wall (Longley and Black 1967). Resin spheres (simulating eggs) coated with radioactive gold have been used also to study transport through the oviduct (Harper *et al.* 1960).

B. *Results on Uterotubal Insufflation*

When gas is forced into the uterus, there is an abrupt rise in pressure due to the closed uterotubal junction. The abruptness is dependent upon both the flow rate and the volume of the uterine lumen. However, once the gas starts to enter the oviduct there is a sharp drop in pressure to some lower plateau (Fig. 19). The pressure which forces the uterotubal

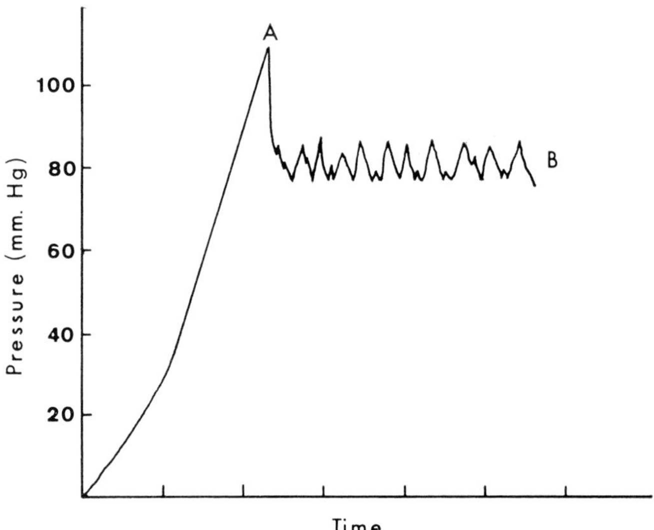

FIG. 19. Pressure fluctuations observed when the rabbit oviduct is insufflated. A composite drawing from several sources; *A*, uterotubal "peak pressure"; *B*, fluctuations due to muscular activity of the uterotubal junction and isthmus.

junction open is commonly referred to as the "peak pressure" or the "uterotubal opening pressure." This pressure depends on both the species and the stage of reproductive cycle (Table 2). Two types of equipment which have been used to determine the "peak pressure" are shown in Figure 20. Although the pressure required to force gas from the uterus into the oviduct is high, gas flows relatively easily in the opposite direction, i.e., from the abdominal ostium of the oviduct into the uterus. The differential resistance to gas flow depending on direction implies the presence of a valve.

1. *Species Differences.* The species-dependent "peak pressure" values are due, in part, to anatomical differences. In some animals such as the pig, guinea pig, and rabbit, elaborate projections are present at the oviductal orifice. As gas is forced into the uterus, these projections could be pressed against one another and act as a valve. The more pressure applied, the tighter the oviduct is closed. It has been suggested also that an anatomical sphincter is present at the uterotubal junction. If the entrance of the oviduct is to one side of the uterine apex, it may likewise act as a valve when lateral pressure is applied to the uterine wall (rat, mouse,

TABLE 2

UTEROTUBAL OPENING PRESSURES OF SEVERAL MAMMALIAN SPECIES

Species	Reproductive Status	Uterotubal Opening Pressure mm Hg	Remarks	Reference
Cat	Anestrus (ovaries without large follicles)	280	Ink injection	Lee 1925*b*
	Immature and late pregnancy	80–400	Uterus ruptured in every case	Andersen 1928
Cattle	Early postestrus (freshly ruptured follicles)	20–38	*In vitro*	Andersen 1928
	Estrogenic phase (large ovarian follicle)	63	*In vitro*	Andersen 1928
	Progestational phase (large corpus luteum)	65	*In vitro*	Andersen 1928
Dog	Nonpregnant	Impossible		Lee 1928
Guinea pig	Nonpregnant	30	*In situ*	Lee 1928
	Unknown	Impossible	No fluid passage at 504 mm Hg pressure	Kelly 1927
Man	Just after menstruation	80–100	*In vitro*	Rubin 1925
	Just before or during menstruation	180–200	*In vitro*	Rubin 1925
Monkey, rhesus		300	*In situ*	Morse and Rubin 1937
Pig	1st 3 days postestrus	360	*In vitro*	Andersen 1928
	During estrous cycle with exception of 1st 3 days postestrus	206	*In vitro*	Andersen 1928
	Pregnant	Pressures over 100 mm rare	*In vitro*	Andersen 1928
	Estrus	200	*In vitro* no gas passage	Whitelaw 1933
	Unknown	100		Lee 1928
	Inter-estrus	40–180	*In vitro*	Whitelaw 1933
Rabbit	Unknown	60, 130	*In situ*	Rubin and Davids 1940
	Late pregnancy	270	*In vitro*	Andersen 1928
	15 days pregnancy	94, 116	*In situ*	Hafez 1962
	8 days pseudopregnant	101–2	*In situ*	Hafez 1962
	15 days pseudopregnant	103–11	*In situ*	Hafez 1962
	Anestrus	50–220	*In situ*	Feresten and Wimpfheimer 1939
	Estrus	50–200	*In situ*	Feresten and Wimpfheimer 1939
	At ovulation	50–200	*In situ*	Feresten and Wimpfheimer 1939
	Pregnancy	50–200	*In situ*	Feresten and Wimpfheimer 1939
	Pseudopregnant	100–220	*In situ*	Feresten and Wimpfheimer 1939
	Castrated	50–200	*In situ*	Feresten and Wimpfheimer 1939
	Unknown	60	Ink injection	Lee 1928
	Unknown	50–200	*In situ*; gas did not pass in some animals at 200 mm Hg pressure	Wimpfheimer and Feresten 1939
Rat	Proestrus and diestrus	51	*In situ*	Alden 1943
	Estrus	100	*In situ*	Alden 1943
	Early pregnancy	100	*In situ*	Alden 1943
	Nonpregnant, nonestrus	250	Ink injection; no passage at 250 mm Hg pressure	Lee 1928
Sheep	Diestrus and late pregnancy	10–83	*In vitro*	Andersen 1928

dog, cat). In the dog and cat, the oviduct enters the uterus at approximately a 120° angle (Andersen 1928), presenting a formidable barrier to the passage of the fluids.

Often data obtained on the same species by different investigators must be compared with caution (Table 2). Too often the flow rates, which are extremely critical, are not reported. This is amply illustrated by the 54 mm Hg pressure required to open the uterotubal junction of a rabbit at a flow rate of 20 cc per min, contrasted to 295 mm Hg required in the same animal at a flow rate of 1,200 cc per min (Stavorski and Hartman 1958).

Animals with a simple uterotubal junction, such as cattle and sheep, have a very low "peak pressure" valve. In these species, the oviduct gradually merges with the uterus (Andersen 1928) at a very slight angle; furthermore, there are no complicated folds in the junction area. There is a flexure in the oviduct just cephalic to the uterus in the sheep (Edgar and Asdell 1960; Hafez, unpublished); the effect on uterotubal insufflation pressures, however, has not been determined.

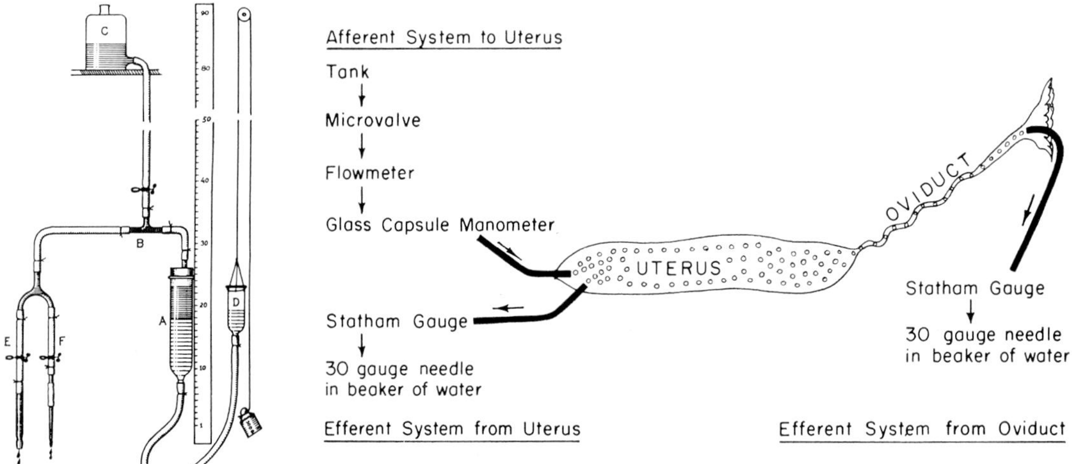

FIG. 20. Equipment used to measure the pressure required to open the uterotubal junction by Andersen 1928 (*left*) and Stavorski and Hartman 1958 (*right*):

Left: *A*, glass tube in which saline and mercury are in contact; *B*, rubber tube of saline connected by T-tube to *C*, a saline reservoir; *D*, mercury bulb to be raised and lowered; *E* and *F* can be controlled by raising and lowering *D* and read on the scale directly in millimeters of mercury as the distance between the mercury levels in *A* and *D*.

Right: Recordings reproduced are manometric, made with the aid of the glass capsule manometer.

2. *Effect of Endocrines*. The opening pressure of the uterotubal junction is dependent to some extent on the endocrinological state of the animal. In the sow, at the time when the eggs are in the oviduct, the villi surrounding the tubal opening are extremely edematous. This is also the period when the insufflation pressures are the greatest. From the 4th to the 8th day after ovulation, there is a decline in edema, and from the 8th to the 13th day, the edematous condition completely disappears (Andersen 1927*a*, 1928).

In man, a pressure of 180 to 200 mm Hg is required to force gas through the uterotubal junction just before or at menstruation while only 80 to 100 mm Hg pressure is required after menses (Rubin 1925). The initial patency of the human uterotubal junction varies so widely that no pattern was consistently related to the menstrual cycle (Davids 1948). The administration of estrogens (Bernstein and Feresten 1940) could increase the initial low insufflation pressure during menopause (average of 57 mm Hg).

In the rabbit, the degree of patency of the uterotubal junction bears no consistent relationship to the estrous cycle and there may be a considerable difference between the two uterotubal junctions in the same animal (Feresten and Wimpfheimer 1939). Peak pressure values are not significantly different during pregnancy from those during pseudopregnancy (Hafez 1962).

3. *Muscular Activity*. After the peak pressure, continued insufflation results in a pressure plateau showing fluctuations. The cause of these fluctuations is outlined in the present section.

Rhythmic activity in the uterotubal junction of the rabbit can be observed and recorded after complete ablation of the oviduct. The bubbles of gas that escape from it do so with a simultaneous decrease in observed pressure (Stavorski and Hartman 1958). However, it has been suggested that a portion of the isthmus must be present for pressure fluctuations to continue (Bonnet 1964).

In man, the musculature of the uterus is arranged around the intramural portion of the oviduct (Vasen 1959). There are also one or two vascular rings with accompanying muscle bundles within the intramural portion. Under normal conditions, the interstitial portion of the oviduct has the smallest diameter (Geist and Goldberger 1925; Lisa, Gioia, and Rubin 1954; Rocker 1964; Rubin 1928; Rubin 1954; Rubin and Novak 1952). This area will be very influential in governing gas pressure unless there is a greater narrowing of the oviduct cephalic to the intramural portion. Even though pressure fluctuations can be observed in the uterus after the oviduct has been removed (Rubin 1954), the ability of the intramural portion of the oviduct to produce manometric fluctuations apparently resides in the outer two-thirds, since removal of this segment stops further fluctuations (Mikulicz-Radecki 1930).

It is not known whether closure of the uterotubal junction is due to the intrinsic contractility of the oviduct or to the uterine musculature around the oviduct. Insertion of minute balloons into the myometrium of the uterine fundus, into the musculature of the uterine cornu near the isthmic portion of the tube, or into the oviductal wall, suggests that the uterine musculature plays an important role. The pressure oscillation changes obtained during insufflation coincide only with the pressure changes in the balloons located in the myometrium near the oviduct (Stabile 1952, 1954).

C. *Oviductal Fluid Flow through the Uterotubal Junction*

For many years the secretory function of oviductal epithelium has been recognized. However, little attention has been given to the direction that this fluid flows, i.e., whether the flow is toward the peritoneal cavity or toward the uterotubal junction. Assuming that fluid flow will be in the direction of least resistance, it is probable that the flow will depend on the physiological state of the uterotubal junction. As pointed out earlier, the junction undergoes changes during the reproductive cycle, which suggest that the direction of flow may alternate during the cycle.

Oviductal distention is present in the sheep, cow, and rabbit for about 3 days after ovulation if the ovarian end is closed. There is little difference in the degree of distention of oviducts ligated at the ovarian end and those ligated at both ends—an indication that during this stage of the reproductive cycle little fluid passes through the uterotubal junction.

In distended oviducts, there is a concomitant edema of the uterotubal junction area.

This edema, in the ewe, causes a sharp flexure of the oviduct immediately cephalic to the uterus, which may occlude the tubal lumen (Edgar and Asdell 1960).

Distention in rabbit oviducts remains about 60 hr *post coitum* and then recedes; at 72 hr distention is no longer evident (Black and Asdell 1958, 1959; Hafez 1963). The reduction of this distention is not merely the result of a reduced secretory rate, for a ligature placed at the uterotubal junction at this time will again result in distention. In the cow distention of the ligated oviduct is evident up to 72 hr after ovulation. The isthmus may be responsible for distention, in this animal, since removal of the uterotubal junction during insufflation does not result in a significant decrease in pressure. The fluid that does flow through the junction appears to be regulated by the endocrinological state of the animal as well as its maturity. Estrogen administration prolongs the distention of the ligated oviducts in both rabbits (Black and Asdell 1959) and cows (Black and Davis 1962). On the other hand, progesterone apparently has no effect on oviductal distention in the rabbit (Black and Asdell 1959).

The oviducts of immature does, unlike those of mature rabbits, do not distend when the ovarian end is ligated. Ligation of both ends, however, is followed by distention. These results may be due to the relaxed condition of the junction in immature animals. Likewise, the uterotubal junction of the immature ewe is relaxed (Edgar 1962). When the endocrine state of the sheep is altered, as with hormonally induced ovulation, distention does not occur. Inadequacy of the uterotubal junction under these circumstances can be overcome by estrogen administration (Edgar 1962).

Fluid flow through the uterotubal junction may be affected also by muscular activity. In sheep there is strong antiperistaltic activity in the isthmus for 48 to 60 hr following ovulation. Later peristaltic movements in this area carry the eggs through the junction and into the uterus (Wintenberger-Torres 1961). Presumably, such activity would be effective in the transport of oviductal fluid.

From the evidence at hand, it appears that fluid normally passes through the oviductal ostium and into the peritoneal cavity. There are, however, certain times during the reproductive cycle when the fluid flow is reversed and passes into the uterus; i.e., when the eggs enter the uterus, and for a brief period before and after parturition. The possible importance of the former movement in the passage of sperm and eggs through the uterotubal junction will be discussed below.

D. *Sperm Transport through the Uterotubal Junction*

It is generally agreed that the uterotubal junction, perhaps by virtue of its extremely small lumen and complex anatomy, is a barrier to sperm transport. The relative importance of this structure as a barrier varies with the species. In the rat, mouse, and horse, in which semen is ejaculated directly into the uterus, the uterotubal junction may provide the major barrier to transport. If sperm are deposited in the vagina, however, as in the rabbit and sheep, both the cervix and the uterotubal junction may act as barriers to movement.

Irrespective of whether or not the uterotubal junction is of primary or secondary importance as a barrier to sperm transport, the number of sperm which enter the oviduct, compared to the number in the uterus, is small. For example, the number of sperm in the uterus of a rat is about 30,000 times more than that found in the oviduct (Blandau and Odor 1949). In the ewe about 1 sperm in every 10 enters the oviducts, whereas in the rabbit 1 per 160 found in the anterior end of the uterus actually enters the oviduct (Braden 1963).

In some instances, perhaps as a result of malfunction of the uterotubal junction, a great number of sperm enter the oviducts (Braden 1963).

Once sperm reach the anterior uterus, they are transported through the uterotubal junction very quickly. In the cow, transport from the cervix to the oviducts requires only 2.5 min (VanDemark and Moeller 1951). In the sheep it requires about 8 min (Mattner and Braden 1963). However, this time can be lengthened by disturbing the animals at the time of copulation or soon thereafter. For example, ewes deliberately disturbed had no sperm in the oviducts within 15 min after copulation (Mattner 1963).

Although sperm may pass the uterotubal junction soon after copulation, there may not be a sufficient number to insure maximum fertilization. In the rabbit, for example, sperm reach the anterior tip of the uterus within 30 min of copulation; however, an additional 2 to 5 hr are required before adequate numbers enter the oviduct to insure maximum fertilization (Adams 1956). In the pig, sperm persist for some time at the uterotubal junction, perhaps due to the diverticulae of the oviduct (Rigby and Glover 1965; Rigby 1966).

Although the uterotubal junction undergoes certain anatomical and physiological alterations during the reproductive cycle, such changes apparently do not prevent sperm transport. Transport occurs at all stages of the estrous cycle in the ewe (Green and Winters 1935; Schott and Phillips 1941), and also during early pregnancy, but not during advanced pregnancy (Green and Winters 1935). Sperm are transported in ovariectomized sheep, but in reduced numbers (Mattner and Braden 1963). Likewise, sperm are transported during any period of the estrous cycle in the rat (Howe and Black 1963*a*). Even in the very young calf (1 month of age), sperm are transported very rapidly into the oviducts after vaginal deposition (Howe and Black 1963*b*).

It has been proposed that one of the functions of the uterotubal junction is to select against foreign and dead sperm (Leonard and Perlman 1949). This proposal is based on the observations that when a mixture of motile rat, guinea pig, and bovine sperm was inseminated into rats, only the rat sperm were found in the oviducts. However, it is clear from other studies that motility is not essential for transport. Nonmotile bovine sperm are transported rapidly in the cow (VanDemark and Moeller 1951); nonmotile bovine, guinea pig, human, and rat sperm are transported in the rat (Howe and Black 1963*b;* Marcus 1965); and nonmotile sperm are transported in the ewe (Mattner 1963) and the rabbit (Braden 1963).

In addition to immobilized sperm, many inert substances are also conveyed through the uterotubal junction. Some examples of such inert materials are carbon particles in man (Egli and Newton 1961), rabbit (Parker 1931), and sheep (Mattner and Braden 1963); radio-opaque oil in the cow (Rowson 1955); and microspheres in the rabbit (Glover and Patterson 1963).

Even though nonmotile sperm as well as other inert substances can be transported through the uterotubal junction, it is doubtful that the effectiveness of transport is the same. Selection of vigorous sperm may be an important function of the uterotubal junction area.

The mechanisms involved with sperm transport through the uterotubal junction are thought to be the same as those involved in transport in other parts of the oviduct. Muscular activity at the time of mating, especially if antiperistalsis in the isthmic portion of the oviduct is occurring, would be extremely important (Wintenberger-Torres 1961). Other equally possible mechanisms are uterine contractions forcing sperm and fluid through the junction (Kuhlmann 1964–65), and flow of oviductal fluid toward the ovarian end.

E. *Egg Transport through the Uterotubal Junction*

Examination of transportation through the oviduct, by locating eggs in the oviduct at various intervals after ovulation, has shown that very few eggs are ever found in the utero-tubal junction area. Rather, they tend to stop their progressive movement some distance (1–2 cm) from the uterus (Greenwald 1961; Longley and Black 1967), from which point they move very rapidly into the uterus. ^{198}Au-coated resin spheres (simulating eggs) behave in a similar manner (Harper *et al.* 1960). Likewise, oil droplets placed in the rabbit oviduct *in vitro* are rapidly shuttled to and fro down the oviduct by the extremely active musculature. Once they reach a position 2–3 cm above the uterotubal junction, they stop movement and are never observed to enter the uterus (Black and Asdell 1959). It should be emphasized also that not all eggs in polytocous animals pass through the uterotubal junction at the same time (Greenwald 1961; Harper *et al.* 1960; Noyes, Adams, and Walton 1959; Longley and Black 1967; Hafez, unpublished data) (Fig. 21).

Other than the fact that eggs pass through the uterotubal junction rapidly, little is known about their transport through this area. The primary obstacle to such studies is the thick oviductal wall which prevents direct observation of egg movement (Alden 1942). The principal forces concerned with this movement are generally assumed to be the same as those responsible for transportation through the ampulla and isthmus; i.e., muscular and ciliary activity, or a combination of the two.

The area most responsible for delaying egg entry into the uterus may depend upon the species or the endocrine state of the animal. For example, estrogen causes retention of ova at the ampullary-isthmic junction in the rabbit (Greenwald 1963). In the mouse, on the other hand, estrogen administration results in the retention of eggs in the lower part of the oviduct. With estrogen treatment, the uterine end of the oviduct is turgid as if it were contracted or distended with lymph (Whitney and Burdick 1936). Under normal conditions, there is relaxation in this area approximately 72 hr after the vaginal plug forms. In addition, there is increased activity in the uterotubal junction at approximately the same time (Burdick, Whitney, and Emerson 1942).

The time at which eggs enter the uterus is under the control of ovarian hormones. Injection of progesterone may (Alden 1942) or may not hasten the rate of transportation (Greenwald 1961; Black and Asdell 1959). Ovariectomy appears to enhance egg transport in rats (Alden 1942) and rabbits (Noyes, Adams, and Walton 1959), but not in mice (Whitney and Burdick 1939). Eggs transferred to the oviducts of pseudopregnant and estrous rabbits without luteal tissue are not transported to the uterus in a normal manner (Austin 1941).

As pointed out previously, oviductal fluid flows toward the cornua at about the time eggs normally enter the cornua. It is likely that relaxation of the uterotubal junction and this flow materially aid transport through this area.

III. Pharmacophysiology of the Uterotubal Junction

At present little work has been reported concerning the pharmacology of the uterotubal junction. It is, however, perhaps safe to extrapolate results obtained from studies on the oviduct, since it appears that all segments respond similarly to certain drugs (Sandberg, Ingleman-Sundberg, and Lindgren 1960). All segments respond similarly to acetylcholine, epinephrine, norepinephrine, or oxytocin. On the other hand, prostaglandin does have differential effects; the uterine one-fourth tends to contract, while the remaining three-fourths undergo relaxation (Sandberg, Ingelman-Sundberg, and Ryden 1963).

Table 3 summarizes the evidence that suggests that the adrenergic nervous system can stimulate motor activity in the oviduct (Hawkins 1964; Davids and Bender 1940*a;* Horton, Main, and Thompson 1965; Brundin 1965; Longley and Black 1967; Rosenblum and Stein 1966), whereas the effect of the cholinergic component is less clear and probably less important in controlling motor activity (Hawkins 1964; Davids and Bender 1940*b;* Horton, Main, and Thompson 1965; Longley and Black 1967). The apparent stimulator of oviductal

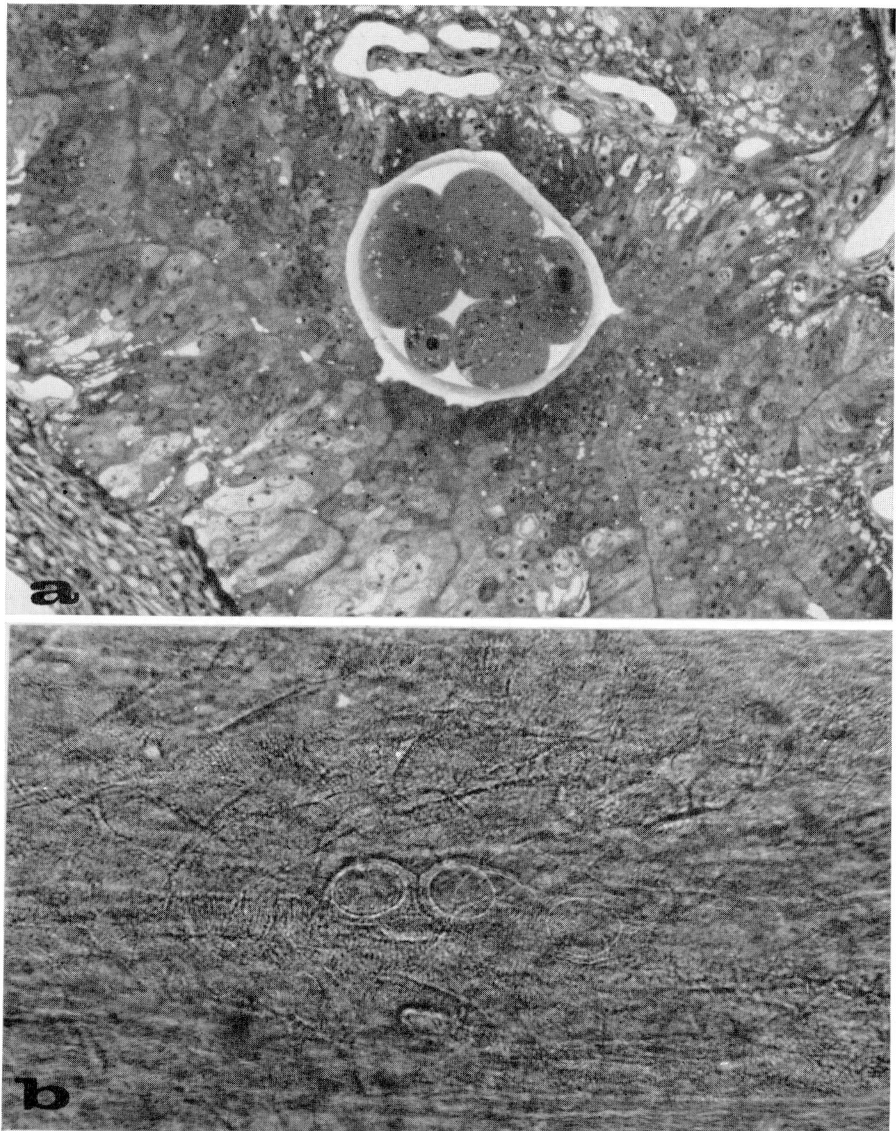

FIG. 21. (*a*) A 4- to 8-cell egg in the uterotubal junction of the mouse. Note the firm contact with the surrounding epithelial cells. Only one cell type can be seen, to some extent resembling the uterine epithelial cell (×375) (from Reinius 1966); (*b*) Eggs in the uterotubal junction area of the rabbit 72 hr *post coitum.* The oviduct has been cleared by the method of Orsini (Black, unpublished data).

TABLE 3

THE EFFECT OF PHARMACOLOGICAL AGENTS ON OVIDUCT MOTILITY

Drug	Dosage	Species	Reaction	Reference
Acetylcholine	5–20 μg/ml	Man†	Stimulatory	Hawkins 1964
	0.5–5.0 μg/ml	Man†	Stimulatory	Sandberg, Ingelman-Sundberg, and Lindgren 1960
	10 μg/ml	Rabbit*	Relaxation (may be preceded by contraction)	Davids and Bender 1940*a*
	....	Rabbit*	No effect	Horton, Main, and Thompson 1965
	2.5–5.0 μg/kg	Rabbit*	No effect	Longley and Black 1967
Amylocaine hydrochloride (high spinal)	50 mg	Man*	Stimulatory	Walker and Stout 1952
Amylocaine hydrochloride (low spinal)	50 mg	Man*	Slightly stimulatory	Walker and Stout 1952
Amylnitrate	....	Man*	Relaxation	Rubin 1947
Atropine	0.5 μg/ml	Man†	No effect	Hawkins 1964
	1/100 gr	Man*	No effect	Walker and Stout 1952
	1 mg	Rabbit*	Blocked action of acetylcholine	Davids and Bender 1940*a*
Atropine	20 mg/kg	Rabbit*	No effect	Brundin 1965
Angiotensin	5 μg/kg	Rabbit*	No effect	Brundin 1965
Barium chloride	0.707.0 mg/ml	Man†	Stimulatory	Hawkins 1964
Bradykinin	....	Rabbit*	Transient stimulation	Horton, Main, and Thompson 1965
Chloroform	Anesthesia	Man*	Slight decrease in tone; slight increase in amplitude	Walker and Stout 1952
Cyclopropane	Anesthesia	Man*	No effect	Walker and Stout 1952
d-Tubo-curarine	11 mg	Man*	Stimulatory to amplitude	Walker and Stout 1952
1,1–dimethyl-4–phenylpiperazinium	1.0–4.0 μg/ml	Man†	Inhibitory	Rosenblum and Stein 1966
Epinephrine	0.1–2.5 μg/ml	Man†	Stimulatory	Hawkins 1964
Epinephrine	0.5–1.0 μg/ml	Man†	Stimulatory	Sandberg, Ingelman-Sundberg, and Lindgren 1960
		Man†	Stimulatory	Gunn 1914
	....	Rabbit*	Stimulatory	Davids and Bender 1940*b*
	0.1–0.3 mg/kg	Rabbit*	Stimulatory	Davids and Bender 1940*b*
	2.0–8.0 μg	Rabbit*	Stimulatory	Davids and Bender 1940*b*
	....	Rabbit*	Stimulatory	Horton, Main, and Thompson 1965
	2.5–7.5 μg/kg	Rabbit*	Stimulatory	Longley and Black 1967
Ergometrine	2 and 8 μg/ml	Man†	Stimulatory	Hawkins 1964
Eserine salicylate	0.6 mg	Rabbit*	No effect alone; potentiates acetylcholine	Davids and Bender 1940*a*
Ether	Anesthesia	Man*	Contractility decreased; frequency increased	Walker and Stout 1952
	Anesthesia	Rabbit*	Inhibitory	Newman 1941
Histamine	3–30 μg/ml	Man†	Stimulatory	Hawkins 1964
Isoproterenol	1–4 μg/kg	Rabbit*	Inhibitory	Longley and Black 1967
	0.5–5.0 μg/ml	Man†	Stimulatory	Rosenblum and Stein 1966
Morphine	1/4 grain	Man*	Stimulatory	Davids and Weiner 1950
Methacholine	10–100 μg/ml	Man†	Stimulatory	Rosenblum and Stein 1966
Nembutal	3 grains	Man*	Stimulatory	Davids and Weiner 1950
	Anesthesia	Rabbit*	No effect	Newman 1941
Nitrous oxide	Anesthesia	Man*	Increased amplitude	Walker and Stout 1952
Norepinephrine	0.1–2.5 μg/ml	Man†	Stimulatory	Hawkins 1964
	0.5–1.0 μg/ml	Man†	Stimulatory	Sandberg, Ingelman-Sundberg, and Lindgren 1960
	0.5–5.0 μg/ml	Man†	Stimulatory	Rosenblum and Stein 1966
Oxytocin	8–800 μl/ml	Man†	Stimulatory	Rorie and Newton 1965
	0.005–0.01 I.E./ml	Man†	Little effect	Sandberg, Ingelman-Sundberg, and Lindgren 1960

* *In vivo* † *In vitro*

TABLE 3—*Continued*

Drug	Dosage	Species	Reaction	Reference
Parasympatholytic	....	Rabbit*	No effect	Joshi and Asdell 1966
		Rabbit†	No effect	
Parasympatho-mimetic	....	Rabbit*	No effect	Joshi and Asdell 1966
		Rabbit†	No effect	
Pentamethonium	4 mg/kg	Rabbit*	No effect	Brundin 1965
Phenoxyhenzamine	3 mg/kg	Rabbit*	Blocks effect of epinephrine	Longley and Black 1967
Phentolamine	0.5–1.0 mg/kg	Rabbit*	Blocked effect of hypogastric nerve stimulation	Brundin 1965
	1.0 mg/kg	Rabbit*	Inhibitory	Brundin 1965
Pilocarpine	5 and 20 μg/ml	Man†	No effect	Hawkins 1964
Propranolol	7 mg/kg	Rabbit*	Blocks effect of isoproterenol	Longley and Black 1967
Prostaglandin	....	Man†	Stimulatory-uterine 1/4; inhibitory-ovarian 3/4	Sandberg, Ingelman-Sundberg, and Lindgren 1960
Prostaglandin E_1	0.5 μg/kg	Rabbit*	Relaxation	Horton, Main, and Thompson 1965
		Sheep*	Variable	Horton, Main, and Thompson 1965
Reserpine	0.25 mg/kg	Rabbit*	No effect on spontaneous activity	Brundin 1965
Scopolamine	....	Man*	Relaxation	Rubin 1947
Sympatholytic drugs	....	Rabbit*	Stimulatory	Joshi and Asdell 1966
		Rabbit†	Stimulatory	
Sympathomimetic	....	Rabbit*	Stimulatory	Joshi and Asdell 1966
		Rabbit†	Stimulatory	
Thiopentone	....	Man*	No effect	Walker and Stout 1952
Tincture belladona demerol	100 mg	Man*	Stimulatory	Davids and Weiner 1950
Trichlorethylene	Anesthesia	Man*	Decreased tone	Walker and Stout 1952

contraction is the alpha and not the beta component of the adrenergic nervous system (Rosenblum and Stein 1966; Longley and Black 1967). Whether or not the beta component has an inhibitory effect has not been determined. Further evidence that an alpha component of the adrenergic nervous system is the stimulant is the abundant amount of norepinephrine present in the uterotubal junction (Brundin 1964; Brundin and Wirsen 1964*a*, *b*, *c*). Results obtained from other experiments in which parasympathetic drugs have been administered are contradictory. For example, administration of acetylcholine may be stimulatory (Hawkins 1964; Sandberg, Ingelman-Sundberg, and Lindgren 1960; Davids and Bender 1940*b*), have no effect (Horton, Main, and Thompson 1965; Longley and Black 1967), or occasionally cause the oviduct to relax prior to contraction (Davids and Bender 1940*b*), a result that may be due to an initial release of endogenous epinephrine.

In studying the action of pharmacological agents, caution must be used in designing the experimental methods. Because anesthetics have a pronounced influence on oviductal motility, great care must be exercised in their use and in interpreting their effects. The endocrine status of the test animal must also be known. Furthermore, the dose level of the drug must be sufficient to be physiologically active, and it must be ascertained whether or not the parameter being measured is the primary or the secondary effect.

The amount of epinephrine required to produce sustained oviductal contractions in rabbits at different endocrine stages is highly variable. The amount required to cause contractions of 20–40 mm Hg is 2–8 mg during anestrus and only 0.2 mg at estrus. Moreover,

4.0–5.0 mg are required after ovariectomy, but if estradiol benzoate is given 0.04 mg are sufficient. After testosterone treatment, higher levels (10–50 mg) are required (Davids and Bender 1940*a*). The effect of acetylcholine, however, is less dependent on the hormonal condition of the animal. In anestrous animals, 10 mg of acetylcholine result in relaxation of the oviducts in 8 out of 20 rabbits, while the same dose causes relaxation in 6 out of 10 estrous animals (Davids and Bender 1940*b*).

The increased sensitivity to drugs at the time of estrus is not apparently a reflection of the accumulation or overproduction of sympathetic amines, since the amount of norepinephrine in the isthmus does not vary with the hormonal state of the animal (Brundin 1965). Rather, the musculature of the oviduct is apparently more reactive to the amine when under the influence of estrogen.

As a guide to the physiological activity of drugs, parameters other than oviduct motility should be determined simultaneously. For example, even though the oviduct may not respond to a given dose of a particular drug, the cardiovascular response, if present, can be used as an effective guide. The respiratory system may be a useful determinant also.

IV. Conclusions

Methods for evaluating uterotubal junction physiology are available, but extreme care must be taken to insure that differences observed are not due to technique alone. Uterotubal insufflation is the most commonly used technique. With this method, differences between species have been found which can often be related to anatomical structure or to the endocrine state of the animal.

Although much has been learned about the physiology of the uterotubal junction, its role in normal reproduction remains obscure. There is evidence which suggests that the uterotubal junction controls the entry of eggs into the uterus of some species. In other species, this evidence is lacking. The junction also presents a barrier to sperm transport, but whether it is really effective in selecting against less vigorous sperm is not known. In the sow, the uterotubal junction functions as a "sperm reservoir." Perhaps one very important function of the uterotubal junction is to direct the flow of oviductal fluid within the oviduct. In this role, the uterotubal junction may assist in the transport of sperm and egg.

To elucidate the function of the uterotubal junction, areas of comparative research which may be beneficial are:

1) neural, vascular, and lymphatic supply of the uterotubal junction;

2) histological and histochemical changes following copulation, at the time of egg transport to the uterus, and following ovariectomy or administration of exogenous steroids;

3) the architecture and fine structure of the muscle fibers in the circular and longitudinal musculature in relation to that of the mucosal folds (cf. Fig. 22);

4) anatomical and physiological function of the ligaments attaching the oviduct to the uterus; the endocrine and pharmacophysiological aspects of these ligaments;

5) relative importance of the uterotubal junction, in comparison with other parts of the oviduct, in controlling entry of eggs into the uterus;

6) mechanisms involved in transport of sperm and eggs through the uterotubal junction;

7) effects of sexual cycle stages, ovarian function, and exogenous hormones on the opening and closing of the junction and on the passage of sperm and eggs through the junction;

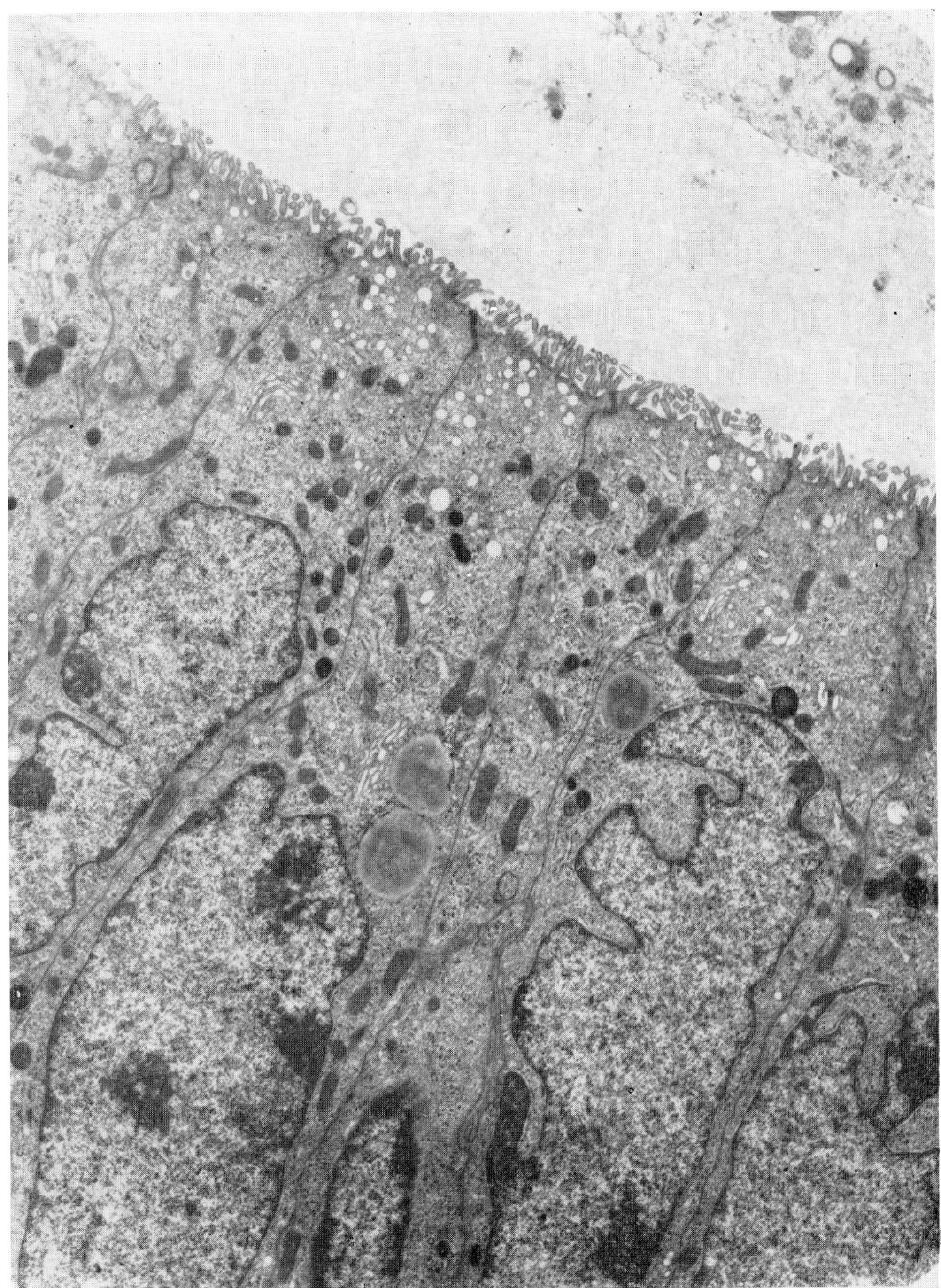

FIG. 22. Electronmicrograph of the egg-epithelium relationship in the uterotubal junction of the mouse. The microvilli are low and regular. Apical vesicles are present at day 3, but are less abundant during days 1, and 2. The endoplasmic reticulum consists of short membrane formations with ribosomes. Lipid granules are numerous in the basal part of the cells (×11,000). (Photo by S. Reinius).

8) neural control, both central and peripheral, of the uterotubal junction;

9) the neuroendocrine control of the uterotubal junction; e.g., the effect of ovariectomy and administration of exogenous steroids on the differential distribution of adrenergic and noradrenergic nerve supply.

Acknowledgment

The section on the anatomy of the uterotubal junction represents unpublished results by E. S. E. Hafez. These results are part of investigations supported by research grants HD-00585-03 and HD-00745, and symposium contract PH-43-67-672 from the National Institute of Child Health and Human Development, United States Public Health Service, and by the Mrs. Lillian Banta Research Fund. Thanks are due to Dr. Philip A. Corfman, United States National Library of Medicine, and to the University of Washington Primate Information Center for providing a list of references, and to Drs. R. M. Brenner, W. L. Herrmann, O. Smith, and F. Young for providing specimens of man and monkeys. Several helpful suggestions and criticisms have been made by Drs. R. J. Blandau, R. E. Click, M. J. K. Harper, and Professor S. J. Hook. Thanks are also due to Mr. Steve Allured and Mr. Walter Carey for preparing some of the drawings, and to Mr. J. A. Lineweaver for the microphotography.

References

1. Adams, C. E. 1956. A study of fertilization in the rabbit: The effect of post-coital ligation of the Fallopian tube or uterine horn. *Endocrinology* 13:296.
2. Alden, R. H. 1942. The oviduct and egg transport in the albino rat. *Anat. Rec.* 84:137.
3. ———. 1943. The uterotubal junction in the albino rat. *Anat. Rec.* 85:290.
4. Allen, E. 1922. The oestrous cycle in the mouse. *Amer. J. Anat.* 30:297.
5. Andersen, D. H. 1927*a*. Lymphatics of the Fallopian tube of the sow. *Contrib. to Embryol. Carnegie Inst.* 19:135.
6. ———. 1927*b*. The rate of passage of the mammalian ovum through various portions of the Fallopian tube. *Amer. J. Physiol.* 82:557.
7. ———. 1928. Comparative anatomy of the tubo-uterine junction. Histology and physiology in the sow. *Amer. J. Anat.* 42:255.
8. Asdell, S. A. 1955. *Cattle fertility and sterility*. Boston: Little, Brown & Co.
9. Austin, C. R. 1941. Fertilization and the transport of gametes in the pseudopregnant rabbit. *J. Endocr.* 6:63.
10. Bernstein, F., and Feresten, M. 1940. Estrogenic effects upon tubal contractility and the vaginal secretion in the menopause. *Endocrinology* 26:946.
11. Black, D. L., and Asdell, S. A. 1958. Transport through the rabbit oviduct. *Amer. J. Physiol.* 192:63.
12. ———. 1959. Mechanism controlling entry of ova into rabbit uterus. *Amer. J. Physiol.* 197 (6):1275.
13. Black, D. L., and Davis, J. 1962. A blocking mechanism in the cow oviduct. *J. Reprod. Fertil.* 4:21.
14. Blandau, R. J., and Odor, D. L. 1949. The total number of spermatozoa reaching various segments of the reproductive tract in the female albino rat at intervals after insemination. *Anat. Rec.* 103:93.

15. Bogli, B. 1959. Das tubo-uterine Ventil beim Goldhamster. *Rev. Suisse Zool.* 66:211.
16. Bonnet, L. 1964. The origin of the oscillations in kymographic tubal insufflation. *Int. J. Fertil.* 9:513.
17. Braden, A. W. H. 1953. Distribution of sperm in the genital tract of the female rabbit after coitus. *Aust. J. Biol. Sci.* 6:693.
18. ———. 1963. Unpublished data as quoted by P. E. Mattner and A. W. H. Braden. Spermatozoa in the genital tract of the ewe. I. Rapidity of transport. *Aust. J. Biol. Sci.* 16:473.
19. Brundin, J. 1964. The distribution of noradrenaline and adrenaline in the Fallopian tube of the rabbit. *Acta Physiol. Scand.* 62:156.
20. ———. 1965. Distribution and function of adrenergic nerves in the rabbit Fallopian tube. *Acta Physiol. Scand.* 66:1.
21. Brundin, J., and Wirsen, C. 1964*a*. The distribution of adrenergic nerve terminals in the rabbit oviduct. *Acta Physiol. Scand.* 61:203.
22. ———. 1964*b*. Adrenergic nerve terminals in the human Fallopian tube examined by fluorescence microscopy. *Acta Physiol. Scand.* 61:505.
23. ———. 1964*c*. The distribution of adrenergic nerve terminals in the rabbit oviduct. *Acta Physiol. Scand.* 61:203.
24. Burdick, H. O.; Whitney, R.; and Emerson, B. 1942. Observations on the transport of tubal ova. *Endocrinology* 31:100.
25. Davids, A. M. 1948. Fallopian tubal motility in relation to the menstrual cycle. *Amer. J. Obstet. Gynec.* 56:655.
26. Davids, A. M., and Bender, M. B. 1940*a*. Effects of adrenaline on tubal contractions of the rabbit in relation to sex hormones (study *in vivo* by Rubin method). *Amer. J. Physiol.* 129:259.
27. ———. 1940*b*. The relaxation effect of acetylcholine on the oviduct of the rabbit in relation to hormonal status. *Amer. J. Physiol.* 131:240.
28. Davids, A. M., and Weiner, I. 1950. The effect of sedation on Fallopian tubal motility. *Amer. J. Obstet. Gynec.* 59:673.
29. Edgar, D. G. 1962. Studies on infertility in ewes. *J. Reprod. Fertil.* 3:50.
30. Edgar, D. G., and Asdell, S. A. 1960. The valve-like action of the utero-tubal junction of the ewe. *J. Endocr.* 21:315.
31. Egli, G. E., and Newton, M. 1961. The transport of carbon particles in the human female reproductive tract. *Fertil. Steril.* 12:151.
32. Falck, B. 1962. Observation on the possibilities of the cellular localization of monoamines by a fluorescence method. *Acta Physiol. Scand.* 56, suppl. 197.
33. Falck, B.; Hillarp, N. A.; Thieme, G.; and Torp, A. 1962. Fluorescence of catecholamines and related compounds condensed with formaldehyde. *J. Histochem. Cytochem.* 10:384.
34. Feresten, M., and Wimpfheimer, S. 1939. Patency of the uterotubal junction of the rabbit: Experimental observations with the aid of CO_2 insufflation (Rubin method). *Endocrinology* 24:510.
35. Geist, S. H., and Goldberger, M. A. 1925. A study of the intramural portion of normal and diseased tubes with special reference to the question of sterility. *Surg. Gynec. Obstet.* 41:646.
36. Glover, T. D., and Patterson, J. A. 1963. The passage of ^{131}I-labeled copolymer poly-

styrene-divinylbenzene microspheres through the reproductive tract of the female rabbit. *J. Endocr.* 26:175.

37. Gough, J. 1960. In *Recent advances in pathology*, ed. C. V. Harrison. London: J. & A. Churchill.
38. Green, W. W., and Winters, L. M. 1935. Studies of the physiology of reproduction in the sheep. III. The time of ovulation and rate of sperm travel. *Anat. Rec.* 61:457.
39. Greenwald, G. S. 1961. A study of the transport of ova through the rabbit oviduct. *Fertil. Steril.* 12:80.
40. ———. 1963. Interruption of early pregnancy in the rabbit by a single injection of oestradial cyclopentylpropionate. *J. Endocr.* 26:133.
41. Gunn, J. A. 1914. The action of certain drugs on the isolated human uterus. *Proc. Roy. Soc. Lond. Ser. B.* 87:551.
42. Hafez, E. S. E. 1962. Pressure fluctuations during uterotubal kymographic insufflation in pregnant rabbits. *Fertil. Steril.* 13:426.
43. ———. 1963. The uterotubal junction and the luminal fluid of the uterine tube of the rabbit. *Anat. Rec.* 145:7.
44. Harper, M. J. K.; Bennett, J. P.; Boursnell, J. C.; and Rowson, L. E. 1960. An autoradiographic method for the study of egg transport in the rabbit Fallopian tube. *J. Reprod. Fertil.* 1:249.
45. Hawkins, D. F. 1964. Some pharmacological reactions of isolated rings of human Fallopian tube. *Arch. Int. Pharmacodyn.* 152:474.
46. Horton, E. W.; Main, I. H.; and Thompson, C. J. 1965. Effect of prostaglandins on the oviduct studied in rabbits and ewes. *J. Physiol.* 180:514.
47. Howe, G. R., and Black, D. L. 1963*a*. Spermatozoan transport and leucocytic responses in the reproductive tract of calves. *J. Reprod. Fertil.* 6:305.
48. ———. 1963*b*. Migration of rat and foreign spermatozoa through the uterotubal junction of the oestrous rat. *J. Reprod. Fertil.* 5:95.
49. Joshi, S. R., and Asdell, S. A. 1966. Pharmacology of the rabbit oviduct. *Fed. Proc.* 25:1427.
50. Kellogg, M. 1945. The postnatal development of the oviduct of the rat. *Anat. Rec.* 93:377.
51. Kelly, G. L. 1927. The uterotubal junction in the guinea pig. *Amer. J. Anat.* 40:373.
52. Kuhlmann, W. 1964–65. Uterotubal valve and the fate of the sperm in the female tract of the mouse. *Jackson Lab. Ann. Rept.* 71.
53. Lee, F. C. 1925*a*. A brief note on the anatomy of the uterine opening of the Fallopian tube. *Proc. Soc. Exp. Biol. Med.* 22:470.
54. ———. 1925*b*. A preliminary note on the physiology of the uterine openings of the Fallopian tube. *Proc. Soc. Exp. Biol. Med.* 22:335.
55. ———. 1928. The tubo-uterine junction in various animals. *Bull. Johns Hopkins Hosp.* 42:335.
56. Leonard, S. L., and Perlman, P. L. 1949. Conditions effecting the passage of spermatozoa through the uterotubal junction of the rat. *Anat. Rec.* 104:89.
57. Lisa, J. R.; Gioia, J. D.; and Rubin, I. C. 1954. Observations on the interstitial portion of the Fallopian tube. *Surg. Gynec. Obstet.* 99:159.
58. Longley, W. J., and Black, D. L. 1967. Unpublished data.
59. Marcus, S. L. 1965. The passage of rat and foreign spermatozoa through the uterotubal junction of the rat. *Amer. J. Obstet. Gynec.* 91:985.

60. Marshall, R., and May, J. W. 1960. Cine (serial) photography of microtome sections. *Med. Biol. Illus.* 10:267.
61. Mastroianni, L. Jr. 1962. The structure and function of the Fallopian tube. *Clin. Obstet. Gynec.* 5:781.
62. Mattner, P. E. 1963. Spermatozoa in the genital tract of the ewe. II. Distribution after coitus. *Austr. J. Biol. Sci.* 16:688.
63. Mattner, P. E., and Braden, A. W. H. 1963. Spermatozoa in the genital tract of the ewe. I. Rapidity of transport. *Austr. J. Biol. Sci.* 16:473.
64. Mikulicz-Radecki, F. V. 1930. Untersuchungen über die Tubenkontraktionen mit Hilfe der Pertubation. *Zbl. Gynaek.* 54:2183.
65. Morse, A. H., and Rubin, I. C. 1937. Uterotubal insufflation in the Macacus rhesus: A method of assaying pharmacologic and hormonal effects on tubal and uterine contractions. *Amer. J. Obstet. Gynec.* 33:1087.
66. Newman, H. F. 1941. Experimental uterotubal insufflation in the rabbit. *J. Lab. Clin. Med.* 26:1129.
67. Novak, J., and Rubin, I. C. 1952. Anatomy and pathology of the Fallopian tubes. *Ciba Clinical Symposia*, vol. 4, no. 6.
68. Noyes, R. W.; Adams, C. E.; and Walton, A. 1959. The transport of ova in relation to the dosage of estrogen in ovariectomized rabbits. *J. Endocr.* 18:108.
69. Parker, G. H. 1931. The passage of sperm and eggs through the oviducts in terrestrial vertebrates. *Phil. Trans. Roy. Soc. Lond. Ser. B.* 219:381.
70. Reinius, S. 1966. Ultrastructure of epithelium in mouse oviduct during egg transport. *Proc. Fifth. Int. Congr. Fertil. Steril.* (*Stockholm*).
71. Rigby, J. P. 1966. Persistence of spermatozoa at the uterotubal junction of the sow. *J. Reprod. Fert.* 11:153.
72. Rigby, J. P., and Glover, T. D. 1965. The structure of the uterotubal junction of the sow. *J. Anat.* 99:416.
73. Rocker, I. 1964. The anatomy of the uterotubal junction area. *Proc. Roy. Soc. Med.* 57 (8):707.
74. Rorie, D. K., and Newton, M. 1965. Response of isolated human Fallopian tube to oxytocin: A preliminary report. *Fertil. Steril.* 16:27.
75. Rosenblum, I., and Stein, A. A. 1966. Autonomic responses of the circular muscles of the isolated human Fallopian tube. *Amer. J. Physiol.* 210:1127.
76. Rowson, L. E. A. 1955. The movement of radio-opaque material in the bovine uterine tract. *Brit. Vet. J.* 111:334.
77. Rubin, I. C. 1925. Most favorable time for transuterine insufflation to test tubal patency. *J.A.M.A.* 84:661.
78. ———. 1928. Clinical study in 650 cases of sterility by the method of peruterine insufflation combined with kymograph. *J.A.M.A.* 90:99.
79. ———. 1947. *Uterotubal insufflation.* St. Louis, Mo.: Mosby.
80. ———. 1954. Manometric oscillations: Discussion of paper by Stabile. *Fertil. Steril.* 5:147.
81. Rubin, I. C., and Davids, A. M. 1940. Pharmacodynamic effects of testosterone propionate on tubal contractions (oviduct of rabbit) determined by CO_2 insufflation. *Endocrinology* 26:523.
82. Rubin, I. C., and Novak, J. 1952. Anatomy and pathology of the Fallopian tube. *Ciba. Clin. Symposia* 4:179.

83. Sandberg, F.; Ingelman-Sundberg, A.; and Lindgren, L. 1960. *In vitro* studies of the motility of the human Fallopian tube. I. The effects of acetylcholine, adrenaline, noradrenaline, and oxytocin on the spontaneous motility. *Acta Obstet. Gynec. Scand.* 39:506.
84. Sandberg, F.; Ingelman-Sundberg, A.; and Ryden, G. 1963. The effect of prostaglandin E_1 on the human uterus and Fallopian tubes *in vitro. Acta Obstet. Gynec. Scand.* 42:269.
85. Schilling, E. 1962. Untersuchungen über den Bau und die Arbeitsweise des Eileiters vom Schaf und Rind. *Z. Veterinärmed.* 9:805.
86. Schneider, P. 1942. The problem of the "tubal sphincter" and of the intramural portion of the Fallopian tube. *Amer. J. Roentgenol.* 48:527.
87. Schott, R. G., and Phillips, R. W. 1941. The rate of sperm travel and time of ovulation in sheep. *Anat. Rec.* 79:531.
88. Stabile, A. 1952. Nuestra interpretación de las alternatives ritmicas en los quimograms de insuflación úterotubaria. *Obstet. Ginec. Lat.-Amer.* 10:40.
89. ———. 1954. Interpretation of manometric ascellations observed during uterotubal insufflation. *Fertil. Steril.* 5:138.
90. Stavorski, J., and Hartman, C. G. 1958. Uterotubal insufflation. A study to determine the origin of fluctuations in pressure. *Obstet. Gynec.* 11:622.
91. Sweeney, William J. 1962. The interstitial portion of the uterine tube—its gross anatomy, course, and length. *Obstet. Gynec.* 19:3.
92. VanDemark, N. L., and Moeller, A. N. 1951. Speed of spermatozoan transport in reproductive tract of estrous cow. *Amer. J. Physiol.* 165:674.
93. Vasen, L. C. 1959. The intramural part of the Fallopian tube. *Int. J. Fertil.* 4:309.
94. Walker, A. H. C., and Stout, R. J. 1952. The effects of anesthesia upon Fallopian tubal motility. *J. Obstet. Gynec. Brit. Emp.* 59:1.
95. Westman, A. 1926. A contribution to the question of the transit of the ovum from the ovary to uterus in rabbits. *Acta Obstet. Gynec. Scand.* 5:104.
96. Whitelaw, M. J. 1933. Tubal contractions in relation to the estrous cycle as determined by uterotubal insufflation. *Amer. J. Obstet. Gynec.* 25:475.
97. Whitney, R., and Burdick, H. O. 1936. Tube-locking of ova by oestrogenic substances. *Endocrinology* 20:643.
98. ———. 1939. Effect of massive doses of an estrogen on ova transport in ovariectomized mice. *Endocrinology* 24:45.
99. Wimpfheimer, S., and Feresten, M. 1939. The effect of castration on tubal contraction of the rabbit as determined by the Rubin test. *Endocrinology* 25:91.
100. Wintenberger-Torres, S. 1961. Movements des trompes et progression des oeufs chez la brebis. *Ann. Biol. Anim. Biochem. Biophys.* 1:121.

Part II Physiology, Endocrinology, and Pharmacology

5

Gamete Transport—Comparative Aspects

R. J. Blandau

Department of Biological Structure
School of Medicine
University of Washington, Seattle

I. Egg Transport

A. *Transport of Ova from the Surface of the Ovary into the Oviductal Ostium*

The manner in which newly ovulated eggs are moved from the site of follicle rupture to the ostium of the oviductal infundibulum is poorly understood. One important reason for this hiatus in our knowledge is the technical difficulties in being able to observe and record ovulation and egg transport simultaneously under normal physiological conditions. The ovulatory mechanism is exceedingly sensitive to environmental conditions and may be inhibited by anesthesia and a variety of other insults to exposed animal tissues. Further, the processes of ovulation and egg transport occur within a remarkably short interval of time during any single reproductive cycle. Species variation in the anatomy of the infundibulum and relationship of its fimbriated end to the surface of the ovary from which the eggs must be captured also presents problems in interpretation of mechanism. These variations may mean that special adaptations exist in effecting egg "pick-up" in different animals. Thus in this area as elsewhere in biology generalizations as to mechanisms of gamete transport are hazardous.

B. *Relationship of Fimbriae to Ovary*

In the *Mustelidae* and *Muridae* the ovaries are enclosed in a membranous periovarial sac. The infundibulum projects into this sac but occupies only a small area of the periovarial space. The fimbriated tip makes very limited contact with the surface of the ovary and the ostium of the oviduct usually lies parallel to its surface (Alden 1942; Wimsatt and Waldo 1945). In contrast, the expansive fimbriae of the oviducts of rabbits and guinea pigs almost completely enclose the ovaries at the time of ovulation (Westman 1926, 1952; Blandau unpublished observations). In primates the relationship between the fimbriae and ovary is somewhat in between the two situations described above. The extent of communication between the fimbriae and ovary varies with the stage of the estrous or menstrual cycle. In the rat the fimbriated collar of the oviduct is quite narrowed during the diestrum but at the time of ovulation it becomes much more expansive and vigorously active. In both the rabbit and the guinea pig not in heat the fimbriae appear somewhat shrunken and only partially cover the ovaries. In contrast, at the time of normal or induced ovulation the fimbriae have expanded to almost completely enclose the ovaries, forming distinct ovarian bursae.

C. *Role of the Fimbriae in Egg Transport*

Various techniques have been devised to observe the behavior of the fimbriae at the time of ovulation and to evaluate their role in egg transport from the surface of the ovary. Westman (1926, 1952) placed windows into the abdominal wall near the oviducts of rabbits and was the first to describe the activity of the fimbriae at ovulation. A disadvantage of this technique is the presence of other viscera within the region of the window, which often obscures adequate visibility of the oviducts. Attempts have been made to observe the movements of human fimbriae by means of abdominal peritoneoscopy and exploratory culdotomy (Elert 1947). Elert described the elongated fimbriae in women grasping the inferior pole of the ovary for several minutes. By using similar techniques Doyle (1951, 1954) failed to observe any kind of sweeping or grasping motion of the fimbriae at the time of follicle rupture in women. He concludes that the initial transport of the egg is by a process in which it simply floats into the cul-de-sac and from there is siphoned into the ampulla. With improved culdotomy techniques, recent observations on fimbrial behavior are similar to those described by Elert (Doyle 1956).

1. *The Vacuum Hypothesis in Egg Transport.* From time to time it has been postulated that a negative pressure is created in the region of the fimbriae of the oviduct by the vigorous peristaltic activity of both the fimbriae and ampulla (Westman 1952; Austin 1963). This assumption is based on the observation that when particulate matter such as dyes, lamp black, spores, and invertebrate eggs are placed in the vicinity of the fimbriae they are drawn into the oviduct. There is no experimental evidence that the fimbriae or ampulla creates a "suction" and it is most unlikely that such a phenomenon exists. Direct observations of the movements of ovulated eggs which have been placed in the vicinity of the fimbriae in rats and rabbits have shown conclusively that the cilia lining the fimbriae are responsible for the movements of such materials. The directional beat of the cilia causes fluid currents to flow toward the oviductal ostium (Blandau, unpublished observations). Furthermore, Clewe and Mastroianni (1958) tied a ligature at the junction of the fimbriae and ampulla in the rabbit and found that eggs were swept into the ostium in the normal manner. Thus, there is no positive evidence for assuming that negative pressures in the region of the infundibulum play any role in egg transport.

D. *Role of Accessory Membranes*

It is stated frequently that ovulated eggs are transported rapidly from the surface of the ovary to the site of fertilization in the ampulla. There are, however, relatively few observations as to just how this transport is accomplished. The variability in the relationship of the fimbriae of the oviduct to the surface of the ovary should make one suspect that the mechanism of egg transport may not be the same in all animals. Too often investigators have directed their attention only to the behavior of the fimbriae and muscular activities of the oviduct itself, disregarding the activities of the various mesenteries and other membranes attached to this system. When one observes the oviduct and ovary *in situ* in the living animal, the contractile activities of the mesovarium, the mesosalpinx, and various other accessory membranes are very impressive indeed. We may emphasize the importance of these membranes by citing the situation in the rabbit. A special chamber may be implanted into the anterior abdominal wall of the anesthetized rabbit and arranged in such a manner that the full extent of the oviduct and ovary may be exposed for viewing without applying traction to the various membranes attached to them. The general arrangement of the oviduct and its membranes and their relationship to the ovary is portrayed in Figure 1. A thin, muscular membrane, the *mesotubarium superius*, spans the entire antimesosalpinxal border. It is attached also to the medial edge of the fimbriae. At the time of ovulation this membrane contracts vigorously at more or less regular intervals, drawing the oviduct into the form of a crescent. More important perhaps is its action on the fimbriae themselves. At each contraction the mesotubarium superius itself slides the fimbriae over the surface of the ovary. Independent and intermittent contractions of the mesosalpinx also continuously change the contour and position of the oviduct. The rhythmic contractions of the mesosalpinx are not as vigorous as those of the mesotubarium superius. Furthermore, the ovary itself undergoes

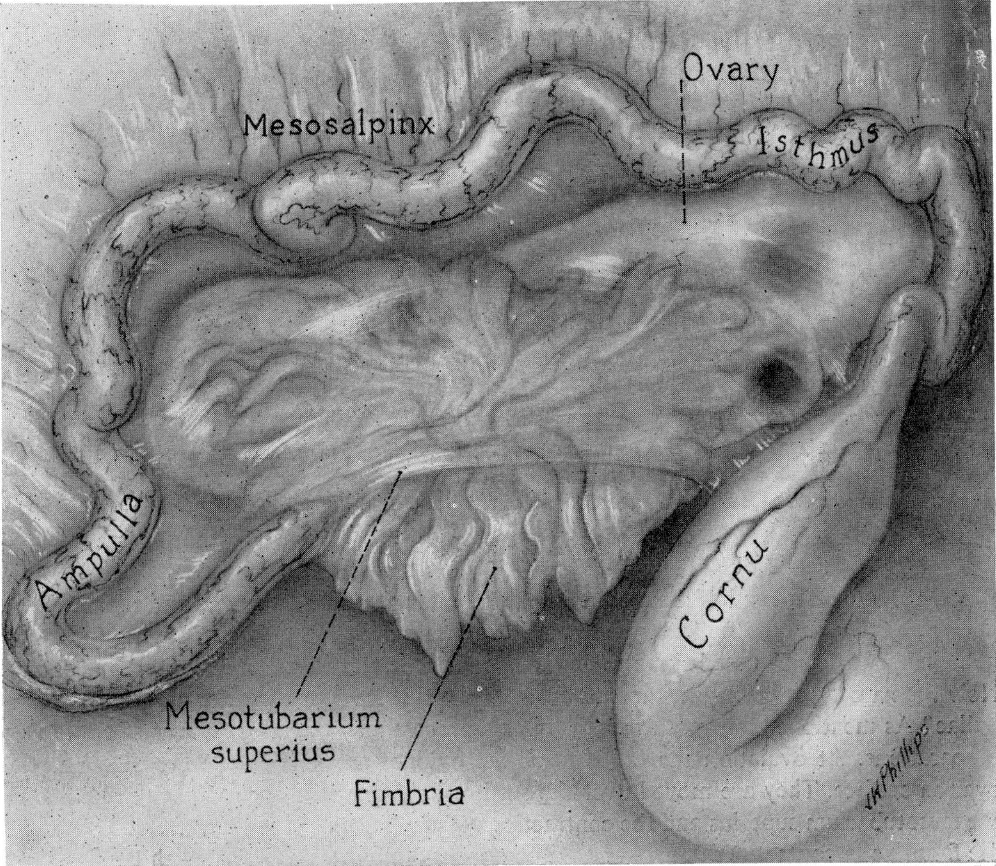

FIG. 1. Relationship of fimbriae and mesotubarium superius to the ovarian surface in the rabbit

a longitudinal rotation as the smooth muscles within the mesovarium contract intermittently. The contractions of the mesovarium change the position of the ovary in relation to the fimbriae. In addition to all the contractile activities of the various membranes that have been mentioned, the fimbriated portion of the infundibulum also undergoes a rhythmic, massage-like contraction which continually changes its position on the surface of the ovary. The vigor with which the various membranes contract at the time of ovulation and the independent rhythms of their contractions are most remarkable. The pattern of movements of the oviducts and ovaries at ovulation are far more complex than had been suspected. All these contractile activities must be analyzed and integrated before we will fully understand egg transport.

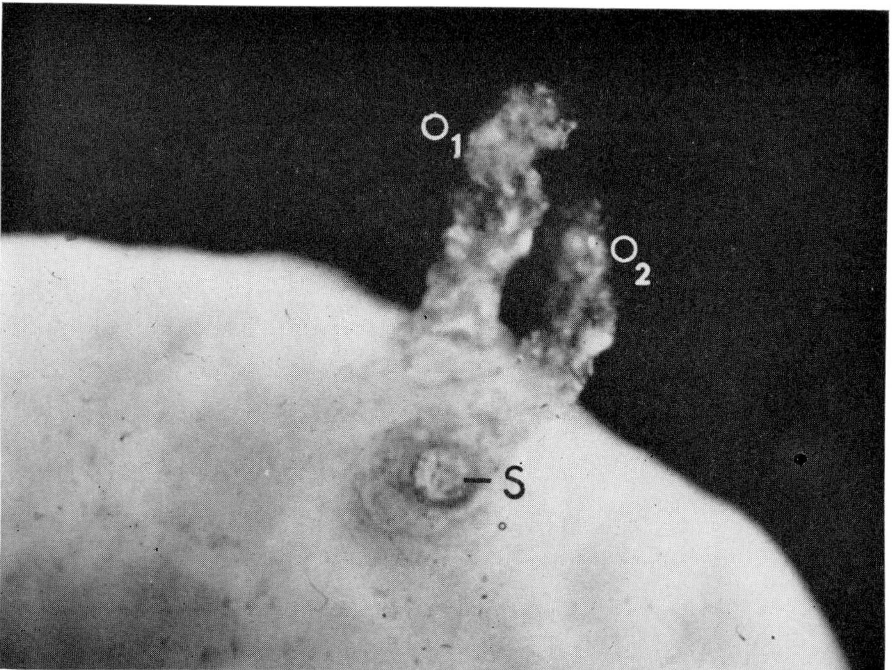

FIG. 2. An enlargement from a 16 mm motion picture film showing fresh ovulations in a rabbit in which the fimbriae had been removed surgically. The extruded follicular contents remain adherent to the stigma for many hours (*S*, stigma; O_1 and O_2, ovulations 1 and 2).

E. *Mechanism of Egg Release and Egg Transport*

If the fimbriated end of the oviduct in the rabbit is surgically removed and the animal is allowed to ovulate, the follicular contents remain adherent to the site of stigma for many hours (Fig. 2). The ovulated cumulus and its matrix are quite highly polymerized and sticky. How then are the rabbit eggs transported from the surface of the ovaries and into the ampullae? As mentioned above, the details of this process may be observed directly in *in vivo* preparations. At ovulation the fimbriae of the oviduct are in intimate contact with the ovarian surface. They are moved over the surface at intervals both by the contractions of the mesotubarium superius and the contractions of the fimbriae themselves. The surfaces of the fimbriae in contact with the ovary are covered by innumerable cilia which beat actively in the direction of the ostium of the oviduct. As mentioned earlier, at ovulation the follicular contents are extruded in the form of an elongated strand. The egg enclosed within its corona radiata usually escapes first and is positioned at the more distal end of the strand (Fig. 3). Almost immediately after ovulation the cilia lining the fimbriae begin to draw the cumulus and matrix toward the ostium. Within 15 to 20 sec the strand is greatly elongated and it may break away, at least partially, from its stigmal attachment. It is then transported rapidly into the ampulla. A stump of the matrix often protrudes from the stigma and may not be completely swept from the surface until several hours after ovulation (Fig. 4). There is no question but that the cilia of the fimbriae, at least in the rabbit, play the primary role in sweeping the ovulated eggs from the ovarian surface. The complex, rhythmic contractions of the oviductal and ovarian membranes assure that the ciliated fimbriae come into intimate contact with all of the surface areas of the ovary. If freshly ovulated rabbit eggs are freed from both the matrix and cumulus cells and then brought onto the folds of the fimbriae, they are transported only very slowly or not at all. The eggs may fall into a fold and be rotated on the spot for long periods of time. If they come into contact with one of the longitudinal ridges they may be moved toward the ostium for short distances, and then fall off into a depression formed by the folds. This implies that the presence of the cumulus and matrix is essential for normal egg transport. The cilia lining the fimbriae are remarkable in their capacity to hold the cumulus and matrix onto the surface of the fimbriae. When a freshly ovulated egg with its cumulus is brought to the fimbriae and transport is initiated, one can pull it away from the surface only with difficulty. The ciliary action of the cells lining the fimbriae is in itself powerful enough to carry the eggs and cumulus into the ampulla.

Stroboscopic observations of oviductal ciliated cells have shown that the cilia beat approximately 1200 times per min (Borell, Nilsson, and Westman 1957). Progesterone increases the rate of beat by approximately 20%. Since there is considerable evidence that progesterone is secreted by the preovulatory follicle it may be that its presence is important in accelerating egg transport at this time (see Chapter 6). Egg transport from the surface of the rabbit ovary is an efficient biological process. Because of the bursa formed by the fimbriae it would seem unlikely that an ovulated egg could escape into the peritoneal cavity.

How is egg transport effected in such animals as the mouse, hamster, and rat in which a much smaller fimbriated tip (Fig. 5) makes only minimal contact with the ovarian surface? If in these animals the antral contents of the follicle remained adherent to the stigma, as has been shown in the rabbit, egg transport would be impossible. Fortunately, in these rodents the eggs are shed free into the periovarial space and are capable of being moved about freely within the fluids filling the space (Fig. 6). The oviducts and ovaries of the rat may be observed *in vivo* by techniques similar to those described for the rabbit. Contractile activity of the various membranes of the ovary and oviduct plays an important role in moving the ovulated egg about in the periovarial space. Intermittent contractions of the mesovarium change the position of the ovary within the periovarial space, thus pivoting it on its axis so that the ovulated eggs are more or less "swished" about. They may drift into the area of the infundibulum where ciliary currents direct them to the ostium of the oviduct. Quite vigorous and intermittent contractions of the ampulla of the oviduct change its position within the periovarial space. In these animals the bursa formed by the periovarial sac insures that the eggs do not escape into the peritoneal cavity.

Observations on egg transport from the ovarian surface to the oviductal ostium in the rabbit and rat indicate that despite large differences in the anatomy and arrangement of the fimbriae there appears to be a basic pattern in the mechanism of egg transport. It is established that cilia play a most important role in the transport of eggs from the ovarian

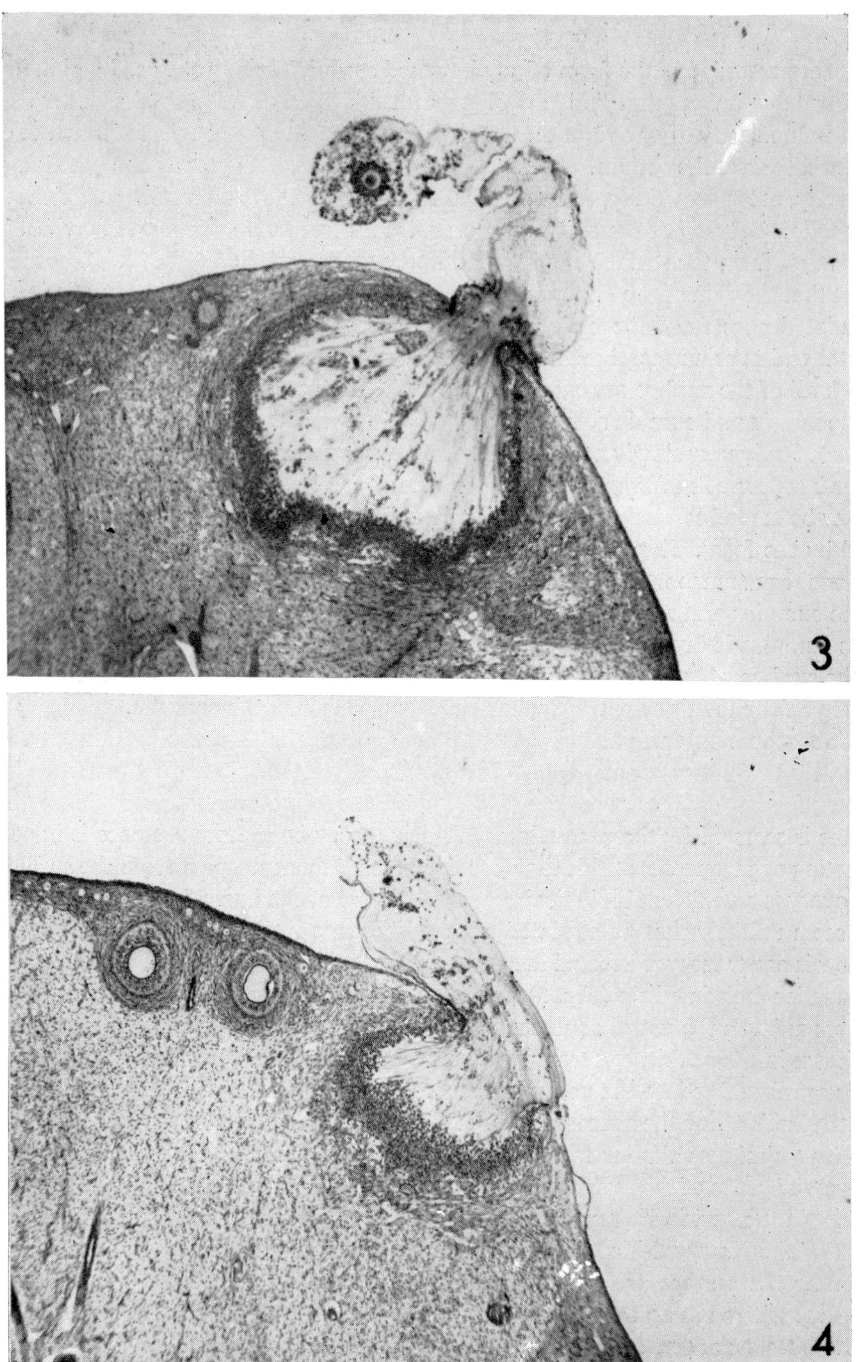

FIG. 3. A section through a follicle of the rabbit in which ovulation has been completed. Note the protrusion of the cumulus and matrix through the stigma and its adherence to the surface (×54).

FIG. 4. Section through an ovulated follicle in the rabbit $\frac{1}{2}$ hr after ovulation. The egg has been pulled away from the protruding matrix. The material remaining will be removed from the surface in approximately 1 hr by the ciliary activity of the fimbriae (×54).

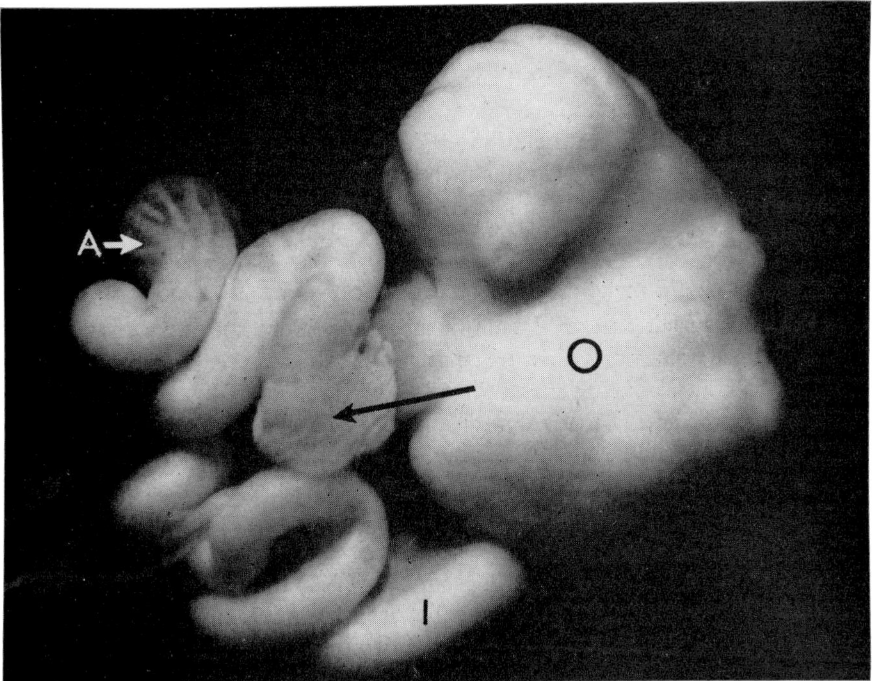

FIG. 5. A fresh preparation of the ovary and oviduct of a rat with the periovarial sac trimmed away. Note the small size of the fimbriated portion of the infundibulum (arrow) in relation to the surface area of the ovary. The dilated loop of ampulla (*A*) contains the ovulated eggs (*O*, ovary; *I*, isthmus).

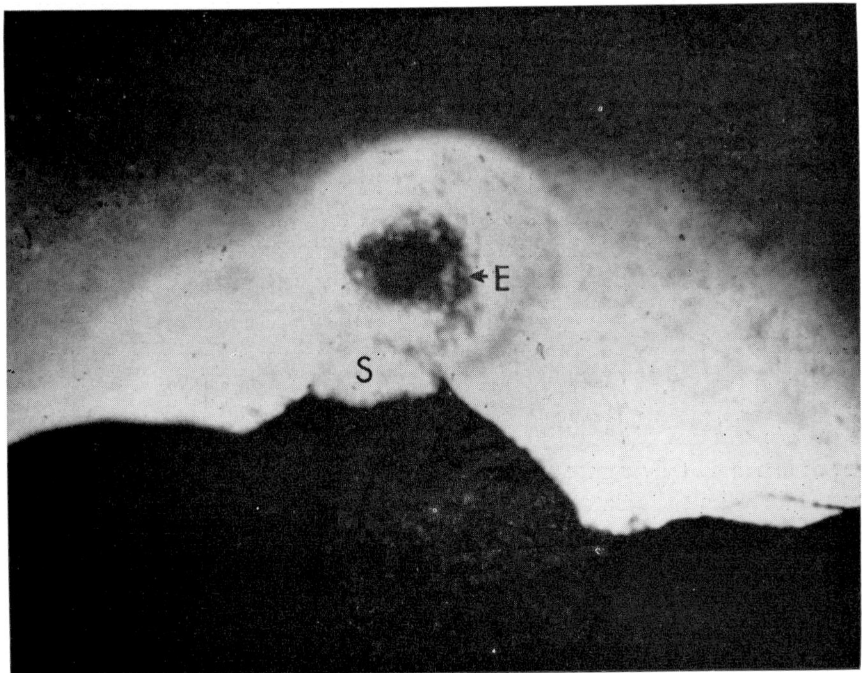

FIG. 6. An enlargement from a 16 mm film showing an ovulated egg and its cumulus freed from the follicle at the moment of ovulation (*S*, stigma; *E*, egg and cumulus).

surface. But if the cilia are to be effective either they must come into intimate contact with the ovulated eggs or the ova must be freed from the follicle so that they can be moved mechanically into the vicinity of the infundibulum. Of great importance are the patterns of the muscular contractions of the various membranes attached to the oviducts and ovaries. The movement of eggs through the first few millimeters of the ampulla is continuous and smooth, apparently the result of ciliary activity of the cells lining the lumen.

F. *Egg Transport through the Ampulla of the Oviduct*

Once the cumulus has entered the ampulla of the oviduct it passes rapidly to the site of fertilization. In most mammals this area lies proximal to the isthmoampullary junction. Exceptions to the ampulla as the site of fertilization are those of the short-tailed shrew and ferret, where sperm penetration apparently begins within the ovarian bursa, and in tenrecs, in which sperm enter the ovarian follicles and fertilize the eggs before ovulation.

Reports as to the rate of transport through the ampulla vary considerably and depend on the techniques used to determine them.

A number of ingenious methods have been devised to explore the mechanism and rate of egg transport through the oviducts. Greenwald (1961) induced ovulation in rabbits by mating them and then removed the oviducts between 12 and 70 hr later. The oviducts were attached to a board and cut into 8 segments by a device consisting of 7 parallel razor blades similar to a guillotine. Each segment was flushed and the eggs searched for. He concluded that the eggs passed through the ampulla within 2 hr after ovulation.

A considerably shorter ampullar transport time has been recorded by Harper (1961*a*, *b*). He recovered fresh cumuli from the oviducts of donor animals and stained them with toluidine blue to make them more visible within the lumina. The eggs were transferred into the oviducts with fine pipettes and a cinematographic film made of their movements.

The movements of the cumuli through the greater part of the ampulla were quite rapid but discontinuous and involved a series of short rushes. Often antiperistaltic contraction waves forced the eggs backward for a short distance. The primary movement, however, was segmental and forward. The time taken for the eggs to reach the isthmoampullar junction ranged from 4 to 12 min. Harper's observations were made by exteriorizing the oviducts and placing them on moistened cotton squares. Very similar ampullar transport times (average 6 min, 26 sec) have been observed by this writer when fresh cumuli were placed on the surface of the rabbit fimbriae and allowed to enter the ostium on their own. The oviducts were examined *in vivo* with minimal handling while being continuously bathed in a tissue culture medium. Considerable variations in the pattern of muscular contractions may be observed even when the environmental conditions appear to be optimal. Under the best physiological conditions attainable, the movements of the egg and cumulus appear to be intermittent and effected by definitive segmental contractions of the tube. The pattern of contractions is exceedingly complex. The presence of the cumulus appears to initiate contractions within a limited area of the tube in which it lies. The rate and amplitude of oviductal contractions can be greatly altered by the injections of exogenous hormones, the details of which will be described in another chapter.

The ampullar portions of the oviducts in most mammals have complex mucosal folds which are heavily ciliated (Fig. 7*a*). When egg transport is observed directly in the living animal cilia apparently play only a minor role in their movement through this segment.

A very similar pattern of egg transport through the ampulla has been described for the rat. When the egg and cumulus is caught on the ciliated fimbriae it is transported into the

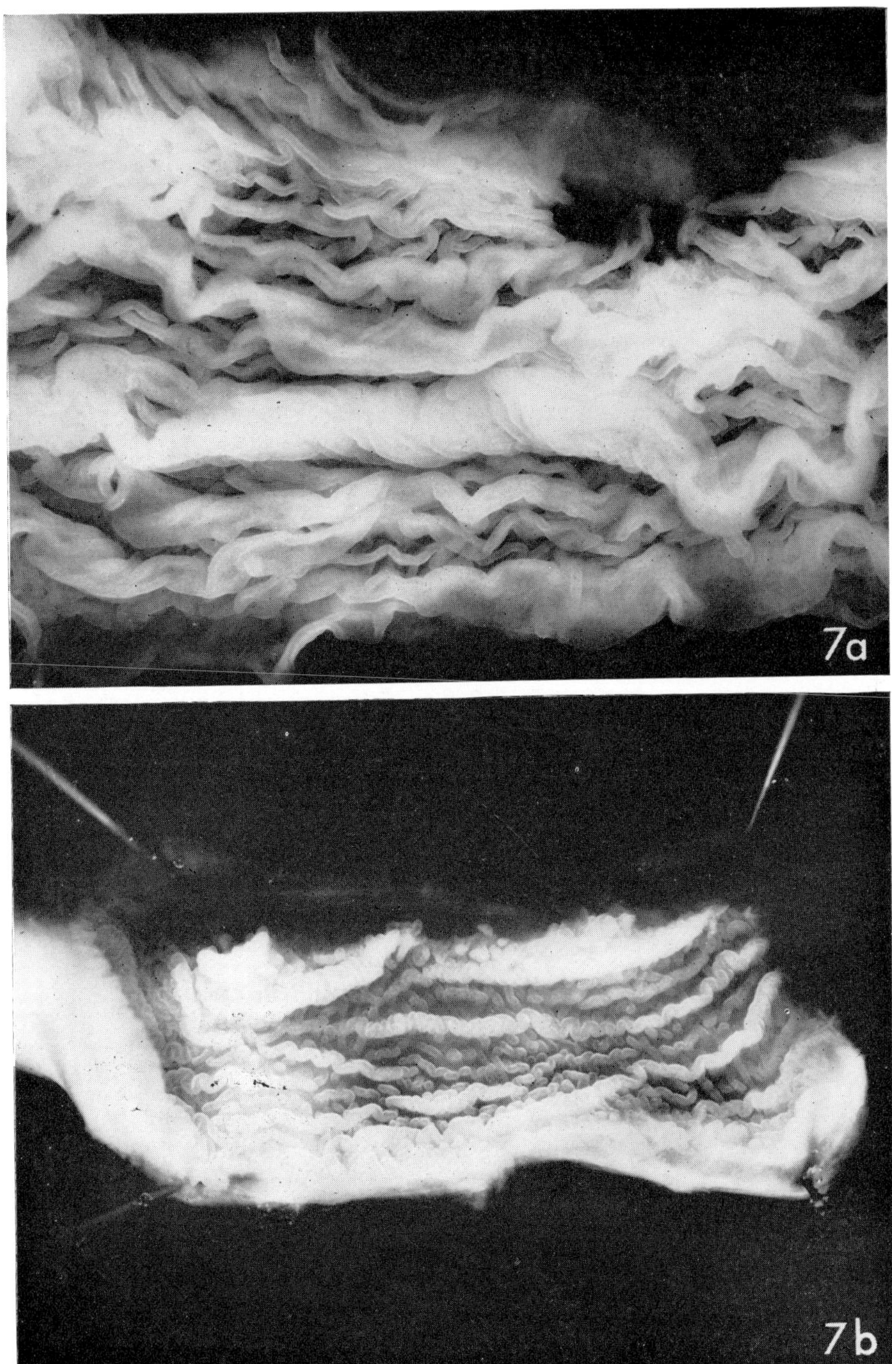

FIG 7. Arrangement and complexity of the mucosal folds of the ampullar and isthmic portions of a freshly opened oviduct of the rabbit: (*a*) ampullar portion (×11); (*b*) isthmic portion (×6.5).

ostium within 8 to 10 secs. It is then moved continuously and smoothly over the ciliated folds of the first loop of the ampulla to enter the more dilated segments. At ovulation the dilated loops of the ampulla exercise vigorous and rhythmic contraction waves which are primarily peristaltic and end at the isthmoampullar junction. Ordinarily as the eggs are moved forward they accumulate in a mass just proximal to the oviductal stricture. Occasionally cumuli may come into contact with the ciliated folds of the ampulla and be rotated about only to be moved forward vigorously during the next contraction.

G. *The Isthmoampullar Junction and Egg Transport*

As mentioned earlier, eggs are retained at the isthmoampullar constriction for a number of hours after ovulation. This would imply that the stricture is sufficient to narrow the lumen of the oviduct and retain the eggs but not so occlusive as to interfere with sperm transport into the ampulla. No structural basis for an isthmic block has been demonstrated, even though various histological methods have been tried (Greenwald 1961).

Intraluminal pressures of both the ampullar and isthmic regions may be recorded simultaneously by sensitive pressure transducers. The maximal pressure changes within the ampulla did not exceed 7 mm Hg while pressure within the isthmic lumen frequently recorded 16–50 mm Hg (Brundin 1964). These pressure differentials were interpreted as indicating the presence of functional occlusive mechanism at the isthmoampullar junction.

In rats, mice and hamsters the isthmoampullar junction is sufficiently constricted to retain considerable quantities of fluid which dilate the ampullar loops and increase the fluids within the periovarial space. Physiological closure of the periovarial-peritoneal slit also aids in fluid retention. The stricture relaxes, usually within 24 hr after ovulation, and the fluids drain from the ampulla.

Nothing is known about the endocrine mechanisms which cause the stricture to form in such a restricted area of the oviduct. Localized narrowing of the oviduct to effectively retain fluids in the ampullar loops would appear to present a significant hindrance to sperm transport into the ampulla. It is not known whether sperm enter the ampulla by their own motility or whether they attain admission by momentary relaxation of the stricture.

H. *Egg Transport through the Isthmus of the Oviduct*

There is very little basic information on the movements of mammalian eggs through the isthmus of the oviduct. In most mammals the eggs are retained in this segment from 24 to 48 hr. Complex patterns of peristaltic and antiperistaltic contractions have been described, but the thickness of the muscular wall of the isthmus makes it difficult to observe the location and movement of eggs within this segment in living animals. Furthermore, when the eggs reach the isthmus, having been freed of the cumulus and corona radiata cells, their identification *in situ* is almost impossible.

In the monotremes (Hill 1933) and marsupials (Hartman 1916) extensive albuminous coatings are deposited on the zonae pellucidae as the eggs move through the isthmus. In the order Lagomorpha a mucus or "albuminous" layer, composed of complex mucopolysaccharides, is deposited on the zonae as the eggs move into the isthmus (Gregory 1930; Pincus 1936). A chemically identical but thinner coating is deposited as tertiary membranes on the eggs of the dog and the horse (Hamilton and Day 1945).

These tertiary membranes are deposited on the zonae so evenly that they form almost perfect spheres. In order to effect such an even coating it would seem that the eggs must be rotated in a very precise manner. In frogs the eggs are evenly coated by rotating approximately every 30 sec as they descend through the oviduct (Yamada 1952). No similar observations have been made in mammals. The specific pattern of the muscular contractions of the isthmus which may cause the eggs to rotate is not known in any mammal.

After rabbit eggs enter the isthmus they are retained in the first segment for a considerable period of time (Greenwald 1961). Deposition of the albuminous coating begins in this segment. When deposition of the albuminous layer is once initiated sperm can no longer penetrate the eggs. The epithelial folds of the isthmus are much less complex than those of the ampulla and are lined by fewer ciliated cells (Fig. 7*b*). It is not known whether the ciliated cells of the isthmus participate in the rotation of the eggs during the coating process or in their forward transport. As one approaches the uterotubal junction almost all of the cells are ciliated; thus egg transport through this segment may be more rapid (Greenwald 1961).

I. *Stages of Development and Location of Eggs within the Oviduct*

In mammals the transport of ova through the oviduct is generally thought to require approximately 3 days. Eggs are transported through the oviducts in swine more rapidly than in other domestic animals. In a series of observations on a significant number of pigs Oxenreider and Day (1965) first recovered ova from the uterus 66 hr following the onset of estrus and all ova were recovered between 90 and 108 hr. Ovulation occurs about 36 hr from the onset of estrus in this animal. In some pigs the eggs were only in the 4-cell stage when they were recovered from the uterus. When reviewing the data in Table 1 one notes that the rates of egg transport in the pig and opossum are remarkably similar.

Data on the rate of cleavage and time that eggs reach the uterus in a number of animals are summarized in Table 1. The precise time of ovulation is difficult to determine, particularly in the sheep, cow, pig, and horse. Thus the stages of egg development in these animals may vary considerably and the times listed in the table may be only approximately accurate. The rate of cleavage is an inherent property of the zygote. In the rabbit the cleavage rate is consistently more rapid in strains producing larger-sized animals than in smaller-sized races. Despite these differences in cleavage rate, embryonic differentiation occurs at the same rate in both breeds (Gregory and Castle 1931).

J. *Fertilizable Life of Ova within the Oviduct*

The reproductive processes in most mammals are timed in such a manner that sperm reach the site of fertilization, have attained capacitation, and are ready to penetrate the eggs almost immediately after they reach the ampullae. If eggs are not penetrated by sperm relatively soon after ovulation they undergo a rapid devitalization and die. Mammalian eggs are among the most short-lived cells in the body. Unfertilized eggs appear to be transported through the oviducts at the same rate as fertilized eggs.

Data available on the fertilizable life of mammalian ova within the oviducts are summarized in Table 2. Again, because of the difficulty in determining the time of ovulation precisely in some animals, only estimations of egg viability can be given. Sperm may penetrate devitalized eggs and undergo transformation into the male pronuclei, but the female nucleus either fails to develop or fragments. Polyspermy increases as eggs are aged before sperm penetration in certain strains of animals (Braden 1958; Odor and Blandau 1956; Piko 1958). In some animals, such as the mouse, polyspermy is not increased after delayed mating (Braden and Austin 1954*b*; Marston and Chang 1964). Overripe eggs penetrated by sperm may not be completely devitalized and may undergo partial development. Such eggs usually develop much more slowly or show distinct abnormalities which are incompatible with continued growth and development.

TABLE 1

Stages of Development of Fertilized Eggs, in Hours, and Time Required to Reach Uterus

Species	1-Cell	2-Cell	3–4-Cell	5–8-Cell	9–16-Cell	Morula	Blastocyst	Time and Stage at Which Egg Reaches Uterus	References
Gerbil	22–24	48	70–72	...	60	...	...	72	Marston and Chang 1966
Opossum	...	60	66	72	...	84	96	24 (1-cell pronuclear stage)	Hartman 1928
Mouse	0–24	24–38	38–50	50–64	60–70	68–80	74–82	72 (morula)	Lewis and Wright 1935
Mouse	0–24	24–38	50	60	70	...	...	...	Sobotta 1895
Rat	0–24	42–70	63–73	89	89–96	...	...	...	Huber 1915
Rat	8.5–27	27–44	60–85	71–95	...	...	...	...	Gilchrist and Pincus 1932
Rat	12–20 (15)	37–61 (45)	57–85 (65)	64–87 (79)	84–92 (90)	...	105–9 (107)	...	Macdonald and Long 1934
Guinea pig	3–30	30–35	30–75	80	...	100–115	115–40	80–85 (8-cell)	Squier 1932
Rabbit	up to 22	22–26	26–32	32–40	40–47	47–68	68–76	70 (blastocyst)	Gregory 1930
Rabbit	12–24	24–28	28	...	...	48	75–96	...	Assheton 1894
Rabbit	11–21	21–24	...	...	...	...	...	72–75 (blastocyst)	Gilchrist and Pincus 1932
Ferret	31–53	...	64–72	64–116	74–120	120–46	146–264	120–40 (32-cell)	Hamilton 1934
Pig	0–51	51–66	66–72	90–110	...	110–14	114	75 (4-cell, morula)	Heuser and Streeter 1929 Oxenreider and Day 1965
Sheep	0–38	38–39	42	44	65–77	96	113–38	77–96 (16-cell)	Clark 1934
Goat	30	30–48	60	85	98	120–40	158	98 (10–13-cell)	Amoroso, Griffiths, and Hamilton 1942
Cattle	34	50–62	...	62–64	110	134	182	110 (16-cell)	Winters, Green, and Comstock 1942
Cattle	23–51	40–55	44–55	46–96	71–141	144	190	96 (8–16-cell)	Hamilton and Laing 1946
Horse	...	24	(3)27–33 (4)30–36	(5)50–60	(15)96±6	98±6	...	...	Hamilton and Day 1945
Monkey (rhesus)	...	0–24	24–36	36–48	48–72	72–96	...	96 (16-cell)	Lewis and Hartman 1941

Source: Blandau 1961.

II. Sperm Transport

A. *Sperm Transport through the Cervix*

At copulation semen is deposited into the female reproductive tract, but the volume of the ejaculate, the site of deposition, and the number of spermatozoa vary greatly. Some data on the number ejaculated, the location of insemination, and the number reaching the site of fertilization are summarized in Table 3. Perhaps the most voluminous ejaculate is that of the pig, in which as many as 500 million sperm are propelled slowly through the cervix and into the uterus during a prolonged coitus. In the rat, by contrast, a very small compacted mass of spermatozoa is catapulted directly into the cornua in a matter of a few seconds. During normal ejaculation semen is deposited in the vagina in the rabbit, cat, cow, ewe, monkey, chimpanzee, and woman. In the mouse, rat, hamster, ferret, dog, sow, and mare the ejaculate passes into the uterus. In the guinea pig semen appears to be deposited both in the region of the cervix and in the body of the uterus (Table 3).

The fluidity of the ejaculate is another variable. Semen is ejaculated in either a liquid or semiliquid form. In such species as dog, pig, and cattle the semen remains liquid; in others it

TABLE 2

THE FERTILIZABLE LIFE OF THE MAMMALIAN OVUM

Species	Length of Fertilizable Life	References
Opossum	24 hr—estimation	Hartman 1916
Mouse	15 hr—experimental	Marston and Chang 1964
Hamster	5 hr—experimental	Yanagimachi and Chang 1961
Dog	24+ hr—estimation	Evans and Cole 1931
Rat	12 hr—experimental	Blandau and Jordan 1941
Guinea pig	20 hr—experimental	Blandau and Young 1939
Ferret	30 hr—experimental	Hammond and Walton 1934
Rabbit	6 to 8 hr—experimental	Hammond 1934; Chang 1952
Sheep	24 hr—estimation	Green and Winters 1935
	12 to 15 hr—experimental	Dauzier and Wintenberger 1952
Horse	4 to 20 hr—estimation	Berliner 1959
Pig	10 hr—experimental	Hunter 1967
Monkey	23 hr—estimation	Lewis and Hartman 1941
Man	6 to 24 hr—estimation	Hartman 1939

TABLE 3

NUMBER OF SPERM EJACULATED, SITE OF INSEMINATION, AND NUMBER REACHING SITE OF FERTILIZATION

Species	Mean Number of Sperm Ejaculated in Millions	Site of Insemination	Number of Sperm in Ampulla	References
Mouse	50	Cornua	17+	Braden and Austin 1954*a*
Rat	58	Cornua	5–100	Austin 1948
				Blandau and Odor 1949
Rabbit	60	Vagina	250–500	Braden 1953
Ferret	....	....	18–1600	Hammond and Walton 1934
Guinea pig	80	Vagina and uterus	25–50	Blandau, unpublished observations
Cattle	3000	Vagina	Few	Austin 1959; Salisbury and VanDemark 1961
Sheep	800	Vagina	600–700	Braden and Austin 1954*a*
			5000+	Warbritton *et al.* 1937

tends to coagulate in varying degrees shortly after ejaculation (Mann 1954). The significance of coagulation, or its role in insemination, is poorly understood in most mammals. In primates the semen ejaculated into the vagina clots within 1 min (Fig. 8), then liquefies partially or completely within approximately 20 min (Huggins and Neal 1942; MacLeod 1946; Hoskins and Patterson 1967). Initially spermatozoa appear to exude from the coagulum. In the human, spermatozoa do not attain their full motility until the seminal clot liquefies. In the pig only about half the entire ejaculate forms a coagulum. Its matrix is derived from the bulbourethral glands and is rich in sialomucoprotein (Mann 1964). The same author has described partial gelatinization of the semen in the rabbit and horse (Mann 1954).

The factors causing liquefication of the coagulum *in vivo* are not fully understood. Indeed, in the rat, mouse, and guinea pig little or no autolysis of the vaginal plug occurs. At

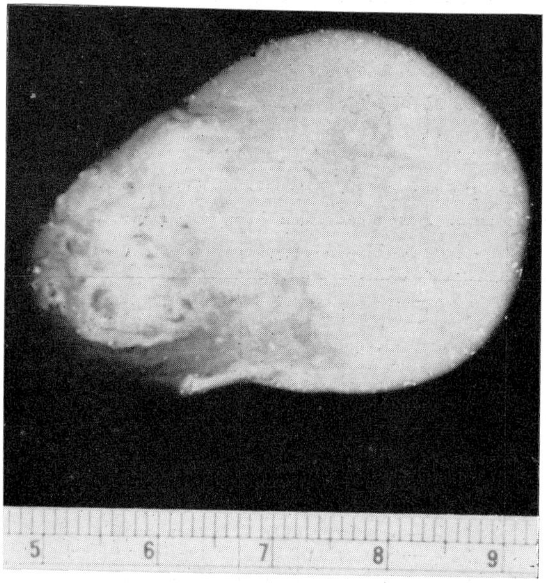

FIG. 8. Appearance of the ejaculated coagulum of a rhesus monkey 4 to 5 min after electroejaculation. (Courtesy of Dr. D. D. Hoskins, Oregon Regional Primate Research Center).

the end of heat it is shed from the vagina largely intact. In the artificially ejaculated rhesus monkey a firm coagulum forms within minutes (Fig. 8) (Hoskins and Patterson 1967). Autolysis under *in vitro* conditions fails to take place even after 30 min. Attempts to render it soluble, by adding various hydrolytic enzymes, while still maintaining spermatozoal motility, were unsuccessful. However, if the ejaculate was placed into a beaker containing 3 ml of 2% alpha-chymotrypsin dissolved in Norman-Johnson's medium, complete solution of the seminal fluid was obtained in approximately 5 min. Usually such freed spermatozoa showed good motility. However, it is not known whether these enzyme-treated spermatozoa retain their capacity for fertilization. The coagulum in the monkey does not always appear to have been completely solubilized *in vivo* since occasionally fragments of plug material may be found in the cage after mating.

As mentioned earlier, coagulation of the seminal plasma is developed to its highest degree in hamsters, rats, mice, and guinea pigs. In these animals relatively firm vaginal plugs form almost immediately after ejaculation (Fig. 9). The plug abuts against the cervix and a portion of it may extend as two delicate prongs into the cervical canals. Occasionally strands of plug material may be found in the fluids of the cornua. In the guinea pig, significant amounts of this material may be found here. This suggests that during normal ejaculation in these animals the semen is injected directly through the cervix and into the body of the uterus. If the vaginal plugs from these animals are sectioned and stained, numerous immotile spermatozoa are found to be trapped. Since there ordinarily is no significant dissolution of the plug material, the trapped spermatozoa are shed from the vagina along with the plug near the end of heat.

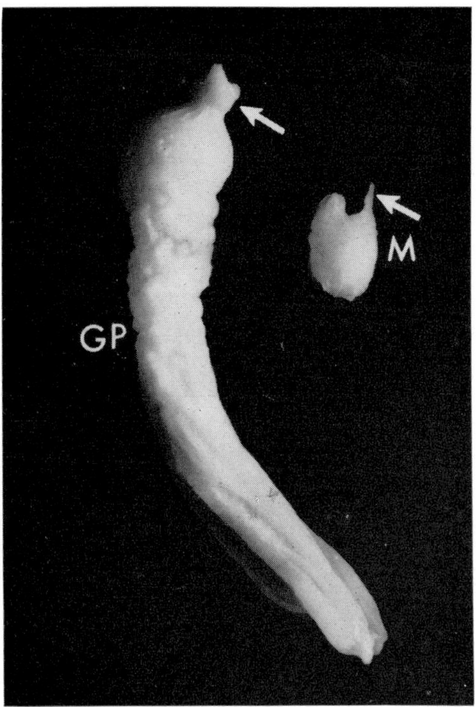

FIG. 9. Copulation plugs from the guinea pig (*GP*) and mouse (*M*). Note the variations in size and extensions of plug material (arrows) into the cervix (×2.5).

If the seminal vesicles and coagulating glands of rats, mice, and hamsters are carefully and completely removed, and the males allowed to mate with females in heat, spermatozoa may be deposited in the vagina but do not enter the cornua. Nor do spermatozoa enter the uterine horns if the contents of the ductus deferens from such males is injected artificially into the vagina of females during heat. These experiments indicate that in these animals, at least, the vaginal plug plays an important role in injecting the semen directly into the cornua at the moment of ejaculation and in then sealing the cervical canals so that the spermatozoa cannot escape into the vagina (Blandau 1945).

Thus, it may be concluded that the cervix is not a formidable barrier to sperm transport in the horse, pig, dog, ferret, rat, mouse, hamster, and guinea pig since insemination occurs directly into the uterus or cornua. But in those animals in which spermatozoa are deposited in the vagina only at the time of ejaculation, transport mechanisms must be significantly more complex. We are then faced with the problem of how spermatozoa pass through the barrier of the cervical canals.

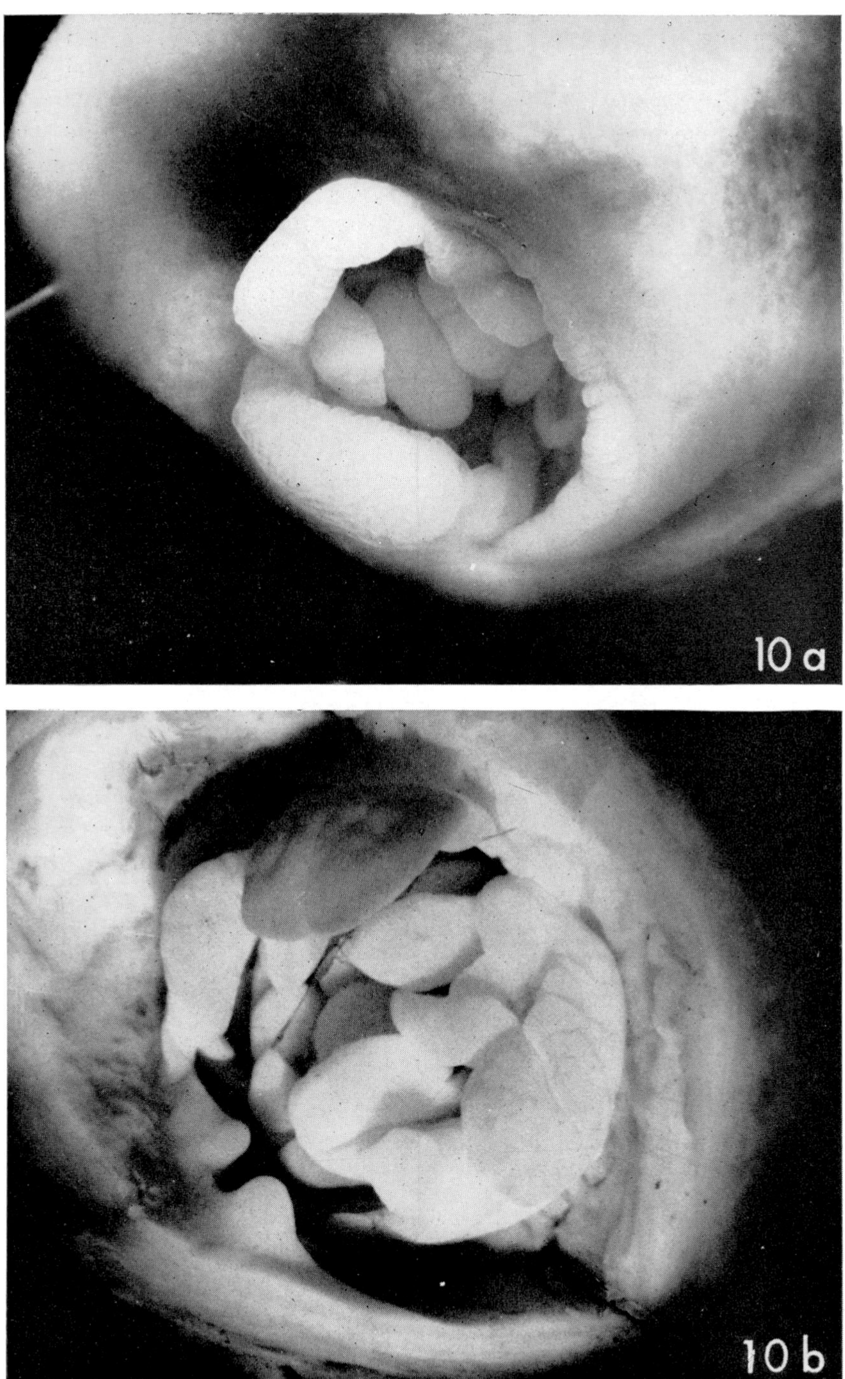

FIG. 10. The appearance of the living uterotubal junctions of rabbits to show the complexity of the folds in this region: (*a*) from a pubertal animal; (*b*) at the time of ovulation in a mature female (×18).

The various factors which must be considered in evaluating the mechanism of transport from the vagina through the cervical canals are as follows: (1) the role of muscular activity of the vagina and cervix; (2) the composition and physical characteristics of the cervical fluids; (3) the directional swimming activity of the spermatozoa themselves; and (4) the role played by the orgasm of the female in altering the mechanical action of the cervix and uterus.

If one examines the modest information available it appears that there is significant species variation as to the role played by any one or all of these factors in the transport of spermatozoa through the cervix (Hartman 1957).

It is well known for man that the first portion of the ejaculate contains most of the spermatozoa (MacLeod and Hotchkiss 1942). Thus, one may assume that at the moment of ejaculation the cervix is bathed by the highest concentration of spermatozoa. Whether spermatozoa move into and through the cervical canal by their own motility or whether their transport is assisted by the contractile muscular activity of the vagina and cervix has not been settled unequivocably. There are remarkably few data on the rate or mechanism of sperm transport through the cervix in women or in those animals in which spermatozoa are ejaculated directly into the vagina.

Rubenstein *et al.* (1951) recovered spermatozoa from the oviducts of a patient within 30 min after coitus. If these data are correct, they would most certainly imply a rapid entrance of spermatozoa into the uterus. Recovery of many motile spermatozoa from the cervical canal within seconds after ejaculation prompted Sims (1869) to postulate a transport en masse through this segment.

Amersbach (1930) placed a cap containing a suspension of lamp black particles over the cervix of a human patient and recovered some of the particles from the uterus after coitus.

Rubenstein (1943) contends that human vaginal contractions play an important role in the entry of spermatozoa into the uterus. Many experiments have been done to test the so-called in-suck theory of sperm passage into and through the cervix in various animals and man. The fact remains that although there are strong proponents of this theory there is insufficient experimental evidence to validate the idea, particularly in man (Hartman 1957; Masters and Johnson 1966).

Krehbiel and Carstens (1939) and later Akester and Inkster (1961) provided conclusive evidence that such a phenomenon may exist, at least in the rabbit. They injected X-ray opaque media into the vagina of a rabbit. Digital stimulation of the vulva during fluoroscopic examination revealed strong vaginal contraction waves which forced the radio-opaque material through the cervix and into the cornua, from where it was quickly transported to the uterotubal junction. The entire transport was completed in a matter of seconds.

There is an increasing body of evidence that a neurohumoral mechanism may activate the rapid transport of semen in some animals. Sexual stimulation effects the release of neurohumoral substances, particularly in the cow and the ewe, causing vigorous contractions of the entire reproductive tract (VanDemark and Hays 1952). In women, also, vigorous contractions of the vagina and uterus during orgasm have been observed and recorded (Masters and Johnson 1966), and there is some evidence that oxytocin is released at the time of orgasm (Campbell and Petersen 1954; Pickles 1953). However, its effect on the cervix and uterus of the human female during insemination has not been evaluated. Experiments in rabbits have shown that administration of oxytocin immediately after mating augments the number of sperm found in the uterus but not in the oviducts (Mroueh 1967).

An inward and outward movement of the mucous column in the cervical canal of the human female has been described by Belonoschkin (1949). At one time the plug could be seen to project into the vagina and then to retreat into the cervical canal. The movements of this mucous column convinced Belonoschkin that the lability of the mucous plug could be a mechanism for the en bloc transport of sperm.

The rate at which spermatozoa travel in cervical mucus in *in vitro* preparations ranges from 19 μ to 50 μ per sec and at this rate of forward movement it would be quite possible for freshly ejaculated sperm to reach the oviducts in 30 min after semination (Harvey 1960; Rubenstein *et al.* 1951).

This may be rendered possible by the fact that the physicochemical properties of cervical mucus in the human are greatly altered during the normal menstrual cycle—presumably by the action of estrogens and progestogens upon the cells lining the cervix. Experimentally, natural and synthetic estrogens increase the fluidity, salt content, and amount of cervical mucus. Indeed, the rates of sperm penetration and survival are enhanced proportional to the dose and estrogenic activity of the steroid used (Zonartu 1966).

The basic component of cervical mucus in the human female is a carbohydrate-rich glycoprotein. Trypsin and chymotrypsin readily hydrolyze this mucoid and cause chemical and physical changes which decrease its viscosity and change its electrophoretic mobility (Neuhaus and Moghissi 1962).

The significance of this is that similar proteases are present in human semen. It is reasonable to assume that enzyme-catalyzed hydrolysis of cervical mucus may play a role in sperm penetration through the cervix (Moghissi *et al.* 1964; Mann 1964).

During *in vitro* experiments sperm appear to move at random when brought into juxtaposition with cervical mucus. There is some evidence that motile spermatozoa have a tendency to migrate along the strands of the cervical mucus. Bull sperm, moving along these strands, appear to be oriented in the longitudinal direction in which the threads are drawn out (Tampion and Gibbons 1962). The number of spermatozoa in the cervix of mated ewes seems to be maximal at about 15 min after coitus (Mattner 1963*b*).

It may be concluded that spermatozoa pass through the cervix in most mammals, including man, at a rapid rate. The mechanism by which this is accomplished varies in different animals. In some the spermatozoa are catapulted directly through the cervix and into the uterus. In others the action of the musculature of the female genital tract plays a major role in the mechanical transport. In still others spermatozoa pass into the cervix by their own motility.

The physiology and pharmacology of the musculature of the vagina and cervix of mammals, particularly at the time of semination, remains largely unexplored. Because of its importance in reproductive biology this is an area that deserves intensive investigation.

B. *Sperm Transport through the Uterus*

It is interesting that Antony van Leeuwenhoeck not only first described the independent motility of spermatozoa in 1677, but 8 years later pioneered the experimental study on the rate of sperm ascent in the reproductive tract of the female rabbit (Leeuwenhoeck 1685). Because of the historical significance of these observations we will quote the pertinent paragraph.

> After this I obtained a rabbit in heat, which was mated with a male in my presence three times, this taking about one minute. I then had it killed and brought to my house, so that, by the time I had removed the womb, about a quarter of an hour had elapsed since the mating. I then opened the horn of the womb, about a finger's breadth from the place where the horns of the womb begin, whence I took a small portion of a liquid substance containing some living animalcules, being the seed of the male rabbit. But when I examined the substance taken from the horn

of the womb, just where it starts from the vagina, I saw an innumerable quantity of these living animalcules. But I could find none of these animalcules in the substance contained by the horn of the womb up to the place where it narrows down, nor in its centre; all I saw was a few globules of blood floating in a small amount of fluid. The reason why I did not find any animalcules from the male seed everywhere in the horn of the womb was, I believe, simply because the male seed had not been in the womb long enough, and that, moreover, the time had been too short for the animalcules to travel right through the horns of the womb.

Leeuwenhoeck's fundamental discovery that spermatozoa moved forward by means of the vibratile activity of their flagella formed the basis for the theory as to the manner by which they traversed the uterus. Recent studies have shown clearly that muscular contractions of the uterus and oviducts are perhaps more important than sperm motility itself in the transport mechanism through the female genital tract.

Several earlier investigations cast doubt as to the significance of the independent motility in sperm transport through the uterus. For example, Hausmann (1840) found spermatozoa throughout the uterus of a dog killed during copulation. This was confirmed by Bischoff in 1845, also in the bitch. Later Leuckart (1853) found spermatozoa in the middle of the guinea pig cornua 20 min after mating.

With few exceptions the independent movements of sperm appear to be a negligible factor in their ascent through the uterus. Rapid transport of sperm through the female genital tract has now been reported for many mammalian species (Parkes 1960; Bishop 1961).

It seems clear that contractile activity of the reproductive tract itself must be primarily responsible for their rapid transport. It has been shown that nonmotile sperm (VanDemark and Moeller 1951) and radio-opaque solutions pass rapidly from the cervix to the oviducts in cows treated with oxytocin (Rowson 1955). Inert particles placed in the vagina of sheep are found in the oviducts within 15 min after coitus (Mattner and Braden 1963). Similarly when radioactive polymer particles of 1 μ in diameter are placed at the base of the rabbit uterus they become more or less rapidly and equally distributed throughout the cornua and some of them pass into the oviduct (Wood 1966). Carbon particles deposited in the anterior vagina of women undergoing hysterectomy are rapidly transported to the oviducts (Egli and Newton 1961).

An interesting series of experiments on sperm transport in the genital tract of the cow has been reported by VanDemark and Hays (1954). Spermatozoa are transported to the ampullae of the oviducts in 2 to 4 min, both after artificial insemination, in which the semen is placed directly into the cervix, and after natural mating. Thus, the speed of transport is far more rapid than could be accounted for by the motility of the spermatozoa themselves. Furthermore, both actively motile and nonmotile spermatozoa are transported at approximately the same rate (VanDemark and Moeller 1951).

The physiological processes involved in this rapid transport have been shown to be related to a significant increase in uterine activity at the moment of insemination (VanDemark and Hays 1952). Various phases of mating usually produce extensive uterine responses. When a balloon is inserted into the uterus of a cow and connected to a recording device, the records show that a variety of sexual stimuli can produce vigorous uterine contractions. Merely bringing the bull near the cow, allowing the bull to nuzzle the vulva and hind quarters or allowing the bull to briefly mount the cow all stimulate uterine contractions. Normal manipulation of the vulva or the cervix during the insertion of an inseminating tube also causes active uterine contractions in cows (Hays and VanDemark 1953*a*). The greatest stimuli for inducing tonic contractions of uterine muscle are copulation and ejaculation. The contractile responses of the uterus to mating and the rapid transport of spermatozoa through the reproductive tract occur in both estrous and proestrous states. Manual stimulation of the genitals and nipples also effects a rapid sperm transport (Ivy, Hartman, and Koff 1931; VanDemark and Hays 1954; VanDemark and Hays 1951).

Detailed pharmacological studies have shown that a variety of sexual stimuli lead to the release of endogenous oxytocin from the neurohypophysis and that it is the local effect of this hormone which causes the uterine contractions (Fig. 11). Intravenous administration of oxytocin at all stages of the estrous cycle increases uterine tone and the rate and amplitude of contractions. The contractions induced by oxytocin injections are remarkably similar in pattern to those recorded during natural mating. Complete obliteration of uterine activity

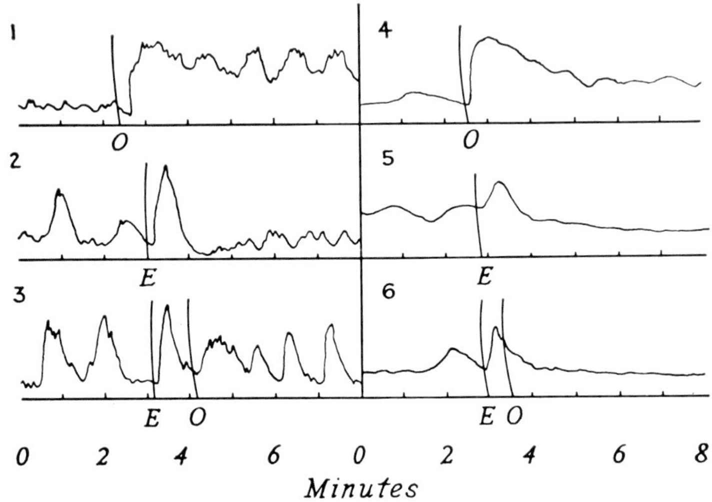

FIG. 11. Kymographic recordings of motility responses of the bovine uterus to oxytocin (*O*) and epinephrine (*E*) injections in the intact cow (1, 2, 3) and in the excised tract *in vitro* (4, 5, 6). Note the remarkable similarity of response to each of the drugs under both *in vivo* and *in vitro* conditions. (From Salisbury and VanDemark 1961.)

can be brought about by frightening the animal or by the intravenous injection of epinephrine (Hays and VanDemark 1953*b*).

The effects of oxytocin on the cow's uterus under *in vitro* conditions have been studied also. Fresh uteri, removed as soon as possible after slaughter, have been perfused through the vascular system so that the effects of hormones added to the perfusate could be evaluated (Fig. 11). Adding oxytocin produces tonic contractions similar to those observed *in vivo*. As the moment of oxytocin is increased the magnitude of contractions increases also. The effectiveness of sperm transport in the *in vitro* preparation is related somewhat to the stage of the estrous cycle at which the reproductive tract is removed. Sperm transport is more rapid in tracts that have been estrogenized. However, the addition of oxytocin to the perfusate enhances sperm transport in tracts from both the follicular and the progestational stages. The addition of epinephrine to the perfusate a few seconds prior to adding oxytocin obliterates all contractions in the *in vitro* preparations.

If semen is deposited into the cervix, in the *in vitro* preparations, and oxytocin is then added to the perfusate, sperm can be recovered from the ovarian portion of the oviducts within 30 min (VanDemark and Hays 1955; Hays and VanDemark 1952).

Thus, it has been demonstrated conclusively that mating or various manipulations of the reproductive tract of the cow causes the release of oxytocin into the blood stream. This agent is responsible for stimulating the vigorous contractions of the reproductive tract. The contraction waves propel the spermatozoa to the site of fertilization in a remarkably short interval of time. It appears also that oxytocin is released in varying amounts depending upon the extent of the stimulus applied. As mentioned previously, the effects of this hormone can be modified significantly by a variety of environmental factors that may cause the release of epinephrine (Hays and VanDemark 1953*b*). It has been suspected for some time that psychosomatic influences may play a significant role in gamete transport within the oviducts of the human female.

TABLE 4

INTERVAL BETWEEN EJACULATION AND APPEARANCE OF SPERM IN THE OVIDUCTS

Species	Time and Place in Oviduct	References
Mouse	15 min—upper regions of oviduct	Lewis and Wright 1935
Rat	15–30 min—ampullae	Blandau and Money 1944
Hamster	2 min—ampullae	Yamanaka and Soderwall 1960
	60 min—ampullae	Yanagimachi and Chang 1963
Rabbit	3 hr—in oviducts	Braden 1953
	5–6 hr—in oviducts sufficient sperm to fertilize all eggs	Adams 1956; Greenwald 1956
Guinea pig	15 min—middle of oviduct few seconds—in oviducts	Leuckart 1853
		Florey and Walton 1932
Dog	2 min—isthmus	Whitney 1937
	Few hours—in oviducts	Bischoff 1845
Sow	15 min—in ampullae	First *et al.* 1965
Cow	2 min—in ampullae	VanDemark and Moeller 1951
	13 min—in ampullae	Howe and Black 1963*a*
Ewe	8–30 min—in ampullae	Mattner and Braden 1963; Mattner 1963*b*
	6 min—in ampullae	Starke 1949
	$2\frac{1}{2}$ hr—in ampullae	Edgar and Asdell 1960
	5 hr—in ampullae	Quinlan, Maré, and Roux 1932; Green and Winters 1935; Kelly 1937
Man	68 min—in oviducts in extirpated tracts *in vitro*	Brown 1944
	30 min—in oviducts—removed at surgery	Rubenstein *et al.* 1951

Spermatozoa are transported rapidly through the reproductive tract of sows and ewes also. Within 15 min after insemination spermatozoa are recovered from the oviducts of estrous and luteal sows. Spermatozoa motility is not significantly affected by the stage of the estrous cycle at which insemination occurs (First *et al.* 1965). At 15 min after insemination 79% of the sperm recovered from the cornua of estrous and luteal sows are motile.

Table 4 summarizes the significant variations in the rate of sperm transport that have been recorded in ewes by several competent investigators. It has been shown that in ewes, as in cows, oxytocin is released following vaginal distension. When oxytocin is present, the ewe's uterine muscle responds by vigorous contractions (Debackere and Peeters 1960; Alexander 1945). Epinephrine inhibits uterine activity in both pregnant and nonpregnant ewes (Alexander 1945). The long intervals of sperm transport reported by the various investigators may be explained by the fact that sheep which are not handled frequently are much more easily disturbed by separation from the flock or restraint. Under these circumstances significant amounts of epinephrine may be secreted which would inhibit contractions of the uterine muscle. Thus the rate of sperm transport could be greatly prolonged. Two processes, therefore, appear to be involved in the transport of sperm from the cervix to the oviducts in the ewe: (1) a rapid transport which results from strong uterine contractions that take place immediately after mating (this phase is of relatively short duration) and (2) a slow transport which persists as long as sperm are present within the cervix (Mattner 1963*a*).

Sperm may be transported through the genital tracts of ewes under widely varying estrogen and progesterone levels. Sperm have been recovered from the oviducts of force-mated ewes during diestrus, early pregnancy, just prior to puberty, and in untreated ovariectomized ewes (Green and Winters 1935; Phillips and Andrews 1937; Warbritton *et al.* 1937; Mattner and Braden 1963).

In the rat and mouse, tonic contractions of the circular muscle of the cervix during heat are responsible for the retention of the uterine fluids. Normally the accumulated uterine fluids distend the cornua. At the moment of ejaculation a small mass of compacted spermatozoa is injected through the momentarily relaxed cervix and directly into the cornua. Strong peristaltic and antiperistaltic contraction waves pass over the cornua, quickly dispersing the sperm throughout the oviduct. Thus, spermatozoa may reach the uterotubal junction within a few seconds after insemination (Blandau and Odor 1949).

When one summarizes the data on the rate of sperm movement through the uterus or cornua it is clear that transport is effected chiefly by the muscular movements of the organ and not by the self-propellant activity of the spermatozoa.

C. *The Uterotubal Junction and Sperm Transport*

The role of the uterotubal junction in sperm transport has not been clarified for any animal.

In the mouse, rat, and hamster the uterotubal junction seems to control the number of sperm that enter the oviduct. For example, in the rat 25 to 30 million spermatozoa may be ejaculated into each cornu. Twelve hours after ejaculation or several hours after ovulation, the average number of spermatozoa found in the various segments of the oviducts is as follows; right isthmus, 488 $\pm$ 628 (range 27 to 2475); left isthmus, 430 $\pm$ 634 (range 19 to 2628); right ampulla, 10 (range 4 to 24); left ampulla 14 (range 1 to 55); right and left periovarian spaces 4 $\pm$ 5 (Blandau and Odor 1949; Huber 1915).

The comparative anatomy of the uterotubal junction has been described by Andersen (1928), Lee (1928) and Hafez and Black (Chapter 4, this volume). Its complexity varies with the animal. In the rabbit (Figs. 10*a*, *b*) and the pig the uterotubal junction has an elaborate arrangement of polyp-like processes which obscure the passage into the oviduct. In the rat and mouse a single small conical papilla or colliculus (0.3 to 0.4 mm in length) extends into the cornu (Kuhlmann 1964). Rhythmic, sphincter-like contractions of the uterotubal junction have been described and photographed only in the rabbit (Hartman, Stravoski, and Rubin 1959).

If physiological saline is injected rapidly into the cornua it is difficult and often impossible to force fluids through the uterotubal junction. Transuterine insufflation often leads to bursting of the uterus before gasses or fluids can be forced through this segment and into the tubes (Sobrero 1959). Apparently, in some animals the uterotubal junction acts as a valve which opens from time to time allowing a few sperm to pass into the oviducts (Kuhlmann 1964).

In the ewe both the cervix and uterotubal junctions appear to act as simple mechanical barriers that are able to maintain a graded concentration of sperm throughout the tract (Mattner 1963*b*).

In the sow, after a large volume of semen has been introduced directly into the uterus during normal coitus, most of the fluid content disappears from the cornua within 2 hr. A high concentration of spermatozoa persists, however, in the region of the uterotubal junctions. This reservoir of sperm remains here for 24 hr and then within the next 48 hr it too disappears. It is believed that the spermatozoa from this reservoir pass continually into the oviducts to reach the site of fertilization (Rigby 1966).

Further investigations are needed to define the role of the uterotubal junction in the transport of sperm. Disagreement remains as to whether or not sperm motility is a prerequisite for migration into the oviducts of animals other than the ungulates. In an experiment in which nonmotile rat, guinea pig, and human sperm were injected into the cornua of estrous rats, small numbers of each were recovered from the oviducts (Marcus 1965). Marcus concludes that motility of spermatozoa is not a prerequisite for their passage into the oviducts. Similar observations have been reported by Howe and Black (1963*a*, *b*) and Phillips and Andrews (1937). However, Leonard and Perlman (1949) report that dead sperm cannot pass the uterotubal junction in rats. It is obvious that the mechanism of sperm transport through the uterotubal junction has not been resolved satisfactorily.

D. *Sperm Transport through the Oviducts*

Although the mode of transport into and through the oviducts has been studied by a number of investigators there is as yet little specific information as to how spermatozoa reach the site of fertilization. Some data as to the time interval between ejaculation and sperm arrival in various segments of the oviducts are summarized in Table 4. The methods used to obtain this information were not the same for the various animals listed. Thus, one must assume that the data given are approximate rather than specific values.

The oviducts possess the remarkable property of conveying spermatozoa and eggs in opposite directions. Ovulated eggs are moved from the ostium of the oviduct to the site of fertilization within the ampullae while at the same time spermatozoa are being transported from the uterotubal junction to the location of the eggs within the ampullae. Most formidable are the technical problems of evaluating the respective roles played by peristalsis, antiperistalsis, ciliary activity of the oviducts, stricture of the isthmoampullar junction, the muscular movements of the mesosalpinx, and the intrinsic activity of the spermatozoa. The mechanism of sperm transport through the oviducts will be known only when the role of each of these activities has been more clearly defined.

It is assumed that transport of sperm within the oviducts is accomplished mainly by muscular contractions of the enclosing walls. There are, however, no specific observations recorded as to just how contractions accomplish transport. When the oviducts of the rabbit are observed at the time of ovulation, the variations and complexity of the muscular contractions are impressive. The patterns of movement in the ampullar and isthmic regions differ from each other significantly. In the isthmic portion peristaltic and antiperistaltic contractions are segmental in nature, vigorous, and almost continuous. In the region of the ampulla strong peristaltic waves move the eggs and their cumuli in a segmental fashion toward the mid-portion of the oviduct. When freshly ovulated eggs are introduced into the ostium of the oviduct, peristaltic contraction waves move them down the oviduct rapidly and in a very precise manner (Harper 1961*a*, *b*; Blandau, unpublished observations). When one observes the vigorous peristaltic contractions that drive the cumulus masses forward it is difficult to believe that spermatozoa could move up into the ampulla by their innate motility.

At the time of ovulation one sees not only the complex muscular movements of the oviducts but the rhythmic contractions of the mesosalpinx as well. These contractions change the position and contour of the tubes continually and it is likely that they play some role in gamete transport.

1. *The Role of Cilia.* Most investigators surmise that the cilia lining the oviducts are of no great importance in sperm transport. Since the original work of Parker (1930, 1931) on the role of cilia in sperm transport in the tortoise, pigeon, and rabbit, no similar investigations have been made in other animals. This is an area where carefully controlled *in vivo* and *in vitro* observations would be instructive in evaluating the role of cilia in gamete transport in different animals.

In the oviducts of the painted tortoise there are two ciliary systems. Most of the luminal surface of the oviduct is lined with cilia that beat toward the cloaca, but on one side of the oviduct there is a band of cilia, 2 to 3 mm wide, in which the cilia beat in the direction of the ovary (Parker 1931). This is well displayed when the oviduct is split open and particulate matter is flushed onto the exposed surface. Then movement in two directions, abovarian and adovarian, is dramatically demonstrated.

The pigeon oviduct also has two ciliary systems which are capable of moving particulate material in opposite directions. The adovarian band is approximately one-fourth the total width of the oviduct and extends from the infundibulum to the uterus (Parker 1931).

It has been assumed that such dual ciliary tracts do not exist in the mammalian oviducts. With the exception of the rabbit the pattern of ciliary beat in the entire oviduct has not been described for any animal. If a fresh oviduct of the rabbit is carefully slit open along the attachment of the mesosalpinx and submerged in a tissue culture medium its elaborate system of longitudinal folds and ciliary pattern may be studied (Blandau, unpublished observations). There is extensive ciliation of the folds and all cilia beat toward the uterus. Parker, also, watched the movement of particulate matter over the surfaces of slit-open oviducts in rabbits and concluded that there was only one direction of ciliary beat; namely, abovarian. When spermatozoa were applied to the surfaces of opened oviducts they were swept along by the ciliary current at such a rate that their own flagellar activity was of little avail against the current. The rate and direction of ciliary beat in the rabbit oviduct appears to present an almost insurmountable handicap for individual sperm movements. Further observations on the opened oviduct showed that the fluid currents in the spaces between the folds do not all move in an abovarian direction. The longitudinal folds often form compartments which are momentarily closed (Figs. 7*a*, *b*). Under these circumstances the abovarian currents would be blocked and as a result fluids would flow in the opposite direction up the center of the compartments. Thus, Parker describes the circulation of fluids as abovarian in direction along the sides of the ciliated ridges and adovarian in the central regions of each compartment. Oviductal contractions would change the configuration of the various compartments momentarily, providing the possibility that fluids and sperm could be moved from one compartment to the next. Parker concluded that these temporary oviductal compartments provide the mechanism by which sperm may be transported in an adovarian direction.

In opposition to this view Black and Asdell (1958) minimize the role of ciliary activity in transporting rabbit sperm. They emphasize the importance of segmental contraction brought about by the circular musculature of the oviduct. Although we tend to agree that probably muscular contractions play the most important role we have repeated the opened tube technique of Parker and have observed the movements of stained lycopodium spores and rabbit spermatozoa along the exposed surfaces. In so doing we have been impressed by Parker's observations and by the length of time ciliary beat and muscular contractions can be maintained in *in vitro* preparations.

E. *The Isthmoampullary Junction and Gamete Transport*

A spasmodic stricture appears in the oviduct at about the time of ovulation in different animals. This is a physiological and not an anatomical sphincter which appears midway between the isthmus (which adjoins the uterus) and the ampulla (which adjoins the infundibulum). For want of a better term it has received the name "isthmoampullary junction." In rats, mice, and hamsters this stricture occludes the lumen sufficiently to allow accumulations of fluids within the loops of the ampulla. In an ovulating rabbit the stricture can be clearly recognized by its median position in the tube and pale color. In this animal as well the constriction is responsible for the retention of considerable amounts of fluid within the ampullae. Ovulated eggs will not pass into the isthmus for some hours after fertilization since they are retained within the ampullae by the strictures. How spermatozoa achieve passage through the narrowed portion of the oviductal lumen is unknown. Although again there is no specific evidence it seems very possible that individual spermatozoa pass this barrier by virtue of their own motility.

The finding of spermatozoa within the oviducts at specific intervals after ejaculation or artificial insemination does not mean that sufficient numbers have entered to fertilize all of the ovulated eggs. If the female rats are mated early in estrus and each isthmus ligated 1 to 5 hr after ejaculation the majority of ova are found to be unfertilized (Leonard 1950). Essentially similar data have been obtained by Yanagimachi and Chang (1963) for the hamster. When female hamsters are mated 5 to 8 hr before ovulation, and one oviduct is ligated 1 hr later, only 17% of the eggs on the ligated side are fertilized. On the control side 100% of the eggs have been penetrated by spermatozoa. Nine out of 12 females have no fertilized eggs in the ligated oviducts. The percentage increases proportionately as the time interval between mating and ligation is prolonged.

It may be significant that oviductal secretions are copious at the time of ovulation (Bishop 1956), but it is not clear whether they serve as nurture for spermatozoa or as a vehicle for their transport. In the ampullar portions of the oviducts the fluids may play a more important role in the transport of eggs than of spermatozoa.

F. *Gamete Transport in the Presence of Intrauterine Devices*

It has been suggested that the mechanism by which intrauterine devices prevent conception is related to alterations in the motility of both the uterus and oviducts. Increased motility is thought to accelerate gamete transport, particularly through the oviducts (Margulies 1962, 1964). Data to substantiate this hypothesis have been difficult to obtain, particularly in women. There is no agreement as to whether women who have had IUD's in place for prolonged periods of time demonstrate altered motility of the reproductive tract.

Analysis of the sites in which spermatozoa have been found has been more productive. Motile spermatozoa have been found in oviductal washings approximately 24 hr after coitus in 4 women in whom IUD's had been inserted 1 to 10 days prior to operation (Malkani and Sujan 1964). Similar results were reported by Morgenstern, Orgebin-Crist, and Clewe (1966) in women in whom IUD's had been in place for variable periods of time and in whom the oviducts were flushed and examined for gametes immediately after surgery. Five of 6 such women with motile spermatozoa had a history of coitus 1 day before surgery. Examination of the sediment from the oviductal flushings revealed the presence of 3 ova—1 fertilized and 2 unfertilized.

In sheep the presence of an IUD appears to interfere with the mechanism involved in sperm transport. Spermatozoa could not be recovered from oviductal flushings of mated females in which IUD's had been inserted 2 to 4 weeks previously (Hawk 1965). Egg transport in these experimental ewes, did not appear to be accelerated. When the unfertilized eggs were transferred into recipient control, mated ewes, they were capable of being fertilized (Hawk 1965).

In the rhesus monkey accelerated egg transport has been reported after insertion of an IUD (Mastroianni and Hongsanand 1964). This variability in observations suggests the possibility that the effects of IUD's on gamete transport may be related directly to the length of time the device has been in place (Noyes *et al.* 1966).

G. *Sperm Survival in the Oviducts*

Table 5 summarizes some data on the maximum length of time sperm may reside in the oviducts and retain their capacities to penetrate and fertilize ova. When spermatozoa have reached the oviducts their fertilizing capacity is relatively limited. It is well known that the ability of a sperm to penetrate an egg is lost more promptly than its motility. In a few animals in which specific information is available (mouse, guinea pig, rat, and sheep) motility

TABLE 5

FERTILIZABLE LIFE OF SPERM IN THE OVIDUCTS OF VARIOUS SPECIES

Species	Duration of Fertility in Hours	References
Mouse	6	Merton 1939
Rat	14	Soderwall and Blandau 1941
Guinea pig	22	Soderwall and Young 1940
Ferret	126	Chang 1965
Rabbit	30–32	Hammond and Asdell 1926
Pig	24–48	du Mesnil du Buisson and Dauzier 1955; Pitkjanen 1960
Sheep	24–48	Green 1947; Dauzier and Wintenberger 1952
Horse	144	Day 1942; Burkhardt 1949
Cow	24–48	Laing 1945; Vandeplassche and Paredis 1948; Tarosz 1961
Man	24–48	Hartman 1939; Farris 1950; Rubenstein *et al.* 1951; Horne and Audet 1958

persists for approximately twice as long as the ability to fertilize (Bishop 1961). Precise data as to the duration of fertilizing capacity of sperm are difficult to obtain except in those animals in which ovulation can be timed precisely.

In contrast to the effects of aging the egg before fertilization, in which abnormal embryos may result, if "aged" sperm are capable of penetrating an egg, development usually proceeds normally (Soderwall and Young 1940; Soderwall and Blandau 1941). The nature of the oviductal fluids and the cells within them no doubt play a vital role in determining the environmental factors for sperm viability. The nature of these fluids will be discussed in greater detail in Chapter 13.

H. *Escape of Spermatozoa into the Peritoneal Cavity*

Relatively little information is available as to whether spermatozoa enter the peritoneal cavity after insemination and in what numbers. In animals possessing a periovarial sac, which almost completely encloses the ovary (rat, mouse, and hamster), there is little possibility that sperm could escape through the periovarial-peritoneal slit.

In those animals in which the fimbriae are not enclosed in a periovarial sac spermatozoa may enter the peritoneal cavity and possibly cross over to enter the ostium of the other oviduct. Mattner and Braden (1963) found that the cranial ends of the oviducts had become occluded in a number of ewes that had been ovariectomized. Estrus was artificially induced and the animals were mated by force. The mean number of sperm recovered from the occluded oviducts was significantly higher than in the patent oviducts. They interpreted this to mean that the spermatozoa move through the patent oviducts and escape into the peritoneal cavity.

Observations of other investigators lend support to the concept that the spermatozoa may pass from the oviducts into the abdominal cavity. Rowlands (1958) tied one uterine horn just above the body of the uterus prior to mating in rabbits and found that eggs were implanted in both uterine horns in 10% of his animals. Horne and Thibault (1962) recovered live sperm in the peritoneal fluids of 5 out of 14 women laparotomized 24 hr *post coitum.*

I. *Concluding Observations*

Although speculation is rampant it is not yet clear how sperm are transported through the oviducts. More precise observations need to be made in a variety of animals in order to resolve the conflicts arising from species variations in the mechanisms involved. Numerous questions remain which require both descriptive and quantitative approaches. What is the role of the segmental peristaltic and antiperistaltic muscular contractions in sperm transport? What controls the specific types of contraction observed in the ampullar and isthmic zones? Are spermatozoa transported by different mechanisms in the isthmic and ampullar segments of the oviduct? What is the arrangement and significance of the oviductal folds in the rotation of eggs and their transport in different animals? What is the arrangement of the ciliated cells and how does their beat affect sperm movement? Does sperm motility itself play any role in transport or is it effective only at various barriers such as the cervix, uterotubal junction, or isthmoampullary junction? Are there local physical or chemical factors that determine the directional movements of sperm within the oviducts? Do the chemical and physical characteristics of the oviductal fluids vary at different times of the cycle and may these differences affect sperm transport? Does the isthmoampullary junction control the number of sperm which reach the site of fertilization? Do rhythmic contractions of the mesosalpinx or other membranes attached to the oviduct play any role in sperm transport? What are the specific effects of various hormones and other pharmacologic agents on the rate, amplitude, and pattern of oviductal contractions?

Because so little specific information is available, these questions are left with the reader as challenging guides to useful research.

Acknowledgment

Unpublished observations referred to in this chapter were supported by grants from the National Institutes of Health, United States Public Health Service.

References

1. Adams, C. E. 1956. A study of fertilization in the rabbit: The effect of post-coital ligation of the Fallopian tube or uterine horn. *J. Endocr.* 13:296.
2. Akester, A. R., and Inkster, I. J. 1961. Cine-radiographic studies of the genital tract of the rabbit. *J. Reprod. Fertil.* 2:507.
3. Alden, R. H. 1942. The periovarial sac in the albino rat. *Anat. Rec.* 83:421.
4. Alexander, F. 1945. Pharmacological studies on the uterus. *J. Comp. Path.* 55:140.

5. Amersbach, R. 1930. Sterilität und Frigidität. *München. Med. Wschr.* 77:225.
6. Amoroso, E. C., Griffiths, W. F. B., and Hamilton, W. J. 1942. The early development of the goat (*Capra hircus*). *J. Anat.* 76:377.
7. Andersen, D. H. 1928. Comparative anatomy of the tubo-uterine junction: Histology and physiology in the sow. *Amer. J. Anat.* 42:255.
8. Assheton, R. 1894. A re-investigation into the early stages of the development of the rabbit. *Quart. J. Microscop. Sci.* 32:113.
9. Austin, C. R. 1948. Number of sperms required for fertilization. *Nature* 162:534.
10. ———. 1959. Fertilization and development of the egg. In *Reproduction in domestic animals*, ed. H. H. Cole and P. T. Cupps, chap. 12. New York: Academic Press.
11. ———. 1963. Fertilization and transport of the ovum. In *Mechanisms concerned with conception*, ed. C. G. Hartman, chap. 6. New York: The Macmillan Co.
12. Belonoschkin, B. 1949. Zeugung beim Menschen im Lichte der Spermatozoenlehre. *S. Jöbergs Forlag, Stockholm.*
13. Berliner, V. R. 1959. The estrous cycle of the mare. In *Reproduction in domestic animals*, vol. 1, chap. 8. New York: Academic Press.
14. Bischoff, T. L. W. von. 1845. *Die Entwicklungsgeschichte des Hunde-Eies.* Braunschweig.
15. Bishop, D. W. 1956. Active secretion in the rabbit oviduct. *Amer. J. Physiol.* 187:347.
16. ———. 1961. Biology of spermatozoa. In *Sex and internal secretions*, ed. W. C. Young, 3d ed. p. 707. Baltimore: Williams & Wilkins Co.
17. Black, D. L., and Asdell, S. A. 1958. Transport through the rabbit oviduct. *Amer. J. Physiol.* 192:63.
18. Blandau, R. J. 1945. On the factors involved in sperm transport through the cervix uteri of the albino rat. *Amer. J. Anat.* 77:263.
19. ———. 1961. The biology of eggs and implantation. In *Sex and internal secretions*, ed. W. C. Young, 3d ed. Baltimore: Williams & Wilkins Co.
20. Blandau, R. J., and Jordan, E. S. 1941. The effect of delayed fertilization on the development of the rat ovum. *Amer. J. Anat.* 68:275.
21. Blandau, R. J., and Money, W. L. 1944. Observations on the rate of transport of spermatozoa in the female genital tract of the rat. *Anat. Rec.* 90:255.
22. Blandau, R. J., and Odor, D. L. 1949. The total number of spermatozoa reaching various segments of the reproductive tract in the female albino rat at intervals after insemination. *Anat. Rec.* 103:93.
23. Blandau, R. J., and Young, W. C. 1939. The effects of delayed fertilization on the development of the guinea pig ovum. *Amer. J. Anat.* 64:303.
24. Borell, U., Nilsson, O., and Westman, A. 1957. Ciliary activity in the rabbit Fallopian tubes during oestrus and after copulation. *Acta Obstet. Gynec. Scand.* 36:22.
25. Braden, A. W. H. 1953. Distribution of sperms in the genital tract of the female rabbit after coitus. *Aust. J. Biol. Sci.* 6:693.
26. ———. 1958. Strain differences in the incidence of polyspermia in rats after delayed mating. *Fertil. Steril.* 9:243.
27. Braden, A. W. H., and Austin, C. R. 1954*a*. The number of sperms about the eggs in mammals and its significance for normal fertilization. *Aust. J. Biol. Sci.* 7:543.
28. ———. 1954*b*. Fertilization of the mouse egg and the effect of delayed coitus and hot shock treatment. *Aust. J. Biol. Sci.* 7:552.
29. Brown, R. L. 1944. Rate of transport of spermia in human uterus and tubes. *Amer. J. Obstet. Gynec.* 47:407.

30. Brundin, Jan 1964. A functional block in the isthmus of the rabbit Fallopian tube. *Acta Physiol. Scand.* 60:295.
31. Burkhardt, J. 1949. Sperm survival in the genital tract of the mare. *J. Agric. Sci.* 39: 201.
32. Campbell, B., and Petersen, W. E. 1954. Milk "let-down" and the orgasm in the human female. *Hum. Biol.* 25:165.
33. Chang, M. C. 1952. Effects of delayed fertilization on segmenting ova, blastocysts and foetuses in rabbits. *Fed. Proc.* 11:24.
34. ———. 1965. Fertilizing life in ferret sperm in the female tract. *J. Exp. Zool.* 158:87.
35. Clark, R. T. 1934. Studies on the physiology of reproduction in the sheep. II. The cleavage stages of the ovum. *Anat. Rec.* 60:135.
36. Clewe, T. H., and Mastroianni, L., Jr. 1958. Mechanisms of ovum pickup. I. Functional capacity of rabbit oviducts ligated near the fimbria. *Fertil. Steril.* 9:13.
37. Dauzier, L., and Wintenberger, S. 1952. Recherches sur la fécondation chez les mammifères: Durée du pourvoir fécondant des spermatozoïdes de bélier dans le tractus génital de la brebis et durée de la période de fécondite de l'oeuf après l'ovulation. *C.R. Soc. Biol. (Par.)* 146:660.
38. Day, F. T. 1942. Survival of spermatozoa in the genital tract of the mare. *J. Agric. Sci.* 32:108.
39. Debackere, M., and Peeters, G. 1960. Milk ejection studied by means of a crossed-circulation technique on sheep. *Naturwissenschaften* 47:189.
40. Doyle, J. B. 1951. Exploratory culdotomy for observation of tubo-ovarian physiology at ovulation time. *Fertil. Steril.* 2:475.
41. ———. 1954. Ovulation and the effects of selective uterotubal denervation. *Fertil. Steril.* 5:105.
42. ———. 1956. Tubo-ovarian mechanism: Observation at laparotomy. *Obstet. Gynec.* 8:686.
43. Du Mesnil du Buisson, F., and Dauzier, L. 1955. Distribution et résorption du sperme dans le tractus génital de la truie des spermatozoïdes. *Ann. Endocr.* 16:413.
44. Edgar, D. G., and Asdell, S. A. 1960. Spermatozoa in the female genital tract. *J. Endocr.* 21:321.
45. Egli, G. E., and Newton, M. 1961. The transport of carbon particles in the human female reproductive tract. *Fertil. Steril.* 12:151.
46. Elert, R. 1947. Der Mechanismus der Eiabnahme im Laparskop. *Zbl. Gynaek.* 69:38.
47. Evans, H. M., and Cole, H. H. 1931. An introduction to the study of the oestrous cycle in the dog. *Mem. Univ. Calif.* 9:65.
48. Farris, E. J. 1950. *Human fertility and problems of the male.* White Plains, New York: The Author's Press.
49. First, N. L., Short, R. E., Peters, J. B., and Stratman, F. W. 1965. Transport of spermatozoa in estrual and luteal sows. *J. Anim. Sci.* 24:917.
50. Florey, H., and Walton, A. 1932. Uterine fistula used to determine the mechanism of ascent of the spermatozoon in the female genital tract. *J. Physiol.* 74:5P.
51. Gilchrist, F., and Pincus, G. 1932. Living rat eggs. *Anat. Rec.* 54:275.
52. Green, W. W. 1947. Duration of sperm fertility in the ewe. *Amer. J. Vet. Res.* 8:299.
53. Green, W. W., and Winters, L. M. 1935. Studies on the physiology of reproduction in the sheep. III. The time of ovulation and rate of sperm travel. *Anat. Rec.* 61:457.

54. Greenwald, G. S. 1956. Sperm transport in the reproductive tract of the female rabbit. *Science* 124:586.
55. ———. 1961. A study of the transport of ova through the rabbit oviduct. *Fertil. Steril.* 12:80.
56. Gregory, P. W. 1930. The early embryology of the rabbit. *Contr. Embryol. Carnegie Inst.* 21:141.
57. Gregory, P. W., and Castle, W. F. 1931. Further studies on the embryological basis of size inheritance in the rabbit. *J. Exp. Zool.* 59:199.
58. Hafez, E. S. E., and Black, D. L. 1968. The mammalian uterotubal junction. In *The mammalian oviduct*, ed. E. S. E. Hafez and R. J. Blandau, chap. 4. Chicago: University of Chicago Press.
59. Hamilton, W. J. 1934. The early stages in the development of the ferret: Fertilization to the formation of the prochordal plate. *Trans. Roy. Soc. Edinb.* 58:251.
60. Hamilton, W. J. and Day, F. T. 1945. Cleavage stages of the ova of the horse with notes on ovulation. *J. Anat.* 79:127.
61. Hamilton, W. J., and Laing, J. A. 1946. Development of the egg of the cow up to the stage of blastocyst formation. *J. Anat.* 80:194.
62. Hammond, J. 1934. The fertilization of rabbit ova in relation to time: A method of controlling litter size, the duration of pregnancy and weight of young at birth. *J. Exp. Biol.* 11:140.
63. Hammond, J., and Asdell, S. A. 1926. The vitality of spermatozoa in the male and female reproductive tract. *J. Exp. Biol.* 4:155.
64. Hammond, J., and Walton, A. 1934. Notes on ovulation and fertilization in the ferret. *J. Exp. Biol.* 11:307.
65. Harper, M. J. K. 1961*a*. The mechanisms involved in the movement of newly ovulated eggs through the ampulla of the rabbit Fallopian tube. *J. Reprod. Fertil.* 2:522.
66. ———. 1961*b*. Egg movement through the ampullar region of the Fallopian tube of the rabbit. *Proc. Fourth Int. Congr. Anim. Reprod., The Hague*, p. 375.
67. Hartman, C. G. 1916. Studies on the development of the opossum (*Didelphis virginiana* L.). I. History of early cleavage. II. Formation of the blastocyst. *J. Morphol.* 27:1.
68. ———. 1928. The breeding season of the opossum (*Didelphys virginiana* L.) and the rate of the intra-uterine and postnatal development. *J. Morphol. Physiol.* 46:143.
69. ———. 1939. Ovulation and the transport and viability of ova and sperm in the female genital tract. In *Sex and internal secretions*, ed. E. Allen, 2d ed., chap. 14. Baltimore: Williams & Wilkins Co., 1939.
70. ———. 1957. How do sperms get into the uterus? *Fertil Steril.* 8:403.
71. Hartman, C. G., Stravoski, J., and Rubin, I. C. 1959. X-ray motion pictures of hydrotubation in rabbit and in man: Film shown before the Amer. Soc. Stud. Steril. Atlantic City, April 5, 1959.
72. Harvey, Clare 1960. The speed of human spermatozoa and the effect on it of various diluents, with some preliminary observations on clinical material. *J. Reprod. Fertil.* 1:84.
73. Hausmann, U. F. 1840. Über die Zeugung und Entstehung des mahren weiblichen Eies bei den Säugethiere, Hannover. Quoted from Bischoff, 1885, p. 14.
74. Hawk, H. W. 1965. Inhibition of ovum fertilization in the ewe by intra-uterine plastic spirals. *J. Reprod. Fertil.* 10:267.
75. Hays, R. L., and VanDemark, N. L. 1952. Effect of hormones on uterine motility and sperm transport in the perfused genital tract of the cow. *J. Dairy Sci.* 35:499.

76. ———. 1953*a*. Effects of oxytocin and epinephrine on uterine motility in the bovine. *Amer. J. Physiol.* 172:557.
77. ———. 1953*b*. Effect of stimulation of the reproductive organs of the cow on the release of an oxytocin-like substance. *Endocrinology*, 52:634.
78. Heuser, C. H., and Streeter, G. L. 1929. Early stages in the development of pig embryos, from the period of initial cleavage to the time of the appearance of limb-buds. *Contr. Embryol. Carnegie Inst.* 20:1.
79. Hill, J. P. 1933. The development of the Monotremata. II. The structure of the eggshell. *Trans. Zool. Soc. Lond.* 21:413.
80. Horne, H. W., Jr., and Audet, C. 1958. Spider cells, a new inhabitant of peritoneal fluid; a preliminary report. *Obstet. Gynec.* 11:421.
81. Horne, H. W., Jr., and Thibault, J. 1962. Sperm migration through the human female reproductive tract. *Fertil. Steril.* 13:444.
82. Hoskins, D. D., and Patterson, D. L. 1967. Prevention of coagulum formation with recovery of motile spermatozoa from rhesus monkey semen. *J. Reprod. Fertil.* 13:337.
83. Howe, G. R., and Black D. L. 1963*a*. Migration of rat and foreign spermatozoa through the utero-tubal junction of the oestrous rat. *J. Reprod. Fertil.* 5:95.
84. ———. 1963*b*. Spermatozoan transport and leucocytic responses in the reproductive tract of calves. *J. Reprod. Fertil.* 6:305.
85. Huber, G. C. 1915. The development of the albino rat, *Mus norvegicus albinus*. I. From the pronuclear stage to the stage of the mesoderm anlage; end of the first to the end of the ninth day. *J. Morphol.* 26:247.
86. Huggins, C., and Neal, W. 1942. Coagulation and liquefaction of semen: Proteolytic enzymes and citrate in prostatic fluid. *J. Exp. Med.* 76:527.
87. Hunter, R. H. F. 1967. The effects of delayed insemination on fertilization and early cleavage in the pig. *J. Reprod. Fertil.* 13:133.
88. Ivy, A. C., Hartman, C. G., and Koff, A. 1931. The contractions of the monkey uterus at term. *Amer. J. Obstet. Gynec.* 22:388.
89. Kelly, R. B. 1937. Studies in fertility of sheep. *Bull. Coun. Sci. Indust. Res. Aust.* No. 112.
90. Krehbiel, R. H., and Carstens, H. P. 1939. Roentgen studies of the mechanism involved in sperm transportation in the female rabbit. *Amer. J. Physiol.* 125:571.
91. Kuhlmann, W. 1964–65. The uterotubal valve and the fate of the sperm in the female tract of the mouse. *Jackson Lab. 36th Annual Report.*
92. Laing, J. A. 1945. Observations on the survival time of the spermatozoa in the genital tract of the cow and its relation to fertility. *J. Agric. Sci.* 35:72.
93. Lee, F. C. 1928. The tubo-uterine junction in various animals. *Johns Hopkins Hosp. Bull.* 42:335.
94. Leeuwenhoeck, A. van 1677. Observationes di Anthonii Leeuwenhoeck de natis è semine genitali animalculis. *Phil. Trans.* 11:1040.
95. ———. 1957. *The collected letters of Antoni van Leeuwenhoeck. 1685*, edited, illustrated and annotated by a committee of noted scientists, vol. 5, p. 187. Amsterdam: Swets & Zeitlinger, Ltd.
96. Leonard, S. L. 1950. The reduction of uterine sperm and uterine fluid on fertilization of rat ova. *Anat. Rec.* 106:607.
97. Leonard, S. L., and Perlman, P. L. 1949. Conditions effecting the passage of spermatozoa through the utero-tubal junction of the rat. *Anat. Rec.* 104:89.
98. Leuckart, R. 1853. Zeugung. In *Wagner*'s *Handwörterbuch der Physiologie* 4:707.

99. Lewis, W. H., and Hartman, C. G. 1941. Tubal ova of the rhesus monkey. *Contr. Embryol. Carnegie Inst.* 29:1.
100. Lewis, W. H., and Wright, E. S. 1935. On the early development of the mouse egg. *Contr. Embryol. Carnegie Inst.* 25:115.
101. Macdonald, E., and Long, J. A. 1934. Some features of cleavage in the living egg of the rat. *Amer. J. Anat.* 55:343.
102. MacLeod, J. 1946. The semen specimen: Laboratory examination. In *Diagnosis in sterility*, ed. E. T. Engle, p. 3. Springfield, Illinois: Charles C Thomas.
103. MacLeod, J., and Hotchkiss, R. 1942. Distribution of spermatozoa and of certain chemical constituents in human ejaculate. *J. Urol.* 48:225.
104. Malkani, P. K., and Sujan, S. 1964. Sperm migration in the female reproductive tract in the presence of intrauterine devices. *Amer. J. Obstet. Gynec.* 88:963.
105. Mann, T. 1954. *The biochemistry of semen.* London: Methuen & Co., Ltd.
106. ———. 1964. *The biochemistry of semen and of the male reproductive tract.* London: Methuen & Co., Ltd.
107. Marcus, S. L. 1965. The passage of rat and foreign spermatozoa through the utero-tubal junction of the rat. *Amer. J. Obstet. Gynec.* 91:985.
108. Margulies, L. C. 1962. Intrauterine contraceptive devices. *Proc. Conf. April 30–May 1, Excerpta Medica Foundation, New York.* P. 61.
109. ———. 1964. Intrauterine contraception—A new approach. *Obstet. Gynec.* 24:515.
110. Marston, J. H., and Chang, M. C. 1964. The fertilizable life of ova and their morphology following delayed insemination in mature and immature mice. *J. Exp. Zool.* 155:237.
111. ———. 1966. The morphology and timing of fertilization and early cleavage in the Mongolian gerbil and deer mouse. *J. Embryol. Exp. Morph.* 15:169.
112. Masters, W. H., and Johnson, V. E. 1966. *Human sexual response.* Boston: Little, Brown & Co.
113. Mastroianni, L., and Hongsanand, C. 1964. Intrauterine contraception. *Proc. II Internat. Conf., October, 1964, Excerpta Medica Foundation, New York.* P. 194.
114. Mattner, P. E. 1963*a*. Spermatozoa in the genital tract of the ewe. II. Distribution after coitus. *Aust. J. Biol. Sci.* 16:688.
115. ———. 1963*b*. Spermatozoa in the genital tract of the ewe. III. Role of spermatozoan motility and of uterine contractions in transport of spermatozoa. *Aust. J. Biol. Sci.* 16:877.
116. Mattner, P. E., and Braden, A. W. H. 1963. Spermatozoa in the genital tract of the ewe. I. Rapidity of transport. *Aust. J. Biol. Sci.* 16:473.
117. Merton, H. 1939. Studies on reproduction in the albino mouse. III. The duration of life of spermatozoa in the female reproductive tract. *Proc. Roy. Soc. Edinb.* 59:207.
118. Moghissi, K. S., Dabich, D., Levine, J., and Neuhaus, O. W. 1964. Mechanism of sperm migration. *Fertil. Steril.* 15:15.
119. Morgenstern, L. L., Orgebin-Crist, M. C., and Clewe, T. H. 1966. Observations on spermatozoa in the human uterus and oviducts in the chronic presence of intrauterine devices. *Amer. J. Obstet. Gynec.* 96:114.
120. Mroueh, A. 1967. Oxytocin and sperm transport in rabbits. *Obstet. Gynec.* 29:671.
121. Neuhaus, O. W., and Moghissi, K. S. 1962. Composition and properties of human cervical mucus. III. A preliminary study of the mucoid component. *Fertil. Steril.* 13:550.

122. Noyes, R. W., Clewe, T. H., Bonney, W. A., Burrus, S. B., DeFeo, V. J., and Morgenstern, L. L. 1966. Searches for ova in the human uterus and tubes. I. Review, clinical methodology and summary of findings. *Amer. J. Obstet. Gynec.* 96:157.
123. Odor, D. L., and Blandau, R. J. 1956. Incidence of polyspermy in normal and delayed matings in rats of the Wistar strain. *Fertil. Steril.* 7:456.
124. Oxenreider, S. L., and Day, B. N. 1965. Transport and cleavage of ova in swine. *J. Anim. Sci.* 24:413, 1965.
125. Parker, G. H. 1930. The passage of the spermatozoa and ova through the oviducts of the rabbit. *Proc. Soc. Exp. Biol. Med.* 27:826.
126. ———. 1931. The passage of sperms and of eggs through the oviducts in terrestrial vertebrates. *Phil. Trans.* 219:381.
127. Parkes, A. S. 1960. Transport, selection and fate of spermatozoa in the female mammal. In *Marshall's physiology of reproduction*, ed. A. S. Parkes, 3d ed., p. 234. London: Longmans & Co.
128. Phillips, R. W., and Andrews, F. N. 1937. The speed of travel of ram spermatozoa. *Anat. Rec.* 68:127.
129. Pickles, V. R. 1953. Blood-flow estimations as indices of mammary activity. *J. Obstet. Gynaec. Brit. Emp.* 60:301.
130. Piko, L. 1958. Étude de la polyspermie chez le rat. *C.R. Soc. Biol.* (*Paris*) 152:1356.
131. Pincus, G. 1936. *The eggs of mammals.* New York: Macmillan Co.
132. Pitkjanen, I. G. 1960. The fate of spermatozoa in the uterus of the sow. *Z. Obsc. Biol.* 21:28.
133. Quinlan, J., Maré, G. S., and Roux, L. L. 1932. The vitality of the spermatozoa in the genital tract of the merino ewe with special reference to its practical application in breeding. *Rep. Div. Vet. Serv. Anim. Ind. S. Afr.* No. 18, p. 831.
134. Rigby, J. P. 1966. The persistence of spermatozoa at the uterotubal junction of the sow. *J. Reprod. Fertil.* 11:153.
135. Rowlands, I. W. 1958. Insemination by peritoneal injection. *Proc. Soc. Stud. Fertil.* 10:150.
136. Rowson, L. E. A. 1955. The movement of radio-opaque material in the bovine uterine tract. *Brit. Vet. J.* 111:334.
137. Rubenstein, B. B. 1943. The transportation and survival of sperm in the vaginal and cervical canals. In *Problems of human fertility*, ed. E. T. Engle, p. 101. Menasha, Wisconsin: George Banta Publishing Co.
138. Rubenstein, B. B., Straus, H., Lazarus, M. L., and Hawkins, H. 1951. Sperm survival in women. *Fertil. Steril.* 2:15.
139. Salisbury, G. W., and VanDemark, N. L. 1961. *Physiology of reproduction and artificial insemination of cattle.* San Francisco: W. H. Freeman Co.
140. Sims, J. M. 1869. On the microscope, as an aid in the diagnosis and treatment of sterility. *N.Y. Med. J.* 8:393.
141. Sobotta, J. 1895. Die Befruchtung und Furchung des Eies der Maus. *Arch. Mikroskop. Anat.* 45:15.
142. Sobrero, A. J. P. 1959. Sperm migration in the female genital tract. In *Mechanisms concerned with conception*, ed. Carl G. Hartman, chap. 4. New York: Macmillan Co.
143. Soderwall, A. L., and Blandau, R. J. 1941. The duration of the fertilizing capacity of spermatozoa in genital tract of the rat. *J. Exp. Zool.* 88:55.

144. Soderwall, A. L., and Young, W. C. 1940. The effect of aging in the female genital tract on the fertilizing capacity of guinea pig spermatozoa. *Anat. Rec.* 78:19.
145. Squier, R. R. 1932. The living egg and early stages of its development in the guinea pig. *Contr. Embryol. Carnegie Inst.* 23:225.
146. Starke, N. C. 1949. The sperm picture of rams of different breeds as an indication of their fertility. II. The rate of sperm travel in the genital tract of the ewe. *Onderstepoort J. Vet. Sci. Anim. Indus.* 22:415.
147. Tampion, D., and Gibbons, R. A. 1962. Orientation of spermatozoa in mucus of the cervix uteri. *Nature* 194:381.
148. Tarosz, S. 1961. Obtaining fertilized and unfertilized ova following spontaneous and induced ovulation in cows. *Zwsz. nuk. Wyisz. Szisz. Szkal. raln. Krkawie. Zootech. Z.* 2:105.
149. VanDemark, N. L., and Hays, R. L. 1951. The effect of oxytocin, adrenalin, breeding techniques and milking on uterine motility of the cow. *J. Anim. Sci.* 10:1083.
150. ———. 1952. Uterine motility responses to mating. *Amer. J. Physiol.* 170:518.
151. ———. 1954. Rapid sperm transport in the cow. *Fertil. Steril.* 5:131.
152. ———. 1955. Sperm transport in the perfused genital tract of the cow. *Amer. J. Physiol.* 183:510.
153. VanDemark, N. L., and Moeller, A. N. 1951. Speed of spermatozoan transport in reproductive tract of the estrous cow. *Amer. J. Physiol.* 165:674.
154. Vandeplassche, M., and Paredis, F. 1948. Preservation of the fertilizing capacity of bull semen in the genital tract of the cow. *Nature* 162:813.
155. Warbritton, V., McKenzie, F. F., Berliner, V., and Andrews, F. N. 1937. Sperm survival in the genital tract of the ewe. *Proc. Amer. Soc. Anim. Prod.* 30:142.
156. Westman, A. 1926. A contribution to the question of the transit of the ovum from ovary to uterus in rabbits. *Acta Obstet. Gynec. Scand.* 5:1.
157. ———. 1952. Investigations into the transport of the ovum. In *Proceedings conference studies on testes and ovary, eggs and sperm*, ed. E. T. Engle, p. 163. Springfield, Illinois: Charles C Thomas.
158. Whitney, L. F. 1937. *How to breed dogs.* New York: Orange Judd Publishing Co.
159. Wimsatt, W. A., and Waldo, C. M. 1945. The normal occurrence of a peritoneal opening in the bursa ovarii of the mouse. *Anat. Rec.* 93:47.
160. Winters, L. M., Green, W. W., and Comstock, R. E. 1942. Prenatal development of the bovine. *Minnesota Agric. Exp. Sta. Tech. Bull.* No. 151.
161. Wood, C. 1966. How spermatozoa move. *Sci. J.* 2:44.
162. Yamada, F. 1952. Studies on the ciliary movements of the oviduct. *Jap. J. Physiol.* 2:194.
163. Yamanaka, H. S., and Soderwall, A. L. 1960. Transport of spermatozoa through the female genital tract of hamsters. *Fertil. Steril.* 11:470.
164. Yanagimachi, R., and Chang, M. C. 1961. Fertilizable life of golden hamster ova and their morphological changes at the time of losing fertilizability. *J. Exp. Zool.* 148:185.
165. ———. 1963. Sperm ascent through the oviduct of the hamster and rabbit in relation to the time of ovulation. *J. Reprod. Fertil.* 6:413.
166. Zonartu, J. 1966. Effect of natural and synthetic sex steroids in cervical mucus, penetration and ascent of spermatozoa. In *Fifth World Congr. Fertil. Steril. Stockholm, June, 1966*, ed. A. Ingelman-Sundberg and B. Westin, Int. Congr. Series No. 109. Excerpta Medica Foundation.

6

Endocrinology of Oviductal Musculature

J. L. Boling
Department of Biology
Linfield College
Linfield Research Institute
McMinnville, Oregon

Attempts to unravel the endocrine basis for egg transport through the oviducts of mammals have been under way ever since Matsumoto and Macht (1919) studied the effect of crude corpus luteum and ovarian extracts on oviductal contractions in the pig. The general concept at present is that estrogens stimulate and progestins repress oviductal muscle contractility (Chang 1966*a*, *b;* Greenwald 1961, 1963, 1967; Harper 1964, 1965*a*, *b*, 1966). Both estrogen and progesterone are recognized as essential factors in normal egg transport (Austin 1949; Harper 1964, 1965*a*, *b*, 1966). However, the literature presenting these concepts contains many contradictory and confusing observations. It has been reported that injected estrogens prevent egg transport through the oviduct in intact rabbits and mice (Burdick and Pincus 1935). This is the phenomenon that was originally described as "tube locking." Later it was reported that large doses accelerated egg transport in the mouse (Burdick and Whitney 1937). Another report indicates that small doses of Depo Estradiol injected at the time of mating accelerated egg transport in rabbits, while large doses of the same hormone caused retention of the eggs in the oviduct for as long as 6 days (Greenwald 1961). Estradiol injections into spayed rabbits restored the rate of egg transport to that seen at estrus (Harper 1966). In another study (Adams 1958) in which progesterone was injected into recently spayed rabbits there was a delay in egg transport. Egg transport was greatly accelerated when intact rabbits were injected with progesterone for 3 days before induction of ovulation with gonadotropins (Chang 1966*a*). It has been suggested that increasing the progesterone level retards egg transport through the ampulla due to its depressing effect on muscular activity (Harper 1965*a*, 1966). Egg transport time appears to be quite normal if the ovaries are removed immediately after ovulation in the rabbit (Adams 1958) and rat (Alden 1942). However, the eggs were retained in the oviducts of mice which were ovariectomized 1 to 12 hr after finding a vaginal plug (Whitney and Burdick 1939). When donor eggs were placed into the ampullae of rabbits ovariectomized 1 month previously, egg transport was significantly accelerated (Noyes, Adams, and Walton 1959); the injection of 1 to 4 mg estradiol benzoate for 5 to 10 days resulted in the retention of eggs in the oviducts. The results of these and other investigations point out clearly the contradictory data that can be obtained by different procedures.

One of the basic problems that confronts the investigator who is studying the endocrinology of contractility is that physiological levels have not been established for the hormones. The injections of unphysiological doses may result in pharmacological reactions which are quite different from the normal.

Recent investigations with radiologically tagged estrogens in the rat have furnished excellent data on physiological levels for that species (Jensen and Jacobson 1960, 1962; Jensen *et al.* 1966). Such data are essential as a basis for experimental design in studying oviductal contractility and egg transport.

Furthermore, the possibility that muscle contractility may be initiated at the time of estrogen withdrawal has received little or no attention. It is conceivable that the reactions attributed to the various estrogens injected may be associated with their withdrawal rather than being the result of their direct action. The attainment of the necessary data for the clarification of the problem depends upon: (1) development of adequate techniques for observing muscle movements without introducing unknown and distorting environmental factors; (2) development of hormone assay methods which permit qualitative evaluations and quantitative determinations of hormones in the various body tissues and fluids; and (3) establishment of a clear relationship between hormone levels and specific events in the reproductive cycle. In this chapter attention will be focused on the endocrine influences that may control oviductal contractility during a very restricted period of the reproductive cycle, namely, just prior to, during, and immediately after ovulation. Also, various experimental procedures which have been designed to reproduce the normal patterns of contractility will be described. Egg transport will be discussed only in instances when it may offer clues to muscular activity.

I. Endogenous Hormones Present during the Period of Rapid Growth of Preovulatory Follicles

A knowledge of the quantities and types of hormones present at each stage of the estrous cycle is essential to an understanding of normal endocrine influences on oviductal musculature. Data on estrogen and progestin concentrations during the preovulatory phase of the cycle in various mammals will be presented in this section.

A. *Estrogens*

The literature contains many observations which indicate that estrogens are present in significant quantities in the body fluids and tissues prior to the time of ovulation (Allen, Hisaw, and Gardner 1939; Brown 1955; Brown, Klopper, and Loraine 1958; Everett 1961; Merrill 1958; Smith 1960). Inasmuch as this point is well established it will not be discussed further. There are, however, obvious needs for more quantitative data on estrogen levels correlated with other events in the estrous cycle.

B. *Progestins*

Originally investigators assumed that the corpus luteum hormone was not secreted until after ovulation. Evidence for the presence of progestins prior to ovulation has now been substantially documented (Everett 1961; Young 1961). This finding will be emphasized, since unpublished data to be presented later indicate a specific relationship between the secretion of progesterone and the increased muscular activity of the oviduct near the time of ovulation.

1. *Physiological Evidence for the Presence of Progestins prior to Ovulation.* The first indication that progestins might be secreted prior to ovulation came from experiments on the induction of mating behavior in the guinea pig (Dempsey, Hertz, and Young 1936). It was found that progesterone given at the proper time after estrogen injection was more effective in inducing normal mating behavior in ovariectomized guinea pigs than the injection of estrogen alone (Dempsey, Hertz, and Young 1936). This same hormonal relationship in the induction of mating behavior was demonstrated in the albino rat (Boling and Blandau 1939) and in the mouse (Ring 1944). At about this same time it was shown that injections of progesterone into estrogen-primed mice reduced the water content of cornual tissues. Astwood (1938, 1939) emphasized that this simulated the loss of cornual fluid during the period of preovulatory swelling of the follicle in the normal animal.

Further evidence that progesterone is secreted prior to the rupture of the follicle was obtained by means of mechanical and electrical recordings of the activity of the vaginal musculature of the living rat. If the muscular activity is recorded at intervals throughout the estrous cycle, it is clear that the contractions of the vaginal musculature are at a minimum near the onset of sexual receptivity. This phase of inactivity is followed by vigorous muscle contractions which appear during the time of rapid follicular growth (Boling and Burr 1941; Boling, unpublished data). Records of vaginal muscle contractions taken in ovariectomized rats show an erratic pattern (Boling and Job 1965). Repeated small injections of estrogen in the castrate cause the muscle to become quiescent and remain this way for as long as the closely spaced injections are continued. This period of estrogen-induced inactivity resembles that observed at the beginning of heat in the intact animal. Twenty-four to 36 hr after the last injection of estrogen, muscle contractions are observed again (Boling and Job 1965). These observations indicate that contractions of the vaginal musculature remain minimal as long as the estrogens are maintained at a physiological level. When the hormone concentration drops below this level, muscle activity is initiated. On the other hand, if injections of estrogen are continued and progesterone is added, the vaginal muscle becomes very active within 24 hr after the initial injection of progesterone (Boling and Job 1965). Progesterone injections in the estrogen-primed animal induce muscle contractility patterns, which are remarkably similar to those displayed in the normal animal near the time of ovulation. When estrogen is withdrawn the muscular activity increases, but the pattern of contractility does not resemble the normal as closely as that induced by progesterone.

2. *Hormone Assays Demonstrating the Presence of Progestins prior to Ovulation.* Progesterone, 20α-hydroxy-pregn-4-en-3-one (20α-OH) and various unidentified progestins have been recovered from ovarian venous and peripheral blood of the rabbit within minutes after mating or the injection of ovulation-inducing gonadotropins (Table 1), (Forbes 1953; Hilliard, Endröczi, and Sawyer 1961; Hilliard, Archibald, and Sawyer 1962; Hilliard, Hayward, and Sawyer 1964). The concentration of 20α-OH increases rapidly after mating (Fig. 1). The site of formation of progestins in ovaries during the preovulatory period has not been established. In ovaries in which the preovulatory follicles have been cauterized, gonadotropins nevertheless effect the release of progestins in amounts comparable to those recovered from normal ovaries (Hilliard, Archibald, and Sawyer 1962). Pregnane-3α:20α-diol was found in rabbit urine "immediately" after mating (Verly, Sommerville, and Marrian

TABLE 1

RECOVERY OF PROGESTINS PRIOR TO OVULATION

Species	Time of Appearance	Tissue	Method of Identification	Hormone	References
Cattle	Follicular phase	Follicular fluid Unruptured mature follicles	Paper chromatography	Progesterone	Edgar 1953
Man	Midcycle days, 15–17	Peripheral blood	Hooker-Forbes assay	Progestins	Forbes 1950
	Follicular phase	Unruptured mature follicles	Paper chromatography	Progesterone and 20α-OH	Zander *et al.* 1958
Monkey	Follicular phase peak near midcycle	Peripheral blood	Hooker-Forbes assay	Progestins	Bryans 1951
	Follicular phase peak near midcycle	Peripheral blood	Hooker-Forbes assay	Progestins	Forbes, Hooker, and Pfeiffer 1950
	Within 1–2 hr (G)[a]	Ovarian venous blood and ovarian tissue	Paper chromatography	Progesterone, 20α-OH and 17α-OH[b]	Hayward, Hilliard, and Sawyer 1963
Mouse (5 day)	Proestrus	Peripheral blood	Hooker-Forbes assay	Progestins	Guttenberg 1961
Pig	Follicular phase	Follicular fluid Unruptured mature follicles	Paper chromatography	Progesterone	Edgar 1953
Rabbit	Within 78 min (G) Within 116 min (M)[c]	Peripheral blood	Hooker-Forbes assay	Progestins	Forbes 1953
	Within 60 min (G) Within 25 min (M)	Ovarian venous blood	Paper chromatography	Progesterone and 20α-OH[d]	Hilliard, Endröczi, and Sawyer 1961; Hilliard, Archibald, and Sawyer 1962
	Within 30 min Through 10 hr period (M)	Ovarian venous blood	Paper chromatography	20α-OH	Hilliard, Hayward, and Sawyer 1964
Rat	Throughout cycle Highest in diestrus and proestrus	Ovarian venous blood	Paper chromatography	Progesterone and 20α-OH	Telegdy and Endröczi 1963
	Throughout cycle Highest in diestrus and proestrus	Ovarian tissue	Paper chromatography	Progesterone and 20α-OH	Telegdy 1963

[a] (G), time of appearance after injection of gonadotropins.

[b] 17α-hydroxyprogesterone.

[c] (M), time of appearance after mating.

[d] 20α-hydroxypregn-4 en-3-one.

1950; Verly 1951). Since the urine was collected during a 24 hr period and pooled, these data, though interesting, are not very helpful in giving specific information as to the hormones secreted before ovulation.

A significant rise in progestin concentration in peripheral blood has been detected during proestrus in CHI mice with 5-day cycles (Table 1) (Guttenberg 1961). Progesterone and 20α-OH have been isolated from both the ovaries and the ovarian venous blood of rats throughout the estrous cycle. The concentrations of both hormones were highest in diestrus and proestrus (Table 1) (Telegdy 1963; Telegdy and Endröczi 1963). Progesterone has been

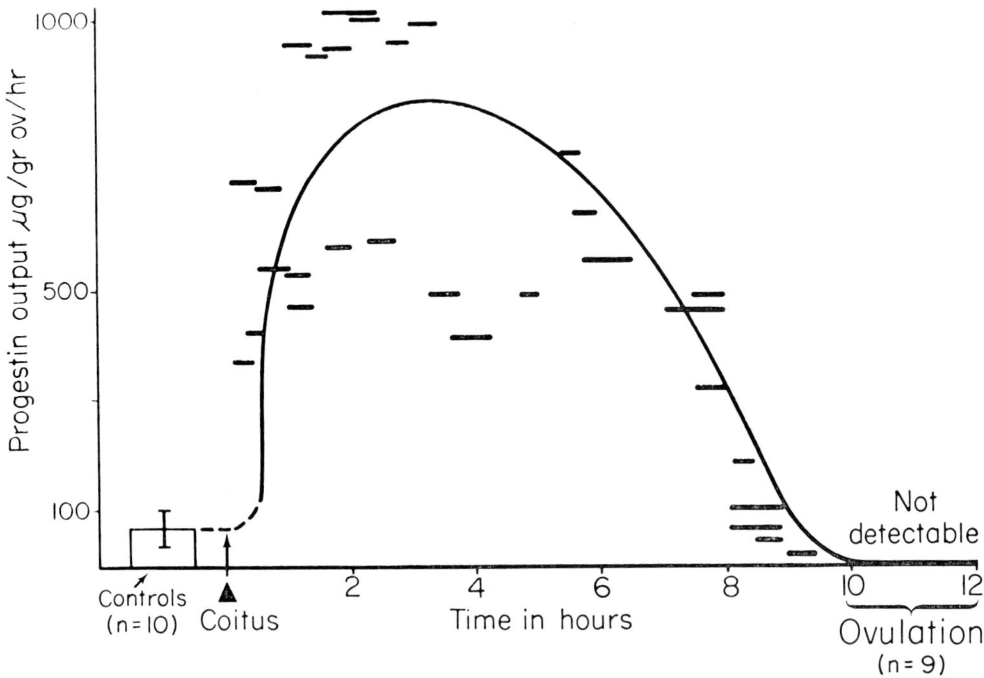

FIG. 1. Progestin (20α-hydroxypregn-4-en-3-one) levels in ovarian venous blood samples collected from 49 individual rabbits before and at varying intervals after coitus. Horizontal lines represent progestin levels obtained in different animals and the length of each line shows the duration of collection (15–60 min). The bars over "Controls" represent the mean level of 10 unmated animals ± standard error of mean. (From Hilliard, Hayward, and Sawyer 1964.)

recovered from large unruptured follicles with stigmas in cattle which had received injections of gonadotropins, as well as from the preovulatory follicles in pigs during the normal period of heat (Table 1) (Edgar 1953).

Progesterone, 17α-OH and 20α-OH were recovered and identified by chromatography from the ovarian venous blood of a monkey following gonadotropic stimulation. Progesterone was recovered also from the whole ovaries removed at the end of the experiment (Table 1) (Hayward, Hilliard, and Sawyer 1963).

Progestins were identified in peripheral blood by means of the Hooker-Forbes bioassay during midcycle in monkey and man (Table 1) (Bryans 1951; Forbes 1950; Forbes, Hooker, and Pfeiffer 1950). The level of progestins decreased immediately after midcycle only to increase again in the lutein phase. Progesterone and Δ^4–3–ketopregnene–20–ol were recovered from both the follicular fluid and the whole follicular wall of mature follicles obtained "one to two days prior to ovulation" in man (Table 1) (Zander *et al.* 1958). The α and β epimers of Δ^4–3–ketopregnene–20–ol were not identified separately.

It is clear from the observations mentioned above that various progestins are secreted during the preovulatory phase of the normal cycle in different animals. This is a significant point when considered in relation to the increase in oviductal contractions which occur at this time (Boling and Blandau, unpublished data; Seckinger 1923; Wislocki and Guttmacher 1924). The effect of progestins on the contractility of oviductal muscle needs to be studied by more carefully timed experiments in all species.

II. Cyclic Changes in the Morphology of the Oviductal Musculature

Brief descriptions of the general arrangement of the circular and longitudinal muscle in the oviducts of mammals are found in articles dealing with various species. Marshall's *Physiology of Reproduction* (Parkes 1966) contains descriptions of the oviducts of many forms. Gross oviductal anatomy and the muscle arrangement for several primates have been reviewed (Wislocki 1932). However there is a distressing lack of studies correlating changes in histological or submicroscopical structure to other events in the estrous cycle. In one such study it has been reported that the length of pig oviductal muscle cells at the time of estrus is nearly twice that at diestrus (Anapolsky 1928). Studies of possible cyclic variations in submicroscopic detail in the oviductal muscle cells are now practicable and should be done.

III. Patterns of Contractility of Oviductal Musculature during Normal Cycle

A summary of the investigations which have related oviductal muscular activity to the various stages of the estrous cycle in mammals is included in Table 2. Although investigators are not in complete agreement as to the pattern of muscular activity, the majority of the reports indicate that the most vigorous contractions occur at about the time of estrus and ovulation. It is agreed also that the vigor of the muscle activity decreases during diestrus. In contrast, there are some reports which indicate that the oviductal muscle is relatively inactive at estrus compared with other times of the cycle (Table 2) (Kok 1926; Whitelaw 1933).

One of the reasons for the confusion in interpretation of the data is that conditions in the ovary at the time of observation have not always been determined accurately. For example, if the stage of development of the follicles is accurately determined, interesting relationships between ovarian and oviductal activity can be demonstrated. This is shown by the work of Seckinger (1923) and Wislocki and Guttmacher (1924) in the pig. Direct quotations from these investigators are included in order to convey accurately the authors' observations. "The physiological change from the slow rhythmical contractions of inter-oestrus to the rapid undulating type of oestrus is coincident with the morphological changes occurring in the theca interna, both changes occurring in follicles that have reached a diameter of between seven and eight millimeters" (Seckinger 1923). "A change occurs in the tubal contractions when the follicles reach a diameter of 7 to 8 millimeters from the 'feeble' ones of the preceding period. The activity of the tubes increases sharply, at first becoming 'moderate,' then 'vigorous' and, finally, at the time of ovulation becoming 'very vigorous.' " (Wislocki and Guttmacher 1924).

These accurate observations made 40 years ago predict the relationship just now being confirmed between the onset of oviductal contractility and secretion of progestins from the ovary prior to ovulation (Boling and Blandau, unpublished data).

TABLE 2

Observations of Normal Activity of Oviductal Musculature

Species	Observation: Method	Observation: Time	Activity of Oviductal Musculature	References
Man	*In vivo*—uterotubal insufflation	At presumed ovulation	Uniform, high amplitude pressure fluctuations 1–3/min	Davids 1948
		Other times of cycle	Irregular, low amplitude pressure fluctuations 4–5/min	
	In vivo—uterotubal insufflation	10th–16th day of cycle	High amplitude pressure fluctuations ≃7/min	Rubin 1939
		Other times of cycle	Slow, lower amplitude pressure fluctuations	
	In vitro—kymography	Mid and late interval	Contractions 5–7/min, varied periodically in amplitude	Seckinger and Snyder 1926
		Other times of cycle	Uniform contractions 4–5/min	
Monkey (Rhesus)	*In vitro*—kymography	0–1 days postovulation	Contractions 8–13/min, varied periodically in amplitude	Seckinger and Corner 1923
		Interestrus	Uniform contractions 3–8/min	
Pig	*In vitro*—kymography and visual observation	0–3 days post ovulation	Nearly complete cessation of activity	Kok 1926
		Other times of cycle	Contractions 7–15/min of varying amplitude	
	In vitro—kymography	21st day and 1st–3d days of cycle	Contractions 13–15/min, varied periodically in amplitude	Seckinger 1923
		Interestrus	Regular contractions 4–6/min	
	In vitro—uterotubal insufflation	Estrus	Uniform, shallow pressure fluctuations ≃15/min	Whitelaw 1933
		3d–5th day of cycle	Stronger pressure fluctuations ≃1/min	
	In vitro—insufflation per infundibulum	Estrus Just before and after estrus	Irregular or fairly smooth pressure fluctuations 13–15/min	
	In vitro—visual observation	21st day and 1st–3d of cycle	Vigorous contractions	Wislocki and Guttmacher 1924
		Interestrus	Feeble to moderate contractions	
Rabbit	*In vivo*—cineradiography	Estrus	Strong contractions 14–18/min	Björk 1959
		Anestrus	Contractions 4–9/min of varying amplitude	
	In vitro—kymography (longitudinal muscle)	0–4 days postovulation	Regular, high amplitude contractions	Black and Asdell 1958
		After 4 days postovulation	Irregular contractions, lower frequency and amplitude	
	In vitro—kymography (circular muscle)	Near ovulation	Contractions ≃10/min	Black and Asdell 1959
		Other times in cycle	Contractions ≃10/min	

TABLE 2—*Continued*

Species	Observation		Activity of Oviductal Musculature	References
	Method	Time		
Rabbit—*Continued*	*In vivo*—intraluminal pressure	Estrus	Uniform, powerful contractions ≃12/min	Greenwald 1963
		By 3 days postovulation	Irregular contractions 3–4/min	
	In vivo—abdominal window	Few days before and after ovulation	Contractions 10–15/min	Mikulicz-Radecki 1925
		Other times of cycle	Contractions 4–6/min	
	In vivo—abdominal window	Estrus	Powerful contractions at 5–12 sec intervals	Westman 1926
		2d day postfertilization	Reduction in amplitude and frequency	
		Sexually resting	Weak, rhythmic contractions 6–25 sec intervals	

The various types and complexity of contractions of the oviductal musculature have been reviewed (Björk 1959; Harper 1961; Hartman 1932; Westman 1926). In analyzing the literature it is difficult to arrive at a clear picture of the patterns of oviductal contractility for any animal. The contractility patterns for the whole oviduct are exceedingly complex and the variations in types of motility described may be a result of examining them at different times in the cycle and under abnormal environmental conditions.

Peristaltic waves traveling for long distances along the oviduct are seldom seen. Localized peristaltic-like contractions originating in isolated segments or loops ordinarily travel only a short distance (Björk 1959; Black and Asdell 1958; Mikulicz-Radecki 1925). Contractions usually proceed in an abovarian direction, adovarian contractions occurring less frequently. Contradictory reports (Rubin and Bendick 1926; Wislocki and Guttmacher 1924) indicate that the predominant form of muscular movement is peristaltic. Other movements that have been described include segmentation contraction (Black and Asdell 1958), and a wormlike writhing of the entire oviduct (Rubin and Bendick 1926). The latter was attributed to the longitudinal muscle fibers.

Some variations in the character of motility between the isthmus and ampulla may be related to the differences in the contractions of the longitudinal or circular muscles. Kymographic records made on pig oviducts show that the longitudinal muscles of the ampullae undergo regular and mild contractions. In contrast, contractions of the isthmus are more vigorous and prolonged (Kok 1926). When contractions of circular muscle rings were observed the type of contractions in ampulla and isthmus were reversed. Slow contractions of high amplitude were recorded in the ampulla. The isthmic rings contracted more rapidly but less vigorously. These results have been confirmed in the rabbit (Black and Asdell 1958, 1959).

Westman (1926) made a comparison of the motility of the isthmus and ampulla immediately after ovulation in the rabbit. By 24 hr after ovulation the intensity and rate of contractions had decreased. The ampulla was less active than the isthmus at all times in the cycle.

IV. Contractility of Oviductal Musculature in the Castrate

All investigators agree that oviductal contractility is reduced in the castrate. However, the musculature does not become completely quiescent (Table 3). A distinct reduction in the amplitude, with no decrease in rate of contraction, is reported for the 27-day castrated rabbit (Greenwald 1963). A decrease in both amplitude and rate following castration in the rabbit has been reported also (Wimpfheimer and Feresten 1939). Ampullary egg transport has been observed in the rabbit 167 days after castration (Boling and Blandau, unpublished data) (Table 4).

V. Contractility of Oviductal Musculature as Influenced by Exogenous Hormones

A summary of experimental studies of the effect of various hormones on the contractility of oviductal musculature is presented in Table 3. Administration of estrogenic substances is followed by an increase in contractility of the oviduct, according to most reports. It should be emphasized that in the majority of the investigations a number of hours had elapsed between the last injection and the observations (Björk 1959; Geist, Salmon, and Mintz 1938*a;* Greenwald 1963; Rubin 1939; Wimpfheimer and Feresten 1939). It is possible, as was pointed out earlier, that these investigators were observing the contractility of the muscle during the period when estrogen was being withdrawn rather than at a time when the hormone was acting directly on the cells. Evidence is accumulating which shows that the pattern of contractility depends upon whether the muscle is under the influence of estrogen at physiological levels or whether the hormone is being withdrawn (Boling and Blandau, unpublished data). Many investigators do not give the exact time that the observations are made after the last injection of estrogen, so it is difficult to make comparisons between the data in the various reports. One investigator observed a very definite decrease in the amplitude of activity after estrogen administration (Ichijo 1960). Likewise, a decrease in activity and tonus of the human oviduct was observed after a large dose of folliculin was administered *in vitro* (Table 3) (Cella and Georgescu 1937).

A limited amount of work was done with extracts of the corpus luteum in the pig before crystalline progestins were isolated. An increase in the rate of contractions, using *in vitro* kymography was observed in the normal animal after administration of the crude extract, however, the time of the cycle at which the material was administered is not known (Matsumoto and Macht 1919).

The influence of progesterone on the activity of the oviduct has not been studied extensively. A decrease in the rate of contractions has been noted in the few cases in which it has been used (Björk 1959; Ichijo 1960).

A few experiments have been performed using estrogen and progestin synergistically. The procedure in all of these has been to administer the progestin to an animal that had been primed previously with an estrogen. Investigators are in agreement that with the administration of progesterone, the activity in the oviduct is reduced to a level below that recorded with estrogen alone (Geist, Salmon, and Mintz 1938*b;* Greenwald 1963; Wimpfheimer and Feresten 1939).

Recent preliminary experiments in rabbits indicate that progestins secreted prior to ovulation "trigger" an increase in oviductal muscular contractility (Boling and Blandau, unpublished data). The experimental procedures and observations supporting this thesis are summarized in Table 4. Rabbits were selected as experimental animals because they are reflex ovulators. Sexually mature females were isolated for at least 1 month. Under these conditions large follicles are continually forming (Hill and White 1934). The animals remain in heat and presumably all tissues of the reproductive tract are under estrogen domination. Such animals are designated as preovulatory in Table 4. Ovulation will occur in a majority of

TABLE 3

EXPERIMENTAL OBSERVATIONS OF OVIDUCTAL MUSCULATURE ACTIVITY

Species	Condition of Animal	Hormone Administration				Observation		Activity of Oviductal Musculature	References
		Hormone	Dose	Time of Admin.	Method of Admin.	Method	Time		
Man	Not specified	Folliculin	4,000 units	During observation	Added to physiological sol.	*In vitro*—kymography	Not specified	Activity and tonus reduced	Cella and Georgescu 1937
	Little or no ovarian function	....	....	....	....	*In vivo*—uterotubal insufflation	Not specified	Low muscle tone, irregular weak pressure fluctuations	Geist, Salmon, and Mintz 1938*a*
	Little or no ovarian function	Progynon-B	120,000–650,000 i.u. divided over a period of 10–14 days	After last preliminary insufflation	Injection	*In vivo*—uterotubal insufflation	0–2 days ALI[a] Progynon	High amplitude, rhythmic pressure fluctuations	Geist, Salmon, and Mintz 1938*a*
							7–16 days ALI Progynon	Irregular, weak, slow pressure fluctuations	
	Postmenopause	Progynon-B only	450,000–900,000 i.u. divided over 10 days	After preliminary record	Intramuscular	*In vivo*—uterotubal insufflation	ALI Progynon-B	Regular, rhythmic pressure fluctuations	Geist, Salmon, and Mintz 1938*b*
		Proluton (after Progynon-B)	20 mg in 4 cc oil	After record taken at end of Progynon-B treatment	Intramuscular	Same[b]	2 hr after proluton inj.	No pressure fluctuations	
	Prolonged secondary amenorrhea	Progynon	50,000–250,000 i.u.	Not specified	Injection	*In vivo*—uterotubal insufflation	2–3 days ALI Progynon	Increased activity and uterotubal tonicity from before treatment	Rubin 1939
Pig	Normal	Ovarian extract	2.0–3.5 cc	During observation	Placed in Tyrode's sol.	*In vitro*—kymography	Not specified	Increased contractions	Matsumoto and Macht 1919
		Corpus luteum extract	0.15 cc	Same	Same	Same	Same	Increased contractions	
	Normal	Corpus luteum extract	0.5–1.0 cc of a 1:5 sol.	During observation	Placed in Locke's sol.	*In vitro*—kymography	Estrus	Resembles interestrous pattern	Seckinger 1924
							Interestrus	Increase in amplitude, decrease in frequency from normal interestrus	

[a] ALI—after last injection.

[b] Same—indicates same method or time as in box immediately above.

TABLE 3—*Continued*

Species	Condition of Animal	Hormone Administration				Observation		Activity of Oviductal Musculature	References
		Hormone	Dose	Time of Admin.	Method of Admin.	Method	Time		
Rabbit	Castrate	H_2O-soluble estrogen	10 mg followed in 6 hr by 20 mg	14–20 days post castration	Intravenous	*In vivo*—cineradiography	During 24 hr AII[c] estrogen	Remained quiet as before estrogen treatment	Björk 1959
		Estradiol valerianate (Progynon Depot)	0.5 mg	14 days after H_2O soluble estrogen	Intramuscular	Same	1 week after injection	Contractions 14–18/min, similar to normal estrus	
	Normal	Progesterone	25 mg	Estrus	Intramuscular	*In vivo*—cineradiography	3 hr after injection	Decreased rate of contractions to 8–9/min	Björk 1959
			0.25 mg	Anestrus	Same	Same	Same	Decreased rate in 2 of 4 cases; no effect in 2 of 4 cases	
	Castrate	....	....	....	....	*In vivo*—intraluminal pressure	27 days postcastration	Reduced amplitude but same frequency as in normal estrus	Greenwald 1963
	Normal	ECP[d]	25 μg or 250 μg	"Immediately" after HCG to induce ovul.	Single injection	*In vivo*—intraluminal pressure	By 3 days after ovulation	Estrous pattern of contractility persists (uniform, powerful contractions ≃12/min)	Greenwald 1963
	Castrate	ECP	25 μg	2 weeks after castration	Same	Same	Same	Estrous pattern of contractility restored	
	Castrate	ECP	25 μg	2 weeks after castration	Single injection	*In vivo*—intraluminal pressure	1 day after prog. treatment	Similar to 3 days after ovulation (i.e., slower than in estrus)	Greenwald 1963
		Progesterone	5 mg	Beginning 3 days after ECP	4 daily 5 mg injections				
	Castrate	Estrogen (ovahormone benzoate susp.)	30 γ/kg/day	Beginning 3 weeks after castration	Injected for 7 consec. days	*In situ*—by O_2 insufflation	After estrogen treatment	High frequency and low amplitude of spontaneous pressure fluctuations	Ichijo 1960

[c] AII—after initial injection.

[d] ECP—estradiol cyclopentylpropionate.

TABLE 3—*Continued*

Species	Condition of Animal	Hormone Administration				Observation		Activity of Oviductal Musculature	References
		Hormone	Dose	Time of Admin.	Method of Admin.	Method	Time		
Rabbit —*Continued*	Castrate	Progesterone	1.0 mg/kg/day	Beginning 3 weeks after castration	Intramuscular for 4 consec. days	*In situ*—O_2 insufflation	"After prog. treatment"	Low frequency and high amplitude of spontaneous pressure fluctuations	Ichijo 1960
	Castrate	....	....	....	....	*In vivo*—uterotubal insufflation	Not specified	Reduced amplitude and frequency from animals in estrus and anestrus	Wimpfheimer and Feresten 1939
	Castrate	Theelin or Progynon-B or Dimenformon benzoate	800–1,200 i.u. / 800–1,200 i.u. / 10,000–13,000 i.u.	1–2 months postcastration	Injected in 4 daily doses	*In vivo*—uterotubal insufflation	Within 24 hr ALI estrogen	Increased contraction rate over castrate	Wimpfheimer and Feresten 1939
	Castrate	Estrogen as above and progesterone	Estrogen as above / 3.5 or 4.0 mg	Estrogen as above / Within 24 hr ALI estrogen	Estrogen as above / Injected in 4 daily doses	Same	Within 24 hr ALI Prog.	Reduced rate in 3 of 6 cases; no change from castrate in 3 of 6 cases	

such animals at approximately 10 to 11 hr after hormone stimulation (Harper 1961, 1963).

The injections of estrogen were made at intervals of 6 hr in an attempt to simulate normal conditions by keeping small amounts of the hormone present at all times. Studies using tritiated estradiol have shown that the hormone concentrations in tissues of the reproductive tract drop significantly between 6 and 16 hr after a single subcutaneous injection of estradiol or estrone in saline or in oil (Figs. 2, 3) (Jensen and Jacobson 1962). These data suggest that when injections of hormones are spaced at one day intervals there is danger of introducing estrogen withdrawal effects.

The data for the injection procedures and the observations are given in Table 4. The preovulatory animals which were assumed to be under endogenous estrogen domination were the only ones which were not subjected to injections or ovariectomy prior to the time of surgical preparation for the observations of oviductal contractility.

The data in Table 4 show that the average egg transport time in preovulatory animals was longer than in animals at the time of an induced ovulation. The contractions of the oviduct and mesotubarium were characteristically gentle. There were some variations in the rate

TABLE 4

NORMAL AND EXPERIMENTAL EGG TRANSPORT IN THE AMPULLA OF RABBIT OVIDUCTS

Condition of Animal	Number of Animals	Days between Ovariectomy and Observation (Range)	Hormone and Amount per Injection[a]	Hours between Injection and Observation	Time of Transport		
					Number of Eggs	Average	Range
Preovulatory[b]	5	....	....	....	19	10 min	3 min 30 sec—28 min
Preovulatory plus progesterone	2	....	Prog. 2 mg	12 AIIP[c] 4–7 ALIP[d]	8	5 min 43 sec	4 min 32 sec—10 min
Induced ovulation	6	....	Follutin 50 i.u.	10–13	46	6 min 12 sec	3 min 53 sec—11 min 55 sec
Castrate	4	10–167	....	....	10	19 min 28 sec	9 min 2 sec—29 min 30 sec
Castrate, estrone only	4	12–21	Estrone 6 μg	72–96 AIIE[e] 1–4 ALIE[f]	17	5 min 42 sec	2 min 12 sec—10 min 54 sec
Castrate, estrone withdrawal	3	13–22	Estrone 6 μg	96–98 AIIE 30–32 ALIE	13	3 min 10 sec	1 min 19 sec—5 min 35 sec
Castrate, estrone plus progesterone	3	14–18	Estrone 6 μg Prog. 2 mg	72–98 AIIE 2–4 ALIE 19–20 AIIP 2–4 ALIP	14	6 min 37 sec	3 min 5 sec—19 min 40 sec
Castrate, estradiol	1	14	Estradiol 6 μg	72 AIIE 3 ALIE	7	4 min 33 sec	4 min 20 sec—5 min 9 sec
Castrate, estradiol withdrawal	1	14	Estradiol 6 μg	96 AIIE 30 ALIE	2	34 sec	29–39 sec

[a] Injections made at intervals of 6 hr after the initial injection. All injections subcutaneous. Estradiol (Progynon Schering), saline suspension, 6 μg/cc. Estrone (Theelin, Parke Davis) saline suspension, 6 μg/cc. Progesterone (Proluton, Schering), sesame oil, 1 mg/cc or 2 mg/0.25 cc or progesterone (Lilly), sesame oil, 1 mg/cc.

[b] Preovulatory designates virgin females which have been isolated for at least 30 days with minimum of handling, 3.5–4.5 kg.

[c] AIIP—After initial injection progesterone.

[d] ALIP—After last injection progesterone.

[e] AIIE—After initial injection estrone or estradiol as indicated.

[f] ALIE—After last injection estrone or estradiol as indicated.

and pattern of contractility between animals. The ovaries are being sectioned for study to see if the stage of development of the follicles is associated with the amount of activity observed.

When the preovulatory females received injections of progesterone beginning 12 hr before observations, the ampullary egg transport time was similar to that observed near the time of an induced ovulation. The contractions of the mesotubarium and oviduct were vigorous, quite like those seen at the time of ovulation. The increase in contractility in the mesotubarium and oviduct was distinct. The evidence is very clear that the progesterone as

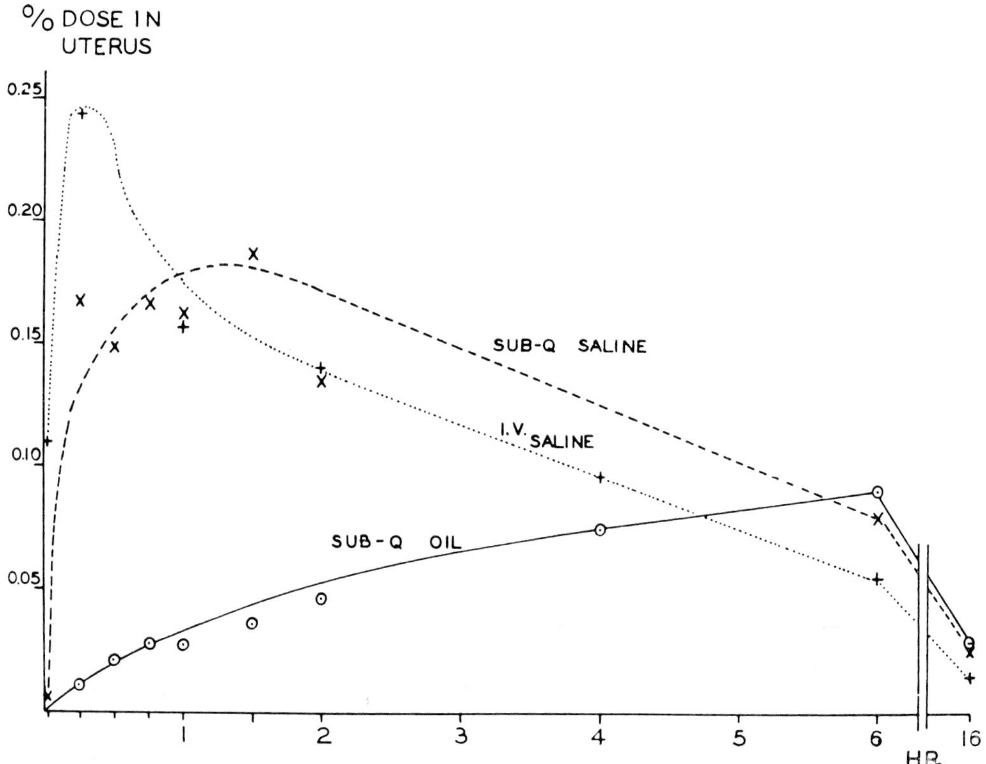

FIG. 2. Radioactivity in uterus of 23-day-old rat after single injection of approximately 0.1 μg of 6,7-tritiated estradiol. Points for intravenous (*I.V.*) injections in 0.5 ml saline are median values of 8 animals; points for subcutaneous (*sub-q*) injections in 0.5 ml saline or 0.2 ml sesame oil are median values for 6 animals. (From Jensen and Jacobson 1962.)

administered was followed by oviductal activity similar to that which occurs at the time of an induced ovulation.

Observations in the castrated animal indicated that egg transport does occur, but at a slower rate (Table 4). Contractions of the oviduct and mesotubarium are very gentle.

Ampullary egg transport times observed in a castrated animal 3 hr after the last estradiol injection were within the range of those receiving estrone. Oviductal muscle activity was characteristically gentle.

The egg transport time observed 30 hr after the last injection of an estradiol series was reduced to approximately $\frac{1}{10}$ of the average transport time observed at the time of induced ovulation. The contractions of the mesotubarium and oviduct were very vigorous.

The contractions of the mesotubarium and oviduct were characteristically gentle in experiments in which the muscle was under estrogen domination. On the other hand, the contractions were vigorous after estrogen withdrawal or progesterone injection in the estrogen-primed animal.

The data presented indicate: (1) that the oviductal muscle remains relatively quiet under the influence of estrogens continuously present in small (physiological) quantities; (2) that oviductal muscle, which has remained relatively quiescent under the influence of estrogens, begins to contract more vigorously when the estrogen is withdrawn; (3) that oviductal

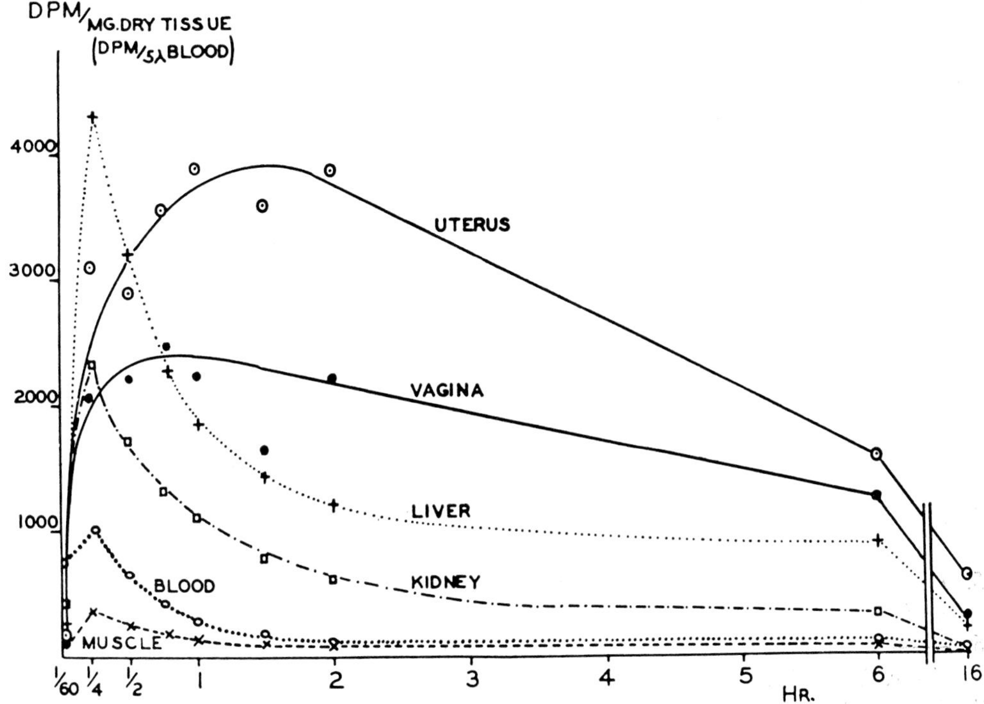

FIG. 3. Concentration of radioactivity in rat tissues after single subcutaneous injection of 0.098 μg (11.5 μc) of estradiol-6,7-H^3 in 0.5 ml saline. Liver and kidney points are mean values of 3 aliquots of dried pooled tissue; other points are median values of individual samples from 6 animals. Muscle is *M. Quadriceps femoris*. (From Jensen and Jacobson 1962.)

muscle, which is quiescent under the influence of an estrogen, begins to contract vigorously a few hours after the injection of progesterone. On the basis of these observations, it is concluded that the increased contractility in the oviduct observed near the time of ovulation in the normal animal is "triggered" in some manner by progestins.

VI. Conclusions

In reviewing the experiments dealing with oviductal contractility it is usually assumed that estrogens are directly and primarily responsible for increased muscle activity. In the normal preovulatory animal in which estrogens are indeed present at physiological levels, the oviducts are in a quiescent state. In attempting to restore the normal endocrine balance in the *castrate* it is necessary to avoid large doses which override normal effects and to maintain hormones at physiological levels. On the other hand, an understanding of the effect of large doses of hormones is important for practical applications.

Progesterone is known as an antagonist of estrogen in several of its actions. The synergistic action of estrogens and progestins in the induction of mating behavior is well established. The data discussed in this chapter indicate that near the time of ovulation they work together in the induction of muscle contractility. It is apparent that the interactions of estrogens and progestins need further study.

Acknowledgment

Unpublished observations referred to in this chapter were supported by grants from the National Institutes of Health, United States Public Health Service. This review could not have been made without considerable aid from my associates Mrs. Nancy Collins and Mr. Lynn Elwell and my students David Dodge, Stuart Markwell, and Sheryl Stafford.

The literature search for references was ended in August, 1967.

References

1. Adams, C. E. 1958. Egg development in the rabbit: The influence of post-coital ligation of the uterine tube and of ovariectomy. *J. Endocr.* 16:283.
2. Alden, R. H. 1942. Aspects of egg-ovary-oviduct relationship in the albino rat. I. Egg passage and development following ovariectomy. *J. Exp. Zool.* 90:159.
3. Allen, E.; Hisaw, F. L.; and Gardner, W. U. 1939. The endocrine functions of the ovaries. In *Sex and internal secretions*, ed. E. Allen, chap. 8. Baltimore: Williams & Wilkins Co.
4. Anapolsky, D. 1928. Cyclic changes in the size of muscle fibers of the Fallopian tube of the sow. *Amer. J. Anat.* 40:459.
5. Astwood, E. B. 1938. A six-hour assay for the quantitative determination of estrogen. *Endocrinology* 23:25.
6. ———. 1939. Changes in weight and water content of the uterus of the normal adult rat. *Amer. J. Physiol.* 126:162.
7. Austin, C. R. 1949. Fertilization and the transport of gametes in the pseudopregnant rabbit. *J. Endocr.* 6:63.
8. Björk, L. 1959. Cineradiographic studies on the Fallopian tubes in rabbits. *Acta Radiol.* Suppl. 176:1.
9. Black, D. L., and Asdell, S. A. 1958. Transport through the rabbit oviduct. *Amer. J. Physiol.* 192:63.
10. ———. 1959. Mechanisms controlling entry of ova into rabbit uterus. *Amer. J. Physiol.* 197:1275.
11. Boling, J. L., and Blandau, R. J. 1939. The estrogen-progesterone induction of mating responses in the spayed female rat. *Endocrinology* 25:359.
12. Boling, J. L., and Burr, H. S. 1941. Factors associated with vaginal electric correlates of the estrous cycle of the albino rat. *Anat. Rec.* 79 (suppl. no. 2):9.
13. Boling, J. L., and Job, D. D. 1965. Studies of the influence of estrogens and progesterone on abdominovaginal electropotential differences and muscular activity of the vagina in the albino rat. *Anat. Rec.* 151:326.
14. Brown, J. B. 1955. Urinary excretion of oestrogens during the menstrual cycle. *Lancet* 268:320.

15. Brown, J. B.; Klopper, A.; and Loraine, J. A. 1958. The urinary excretion of oestrogens, pregnanediol and gonadotrophins during the menstrual cycle. *J. Endocr.* 17:401.
16. Bryans, F. E. 1951. Progesterone of the blood in the menstrual cycle of the monkey. *Endocrinology* 48:733.
17. Burdick, H. O., and Pincus, G. 1935. The effect of oestrin injections upon the developing ova of mice and rabbits. *Amer. J. Physiol.* 111:201.
18. Burdick, H. O., and Whitney, R. 1937. Acceleration of the rate of passage of fertilized ova through the Fallopian tubes of mice by massive injections of an estrogenic substance. *Endocrinology* 21:637.
19. Cella, C., and Georgescu, I. D. 1937. Experimentelle Untersuchungen über die Physiologie und Pharmakodynamik des Eileiters. *Arch. Gynaek.* 165:36.
20. Chang, M. C. 1966*a*. Effects of oral administration of medroxyprogesterone acetate and ethinyl estradiol on the transportation and development of rabbit eggs. *Endocrinology* 79:939.
21. ———. 1966*b*. Transport of eggs from the Fallopian tube to the uterus as a function of oestrogen. *Nature* 212:1048.
22. Davids, A. M. 1948. Fallopian tubal motility in relation to the menstrual cycle. *Amer. J. Obstet. Gynec.* 56:655.
23. Dempsey, E. W.; Hertz, R.; and Young, W. C. 1936. The experimental induction of oestrous (sexual receptivity) in the normal and ovariectomized guinea pig. *Amer. J. Physiol.* 116:201.
24. Edgar, D. G. 1953. The progesterone content of body fluids and tissues. *J. Endocr.* 10:54.
25. Everett, J. W. 1961. The mammalian female reproductive cycle and its controlling mechanisms. In *Sex and internal secretions*, ed. W. C. Young, vol. 1, chap. 8. Baltimore: Williams & Wilkins Co.
26. Forbes, T. R. 1950. Systemic plasma progesterone levels during the human menstrual cycle. *Amer. J. Obstet. Gynec.* 60:180.
27. ———. 1953. Pre-ovulatory progesterone in the peripheral blood of the rabbit. *Endocrinology* 53:79.
28. Forbes, T. R.; Hooker, C. W.; and Pfeiffer, C. A. 1950. Plasma progesterone levels and the menstrual cycle of the monkey. *Proc. Soc. Exp. Biol. Med.* 73:177.
29. Geist, S. H.; Salmon, U. J.; and Mintz, M. 1938*a*. The effect of estrogenic hormone upon the contractility of the Fallopian tube. *Amer. J. Obstet. Gynec.* 36:67.
30. ———. 1938*b*. Effect of progesterone on Fallopian tube contractility. *Proc. Soc. Exp. Biol. Med.* 38:783.
31. Greenwald, G. S. 1961. A study of the transport of ova through the rabbit oviduct. *Fertil. Steril.* 12:80.
32. ———. 1963. *In vivo* recording of intraluminal pressure changes in the rabbit oviduct. *Fertil. Steril.* 14:666.
33. ———. 1967. Species differences in egg transport in response to exogenous estrogen. *Anat. Rec.* 157:163.
34. Guttenberg, I. 1961. Plasma levels of "free" progestin during the estrous cycle of the mouse. *Endocrinology* 68:1006.
35. Harper, M. J. K. 1961. The time of ovulation in the rabbit following the injection of luteinizing hormone. *J. Endocr.* 22:147.
36. ———. 1963. Ovulation in the rabbit: The time of follicular rupture and expulsion of the eggs, in relation to injection of luteinizing hormone. *J. Endocr.* 26:307.

37. ———. 1964. The effects of constant doses of oestrogen and progesterone on the transport of artificial eggs through the reproductive tract of ovariectomized rabbits. *J. Endocr.* 30:1.
38. ———. 1965*a*. Transport of eggs in cumulus through the ampulla of the rabbit oviduct in relation to day of pseudopregnancy. *Endocrinology* 77:114.
39. ———. 1965*b*. The effects of decreasing doses of oestrogen and increasing doses of progesterone on the transport of artificial eggs through the reproductive tract of ovariectomized rabbits. *J. Endocr.* 31:217.
40. ———. 1966. Hormonal control of transport of eggs in cumulus through the ampulla of the rabbit oviduct. *Endocrinology* 78:568.
41. Hartman, C. G. 1932. Ovulation and the transport and viability of ova and sperm in the female genital tract. In *Sex and internal secretions*, ed. E. Allen, chap. 14. Baltimore: Williams & Wilkins Co.
42. Hayward, J. N.; Hilliard, J.; and Sawyer, C. H. 1963. Preovulatory and postovulatory progestins in monkey ovary and ovarian vein blood. *Proc. Soc. Exp. Biol. Med.* 113:256.
43. Hill, M., and White, W. E. 1934. The growth and regression of follicles in the oestrous rabbit. *J. Physiol.* 80:174.
44. Hilliard, J.; Archibald, D.; and Sawyer, C. H. 1962. Gonadotropic activation of preovulatory synthesis and release of progestin in the rabbit. *Endocrinology* 72:59.
45. Hilliard, J.; Endröczi, E.; and Sawyer, C. H. 1961. Stimulation of progestin release from rabbit ovary *in vivo*. *Proc. Soc. Exp. Biol. Med.* 108:154.
46. Hilliard, J.; Hayward, J. N.; and Sawyer, C. H. 1964. Postcoital patterns of secretion of pituitary gonadotropin and ovarian progestin in the rabbit. *Endocrinology* 75:957.
47. Ichijo, M. 1960. Studies on the motile function of the Fallopian tube. I. Analytic studies on the motile function of the Fallopian tube. *Tohoku J. Exp. Med.* 72:211.
48. Jensen, E. V., and Jacobson, H. I. 1960. Fate of steroid estrogens in target tissues. In *Biological activities in relation to cancer*, ed. V. Pincus. New York: Academic Press.
49. ———. 1962. Basic guides to the mechanism of estrogen action. *Recent Prog. Horm. Res.* 18:387.
50. Jensen, E. V.; Jacobson, H. I.; Flesher, J. W.; Saha, N. N.; Gupta, G. N.; Smith, S.; Colucci, V.; Shiplacoff, D.; Neumann, H. G.; DeSombre, E. R.; and Jungblut, P. W. 1966. Estrogen receptors in target tissue. In *Steroid dynamics*, ed. E. G. Pincus *et al.* New York: Academic Press.
51. Kok, F. 1926. Bewegungen des muskulösen Rohres der Fallopischen Tube. *Arch. Gynaek.* 127:384.
52. Matsumoto, S., and Macht, D. I. 1919. A pharmacological study of ovarian and corpus luteum extracts, with a special reference to the contractions of the genito-urinary organs. *J. Urol.* 3:63.
53. Merrill, R. C. 1958. Estriol: A review. *Physiol. Rev.* 38:463.
54. Mikulicz-Radecki, F. v. 1925. Zur Physiologie der Tube. I. Mitteilung. Experimentelle Studien über die Spontanbewegungen der Kaninchentube *in situ*. *Zbl. Gynaek.* 49:1655.
55. Noyes, R. W.; Adams, C. E.; and Walton, A. 1959. The transport of ova in relation to the dosage of oestrogen in ovariectomized rabbits. *J. Endocr.* 18:108.
56. Parkes, A. S., ed. 1966. *Marshall's physiology of reproduction*, vol. 2. Boston: Little, Brown & Company.
57. Ring, J. R. 1944. The estrogen-progesterone induction of sexual receptivity in the spayed female mouse. *Endocrinology* 34:269.

58. Rubin, I. C. 1939. The influence of hormonal activity of the ovaries upon the character of tubal contractions as determined by uterine insufflation. *Amer. J. Obstet. Gynec.* 37:394.
59. Rubin, I. C., and Bendick, A. J. 1926. Fluoroscopic visualization of tubal peristalsis in women. *J. Amer. Med. Ass.* 87:657.
60. Seckinger, D. L. 1923. Spontaneous contractions of the Fallopian tube of the domestic pig with reference to the oestrous cycle. *Bull. Johns Hopkins Hosp.* 34:236.
61. ———. 1924. The effect of ovarian extracts upon the spontaneous contractions of the Fallopian tube of the domestic pig with reference to the oestrous cycle. *Amer. J. Physiol.* 70:538.
62. Seckinger, D. L., and Corner, G. W. 1923. Cyclic variations in the spontaneous contractions of the Fallopian tube of *Macacus rhesus*. *Anat. Rec.* 26:299.
63. Seckinger, D. L., and Snyder, F. F. 1926. Cyclic changes in the spontaneous contractions of the human Fallopian tube. *Bull. Johns Hopkins Hosp.* 39:371.
64. Smith, O. W. 1960. Estrogens in the ovarian fluids of normally menstruating women. *Endocrinology* 67:698.
65. Telegdy, G. 1963. Die Veränderungen des Progesteron- und Δ^4-Pregnenol-(20α)-on-(3)-Gehaltes im Ovargewebe von Ratten während des Östruszyklus. *Endokrinologie* 44:29.
66. Telegdy, G., and Endröczi, E. 1963. The ovarian secretion of progesterone and 20α-hydroxypregn–4–en–3–one in rats during the estrous cycle. *Steroids* 2:119.
67. Verly, W. G. 1951. The urinary excretion of pregnane-3α:20α-diol in the female rabbit immediately following mating. *J. Endocr.* 7:258.
68. Verly, W. G.; Sommerville, I. F.; and Marrian, G. F. 1950. The quantitative determination and identification of pregnane-3α:20α-diol in the urine of the pregnant rabbit. *Biochem. J.* 46:186.
69. Westman, A. 1926. A contribution to the question of the transit of the ovum from ovary to uterus in rabbits. *Acta Obstet. Gynec. Scand.* 5 (suppl. no. 3):1.
70. Whitelaw, M. J. 1933. Tubal contractions in relation to the estrous cycle as determined by uterotubal insufflation. *Amer. J. Obstet. Gynec.* 25:475.
71. Whitney, R., and Burdick, H. O. 1939. Effect of massive doses of an estrogen on ova transport in ovariectomized mice. *Endocrinology* 24:45.
72. Wimpfheimer, S., and Feresten, M. 1939. The effect of castration on tubal contractions of the rabbit, as determined by the Rubin test. *Endocrinology* 25:91.
73. Wislocki, G. B. 1932. On the female reproductive tract of the gorilla with a comparison of that of other primates. *Contr. Embryol. Carnegie Inst.* 23:163.
74. Wislocki, G. B., and Guttmacher, A. F. 1924. Spontaneous peristalsis of the excised whole uterus and Fallopian tubes of the sow with reference to the ovulation cycle. *Bull. Johns Hopkins Hosp.* 35:246.
75. Young, W. C. 1961. The mammalian ovary. In *Sex and internal secretions*, ed. W. C. Young, vol. 1, chap. 7. Baltimore: Williams & Wilkins Co.
76. Zander, J.; Forbes, T. R.; von Münstermann, A. M.; and Neher, R. 1958. Δ^4-keto-pregnene-20α-ol and Δ^4-3-ketopregnene-20β-ol, two naturally occurring metabolites of progesterone: Isolation, identification, biologic activity and concentration in human tissues. *J. Clin. Endocr. Metab.* 18:337.

7

Endocrinology of Oviductal Secretions

G. S. Greenwald

Departments of Obstetrics, Gynecology, and Anatomy
University of Kansas Medical Center, Kansas City

Secretion is defined by *Webster's International Dictionary* as "material separated, elaborated, and discharged by a cell or cells, especially (in animals) by the epithelial cells of glands." This definition, which at first glance seems unduly broad, reflects the current biological usage of the term. Various aspects of oviductal secretion will be considered in other sections of this book and of necessity this chapter will duplicate some of this work. The major aim of this article is to concentrate on the cyclicity of oviductal changes which are regulated by hormones. Specifically, this chapter will include a survey of the pertinent literature and my own work on the mucin layer of the rabbit egg.

I. Endocrine Control of Oviductal Secretion

A. *Mouse*

A comprehensive account of the microanatomy of the oviduct of the mouse is provided by Agduhr (1927). Secretory changes in the oviductal epithelium are apparent even during the fetal and early postnatal stages in the form of bulging or filiform processes jutting into the lumen. It is possible that these early changes represent a response to maternal hormone levels. Agduhr confirmed the observations of Allen (1922), who noted cyclic changes in the oviductal epithelium of the mouse. During proestrus and estrus the nuclei are arranged in a uniform row. However, during metestrus some of the nuclei begin to migrate to the free ends of the cells and are then extruded. These cells constitute the "peg" cells which have been described by numerous authors. The height of the epithelium decreases during metestrus and early diestrus and the cytoplasm becomes highly vacuolated (Allen 1922).

Secretory granules have been observed in both the ciliated and nonciliated cells of the oviduct of the mouse (Espinasse 1935). Some of the nonciliated cells gradually bulge and herniate through the ciliated surfaces. Subsequently, the nucleus is compressed and moves toward the lumen. In contrast to the previous authors, Espinasse believed that the cells do not become completely detached but eventually move back to their original position in the epithelium.

The cytoplasmic contents of the ruptured cells may be of physiological significance. Mouse eggs cannot be cultured as zygotes, but from the 2-cell stage on, they will mature *in vitro* into viable blastocysts. It has been suggested that the PAS-positive peg cells—which are most conspicuous at estrus—may somehow provide a factor which is essential for early embryonic development (Biggers, p. 397 in Wolstenholme and O'Connor 1965). It is of interest that the culturing of mouse oviducts containing fertilized ova for 4 days results in the differentiation of normal blastocysts (Gwatkin and Biggers 1963).

The uptake of macromolecular substances produced by the adult and transferred to the embryo has been studied in the mouse by use of the fluorescent antibody technique (Glass 1963). The ampullary epithelium gives an intense reaction at about the time of ovulation; the fluorescence is unrelated to the major cell types. Epithelial fluorescence decreases as pregnancy progresses. There is appreciable transfer of serum antigens to 2-cell embryos and little or none to 8-cell and older oviductal stages. A subsequent study revealed that animals of different ages have different patterns of heterosynthetic transfer, suggesting the influence of sex steroids (Glass and McClure 1965). Moreover, preferential regions of oviductal transfer exist, with the ampulla uniformly exhibiting a much more intense response than the isthmus.

The *in vivo* incorporation of ^{35}S methionine by oviductal epithelium and ova of the mouse has been determined (Greenwald and Everett 1959; Weitlauf and Greenwald 1965). The oviductal epithelium, throughout the period of egg transport, shows a pronounced response, as judged by radioautography. The oviductal eggs, however, show negligible uptake of ^{35}S methionine, whereas the blastocysts *in utero* incorporate ^{35}S methionine to a significant degree, indicating a flurry of protein synthesis at this critical stage of embryogenesis. Blastocysts transferred into the oviducts of mice on day 2 of pregnancy *do* incorporate methionine, whereas the recipient's own ova show only limited uptake (Weitlauf and Greenwald 1967). This indicates that maturational changes in the mouse egg are of greater significance in triggering protein synthesis than environmental differences between oviduct and uterus.

Some insight into the endocrine regulation of oviductal secretion may be provided by studying the effects of early bilateral ovariectomy on viability of ova. Following castration of mice on the day of mating or during the next 2 days, blastocysts can be recovered 30 to 46 days later in approximately one-third of all animals (Smithberg and Runner 1960). Conversely, in intact mice, blastocysts retained in the oviduct by a ligature for 7 days appear morphologically normal (Bloch 1952). The intriguing findings of Kirby (1965) reveal that transfer of oviductal stages of mouse eggs to an extrauterine site leads to differentiation of trophoblasts only whereas uterine blastocysts similarly transplanted develop into embryos.

Mouse eggs *in vitro* are capable of normal cleavage and development into viable blastocysts (for references see Brinster 1965). It should be obvious that development *in vitro* is not proof of the independence of the egg from the influence of the oviduct but merely indicates that optimal conditions have been produced which may simulate the normal oviductal environment.

B. *Rat*

Descriptions of the oviduct of the rat are given by Alden (1942*a*) and Kellogg (1945). The third division, which includes the isthmus proper, is highly secretory with mucoid granules in both the epithelium and lumen (Kellogg 1945). The mucoid secretion begins to accumulate in the 2-week-old animal (Kellogg 1945).

During estrus in the rat, two types of cells are present—ciliated and secretory. At this stage, numerous long processes protrude from the secretory cells into the lumen (Borell *et al.* 1959). As in most species in which a closed periovarial sac exists, estrus in the rat is accompanied by a fluid-filled distension of the ampulla. This is preceded during proestrus by a hyperemic response of the ampulla, suggesting that the fluid contents represent oviductal secretion rather than appreciable contributions of follicular or periovarial fluid. The pH of the dilated ampulla is 8.04 (Blandau, Jensen, and Rumery 1958); cyclic variations in oviductal pH have not been evaluated.

The peg cells which are characteristic of the oviduct of the mouse are rarely encountered in the rat (Alden 1942*a*). In fact, cyclic changes cannot be demonstrated clearly in the rat oviduct by either histologic (Alden 1942*a*) or histochemical (Deane 1952) techniques.

The cells of the fimbriated end contain lipid droplets, and in the isthmus the cell surfaces are positive for alkaline phosphatase and periodic acid–Schiff reactions (Deane 1952). Alkaline phosphatase activity is intense in the epithelium of the fimbriae and ampulla and weak in the isthmus (Augustin and Moser 1955). The activity is maximal during estrus and metestrus and minimal at diestrus. These histochemical findings agree with biochemical determinations for hydrolytic enzymes. Thus, alkaline phosphomonesterases are greater in the oviduct of the rat during proestrus and estrus. Following ovariectomy, the levels decline and they can be restored to normal values by the administration of estrogen (Robboy and Kahn 1964).

Following ovariectomy and ligation of the oviduct on the day of mating, rat eggs develop at the normal rate and can be recovered from the ampulla as blastocysts on day 4 (Alden 1942*b*). Hence, in the rat, ovarian steroids do not appear to be necessary for preimplantation embryogenesis.

C. *Rabbit*

Oviductal secretions have been more thoroughly investigated in the rabbit than in any other species. This can be attributed to the ease of dating the time of ovulation, the relatively straight, uncoiled nature of the oviduct, and the large quantities of fluid which can be recovered from the ligated oviduct.

The oviductal epithelium of the rabbit consists of two principal cell types: ciliated and secretory. It is generally believed that the ciliated cells do not undergo any structural changes during the reproductive cycle (Borell *et al.* 1956; Hashimoto *et al.* 1959*b*). A controversy which dates from the initial studies of the oviduct is whether the ciliated and secretory cells represent stages in the life cycle of a single cell; this viewpoint is represented by a classic paper of Moreaux (1913). The transformation theory appears to have been refuted by recent work which demonstrated that two discrete populations of cells exist in the epithelium (Fredricsson 1959*a*). The number of secretory cells which regenerate in the epithelium of the castrated rabbit is dependent on the dose of estrogen administered.

The secretory cells of the ampulla and the isthmus differ both morphologically and biochemically. The rod-shaped ampullary cells react positively to both the PAS technique and Mayer's mucicarmine stain, but the secretion released into the lumen does not adhere to eggs—unlike the secretory product of the isthmic secretory cells (Greenwald, unpublished). In the fimbriated region of the rabbit oviduct, nonciliated nucleated cells, which lack secretory droplets, escape into the lumen in both the pre- and postovulatory stages (Hashimoto *et al.* 1959*a*).

In the intact anestrous rabbit, the nonciliated cells contain only a few secretory granules (Borell *et al.* 1956). However, during estrus the secretory cells bulge into the lumen; the cytoplasm sometimes appears constricted and ready to be released into the lumen. This suggests an apocrine secretion before ovulation (Hashimoto *et al.* 1959*a*).

Following coitus, finger-like protrusions extend from the secretory cells far beyond the free surfaces of the ciliated cells (Borell *et al.* 1956). After ovulation a change occurs in the secretory granules. At this time the granules are most numerous (Nilsson and Rutberg 1960; Hashimoto *et al.* 1959*a*) but stain less intensely.

At the time of ovulation in the rabbit, the ciliated cells show a decrease in supranuclear glycogen and there is decreased alkaline phosphatase activity at the secretory cell surfaces (Fredricsson 1959*c*). PAS-reactive and diastase-resistant material is the most characteristic product discharged after ovulation in the rabbit (Fredricsson 1959*c*). This substance is probably an acid sulphuric mucopolysaccharide. Under the influence of estrogen, large quantities of radioactive sulphate accumulate in the oviductal epithelium of ovariectomized rabbits (Zachariae 1958). Cyclic variations in the uptake and distribution of $Na_2{}^{35}SO_4$ have been noted also in intact rabbits. Thus, 24 hr after ovulation high activity is concentrated in the epithelium, whereas 36 hr later an intense reaction is localized in the lumen (Friz 1959). Similarly, the oviductal secretion contains sulfur which, at the time of fertilization, adheres to the zona pellucida of the cleaving egg (Gothié and Moricard 1955).

It is apparent that the epithelium of the ovariectomized rabbit responds very rapidly to estrogen by renewing its growth and secretory activity. This is demonstrated also by the study of Flerkó (1954) as well as by the numerous observations on rates of secretion in catheterized oviducts.

The modern era of investigations on secretory rates was initiated by Bishop (1956), who utilized ligated oviducts of anesthetized rabbits. The secretory rates were definitely influenced by the hormonal status of the animal. Secretion was high during estrus and low during late pregnancy. The secretion was stimulated by the administration of pilocarpine, which indicates that it is an active process.

A subsequent method for the monitoring of volumetric collections of oviduct secretion in unanesthetized rabbits was devised by Clewe and Mastroianni (1960). Seventy hr after castration, secretion fell to 1/3 of the estrous values. After estradiol administration precastration levels were restored. Further, the injection of intact, estrous rabbits with progesterone caused a sharp drop in the secretion rate (Mastroianni *et al.* 1961). Another study indicated that 3 days after mating, the secretion rate fell to 50% of the estrous levels (Mastroianni and Wallach 1961).

Cyclic variations in the biochemical constituents of oviductal fluid have been demonstrated also in the rabbit. Exogenous progesterone results in a twofold increase of glycine and serine content in oviductal fluid (Gregoire, Gongsakdi, and Rakoff 1961). It was suggested that these amino acids may be utilized by the embryo directly during its preimplantation phase of development. Progesterone administration also increased inositol content in oviductal fluid (Gregoire, Gongsakdi, and Rakoff 1962). It was suggested that inositol may serve as a reserve carbohydrate. During the first 3 days of pregnancy, the concentration of lactic acid is significantly increased in the rabbit oviduct (Mastroianni and Wallach 1961).

The pH of oviductal fluid of estrous rabbits is 7.9 and the content of bicarbonate is approximately twice that of plasma, indicating an active transport system (Vishwakarma 1962). These findings were confirmed and extended by Hamner and Williams (1965), who postulated that the bicarbonate content may be involved in the capacitation of sperm.

The metabolism of endometrium and oviduct has been compared recently for the estrous and pseudopregnant rabbit (Mounib and Chang 1965). In contrast to the definite trend for increased oxygen uptake by the endometrium, in the Fallopian tube the consumption of oxygen did not vary between estrus and days 1 to 19 of pseudopregnancy.

The cumulative evidence presented in this section indicates that striking differences exist in the pattern of oviductal secretions in the rabbit which are correlated with phases of follicular or luteal dominance. Whether these changes are of functional significance is, of course, the critical question. Westman (1930), among others, ovariectomized pregnant rabbits shortly after ovulation and observed that the oviductal eggs soon became abnormal. He concluded that "the secretory activity of tubal mucous membrane is of decided importance for the life of the ovum."

An interesting difference exists in the ability to culture rabbit and mouse eggs. Unlike those of the mouse, rabbit ova can be fertilized *in vitro* (Chang 1959; Bedford and Chang 1962). However, rabbit eggs fertilized *in vivo* and then grown *in vitro* fail to develop past the morula stage (see appendix II, Austin 1961). This contrasts with the situation in the mouse.

It is noteworthy that rabbit blastocysts retained in a ligated oviduct degenerate at about 84 hr after mating (Adams 1958). Thus, the rabbit oviduct which appears to be essential for development during the first 3 days of embryonic life is an inhospitable environment thereafter.

D. *Man and Rhesus Monkey*

The oviductal epithelium in the human female is at its maximal height during the follicular stage. The postovulatory period is characterized by an epithelium not more than 20 μ in height, with long processes projecting from the nonciliated cells (Snyder 1924). The ciliated cells retain their integrity throughout the cycle, although changes in height and width do occur (Novak and Everett 1928). Novak and Everett noted the presence of peg cells—wedged between others—which they believed represented a phase of the secretory cells. Extrusion of nuclei and cytoplasm was observed in the premenstrual oviduct. The nuclear volume of the oviductal epithelial cells shows a steady increase throughout the menstrual cycle with a peak at the end of the secretory phase on days 25–27. There is a significant difference between the average nuclear volumes of the proliferative and secretory stages (Lehto 1963).

In the newborn infant, the differentiation into ciliated and nonciliated cells is already apparent (Stegner 1962). In the postovulatory phase, the Golgi complex is accentuated and located in an infranuclear position (Stegner 1962).

The ampullary secretory cells have the ultrastructural characteristics of both protein and carbohydrate secreting complexes (Björkman and Fredricsson 1962). Another electron microscope study indicated significant pinocytosis in both secretory and ciliated cells during the follicular phase but not during the luteal phase (Fredricsson and Björkman 1962). Furthermore, the appearance of the endoplasmic reticulum varies between the follicular phase, when the membranes are studded with ribosomes, and the luteal phase, when there are numerous free ribosomes (Fredricsson and Björkman 1962).

In man, on the basis of electron microscopy, there seems to be more secretory activity during the follicular than luteal stage.

Extensive histochemical studies have been carried out on the oviduct of man (for references see Fredricsson 1959*b*). A PAS-positive secretion which is diastase resistant is found on the surface of the secretory cells and in slight amounts within the lumen at about the time of ovulation, the maximal secretion occurring on days 22 to 24 (Fredricsson 1959*b*). The latter period of luminal secretion coincides with a peak of endometrial activity. It has been postulated that alpha-amylase, which is secreted by the oviduct, plays a vital role in the degradation of uterine glycogen to glucose (Green 1957).

The pattern of secretion of fluid in the oviduct of the rhesus monkey parallels the findings in the rabbit. Maximal fluid production usually occurs within 1 day of the height of vaginal cornification. It appears likely that the secretion is the result of the action of estrogen (Mastroianni, Shah, and Abdul-Karim 1961). No increase in fluid secretion was observed during anovulatory cycles. Most of the protein components of the oviductal fluid of the rhesus monkey closely parallel in percentage ranges the concentrations of these components in the animal's serum. It appears that in this species oviductal fluid represents a combination of secretion by epithelial cells and a transudation from the vascular system (Marcus and Saravis 1965).

E. *Pig, Cow, and Sheep*

In the estrous pig, the oviductal epithelium is about 25 μ in height. On day 12 after ovulation, the epithelium is reduced to 10 μ and the most conspicuous feature is a set of long finger-like processes protruding from the nonciliated cells. Pregnancy leads to the persistence of the low epithelium and long cytoplasmic projections (Snyder 1923). There is no demonstrable carbonic anhydrase in the oviduct of the pig (Lutwak-Mann 1955).

In the bovine oviduct, the ciliated cells are fairly constant throughout the cycle. During the follicular phase, the apical portions of the secretory cells bulge into the lumen and empty the contents of the granules into the oviductal lumen. The cells have abundant RNA, sparse lipid and PAS-positive—diastase-resistant material. The process of secretion appears to be only slightly influenced by the phases of the cycle (Björkman and Fredricsson 1961). The secretory cells have long and slender protrusions during the follicular phase, but shorter and blunter processes after the formation of the corpus luteum; the cells contain no alkaline phosphatase (Björkman and Fredricsson 1960). Alkaline phosphatase activity is present at the surfaces of the ciliated cells.

The composition of oviductal luminal fluids in the cow has been analyzed and there is no real evidence for cyclic changes (Olds and VanDemark 1957). The fimbriated portion of the oviduct contains carbonic anhydrase which is especially enhanced following ovulation (Lutwak-Mann 1955).

A description of the histology of the sheep oviduct is provided by Hadek (1955). Maximal cell height occurs during estrus and metestrus. PAS-positive granules are present during proestrus, estrus, and metestrus but are absent in anestrus and late diestrus. At the beginning of diestrus, cytoplasmic projections appear to become detached and lie free in the oviductal lumen. Cyclic variations occur in the amount of acid mucopolysaccharides and alkaline phosphatase in the secretory cells. At estrus and metestrus, the PAS-positive material appears in the lumen and simultaneously the number of granules decreases in the epithelium.

The fluid volume of the ovine oviduct is low during the luteal phase of the cycle and begins to rise at the onset of estrus. Maximum amounts of fluid are secreted on the day after the onset of estrus and the values thereafter decline rapidly (Perkins *et al.* 1965). A chemical analysis of the fluid from the sheep oviduct indicates that estrogen increases the secretion of potassium, bicarbonate, and lactate, whereas progesterone decreases lactate secretion. The lactate could serve as a substrate for sperm metabolism (Restall and Wales 1966*a*). It is of interest that the oviductal fluids harvested at different stages of the estrous cycle uniformly depress the respiratory activity of ram spermatozoa (Restall and Wales 1966*b*). This differs from the response observed in the cow and rabbit.

Cyclical changes occur in the pH of the sheep's oviduct. The lowest values occur during diestrus (6–6.4); the highest during estrus and metestrus (6.8–7.0) (Hadek 1953). The levels of carbonic anhydrase in the ovine oviduct do not correlate with the stages of the cycle; in fact, 3 weeks after ovariectomy the values are unchanged. The ampulla is inactive in comparison to the isthmus (Lutwak-Mann and Averill 1954).

II. Secretion of Mucin by the Rabbit Oviduct

The preceding portion of this article has surveyed the literature on endocrine regulation of oviductal secretions. The purpose of this section is to summarize observations on the secretion of mucin by the rabbit oviduct. As previously mentioned, the reproductive tract of the rabbit is rich in secretory granules which are PAS positive and diastase resistant. In the case of the oviduct, these compounds are deposited around the ovum as a tertiary membrane known as the mucin layer. The mucin layer consists primarily of a mucopolysaccharide (Braden 1952; Bacsich and Hamilton 1954).

Various theories have been proposed to account for the extensive deposition of mucin around the rabbit egg. Gregory (1930) believed that the mucin layer was responsible for the *expansion* of the rabbit blastocyst by releasing its water content. Assheton (1895) assigned a purely *protective* role to the mucin layer. According to this theory, the rigid mucin layer prevented the rupture of the rapidly expanding blastocyst. On the other hand, Hartman (1925) proposed that the coating was a primitive, nonfunctional membrane commonly found in marsupials.

My interest in the possible functional significance of the mucin layer was aroused by the finding that after the administration of estrogen the thickness of the layer around oviductal eggs was considerably reduced (Greenwald 1957). By 96 hr after mating dead eggs were recovered *in utero* and it was presumed that the deficient mucin layer accounted for the embryonic mortality.

The difference in thickness of the mucin layer between control and estrogen-treated rabbits is first apparent by day 2 *post coitum* and is most striking by day 3 (Fig. 1*a*, *b*). Despite the reduced deposition of mucin after injection of estrogen, the eggs cleave normally and develop into morulae (Fig. 1*a*, *b*).

It has been suggested that the thickness of the mucin layer depends on the time spent in the mucin-rich regions of the oviduct (Noyes, Adams, and Walton 1959). Thus, acceleration of egg transport would reduce the length of time that eggs would be exposed to the secretory cells of the oviduct and presumably result in a diminished mucin layer. Conversely, a delay in egg transport would prolong the period of contact between ova and epithelium and hence increase the thickness of its mucin layer. Arguing against this theory is the observation that high doses of estrogen lead to prolonged retention of rabbit ova at the ampullary-isthmic junction (tube locking) (Greenwald 1961, 1963). Under these circumstances, the trapped ova have a considerably diminished coating of mucin, although they lie in a region of secretory cells which are laden with PAS-positive granules.

In view of the preceding observations, attempts were made to determine which hormones are responsible for the regulation of the secretion of mucin in the oviductal epithelium of the rabbit (Greenwald 1958). Originally, Mayer's mucicarmine stain was used to identify the mucigenous cells; more recently, a number of the observations have been repeated and verified by the use of the periodic acid-Schiff procedure (Greenwald, unpublished).

As a baseline for the experimental studies, it is necessary to describe the normal sequence of events in the oviductal epithelium of untreated New Zealand giant rabbits. During estrus, two kinds of epithelial cells are present: ciliated cells and nonciliated mucigenous cells (Fig. 2*a*). As one approaches the isthmus, the mucin cells become increasingly abundant. The most conspicuous feature at estrus is that the mucin granules are confined to the cells and none are found in the lumen.

Twenty-four hr after coitus, the mucin secretion begins to be extruded into the lumen or on the surface of the secretory cells (Fig. 2*b*). The mucin appears to stain darker with mucicarmine, indicating a possible change in chemical composition. Figure 2*c* shows the layer of mucin investing a 2-cell egg. At 48 hr there is considerable variation in the amount of mucin discharged without any evidence for replenishment. There is a correlation between the amount of mucin found upon the ova and stored in the secretory cells. The ciliated cells become increasingly evident at 3 or more days after coitus, due to the reduced content of mucin

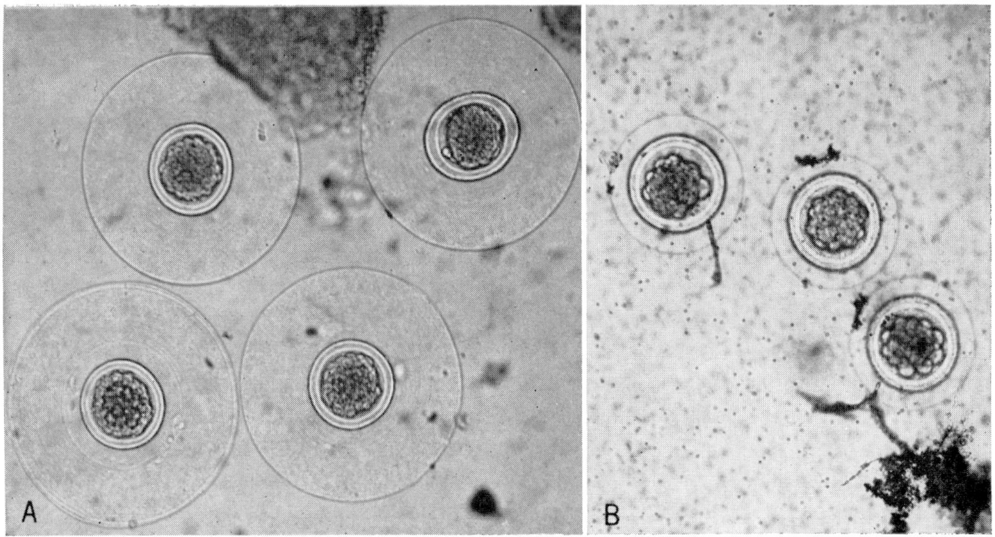

FIG. 1. (*a*) Normal rabbit egg, 72 hr *p.c.* (*b*) Eggs at 72 hr *p.c.* after injection of 5 μg estradiol benzoate at 24 and 48 hr *p.c.* (From Greenwald 1957.)

in the secretory cells (Fig. 2*d*). By the 10th day only a fringe of old mucin clings to the surface of the epithelium. New granules of mucin appear in the apical portions of the secretory cells (Fig. 2*e*). The secretory cycle is completed between the 15th and the 30th days and the cells once more contain their normal complement of mucin (Fig. 2*f*).

The next step was to determine the effect of exogenous estrogen on mucin secretion. Little change can be detected in the oviductal epithelium 24 hr after the injection of 5 μg of estradiol benzoate (injected immediately after mating). In the experimental animals the secretory cells are still loaded with granules 3 days later, in contrast to the depleted appearance in the control group. Thus, the injection of estrogen largely inhibits the release of mucin, which accounts for the meager coating on the oviductal eggs.

Collectively, the sequence of events described above suggests that following coitus mucin is released through the action of progesterone but that the injection of estrogen inhibits the secretory discharge. What regulates the synthesis of mucin within the cell? The following experiments were designed to shed light on this problem.

Rabbits were ovariectomized and left untreated for 2 weeks, and then injected intra-

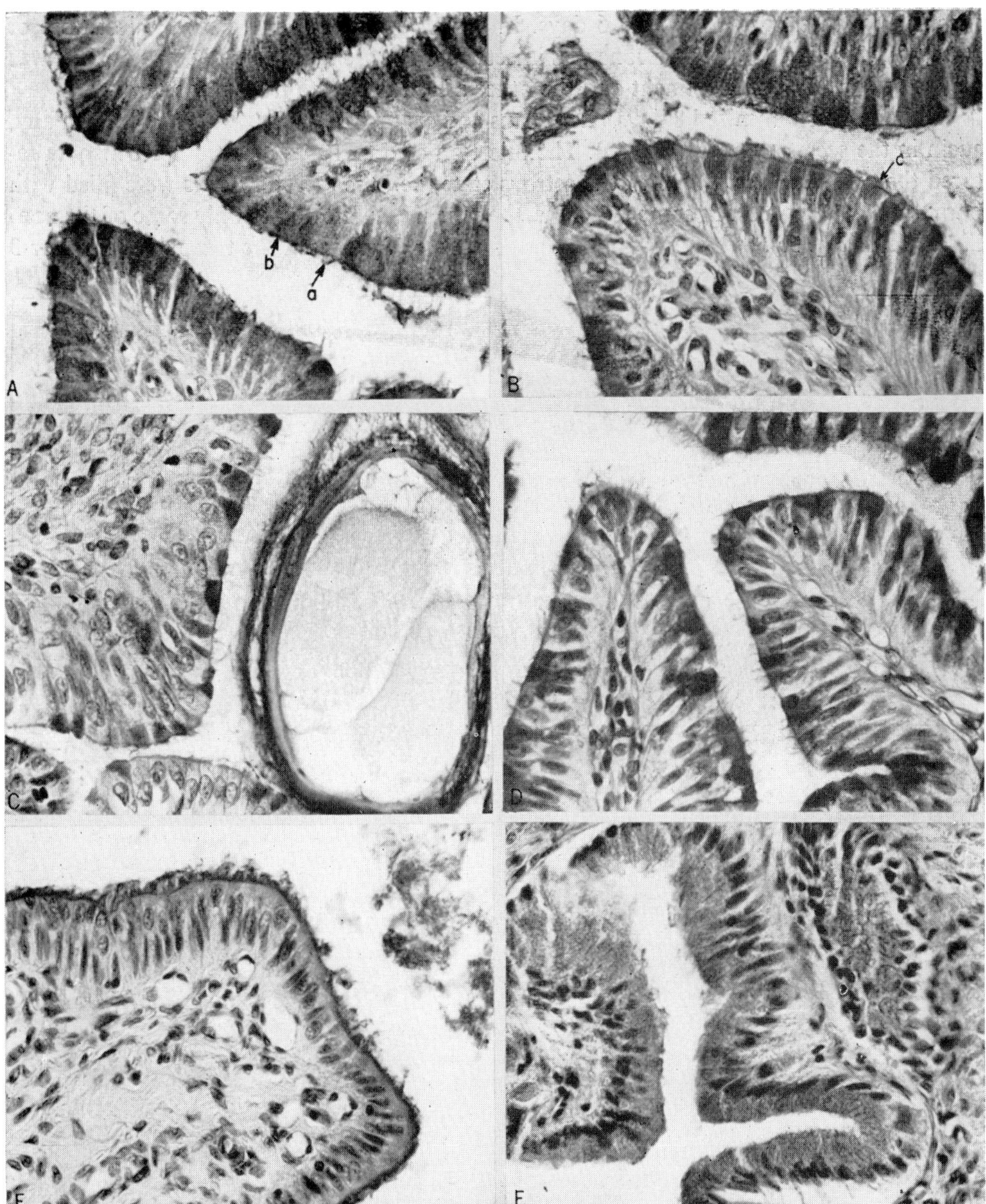

FIG. 2. (*a*) Estrous oviductal epithelium. Note ciliated nonmucin cells (a) and mucin stored in secretory cells (b) (×169). (*b*) Oviductal epithelium, 1 day *p.c.* Note mucin discharged from secretory cells (c) (×169). (*c*) Ampullary epithelium, 1 day *p.c.* The 2-cell egg is surrounded by mucin identical with that of epithelium (×169). (*d*) Oviductal epithelium, 3 days *p.c.* Some secretory cells are depleted of mucin (×169). (*e*) Oviductal epithelium, 10 days *p.c.* Mucin discharged; beginning of new cycle of synthesis and storage (×169). (*f*) Oviductal epithelium, 23 days *p.c.* Mucin restored to secretory cells (×169). (From Greenwald 1958.)

muscularly with estradiol for 5 or 6 days. In some rabbits, the day after the last dose of estradiol, 2.5 to 5 mg of progesterone was injected for 3 more days. Usually, one oviduct was removed after estradiol treatment and the other after the injection of progesterone.

Rabbits ovariectomized 8 days after coitus and examined 14 days later had little, if any, mucin in the secretory cells (Fig. 3*a*). The injection of 5 to 10 μg of estradiol benzoate restored the oviductal epithelium to the estrous state, i.e., the secretory cells were filled with coarse granules (Fig. 3*b*). In some instances, the mucin was restored to only some of the nonciliated cells (Fig. 3*c*). Subsequent injections of 2.5 mg of progesterone twice daily for 2 or 3

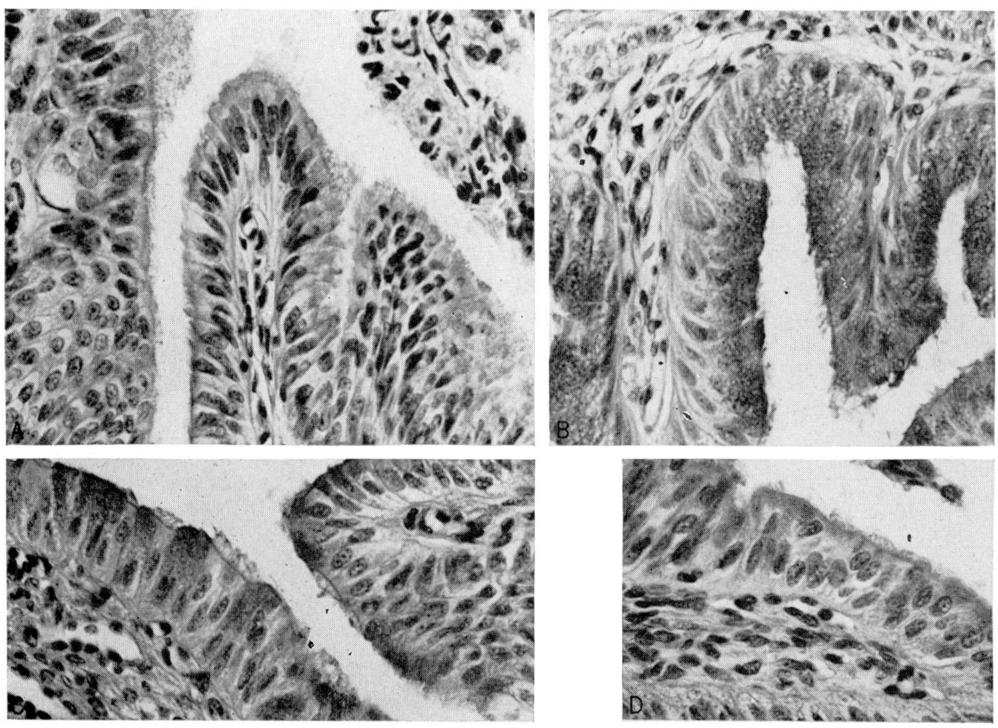

FIG. 3. (*a*) Oviductal epithelium 14 days after ovariectomy. Right oviduct of animal 6 (×185). (*b*) Oviductal epithelium after 5 μg of estradiol benzoate for 5 days. Left oviduct of animal 6 (×185). (*c*) Oviductal epithelium of castrate after 5 μg of estradiol benzoate for 5 days. Mucin restored to only some of the nonciliated cells (×185). (*d*) Oviductal epithelium of castrate after treatment with estrogen followed by progesterone. Mucin released from epithelium (×185). (From Greenwald 1958.)

days depleted the epithelium of mucin (Fig. 3*d*). In summary, estrogen is essential for the synthesis of mucin and progesterone effects its release.

Spheres of pure mucin, measuring up to 500 μ in diameter, are occasionally found in the oviducts of rabbits. These solid globules represent accretions around sloughed epithelial cells or other detritus. It seemed likely that mucin may coat any foreign object introduced into the oviducts after the proper hormonal sequence of estrogen and progesterone.

To test this possibility the following experiment was carried out: 10 rabbits were ovariectomized and injected 2 weeks later with 5 μg of estradiol benzoate; the injections were continued daily for the next 5 to 7 days. Glass beads were then introduced into the oviducts via either the ostium tubae abdominale or the uterotubal junction. The animals were divided into two groups; 3 rabbits continued to receive 5 μg of estradiol for three days (n = 3); the remaining 7 animals were injected with 5 mg of progesterone. On flushing the oviducts it was observed that mucin was deposited on the glass beads only when the ovariectomized rabbits were treated first with estradiol and then with progesterone (Fig. 4*b*). When 0.05% aqueous toluidine blue was added to the beads a metachromatic reaction occurred, indicating the acid mucopolysaccharide nature of the mucin.

As mentioned previously, the thickness of the mucin layer on oviductal eggs varied in different animals. In view of the above experiments, it appeared plausible to attribute this variability to differences in the level of circulating progesterone. When 2.5 mg of progesterone were injected twice daily for 3 days following coitus, there was a considerable increase in the thickness of the mucin envelope (Fig. 4*a*, *c*). As anticipated, the oviductal epithelium of such animals shows more depletion of mucin than that of untreated rabbits of the same postovulatory age (Fig. 4*d*).

There remains the question, What is the functional significance of the mucin layer? It

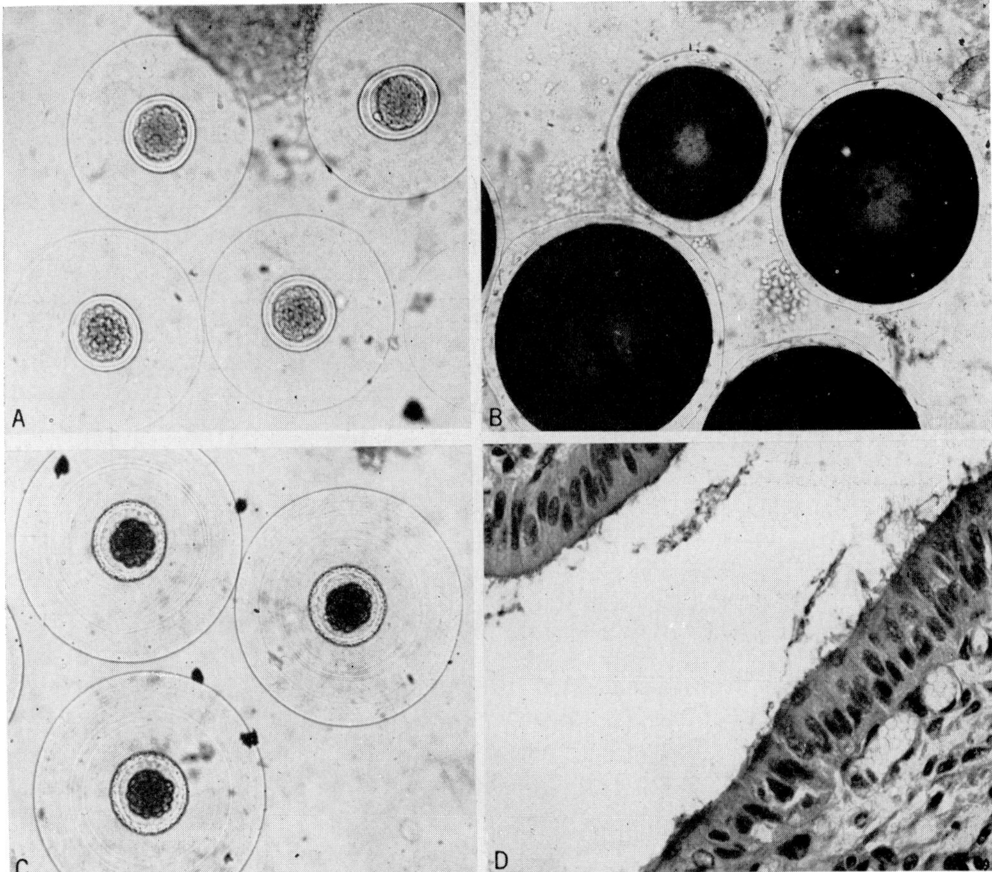

FIG. 4. (*a*) Normal rabbit eggs at 3 days *p.c.* Mucin volume is 0.03 mm³; mean diameter of eggs including mucin layer is 411 μ (×32). (*b*) Glass beads covered with mucin after insertion into castrate treated with estrogen followed by progesterone (×20). (*c*) Rabbit eggs at 3 days *p.c.*, after 2.5 mg progesterone twice daily for 3 days. Mucin volume is 0.06 mm³; mean diameter is 500 μ (× 32). (*d*) Oviductal epithelium of animal whose eggs are shown in C. Epithelium is depleted of mucin (×190). (From Greenwald 1958.)

became apparent that the administration of estrogen to rabbits in the first few days after mating altered the rate of egg transport, thus exerting a deleterious effect on the course of pregnancy. Small doses of exogenous estrogen accelerated egg transport, whereas large doses caused the eggs to be retained in the oviducts for longer periods than normal (Greenwald 1961, 1963). The effects of a greatly reduced layer of mucin on embryonic survival could not be evaluated because of the overriding disturbances in the role of egg transport.

This problem was circumvented by recovering 3-day-old rabbit embryos from animals which had been treated with estrogen and transplanting them to the cornua of animals that were 3 days pseudopregnant (Greenwald 1962). The experimental design is illustrated in

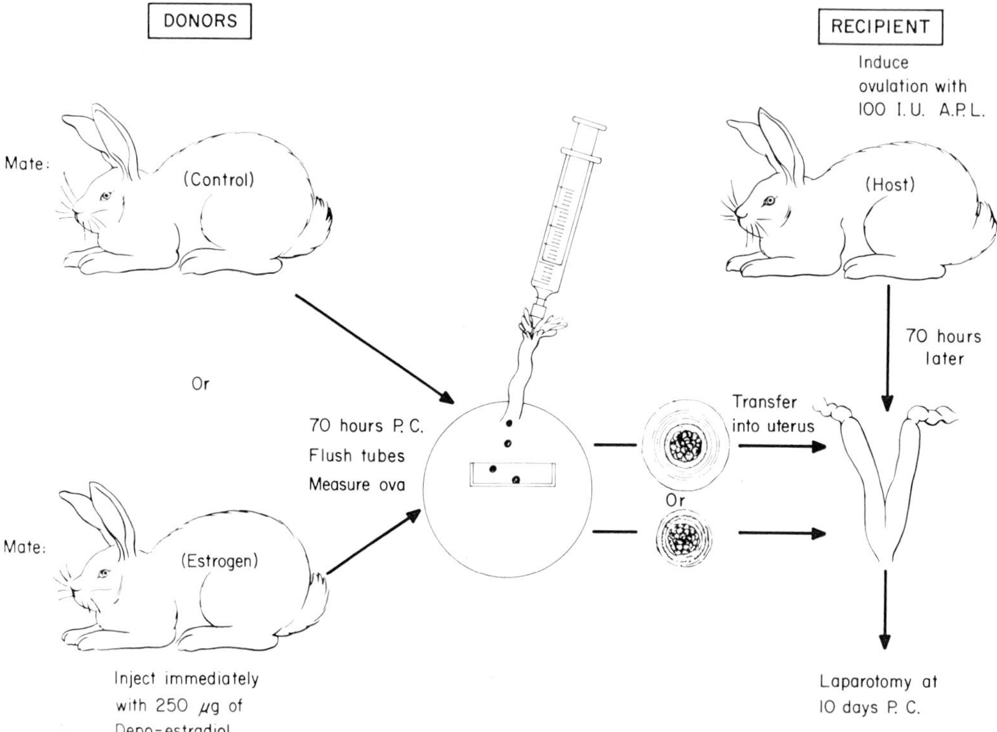

FIG. 5. Design of the transplantation experiments. Depo-estradiol is the trade name for estradiol cyclopentylpropionate (ECP). (From Greenwald 1962.)

Figure 5. The donors were either injected intramuscularly with 250 μg of estradiol cyclopentylpropionate (ECP) or left untreated. Ovulation was induced in the recipients by the intravenous injection of 100 i.u. of human chorionic gonadotropin. At 70 hr *p.c.* the donors were killed by the intravenous injection of nembutal. The oviducts were removed and flushed with 5 ml of freshly prepared normal saline to which penicillin (20,000 i.u. per ml) was added. The eggs were examined at a magnification of 25$\times$ with a dissecting microscope and the diameter of each, including the mucin layer, was measured with an ocular micrometer. It was determined previously that in this strain of rabbits the diameter of the eggs to the outer limit of the zona pellucida was relatively constant at 172.0 μ. By subtracting this figure from the total diameter, the diameter of the mucin layer was determined; the thickness of the mucin layer was calculated by dividing the two.

The ova were transferred into the cornua of the recipient by a braking pipette. A laparotomy was performed 1 week later and the number of implantation sites counted and compared with the number of ova that had been transferred. Seventeen transfers were made of normal ova and 40 transfers of ova with a reduced layer of mucin. Ova collected from control donors had a thickness of mucin ranging from 65 to 129 μ, compared with 12 to 43 μ in the ova recovered from the estrogen-treated series. Of the 148 normal eggs transferred at 3 days *p.c.*, 102 were implanted 1 week later (or 68.9%). The success of implantation had no relationship to the thickness of the mucin layer. Five females were examined at term or shortly before delivery. Twenty-three of the original 31 implantations contained living young.

Of 332 transplanted ova with a reduced coating of mucin, 97 of them (or 29.2%) implanted successfully. The relationship between the thickness of mucin on eggs subjected to estrogen and successful implantation is shown in Table 1. Transplanted ova with a mucin

TABLE 1

RELATIONSHIP BETWEEN THE THICKNESS OF MUCIN OF TRANSFERRED ECP[a]-TREATED OVA AND SUCCESSFUL IMPLANTATION

Total Diameter of Ova at Transfer (μ)	Mucin Thickness (μ)	Number of Females (Recipients)	Number of Pregnant Females	Number of Eggs Transferred	Number Implanted	Per Cent
258	43	3	2	24	9	37.5
236	32	5	4	49	19	38.8
215	22	10	7	75	30	40.0
195	12	11	11	89	39	37.1
195	12	11	0	95	0	....
Total	....	40	24	332	97	29.2

SOURCE: Greenwald 1962.
[a] ECP—estradiol cyclopentylpropionate.

thickness which varied from 22 to 43 μ had a similar implantation rate. When ova with the minimal mucin coating (12 μ) were transplanted, only 11 out of 22 recipients became pregnant. Despite this heavy loss of ova, in one transfer 3 normal living young were born at 34 days *p.c.*

In another experiment, recipients were killed 1 to 3 days after the transfer of normal eggs or eggs with reduced mucin and the blastocysts flushed from the uterus. The diameters of the blastocysts were measured with an ocular micrometer and compared with those recovered from untreated animals killed on days 4, 5, or 6 *post coitum* (Fig. 6). The results indicate that at 24 hr after transfer, the diameters of normal ova fell within the size range of the control blastocysts (i.e., nontransferred group) while the reduced mucin ova were slightly smaller. However, by day 5, there was no significant difference in the size of the blastocysts of the three groups.

By 6 days *p.c.*, there was considerable variation in the size of the control blastocysts. In one female the blastocysts' size varied from 2.3 to 3.6 mm in diameter and in another from 1.5 to 2.6 mm. Ova transferred with a normal or diminished layer of mucin reached the same size as the control eggs by day 6 (Fig. 6). Of 47 normal ova that were transferred, 35 (74%) were recovered 3 days later while out of 52 reduced mucin ova, 41 (79%) were retrieved. Thus, the recovery rate at day 6 was essentially the same for both groups.

On the 6th day, 61% of the transferred control ova recovered were large and viable blastocysts. Only 30.7% of the blastocysts from the reduced mucin series appeared normal. The remaining eggs with the reduced mucin layer were degenerating at day 6. Apparently they expanded to the size of normal 5-day blastocysts and then collapsed.

It should be stressed that some of the blastocysts developing from the reduced mucin ova were as well developed as control embryos, although most had degenerated by day 6. For example, the ova of one female at the time of transfer had a mucin thickness of only 12 μ, but

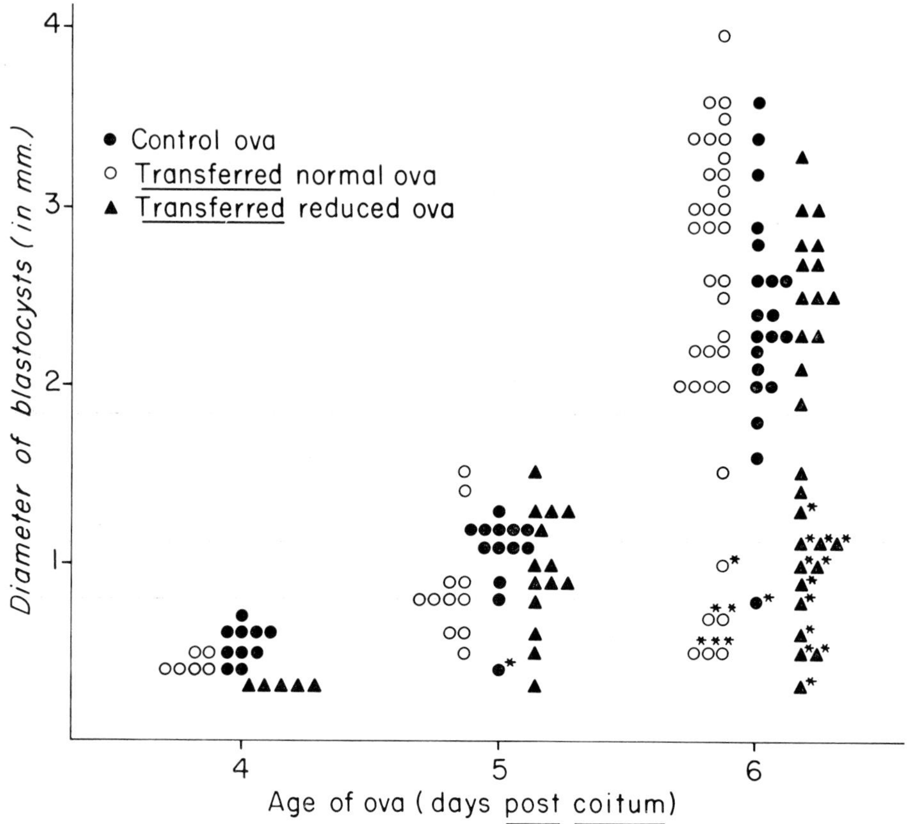

FIG. 6. The size of individual blastocysts at 4–6 days *p.c.* The asterisks designate ova that were degenerating. (From Greenwald 1962.)

5 living blastocysts recovered from the recipient measured 2.3 to 2.7 mm in diameter. There appeared to be no positive correlation between the initial thickness of the mucin layer and the ultimate size of the blastocysts.

III. Discussion

In 1930 Westman wrote, "Although the anatomical changes in the tube during different phases of the sexual cycle have been clarified, no attempt has been made to clarify what is the physiological function of the tubal secretion." Despite a considerable amount of work in the past 37 years, it must be conceded that Westman's statement is still true.

A relatively recent subject of interest is the variation in oviductal secretion rates during the reproductive cycle. In the species which have been studied, the peak of fluid secretion occurs at the time of ovulation. Experiments with ovariectomized animals indicate that exogenous estrogen stimulates secretion whereas progesterone inhibits it. These effects are similar to the action of the hormones on fluid imbibition by the uterus. In the case of the estrogen-treated uterus, the fluid uptake precedes a more important phase of carbohydrate metabolism and protein synthesis. It is quite possible that the increase in oviductal fluid following the administration of estrogen is also only a manifestation of much more pronounced changes in growth and differentiation of the oviduct. Hence, detailed biochemical studies of the changes in the composition of oviductal fluid—and of the oviduct itself—are of considerable importance in elucidating the functions of oviductal secretions.

Differences can be demonstrated in oviductal activity between pre- and postovulatory stages in intact animals. It is tempting to correlate these differences to periods of follicular and luteal dominance. However, at times this may be a dangerous assumption, since in most species there appears to be an abrupt but transitory drop in estrogen secretion following ovulation. Are the observed changes in the oviduct due to reduced estrogen levels or to rising titers of progesterone? It is obvious that baseline studies with intact, cycling animals are indispensable, but further investigations require suitably treated ovariectomized animals as well as the administration of exogenous hormones to intact animals (e.g., progesterone injected into preovulatory rabbits).

It is apparent from the review of the literature on oviductal secretion that species differences exist in the magnitude of the cyclic changes. In the rabbit, the most pronounced histochemical and histologic changes occur after ovulation, while the egg is passing through the oviduct. On the other hand, in man there seems to be more secretory activity during the follicular stage than the luteal stage. In the rat, cyclic changes in the oviduct have so far not been clearly demonstrated at any stage of the reproductive cycle.

The functional significance of oviductal secretion can obviously not be understood without considering its relationship to the gametes and the developing embryo. This has probably been the principal stumbling block in elucidating the role of oviductal secretions, since the energy requirements of oviductal eggs have only recently begun to be clarified.

Oviductal secretions may function in several ways. (1) They may be necessary for the final maturation of sperms and ova, the essential preludes to fertilization. (2) Oviductal secretions may provide nutrients or protection for oviductal embryos. (3) The effects of oviductal secretions may extend to the endometrium rather than exclusively influencing the environment of the oviduct. (4) The disquieting possibility exists also that oviductal secretions may be unessential, to a great extent, for embryonic development and merely represent changes in a secondary target organ to estrogen and progesterone. Only further research will reveal which parameters of oviductal secretions are of primary importance.

The hormonal regulation of the synthesis and release of mucin in the Fallopian tube of the rabbit (Greenwald 1958, 1962) is an example of the difficulty of assessing the significance of oviductal secretions. The fact that some rabbit eggs with a thin shell of mucin normally expand *in utero*, implant, and go to term raises the fundamental question whether or not the mucin layer is essential for normal development. It must be recollected that the majority of transferred ova with a reduced thickness of mucin fail to develop into fetuses. The dilemma of any inconclusive experiment is whether one stresses the positive or negative findings.

Adams (1960) approached the problem of the significance of the mucin layer in still another way. Two-cell rabbit ova were recovered and cultured *in vitro* for 24 hr; morulae with a mucin layer of 10 μ or less were then transferred to 7 recipients. At laparotomy 10 days *p.c.*, 44% of the eggs had implanted successfully in 5 of the recipients. These results agree with the findings reported by this author.

It is possible that rabbit ova can be recovered immediately after fertilization (*before any mucin is deposited*) and grown *in vitro* for 2 or 3 days. The ability or inability of these ova to develop into normal fetuses on transfer into the cornua of a suitably synchronized recipient appears to be the most direct method of resolving the role of the mucin layer. However, this approach will depend on the development of techniques for long-term culture of rabbit ova.

Acknowledgment

This is a contribution from the Research Professorship in Human Reproduction. The work of the author included in this paper has been supported by grants from The Population Council, National Institutes of Health (HD 00596) and The Ford Foundation.

References

1. Adams, C. E. 1958. Egg development in the rabbit: The influence of post-coital ligation of the uterine tube and of ovariectomy. *J. Endocr.* 16:283.
2. ———. 1960. Development of the rabbit eggs with special reference to the mucin layer. *Adv. Abstr. First Int. Congr. Endocr.* Abstr. no. 345, p. 687.
3. Agduhr, E. 1927. Studies on the structure and development of the bursa ovarica and the tuba uterina in the mouse. *Acta Zool.* 8:1.
4. Alden, R. H. 1942*a*. The oviduct and egg transport in the albino rat. *Anat. Rec.* 84:137.
5. ———. 1942*b*. Aspects of the egg-ovary oviduct relationship in the albino rat: II. Egg development within the oviduct. *J. Exp. Zool.* 90:171.
6. Allen, E. 1922. The oestrus cycle in the mouse. *Amer. J. Anat.* 30:297.
7. Assheton, R. 1895. A reinvestigation into the early stages of development of the rabbit. *Quart. J. Micr. Sci.* 37:113.
8. Augustin, E., and Moser, A. 1955. Vorkommen und Aktivität von alkalischer Phosphatase im Eileiter der Ratte und in unbefruchteten und befruchteten Eiern während der Tubenwanderung. *Arch. Gynaek.* 185:759.
9. Austin, C. R. 1961. *The mammalian egg.* Springfield, Illinois: Charles C Thomas.
10. Bacsich, P., and Hamilton, W. J. 1954. Some observations on vitally stained rabbit ova with special reference to their albuminous coat. *J. Emb. Exp. Morph.* 2:81.
11. Bedford, J. M., and Chang, M. C. 1962. Fertilization of rabbit ova *in vitro*. *Nature* 193:898.
12. Bishop, D. W. 1956. Active secretion in the rabbit oviduct. *Amer. J. Physiol.* 187:347.
13. Björkman, N., and Fredricsson, B. 1960. The ultrastructural organization and the alkaline phosphatase activity of the epithelial surface of the bovine Fallopian tube. *Z. Zellforsch.* 51:589.
14. ———. 1961. The bovine oviduct epithelium and its secretory process as studied with the electron microscope and histochemical tests. *Z. Zellforsch.* 55:500.
15. ———. 1962. Ultrastructural features of the human oviduct epithelium. *Int. J. Fertil.* 7:259.
16. Blandau, R.; Jensen, L.; and Rumery, R. 1958. Determination of the pH values of the reproductive-tract fluids of the rat during heat. *Fertil. Steril.* 9:207.
17. Bloch, S. 1952. Untersuchungen über Superfetation an der Maus. *Schweiz. Med. Wsch.* 82:1.

18. Borell, U.; Gustavson, K. H.; Nilsson, O.; and Westman, A. 1959. The structure of the epithelium lining the Fallopian tube of the rat in oestrus. *Acta Obstet. Gynec. Scand.* 38:203.
19. Borell, U.; Nilsson, O.; Wersäll, J.; and Westman, A. 1956. Electron-microscope studies of the epithelium of the rabbit Fallopian tube under different hormonal influences. *Acta Obstet. Gynec. Scand.* 35:35.
20. Braden, A. W. H. 1952. Properties of the membranes of rat and rabbit eggs. *Austral. J. Sci. Res. Series B.* 5:460.
21. Brinster, R. L. 1965. Studies of the development of mouse embryos *in vitro*: Energy metabolism. In *Preimplantation stages of pregnancy*, ed. G. E. W. Wolstenholme, pp. 60–74. Boston: Little, Brown and Co.
22. Chang, M. C. 1959. Fertilization of rabbit ova in vitro. *Nature* 184:466.
23. Clewe, T. H., and Mastroianni, L., Jr. 1960. A method for continuous volumetric collection of oviduct secretions. *J. Reprod. Fertil.* 1:146.
24. Deane, H. W. 1952. Histochemical observations on the ovary and oviduct of the albino rat during the estrous cycle. *Amer. J. Anat.* 91:363.
25. Espinasse, P. G. 1935. The oviductal epithelium of the mouse. *J. Anat.* 69:363.
26. Flerkó, B. 1954. Die Epithelien des Eileiters und ihre hormonalen Reaktionen. *Z. Mikroskopischanat. Forsch.* 61:99.
27. Fredricsson, B. 1959*a*. Proliferation of rabbit oviduct epithelium after estrogenic stimulation, with reference to the relationship between ciliated and secretory cells. *Acta Morph. Neerl. Scand.* 2:193.
28. ———. 1959*b*. Histochemical observations on the epithelium of human Fallopian tubes. *Acta Obstet. Gynec. Scand.* 38:109.
29. ———. 1959*c*. Studies on the morphology and histochemistry of the Fallopian tube epithelium. *Acta Anat. (Basel)* 38 (Supp. 37): 1–23.
30. Fredricsson, B., and Björkman, N. 1962. Studies on the ultrastructure of the human oviduct epithelium in different functional states. *Z. Zellforsch.* 58:387.
31. Friz, M. 1959. Tierexperimentelle Untersuchungen zur Frage der Tubensekretion. *Z. Geburtsh. Gynaek.* 153:285.
32. Glass, L. E. 1963. Transfer of native and foreign serum antigens to oviducal mouse eggs. *Amer. Zool.* 3:135.
33. Glass, L. E., and McClure, T. R. 1965. Post natal development of the mouse oviduct: transfer of serum antigens to the tubal epithelium. In *Preimplantation stages of pregnancy*, ed. G. E. W. Wolstenholme, pp. 294–321. Boston: Little, Brown and Co.
34. Gothié, S., and Moricard, R. 1955. Étude sur la répartition du S^{35} dans la trompe: Perméabilité de la membrane pellucide de l'oeuf fécondé intratubaire. *C.R. Soc. Biol.* 149:2084.
35. Green, C. L. 1957. Identification of alpha-amylase as a secretion of the human Fallopian tube and "tubelike" epithelium of Mullerian and mesonephric duct origin. *Amer. J. Obstet. Gynec.* 73:402.
36. Greenwald, G. S. 1957. Interruption of pregnancy in the rabbit by the administration of estrogen. *J. Exp. Zool.* 135:461.
37. ———. 1958. Endocrine regulation of the secretion of mucin in the tubal epithelium of the rabbit. *Anat. Rec.* 130:477.
38. ———. 1961. A study of the transport of ova through the rabbit oviduct. *Fertil. Steril.* 12:80.

39. Greenwald, G. S. 1962. The role of the mucin layer in development of the rabbit blastocyst. *Anat. Rec.* 142:407.
40. ———. 1963. Interruption of early pregnancy in the rabbit by a single injection of oestradiol cyclopentylpropionate. *J. Endocr.* 26:133.
41. Greenwald, G. S., and Everett, N. B. 1959. The incorporation of S^{35} methionine by the uterus and ova of the mouse. *Anat. Rec.* 134:171.
42. Gregoire, A. T.; Gongsakdi, D.; and Rakoff, A. E. 1961. The free amino acid content of the female rabbit genital tract. *Fertil. Steril.* 12:322.
43. ———. 1962. The presence of inositol in genital tract secretions of the female rabbit. *Fertil. Steril.* 13:432.
44. Gregory, P. W. 1930. The early embryology of the rabbit. *Contrib. to Embryol. Carnegie Inst.* 21:141.
45. Gwatkin, R. B. L., and Biggers, J. D. 1963. Histology of mouse Fallopian tubes maintained as organ cultures on a chemically defined medium. *Int. J. Fertil.* 8:435.
46. Hadek, R. 1953. Alteration of pH in the sheep's oviduct. *Nature* 171:976.
47. ———. 1955. The secretory process in the sheep's oviduct. *Anat. Rec.* 121:187.
48. Hamner, C. E., and Williams, W. L. 1965. Composition of rabbit oviduct secretions. *Fertil. Steril.* 16:170.
49. Hartman, C. G. 1925. On some characters of taxonomic value pertaining to the egg and the ovary of rabbits. *J. Mamm.* 6:114.
50. Hashimoto, M.; Shimoyama, T.; Mori, Y.; Komori, A.; Kosaka, M.; and Akashi, K. 1959*a*. Electronmicroscopic observations on the secretory process in the Fallopian tube of the rabbit (Report II). *J. Jap. Obstet. Gynec. Soc.* 6:384.
51. Hashimoto, M.; Shimoyama, T.; Mori, Y.; Komori, A.; Tomita, H.; and Akashi, K. 1959*b*. Electronmicroscopic observations on the secretory process in the Fallopian tube of the rabbit (Report I). *J. Jap. Obstet. Gynec. Soc.* 6:235.
52. Kellogg, M. 1945. The postnatal development of the oviduct of the rat. *Anat. Rec.* 93: 377.
53. Kirby, D. R. S. 1965. The role of the uterus in the early stages of mouse development. In *Preimplantation stages of pregnancy*, ed. G. E. W. Wolstenholme, pp. 325–39. Boston: Little, Brown and Co.
54. Lehto, L. 1963. Cytology of the human Fallopian tube: Observations on the epithelial cells of the human Fallopian tube during foetal and menstrual life and on tubal cancer. *Acta Obstet. Gynec. Scand.* 42 (supp. 4):3–95.
55. Lutwak-Mann, C. 1955. Carbonic anhydrase in the female reproductive tract. Occurrence, distribution and hormonal dependence. *J. Endocr.* 13:26.
56. Lutwak-Mann, C., and Averill, R. L. W. 1954. Carbonic anhydrase activity in the uterus and Fallopian tubes of the ewe. *J. Endocr.* 11:12.
57. Marcus, S. L., and Saravis, C. A. 1965. Oviduct fluid in the rhesus monkey: A study of its protein components and its origin. *Fertil. Steril.* 16:785.
58. Mastroianni, L., Jr.; Beer, F.; Shah, U.; and Clewe, T. H. 1961. Endocrine regulation of oviduct secretions in the rabbit. *Endocrinology* 68:92.
59. Mastroianni, L., Jr.; Shah, U.; and Abdul-Karim, R. 1961. Prolonged volumetric collection of oviduct fluid in the rhesus monkey. *Fertil. Steril.* 12:417.
60. Mastroianni, L., Jr., and Wallach, R. C. 1961. Effect of ovulation and early gestation on oviduct secretions in the rabbit. *Amer. J. Physiol.* 200:815.

61. Moreaux, R. 1913. Recherches sur la morphologie et la fonction glandulaire de l'épithélium de la trompe utérine chez les Mammifères. *Arch. Anat. Micr. Morph. Exp.* 14:515.
62. Mounib, M. S., and Chang, M. C. 1965. Metabolism of endometrium and Fallopian tube in the estrous and the pseudopregnant rabbit. *Endocrinology* 76:542.
63. Nilsson, O., and Rutberg, U. 1960. Ultrastructure of secretory granules in postovulatory rabbit oviduct. *Exp. Cell Res.* 21:622.
64. Novak, E. and Everett, H. S. 1928. Cyclical and other variations in the tubal epithelium. *Amer. J. Obstet. Gynec.* 16:499.
65. Noyes, R. W.; Adams, C. E.; and Walton, A. 1959. The transport of ova in relation to the dosage of oestrogen in ovariectomized rabbits. *J. Endocr.* 18:108.
66. Olds, D., and VanDemark, N. L. 1957. Composition of luminal fluids in bovine female genitalia. *Fertil. Steril.* 8:345.
67. Perkins, J. L.; Goode, L.; Wilder, W. A., Jr.; and Henson, D. B. 1965. Collection of secretions from the oviduct and uterus of the ewe. *J. Anim. Sci.* 24:383.
68. Restall, B. J., and Wales, R. G. 1966*a*. The Fallopian tube of the sheep. III. The chemical composition of the fluid from the Fallopian tube. *Aust. J. Biol. Sci.* 19:687.
69. ———. 1966*b*. The Fallopian tube of the sheep. IV. The metabolism of ram spermatozoa in the presence of fluid from the Fallopian tube. *Aust. J. Biol. Sci.* 19:883.
70. Robboy, S. J., and Kahn, R. H. 1964. Electrophoretic separation of hydrolytic enzymes of the female rat reproductive tract. *Endocrinology* 75:97.
71. Smithberg, M., and Runner, M. N. 1960. Retention of blastocysts in nonprogestational uteri of mice. *J. Exp. Zool.* 143:21.
72. Snyder, F. F. 1923. Changes in the Fallopian tube during the ovulation cycle and early pregnancy. *Bull. Johns Hopkins Hosp.* 34:121.
73. ———. 1924. Changes in the human oviduct during the menstrual cycle and pregnancy. *Bull. Johns Hopkins Hosp.* 35:141.
74. Stegner, H. E. 1962. Elektronenmikroskopische Untersuchungen über die Sekretionsmorphologie des menschlichen Tubenepithels. *Arch. Gynaek.* 197:351.
75. Vishwakarma, P. 1962. The pH and bicarbonate-ion content of the oviduct and uterine fluids. *Fertil. Steril.* 13:481.
76. Weitlauf, H. M., and Greenwald, G. S. 1965. A comparison of ^{35}S methionine incorporation by the blastocysts of normal and delayed implanting mice. *J. Reprod. Fertil.* 10:203.
77. ———. 1967. A comparison of the *in vivo* incorporation of S^{35} methionine by two-celled mouse eggs and blastocysts. *Anat. Rec.* 159:249.
78. Westman, A. 1930. Studies of the function of the mucous membrane of the uterine tube. *Acta Obstet. Gynec. Scand.* 10:288.
79. Wolstenholme, G. E. W., and O'Connor, M., eds. 1965. *Preimplantation stages of pregnancy.* Boston: Little, Brown and Co.
80. Zachariae, F. 1958. Autoradiographic (S^{35}) and histochemical studies of sulphomucopolysaccharides in the rabbit uterus, oviducts and vagina. *Acta Endocr.* 29:118.

8

The Biology of Oviductal Cilia

R. M. Brenner

Department of Electron Microscopy
Oregon Regional Primate Research Center
Beaverton, Oregon

The literature on the structure and function of cilia and flagella has been the subject of a number of recent reviews. The basic ultrastructural pattern and the variations in ciliary structure throughout the biological world are well understood and thoroughly documented (Fawcett 1961; Faure-Fremiet 1961; Grimstone 1961; Pitelka 1963; Sleigh 1962; Satir 1965). The mechanism of ciliary motion is not so well understood, and the problem of the control of ciliary motility remains an important research area for the future (Kinosita and Murakami 1967; Holwill 1966; Sleigh 1962; Rivera 1962; Satir 1965).

Kinocilia (motile cilia) play diverse roles in the economy of living organisms. Locomotion, respiration, circulation, alimentation, sensory reception, and reproduction are all processes in which cilia and flagella play important parts. There is a considerable body of literature on each of these phases of ciliary activity (Fawcett 1961) but, in this review, we shall discuss only those cilia that play a reproductive role. More specifically, this review will cover the cilia of the mammalian oviduct—their structure, growth patterns, and response to ovarian steroids. For purposes of illustration, diagrams and photographs of the cilia of the rhesus monkey oviduct will be used.

I. The Ultrastructure of Oviductal Cilia

The cilia of the mammalian oviduct are essentially the same as those found in the vast majority of ciliated organisms and are identical with those in the respiratory tract. Each cilium is a slender, hairlike process that extends from the free surface of the cell. An extension of the cell membrane forms the ciliary membrane. In the rhesus monkey, the cilia are 5–6 μ in length and 0.2–0.3 μ in diameter (Fig. 6*a*, *c*). Within the ciliary membrane is the "axoneme" or "axial filament complex," a bundle of 11 fibrils consisting of 2 central single fibrils and 9 peripheral double fibrils. The fibrils are approximately 200 Å in diameter. One of the subfibrils of each peripheral doublet has 2 short projections or "arms," which are approximately 50–70 Å long in cross section. The fibrils are embedded in a matrix material which, although it is difficult to preserve, appears to contain fibers radiating out to the peripheral fibrils like spokes in a wheel (Fig. 1).

In some organisms there is a distinct basal plate separating the cilium proper from the basal body, but mammalian oviductal cilia lack such a plate (Fawcett 1961) and the fibrils of the cilium join directly with the fibrils of the basal body (Figs. 1, 11*a*, *b*).

The basal bodies are lined up in orderly rows immediately beneath the cell membrane (Fig. 6*a*). In the rhesus monkey oviduct, each basal body is a hollow, cylindrical structure about 0.2 μ wide and 0.4–0.5 μ long. Its internal structure is identical with the centriole of the

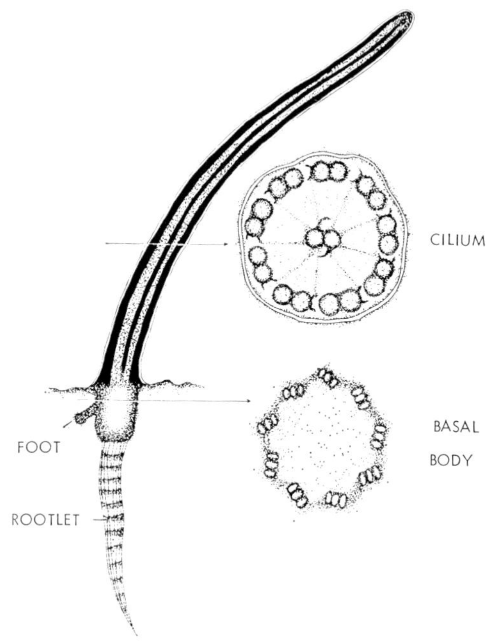

FIG. 1. Ultrastructure of cilium and basal body in rhesus monkey oviduct. The axoneme of each cilium consists of 9 peripheral double fibrils and 2 central single fibrils. One subfibril of each peripheral doublet has 2 short projections, or "arms." The central fibrils also have arms. The delicate matrix material appears to radiate in fibers toward the peripheral fibrils. The basal bodies have 9 peripheral triplet fibrils and no central fibrils.

cell. Embedded within the wall of the basal body are 9 sets of triplet fibrils with each triplet set at an angle to the axis of the basal body so that the cross-sectional appearance resembles a pinwheel (Fig. 6*d*). In many organisms, including the rhesus monkey, a 100 μ lateral projection of dense material known as the basal foot is present on the wall of the basal body (Figs. 1, 11*b*, 12). Fibrous rootlets with periodic cross-striations extend downward into the cytoplasm from the proximal end of each basal body (Figs. 1, 11*b*). These rootlets are not always seen in mammalian cilia, but they are present in rhesus monkey and human oviducts (Clyman 1966) though they are not so well developed as in invertebrates (Fawcett 1961). Presumably the rootlets provide a cytoplasmic anchorage for the motile cilia.

The secretory cells of the rat oviduct bear many filamentous projections that presumably serve to increase the cell surface area (Odor 1953; Nilsson 1957). These projections are nonmotile stereocilia and lack the axial filament complex of true kinocilia. They are not extensively developed in rhesus monkey oviduct.

II. Hormonal Control of Growth of Oviductal Cilia

The literature on the hormonal control of the growth of oviductal ciliated cells can be considered in two categories. First are studies of the effects of such experimental procedures as hypophysectomy and ovariectomy with and without subsequent replacement therapy, and also the hormonal treatment of intact animals. Second are investigations of the variation in the number and morphology of ciliated cells during the changes in hormone levels that normally occur during fetal life, prepubertal life, the estrous and menstrual cycles, pregnancy, and the period of senescence.

By far the greatest number of reports in the literature concerns the variation in number and morphology of ciliated cells during various phases of normal life. This literature contains many conflicting reports and it seems clear that there are considerable species differences.

The data on experimentally induced hormonal imbalances are much more meager but far more harmonious. Most authors, including myself, find that estrogens stimulate the growth of ciliated cells. For purposes of clarity then, it will be best to begin with a discussion of the various experimental studies.

A. *Experimental Studies of Estrogen Effects on Oviductal Cilia*

The oviducts of untreated, immature rhesus monkeys have very few ciliated cells, but immature estrogen-treated monkeys develop a completely ciliated epithelium. Also, ciliation is stimulated by estrogen in ovariectomized immature monkeys (Allen 1928*a*).

In adult ovariectomized monkeys, estrogen treatment produces patches of ciliated cells in the oviducts, and clippings of the oviducts placed in Ringer's solution show motile cilia (Allen 1928*b*).

Surprising as it may seem, there are no further reports in the literature on the effects of castration and hormone replacement therapy on the ciliated cells of the oviduct of any mammal until 1951. A complete loss of cilia from the rabbit oviduct follows castration, hypophysectomy, stalk-section, or hypothalamic lesioning, provided sufficient time has elapsed (3–4 months) subsequent to the procedure (Flerkó 1951, 1954). When such animals are treated with estrogen (800 "units" over 2 weeks) there is complete restoration of the ciliated epithelium regardless of the previous experimental procedure. Treatment with estrogen also stimulates development of a mature ciliated epithelium in the oviducts of infantile rabbits, which normally possess a nonciliated epithelium. Unfortunately, the hypophysectomized rabbits could not be maintained long enough to test them with estrogen, but the evidence indicates that the ciliated epithelium of the oviduct is under direct ovarian control and that estrogens are specifically responsible for the growth and maintenance of the ciliated cells (Flerkó 1954).

One study of the rabbit oviduct, however, conflicts with these data. No change was found in the oviductal ciliated cells after castration in rabbits, though the secretory cells did atrophy (Borell *et al.* 1956). The elapsed time after castration was not stated, however, and, according to Flerkó (1954) the loss of the ciliated cells requires 3–4 months, whereas the secretory cells atrophy within a few weeks of castration.

In postparturient women at the time of delivery, the oviductal epithelium is low (average height, 16 μ) and becomes increasingly lower during the postpartum period (average height,

10 μ), and the ciliated cells decrease in number (Andrews 1951). Women who have received stilbestrol alone, starting on the day of delivery (5 mg/day for 5–9 days), show an intense proliferation of the ciliated epithelium. The cells reach 20–25 μ in height, as compared with the average height of approximately 10 μ in untreated cases. Women who have received stilbestrol in combination with progesterone show an inhibition of the estrogen effect, and the ciliated epithelium resembles that found at the time of delivery. Treatment with stilbestrol 5–10 days before delivery produces no stimulation of the ciliated epithelium, and progesterone treatment alone has no effect.

Because exogenous progesterone inhibits the effect of estrogen upon ciliated cell growth in postpartum women, a similar process is presumed to occur during pregnancy (Andrews 1951), and the loss of ciliated cells and diminished height of the oviductal epithelium during the puerperium is presumably due to the low levels of circulating estrogen present at that time. Recently, we have found (Brenner 1967*a*) that there is almost complete loss of cilia from the oviduct of the rhesus monkey subsequent to ovariectomy or hypophysectomy, and that treatment with estrogen restores a fully ciliated epithelium (see this chapter, part IV A).

It seems clear that growth of the oviductal ciliated epithelium in man, monkey, and rabbit can be stimulated by exogenous estrogen and that exogenous progesterone antagonizes the estrogen-driven ciliated cell growth (Allen 1928*a, b;* Flerkó 1951, 1954; Andrews 1951; Brenner 1967*a*).

B. *Cyclical Variations in the Ciliated Epithelium of the Oviduct*

The literature concerning the question of cyclical change in oviductal epithelium abounds with disagreements. Many authors maintain that during the menstrual cycle, the ciliated cells wax and wane in number as well as in height; and, moreover, that ciliated cells transform into secretory cells during the last portion of the cycle. Others find that the ciliated cells wax and wane only in height, and that there is no transformation of ciliated cells into secretory cells. Some find no cyclic change at all in the ciliated cells. All authors seem to agree that the activity of the secretory cells is at its maximum during the premenstrual phase. Details of these discrepancies can be found in several of the available comprehensive reviews (Schaffer 1908; Novak and Everett 1928; Balboni 1954; Schultka 1963; Hashimoto *et al.* 1964; Horstmann and Stegner 1966).

Table 1 summarizes some of these discordancies in the literature, and the following is a brief review of some of the major conflicting reports.

1. *The Cell Types.* The mammalian oviductal epithelium consists of two major types of cells, one ciliated and the other secretory. A third type, a rodlike cell with a flat compressed nucleus and practically no cytoplasm, is known as the "peg cell," "intercalary cell," or "Stiftchenzellen," and a fourth type, the basal cell, is also present (Horstmann and Stegner 1966). There is some recent evidence that basal cells, also called "indifferent" cells, may form a population of reserve cells for oviductal growth (Pauerstein and Woodruff 1967). The number and character of the ciliated and secretory cells varies depending on the species studied, the portion of the oviduct sampled, and the phase of the sexual cycle (Schaffer 1908). It has been suggested (Schaffer 1908) that the ciliated cells become transformed into secretory cells as the sexual cycle progresses and that a ciliated phase alternates regularly with a secretory phase. The "peg cells" have been presumed to be either independent elements (Schaffer 1908) or emptied secretory cells (Hörmann 1909). Most authors believe "peg cells" to be the latter.

2. *The Problem of Cyclic Change.* In the older literature, the first 4–5 days after menstruation are referred to as the "postmenstrual" phase, and the ensuing period up to 5–6 days before menstruation is called the "interval" phase. The "premenstrual" phase is the period 5–6 days before menstruation. In the first thorough correlative study of the changes in the oviduct and the uterine endometrium during all the phases of the menstrual cycle in man, the ciliated cells have been found to predominate in number in the postmenstrual and interval phases, whereas the secretory cells predominate in number during the premenstrual phase. Thus a periodic variation in the oviductal epithelium involving an alternation of a ciliated and a nonciliated (secretory) phase is presumed to occur (Tröscher 1917).

However, active ciliated cells have been seen at all stages of the cycle, without any significant alteration in their distribution, number, or activity, in the oviductal epithelium of

TABLE 1

PARTIAL LISTING OF DISCREPANCIES IN THE LITERATURE ON THE CYCLIC CHANGES IN OVIDUCTAL CILIATED CELLS

Species	Author	Region of Oviduct	Type of Microscopy	Change in Height	Change in Number
Mouse	Allen 1922	All regions	Light	No	Yes
"	Espinasse 1935	Isthmus	Light	No	No
Rat	Deane 1952	All regions	Light	No	No
Pig	Snyder 1923	Fimbriae, ampulla, isthmus	Light	Yes	No
Rhesus monkey	Westman 1932, 1934	Not stated	Light	Yes	Yes
Java monkey	Joachimovits 1935	Not stated	Light	Yes	Yes
Man	Balboni 1954	Not stated	Light	Yes	Yes (slight)
"	Clyman 1966	Fimbriae, ampulla	Electron	Yes	No
"	Fredricsson 1959*a*, *b*	Fimbriae, ampulla, isthmus	Light	Yes	No
"	Hashimoto *et al.* 1964	Fimbriae, ampulla, isthmus	Electron	No	No
"	Hølund 1946	Not stated	Light	No	No
"	Novak and Everett 1928	Isthmus	Light	Yes	No
"	Pernkopf and Pichler 1953	Not stated	Light	Yes	Yes
"	Scheyer 1926	Not stated	Light	No	No
"	Schröder 1930	Not stated	Light	Yes	Yes
"	Schultka 1963	Not stated	Light	Yes	Yes
"	Snyder 1924	Ampulla, isthmus	Light	Yes	No
"	Tröscher 1917	Not stated	Light	Yes	Yes

domestic pigs. Only a change in height has been observed, with a maximum (25 μ) at 1–3 days after ovulation and a minimum (10 μ) at 2 weeks after ovulation. During the 3d week, a rise in the height of the epithelium occurs, with a gradual increase to a maximum at the time of ovulation (Snyder 1923).

A similar variation, restricted to the height of the epithelium during the cycle, has been found in the oviductal epithelium of man (Snyder 1924; Novak and Everett 1928). The ciliated cells are tallest during the interval phase (30 μ) and lowest during the premenstrual phase (20 μ) as well as during pregnancy. Ciliated cells are seen throughout the cycle without any change in number. The height of the ciliated cells is at a maximum during the interval phase and at a minimum during menstruation and pregnancy, and the postmenstrual phase is characterized by a rapid increase in height of the ciliated cells. There is no evidence of transformation of ciliated cells into secretory cells. "Peg cells" are thought to be the remnants of emptied secretory cells. The only oviducts showing any evidence of an increase in the number of ciliated cells are from women with endometrial hyperplasia, a condition known to be associated with hyperestrogenism. These observations do not support the hypothesis of a normal alternation of ciliated and nonciliated phases in the human oviductal epithelium (Novak and Everett 1928).

Some recent electron microscopic studies also fail to show cyclic changes in the ciliated cells of man (Shimoyama 1963*a;* Clyman 1966). One study of mouse oviductal cilia fails to show cyclic changes (Espinasse 1935), while another indicates a small degree of loss of ciliated cells during metestrus (Allen 1922). A study of the rat oviduct shows no cyclic change (Deane 1952), while another reports cyclic loss and regeneration of cilia (Flerkó 1954).

A large number of reports show that cyclic changes do occur in human oviductal ciliated cells (Schröder 1930; Westman 1930, 1934; Tietze 1929, 1932; Cotte 1949; Bruni 1950; Balboni 1954; Pernkop and Pichler 1953; Schultka 1963; Schultka and Scharf 1963). In the oviduct of the rhesus monkey (Westman 1932, 1934) and Java monkey (Joachimovits 1935) the ciliated cells increase in number during the follicular phase and decrease in number during the luteal phase. We have found recently, in rhesus monkey oviduct, that the ciliated cells of the fimbriae and the ampulla show marked differences in cyclic behavior (Brenner 1967*b* and this chapter, part IV B). The fimbriae shed and regenerate their ciliated cells cyclically, but the ciliated cells of the ampulla fluctuate only in height. The fact that different authors have sampled different regions of the oviduct may account for some of the discordant findings in the literature (Table 1).

3. *The Fate of the Deciliated Cells.* Many authors who have found changes in the number of ciliated cells during the menstrual cycle have described cytological processes which they interpret as the transformation of ciliated cells to secretory cells. During the late luteal phase, the entire distal third of the ciliated cell becomes constricted from the rest of the cell and is cast off into the lumen (Balboni 1954). This process of deciliation also occurs in the oviduct of both the rabbit and the pig (Flerkó 1954), in the human uterus (Hamperl 1950), and in monkey oviduct (Brenner 1967*b*). Two other mechanisms exist by which cilia may be lost from the oviduct. Whole ciliated cells may fall out of the epithelium, and individual cilia may fall out of the ciliated cells (Flerkó 1954).

There is disagreement as to the ultimate fate of the deciliated cells (Schultka 1963; Flerkó 1954). One view is that the deciliated cells fill up with secretory granules and become active secretory cells (Schultka 1963). The secretory cells then become "peg cells" which are eventually lost from the oviduct. New ciliated cells form again, at the beginning of the next cycle from another cell type, which is found in the oviduct only during the first postmenstrual week. This latter cell type is large, has very clear cytoplasm and has variously been referred to as a "clear cell," "light cell," or "Ersatzzelle." "Ersatzzellen" are considered to be undifferentiated cells whose only function is to regenerate cilia (Schultka 1963).

However, "Ersatzzellen" have been considered as universal replacement cells that can become either ciliated cells or secretory cells. Deciliated cells at the end of the cycle are either lost from the oviduct or remain and become "Ersatzzellen" at the beginning of the next cycle. The depleted secretory cells become "peg cells" and are eventually lost. This scheme applies to the oviductal epithelium of a variety of mammals—man, rabbit, rat, domestic pig, cow, horse, sheep, dog, and cat (Flerkó 1954).

We have found no evidence for a transformation of ciliated cells to secretory cells in monkey oviduct. The deciliated cells either fall out or remain to become new ciliated cells in the next cycle. There is not sufficient evidence at present to evaluate the concept of a universal replacement cell (Brenner 1967*a*, *b*, and this chapter, part IV A, B).

4. *Conclusions concerning the Problem of Cyclic Change.* It is indeed difficult to evaluate the conflicting reports concerning the cyclic changes in the oviductal ciliated cells. In primates, the rising blood level of estrogen during the follicular phase seems to stimulate the production and growth of ciliated cells. As progesterone levels increase and estrogen levels decrease during the luteal phase, ciliated cells diminish in number and height. The progesterone-dominated state of pregnancy is characterized by a low oviductal epithelium (Andrews 1951; Fredricsson 1959*a*, *b;* Snyder 1924) and in the late menopausal state, with minimal levels of estrogen, the epithelium loses almost all its ciliated cells (Novak and Everett 1928; Naumann 1931; Shimoyama 1963*b*).

Perhaps one reason for the conflicting findings in man lies with the inherent difficulty of obtaining tissue samples from normal patients. Most of the women who come to surgery suffer from uterine myomas, uterine cancer, ovarian cancer, and other diseases. Ovarian function is undoubtedly disturbed in some of these patients and the normal estrogen-progestin cyclicity is probably interrupted. Further, women facing surgery normally undergo varying degrees of anxiety, which may be sufficient to induce ovarian malfunction. If corpora lutea are not formed, progesterone levels may not rise high enough and estrogen levels fall low enough to induce the loss of ciliated cells from the oviduct. In rats and mice, with their very short estrous cycles, sufficient time does not elapse for an extreme loss of ciliated cells from the oviduct, so that these common laboratory animals do not show a dramatic deciliation. Sufficient data are not present in the literature on many mammals with long luteal phases, although it has been stated that ciliated cells are regenerated cyclically in the oviducts of sheep, horse, pig, cow, dog, and cat (Flerkó 1954).

Surprisingly enough, there are no reports of the effects of castration and estrogen-replacement therapy on the oviducts of mice, rats, or any of the mammalian species listed above other than the rabbit. Such information would be invaluable in unifying our ideas on the sensitivity of ciliated cells to estrogen.

In summary, the data on castration and estrogen-replacement therapy in rabbits and monkeys and those on the effects of estrogen during the puerperium in man, as well as the observation of a correlation between endometrial hyperplasia and extensive ciliation in man, all serve as convincing evidence that estrogen controls the growth of oviductal ciliated cells. Further, the large number of reports on primates in which a normal cyclic regeneration of the ciliated epithelium was found, as well as our own findings, indicates that, at least in primates, normal ovarian cycles can produce a normal oviductal cycle of ciliary loss and repair.

III. The Process of Ciliogenesis

A. *The "Flimmerblase"*

The earlier literature on oviductal ciliogenesis contains many descriptions of an unusual cytological structure known as a "Flimmerblase" or ciliary vacuole (Mihálik 1934). The ciliary vacuole is presumed to consist of an intracellular sphere, the surface of which is covered by numerous basal bodies, and which develops deep in the cytoplasm of nonciliated cells. Each basal body eventually grows a cilium which points toward the center of the sphere. Eventually this ciliary vacuole migrates to the cell surface and opens up in such a manner that all the cilia eventually point toward the oviductal lumen.

This ciliary vacuole has been observed by light microscopy in the oviducts of several mammalian species including man (Flerkó 1951; Schultka 1963) and in human uterine ciliated cells (Hamperl 1950). Other workers have been unable to find ciliary vacuoles in the human Fallopian tube (Fredricsson 1959*b*), and the "Flimmerblase" is thought to be an artifact observed only in tangential sections (Hølund 1946). Most electron microscopists have not seen ciliary vacuoles in the oviducts of any species (Stegner 1961, 1962; Shimoyama 1963*a, b;* Hashimoto *et al.* 1964; Clyman 1966; Brenner 1967*a, b*) although there is one report of a presumed ciliary vacuole in fetal human oviduct (Overbeck 1967). However, the only electron micrograph in the above work is clearly a tangential section of the epithelium. The so-called ciliary vacuole could easily be the cell surface seen from below. Until more evidence is forthcoming, the reality of the "Flimmerblase" and its role in oviductal ciliogenesis must remain equivocal.

B. *Mechanisms of Ciliogenesis*

Cilium formation occurs throughout the taxonomic spectrum, and the pattern of ciliogenesis in mammalian oviduct differs from that seen in some other organisms. Throughout the biological world there are at least three different ciliogenic pathways with several common denominators (Satir 1965). These are, first, that ciliary morphogenesis always requires previous basal body morphogenesis. If basal bodies are not present, the cell first synthesizes them. Second, filament morphogenesis takes place within the cytoplasmic matrix, and third, smooth membrane-bound elements, either from the Golgi zone or the cell membrane, are involved in the growth of the organelle.

One developmental pathway occurs in fibroblasts from both birds and mammals (Sorokin 1962) and in the water mold, *Allomyces* (Renaud and Swift 1964). In this process, a cytoplasmic vesicle forms and migrates toward an existing centriole. The vesicle fuses with the centriole and forms an acorn-shaped cap. The ciliary fibrils develop within the well formed by the cap and the centriole. As the cilium elongates, secondary vesicles fuse with the cap and the cell membrane. Eventually the cilium is liberated from the cell.

Another pathway is illustrated by the ameboflagellate organism, *Naegleria gruberi* (Manton 1959; Satir 1965). This organism has no centriole during the amoeboid phase of its life cycle (Schuster 1963), but it synthesizes a centriole when it enters the flagellated phase. Once formed, the centriole sprouts a flagellum within the cytoplasm. This flagellum lacks a membrane covering at first, but a series of vesicles eventually coat the flagellum and these fuse to form the flagellary membrane. Fusion of the flagellar membrane and the cell membrane finally brings the flagellum to the exterior.

In a third developmental pathway, a centriole migrates to the cell surface and attaches there. Above this attachment point, a ciliary bud forms, and the fibrils of the cilium develop and elongate within this bud. This form of ciliogenesis occurs in the vertebrate rod cell (Tokuyasu and Yamada 1959), in neural epithelium (Sotelo and Trujillo-Cenoz 1958) and in various protozoa (Roth and Shigenaka 1964). In mammalian oviducts, cilia grow by means of a similar ciliary bud formation (see Fig. 11*a*).

C. *Mechanisms of Basal Body Replication*

The mechanism of basal body replication in cells undergoing ciliogenesis or flagellogenesis has been a matter of intense interest to cytologists for many years. Early studies of ciliated epithelia lead to the conclusion that the basal bodies are derived from the original centrioles of the cell by a process of repetitive self-replication (Henneguy 1898; Lenhossek 1898; Jordan and Helvestine 1923). However, recent electron microscopic studies indicate that, as with cilium formation, multiple pathways exist for basal body replication. For example, in chick embryo neural epithelium the centrioles themselves migrate to the cell surface and then sprout cilia, thus becoming basal bodies by virtue of a change in function (Sotelo and Trujillo-Cenoz 1958). But in ciliated protozoans, new kinetosomes (basal bodies) always arise from preexisting kinetosomes (Lwoff 1950; Bradbury and Pitelka 1965). Sometimes pericentriolar dense particles serve as precursors of the new centrioles (Bernhard and Deharven 1960). Also, during development of the multiflagellated sperm of viviparid snails, the original centrioles become surrounded by miniature centrioles, or "procentrioles," which later enlarge, develop further, and become the mature basal bodies (Gall 1961).

In *Allomyces arbusculus* (the water mold) a small centriole near the nucleus, similar to a procentriole, enlarges and develops into the mature basal body of the uniflagellated gametangium (Renaud and Swift 1964).

The developing multiflagellated spermatozoa of the fern (*Marsilea*) and the cycad (*Zamia*) contain cytoplasmic structures known as blepharoplasts, which are spheres containing a mass of tubules. At the time of the last spermatogenous mitosis, the blepharoplast becomes hollow and its surface becomes studded with a large number of procentrioles. Each small procentriole then leaves the surface of the blepharoplast and develops into a large, mature basal body which sprouts flagella at the cell surface (Mizukami and Gall 1966).

D. *Basal Body Replication in Mammalian Oviduct and Trachea*

In fetal rat trachea and oviduct, each future ciliated cell is destined to synthesize approximately 300 cilia (Rhodin and Dalhamn 1956), and since these cells possess only 2 centrioles initially, approximately 300 centrioles have to be synthesized before cilia formation. A precise series of stages occur during the process of centriole replication in fetal rat trachea and oviduct, and these stages are different from any of those previously described (Dirksen and Crocker 1965).

First, the parent centrioles become associated with large masses of fibrogranular material referred to as "proliferative elements." Within the proliferative elements, some of the granules become denser and eventually separate from the main mass. Around these isolated dense granules, called "condensation forms," and in contact with them, small nascent centrioles appear. The combination of the condensation forms with the nascent centrioles is referred to as a "generative complex." Each generative complex consists of 5–6 nascent centrioles distributed radially around a condensation form. Eventually, the nascent centrioles increase in length, break away from the generative complex, and migrate toward the cell surface. At the surface, each centriole becomes associated with the cell membrane, a ciliary bud forms, and a cilium grows out above the cell (Figs. 2, 3).

Similar structures appear in the developing ciliated cells of the nasal mucosa in 14- to 17-day-old fetal mice, but no association between the parent centriole and the proliferative element is seen (Frisch 1967). Structures that appear identical with the proliferative elements, condensation forms, and generative complexes are present during ciliogenesis in the monkey oviduct (Fig. 9). However, no evidence has been found yet of an association between the original centrioles and the proliferative elements (Brenner 1967*a*, *b* and this chapter, part IV A).

E. *Control of Ciliogenesis*

Considerable interest has centered on the control mechanisms that regulate these complex replicative processes. There is good evidence that the process of ciliogenesis is under feedback control and that the cell nucleus plays an important role in regulating ciliary and flagellary growth. For example, if the flagella of the organism *Ochromonas* are amputated, a lag period occurs followed by a period of maximal growth, which is then succeeded by a period of deceleration, so that the flagella always reach a length of 10 μ within 5 hr (Dubnau 1961).

In *Peranema trichophorum* (Tamm 1965) and *Euglena gracilis* (Rosenbaum 1965) flagellar regeneration after mechanical removal also shows precise lag periods, maximal growth periods, and a deceleration of growth which results in flagella of definitive lengths characteristic for the organism.

If actinomycin-D is added to a culture of *Naegleria gruberi* organisms just before induced flagellar regeneration, there is considerable inhibition of flagellar growth (Satir 1965). Since actinomycin-D blocks DNA-dependent RNA synthesis (Reich *et al.* 1961), it is clear that the cell nucleus plays an important role in controlling flagellary growth. Compounds that block protein synthesis at the ribosomal level, such as cycloheximide, have been shown also to inhibit flagellary growth (Rosenbaum 1965). Studies of ciliary regeneration with puromycin, cycloheximide, actinomycin-D, and DNP show that the process requires oxidative phosphorylation and protein synthesis as well as DNA-dependent RNA synthesis (Child 1965).

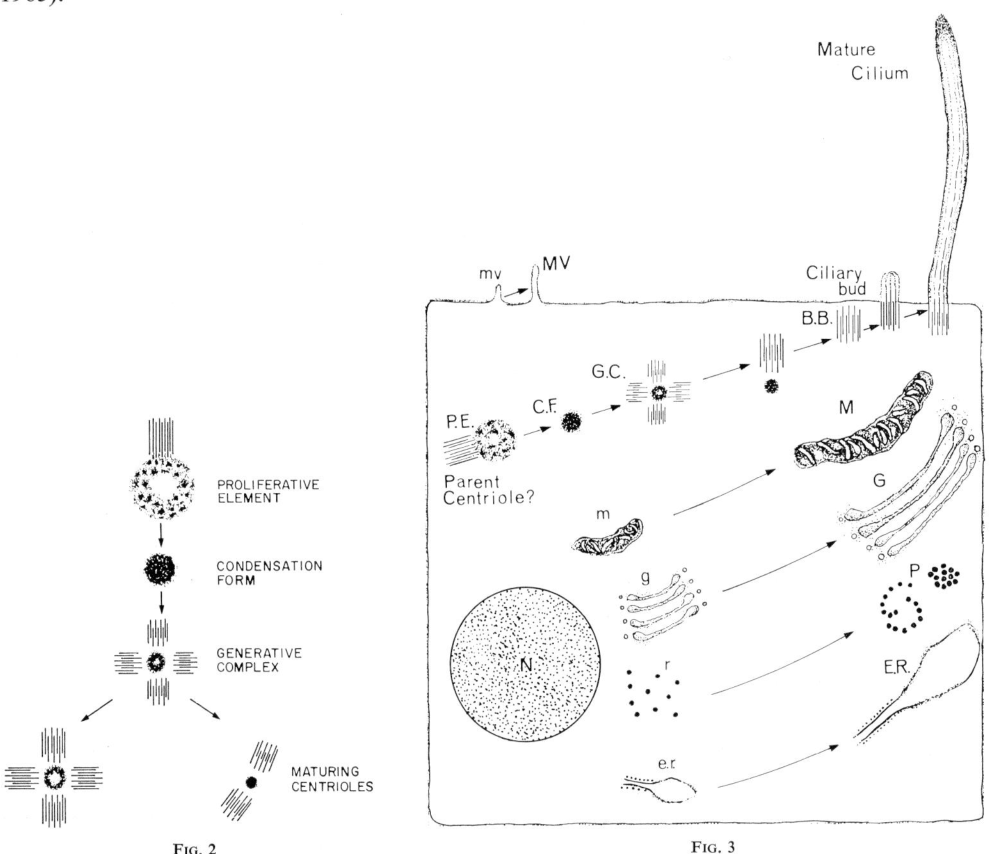

Fig. 2 Fig. 3

Fig. 2. Hypothetical scheme for development of basal bodies. The mature centriole presumably gives rise to proliferative elements which in turn develop condensation forms. The latter give rise to nascent centrioles in a grouping referred to as a generative complex. As the basal bodies mature, the condensation forms shrink. (From Dirksen and Crocker 1965.)

Fig. 3. Summary diagram of estrogen-driven ciliogenesis in rhesus monkey oviduct. The drawing indicates the various cellular processes which occur over a 5-day period of estrogen treatment in ovariectomized monkeys. Microvilli lengthen, the mitochondria elongate, the Golgi material hypertrophies, the endoplasmic reticulum becomes more extensive, and the ribosomes become polysomes. The basal bodies replicate, migrate to the cell surface, and sprout ciliary buds from which the cilia grow.

A similar DNA-dependent RNA synthesis may occur during estrogen-driven oviductal ciliogenesis, although as yet such evidence is not at hand. There is considerable evidence, however, that estrogens do stimulate RNA synthesis in various target organs, such as the uterus of immature or castrated animals, and indeed that an important locus of steroid hormone action may be the genome of the cell (Karlson 1963; Williams-Ashman 1965). For example, estradiol will stimulate RNA synthesis in the castrate rat uterus, and actinomycin-D will block such synthesis, as well as prevent the estrogenic acceleration of both phospholipid and protein synthesis (Ui and Mueller 1963). In the immature rat uterus, estradiol initially enhances the synthesis of transfer and template RNA, and enhancement of ribosomal structural RNA is a later effect (Wilson 1963). In the castrate rat uterus, estradiol stimulates the production of uterine ribosomes, and either actinomycin-D or puromycin will inhibit this estrogen effect (Moore and Hamilton 1964). It seems clear that protein synthesis follows an antecedent DNA-dependent RNA synthesis in estrogen-stimulated mammalian target tissue, just as it does in other organisms (Volkin and Astrachan 1956; Nomura, Hall, and Spiegelman 1960; Brenner, Jacob, and Meselson 1961).

Exactly what role DNA-dependent RNA synthesis plays in oviductal ciliogenesis is a problem for the future, and such studies are certain to be tremendously exciting.

IV. Ciliogenesis in Rhesus Monkey Oviduct

Recently, in our laboratory, we have been studying ciliogenesis in rhesus monkey oviduct with the electron microscope. Normal animals during the menstrual cycle, as well as ovariectomized animals treated with estrogen, have been examined. The results of this work have been alluded to in the previous discussion, but a more complete description of our findings is presented below. Our results are illustrated in Figures 4–12, which are all previously unpublished photographs.

A. *Effects of Ovariectomy and Estrogen Replacement*

1. *Methods.* Twelve adolescent female rhesus monkeys (average weight, 3.0 kg) were bilaterally ovariectomized, and at the time of ovariectomy a biopsy of the fimbriae and the ampulla of the oviduct was performed. After a period of 7 weeks, a second biopsy was performed to determine if the cilia had been lost. Nine weeks later, the animals received a series of intramuscular injections of estradiol benzoate in sesame oil at a rate of 10 μg/day and were serially biopsied on days 1, 2, 3, 4, 5, 6, and 9 after the injections had begun. Two monkeys were biopsied on each of days 3, 4, and 5, one in the morning and one in the afternoon; and one animal was biopsied on each of the other days. Two control animals were injected with 1 cc sesame oil/day and biopsied on the 9th day of treatment. Before fixation, a sample of the biopsy tissue was placed in a drop of Lactate-Ringer solution on a slide and examined with the light microscope for the presence of motile cilia. The remainder of the biopsy tissues was prepared for electron microscopy (Brenner 1966).

2. *Response to Estrogen.* The initial biopsies showed that 8 of the 12 animals had a fully ciliated oviductal epithelium (Fig. 4*b*). Seven weeks after ovariectomy, all the animals lacked cilia (Fig. 4*c*). Motile cilia developed after 4–5 days of estrogen treatment and the number of ciliated cells increased through the 9th day (Fig. 5*a–d*). The oviducts of the control animals did not develop cilia.

3. *Electron Microscope Findings.* After 1 day of estrogen treatment, there are no visible cytological changes in the atrophied oviductal epithelium at the electron microscope level

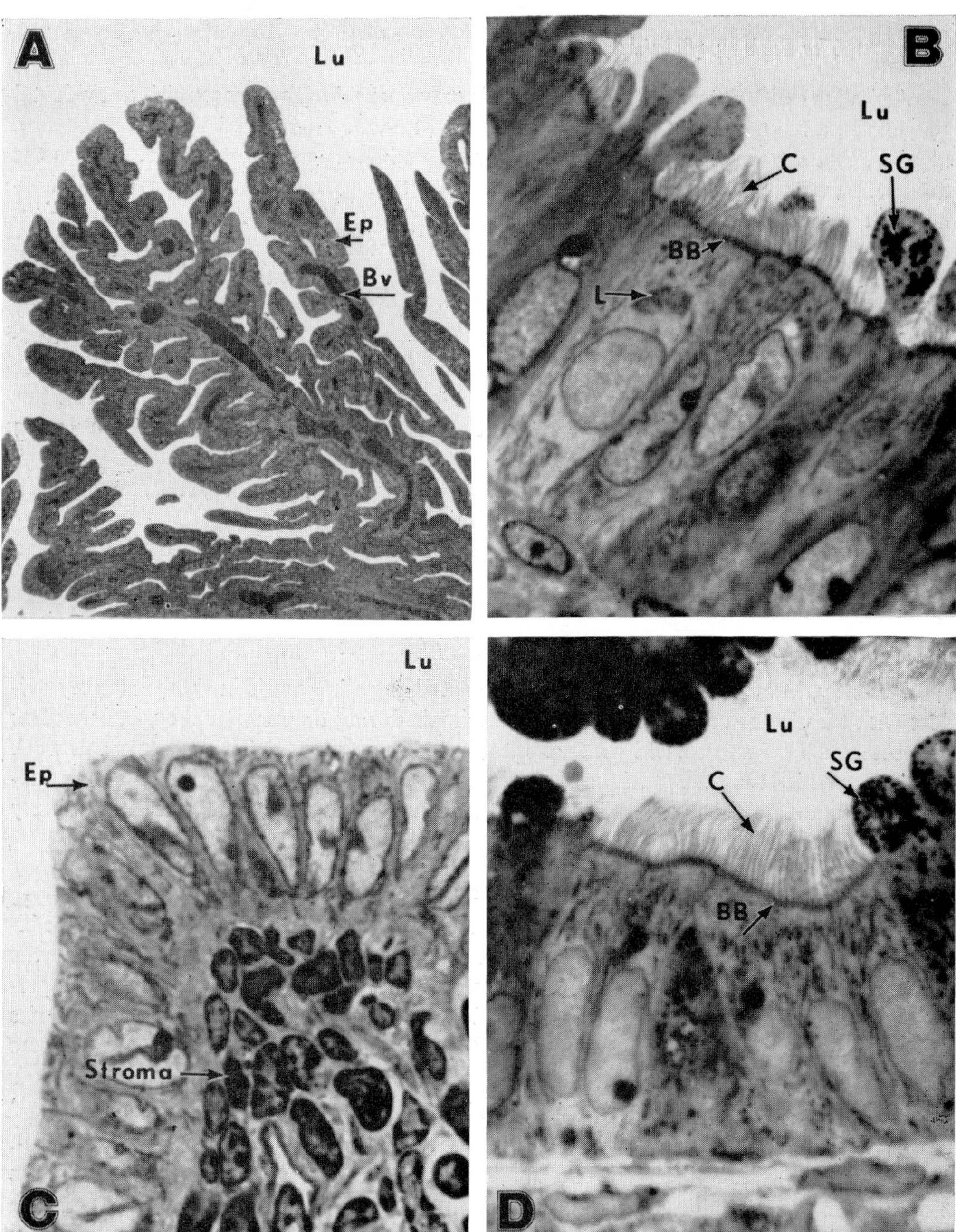

FIG. 4. (*a*) Fimbriated end of the oviduct from an intact rhesus monkey in the follicular phase of the menstrual cycle. Note the extreme folding of the epithelial surface (*Ep*) and the large number of blood vessels (*Bv*) ×39.

(*b*) Ciliated and secretory cells of the fimbriae shown in *a*. The ciliated cells are characterized by rows of basal bodies (*BB*) at the surface of the cell. Each basal body is attached to a cilium (*C*). Ciliated cells are also characterized by a large lipid complex (*L*). The secretory cells are characterized by small, dark granules (*SG*); *Lu* = Lumen (×990).

(*c*) Fimbriae from a monkey 6 weeks after ovariectomy. The epithelium is atrophied and the cilia, basal bodies, and secretory granules have been lost. The nuclei of stromal cells have become shrunken and hyperchromatic (×990).

(*d*) Fimbriae from an ovariectomized monkey after 2 weeks treatment with 10 μg estradiol benzoate/day. Estrogen treatment completely restores a normal epithelium with both ciliated and secretory cells (×990).

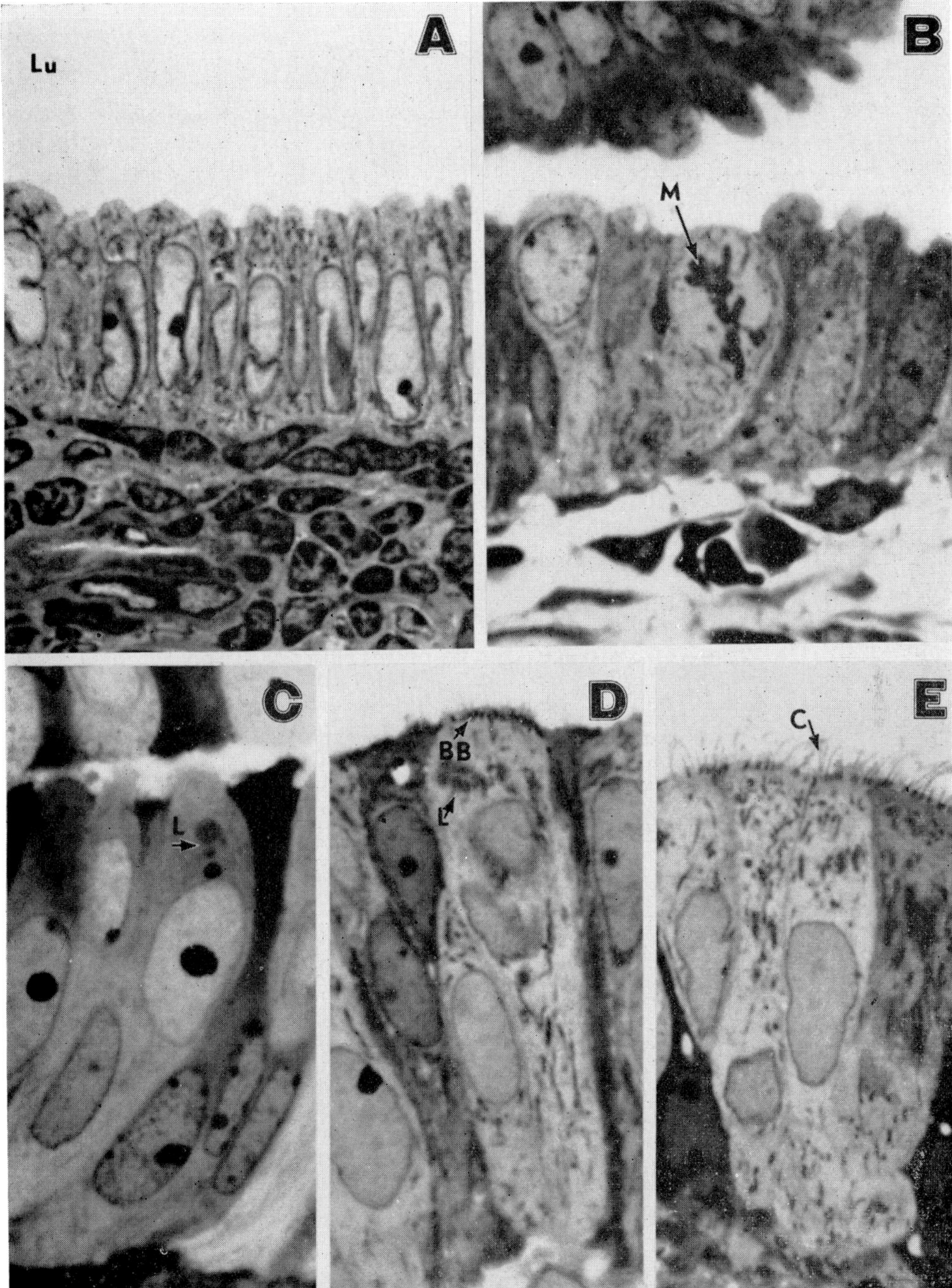

FIG. 5. Note: Photographs *a–e* are all of fimbriae from monkeys which had been ovariectomized 6 weeks previous to estrogen treatment. All the micrographs are at the same magnification:

(*a*) Fimbriae after 1 day of treatment with estradiol benzoate at 10 μg/day. There are no visible changes (×990). Compare with Figure 4*c*.

(*b*) Fimbriae after 2 days of treatment with estradiol benzoate at 10 μg/day. The epithelium shows both hypertrophy and hyperplasia. Mitotic figures (*M*) are common (×990).

(*c*) Fimbriae after 3 days of treatment with estrogen. The epithelial cells are maximally hypertrophied. Clear cells and dark cells are present. Only the clear cells contain the lipid complexes; these are the future ciliated cells (×990).

(*d*) Fimbriae after 4 days treatment with estrogen. The clear cells develop basal bodies which line up at the luminal surface of the cell (×990).

(*e*) Fimbriae after 5 days treatment with estrogen. The basal bodies sprout cilia and the cells attain a mature appearance (×990).

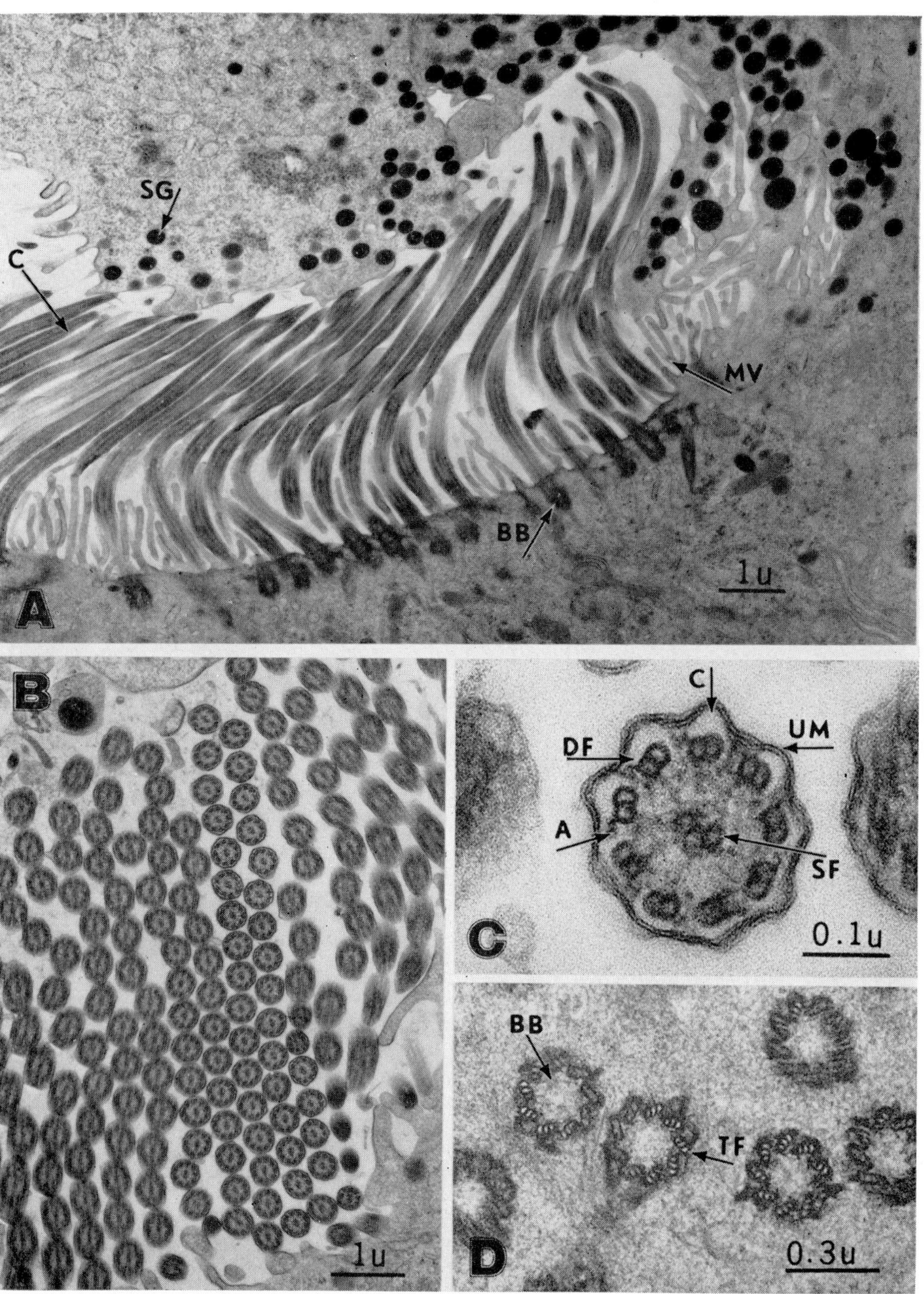

Fig. 6. (*a*) Cilia of fimbriae from an intact rhesus monkey. This cell was favorably fixed to illustrate the different stages of flexion characteristic of each cilium during the ciliary beat. Between the cilia, the cell surface has numerous microvilli. Secretory granules are present in a secretory cell adjacent to the ciliated cell (×10,030).

(*b*) A cross section through a group of cilia from the fimbriae. At any one time, each cilium is in a different phase of the beat. Thus the different cilia pass through the plane of the section at slightly different angles (×11,040).

(*c*) Cross section through a single cilium. The orderly substructure of the cilium can be seen. The cilium contains 9 peripheral double-fibrils (*DF*) and 2 central single-fibrils (*SF*). Small projections or arms (*A*) are present on one subfibril of each peripheral doublet. A delicate matrix material is present between the central and peripheral fibrils. The cilium is bounded by a unit membrane (*UM*) (×132,000).

(*d*) Cross section through a group of basal bodies. There are 9 peripheral triplet fibrils and no central fibrils (×44,000).

(Fig. 5*a*). The organelles are small and the ribosomes are in a monosomal state (Fig. 7*b*). The first sign of estrogen stimulation is an increase in mitotic activity in the oviductal epithelium which begins after 2 days of estrogen treatment (Fig. 5*b*). On the 3d day two cell types become evident, a "clear" cell and a dense cell (Fig. 5*c*). In the clear cells, particularly, the organelles show evidence of hypertrophy. Polyribosomes appear where only ribosomes had been present, mitochondria become longer, the Golgi membranes increase in amount, the endoplasmic reticulum becomes more prominent, and a large complex lipid aggregate appears distal to the nucleus (Fig. 8*a*, *b*). On the 4th day, aggregates of small, dense granules interwoven by a network of fine filaments appear at random in the cytoplasm of the clear cells (Fig. 9*a*). These aggregates resemble the structures referred to as proliferative elements (Dirksen and Crocker 1965). In many cells, the dense granules are enlarged and distinct from the filamentous network. These enlarged, dense granules are the so-called condensation forms (Fig. 9*b*). Scattered throughout the cytoplasm of many clear cells are clusters of very small centrioles, radially arranged around dense granules, which are the structures known as generative complexes (Fig. 9*b*).

In other clear cells, the radial clusters of nascent centrioles disappear and the cytoplasm is filled with enlarged basal bodies of mature appearance. These basal bodies eventually arrange themselves linearly at the luminal surface of the cell. Some of them have condensation forms still attached to the proximal end (Figs. 9*c*; 10*a*, *b*). Small ciliary buds then appear above the cell surface in contact with the centrioles and ciliary filamentogenesis occurs within this ciliary bud (Fig. 11*a*, *b*). From the 5th to the 9th day, the number of fully ciliated cells increases (Figs. 11*b*; 12).

Mature ciliated cells contain a large lipid complex (Fig. 4*b*) which serves as an identifying marker for the potential or actual ciliated cell. The dense cells lack the lipid aggregate and these cells develop only into secretory cells (Fig. 11*b*).

As noted previously, contact regions have been observed between the old centriole of the cell and the proliferative elements in fetal rat trachea and oviduct (Dirksen and Crocker 1965). Such connections have not been observed by us in monkey oviduct, but our failure to find such connections may be due only to the rarity of their occurrence. In general, the Dirksen-Crocker hypothetical scheme seems to fit the process we observed. Proliferative elements appear first, followed by condensation forms and generative complexes with their radially arranged clusters of nascent centrioles. The centrioles then enlarge, migrate to the cell surface, and develop ciliary buds from which the cilia grow (Fig. 3).

B. *Ciliogenesis during the Menstrual Cycle*

1. *Methods.* Thirty-five cycling, mature rhesus females were bilaterally ovariectomized, and concurrent biopsies of the fimbriae and the ampulla of the oviduct were obtained at various times during the menstrual cycle. The distribution of animals throughout the cycle is shown in the following tabulation.

Day in cycle	2	3	4	5	6	7	9	11	12	13	15	16	18	19	21	23	24	25	27
Number of monkeys	1	1	2	1	1	2	3	3	4	2	1	1	2	2	1	3	2	1	2

The fimbriae and ampullae were processed for electron microscopy, and the ovaries were embedded in paraffin and serially sectioned for light microscopic analysis. This work was done in collaboration with Koering, Resko, Goy, and Phoenix (1967).

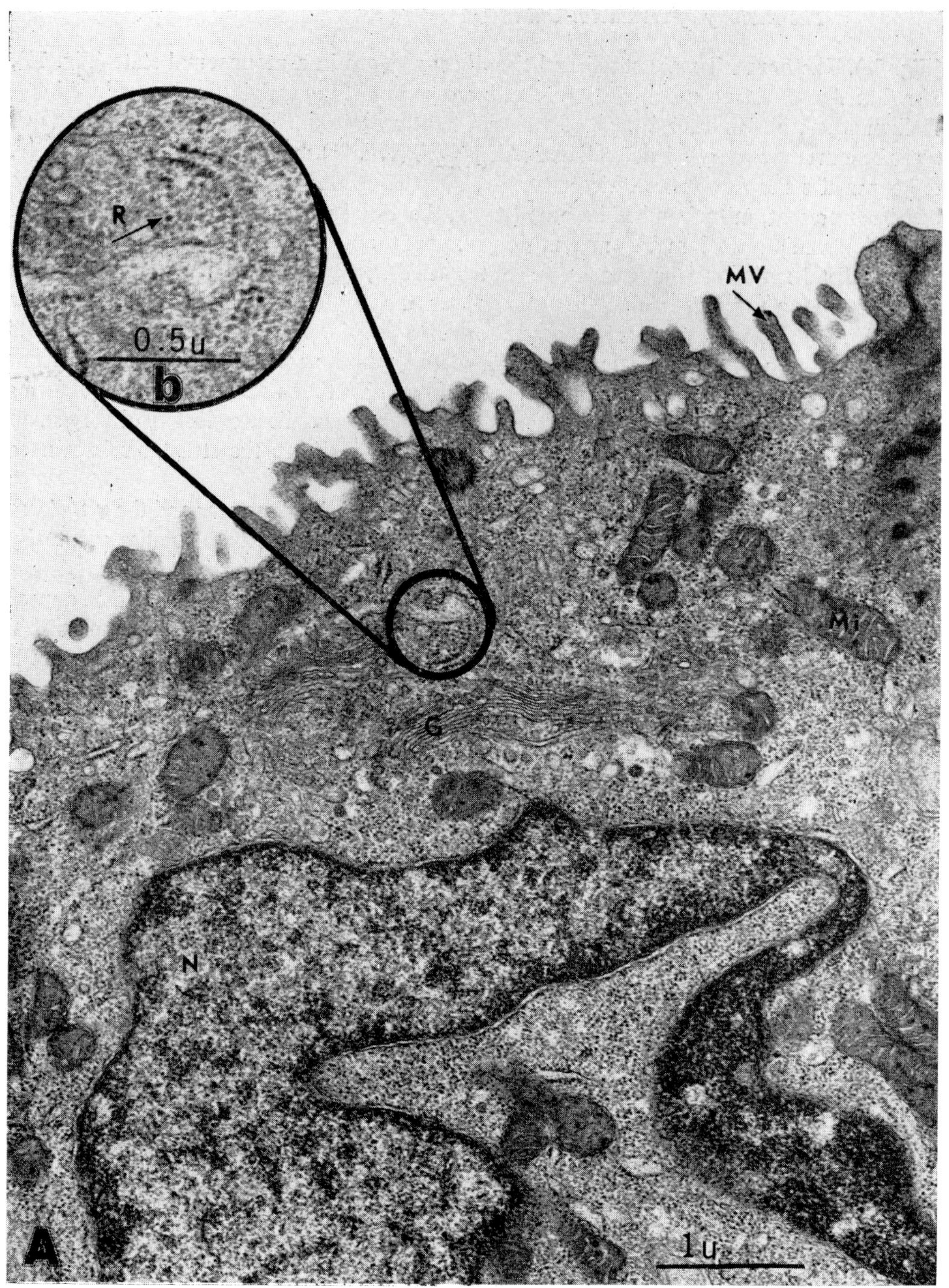

FIG. 7. (*a*) Cell from fimbriae of monkey 6 weeks after ovariectomy. All the cilia are gone, although 1 basal body (or centriole) remains at the surface of the cell. The mitochondria are small and dense, and the Golgi apparatus is relatively inactive. The microvilli are present but appear shorter than in the intact animal (×20,900).

(*b*) An enlargement of a cytoplasmic region from *a*. The ribosomes are in a single or monosomal state, which indicates a low level of protein synthesis (×41,800).

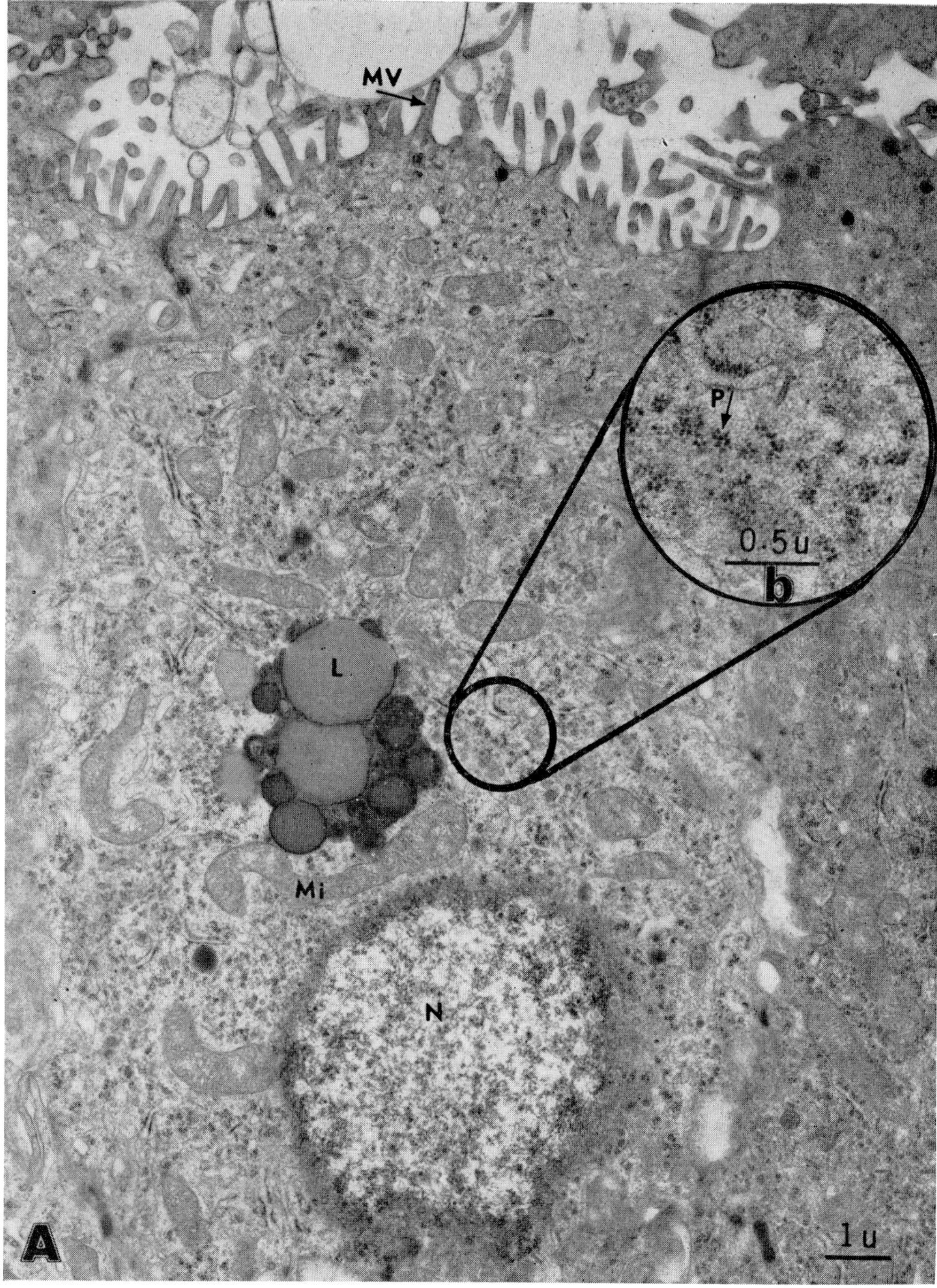

FIG. 8. (*a*) Fimbriae of ovariectomized monkey after 3 days treatment with estradiol benzoate, 10 μg/day. This cell is marked as a future ciliated cell by the presence of the large lipid complex (*L*) which is found only in ciliated cells and never in secretory cells. The microvilli and the mitochondria are enlarged and the cytoplasm has an abundance of polysomes. The endoplasmic reticulum is more extensive ($\times$9,450).

(*b*) The ribosomes are in a polysomal state, which indicates a high level of protein synthesis ($\times$28,350). (Compare with Fig. 7*b*.)

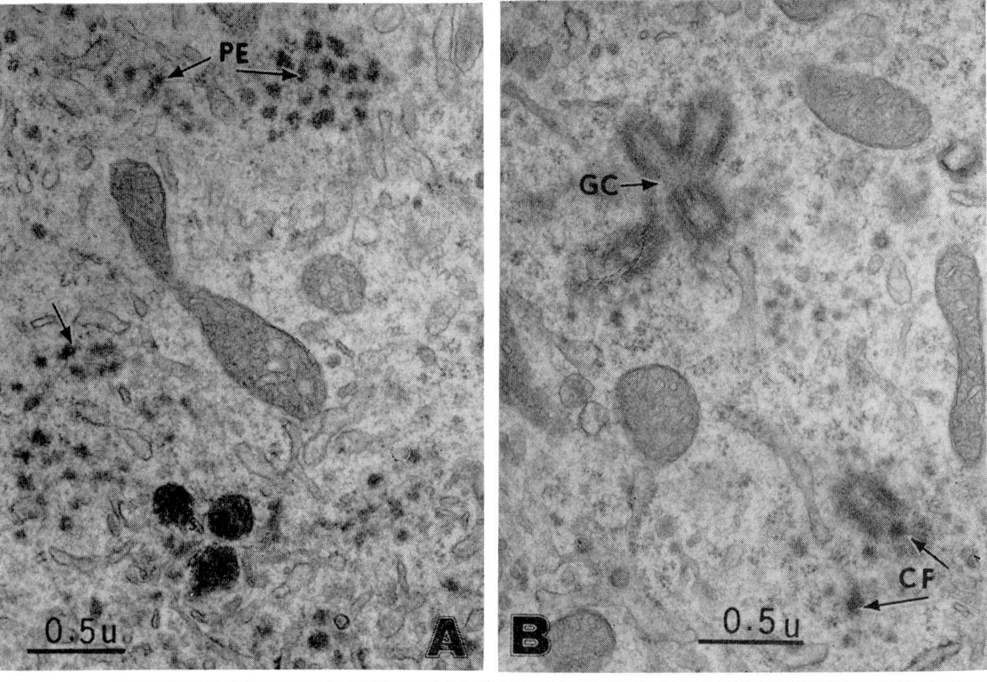

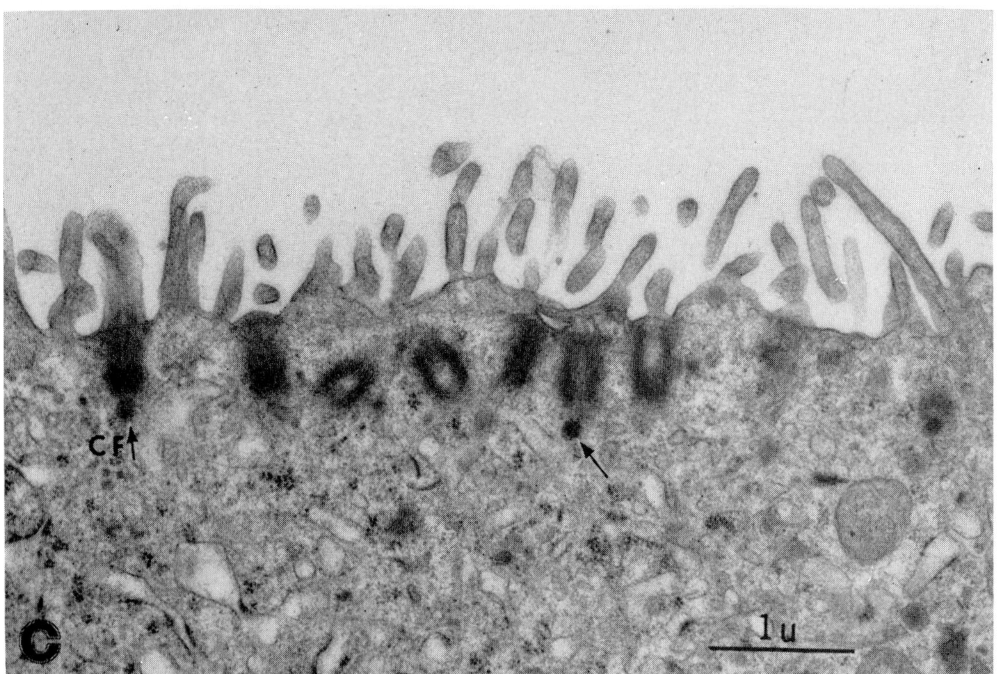

FIG. 9. (*a*) Fimbriae from ovariectomized monkey after 4 days estrogen treatment. The micrograph shows several clusters of small dense granules with indistinct borders associated with a delicate filamentous meshwork. These clusters are referred to as proliferative elements (*PE*) (×27,650).

(*b*) Fimbriae after 4 days estrogen treatment. The micrograph shows small groups of nascent centrioles which appear to radiate from a central region. In the lower right, nascent centrioles are associated with a dense granule which is referred to as a condensation form (*CF*). The combination of condensation forms with nascent centrioles is referred to as a generative complex (*GC*). In the generative complex at the upper left the central dense granule is not in the plane of the section (×27,650).

(*c*) Fimbriae after 4 days estrogen treatment. The basal bodies break away from the generative complexes and migrate to the surface of the cell. Several condensation forms are still attached to the basal bodies. Early cilia formation can be seen at the extreme left of the micrograph (×19,470).

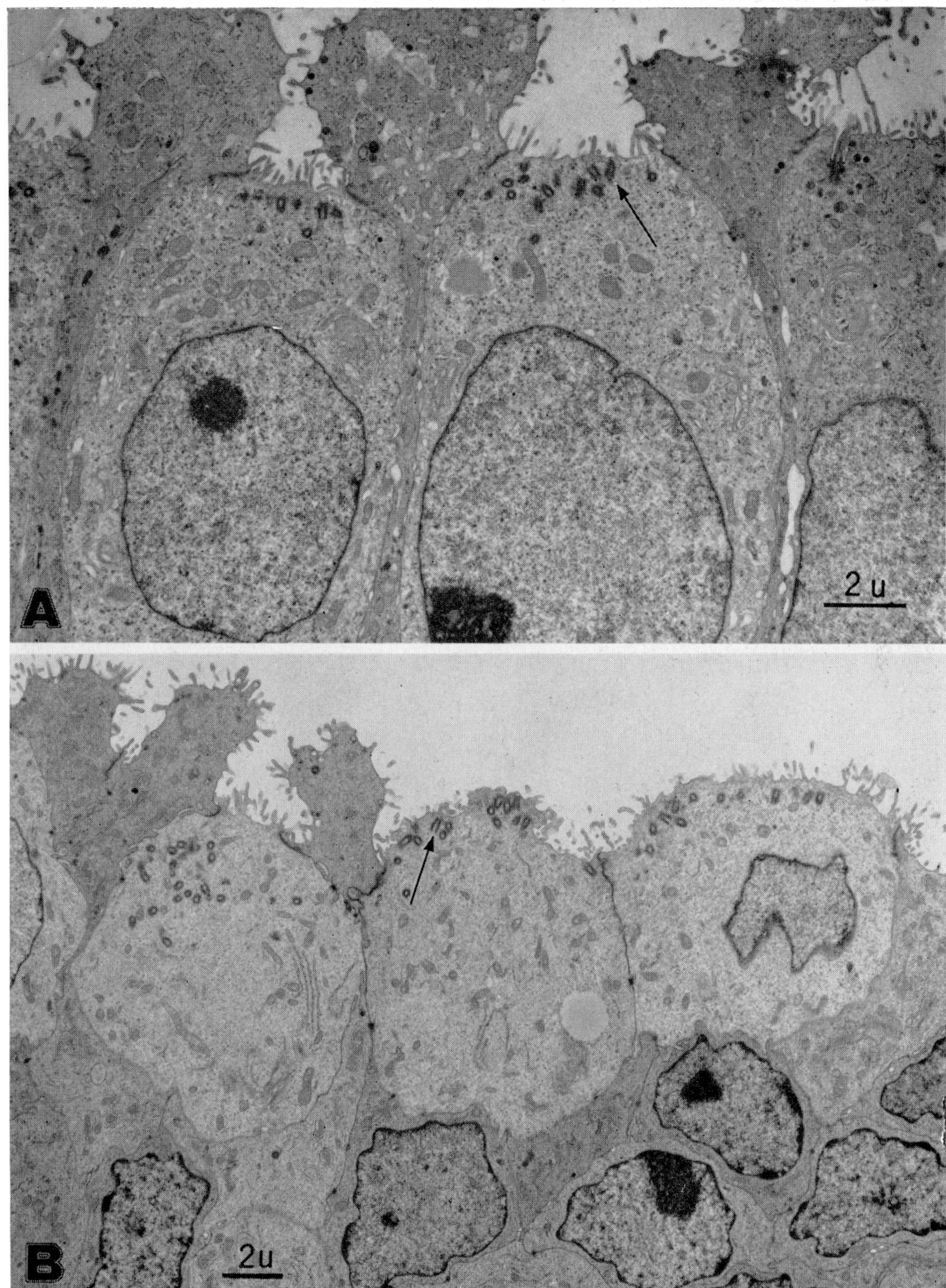

FIG. 10. (*a*) Fimbriae after 4 days of estrogen treatment. The basal bodies are migrating to the cell surface (×6,080).

(*b*) Fimbriae after 4 days of estrogen treatment. Another illustration of basal body migration. This micrograph, as well as *a*, illustrates that neighboring cells show a considerable degree of developmental synchrony (×4,140).

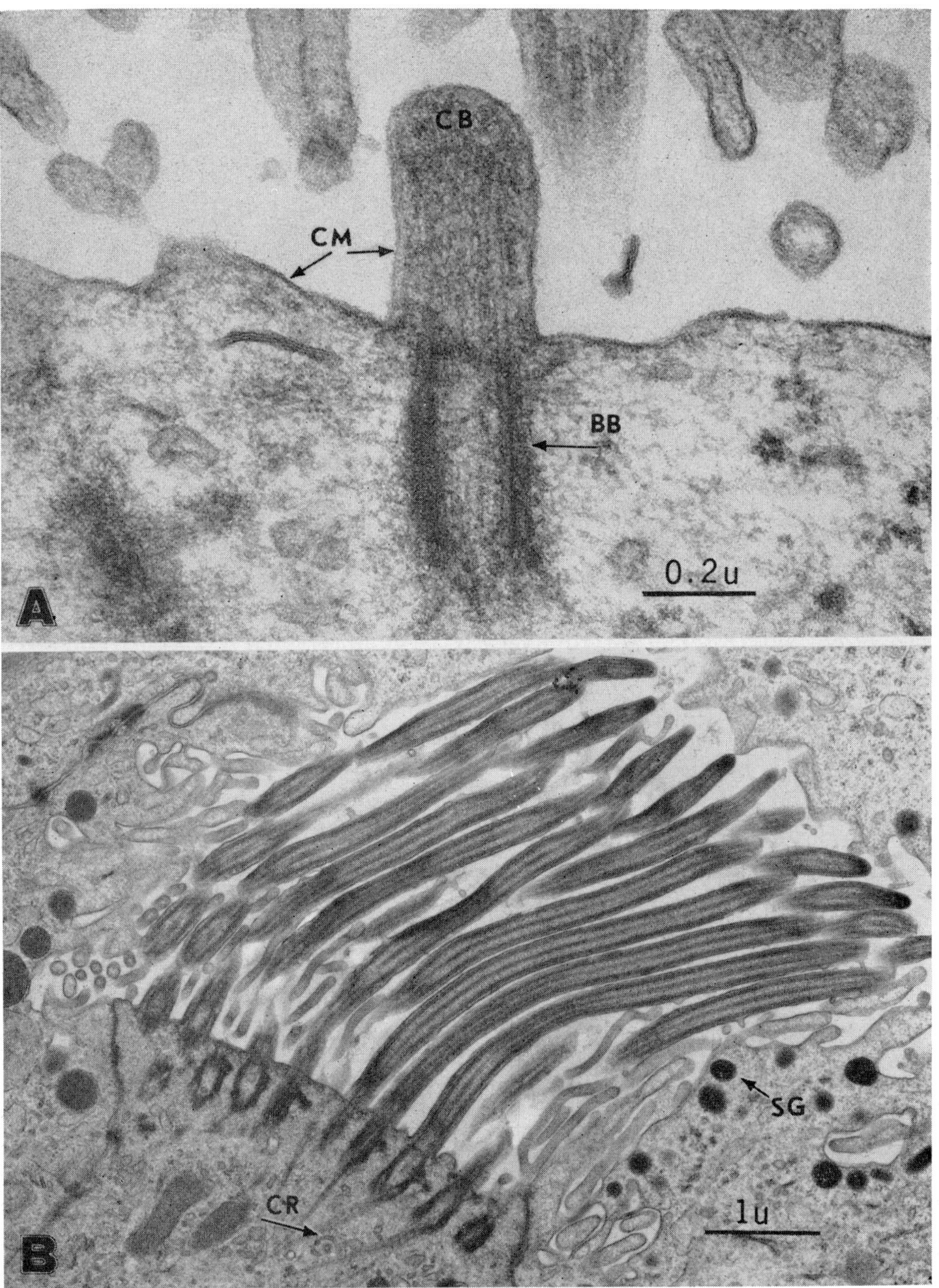

FIG. 11. (*a*) Fimbriae after 4 days of estrogen treatment. A ciliary bud (*CB*) develops once the basal body reaches the cell surface. The cell membrane becomes continuous with the membrane surrounding the cilium. Longitudinal sections of the ciliary fibrils are visible at the right of the bud (×81,090).

(*b*) Fimbriae after 5 days of estrogen treatment. Cilia reach their full length by the 5th day of treatment. Ciliary rootlets (*CR*) are present in this cell, though their occurrence is variable (×16,900).

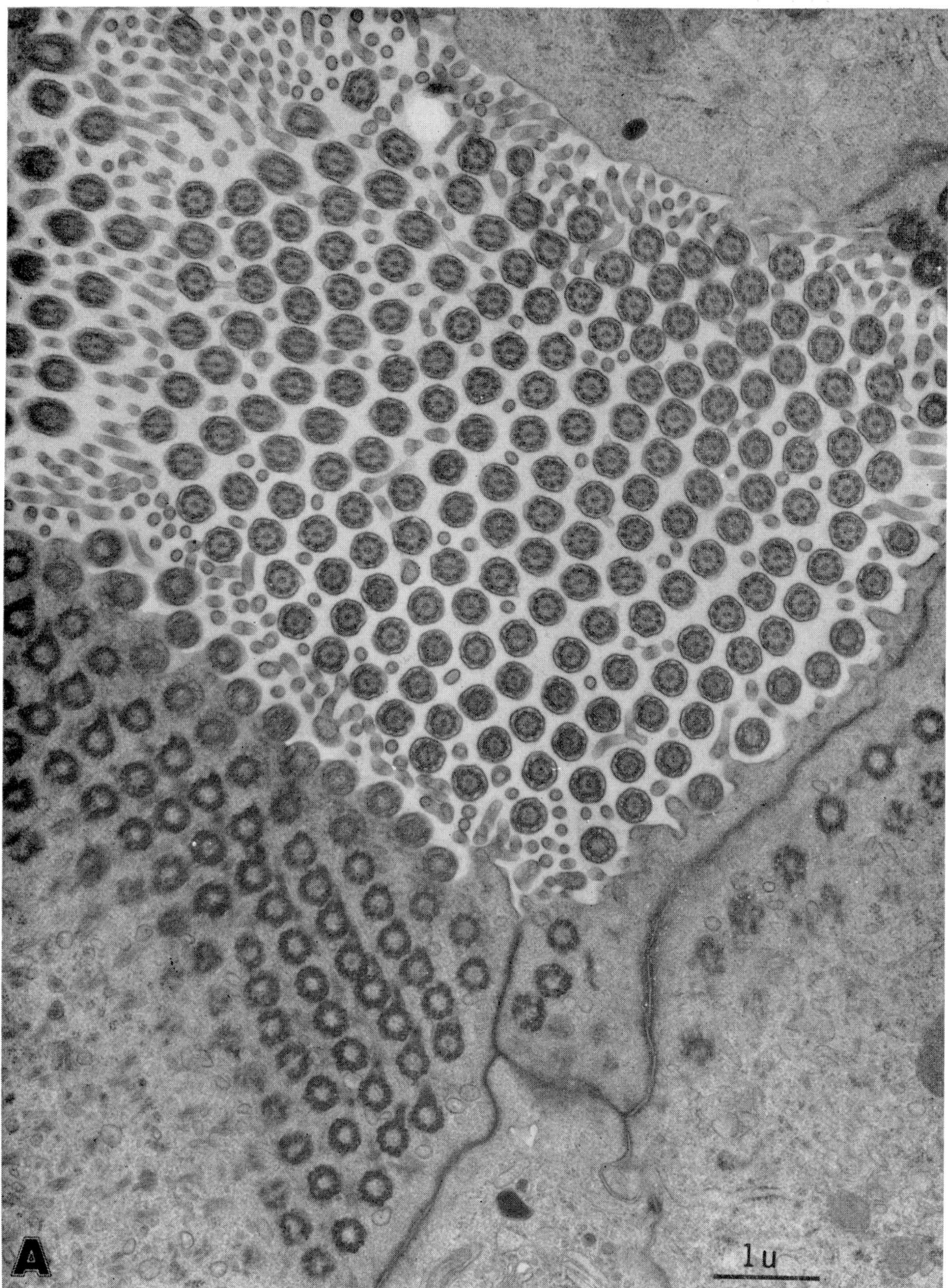

FIG. 12. Fimbriae after 5 days estrogen treatment. This micrograph is a very fortunate tangential section of a ciliated cell which effectively serial-sections the entire ciliary complex. On the left one sees the most basal regions of the basal bodies, then higher and higher regions, then the microvilli, and finally the ciliary shafts (×15,390).

2. *Observations.* The cyclic changes in the ciliated epithelium are strikingly different in the fimbriae as compared with the ampulla of the oviduct.

There are practically no mature ciliated cells in the fimbriae on day 2 of the cycle, but there are large numbers of clear cells in various phases of ciliogenesis. The cytoplasm of these cells contains all the various basal body precursor structures previously described. In many cells the basal bodies are lined up at the cell margins and some cells have sprouted tiny cilia. From days 3 through 9, the number of cells in ciliogenesis decreases while the number of tall, mature ciliated cells increases. The ciliated cells remain tall and normal in appearance through day 18, at which time they begin to diminish in both height and number. Many of them become dense and shrunken; and in others the distal third of the cell, including the entire ciliary apparatus, pinches off and is shed into the lumen in a manner similar to an apocrine secretion. On days 21 through 27, approximately 95% of the epithelial cells are completely deciliated. Only an occasional, short ciliated cell remains. At this time the fimbriae resemble those from ovariectomized animals.

In the ampulla, however, there is no dramatic loss of ciliated cells, though there is a definite change in cell height. On days 2 through 9, there are some ciliogenic cells, but not as many as in the fimbriae. The cells are tallest through the mid-portion of the cycle and by day 18 they begin to shrink. Some ciliated cells become very dark and atrophied, and a number of them shed their ciliated tips, but the general degree of ciliary loss is minimal compared with that which occurs on the fimbriae. On days 21–27 the ampullary epithelium is characteristically very low but fully ciliated. It is indeed surprising that such a difference in behavior should occur in tissues so similar in nature.

One very interesting observation occurred entirely by chance in this study. The ovaries of two animals, which had been obtained on day 23 and day 27 of the menstrual cycle, had large follicles and no fresh corpora lutea, which indicates that ovulation had not occurred. The fimbriae of these two animals had a fully ciliated epithelium resembling the epithelium of animals in the mid-portion of the cycle. Evidently, the estrogen levels had not fallen low enough or the progesterone levels risen high enough to provoke the normal deciliation process in the fimbriae of these two animals. The ovaries from the animals in which deciliation of the fimbriae had occurred by days 21–27 all showed corpora lutea. It seems evident, therefore, that development of corpora lutea and an increase in the blood level of progesterone are prerequisites for the loss of ciliated cells from the fimbriae.

This study of the menstrual cycle supports the contention that the ciliated epithelium of both the ampullary and fimbriated portions of the oviduct is sensitive to the varying levels of estrogen and progesterone found during the cycle but that only the fimbriated portion shows a dramatic cyclic loss and regeneration of ciliated cells. After ovariectomy, however, the ampullary region of the oviduct eventually loses most of its ciliated cells. It seems reasonable to conclude, therefore, that the real difference between fimbriae and ampulla is one of sensitivity, and that the fimbriae are more responsive to changes in estrogen-progestin blood levels than the ampulla.

C. *Future Problems*

From these findings a series of questions immediately arises. Is the rate of ciliogenesis dose-dependent? What is the minimal dose that will provoke ciliogenesis? Will this minimal dose be different for the fimbriae versus the ampulla? Will an extremely high dose cause each cilium to grow longer, or will more ciliated cells regenerate? How do cilia grow—do they add new proteins at their growing tips or at their bases? If they add new proteins at the base, how do the old proteins migrate up the growing shaft? What is the role of the old centriole in the replication of the new crop of basal bodies? Will inhibitors of RNA synthesis and protein synthesis prevent cilia regeneration in mammalian oviducts, as they do in protozoa? Can we find a way of specifically inhibiting ciliogenesis without interfering with other cell functions? If so, would it be of any contraceptive value? What are the biochemical changes associated with each phase of ciliogenesis? Are there key enzymes which are synthesized at specific times during the growth process? Is there a characteristic pattern of enzymogenesis? These and other questions point the way to an exciting period of research on the control of ciliogenesis in the mammalian oviduct.

Acknowledgment

This work was supported in part by Public Health Service Grant HD-02753-01 and Population Council Grant M66.110. Publication Number 254 of The Oregon Regional Primate Research Center, Beaverton, Oregon. I would like to express my appreciation to Miss Mary Jane Kamm and Miss Carol Kurilo for their excellent technical assistance and to Miss Betty Kohl for her superb secretarial work.

References

1. Allen, E. 1922. The oestrous cycle in the mouse. *Amer. J. Anat.* 30:297–372.
2. ———. 1928*a*. Reactions of immature monkey (*Macacus rhesus*) to injections of ovarian hormone. *J. Morphol.* 46:479–520.
3. ———. 1928*b*. Further experiments with an ovarian hormone in the ovariectomized adult monkey, *Macacus rhesus*, especially the degenerative phase of the experimental menstrual cycle. *Amer. J. Anat.* 42:467–87.
4. Andrews, M. C.: 1951. Epithelial changes in the puerperal Fallopian tube. *Amer. J. Obstet. Gynec.* 62:28–37.
5. Balboni, G. 1954. Ricerche istochimischo sull'epitelio tubarico della donna. *Riv. Ostet. Ginec.* 9:164–94.
6. Bernhard, W., and Deharven, E. 1960. L'ultrastructure du centriole et d'autres éléments de l'appareil achromatique. *Quatrième Congr. Int. Micr. Electr., Berlin* 2:217–27.
7. Borell, U.; Nilsson, O.; Wersäll, J.; and Westman, A. 1956. Electron-microscope studies of the epithelium of the rabbit Fallopian tube under different hormonal influences. *Acta Obstet. Gynec. Scand.* 35:36–41.
8. Bradbury, P., and Pitelka, D. R. 1965. Observations on kinetosome formation in an apostome ciliate. *J. Microscopie* 4 (6):805–10.
9. Brenner, R. M. 1966. Adrenocortical fine structure in the rhesus monkey. *Anat. Rec.* 154:321.
10. ———. 1967*a*. Electron microscopy of estrogen effects on ciliogenesis and secretory cell growth in rhesus monkey oviduct. *Anat. Rec.* 157:218.
11. ———. 1967*b*. Ciliogenesis during the menstrual cycle in rhesus monkey oviduct. *J. Cell Biol.* 35:16A.
12. Brenner, S.; Jacob, F.; and Meselson, M. 1961. An unstable intermediate carrying information from genes to ribosomes for protein synthesis. *Nature* 190:576–81.
13. Bruni, A. C. 1950. L'intima struttura delle trombe uterine e il transito dell'ovulo. *Monit. Zool. Ital.* 69: Suppl. 1–36.

14. Child, F. M. 1965. Ciliary co-ordination in glycerinated mussel gills. In *Progress in protozoology*, p. 110. Amsterdam: Excerpta Med. Found.
15. Clyman, J. J. 1966. Electron microscopy of the human Fallopian tube. *Fertil. Steril.* 17:281–301.
16. Cotte, G. 1949. *Troubles fonctionelles de l'appareil génitale de la femme*. Paris: Masson & Cie.
17. Deane, H. W. 1952. Histochemical observations on the ovary and oviduct of the albino rat during the estrous cycle. *Amer. J. Anat.* 91:363–413.
18. Dirksen, E. R., and Crocker, T. T. 1965. Centriole replication in differentiating ciliated cells of mammalian respiratory epithelium: An electron microscopic study. *J. Microscopie* 5:629–44.
19. Dubnau, D. A. 1961. The regeneration of flagella by *Ochromonas danica*. Thesis, Columbia Univ., N.Y.
20. Espinasse, P. G. 1935. The oviducal epithelium in the mouse. *J. Anat. Lond.* 69:363–68.
21. Faure-Fremiet, E. 1951. Cils vibratiles et flagelles. *Biol. Rev.* 36:464–536, 1961.
22. Fawcett, D. W. 1961. Cilia and flagella. In *The cell*, ed. J. Brachet and A. E. Mirsky, vol. 2, pp. 217–97. New York: Academic Press.
23. Flerkó, B. 1951. Einfluss experimenteller Hypothalamus-Läsion auf das Eileiterepithel. *Acta Morph. Acad. Sci. Hung.* 1:5–14.
24. ———. 1954. Die Epithelien des Eileiters und ihre hormonalen Reaktionen. *Z. Mikroskopisch-anat. Forsch.* 61:99–118.
25. Fredricsson, B. 1959*a*. Histochemical observations on the epithelium of human Fallopian tubes. *Acta Obstet. Gynec. Scand.* 58:109–34.
26. ———. 1959*b*. Studies on the morphology and histochemistry of the Fallopian tube epithelium. *Acta Anat. Basel*, Suppl. 37 38:3–23.
27. Frisch, D. 1967. Fine structure of the early differentiation of ciliary basal bodies (abstr.). *Anat. Rec.* 157 (2):245.
28. Gall, J. G. 1961. Centriole replication: A study of spermatogenesis in the snail *Viviparus*. *J. Biophys. Biochem. Cytol.* 10:163–93.
29. Grimstone, A. V. 1961. Fine structure and morphogenesis in Protozoa. *Biol. Rev.* 36: 97–150.
30. Hamperl, H. 1950. Über die "hellen" Flimmerepithelzellen de menschlichen Uterusschleimhaut. *Virchows Arch. Path. Anat.* 319:265–81.
31. Hashimoto, M.; Shimoyama, T.; Kosaka, M.; Komori, A.; Hirasawa, T.; Yokoyama, Y.; Kawase, N.; and Nakamura, T. 1964. Electron microscopic studies on the epithelial cells of the human Fallopian tube (Report II). *J. Jap. Obstet. Gynec. Soc.* 11 (2):92–100.
32. Henneguy, L. F. 1898. Sur les rapports des cils vibratiles avec les centrosomes. *Arch. Anat. Micr.* 1:482–95.
33. Hølund, T. 1946. Tuba-og Uterinslimkindens Epithel morfologiske og cytologiske undersøgelser. Copenhagen: A. Busek.
34. Hörmann, E. 1909. Über das Bindegewebe der weiblichen Geschlechtsorgane. *Arch. Gynaek.* 86:404–33.
35. Holwill, M. E. J. 1966. Physical aspects of flagellar movement. *Physiol. Rev.* 46 (4):696–785.
36. Horstmann, E., and Stegner, H. 1966. Tube, Vagina und äußere weibliche Genitalorgane. In *Handbuch der mikroskopischen Anatomie des Menschen*, vol. 7/4, 35–85. Berlin: Springer.

37. Joachimovits, R. 1935. Studien zur Menstruation, Ovulation, Aufbau und Pathologie des weiblichen Genitales bei Mensch und Affe (*Pithecus fascicularis mordax*), Eileiter und Ovar. *Biol. Generalis* (*Wien*) 11:281–38.
38. Jordan, H. E., and Helvestine, F. 1923. Ciliogenesis in the epididymis of the white rat. *Anat. Rec.* 25:7–18.
39. Karlson, P. 1963. New concepts on the mode of action of hormones. 1963. *Perspect. Biol. Med.*, 6 (2):203–13.
40. Kinosita, H., and Murakami, A. 1967. Control of ciliary motion. *Physiol. Rev.* 47 (1): 53–82.
41. Koering, M.; Resko, J.; Goy, R.; Phoenix, C. 1967. Personal communication.
42. Lenhossek, M. von. 1898. Über Flimmerzellen. *Verhandl. Anat. Gesellsch. Kiel* 12:106.
43. Lwoff, A. 1950. *Problems of morphogenesis in ciliates.* New York: John Wiley & Sons.
44. Manton, I. 1959. Electron microscopical observations on a very small flagellate: the problem of *Chromulina pusilla* Butcher. *J. Marine Biol. Assoc. U.K.* 38:319–33.
45. Mihálik, P. von. 1934. Die Bildung des Flimmerapparates im Eileiterepithel. *Anat. Anz.* 79:259–65.
46. Mizukami, I., and Gall, J. 1966. Centriole replication. II. Sperm formation in the fern, *Marsilea*, and the Cycad, *Zammia. J. Cell Biol.* 29 (1):97–111.
47. Moore, R. J., and Hamilton, T. H. 1964. Estrogen-induced formation of uterine ribosomes. *Proc. Natl. Acad. Sci.* 52 (2):439–46.
48. Naumann, K. 1931. Schwangerschaftsveränderungen am menschlichen Eileiter. *Zbl. Gynaek.* 55:3618–23.
49. Nilsson, O. 1957. Observations on a type of cilia in the rat oviduct. *J. Ultrastruct. Res.* 1:170–77.
50. Nomura, M.; Hall, B. D.; and Spiegelman, S. 1960. Characterization of RNA synthesized in *Escherichia coli* after bacteriophage T2 infection. *J. Mol. Biol.* 2:306.
51. Novak, E., and Everett, H. S. 1928. Cyclical and other variations in the tubal epithelium. *Amer. J. Obstet. Gynec.* 16:499–530.
52. Odor, D. L. 1953. Electron microscopy of rat oviduct. *Anat. Rec.* 115:434–35.
53. Overbeck, L. 1967. Entwicklung der Kinocilien im Tubenepithel des Menschen. *Naturwissenschaften* 54 (9):229.
54. Pauerstein, C. J., and Woodruff, J. D. 1967. The role of the "indifferent" cell of the tubal epithelium. *Amer. J. Obstet. Gynec.* 98 (1):121–25.
55. Pernkopf, E. and Pichler, A. 1953. Systematische und topographische Anatomie des weiblichen Beckens. In *Biologie und Pathologie des Weibes*, ed. L. Seitz and A. J. Amreich, pp. 83–211. Berlin: Urban & Schwarzerberg.
56. Pitelka, D. R. 1963. *Electron microscopic structure of protozoa.* London: Pergamon.
57. Reich, E.; Franklin, R. M.; Shatkin, A. J.; and Tatum, E. L. 1961. Effect of actinomycin D on cellular nucleic acid synthesis and virus production. *Science* 134: 556–57.
58. Renaud, F., and Swift, H. 1964. The development of basal bodies and flagella in *Allomyces arbusculus. J. Cell Biol.* 23:339–54.
59. Rhodin, J., and Dalhamn, T. 1956. Electron microscopy of the tracheal ciliated mucosa in the rat. *Z. Mikroskopischanat. Forsch.* 62:345-412.
60. Rivera, J. A. 1962. *Cilia, ciliary activity and ciliated epithelium.* Oxford: Pergamon.
61. Rosenbaum, J. L. 1965. Effect of cycloheximide on flagellar regeneration in *Euglena gracilis.* (abstr.) *J. Cell Biol.* 27:89A–90A.

62. Roth, J. E., and Shigenaka, Y. 1964. The structure and formation of cilia and filaments in rumen protozoa. *J. Cell Biol.* 20:249–70.
63. Satir, P. 1965. Structure and function in cilia and flagella: Protoplasmatologia. In *Handbuch der Protoplasmaforschung*, vol. 3/E, 1–52. New York: Springer.
64. Schaffer, J. 1908. Über Bau und Funktion des Eileiterepithels beim Menschen und bei Säugetieren. *Mschr. Geburtsh. Gynaek.* 28:526–42.
65. Scheyer, H.-E. 1926. Über die Lipoide der Tube. *Virchows Arch. Path. Anat.* 262:712–34.
66. Schröder, R. 1930. Der Eileiter. In *Handbuch der mikroskopischen Anatomie des Menschen.* Berlin: Springer.
67. Schultka, R. 1963. Der Sekretionscyklus der Flimmerzellen der menschlichen Tuba uterina auf Grund cytologischer und cytotopochemischer Untersuchungen. *Acta Histochem.* 15:285–315.
68. Schultka, R., and Scharf, J. H. 1963. Sekretionszyklus der Tubenepithelzelle in Abhängigkeit vom ovariellen Zyklus. *Zbl. Gynaek.* 85:1601–1906.
69. Schuster, F. 1963. An electron microscopic study of the amoeboflagellate *Naegleria gruberi* (Schardinger). I. The amoeboid and flagellate stages. *J. Protozool.* 10:297–312.
70. Shimoyama, T. 1963*a*. Electron microscopic study of the epithelial cells of the Fallopian tube mucosa in the mature woman. *J. Jap. Obstet. Gynec. Soc.* 15:1237.
71. ———. 1963*b*. Electron microscopic studies of the epithelial cells of the mucosa membrane of the human Fallopian tube in the foetus, pregnancy and menopause. *J. Jap. Obstet. Gynec. Soc.* 15:1307.
72. Sleigh, M. A. 1962. *The biology of cilia and flagella.* London: Pergamon.
73. Snyder, F. F. 1923. Changes in the Fallopian tube during the ovulation cycle and early pregnancy. *Bull. Johns Hopkins Hosp.* 34:121–25.
74. ———. 1924. Changes in the human oviduct during the menstrual cycle and pregnancy. *Bull. Johns Hopkins Hosp.* 35:141–46.
75. Sorokin, S. 1962. Centrioles and the formation of rudimentary cilia by fibroblasts and smooth muscle cells. *J. Cell Biol.* 15 (2):363–77.
76. Sotelo, J. R., and Trujillo-Cenoz, O. 1958. Electron microscope study of the development of ciliary components of the neural epithelium of the chick embryo. *Z. Zellforsch.* 49:1–12.
77. Stegner, H.-E. 1961. Das Epithel der Tuba uterina des Neugeborenen, elektronenmikroskopische Befunde. *Z. Zellforsch.* 55:247–62.
78. ———. 1962. Elektronenmikroskopische Untersuchungen über die Sekretionsmorphologie des menschlichen Tubenepithels. *Arch. Gynaek.* 197:351–63.
79. Tamm, S. L. 1965. Flagellar development in the Protozoan *Peranema trichophorum* (abstr.) *J. Cell Biol.* 27:104A.
80. Tietze, K. 1929. Zur Frage nach den zyklischen Veränderungen des menschlichen Tubenepithels. *Zbl. Gynaek.* 53:32–38.
81. ———. 1932. Histologische Tubenveränderungen in den einzelnen Lebensphasen des Weibes und bei Ovarialtumoren. *Arch. Gynaek.* 148:724–37.
82. Tokuyasu, K., and Yamada, E. 1959. The fine structure of the retina studied with the electron microscope. IV. Morphogenesis of the outer segments of the retinal rods. *J. Biophys. Biochem. Cytol.* 6:225–30.
83. Treche, A. 1893. Essai sur la morphologie de l'épithélium tuboutérine chez la femme et dehors de la grossesse et de la menstruation. Thesis, Nancy, France.

84. Tröscher, H. 1917. Über den Bau und die Funktion des Tubenepithels beim Menschen. *Mschr. Geburtsh. Gynaek.* 45:205–20.
85. Ui, H., and Mueller, G. C. 1963. The role of RNA synthesis in early estrogen action. *Proc. Natl. Acad. Sci.* 50 (2):256–60.
86. Volkin, E., and Astrachan, L. 1956. Phosphorus incorporation in *Escherichia coli* ribonucleic acid after infection with bacteriophage T2. *Virology* 2:149.
87. Westman, A. E. 1930. Studies of the function of the mucous membrane of the uterine tube. *Acta Obstet. Gynec. Scand.* 10:288–98.
88. ———. 1932. Studien über den Sexualzyklus bei Makakus-Rhesus-Affen, nebst einigen Bemerkungen über den menstruellen Blutungsmechanismus. *Acta Obstet. Gynec. Scand.* 12:282–328.
89. ———. 1934. Einige Bemerkungen aus Anlass des Aufsatzes von Jägeroos: Die sexualzyklischen Umwandlungen in der Tuba uterina beim Menschen und bei den niedrigen Primaten. *Acta Obstet. Gynec. Scand.* 13:263–68.
90. Williams-Ashman, H. G. 1965. New facets of the biochemistry of steroid hormone action. *Cancer Res.* 25:1096–1120.
91. Wilson, J. D. 1963. The nature of the RNA response to estradiol administration by the uterus of the rat. *Proc. Natl. Acad. Sci.* 50 (1):93–100.

9

Sperm Physiology in Relation to the Oviduct

D. W. Bishop

Department of Animal Science
Cornell University
Ithaca, New York

In light of less than a comfortable body of evidence as to precisely what transpires within the mammalian oviduct *in situ* after insemination, any contribution to an understanding of sperm function, vis-à-vis the internal environment, is immediately confronted by a two-pronged dilemma. To what extent can *in vitro* data on sperm behavior be safely applied to processes presumed to occur within the genital tract? What are the normal, if changing, conditions in the interior of the oviduct, particularly with respect to its successive segments and micro-environments? It is tempting to speculate, and easier to jump to conclusions, but in the following pages a conservative posture will be assumed, if for no other reason than to indicate wherein lie the areas of ignorance and, therefore, the areas of possibly fruitful new research effort. If all the answers are not yet in, this symposium volume should be instrumental in suggesting new ways, means, and approaches to oviductal physiology—timely and overdue, since 400 years have already lapsed since the age of Gabriellus Fallopius, whose observations and discoveries graced the appellation of the human female oviduct.

I. The Problem

The oviduct must perform certain functions and provide an environment amenable to spermatozoa which engage in internal fertilization, a luxury enjoyed by higher forms of terrestrial, air-breathing animals. The spermatozoa, on their part, display a remarkable capacity for adaptation and are well equipped to adjust to a new, essentially foreign environment, but they reflect the complicated history of their phylogenetic and developmental endowment (Bishop 1961). Sperm function can be approached from a variety of angles and may be viewed not only in physiological, biochemical, and ultrastructural terms, but also in light of

This contribution is dedicated with affection and appreciation to Carl G. Hartman. The author is presently at the Department of Physiology, Medical College of Ohio at Toledo.

comparative and differentiative processes—with fertilizing capacity being the final voice of judgment.

We have elected to pursue the subject in the most direct, if pedestrian, manner possible; if erudition is to suffer, perhaps elucidation may prevail. First, there is the problem of *transport* through the female tract: the mechanical activity of the oviducts, the possible role of sperm motility itself, and the rate and duration of migration, since once sperm activation is initiated, the countdown to fertilization is generally set into motion. Second, the processes associated with sperm *maturation*, including prefertilization changes, begin in the testes but seem to continue throughout passage of the gametes through both male and female genital tracts; maturation of spermatozoa, short of overaging, is essential for fertilization. Third, the *metabolism* of spermatozoa within the oviduct, including the questions of patterns, rates, and controls, underlies the entire framework of gametic welfare and survival and is, perhaps, the most difficult to assess. Finally, there is that area of miscellany which includes such phenomena as the terminal cell-to-cell specific interaction of sperm with egg in lieu of attachment to any other cell, the fate of excess spermatozoa, and the general failure of sperm to induce any significant degree of sensitization, introduced as they are as "foreign bodies" into a recipient host.

II. Oviductal Fluid

Although more keenly detailed and recent studies of the nature of the oviductal contents are presented elsewhere in this book, one may be forgiven for a brief discussion here of certain general aspects of oviductal physiology, since it serves as a frame of reference for consideration of sperm function. Surprisingly few species have been investigated in this regard and none has been explored with thoroughness. The most extensive studies of oviductal fluid have been confined to the rabbit (Bishop 1956*a*, *b*, 1957; Clewe and Mastroianni 1960; Mastroianni and Wallach 1961; Mastroianni *et al.* 1961; Mastroianni and Jones 1965; Hamner and Williams 1965), Rhesus monkey (Mastroianni, Shah, and Abdul-Karim 1961), ewe (Black, Duby, and Riesen 1963; Perkins *et al.* 1965; Perkins and Goode 1966; Restall 1966*a*, *b;* Restall and Wales 1966), and cow (Olds and VanDemark 1957*a*, *b*), and in all of these the endocrine control or modulation of oviductal secretion has been demonstrated. It is noteworthy that virtually no investigations have as yet been carried out on human oviductal fluid.

Oviductal fluid is the product of active secretion and is not merely a passive transudate or diffusable filtrate from the blood-vascular system. The cellular and histological makeup of the oviductal epithelium would indicate that this would be so (Novak and Everett 1928; Hadek 1955; Restall 1966*c*). Active secretion against a pressure gradient has been demonstrated in the rabbit (Fig. 1) and this proved to be a pilocarpine-sensitive (parasympathetic) process (Bishop 1956*a*). Consistent rates of hormone-dominated oviductal fluid production strongly indicate an active rather than passive mechanism in the rabbit, monkey, and sheep; secretion rate is elevated during estrus and is depressed during the luteal phase (Bishop 1956*a;* Mastroianni *et al.* 1961; Mastroianni, Shah, and Abdul-Karim 1961; Restall 1966*b*). The chemical composition of the luminal fluid itself bespeaks a selective secretory process, or at least a highly modified and refined system of transudation, if blood plasma is to be considered the source material. The distinctive, generally alkaline, pH values of oviductal fluid (Bishop 1957; Blandau, Jensen, and Rumery 1958; Vishwakarma 1962; Hamner and Williams 1965), a high K^+ concentration (Olds and VanDemark 1957*b*) the presence of large molecules—including polysaccharide, biuret-protein, and enzymes such as glycerylphosphorylcholine diesterase (Hamner and Williams 1965; Wallace and White 1965)—and the occurrence of relatively exotic reproductive constituents like inositol (Gregoire, Gongsakdi, and Rakoff, 1962), all lend credence to the conclusion that the fluids present in the mammalian oviduct, of some species at least, are actively elaborated and transported into the lumen. In addition, there may, of course, be instances of "excretory" overflow, analogous to the drainage of sulfonamides and alcohol into the seminal plasma in man (Mann 1964), as well as some contribution to the oviductal fluid from the peritoneal cavity, but these sources of supply are not regarded as significant in the normally breeding mammal.

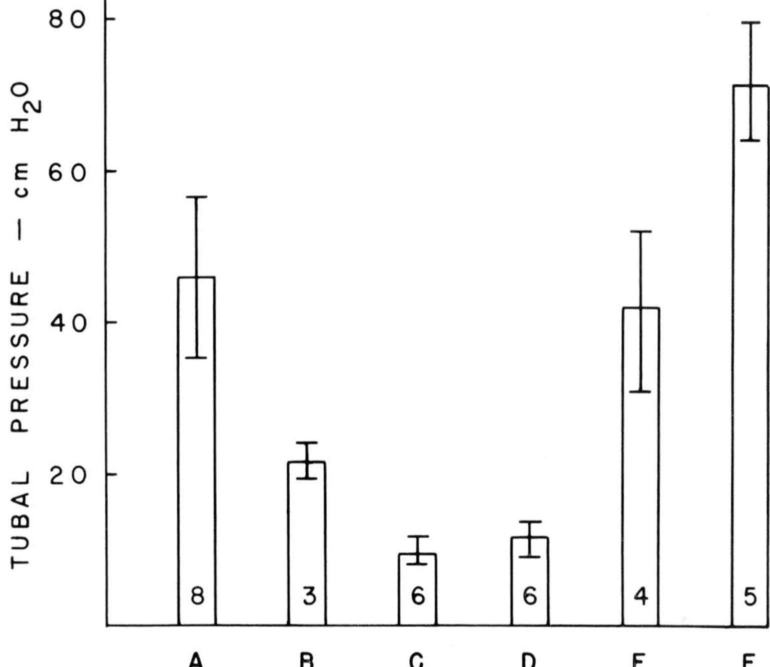

FIG. 1. Oviductal secretion pressures determined by means of water manometer in rabbits under Dial-urethane anesthesia; *a*, estrous; *b*, 9–14 days pregnant; *c*, 18–24 days pregnant; *d*, castrate; *e*, castrate plus estradiol; *f*, estrous plus pilocarpine. Range of values and number of animals shown. (From Bishop 1956*a*.)

There is, on the other hand, the likelihood of important contributions to the female genital tract by the seminal plasma, introduced at the time of insemination, particularly in those animals in which ejaculation occurs directly into the uterus, for example, the mouse, rat, sow, and mare (Bishop 1961). The transport of such materials has been tested. By observing the movement of radio-opaque material (Neo-hydriol) through the bovine reproductive tract, Rowson (1955) demonstrated penetration into the oviduct of the estrous heifer within 30 min of deposition of the substance in the uterus; migration time was shortened to 2 min by the addition of oxytocin (Pitocin). Mann and co-workers (1956) investigated the passage of semen into the oviduct of the pig and horse after mating by employing a system of naturally occurring markers, namely, fructose, citric acid, and ergothionine. They found in the mare, but not in the gilt, an indication that seminal citric acid and ergothionine reach the oviduct some 40 min after mating. In experiments on rabbits, the evidence suggests that very little seminal fructose or citric acid appears even in the uterine fluid within 24 hr after copulation (Lutwak-Mann 1962). Species differences are to be expected, depending on the site of insemination, activity of the female genital tract, concentration of biochemical indicator, and the like. It is quite possible that, at best, seminal plasma constituents are of relatively little consequence to sperm welfare and longevity within the oviduct, because of either extreme dilution, neutralization, or competition with other components of female origin. Nevertheless, in the light of reports of very rapid sperm transport in some mammals (see below), this question of the penetration into the uterus and oviduct of biochemically active components and their potential use by, or effect on, the spermatozoa would seem to warrant additional attention, preferably from a broad, comparative approach.

What metabolically active constituents, one may ask, are normally present in the mammalian oviduct? Glycolytic substrate, assayed as reducing sugar or glucose, has been demonstrated in rabbit oviductal fluid, as well as in adjacent regions of the tract in these and other mammals. In the oviductal fluid of either estrogen- or progesterone-dominated rabbits, Bishop (1957) found detectable quantities (1–2 mg/100 ml) of glucose, assayed colorimetrically by the benzidine method. Hamner and Williams (1965) also demonstrated traces of glucose, but no other free sugars from a number of hexoses and pentoses tested. Rabbit uterine secretion, on the other hand, was found to contain rather large quantities of glucose (12–30 mg/100 ml) after gonadotrophin-induced ovulation or copulation (Lutwak-Mann 1962). In the oviductal fluid of the ewe, little or no free glucose is present, but large amounts (averaging 63 mg/100 ml) of orcinol-reactive carbohydrate can be demonstrated, regardless of the endocrine state of the animal (Perkins and Goode 1966; Restall and Wales 1966). In contrast to that of sheep, bovine oviductal fluid is reported to have about 90 mg/100 ml of reducing sugar, as determined photometrically by either the Somogyi or Nelson-modified procedure (Olds and VanDemark 1957*b*). It is a matter of interest also that the follicular fluid of both the bovine and human ovary gives a positive reaction for glucose, assayed enzymatically (Lutwak-Mann 1954; Birnberg and Gross 1958), but it remains an open question how much, if any, of this substrate is made available for sperm metabolism.

The presence of metabolic substrates which can sustain oxidative respiration in the oviduct is somewhat more consistent. Lactate, for example, is found frequently and in significant quantities; it appears to undergo some degree of elevation in the progestational rabbit (Bishop 1957; Mastroianni and Wallach 1961; Hamner and Williams 1965). Mastroianni and colleagues have concluded from their impressive metabolic studies of human and rabbit oviductal tissue slices *in vitro* that the mucosal cells are responsible for lactic acid production even under aerobic conditions (Mastroianni, Winternitz, and Lowi 1958; Mastroianni, Forrest, and Winternitz 1961). Lactate concentration in the ewe oviductal fluid may attain values as high as 62 mg/100 ml, but do not appear to reflect differences in the endocrine balance (Perkins and Goode 1966; Restall and Wales 1966). Other respiratory substrates, of potential use to inseminated spermatozoa, have been demonstrated in the oviductal secretions of the rabbit: inositol (Gregoire, Gongsakdi, and Rakoff 1962), several free amino acids (Gregoire, Gongsakdi, and Rakoff 1961; Hamner and Williams 1965), and possibly phospholipid (Bishop 1957).

In reviewing past investigations and sometimes equivocal demonstrations of metabolic substrates, as well as other biologically active components, in the genital tract, one is provoked to inquire as to the possible significance of the microenvironment(s) immediately surrounding the gametes. In contrast to concentrations in the total washing medium generally analyzed, local concentrations of key constituents might be quite high, as adjacent to the oviductal mucosa, for example. Such a condition would seem to apply with respect to a lactic acid gradient noted above, even though the oviduct mobility, ciliary activity, and possibly motility of the sperm themselves would tend to mix the luminal contents rapidly. Indeed, some evidence for this type of local substrate elaboration and distribution is suggested through efforts to demonstrate hexose secretion by the oviductal epithelium. Microquantities of substrate, too small to detect in the total volume of luminal fluid, could exert an influence in relation to a small, relatively confined space occupied by one or more spermatozoa.

It is essential to know whether conditions within the oviduct *in situ* are aerobic or anaerobic, and whether they change during the cycle, in order to appreciate the utilization of either endogenous or exogenous substrate by spermatozoa and the relative importance of glycolysis versus respiration. This problem has been approached in the rabbit and rat. Campbell (1932) first indicated that aerobic conditions do prevail and that the oxygen tension of the rabbit uterine mucosa is in equilibrium with that of the vascular supply. By direct,

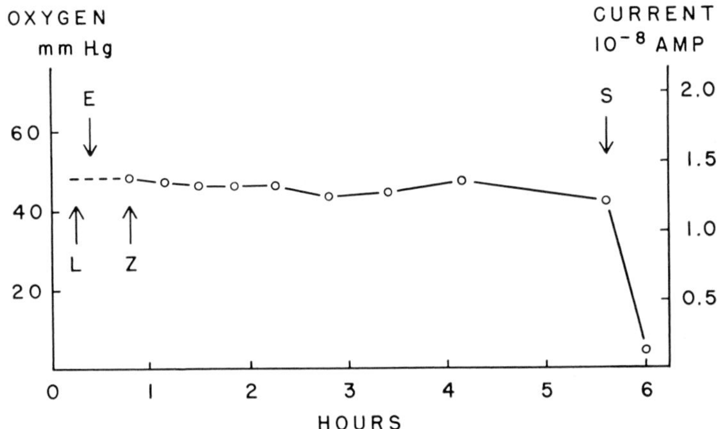

FIG. 2. Steady-state oxygen tension in oviduct of Dial-urethane anesthetized adult estrous rabbit recorded by oxygen electrode; *l*, laparotomy; *e*, electrode inserted; *z*, galvanometer circuit zeroed; *s*, sacrifice of animal by ether. (After Bishop 1956*b*.)

potentiometric measurement with the oxygen electrode, Bishop (1956*b*) and Mastroianni and Jones (1965) have found the luminal oxygen tension of the rabbit oviduct to average 45 and 60 mm Hg pressure in anesthetized and unanesthetized does, respectively (Fig. 2). Aerobic conditions exist also in the rat uterus; oxygen tension varies with stage of estrus (Mitchell and Yochim 1967). What would seem to be a conflicting report, in part perhaps attributable to species difference, is the claim of Birnberg and Gross (1958) that the upper level of the human female tract is anaerobic, a conclusion based on their finding of an effective enzyme reductase system in isolated follicular fluid. It remains to be seen whether this interpretation will be substantiated, but it is nevertheless obvious that many more such studies, on a wider variety of animals and employing different techniques, are desirable, if not mandatory, for the acquisition of any real understanding of the particular conditions which exist in oviducts of various species and which may affect and regulate the activities of the germ cells within.

The fluids of the mammalian oviduct may contain, in different species and under various hormonal and breeding conditions, a number of other biologically active ingredients with physiological, pharmacodynamic, or biochemical effects on the gametes or the oviductal tissue itself. In the sections which follow these luminal constituents will be considered where appropriate. For further clarification, two excellent sources of information are recommended: the monograph by Mann (1964) and the review of Restall (1967).

III. Sperm Transport and Survival

The principal, and in most mammalian species the entire burden for sperm passage through the female genital tract rests upon the duct itself (Bishop 1961; Mann 1964; Restall 1967). Active swimming movement of the gametes may be regarded generally as of little consequence in transport, although some evidence does indicate that sperm migration through the human (Hartman 1957) and rabbit (Braden 1953; Noyes, Adams, and Walton 1958) cervix and across the uterotubal junction in the rat (Leonard and Perlman 1949) and rabbit (Braden 1953) may be facilitated by flagellar activity of the gametes. The extraordinarily rapid rates of sperm ascent of the female tract after insemination (Table 1) certainly support a transport mechanism dependent upon muscular activity of the uterine and oviductal tissues, as does

TABLE 1

SPERM TRANSPORT TIME AFTER NATURAL MATING

Species	Site of Insemination	Appearance in Oviduct (Min)	Reference
Rat	Uterus	2	Hartman and Ball 1930; Warren 1938
Mouse	Uterus	15	Lewis and Wright 1935
Hamster	Uterus	2; 30	Yamanaka and Soderwall 1960; Chang and Sheaffer 1957; Yanagimachi and Chang 1963
Rabbit	Vagina	60	Chang 1952; Adams 1956
Dog	Vagina	20	Evans 1933; Whitney 1937
Cow	Vagina	4	VanDemark and Moeller 1951; VanDemark and Hays 1954
Sheep	Vagina	8	Schott and Phillips 1941; Starke 1949; Mattner and Braden 1963
Man	Vagina	30	Rubenstein *et al.* 1951

the passage of seminal plasma constituents, dead spermatozoa (VanDemark and Moeller 1951; Mattner 1963*b*), and inert material introduced into the vagina (Rowson 1955; Akester and Inkster 1961; Egli and Newton 1961). Both oviductal activity and gametic migration rates are influenced by hormonal factors (see Restall 1967), particularly oxytocin (VanDemark and Hays 1952, 1954; Hays and VanDemark 1953; Egli and Newton 1961; Mrouch 1967), and may be modified *in vivo* by such seminal pharmacodynamic agents as prostaglandin (von Euler 1937; Eliasson 1963).

The data presented in Table 1 indicate the times of first appearance of spermatozoa in the oviducts and suggest rapid rates of migration, but they should not be overdramatized. In no species are many viable spermatozoa found in the ampulla at the time of fertilization (Bishop 1961), but the first spermatozoa to arrive need not be those that engage in successful sperm-egg interaction (Adams 1956; Mattner 1963*a;* Turnbull 1966). Gametes continue to pass up the genital tract, and fertilization must depend on a variety of factors, including the time of ovulation with respect to that of insemination and the necessity for sperm capacitation in some species. An interesting observation which is clear from the accompanying table is that transport may be quite rapid regardless of whether or not sperm are naturally inseminated into either the vagina or uterus. There would seem to be considerable room for further investigation by physiological, pharmacodynamic, and immunological methods of the problems of sperm migration, the factors which modify and control transport, and the relation between rate of passage and success in fertilization.

The survival of spermatozoa in the female tract may be expressed in terms of the duration of their motility or their fertilizing capacity (Table 2). The former may give useful data as to the viability of the gametes under various conditions, such as the endocrine domination of the female, but relatively little is known about why spermatozoa persist so much longer in some species than in others. Moreover, it is likely that motility of sperm from the tract, *per se*, is not necessarily a good criterion of fertilizing capacity, since the gametes are probably

TABLE 2

SURVIVAL TIMES AND RETENTION OF FERTILIZING CAPACITY BY SPERMATOZOA IN THE FEMALE REPRODUCTIVE TRACT

Animal	Maximal Duration of Fertility (Hr)	Maximal Duration of Motility (Hr)	Reference
Rabbit	28–32	....	Hammond and Asdell 1926; Chang and Pincus 1964
Mouse	6	13	Merton 1939
Rat	14	17	Soderwall and Blandau 1941
Guinea pig	21–22	41	Yochem 1929; Soderwall and Young 1940
Ferret	36–48; 126	....	Hammond and Walton 1934; Chang 1965
Dog	134 est.	268	Doak, Hall, and Dale 1967
Sheep	30–48	48	Green 1947; Dauzier and Wintenberger 1952
Cow	28–50	96	Laing 1945; Vandeplassche and Paredis 1948; Gibbons 1959
Horse	144	144	Day 1942; Burkhardt 1949
Man	28–48	48–60	Farris 1950; Rubenstein *et al.* 1951; Horne and Audet 1958
Bat	135–50 days	149–56 days	Wimsatt 1942, 1944

SOURCE: After Restall (1967), with modification.

quiescent in those forms in which the duration of fertility persists for many days, e.g., the ferret, dog, fowl, and hibernating bat. Presumably, these spermatozoa must be reactivated to engage in the fertilization process. On the other hand, the duration of the capacity for fertilization displayed by inseminated sperm is a more meaningful biological index of livability, but its determination, in some species, is beset with difficulties, such as pinpointing the precise time of ovulation, and therefore of fertilization, in man.

The data in Table 2 do represent some interesting features of physiological adaptation with respect to sperm longevity and reproductive pattern. As noted by Chang (1965), the extended duration of fertilizing capacity of ferret sperm coincides with the fact that, in this animal, ovulation occurs between 30 and 40 hr after copulation. In the rabbit, the copulation-ovulation interval is roughly 10 hr, a period during which sperm viability must be assured and sperm capacitation may be achieved. In the horse and the dog, extreme survival times of spermatozoa in the female tracts would seem to correlate with the long estrous periods in these animals. The bat, a hibernating mammal, is unique among those forms listed; reduced temperature must play some role in sperm survival in the female, but special anatomical and physiological features of the genital tract also are significant (Wimsatt, Krutzsch, and Napolitano 1966). Sperm survival in the oviduct of the fowl, an animal with an elevated body temperature, is eloquently discussed by Lake (1967).

In the light of the foregoing evidence, sperm motility itself is relegated to an ancillary position in accounting for transport through the female tract. The process of sperm flagellation, a spectacular form of cellular movement, has considerable intrinsic and theoretical interest in its own right and has been extensively investigated and frequently reviewed (Bishop 1962*a*, *b*; Sleigh 1962; Holwill 1966; Nelson 1967). However, the question may be asked, What good is it (in the mammal) or why does it occur? There are indeed some grounds for skepticism. Whereas sperm motility is generally regarded as essential for fertilization, its role even in this process in birds and mammals has been questioned (Casida and Murphree 1942; Allen and Grigg 1957; Chang and Thorsteinsson 1958), although not as a result of experimental conditions which rule out remobilization of spermatozoa in the female tract. In an extreme

TABLE 3

ACTIVATION OF ELECTRICALLY SPAWNED SPERMATOZOA OF THE HORSESHOE CRAB

Agent	Motility	Acrosome Reaction
Sea water (SW)	−	< 1%
SW, pH 4.9	−	−
SW, pH 9.0	−	−
Oviduct exudate	+	< 1%
Egg water	+	< 1%
Egg water dialyzed	−	−
Acid-alcohol egg extract	+	10%
SW+$CaCl_2$ (0.05 M)	−	6%
SW+EDTA (0.01 M)	+	−
Ca^{++} free	+	......
Mg^{++} free	−	......
Ca^{++} and Mg^{++} free	−	......

SOURCE: Data after Shoger and Bishop 1967.

specialization of the process, however, fertilization readily occurs in many invertebrate animals, such as the decapod crustaceans, even though the sperm are both nonflagellated and nonmotile. Moreover, if the process of fertilization is observed carefully, *in vitro*, in an animal like the sandworm (*Nereis*), the fertilizing spermatozoon appears to be quite motionless immediately upon making contact with the egg.

In general, the motility of spermatozoa is assured at the time of spawning or ejaculation. In some species, however, what would appear as possibly a more efficient mechanism has been evolved where a shortage of energy substrate prevails and motility is of very short duration. In the fish, *Clupea*, for example, the motionless sperm are activated only when they come into the vicinity of the egg and particularly of the egg micropyle; the activating factor is believed to be calcium ion (Yanagimachi and Kanoh 1953). Similarly, in the horseshoe crab, *Limulus*, the flagellated sperm are immotile when spawned but become extremely active (see Table 3) when stimulated by oviductal fluid extruded when the eggs are shed (Shoger and Bishop 1967). In the higher forms, mammals included, no such female motility-triggering device has been evolved or, at least, discovered. The possibility would seem to exist, however, that in those species with long sperm-survival times, the dog, ferret, bat, and fowl, for example, the sperm become motionless after ejaculation and are reactivated prior to fertilization.

These gametes are thus somehow protected from premature aging and retain their capacity for fertilization for days, weeks, or months. Seen in this light, constant or excessive sperm motility in the female genital tract of most mammals may be regarded as a disadvantage, a process of physiological aging, since once spermatozoa are activated, the inevitability of fertilization or death is assured. Much remains to be learned with regard to the factors which affect sperm motility, its control, and the nature of the processes associated with biochemical senescence of the gametes *in vivo*.

IV. Sperm Maturation

Cellular changes of both an ultrastructural and a biochemical nature continue in spermatozoa after their departure from the testes (Henle and Zittle 1942; Bishop 1955; Ånberg 1957; Mann 1959; Fawcett and Hollenberg 1963; Bedford 1965*a;* Gledhill 1966). These modifications may be regarded as normal processes of "ripening" and as necessary for the attainment of full fertilizing capacity, to be followed by destructive and deteriorating reactions associated with cell senescence, aberrant fertilizing capacity, and death (Soderwall and Young 1940; Nalbandov and Card 1943; Bedford 1963; Anand, Hoekstra, and First 1967). Most of the final developmental changes in spermatozoa occur while the gametes pass through or remain in the male reproductive tract; others may be thought of as continuing in or being initiated after insemination into the female tract, i.e., "prefertilization" alterations. A good deal might be gained from studies of the minute and subtle changes which may occur in spermatozoa during their residence within the uterus and oviduct whether or not fertilization is to occur—particularly comparative studies of different species with long and short sperm-survival times, under different endocrinological states, and with different pharmacodynamic treatments.

Sperm *activation*, expressed as increased oxygen uptake, has been demonstrated in suspensions of washed cells, *in vitro*, in the presence of fluids from the female genital tract (Olds and VanDemark 1957*a;* Hamner and Williams 1963, 1964; Mounib and Chang 1964; Black *et al.* 1968). In experiments on rabbits, spermatozoa which had been incubated for periods up to 6 hr in the uterus showed a fourfold increase in oxygen consumption (Hamner and Williams 1963). Employing radiolabeled substrate, Mounib and Chang (1964) found evidence which was interpreted as suggesting a shift in metabolic pattern by these spermatozoa, greater utilization of endogenous substrate, an enhancement of the hexose monophosphate shunt, and rapid aging of the cells. In rabbit oviductal fluid a dialyzable, heat-stable factor which increased respiratory activity of sperm was identified as bicarbonate ion (Hamner and Williams 1964). In contrast, in the oviductal fluid of the fowl a high molecular weight protein component has been observed to increase oxygen consumption up to twofold (Ogasawara and Lorenz 1964). The metabolic activity of mammalian spermatozoa in the presence of female reproductive fluids has been implicated in the phenomenon of capacitation and interpreted as the sperm's ability to produce usable energy at a higher level (Hamner and Williams 1963; Mounib and Chang 1964). Such provocative findings suggest that much more work should be done on other species, that specific respiratory stimulators be sought, and that the seemingly enhanced energy production be reevaluated in terms of the possibility of uncoupling of phosphorylation and a general *loss* in efficiency. It may be noted that the stimulation of spermatozoa, both in terms of motility and oxygen consumption, is an old story with respect to invertebrate spermatozoa (Rothschild 1951). Recent work in the author's laboratory on the invertebrate horseshoe crab, that "living fossil," *Limulus poly-*

phemus, shows that it is possible to separate the motility-stimulating action from other effects on the spermatozoa, such as the acrosome reaction, and that activation is a magnesium-dependent response sensitive to toxic, chelatable ions in the seawater into which the gametes are normally shed (Table 3).

The phenomenon of *capacitation*, the changes brought about in spermatozoa by several hours' sojourn in the female tract prior to attainment of fertilizing capacity, was first demonstrated in the rabbit (Austin 1951; Chang 1951) and has been extended to a few other species, including the rat, ferret, and possibly sheep (Austin 1951; Noyes 1959; Chang and Yanagimachi 1963; Mattner 1963*c*). The nature of the cell changes has not been established; over the years, structural modifications involving the acrosome and cell cap, enzymatic alterations, and surface changes have been implicated by various investigators. Although the uterus and oviduct are the logical and normal sites for capacitation, the reaction also is claimed to occur in the isolated bladder, colon, and anterior chamber of the eye (Noyes, Walton, and Adams 1958). Moreover, capacitation appears to be reversible; Williams and co-workers (1967) have isolated a decapacitation factor from seminal plasma after treatment with which the sperm can be "recapacitated." The area of capacitation research is an active one for investigation and has amassed a sizable background of literature, much of which has been brought together for the interested specialist in the *Bibliography of Reproduction* (vol. 10, no. 3, 1967) and the *Journal of Reproduction and Fertility* (Supplement 2, 1967), which contains the proceedings of the Third Brook Lodge Workshop concerning "Capacitation of Spermatozoa and Endocrine Control of Spermatogenesis." Uneducated opinions are always dangerous, but it seems safe to suggest that if capacitation proves to be a significant phenomenon in a wide variety of mammals, the changes involved in the sperm, and the interplay between oviductal and gametic systems, will probably be found to concern subtle alterations, such as the activation of an enzyme system essential for fertilization or the counteraction of an epididymal-seminal plasma block, which is restricted to higher forms which engage in internal fertilization. No specific oviductal constituents have yet been implicated in the capacitation reaction.

The *acrosome reaction* involves the breakdown or modification of this organelle on the sperm head with the release of components which effect initial contact with the egg; until recently, the subject has been the legacy of investigators concerned with invertebrate fertilization (Dan 1967). Acrosomal changes in oviductal sperm have now been described, however, which indicate that this phenomenon, too, may be a natural maturation or prefertilization modification of mammalian sperm (Austin 1965).

It is generally accepted that a form of acrosomal reaction occurs in mammalian sperm similar to the well-known event in invertebrates (Austin, 1965). The process has now been demonstrated in spermatozoa of the rabbit, hamster, and guinea pig. A loosening and wrinkling of the acrosome occurs which permits the release of hyaluronidase, which facilitates penetration of the corona radiata by the sperm. A second enzyme, "zona lysin," becomes effective, allowing penetration of the zona pellucida (Austin and Bishop 1958; Srivastava, Adams, and Hartree 1965). Bedford (1967) has presented beautiful electron micrographic evidence to account for the disappearance of the acrosome during sperm entry of the rabbit egg. Most of the reactions involve sperm and egg, but they occur within an oviductal microenvironment, the nature of which can only be approximated.

Certain other kinds of prefertilization reactions of mammalian spermatozoa have been suggested, such as the possibility of a fertilizin-antifertilizin recognition system between the gametes of rabbits (Bishop and Tyler 1956; Thibault and Dauzier 1960) and the plausibility of sperm chemotaxis in the same animal (Dickmann 1963). But, despite the fact that the latter concept has recently been made respectable among animal gametes by Miller's (1966) convincing demonstration of chemotaxis in the coelenterate, *Campanularia*, the role of the oviduct in such reactions seems minimal at best, and these controversial issues need not be aired here.

V. Sperm Metabolism

This is not an appropriate place for an extensive survey of the wide range of metabolic patterns and substrates which have been described for mammalian spermatozoa, *in vitro*, particularly since several very knowledgeable and thorough reviews of the subject have appeared recently (Salisbury and Lodge 1962; Mann 1964, 1967). Two points may be stressed, however, since they must permit spermatozoa to adjust to the luminal environment of the oviduct, whatever it may be, and so enhance the chances of the inseminating spermatozoa to survive the passage and sojourn in the female tract. On the one hand, the wide metabolic adaptability of mammalian spermatozoa is impressive and, on the other hand, the metabolic patterns adopted are, with certain exceptions, comparable to, or identical with, those developed by other cells and tissues. Nature has proved conservative in her evolution of new molecules to effect established processes.

The realm of metabolic patterns which mammalian spermatozoa can encompass is depicted in Figure 3 and includes both respiratory and glycolytic schemes, with the option of pursuing the latter under either aerobic or anaerobic conditions. The glycolytic enzymes (Fig. 3) are relatively freely distributed throughout the cytoplasm; the respiratory portions of the system appear to be associated with the mitochondria, the citric acid cycle components with the outer membranes, and the cytochrome-cytochrome oxidase system with the inner membranes (Racker 1968). Since oxidative respiration, coupled as it is to phosphorylation, is the more efficient system in terms of ATP regeneration, this presumably should suffice to provide for the needs of the gametes. Indeed, such is the case predominantly among invertebrate species, or those that engage in external fertilization. But this imposes three restrictions: a guaranteed oxygen supply, a respiratory substrate such as the sperm's endogenous phospholipid, and motility of short duration. When, on the other hand, exogenous energy substrate can be made available to the spermatozoa, as is the case if fertilization occurs internally, these restrictions are, in part, removed. In addition to their own endogenous plasmalogen, mammalian spermatozoa can be supplied with, and can efficiently utilize, a variety of respiratory substrates, including lactate, pyruvate, and glycerol (Morton and Lardy 1967). They are, moreover, equipped to handle glycolytic substrates, fructose, glucose, and mannose, the first of which is the principal mammalian seminal component. The initiation of mammalian sperm motility is assured by two physiological adaptations to internal fertilization: (*a*) substrate sugars, and (*b*) the components of a glycolytic pathway which permits their utilization under anaerobic as well as aerobic conditions. After insemination, there is no certainty that glycolysis continues to any great extent—the issue is in question—and, in fact, intratubal conditions are favorable for oxidative respiration. It is rash, perhaps, to suggest that in the oviduct the spermatozoa themselves become modified and are increasingly dependent on aerobic processes, and, like the developing egg, in a sense revert to an earlier phylogenetic biochemical pattern. An interesting array of correlations may be seen if one compares the biochemical and ultrastructural characteristics of the two types of motile spermatozoa, in this instance the sea-urchin and the bovine sperm (Table 4).

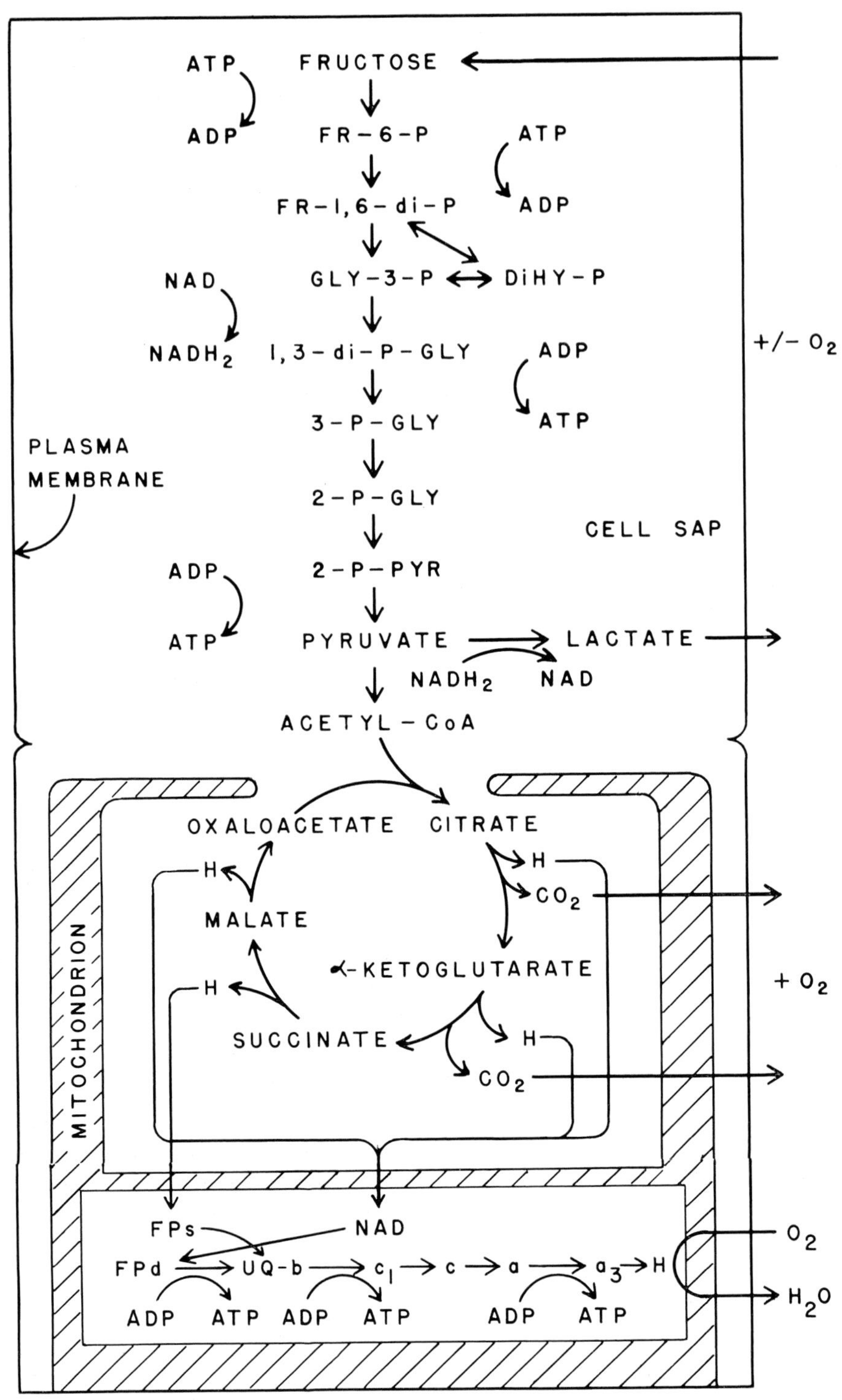

ATP
FRUCTOSE
ADP
FR-6-P
ATP
ADP
FR-1,6-di-P
NAD
GLY-3-P
DiHY-P
$NADH_2$
1,3-di-P-GLY
ADP
ATP
+/- O_2
3-P-GLY
PLASMA MEMBRANE
2-P-GLY
CELL SAP
ADP
2-P-PYR
ATP
PYRUVATE
LACTATE
$NADH_2$
NAD
ACETYL-CoA
OXALOACETATE
CITRATE
H
H
CO_2
MALATE
MITOCHONDRION
α-KETOGLUTARATE
+ O_2
H
SUCCINATE
H
CO_2
FPs
NAD
O_2
FPd
UQ-b
c_1
c
a
a_3
H
ADP
ATP
ADP
ATP
ADP
ATP
H_2O

Finally, a word of caution in regard to the supposed function of spermatozoan components in the mature cell, vis-à-vis their role during development. A case in point is the enzyme, sorbitol dehydrogenase (SDH), present in the spermatozoa of mammals (see Mann 1964). It has been regarded as playing a significant role in the genital tract with respect to the sorbitol-fructose oxidoreduction system and the regulation of the NAD-$NADH_2$ balance. Curiously enough, however, testicular SDH activity has been demonstrated in significant quantities in a wide variety of vertebrate and invertebrate species, many of which breed by external fertilization and under conditions where no exogenous substrate, sorbitol or fructose, can be made available. The significant role of the enzyme under these circumstances seems to involve the differentiation of the gametes during spermatogenesis, rather than the metabolic activity of the mature spermatozoa (Bishop 1968).

TABLE 4

METABOLIC ADAPTATION CORRELATES

PHYSIOLOGICAL CHARACTERISTICS	REPRESENTATIVE SPERMATOZOON	
	Sea Urchin and Bull	Bull
Metabolic pattern	Oxidative respiration	Glycolysis, $\pm O_2$
Energy substrate	Respiratory	Glycolytic
Substrate supply	Endogenous	Exogenous
Glycolytic enzymes	Absent[a]	Present
Cellular compartment	Mitochondria	Cell sap
Surface layer	Permeable plasma membrane	Semipermeable lipoprotein coat
Type of reproduction	External fertilization[a]	Internal fertilization

[a] Does not apply to bull spermatozoa.

VI. Fate of Nonfertilizing Sperm

Of the many millions of spermatozoa deposited in the female genital tract, relatively few persist, generally, for any great length of time. The earlier literature concerning this oviductal expurgation and the suggested sensitizing effects of sperm has been previously reviewed (Bishop 1961). Despite technical objections and claims to the contrary, there would seem to be evidence to indicate that some sperm, at least, in a few species, become incorporated into the uterine and oviductal mucosa (Austin 1960). On the other hand, the major sperm-removal mechanism is made available by the intense leucocytic invasion which makes its appearance after copulation (Austin 1957; Bedford 1965*b*; Howe 1967). In the mouse and rat uterus a great increase in phagocytic polymorphonuclear leucocytes, 15 to 20 hr after coitus, coincides with a decrease and eventual disappearance of spermatozoa in the lumen. In the rabbit, spermatozoa are removed from the oviduct, uterus, and vagina by phagocytic engulfment within 24 hr after their injection into ligated segments of the tract (Howe 1967). This reaction is not sperm-specific, since injected bacteria undergo similar phagocytosis. The leucocytic

FIG. 3. Metabolic pathways utilized by mammalian spermatozoa according to availability of substrate, oxygen, and necessary cofactors. The upper, glycolytic, pathway is indicated as occurring in the cell sap, the citric acid cycle associated with the outer membrane, and the cytochrome system with the inner membrane of the mitochondrion. This schematized plan conforms generally with evidence cited by Mann (1964, 1967) and Racker (1968).

influx depends, to a degree, on the endocrine balance of the doe. After this cellular cleaning, the oviductal contents are presumably evacuated, and little or no debris is conducted into the mucosa.

It is perhaps unwise to generalize from these instances based on a few species, a statement which is singularly appropriate for *any* discussion of phenomena associated with reproductive physiology! But in this situation one wonders what condition prevails in animals like the horse, dog, and ferret (Table 2) where sperm survival is inordinately long; perhaps here, as well as in the bat, leucocytic invasion is simply delayed, or it fails to occur altogether, as has been claimed for the domestic fowl.

VII. Conclusion

A review of this kind appears to raise more questions rather than to provide answers. Perhaps we are just on the verge of asking meaningful questions. A decade ago, VanDemark (1958) published a provocative paper in which he stated:

> Information concerning the environment afforded the spermatozoa by the female genitalia is meagre. More complete data on the composition and changes that occur in the luminal fluids should lead to a clearer understanding of the factors that affect sperm survival in the female reproductive organs.

In this light, the present paper has been a disappointing assignment because one cannot say, today, that we have a very clear understanding of the factors which affect, modify, or control sperm survival or behavior *in situ*. For those about to launch a research career in reproductive physiology—with new ideas and new tools—there need be, however, no disappointment or lack of challenge. There is still much to be done, as recently proclaimed by Mann (1967):

> What, however, still remains to be elucidated is the relative importance to the spermatozoa of anaerobic glycolysis, aerobic glycolysis, exogenous respiration, and endogenous respiration, not merely during their ascent along the reproductive tract but also at the site of fertilization. Further study is also needed on the relations between sperm metabolism, motility, and the energy changes associated with alterations in the adenine nucleotide pool of spermatozoa, which are associated with their activity, and, ultimately, senescence.

Acknowledgment

I wish to take this opportunity to acknowledge sincere gratitude to my colleagues at Cornell University for their many kindnesses and cooperation during the preparation of this review. In particular, I deeply appreciate the stimulation received from, and tolerance shown by, Professors van Tienhoven, Hansel, Foote, and Visek and their many able students.

References

1. Adams, C. E. 1956. A study of fertilization in the rabbit: The effect of post-coital ligation of the Fallopian tube or uterine horn. *J. Endocr.* 13:296–308.
2. Akester, A. R., and Inkster, I. J. 1961. Cine-radiographic studies of the genital tract of the rabbit. *J. Reprod. Fertil.* 2:507–8.
3. Allen, T. E., and Grigg, G. W. 1957. Sperm transport in the fowl. *Aust. J. Agric. Res.* 8:788–99.
4. Anand, A. S.; Hoekstra, W. G.; and First, N. L. 1967. Effect of aging of boar spermatozoa on cellular loss of DNA. *J. Anim. Sci.* 26:171–73.

5. Ånberg, Å. 1957. The ultrastructure of the human spermatozoon. *Acta Obstet. Gynecol. Scand.* 36 (Suppl. 2):1–133.
6. Austin, C. R. 1951. Observations on the penetration of the sperm into the mammalian egg. *Aust. J. Sci. Res., Ser. B* 4:581–96.
7. ———. 1957. Fate of spermatozoa in the uterus of the mouse and rat. *J. Endocr.* 14: 335–42.
8. ———. 1960. Fate of spermatozoa in the female genital tract. *J. Reprod. Fertil.* 1:151–56.
9. ———. 1965. *Fertilization.* New York: Prentice-Hall, Inc.
10. Austin, C. R., and Bishop, M. W. H. 1958. Role of the rodent acrosome and perforatorium in fertilization. *Proc. Roy. Soc., Ser. B* 149:241–48.
11. Bedford, J. M. 1963. Morphological reaction of spermatozoa in the female reproductive tract of the rabbit. *J. Reprod. Fertil.* 6:245–55.
12. ———. 1965*a*. Changes in fine structure of the rabbit sperm head during passage through the epididymis. *J. Anat.* 99:891–906.
13. ———. 1965*b*. Effect of environment on phagocytosis of rabbit spermatozoa. *J. Reprod. Fertil.* 9:249–56.
14. ———. 1967. Experimental requirement for capacitation and observations on ultrastructural changes in rabbit spermatozoa during fertilization. *J. Reprod. Fertil.* Suppl. 2, pp. 35–48.
15. Birnberg, C. H., and Gross, M. 1958. Enzymatic activities of follicular fluid. *Int. J. Fertil.* 3:374–81.
16. Bishop, D. W. 1955. Sperm maturescence. *Scientific Monthly* 80:86–92.
17. ———. 1956*a*. Active secretion in the rabbit oviduct. *Amer. J. Physiol.* 187:347–52.
18. ———. 1956*b*. Oxygen concentrations in the rabbit genital tract. *Proc. Third Int. Congr. Anim. Reprod.* pp. 53–55.
19. ———. 1957. Metabolic conditions within the oviduct of the rabbit. *Int. J. Fertil.* 2:11–22.
20. ———. 1961. Biology of spermatozoa. In *Sex and internal secretions*, 3d ed., ed. W. C. Young, pp. 707–96. Baltimore: Williams & Wilkins Co.
21. ———. 1962*a*. *Spermatozoan motility*, ed. D. W. Bishop. Washington, D. C.: American Association for the Advancement of Science.
22. ———. 1962*b*. Sperm motility. *Physiol. Rev.* 42:1–59.
23. ———. 1968. Testicular enzymes as fingerprints in the study of spermatogenesis. In *Perspectives in reproduction and sexual behavior*, ed. M. Diamond. Bloomington, Ind.: Indiana University Press. (In press.)
24. Bishop, D. W., and Tyler, A. 1956. Fertilizin of mammalian eggs. *J. Exp. Zool.* 132: 575–601.
25. Black, D. L.; Crowley, L. V.; Duby, R. T.; and Spilman, C. H. 1968. Oviduct secretion in the ewe and the effect of oviduct fluid on oxygen uptake by ram spermatozoa *in vitro*. *J. Reprod. Fertil.* 15:127–30.
26. Black, D. L.; Duby, R. T.; and Riesen, J. 1963. Apparatus for the continuous collection of sheep oviduct fluid. *J. Reprod. Fertil.* 6:257–60.
27. Blandau, R.; Jensen, L.; and Rumery, R. 1958. Determination of the pH values of the reproductive-tract fluids of the rat during heat. *Fertil. Steril.* 9:207–14.
28. Braden, A. W. H. 1953. Distribution of sperms in the genital tract of the female rabbit after coitus. *Aust. J. Biol. Sci.* 6:693–705.

29. Burkhardt, J. 1949. Sperm survival in the genital tract of the mare. *J. Agric. Sci.* 39: 201–3.
30. Campbell, J. A. 1932. Normal gas tensions in the mucus membrane of the rabbit's uterus. *J. Physiol.* 76:13P.
31. Casida, L. E., and Murphree, R. L. 1942. Fertility and sex ratios in the rabbit. *J. Hered.* 33:434–38.
32. Chang, M. C. 1951. The fertilizing capacity of spermatozoa deposited into the Fallopian tubes. *Nature (Lond.)* 168:697–98.
33. ———. 1952. Fertilizability of rabbit ova and the effects of temperature *in vitro* on their subsequent fertilization and activation *in vivo*. *J. Exp. Zool.* 121:351–81.
34. ———. 1965. Fertilizing life of ferret sperm in the female tract. *J. Exp. Zool.* 158:87–100.
35. Chang, M. C., and Pincus, G. 1964. Fertilizable life of rabbit sperm deposited into different parts of the female tract. *Proc. Fifth Int. Congr. Anim. Reprod. A.I. (Trento)* IV:377–80.
36. Chang, M. C., and Sheaffer, D. 1957. Number of spermatozoa ejaculated at copulation, transported into the female tract, and present in the male tract of the golden hamster. *J. Hered.* 48:107–9.
37. Chang, M. C., and Thorsteinsson, T. 1958. Effects of urine on motility and fertilizing capacity of rabbit spermatozoa. *Fertil. Steril.* 9:231–37.
38. Chang, M. C., and Yanagimachi, R. 1963. Fertilization of ferret ova by deposition of epididymal sperm into the ovarian capsule, with special reference to the fertilizable life of ova and capacitation of sperm. *J. Exp. Zool.* 154:175–87.
39. Clewe, T. H., and Mastroianni, L. 1960. A method for continuous volumetric collection of oviduct secretions. *J. Reprod. Fertil.* 1:146–50.
40. Dan, J. C. 1967. Acrosome reaction and lysins. In *Fertilization: Comparative morphology, biochemistry, and immunology*, ed. C. B. Metz and A. Monroy, 1:237–93. New York: Academic Press.
41. Dauzier, L., and Wintenberger, S. 1952. La vitesse de remontée des spermatozoïdes dans le tractus génital de la brebis. *Ann. Inst. Nat. Rech. Agron.* no. 1, pp. 13–22.
42. Day, F. T. 1942. Survival of spermatozoa in the genital tract of the mare. *J. Agric. Sci.* 32:108–11.
43. Dickmann, Z. 1963. Chemotaxis of rabbit spermatozoa. *J. Exp. Biol.* 40:1–5.
44. Doak, R. L.; Hall, A.; and Dale, H. E. 1967. Longevity of spermatozoa in the reproductive tract of the bitch. *J. Reprod. Fertil.* 13:51–58.
45. Egli, G. E., and Newton, M. 1961. The transport of carbon particles in the human female reproductive tract. *Fertil. Steril.* 12:151–55.
46. Eliasson, R. 1963. Prostaglandin—properties, actions and significance. *Biochem. Pharmacol.* 12:405–12.
47. Euler, U. S. von. 1937. On the specific vaso-dilating and plain muscle stimulating substances from accessory genital glands in man and certain animals (prostaglandin and vesiglandin). *J. Physiol.* 88:213–34.
48. Evans, E. I. 1933. The transport of spermatozoa in the dog. *Amer. J. Physiol.* 105:287–93.
49. Farris, E. J. 1950. *Human fertility and problems of the male*. White Plains, N. Y.: The Author's Press.

50. Fawcett, D. W., and Hollenberg, R. D. 1963. Changes in the acrosome of guinea-pig spermatozoa during passage through the epididymis. *Z. Zellforsch.* 60:276–92.
51. Gibbons, R. A. 1959. Physical and chemical properties of mucoids from bovine cervical mucin. *Biochem. J.* 72:27P–28P.
52. Gledhill, B. L. 1966. Studies on the DNA content, dry mass, and optical area of bull spermatozoal heads during epididymal maturation. *Acta Vet. Scand.* 7:131–42.
53. Green, W. W. 1947. Duration of sperm fertility in the ewe. *Amer. J. Vet. Res.* 8:299–300.
54. Gregoire, A. T.; Gongsakdi, D.; and Rakoff, A. E. 1961. The free amino acid content of the female rabbit genital tract. *Fertil. Steril.* 12:322–27.
55. ———. 1962. The presence of inositol in genital tract secretions of the female rabbit. *Fertil. Steril.* 13:432–35.
56. Hadek, R. 1955. The secretory process in the sheep's oviduct. *Anat. Rec.* 121:187–205.
57. Hammond, J., and Asdell, S. A. 1926. The vitality of the spermatozoa in the male and female reproductive tract. *J. Exp. Biol.* 4:155–85.
58. Hammond, J., and Walton, A. 1934. Notes on ovulation and fertilisation in the ferret. *J. Exp. Biol.* 11:307–19.
59. Hamner, C. E., and Williams, W. L. 1963. Effect of the female reproductive tract on sperm metabolism in the rabbit and fowl. *J. Reprod. Fertil.* 5:143–50.
60. ———. 1964. Identification of sperm stimulating factor of rabbit oviduct fluid. *Proc. Soc. Exper. Biol. Med.* 117:240–43.
61. ———. 1965. Composition of rabbit oviduct secretions. *Fertil. Steril.* 16:170–76.
62. Hartman, C. G. 1957. How do sperms get into the uterus? *Fertil. Steril.* 8:403–27.
63. Hartman, C. G., and Ball, J. 1930. On the almost instantaneous transport of spermatozoa through the cervix and the uterus of the rat. *Proc. Soc. Exp. Biol. Med.* 28:312–14.
64. Hays, R. L., and VanDemark, N. L. 1953. Effects of oxytocin and epinephrine on uterine motility in the bovine. *Amer. J. Physiol.* 172:557–60.
65. Henle, G., and Zittle, C. A. 1942. Studies of the metabolism of bovine epididymal spermatozoa. *Amer. J. Physiol.* 136:70–78.
66. Holwill, M. E. J. 1966. Physical aspects of flagellar movement. *Physiol. Rev.* 46:696–785.
67. Horne, H. W., Jr., and Audet, C. 1958. Spider cells, a new inhabitant of peritoneal fluid: A preliminary report. *Obstet. Gynec.* 11:421–23.
68. Howe, G. R. 1967. Leucocytic response to spermatozoa in ligated segments of the rabbit vagina, uterus and oviduct. *J. Reprod. Fertil.* 13:563–66.
69. Laing, J. A. 1945. Observations on the survival time of the spermatozoa in the genital tract of the cow and its relation to fertility. *J. Agric. Sci.* 35:72–83.
70. Lake, P. E. 1967. The maintenance of spermatozoa in the oviduct of the domestic fowl. In *Reproduction in the female mammal*, ed. G. E. Lamming and E. C. Amoroso, pp. 254–66. London: Butterworth and Co., Ltd.
71. Leonard, S. L., and Perlman, P. L. 1949. Conditions effecting the passage of spermatozoa through the utero-tubal junction of the rat. *Anat. Rec.* 104:89–102.
72. Lewis, W. H., and Wright, E. S. 1935. On the early development of the mouse egg. *Contrib. Embryol. Carnegie Inst.* 25:113–44.
73. Lutwak-Mann, C. 1954. Note on the chemical composition of bovine follicular fluid. *J. Agric. Sci.* 44:477–80.

74. Lutwak-Mann, C. 1962. Some properties of uterine and cervical fluid in the rabbit. *Biochim. Biophys. Acta* 58:637–39.
75. Mann, T. 1959. Biological changes associated with spermatogenesis, sperm maturation and sperm senescence. *Proc. Sixth Int. Conf. Planned Parenthood* (New Delhi), pp. 122–27.
76. ———. 1964. *The biochemistry of semen and of the male reproductive tract.* London: Methuen & Co.
77. ———. 1967. Sperm metabolism. In *Fertilization: Comparative morphology, biochemistry, and immunology*, ed. C. B. Metz and A. Monroy, 1:99–116. New York: Academic Press.
78. Mann, T.; Polge, C.; and Rowson, L. E. A. 1956. Participation of seminal plasma during the passage of spermatozoa in the female reproductive tract of the pig and horse. *J. Endocr.* 13:133–40.
79. Mastroianni, L.; Beer, F.; Shah, U.; and Clewe, T. H. 1961. Endocrine regulation of oviduct secretions in the rabbit. *Endocrinology* 68:92–100.
80. Mastroianni, L.; Forrest, W.; and Winternitz, W. W. 1961. Some metabolic properties of the rabbit oviduct. *Proc. Soc. Exp. Biol. Med.* 107:86–88.
81. Mastroianni, L., and Jones, R. 1965. Oxygen tension within the rabbit Fallopian tube. *J. Reprod. Fertil.* 9:99–102.
82. Mastroianni, L.; Shah, U.; and Abdul-Karim, R. 1961. Prolonged volumetric collection of oviduct fluid in the Rhesus monkey. *Fertil. Steril.* 12:417–24.
83. Mastroianni, L., and Wallach, R. C. 1961. Effect of ovulation and early gestation on oviduct secretions in the rabbit. *Amer. J. Physiol.* 200:815–18.
84. Mastroianni, L.; Winternitz, W. W.; and Lowi, N. P. 1958. The *in vitro* metabolism of the human endosalpinx. *Fertil. Steril.* 9:500–509.
85. Mattner, P. E. 1963*a*. Spermatozoa in the genital tract of the ewe. II. Distribution after coitus. *Aust. J. Biol. Sci.* 16:688–94.
86. ———. 1963*b*. Spermatozoa in the genital tract of the ewe. III. The role of spermatozoan motility and of uterine contractions in transport of spermatozoa. *Aust. J. Biol. Sci.* 16:877–84.
87. ———. 1963*c*. Capacitation of ram spermatozoa and penetration of the ovine egg. *Nature (Lond.)* 199:772–73.
88. Mattner, P. E., and Braden, A. W. H. 1963. Spermatozoa in the genital tract of the ewe. I. Rapidity of transport. *Aust. J. Biol. Sci.* 16:473–81.
89. Merton, H. 1939. Studies on reproduction in the albino mouse. III. The duration of life of spermatozoa in the female reproductive tract. *Proc. Roy. Soc., Edinburgh* 59: 207–18.
90. Miller, R. L. 1966. Chemotaxis during fertilization in the hydroid Campanularia. *J. Exper. Zool.* 162:23–44.
91. Mitchell, J. A., and Yochim, J. M. 1967. Intrauterine oxygen tension in the rat during pseudopregnancy. *Anat. Rec.* 157:288.
92. Morton, B. E., and Lardy, H. A. 1967. Cellular oxidative phosphorylation. II. Measurement in physically modified spermatozoa. *Biochemistry* 6:50–56.
93. Mounib, M. S., and Chang, M. C. 1964. Effect of *in utero* incubation on the metabolism of rabbit spermatozoa. *Nature (Lond.)* 201:943–44.
94. Mrouch, A. 1967. Oxytocin and sperm transport in rabbits. *Obstet. Gynec.* 29:671–73.
95. Murdoch, R. N., and White, I. G. 1967. "The metabolism of labelled glucose by rabbit spermatozoa after incubation *in utero*." *J. Reprod. Fertil.* 14:213–23.

96. Nalbandov, A., and Card, L. E. 1943. Effect of stale sperm on fertility and hatchability of chicken eggs. *Poultry Sci.* 22:218–26.
97. Nelson, L. 1967. Sperm motility. In *Fertilization: Comparative morphology, biochemistry, and immunology*, ed. C. B. Metz and A. Monroy, 1:27–97. New York: Academic Press.
98. Novak, E., and Everett, H. S. 1928. Cyclical and other variations in the tubal epithelium. *Amer. J. Obstet. Gynec.* 16:499–530.
99. Noyes, R. W. 1959. The capacitation of spermatozoa. *Obstet. Gynec. Survey* 14:785–97.
100. Noyes, R. W.; Adams, C. E.; and Walton, A. 1958. Transport of spermatozoa into the uterus of the rabbit. *Fertil. Steril.* 9:288–99.
101. Noyes, R. W.; Walton, A.; and Adams, C. E. 1958. Capacitation of rabbit spermatozoa. *J. Endocr.* 17:374–80.
102. Ogasawara, F. X., and Lorenz, F. W. 1964. Respiratory rate of cock spermatozoa as affected by oviduct extracts. *J. Reprod. Fertil.* 7:281–88.
103. Olds, D., and VanDemark, N. L. 1957*a*. The behavior of spermatozoa in luminal fluids of bovine female genitalia. *Amer. J. Vet. Res.* 18:603–7.
104. ———. 1957*b*. Composition of luminal fluids in bovine female genitalia. *Fertil. Steril.* 8:345–54.
105. Perkins, J. L., and Goode, L. 1966. Effects of the stage of the estrous cycle and exogenous hormones upon the volume and composition of oviduct fluid in ewes. *J. Anim. Sci.* 25:465–71.
106. Perkins, J. L.; Goode, L.; Wilder, W. A., Jr.; and Henson, D. B. 1965. Collection of secretions from the oviduct and uterus of the ewe. *J. Anim. Sci.* 24:383–87.
107. Racker, E. 1968. The membrane of the mitochondrion. *Sci. Amer.* 218:32–39.
108. Restall, B. J. 1966*a*. The Fallopian tube of the sheep. I. Cannulation of the Fallopian tube. *Aust. J. Biol. Sci.* 19:181–86.
109. ———. 1966*b*. The Fallopian tube of the sheep. II. The influence of progesterone and oestrogen on the secretory activities of the Fallopian tube. *Aust. J. Biol. Sci.* 19:187–97.
110. ———. 1966*c*. Histological observations on the reproductive tract of the ewe. *Aust. J. Biol. Sci.* 19:673–86.
111. ———. 1967. The biochemical and physiological relationships between the gametes and the female reproductive tract. *Adv. Reprod. Physiol.* 2:181–212.
112. Restall, B. J., and Wales, R. G. 1966. The Fallopian tube of the sheep. III. The chemical composition of the fluid from the Fallopian tube. *Aust. J. Biol. Sci.* 19:687–98.
113. Rothschild, Lord. 1951. Sea-urchin spermatozoa. *Biol. Rev.* 26:1–27.
114. Rowson, L. E. 1955. The movement of radio opaque material in the bovine uterine tract. *Brit. Vet. J.* 111:334–42.
115. Rubenstein, B. B.; Strauss, H.; Lazarus, M. L.; and Hankin, H. 1951. Sperm survival in women: Motile sperm in the fundus and tubes of surgical cases. *Fertil. Steril.* 2:15–19.
116. Salisbury, G. W., and Lodge, J. R. 1962. Metabolism of spermatozoa. *Adv. Enzymol.* 24:35–104.
117. Schott, R. G., and Phillips, R. W. 1941. Rate of sperm travel and time of ovulation in sheep. *Anat. Rec.* 79:531–40.
118. Shoger, R. L., and Bishop, D. W. 1967. Sperm activation and fertilization in *Limulus polyphemus*. *Biol. Bull.* 133:485.
119. Sleigh, M. A. 1962. *The biology of cilia and flagella*. London: Pergamon Press.

120. Soderwall, A. L., and Blandau, R. J. 1941. The duration of the fertilizing capacity of spermatozoa in the female genital tract of the rat. *J. Exp. Zool.* 88:55–63.
121. Soderwall, A. L., and Young, W. C. 1940. The effect of aging in the female genital tract on the fertilizing capacity of guinea pig spermatozoa. *Anat. Rec.* 78:19–29.
122. Srivastava, P. N.; Adams, C. E.; and Hartree, E. F. 1965. Enzymatic action of lipoglycoprotein preparations from sperm-acrosomes on rabbit ova. *Nature (Lond.)* 205:498.
123. Starke, N. C. 1949. The sperm picture of rams of different breeds as an indication of their fertility. II. The rate of sperm travel in the genital tract of the ewe. *Onderstepoort J. Vet. Sci. Anim. Husb.* 22:415–525.
124. Thibault, C., and Dauzier, L. 1960. "Fertilisines" et fécondation *in vitro* de l'oeuf de la Lapine. *C.R. Acad. Sci.* 250:1358.
125. Turnbull, K. E. 1966. The transport of spermatozoa in the rabbit doe before and after ovulation. *Aust. J. Biol. Sci.* 19:1095–99.
126. VanDemark, N. L. 1958. Spermatozoa in the female genital tract. *Int. J. Fertil.* 3:220–30.
127. VanDemark, N. L., and Hays, R. L. 1952. Uterine motility responses to mating. *Amer. J. Physiol.* 170:518–21.
128. ———. 1954. Rapid sperm transport in the cow. *Fertil. Steril.* 5:131–37.
129. VanDemark, N. L., and Moeller, A. N. 1951. Speed of spermatozoan transport in reproductive tract of estrous cow. *Amer. J. Physiol.* 165:674–79.
130. Vandeplassche, M., and Paredis, F. 1948. Preservation of the fertilizing capacity of bull semen in the genital tract of the cow. *Nature (Lond.)* 162:813.
131. Vishwakarma, P. 1962. The pH and bicarbonate-ion content of the oviduct and uterine fluids. *Fertil. Steril.* 13:481–85.
132. Wallace, J. C., and White, I. G. 1965. Studies of glycerylphosphorylcholine diesterase in the female reproductive tract. *J. Reprod. Fertil.* 9:163–76.
133. Warren, M. R. 1938. Observations on the uterine fluid of the rat. *Amer. J. Physiol.* 122: 602–8.
134. Whitney, L. F. 1937. *How to breed dogs.* New York: Orange Judd Publishing Co.
135. Williams, W. L.; Abney, T. O.; Chernoff, H. N.; Dukelow, W. R.; and Pinsker, M. C. 1967. Biochemistry and physiology of decapacitation factor. *J. Reprod. Fertil.* Suppl. 2:11–21.
136. Wimsatt, W. A. 1942. Survival of spermatozoa in the female reproductive tract of the bat. *Anat. Rec.* 83:299–305.
137. ———. 1944. Further studies on the survival of spermatozoa in the female reproductive tract of the bat. *Anat. Rec.*, 88:193–204.
138. Wimsatt, W. A.; Krutzsch, P. H.; and Napolitano, L. 1966. Studies on sperm survival mechanisms in the female reproductive tract of hibernating bats. I. Cytology and ultrastructure of intra-uterine spermatozoa of *Myotis lucifugus*. *Amer. J. Anat.* 119:25–59.
139. Yamanaka, H. S., and Soderwall, A. L. 1960. Transport of spermatozoa through the female genital tract of hamsters. *Fertil. Steril.* 11:470–74.
140. Yanagimachi, R., and Chang, M. C. 1963. Sperm ascent through the oviduct of the hamster and rabbit in relation to the time of ovulation. *J. Reprod. Fertil.* 6:413–20.
141. Yanagimachi, R., and Kanoh, Y. 1953. Manner of sperm entry in herring egg, with special reference to the role of calcium ions in fertilization. *J. Faculty Sci., Hokkaido Univ., Sect. VI*, 11:487–94.
142. Yochem, D. E. 1929. Spermatozoön life in the female reproductive tract of the guinea pig and rat. *Biol. Bull.* 56:274–97.

10

Pharmacology of the Oviduct

J. Brundin
Department of Physiology
Karolinska Institutet
Stockholm, Sweden

I. Historical Review

A historical review of the scientific study of the oviduct should start with Giuseppe Fallopius, who in 1561 first described this organ in *Observationes anatomicae*. During the subsequent centuries, studies on the anatomy of the genital tract advanced rapidly, but studies of its function were few. In 1819 Blundell was the first to publish a study on the physiology of the reproductive tract, where he described the muscular movements of the uterus and oviducts. He noted further that fertilization did not take place in a transected uterine horn. Later Bischoff (1842) stated that muscular contractions of the oviduct and broad ligament, as well as ciliary activity of the epithelial cells, were necessary if the eggs were to be conveyed from the ovary to the uterus. In 1862 Thiry concluded that ciliary activity was responsible for the movements of eggs from the ovary to the oviduct. Even today, there are contradictory opinions as to the role of muscular contractions and ciliary beat in the movement of eggs to the uterus.

Martin Barry (1842) was the first to demonstrate that the eggs of rabbits were penetrated by sperm within the oviducts. Assheton (1894) reported that the passage of eggs from the fimbriated end of the oviduct to the cornu required 3 days in the rabbit. Experimental studies on the rate of egg passage through the rabbit genital tract were performed by Pinner (1880), who introduced granules of different sizes, mainly India ink, in the ovarian end. These particles passed to the vagina within 2.5 hr. Lode (1894) used Ascaris eggs in the same way and found them in the middle of the rabbit oviduct after 10 to 36 hr. Sobotta (1922) made the generalization that the time for the journey of ova from the ovary to the uterus was fairly constant in all mammals regardless of the length of the oviduct and the size of the egg. This conclusion was modified later by Andersen (1927), who emphasized that this time schedule was constant only for each family of mammalian species.

In vitro studies of the motility of oviducts were described by Kehrer (1907) and opened the way for further studies in this field. Seckinger (1923) made the first physiological study of oviductal motility which correlated this function with the endocrine status of the animal. In his experiments circular rings of the oviduct of the sow showed different motility patterns during estrus compared with interestrous periods. The frequency of the contractions was reported to be rapid (13–15/min) in estrus and slow (4–6/min) during interestrous periods, though the amplitude was the same in both. Wislocki and Guttmacher (1924) verified these observations in their study of the excised uterus and oviducts of the sow. The work of Seckinger was followed by similar *in vitro* studies by Corner (1923) and Keye (1923). *In vivo* recording of circular muscle activity was introduced later by Mikulicz-Radecki (1926) by the use of a method based on mechanical photokymography. He studied four different levels of the rabbit oviduct under urethane anesthesia.

Pharmacological studies on oviducts were introduced by Kok (1927*b*). He studied the oviducts of man and sow *in vitro* and found that in both species the isthmus and ampulla reacted differently to drugs. He reported that during follicular growth the isthmus was stimulated but the ampulla was inhibited by adrenaline; on the other hand, during the luteal phase both the isthmus and the ampulla contracted upon the addition of adrenaline to the organ bath. He reported also that pilocarpine stimulated all parts of the oviduct regardless of the endocrine phase of the reproductive cycle. Kok was convinced that the nerve supply to the oviduct played a dominant role in the control of motility. This led to rather extensive denervation experiments in the rabbit (Kok 1927*a*). He observed an intensified peristalsis after sympathectomy and concluded that the parasympathetic division of the autonomic nervous system played the major role in maintaining peristalsis. Unfortunately, he did not record the effects of denervation objectively. Furthermore, Kok (1927*a*) discussed the hypothesis concerning presence of an oviductal sphincter in the uterine end of the oviduct which had been suggested by Henle (1866) and had received further support by X-ray studies in the 1920's (cf. Hermstein 1928) and by perfusion studies on sow oviducts (Kok 1929). Since then histological evidence has not verified the presence of any structural sphincter in the mammalian oviduct (cf. Lisa, Gioia, and Rubin 1954; Greenwald 1961*a*).

In 1920 Rubin described the air-insufflation method for the *in vivo* studies of oviduct patency. Data on patency and muscular activity have been obtained by this method. The validity of the method for precise studies of isolated oviductal contractions is still a matter of discussion, although the method has been used frequently in many species (e.g., Morse and Rubin 1937; Geist, Salmon, and Mintz 1938; Wimpfheimer and Feresten 1939; Feresten and Wimpfheimer 1939; Hafez 1962).

During the last 30 years concepts of oviductal functions have changed from the earlier opinion that the oviduct was simply a transporting channel for ova from the ovary to the uterus. It is now known to be an organ where fundamentally important events for the evolution, including fertilization and early development of the ova, take place. Coating of the ova by epithelial secretory products and the retention of the eggs in the uterine part of the ampulla for a considerable period are also important functions of the oviduct.

For the pharmacological studies on oviduct motility, the retention of eggs has been the object of increasing attention during these years. The original observation of Burdick and Pincus (1935) that estrogenic substances could prolong the retention or "tube lock" the eggs of rabbits and mice at the isthmoampullary junction and cause destruction of the blastocyst marked a new era in oviduct physiology. This retention of the recently ovulated eggs in the ampulla has been verified in the rabbit by Pincus and Kirsch (1936); Black and Asdell (1959); Noyes, Adams, and Walton (1959); Greenwald (1961*a*, *b*); Harper (1964, 1965*a*, *b*); and others. The same retention has also been described in the rat (Alden 1942) and cow (Black and Davis 1962).

II. Methods Used and Responses Obtained in the Study of the Effects of Drugs on Oviducts

Smooth muscle of the oviducts has been studied both *in vitro* and *in vivo* and on anesthetized and nonanesthetized animals. Visual observations reported especially in earlier literature may be excluded as nonobjective. These include methods such as the abdominal window technique and some earlier techniques which preceded the more modern salpingography. The more objective methods include kymography, either direct or by photoregistration, and the measurements of the intraluminal pressure, either by the open-end technique (fluid or gas) or in a closed system.

Various drugs belonging to the groups of sympathomimetics, parasympathomimetics, parasympatholytics, and oxytocics which have been tested *in vitro* are listed in Table 1 with

TABLE 1

In Vitro Effects of Substances Studied on Various Oviduct Preparations

Substance	Effect		Species
	Longitudinal Muscles	Circular Muscles	
Sympathomimetics			
Noradrenaline	+	+	Man
"	+	+	Rabbit
Adrenaline	+	+	Man
"	+	....	Monkey
"	+	....	Sow
"	+	....	Cow
Ephedrine	−	....	Sow
Isoproterenol	....	+	Man
Parasympathomimetics			
Acetylcholine	+	+	Man
"	+	+	Rabbit
"	+	....	Sow
Methacholine	....	+	Man
Carbacholine	+	....	Sow
Pilocarpine	−+	−+	Man
"	−+	−+	Sow
Physostigmine	+	....	Sow
Parasympatholytics			
Atropine	0−+	....	Sow
"	0	+	Man
Hyoscyamine	+	....	Sow
Scopolamine	+	....	Man
Oxytocics			
Ergonovine	0	+	Man
"	+	....	Sow
Oxytocin	....	−	Man
"	0	....	Rabbit
Neurohypophyseal extract	....	+	Monkey

Source: Data from many authors.

their recorded effects. A heterogenous group of other substances that have been tested *in vitro* on oviductal preparations are listed in Table 2. As is evident from the tables, only relatively few such substances and drugs have been tested on oviducts. The vast majority of experiments which have been performed on oviducts have been endocrine in nature (see Chapter 7). Since *in vivo* experiments carried out on oviductal preparations are scarce, the pharmacology of the oviduct in the functioning organism is essentially unexplored. It must be emphasized that the physiology and pharmacology of an organ are to be considered incompletely understood if the studies are not made on the unrestrained animal.

Nevertheless, *in vitro* experiments have shown some of the basic properties of the oviduct as an organ with smooth muscle as a major component. We have learned that certain spasmolytic drugs such as papaverine, aminophylline, and sodium nitrite inhibit spontaneous contractions. The contractility of the organ can be stimulated by both sympathomimetic and

TABLE 2

In Vitro EFFECT OF SUBSTANCES TESTED ON OVIDUCT SMOOTH MUSCLE PREPARATIONS

SUBSTANCE	EFFECT		SPECIES
	Longitudinal Muscles	Circular Muscles	
Cocaine	−	....	Man
"	+−	....	Rabbit
Nicotine	+	....	Sow
DMPP	....	−	Man
Sympathomimetics in presence of PBA	....	−	Man
Papaverine	−	....	Sow
"	−	−	Man
Aminophylline	−	−	Man
Histamine	+	+	Man
Sodium nitrite	−	−	Man
Ethanol	+−	....	Sow
Chlorpromazine	−	−	Man

SOURCE: Data from many authors.

parasympathomimetic drugs. For detailed information on the experimental approaches and the effects of various drugs on oviductal smooth muscle *in vitro* see the following authors: Gohara 1919; Kok 1926, 1927*b;* Seckinger and Snyder 1926; Li 1935; Cella and Georgescu 1937; Euler and Hammarström 1937; Donner 1954; Black and Asdell 1958; Bielecki and Kurzepa 1960; Sandberg *et al.* 1960; Sandberg, Ingelman-Sundberg, and Rydén 1963; Freund, Wiederman, and Saphier 1963; Brundin 1964*b;* Hawkins 1964; Rosenblum and Stein 1966.

In vivo measurements of the activity in the smooth muscles of the oviduct of different species have been performed either by the method for direct kymography (Mikulicz-Radecki 1926; Ichijo 1960), direct perfusion with fluid (Brundin 1965; Horton, Main, and Thompson 1965) or by the Rubin air-insufflation technique (Morse and Rubin 1937; Davids and Bender 1940; Artner and Tulzer 1957; Hafez 1962). All experiments except those of Ichijo were carried out under anesthesia. A good method for studies on oviductal activity would be to use a thin catheter placed in the lumen and record the intraluminal pressure in the unrestrained, nonanesthetized animal. No pharmacological studies have been published where this method has been used. The observations that have been published on oviductal pharmacology *in vivo* are listed in Table 3.

III. Neural Effects on Oviduct Activity

The endocrine influences upon genital functions have been well established in many reports. This is true also for the oviduct (cf. Burdick and Pincus 1935; Alden 1942; Greenwald 1957, 1958, 1959*a*, *b*, 1961*a*, *b*, 1963*a*, *b*; Black and Asdell 1958, 1959; Hafez 1962, 1963; Harper 1961, 1964, 1965*a*, *b*; Harrington 1964, and others). While the endocrine events are governed by the pituitary, the pituitary in turn is controlled by the neuronal structures in the hypothalamus (for ref. see Brown-Grant and Cross 1966). In recent years hypothalamic control has been considered more and more important. Similarly there has been an increasing interest also in the peripheral innervation of the genital organs. It has been known for some time that peripheral endocrine organs may have direct neuronal connections to the hypothalamus (Folkow and Euler 1954). There is increasing evidence for the importance of the autonomic nervous system in the control of the genital organ functions.

TABLE 3

In Vivo Effects of Tested Drugs on Oviductal Motility

Substance	Method	Effect	Species
Noradrenaline	Perfusion	Stimul.	Rabbit
Adrenaline	Rubin	Stimul.	Man
"	"	Stimul.	Rabbit
Acetylcholine	"	Inhib.	Man
Oxytocin	Perfusion	No eff.	Rabbit
Angiotensin	"	No eff.	Rabbit
Bradykinin	"	No eff.	Rabbit

Source: Data from many authors.

A. *Morphological Basis for Neural Supply to the Oviduct*

Fundamental observations on the innervation of the genital organs of the cat and rabbit were published by Langley and Anderson (1894, 1895, 1896). In the rabbit oviduct all the nerve fibers originating from the sympathetic division of the autonomic nervous system are conveyed in the hypogastric nerve bundles. In the human female as well, sympathetic nerve fibers which innervate the oviduct originate from the lumbar sympathetic chain and pass via the hypogastric nerve (Mitchell 1938). In man the hypogastric nerve contains these adrenergic, sympathetic fibers and even cholinergic parasympathetic fibers of sacral origin (Learmonth 1931). Most fibers of the hypogastric trunk are nonmyelinated, indicating that they are postganglionic (Langley and Anderson 1894, 1895, 1896).

By the use of electrophysiological technique slow-conducting, nonmyelinated C-fibers have been demonstrated in the hypogastric nerve bundle (Adrian, Bronk, and Phillips 1932). A small fraction of the fibers in the hypogastric nerve have been identified as preganglionic B-fibers (Lloyd 1937; Grundfest and Gasser 1938). These findings suggested that some of the hypogastric nerve fibers have a peripheral synapse. This was recently demonstrated to be true for both the male (Sjöstrand 1965) and female (Brundin 1965).

1. *Histochemical Observations.* Classical histological techniques cannot adequately separate the peripheral autonomic nerve endings into sympathetic and parasympathetic di-

visions. The introduction of the fluorescence method for identifying the peripheral sympathetic nerve endings is an important technique for mapping out the different parts of the autonomic nervous system in the mammalian organism (Falck *et al.* 1962). A choline esterase staining technique is of value also for the visualization of parasympathetic nerves (Jacobowitz and Koelle 1965).

The sympathetic innervation of the oviduct was first studied in the rabbit and in man by means of the fluorescence method (Brundin and Wirsén 1964*a*, *b*). The circular musculature of the isthmus is very richly innervated by adrenergic nerve terminals intermingling with the circular muscle fibers. On the other hand, the adrenergic innervation of the ampulla and infundibulum is poor, as also in the uterine cornua, where the adrenergic innervation is almost

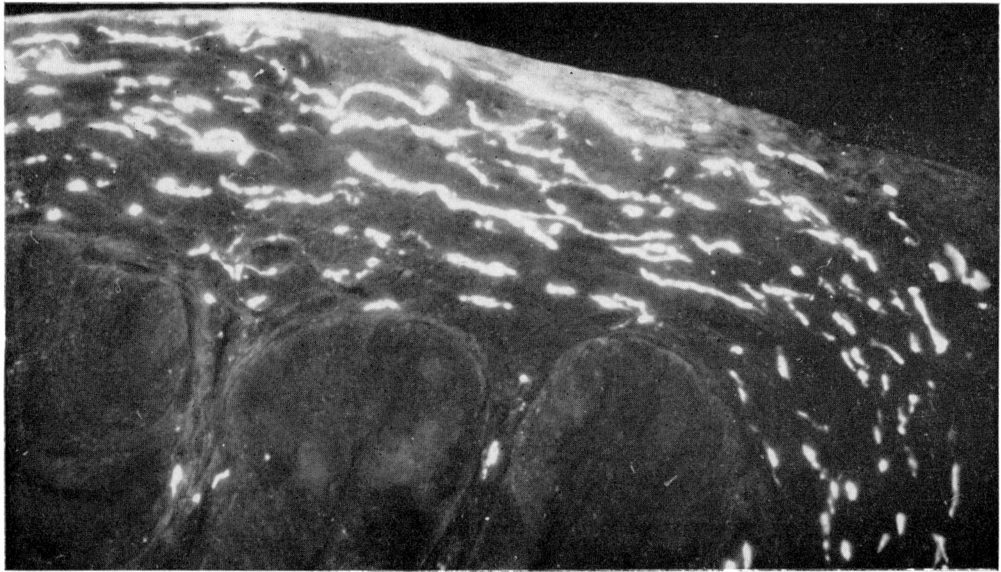

FIG. 1. Isthmic wall of rabbit oviduct. Numerous fluorescent adrenergic nerve terminals mainly in circular muscle layer. Slightly oblique section (×134).

entirely restricted to the walls of the blood vessels. This unique discrepancy between the degrees of innervation in different parts of the same organ has been verified for the rabbit oviduct (Owman and Sjöberg 1966). Figure 1 shows the rich net of fluorescent nerve fibers in the isthmic part of the rabbit oviduct. Figure 2 illustrates the poor innervation of the ampulla. Owman and Sjöberg report a denser net of nerve fibers in the isthmoampullary junction than in the rest of the isthmus. The similar pattern of adrenergic innervation was observed in the human oviduct (Brundin and Wirsén 1964*b*). This was confirmed in later studies by Owman, Rosengren, and Sjöberg 1967. The same distribution of adrenergic nerve supply is reported also for the rat (Norberg and Fredricsson 1966) and cat oviduct (Rosengren and Sjöberg 1967).

If the adrenergic nerve fibers to a peripheral organ are cut, the noradrenaline content of the innervated organ is diminished or abolished (Cannon and Lissák 1939; Euler and Purkhold 1951; Goodall 1951). As the noradrenaline-containing neurons are reduced, there is a reduction in the content of noradrenaline itself which may be measured by the fluorimetric technique. Thus, if the inferior mesenteric ganglion is extirpated, one would expect that the noradrenaline content of the oviduct would be reduced to zero. After this operation on the rabbit 1/3 of the neurons are left, which means that 1/3 of the nerve fibers are relayed to post-ganglionic neurons distal to the inferior mesenteric ganglion (Brundin 1965). No ganglia have been demonstrated in the oviductal wall of various species (Dahl 1916; Beaufays 1937; Chiara 1959; Damiani and Capodacqua 1961). This would indicate that the actual ganglia ought to be situated somewhere along the course of the hypogastric nerve. In the rabbit these ganglia are situated within and outside the middle third of the vaginal wall (Owman, Rosengren, and Sjöberg 1966). In the human female a dense net of adrenergic nerve cell bodies is visible at the uterovaginal junction. It is suggested that these are the ganglia of the short post-ganglionic adrenergic neurons to the oviduct in this species (Owman, Rosengren, and Sjöberg 1967). In 1965 Jacobowitz and Koelle published a histochemical study of the autonomic

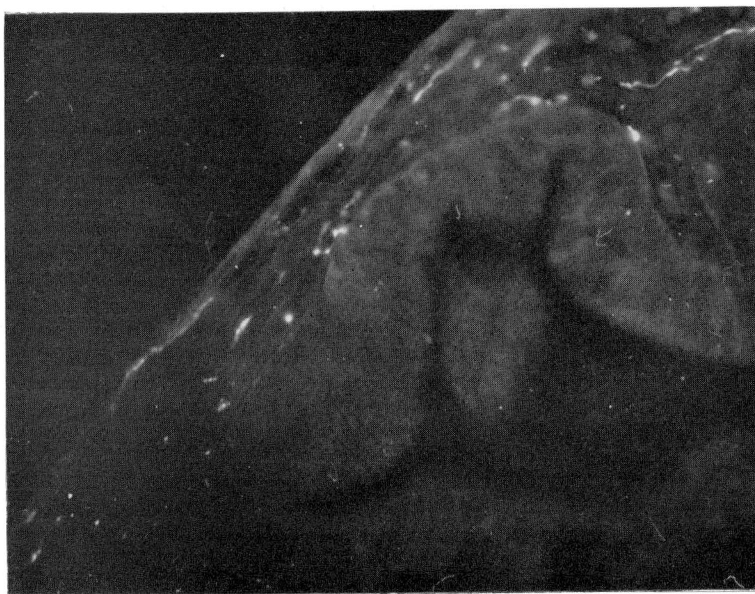

FIG. 2. Ampullary wall of rabbit oviduct. Scarce adrenergic nerve terminals fluorescing in circular muscle layer. Transverse section (×140).

innervation in the cat, rabbit, and guinea pig. They combined the catecholamine staining method of Falck *et al.* (1962) with a choline esterase staining method (Koelle 1955), and their interpretations implied that a great number of the catecholamine-containing nerves also exhibit the staining properties of acetylcholine esterase. This should be seen in the oviduct of the cat and in the vas deferens of cats and guinea pigs. Their hypothesis has not been confirmed (cf. Owman and Sjöberg 1966).

2. *Biochemical Studies.* It is well known from the pioneer works of Euler (1948) and from the extensive subsequent literature on noradrenergic mechanisms (cf. Euler 1956, 1966) that the noradrenaline content of an organ is directly proportional to the degree of adrenergic innervation. Biochemical analysis of the noradrenaline content of oviducts from different mammalian species has revealed large amounts of noradrenaline in the isthmus and a moderate or low content in the ampullary part of the organ. This was first observed in the rabbit (Brundin 1964*c*). The uterine horns also show small amounts of noradrenaline.

Thus, these data confirm the findings obtained later by histochemical techniques. In the species studied the noradrenaline content in the isthmus exceeds that of the ampulla. It may thus be concluded that the degree of adrenergic innervation to the different parts of the oviduct varies. The adrenergic innervation of the isthmus is rich, whereas only a few adrenergic nerves are found in the ampulla.

B. *Neuropharmacology of the Oviduct*

Since the autonomic, sympathetic innervation varies from one part of the oviduct to the next, one could presume that the effects of autonomic transmitter substances would vary in different parts of the oviduct.

As mentioned previously, the effects of adrenaline and noradrenaline and acetylcholine have been studied. Acetylcholine is reported to stimulate the tone and amplitude of both the human (Sandberg *et al.* 1960) and rabbit oviduct *in vitro* (Brundin 1964*b*). On the contrary, acetylcholine reduces the tone and amplitude of oviductal contractions in *in vivo* experiments in man (Artner and Tulzer 1957). The question has been raised whether the circular and longitudinal muscles of the oviduct react differently to various drugs. In the human oviduct these two muscular systems react in the same way to sympathomimetic and parasympathomimetic drugs (Sandberg *et al.* 1960; Hawkins 1964). In their *in vitro* experiments both groups of substances stimulated either the circular or the longitudinal musculature. On the other hand, the circular musculature of the rabbit isthmus is stimulated by noradrenaline whereas that of the ampulla is inhibited. Acetylcholine stimulates the circular muscles of either subdivision of the rabbit oviduct (Brundin 1964*b*).

The effect of noradrenaline has been tested in *in vivo* experiments in rabbits. The animals were anesthetized with pentobarbitol. Noradrenaline was injected intravenously and invariably elicited a contraction of the circular muscles of the oviduct. In Figure 3, the perfusion pressures in the isthmus were recorded after decreasing the intravenous doses of noradrenaline. The dose of 0.25 μg/kg still elicits a detectable response by this method of recording.

Since noradrenaline is a potent excitor of α-receptors, it is evident that this type of receptor predominates in the isthmus of the rabbit oviduct. The enhanced systemic arterial blood pressure following an intravenous injection of noradrenaline is not responsible per se for the enhanced perfusion pressure recorded. Thus, if angiotensin is given (Fig. 4), no perfusion pressure response can be recorded. Noradrenaline, on the other hand, still has an effect on both the systemic arterial blood pressure and the perfusion pressure. Beta-receptors have not been demonstrated in the rabbit oviduct. In human oviducts α-receptors are predominant although a small number of β-receptors can be demonstrated (Rosenblum and Stein 1966). It is likely that the species variations in this respect are only quantitative even in other groups of mammals.

C. *Effects of Stimulation of Adrenergic Nerves* in Vivo

A presumably continuous outflow of adrenergic nerve impulses to the oviduct ought to cause a constriction of the circular muscles of the isthmus since noradrenaline is then released from the nerve endings. By use of a perfusion technique, originally described by Horton and Main (1963) the effects of adrenergic nerve impulses have been studied in rabbit oviducts.

FIG. 3. Perfusion pressure response of rabbit oviduct isthmus to decreasing systemic doses of noradrenaline (NA). Perfusion in ovarian direction from 0.5 cm from the uterotubal junction. Saline: 28 μl/min. *Vaginal smear*, atrophy.

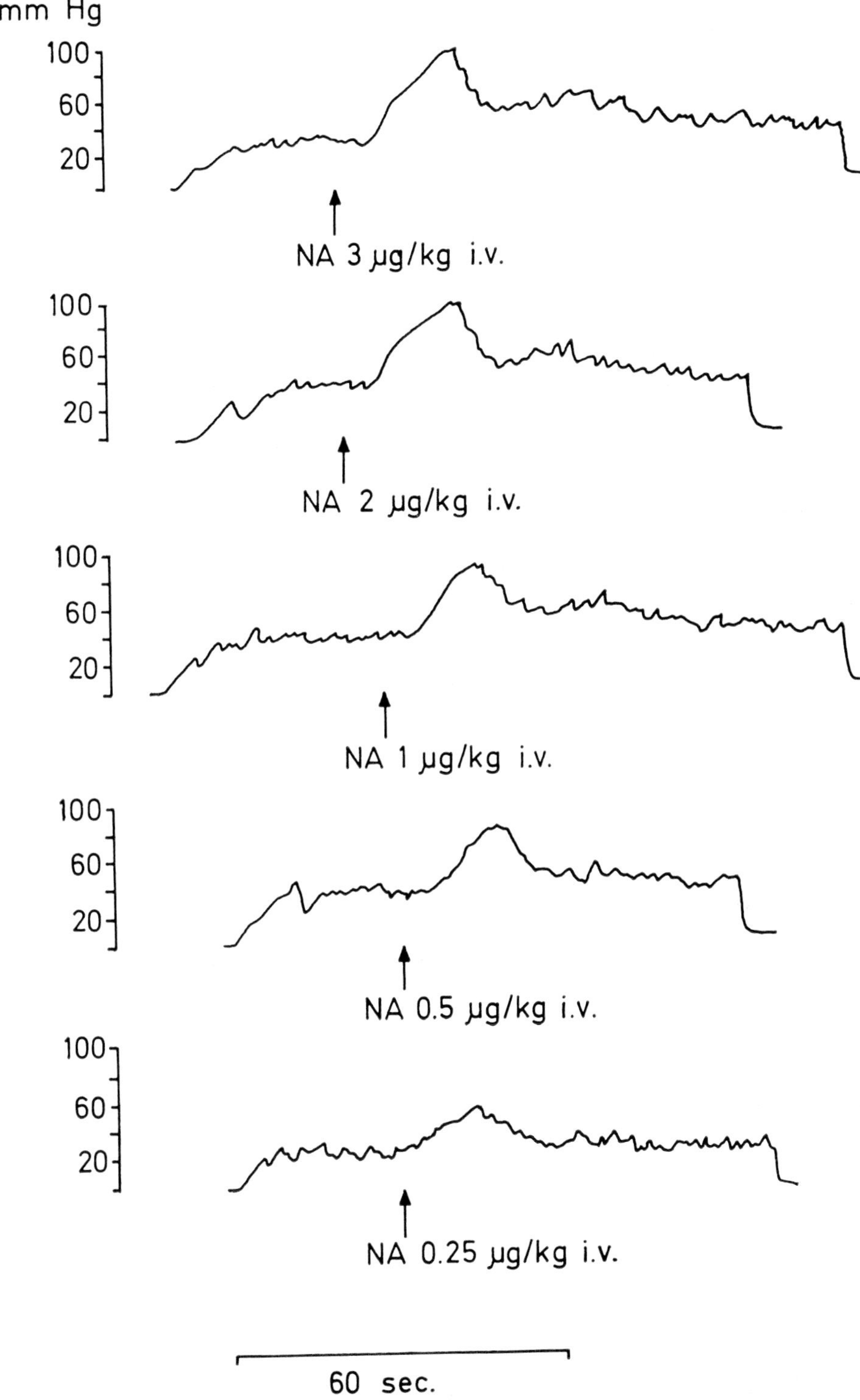
mm Hg
100
60
20
NA 3 µg/kg i.v.
100
60
20
NA 2 µg/kg i.v.
100
60
20
NA 1 µg/kg i.v.
100
60
20
NA 0.5 µg/kg i.v.
100
60
20
NA 0.25 µg/kg i.v.
60 sec.

The method used is shown in Figure 5. The hypogastric nerves are cut above the stimulating electrodes, which are placed just below the inferior mesenteric ganglia. The experimental animal was kept under general anesthesia (pentobarbital sodium). Atropine (20 mg/kg) was given before the experiment to facilitate the anesthesia. When a stimulus of supramaximal strength was applied to the nerve bundle, the perfusion pressure rose. The slope of the curve was then parallel to the pressure rise obtained with perfusion versus closed perfusion catheter which was always performed before and after each experiment (Fig. 6) (cf. Brundin 1965). The occlusive rate of rise is indicated by the dotted line. The parallel rise of the curves allows

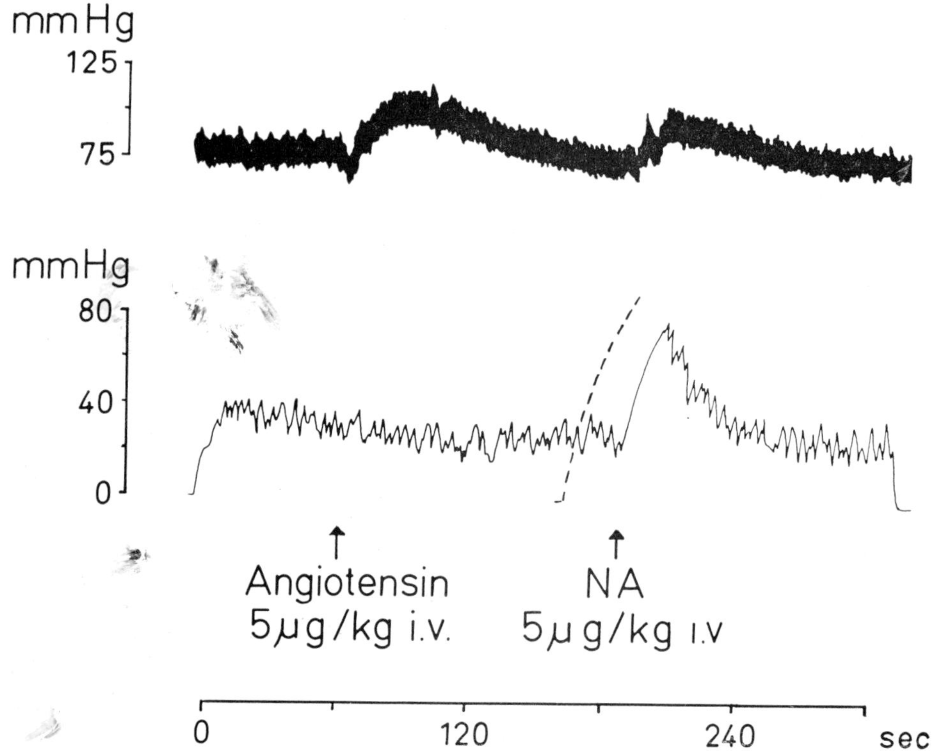

FIG. 4. Comparison of perfusion pressure curves from rabbit oviduct isthmus after angiotensin and after noradrenaline (NA). *Top tracing*, systemic arterial blood pressure. Saline perfusion in ovarian direction from 0.5 cm from uterotubal junction. Perfusion rate: 28 µl/min. *Vaginal smear*, estrus.

one to conclude that the isthmic lumen was closed during the nerve stimulation. The frequency of the square wave pulses applied to the nerves determined the degree of resistance to perfusion of the isthmic lumen as illustrated in Figure 7. The occlusive rate of rise is obtained at a frequency of 10 p/s.

The adrenergic innervation of the ampulla is, as mentioned before, very scarce and confined almost entirely to a vascular supply. This low degree of direct muscular innervation can be illustrated also in the perfusion experiments. If only the ampulla is perfused, there is but a moderate increase in the perfusion pressure recorded during hypogastric nerve stimulation (Fig. 8). A rate of rise of the pressure curve as steep as the occlusive rate of rise was never recorded during hypogastric nerve stimulation.

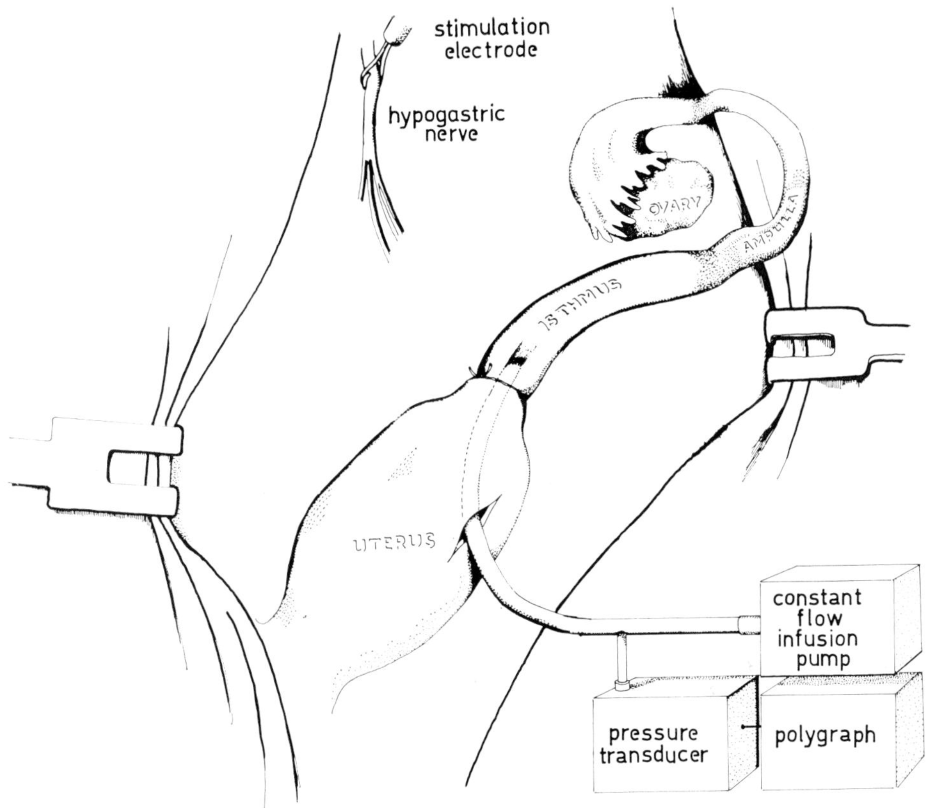

FIG. 5. Experimental device for combined oviduct perfusion and hypogastric nerve stimulation

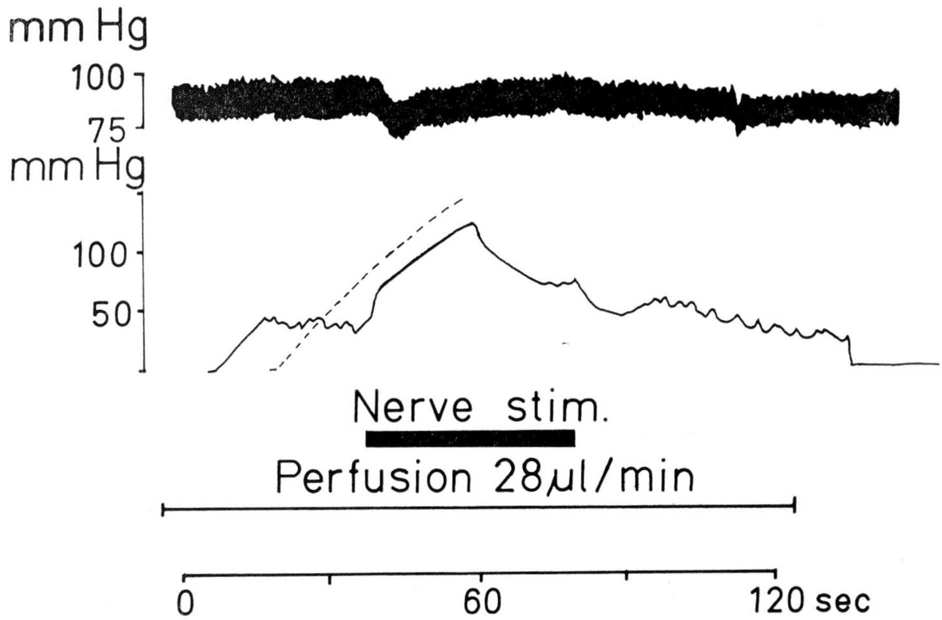

FIG. 6. Hypogastric nerve stimulation (17.5 p/s) during perfusion of rabbit isthmus in ovarian direction starting 0.5 cm from uterotubal junction. *Dotted line*, occlusive pressure rise. *Top tracing*, systemic arterial blood pressure. *Vaginal smear*, atrophy.

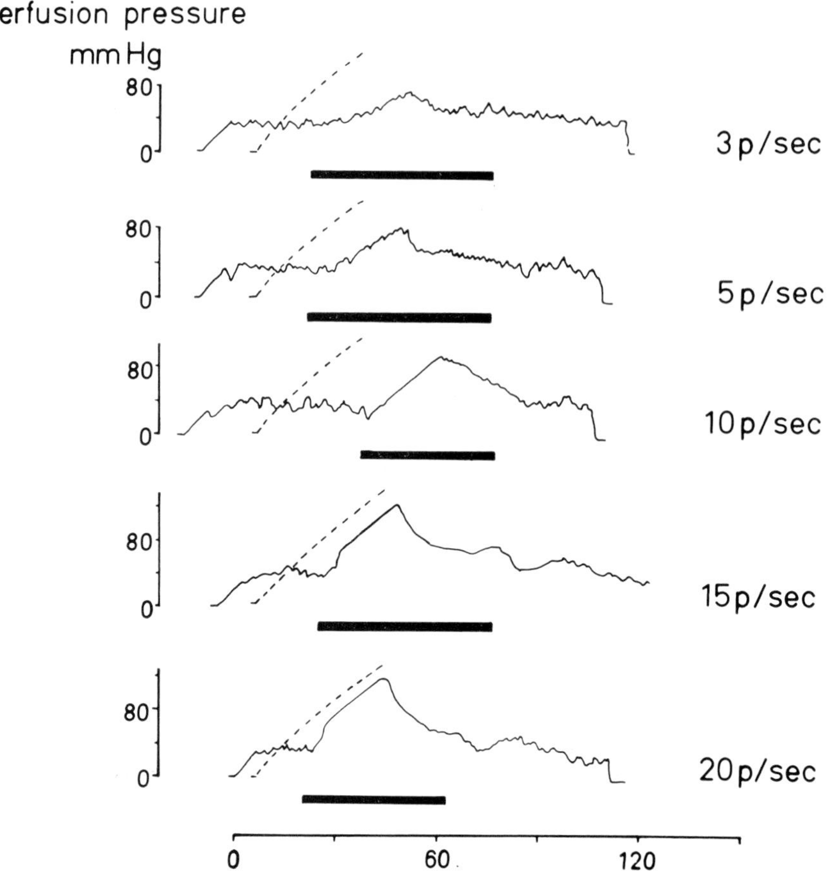

FIG. 7. Perfusion pressure from rabbit oviduct isthmus during increasing frequencies of hypogastric nerve stimulation (*bars*). Saline perfusion (28 μl/min) starts at 0.5 cm from uterotubal junction. *Dotted lines*, occlusive pressure rise. *Vaginal smear*, atrophy.

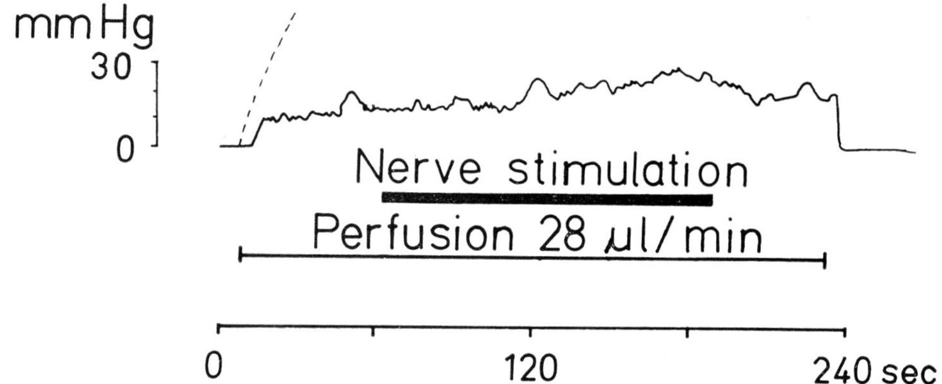

FIG. 8. Perfusion pressure of rabbit oviduct ampulla before and during hypogastric nerve stimulation (20 p/s). *Vaginal smear*, estrus.

The isthmic perfusion pressure response, if elicited by α-receptors during the stimulation, should be blocked by α-receptor blocking drugs. In Figure 9 phentolamine (0.5 mg/kg i.v.) completely abolished the effect of supramaximal nerve stimulation.

IV. The Isthmus as a Noradrenergic Sphincter

In the series of perfusion experiments, the spontaneous resistance to flow through the oviductal isthmus exceeded 30 mm Hg before the lumen opened and the perfusion pressure curve assumed a horizontal course. The spontaneous opening pressure was not significantly different in estrus and anestrus. This initial resistance to saline flow must be due to an obliteration of the isthmic lumen at the beginning of the perfusion. After either α-receptor blocking

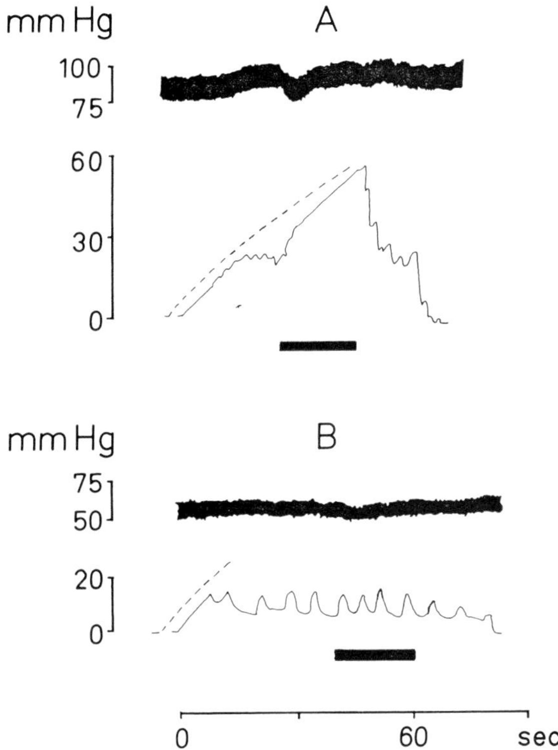

FIG. 9. Effect of α-receptor blocking drug on isthmic perfusion pressure response to hypogastric nerve stimulation in rabbit. *a*, hypogastric nerve stimulation (25 p/s) before phentolamine; *b*, hypogastric nerve stimulation (25 p/s) 3 min after phentolamine (0.5 mg/kg i.v.). *Top tracings*, systemic arterial blood pressure; *dotted lines*, occlusive pressure rise; *bars*, nerve stimulation; *vaginal smear*, atrophy.

or reserpine pretreatment or during experiments performed 72–98 hr *p.c.* the spontaneous opening pressure was reduced to 4–10 mm Hg, which is 1/3 of the normal mean value. The force of contraction of the rabbit oviduct never exceeds 7 mm Hg (Brundin 1964*a*). The force of contraction in the nonpregnant rabbit uterus *in vivo* never exceeds 37 mm Hg even during intense hypogastric nerve stimulation (Setekleiv 1964*a*, *b*, *c*). Thus, the isthmus constitutes an efficient barrier between the uterine and ampullary lumina. The efficiency of the sphincteric function is still more enhanced by a "physiological" stimulation frequency applied to the hypogastric nerves. In view of the fact that the sphincteric function of the isthmus must be "timed" to release the eggs for implantation, the role of the mechanisms associated with sphincteric function becomes even more important.

V. Conclusions

It may be concluded that the isthmus of many mammalian species has a noradrenergic innervation which allows this part of the organ to act as a sphincter. The function of this sphincter has been studied only in rabbits, but the muscular and neuronal prerequisites for a sphincter have been demonstrated also in the cat, dog, rat, and man. Nothing is known of the possible endocrine-dependent fluctuations in adrenergic nervous activity. It has been reported recently that certain antiestrogenic substances speed up the transport of eggs through the oviduct. This implies that the endocrine influence that has been demonstrated so clearly upon various oviductal functions may not be entirely direct upon the smooth muscle fibers. The changes that are induced by alterations in the endocrine environment also may affect the activity of the adrenergic nerves to the oviduct, as well as the effect of adrenergic nerve impulses on the smooth muscles. This constitutes one of the most important gaps in our present knowledge on reproduction physiology. Attention to nervous supply has been neglected by interests in the endocrine control of reproductive functions.

References

1. Adrian, E. D.; Bronk, D. W.; and Phillips, G. 1932. Discharges in mammalian sympathetic nerves. *J. Physiol.* (*Lond.*) 74:115.
2. Alden, R. H. 1942. The oviduct and egg transport in the albino rat. *Anat. Rec.* 34:137.
3. Andersen, D. H. 1927. The rate of passage of the mammalian ovum through various portions of the Fallopian tube. *Amer. J. Physiol.* 82:557.
4. Artner, J., and Tulzer, H. 1957. Über vegetativ bedingte Änderungen der Tubenmotilität. *Arch. Gynaek.* 188:364.
5. Assheton, R. 1894. A reinvestigation into the early stages of the development of the rabbit. *Quart. J. Micr. Sci.* 146:113.
6. Barry, M. 1842. Spermatozoa observed within the mammiferous ovum. *Phil. Trans. of the Royal Society*, *Ser. B* 1:33.
7. Beaufays, J. 1937. Die Endausbreitung des vegetativen Nervengewebes in der gesunden Tube und seine Veränderungen bei Entzündungen der Tube. *Arch. Gynaek.* 164:624.
8. Bielecki, A., and Kurzepa, S. 1960. The action of chlorpromazine (largactil), bromopromazine and fevoropromazine on the Fallopian tubes. *Ginek. Pol.* 31:275.
9. Bischoff, T. L. W. 1842. *Entwicklungsgeschichte des Kaninchenseies.* Braunschweig.
10. Black, D. L., and Asdell, S. A. 1958. Transport through the rabbit oviduct. *Amer. J. Physiol.* 192:63.
11. ———. 1959. Mechanism controlling entry of ova into rabbit uterus. *Amer. J. Physiol.* 197:1275.
12. Black, D. L., and Davis, J. 1962. A blocking mechanism in the cow oviduct. *J. Reprod. Fertil.* 4:21.
13. Blundell, J. 1819. Experiments on a few controverted points respecting the physiology of generation. *Med. Chir. Soc. Trans.* 10:245.

14. Brown-Grant, K., and Cross, B. A. 1966. Recent studies on the hypothalamus. *Brit. Med. Bull.* 22:195.
15. Brundin, J. 1964*a*. A functional block in the isthmus of the rabbit Fallopian tube. *Acta Physiol. Scand.* 60:295.
16. ———. 1964*b*. An occlusive mechanism in the Fallopian tube of the rabbit. *Acta Physiol. Scand.* 61:219.
17. ———. 1964*c*. The distribution of noradrenaline and adrenaline in the Fallopian tube of the rabbit. *Acta Physiol. Scand.* 62:156.
18. ———. 1965. Distribution and function of adrenergic nerves in the rabbit Fallopian tube. *Acta Physiol. Scand.* 66 (suppl. 259):1.
19. Brundin, J., and Wirsén, C. 1964*a*. The distribution of adrenergic nerve terminals in the rabbit oviduct. *Acta Physiol. Scand.* 61:203.
20. ———. 1964*b*. Adrenergic nerve terminals in the human Fallopian tube examined by fluorescence microscopy. *Acta Physiol. Scand.* 61:505.
21. Burdick, H. O., and Pincus, G. 1935. The effect of estrin injections upon the development of ova of mice and rabbits. *Amer. J. Physiol.* 111:201.
22. Cannon, W. B., and Lissák, K. 1939. Evidence for adrenaline in adrenergic neurones. *Amer. J. Physiol.* 125:765.
23. Cella, C., and Georgescu, I. D. 1937. Experimentelle Untersuchungen über die Physiologie und Pharmakodynamik des Eileiters. *Arch. Gynaek.* 165:36.
24. Chiara, F. 1959. Studio sulla fine innervazione dei genitali femminili tube. *Ann. Ostet. Ginec.* 81:1161.
25. Corner, G. W. 1923. Cyclic variation in uterine and tubal contraction waves. *Amer. J. Anat.* 32–33:345.
26. Dahl, W. 1916. Die Innervation der weiblichen Genitalien. *Z. Geburtsh. Gynaek.* 78: 539.
27. Damiani, N., and Capodacqua, A. 1961. Sull'inervazione intrinseca della tuba. *Ann. Ostet. Ginec.* 83:436.
28. Davids, A. M., and Bender, M. B. 1940. Effects of adrenaline on tubal contractions of the rabbit in relation to sex hormones (study *in vivo* by Rubin method). *Amer. J. Physiol.* 129:259.
29. Donner, H. 1954. Die Wirkung des Buscopan auf Tube und Uterus. *Zbl. Gynaek.* 23: 894.
30. Euler, U. S. v. 1948. Identification of the sympathomimetic ergone in adrenergic nerves of cattle (Sympathin N) with laevonoradrenaline. *Acta Physiol. Scand.* 16:63.
31. ———. 1956. *Noradrenaline.* Springfield, Ill.: Charles C Thomas.
32. ———. 1966. Twenty years of noradrenaline. *Pharmacol. Rev.* 18:29.
33. Euler, U. S. v., and Hammarström, S. 1937. Über Vorkommen und Wirkung von Adrenalin in Ovarien. *Skand. Arch. Physiol.* 77:163.
34. Euler, U. S. v., and Purkhold, A. 1951. Effect of sympathetic denervation on the noradrenaline and adrenaline content of the spleen, kidney and salivary glands in the sheep. *Acta Physiol. Scand.* 24:212.
35. Falck, B.; Hillarp, N.-Å.; Thieme, G.; and Torp, A. 1962. Fluorescence of catecholamines and related compounds condensed with formaldehyde. *J. Histochem. Cytochem.* 10:348.
36. Fallopius, G. 1561. *Observationes anatomicae.* Padua.

37. Feresten, M., and Wimpfheimer, S. 1939. Patency of the uterotubal junction of the rabbit: Experimental observations with the aid of CO_2 insufflation (Rubin method). *Endocrinology* 24:510.
38. Folkow, B., and Euler, U. S. v. 1954. Selective activation of noradrenaline and adrenaline producing cells in the suprarenal gland of the cat by hypothalamic stimulation. *Circulat. Res.* 2:191.
39. Freund, M.; Wiederman, J.; and Saphier, A. 1963. A method for the simultaneous recording *in vitro* of the motility of the vagina, of the body of the uterus and of both uterine horns in the guinea pig. *Fertil. Steril.* 14:416.
40. Geist, S. H.; Salmon, U. J.; and Mintz, M. 1938. The effect of estrogenic hormone upon the contractility of the Fallopian tubes. *Amer. J. Obstet. Gynec.* 36:67.
41. Gohara, A. 1919. Über die Wirkung des Kokains: Das Kokain ist nicht ein Sympathicusgift, sondern ein Muskelgift für glattmusklige Organe. *Acta Sch. Med. Univ. Kioto* 3:321.
42. Goodall, McCh. 1951. Studies of adrenaline and noradrenaline in mammalian heart and suprarenals. *Acta Physiol. Scand.* 24 (suppl. 85).
43. Greenwald, G. S. 1957. Interruption of pregnancy in the rabbit by the administration of estrogen. *J. Exp. Zool.* 135:461.
44. ———. 1958. Endocrine regulation of the secretion of mucin in the tubal epithelium of the rabbit. *Anat. Rec.* 130:477.
45. ———. 1959*a*. Tubal transport of ova in the rabbit. *Anat. Rec.* 133:386.
46. ———. 1959*b*. The comparative effectiveness of estrogens in interrupting pregnancy in the rabbit. *Fertil. Steril.* 10:155.
47. ———. 1961*a*. A study of the transport of ova through the rabbit oviduct. *Fertil. Steril.* 12:80.
48. ———. 1961*b*. The anti-fertility effects in pregnant rats of a single injection of estradiol cyclopentylpropionate. *Endocrinology* 69:1068.
49. ———. 1963*a*. Interruption of early pregnancy in the rabbit by a single injection of oestradiol cyclopentylpropionate. *J. Endocr.* 26:133.
50. ———. 1963*b*. *In vivo* recording of intraluminal pressure changes in the rabbit oviduct. *Fertil. Steril.* 14:666.
51. Grundfest, H., and Gasser, H. S. 1938. Properties of mammalian nerve fibers of slowest conduction. *Amer. J. Physiol.* 123:307.
52. Hafez, E. S. E. 1962. Pressure fluctuations during uterotubal kymographic insufflation in pregnant rabbits. *Fertil. Steril.* 13:426.
53. ———. 1963. The uterotubal junction and the luminal fluid of the uterine tube in the rabbit. *Anat. Rec.* 145:7.
54. Harper, M. J. K. 1961. The mechanisms involved in the movement of newly ovulated eggs through the ampulla of the rabbit Fallopian tube. *J. Reprod. Fertil.* 2:522.
55. ———. 1964. The effects of constant doses of oestrogen and progesterone on the transport of artificial eggs through the reproductive tract of ovariectomized rabbits. *J. Endocr.* 30:1.
56. ———. 1965*a*. Transport of eggs in cumulus through the ampulla of the rabbit oviduct in relation to day of pseudopregnancy. *Endocrinology* 77:114.
57. ———. 1965*b*. The effects of decreasing doses of oestrogen and increasing doses of progesterone on the transport of artificial eggs through the reproductive tract of ovariectomized rabbits. *J. Endocr.* 31:217.

58. Harrington, F. E. 1964. Effect of estradiol benzoate on ova transport in superovulated immature mice. *Endocrinology* 75:461.
59. Hawkins, D. F. 1964. Some pharmacological reactions of isolated rings of human Fallopian tube. *Arch. Int. Pharmacodyn.* 152:474.
60. Henle, J. 1866. *Handbuch der systematischen Anatomie des Menschen.* Braunschweig.
61. Hermstein, A. 1928. Zur Frage des Tubensphinkters. *Zbl. Gynaek.* 52:1823.
62. Horton, E. W., and Main, I. H. M. 1963. Comparison of the biological activities of four prostaglandins. *Brit. J. Pharmacol.* 21:182.
63. Horton, E. W.; Main, I. H. M.; and Thompson, C. J. 1965. Effects of prostaglandins on the oviduct, studied in rabbits and ewes. *J. Physiol.* (*Lond.*) 180:514.
64. Ichijo, M. 1960. Studies on the motile function of the Fallopian tube. Report 1. Analytic studies on the motile function of the Fallopian tube. *Tohoku J. Exp. Med.* 72:211.
65. Jacobowitz, D., and Koelle, G. B. 1965. Histochemical correlations of acetylcholinesterase and catecholamines in postganglionic autonomic nerves of the cat, rabbit, and guinea pig. *J. Pharmacol. Exp. Ther.* 148:225.
66. Kehrer, E. 1907. Physiologische und pharmakologische Untersuchungen an den überlebenden und lebenden inneren Genitalien. *Arch. Gynaek.* 81:160.
67. Keye, J. D. 1923. Periodic variations in spontaneous contractions of uterine muscle, in relation to the oestrous cycle and early pregnancy. *Bull. Johns Hopkins Hosp.* 34:60.
68. Koelle, G. B. 1955. The histochemical identification of acetylcholinesterase in cholinergic, adrenergic and sensory neurons. *J. Pharmacol. Exp. Ther.* 114:167.
69. Kok, F. 1926. Bewegungen des muskulösen Rohres der Fallopischen Tube. *Arch. Gynaek.* 127:384.
70. ———. 1927*a*. Über die Versorgung der Fallopischen Tube mit motorischen Nerven. *Arch. Gynaek.* 130:173.
71. ———. 1927*b*. Experimentelle Untersuchungen über die pharmakologische Beeinflussung der Eileitermuskulatur als Beitrag zur Klärung der Frage nach dem Mechanismus des Eitransportes. *Zbl. Gynaek.* 51:2650.
72. ———. 1929. Über den Einfluss der Tubenmuskelbewegungen auf den Tubeninhalt; zugleich zur Frage des Tubensphinkters. *Zbl. Gynaek.* 53:26.
73. Langley, J. N., and Anderson, H. K. 1894. The constituents of the hypogastric nerves. *J. Physiol.* (*Lond.*) 17:177.
74. ———. 1895. The innervation of the pelvic and adjoining viscera. Part IV. The internal generative organs. Part V. Position of the nerve cells on the course of the efferent nerve fibres. *J. Physiol.* (*Lond.*) 19:71.
75. ———. 1896. The innervation of the pelvic and adjoining viscera. Part VI. Anatomical observations. *J. Physiol.* (*Lond.*) 20:372.
76. Learmonth, J. R. 1931. A contribution to the neurophysiology of the urinary bladder in man. *Brain* 54:147.
77. Li, R. C. 1935. The effect of posterior pituitary extract, epinephrine and acetylcholine on the isolated Fallopian tube of the macaque at different stages of the menstrual cycle. *Chin. J. Physiol.* 9:315.
78. Lisa, J. R.; Gioia, J. D.; and Rubin, I. C. 1954. Observations on the interstitial portion of the Fallopian tube. *Surg. Gynec. Obstet.* 99:159.

79. Lloyd, D. P. C. 1937. Transmission of impulses through inferior mesenteric ganglia. *J. Physiol. (Lond.)* 91:296.

80. Lode, A. 1894. Experimentelle Beiträge zur Lehre der Wanderung des Eies vom Ovarium zur Tube. *Arch. Gynaek.* 45:293.

81. Mikulicz-Radecki, F. v. 1926. Experimentelle Untersuchungen über Tubenbewegungen. *Arch. Gynaek.* 128:318.

82. Mitchell, G. A. G. 1938. The innervation of the ovary, uterine tube, testis and epididymis. *J. Anat. (Lond.)* 72:508.

83. Morse, A. H., and Rubin, I. C. 1937. Uterotubal insufflation in the *Macacus rhesus:* A method of assaying pharmacologic and hormonal effects on tubal and uterine contractions. *Amer. J. Obstet. Gynec.* 33:1087.

84. Norberg, K.-A., and Fredricsson, B. 1966. Cellular distribution of monoamines in the uterine and tubal walls of the rat. *Acta Physiol. Scand.* 68 (suppl. 277):149.

85. Noyes, R. W.; Adams, C. E.; and Walton, A. 1959. The transport of ova in relation to the dosage of oestrogen in ovariectomized rabbits. *J. Endocr.* 18:108.

86. Owman, Ch.; Rosengren, E.; and Sjöberg, N.-O. 1966. Origin of the adrenergic innervation to the female genital tract of the rabbit. *Life Sci.* 5:1389.

87. ———. 1967. Adrenergic innervation of the human female reproductive organs: a histochemical and chemical investigation. *Obstet. Gynec.* 30:763.

88. Owman, Ch., and Sjöberg, N.-O. 1966. Adrenergic nerves in the female genital extract of the rabbit: With remarks on cholinesterase-containing structures. *Z. Zellforsch.* 74:182.

89. Pincus, G., and Kirsch, R. E. 1936. The sterility in rabbits produced by injections of oestrone and related compounds. *Amer. J. Physiol.* 115:219.

90. Pinner, O. 1880. Über den Übertritt des Eies aus dem Ovarium in die Tube beim Säugetier. *Arch. Physiol.* 241.

91. Rosenblum, I., and Stein, A. A. 1966. Autonomic responses of the circular muscles of isolated human Fallopian tube. *Amer. J. Physiol.* 210:1127.

92. Rosengren, E., and Sjöberg, N.-O. 1967. The adrenergic nerve supply to the female reproductive tract of the cat. *Amer. J. Anat.* 121:271.

93. Rubin, I. C. 1920. Nonoperative determination of patency of Fallopian tubes in sterility; intrauterine inflation of oxygen, and production of an artificial pneumoperitoneum; preliminary report. *J.A.M.A.* 74:1017.

94. Sandberg, F.; Ingelman-Sundberg, A.; Lindgren, L.; and Rydén, G. 1960. *In vitro* studies of the motility of the human Fallopian tube. Part I. The effects of acetylcholine, adrenaline, noradrenaline and oxytocin on the spontaneous motility. *Acta Obstet. Gynec. Scand.* 39:506.

95. Sandberg, F.; Ingelman-Sundberg, A.; and Rydén, G. 1963. *In vitro* studies of the motility of the human Fallopian tube. Part II. The effects of methylergometrine and papaverine. *Acta Obstet. Gynec. Scand.* 42:1.

96. Seckinger, D. L. 1923. Spontaneous contractions of the Fallopian tube of the domestic pig with reference to the oestrous cycle. *Amer. J. Physiol.* 34:236.

97. Seckinger, D. L., and Snyder, F. F. 1926. Cyclic changes in the spontaneous contractions of the human Fallopian tube. *Bull. Johns Hopkins Hosp.* 39:371.

98. Setekleiv, J. 1964*a*. Uterine motility of the estrogenized rabbit. I. Isotonic and isometric recording *in vivo*. Influence of anesthesia and temperature. *Acta Physiol. Scand.* 62:68.

99. ———. 1964*b*. Uterine motility of the estrogenized rabbit. II. Response to distension. *Acta Physiol. Scand.* 62:79.
100. ———. 1964*c*. Uterine motility of the estrogenized rabbit. III. Response to hypogastric and splanchnic nerve stimulation. *Acta Physiol. Scand.* 62:137.
101. Sjöstrand, N. O. 1965. The adrenergic innervation of the vas deferens and the accessory male genital glands. *Acta Physiol. Scand.* 65 (suppl. 257).
102. Sobotta, J. 1922. Mechanismus der Wanderung des Eies durch den Eileiter. *Dtsch. Med. Wschr.* 48:1088.
103. Thiry, L. 1862. Über das Vorkommen eines Flimmerepithelium auf dem Bauchfell des weiblichen Frosches. *Göttinger Nachrichten* 171.
104. Wimpfheimer, S., and Feresten, M. 1939. The effect of castration on tubal contractions of the rabbit as determined by the Rubin test. *Endocrinology* 25:91.
105. Wislocki, G. B., and Guttmacher, A. F. 1924. Spontaneous peristalsis of the excised whole uterus and Fallopian tubes of the sow with reference to the ovulation cycle. *Bull. Johns Hopkins Hosp.* 35:246.

11

Pathologic Processes of the Oviduct

K. Benirschke

Department of Pathology
Dartmouth Medical School
Hanover, New Hampshire

Diseases of the oviduct constitute an important aspect of illness in general, and the obstetrician is constantly concerned over the possibility of oviductal ectopic pregnancy and salpingitis, to name the two most common conditions. The latter often results in occlusion and, therefore, investigation of oviductal patency is usually the first study performed when an infertile couple is investigated. In striking contrast is the relative paucity of oviductal conditions in domestic and wild animals.

We must ask then the question why diseases of the oviduct are of such importance in man and why most animals suffer much less from diseases of the oviducts. Is it different reproductive mechanisms, different age groups, lack of intensive study, or other unexplored factors? In the following the principal anomalies and diseases of the oviduct will be described and differences in incidence compared as we find them between man and animals. While man will be given most emphasis this is merely to reflect that most knowledge exists of his diseases. More detailed studies of oviductal pathology of animals are needed.

I. Congenital and Structural Anomalies

A. *Aplasia; Hypoplasia; Ectopia; Atresia*

A wide spectrum of congenital anomalies affects the oviducts and, because they are but part of the Müllerian system, often the uterus and vagina are involved as well. Indeed, presumably because of a local embryonic developmental disturbance in the region of the renal and Müllerian duct anlage, and because of the close proximity of these structures, it is not infrequent that both are simultaneously involved. This has considerable implications. It is known that malformed uteri yield a poorer reproductive performance than fused uteri, and, in such patients, unilateral renal aplasia often comes as a surprise to the physician. This diagnosis is often made too late for adequate clinical management of urinary tract diseases. A patient with an absent kidney can occasionally be helped in her obstetric performance if a uterine malformation is suspected early, and Felding (1965) advocates that "hysterosalpingography should be undertaken in all fertile women with congenital solitary kidney and urography in all patients with genital malformations."

The incidence of genital anomalies in women has been reviewed (Greiss and Mauzy 1961). Uterine malformations are common, occurring perhaps as often as in 3% of postpartum uteri examined closely. Many of these malformations have a poor reproductive performance, presumably because some of the anomalies really involve what Jones (1957) chose to call hemiuteri. Anomalies of the oviducts with functional significance are principally: hypoplasia, aplasia (associated with and without uterus unicornis) and ectopia (with and without ectopia of ovary or kidney). Felding (1965), who describes these problems in 13 women with solitary kidney finds unilateral renal aplasia in 1:1275; 39 cases of renal aplasia were found in 53 patients with uterus unicornis (see also Kamm and Beernink 1962). Renal agenesis is reviewed by Lenz (1964) who mentions the higher frequency in males and also its association with absence of one umbilical artery. He is particularly concerned with the genetic and familial aspects of renal malformations. The veterinary literature contains similar reports but fewer systematic reviews. Heinze (1964) presents a case of ovarian and oviductal agenesis in a calf and rules out the possibility of freemartinism. He quotes Lehmann to have found aplasia of oviducts in sheep with the astonishing frequency of 1.5%. Ovarian and Müllerian duct agenesis in association with homolateral (left) renal anomalies (agenesis, hypoplasia) in the rat has been observed in association with eye defects, possibly on a hereditary basis (Nachtsheim 1958). The correlation of solitary kidney with uterine or Wolffian duct anomaly has also been reported in the A × C strain of rats (Deringer and Heston 1956). Hutt (1967) finds atresia of the oviduct in fowl to be a dominant heritable trait. Males transmit the gene to half their daughters. The hens lay eggs but eventually develop a fatal peritonitis from the accumulation of eggs in the body cavity. Roberts (1956) quotes McEntee and Moberg each as having found a case of oviductal aplasia in the cow. The relatively frequent persistence of the normally degenerating right oviduct in pigeons and parrots, occasionally leading to retention cysts, has been described by Köhler (1958).

Other sporadic anomalies are depicted in newborns by Potter (1961) and the relationship to torsion-atrophy of what is considered to be the most frequent anomaly, namely, absence of the ampullary portion, is discussed by Kistner (1965; see also Frankl 1930). These anomalies are not anatomical curiosities, but often demand clinical attention. At times they cause dramatic clinical pictures because of ectopic pregnancies, as in the case of Bollack, Kuhlmann, and Papaevangelos (1964). In their 13-year-old patient, urinary retention and an abdominal mass necessitated laparotomy. There were bifid uterus and vagina and, due to the unilateral atresia of the vaginal opening, there were hematocolpos and hematosalpinx, containing 3000 ml of bloody fluid. The cystically dilated oviduct contained much hemosiderin. While no hematometra was found, such is not infrequent in hymeneal atresia, whether it is associated with hematosalpinx or not.

Displacement of the oviduct has frequently been described (Frankl 1930). This displacement often takes place with simultaneous ectopia of ovary and kidney, and not infrequently such an ectopic appendage is found in hernia sacs. Thus, in the case of Kent (1963) the oviduct and ovary were retroperitoneally dislocated into an inguinal hernia and the homolateral kidney was also displaced. He postulates a failure of the gubernaculum ovarii which normally carries the appendage medially with the uterus.

B. *Intersex States*

Anomalies of the oviducts and the remainder of the urogenital apparatus are a frequent finding in various intersex states (Overzier 1961; Lenz 1964). Ductal differentiation is largely the result of fetal gonadal activity. The latter is presumably of endocrine nature and, in some animals such as the rat and rabbit, Jost's experiments (1948, 1963) have shown that the fetal gonads are under fetal pituitary control. This latter assumption probably does not hold for all animals for, in man and cattle, gonadal differentiation takes place before pituitary activity commences and even in the congenital absence of this gland. Whether placental hormones "direct" gonadal activity in such species is not yet established and it is possible that the genetic makeup of the gonad (either its germ cells or, more likely, its stroma) plays an important role in this early activity. Thus, in a variety of genetic mosaics or chimeras, particularly the true hermaphrodites, some correlation exists between the type of gonadal differentiation and the degree of homolateral duct development. It is suggested that internal ductal differentiation is mediated, in part, by local permeation of differentiating substances (steroid or otherwise). For instance, the true hermaphrodite mink recently described by Nes (1966) with a chimeric diploid XX/triploid XXY chromosome makeup showed a greater degree of Müllerian duct reduction on the side which contained the more testicular gonad. At the moment generalizations cannot be made from the study of these chromosomal mosaics and chimeras as to details of sex differentiation and its relation to percentage of cell lines (Benirschke 1967). True hermaphroditism has been found in man with variable ductal development and in the apparent absence of a cell line containing a Y chromosome; on the other hand, unilateral testicular structure (and ducts) with ovarian development (and ducts) has been described in which only local gonadal effect on ductal differentiation explains the picture satisfactorily (see Grumbach, Wyk, and Wilkins 1955). The variables can be appreciated by a comparison of well-studied calico cats: our male calico cat with XX/XY chimerism had no female external or internal structures (Malouf, Benirschke, and Hoefnagel 1967), while a similar chimera studied by Thuline and Norby (1965) was a true hermaphrodite. We are currently studying an intersex horse (XX/XY chimera) with hypospadias, rudimentary vagina, and atrophic testes in whom no ovaries are present either. In beagles, Murti, Gilbert, and Borgmann (1966) report absence of oviducts in the presence of uteri and ovotestes. Not even the notorious ovarian change in freemartins with the frequent hypoplasia (or atrophy) of Müllerian ducts (Figs. 1 and 2) is really understood. Although there is little doubt concerning the relationship of the freemartin condition to placental anastomoses between heterosexual twins and ensuing blood chimerism, the inability to reproduce the condition experimentally (Jost, Chodkiewicz, and Mauleon 1963; Jainudeen and Hafez 1965) has raised doubt as to the steroid-induction hypothesis.

All these aspects are currently under investigation, both biochemically and cytogenetically. It must be emphasized, though, that this is a most complex area of study, and careless comparison among quite divergent species of mammals must be avoided. (For complete reviews in domestic animals see Bertrand 1965; Biggers and McFeely 1966; Hafez and Jainudeen 1966).

C. *Walthard's Cell Nests*

It is normal to find on the serosal surface of almost every oviduct minute white granules (Fig. 3) which upon histologic examination are either solid or cystic (serous or mucoid) nests of cells (Fig. 5), whose origin has been debated at length. Their histogenetic relation-

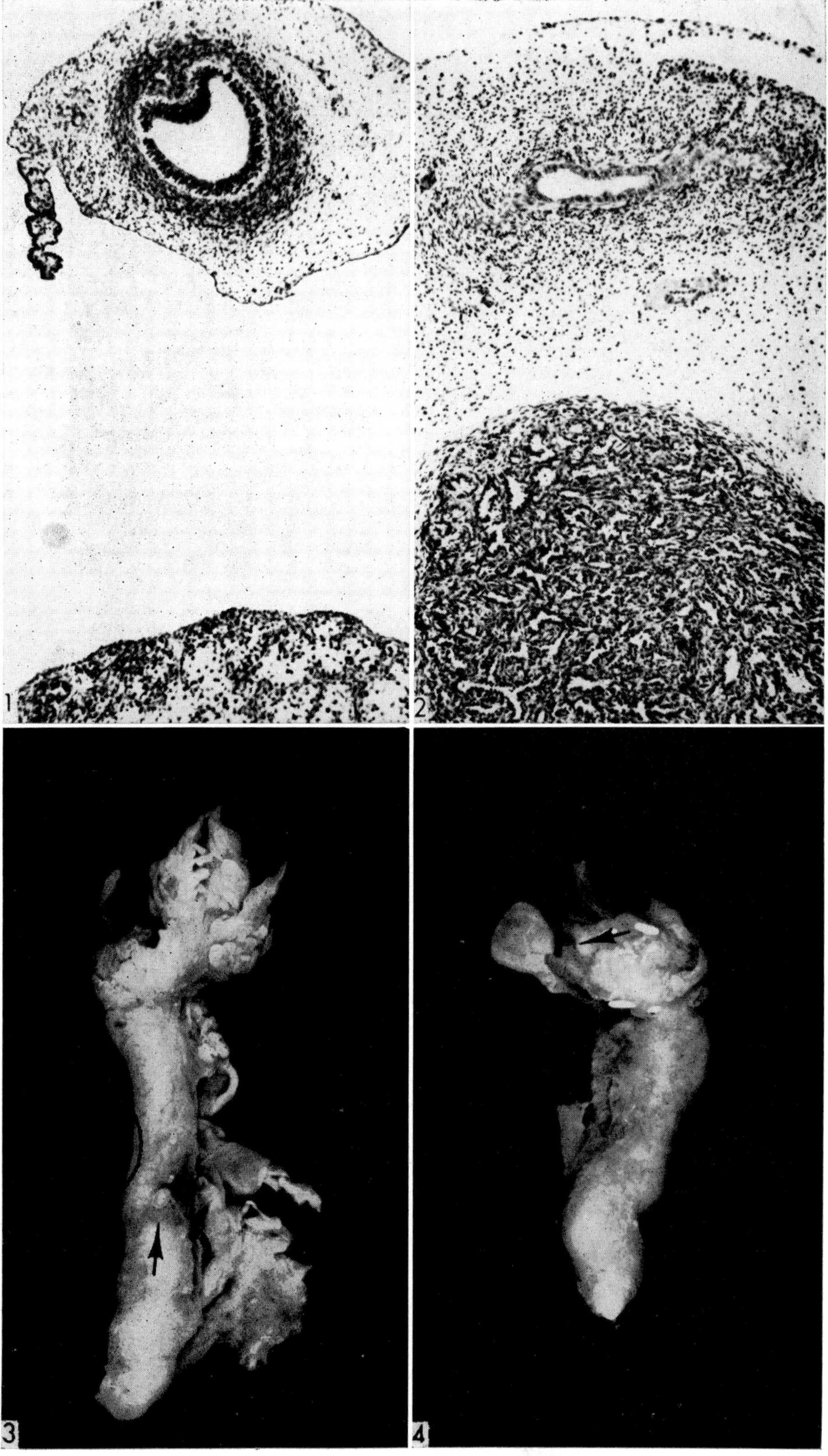

ship to Brenner tumors of the ovary (Fig. 6) is fully discussed by Novak and Woodruff (1962). A critical analysis of these similarities has been undertaken by Arey (1943). Willis (1958) considers the nests of cells to represent metaplasia of the peritoneum and Teoh (1951*a*), in his department, found that these cells never produce keratin, although superficially they resemble squamous epithelium. Similar nests occur in testicular appendages as well, and he refers to cases in which similar metaplastic changes were seen on gut surfaces. The two studies by Teoh (1951*a*, *b*) are exemplary and negate the relationship to Brenner tumors. The reader is also referred to the editorial comments following the abstract of Lauchlan's (1966) paper. The subject should better be discussed henceforth with the other and more important metaplastic event, endometriosis (see also Rewell 1960).

D. *Hydatid Cysts; Duplications*

In man and cattle the fimbriated end of the oviduct frequently contains cystic structures of various sizes, covered by a smooth serosa and containing a watery fluid (Lombard, Morgan, and McNutt 1951). The cysts are usually pedunculated; commonly they are single, but occasionally 2 or 3 cysts have been found and they are frequent in rabbits (Hafez, personal communication). They usually attain a size of only 1 cm (Fig. 4) and are composed of a wall which simulates a miniature oviduct (Figs. 7 and 8). They contain ciliated and columnar epithelium, and have a lamina propria and a thin layer of smooth muscle. Monroe and Spector (1962) describe the presence of plicae with fusion and the production of cysts within cysts. These authors consider their origin the most cephalic portion of the Müllerian duct, while Novak and Rubin (1952) believe them to be "harmless remnants of mesonephric tubules." These structures are often called the hydatids of Morgagni after the first author to consider their possible clinical importance. In males, analogous cysts, sessile and pedunculated, are commonly present at the upper pole of the testes and, according to Zuckerman and Krohn (1938), who review the literature extensively, Morgagni believed hydroceles to have formed from the rupture of these cysts. They describe serial sections of 108 such male structures from man and 9 species of monkeys and provide abundant evidence that they have a Müllerian duct origin. On rare occasions, a direct connection to the mesonephric duct was found. Nevertheless, from their findings, their review of all pertinent literature, and the histologic structure, the Müllerian origin can no longer be doubted. In males the cysts respond to estrone stimulation by growth of connective tissue and epithelium. Frankl (1930), who discusses oviductal anomalies exhaustively, finds that a variety of cysts may be attached to the surface of the oviduct, some arising from misplaced coelomic (Müllerian) epithelium, others from the mesonephric epoophoron. For the cysts at the fimbriated end of the oviduct he assumes as the origin lymphangiectasia, a view which we do not share.

Duplications of the oviducts, partial or complete, are relatively rare. One of the more intriguing cases is that by Marin-Padilla, Hoefnagel, and Benirschke (1964) in human trisomy D_1 (13–15). Their thesis is that in these cases it is not so much a failure of fusion

---

FIG. 1. Normal ovary (*below*) and Müllerian duct (*above*) of cow embryo at 12 cm CR length (H and E ×95).

FIG. 2. Freemartin gonad of cow (*below*) and degenerating Müllerian duct (*above*) at same age as Fig. 1 (H and E ×95).

FIG. 3. Normal human oviduct with cystic Walthard cell nest at arrow (see Fig. 5).

FIG. 4. Contralateral oviduct with typical hydatid of Morgagni attached to fimbriae at arrow.

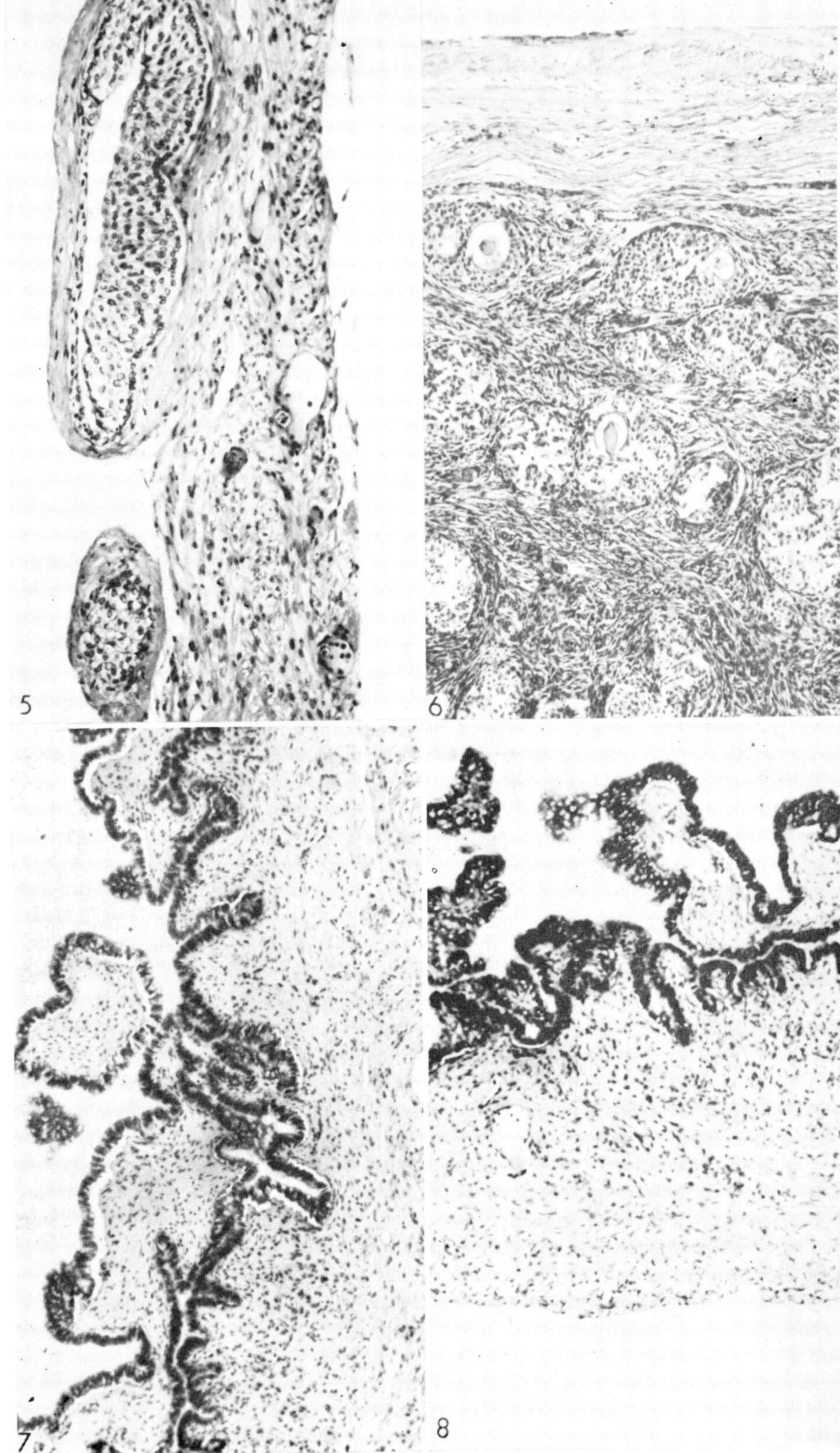
5
6
7
8

which gives rise to a bicornuate uterus but, perhaps more often, excess formation of tissues. They found not only duplication of oviducts, but also the remainder of Müllerian ducts and other tissues, e.g., pancreas, shared this excess formation of structures. Both Frankl (1930) and Zuckerman and Krohn (1938) refer to Richard, who described in 1851 up to 13 accessory openings of the oviduct. These may or may not open into the main oviductal lumen, as is discussed by Kistner (1965).

E. *Hematosalpinx; Torsion; Infarction*

Hematosalpinx may accompany hematometra and hematocolpos in the case of imperforate hymen or other lower Müllerian duct obstruction or, indeed, it may occur without hematometra for the same reason, as described previously. Hematosalpinx is rarely the sequel of trauma, as described in a 14-year-old child after gymnastics by Peter and Veselý (1966).

Much more commonly hemorrhage and infarction occur in the oviduct because of torsion, a relatively frequent condition. It may lead to complete infarction of the oviduct. Only a few studies will be cited from which further references may be obtained in addition to those given by Frankl (1930).

The subject is adequately reviewed by Youssef, Fayad, and Shafeek (1962) who present 6 cases of their own. Bilateral torsion has been described only 13 times. Although occasionally disputed, torsion may occur in previously entirely normal adnexa. Hydrosalpinx, ovarian cysts, and tumors are predisposing factors. After reduction of blood supply to the oviduct by this event, hemorrhage and fluid secretion may mimic a preexisting hydrosalpinx which has been overdiagnosed, according to these authors. An enlarged uterus (e.g, by pregnancy) may also predispose and, because of the association with the menarche, it is suggested that the same is true of pelvic congestion. Hypermotility, unusually long mesosalpinx, and adhesions have been named as further culprits. Severe, acute pain is the commonest symptom and gangrene follows quickly. Because the obstruction to circulation is not always complete or of long duration, these authors plead that restoration of circulation and conservative management should be tried more often. In about two-thirds of cases the ovary is also infarcted. Schultz, Newton, and Clatworthy (1963) found the condition in 5 children within a 10-year period and they needed to remove the adnexa because of infarction in all. The children were between 4½ and 12 years of age and up to 5 twists were seen at operation. They state that the adnexa are very mobile in children when examined at autopsy. Nevertheless, these authors find no real etiology, either at surgery or in examining the specimen, to account for the torsion. During the same time interval 6 adnexal torsions were treated which were associated with, if not due to, ovarian cysts or neoplasms. On the other hand, Braunstein, Ryan, and McCormick (1961) reviewed the literature and found 25% of cases in which normal adnexa are claimed and disagree with this interpretation. They discuss two of their own cases and emphasize the difficulty of differential diagnosis.

FIG. 5. Solid and cystic Walthard cell nests of normal human oviduct shown in Figure 3 (H and E ×95).

FIG. 6. Brenner tumor of ovary to compare structure with Walthard cell nests in Figure 5. Age 83, well-encapsulated tumor (H and E ×95).

FIGS. 7 and 8. Different areas of hydatid of Morgagni shown in Figure 4. Typical plical structure. ciliated epithelium, very thin wall with few smooth muscle cells (H and E ×95).

II. Inflammation and Infections

Salpingitis and its various sequelae are the most frequent diseases affecting the oviduct. Not only are symptoms often acute and in demand of treatment, but the obstructive consequences, with sterility so frequently ensuing, make oviductal inflammation an important condition. Moreover, salpingitis is almost the only disease of the oviduct described in the veterinary literature.

A. *Physiological Salpingitis*

The epithelium of the oviduct participates to a minor degree in the cyclical changes of the menstrual cycle as demonstrated, i.e., by electron-microscopic studies (Clyman 1966; Rewell 1960). Similar changes occur in a variety of domestic animals (Hafez 1968). The frequency with which actual inflammatory changes occur in the menstrual and puerperal phase is surprising. The first authors to draw attention to this physiological salpingitis were Nassberg, McKay, and Hertig (1954). In 62% of uteri and oviducts removed during the period of menstruation an acute inflammatory process was found in the lumina, often bilaterally, and usually limited to the plicae. Leukostasis, edema, leukocyte exudation, red cell debris, and a remarkable degree of dilatation of lymphatics were found sequentially. The process is presumably without sequelae and may give rise to symptoms of dysmenorrhea, which was associated in 23% of cases studied. These authors conjecture that the reflux of "menstrual toxin" is responsible for the histologic changes, particularly since injection of this material into oviducts of a rabbit led to a similar histologic picture. Clyman (1966) shows electron micrographs and states that on the first day of the normal menstrual cycle there is an "influx of many plasma cells and scattered lymphocytes with occasional polymorphonuclear leukocytes in the subepithelial stroma." Hellman (1949) studied segments of oviducts after postpartum sterilization and found an acute nonbacterial salpingitis with edema and lymphangiectasia in one-third, usually between the 6th and 8th postpartum days.

B. *Foreign Bodies*

Foreign bodies of an amazing variety have been found in human oviducts (Frankl 1930). Köhler (1958) reports inflammatory changes occurring in pigeons secondary to seeds. Nonspecific granulomata with foreign body giant cells are found on occasion without etiologic clues. Such a case is shown in Figure 32 in which neither tubercle bacilli nor foreign bodies were found, although they were extensively searched for. In another case, similar granulomata were traced to the content of a leaking cystic teratoma of the ovary. Hertig and Gore (1966) describe such lesions as due to talc, fungi, sarcoidosis, and lipiodol from hysterosalpingography ("Strawberry gallbladder"; xanthomatosis). The latter substance has been incriminated to cause a mucosal macrophage accumulation with abundant lipid, occasionally with foreign body giant cells, granulomata, and adhesions. Other contrast media have been employed consequently and Proust, deBrux, and Palmer (1964) make a searching inquiry into the fate of one of these, an organic iodine complex in a polyvinylpyrrolidone vehicle. The material can be detected occasionally as basophilic inclusions in macrophages. Of 80 women with hysterosalpingography, unresorbed material was found in 4, and in only 1 was there a local reaction to this material 4 years later. A comprehensive review of all types of complications of hysterosalpingography (Siegler 1967) includes the hazard of granuloma formation. These are found even after application of a water-soluble medium, and adhesions are thought to occur rarely as the consequence. It is suggested that normal oviductal epithelium is perhaps less frequently injured than that already previously altered. He concludes that granulomas are an infrequent sequel of salpingography. As a final curiosity in this context it may be mentioned that oviducts have been seen filled with masses of petroleum jelly subsequent to continual use of this coital lubricant. No demonstrable effect on oviductal patency, despite menorrhagia and abdominal pain, has been found in women using contraceptive coils (Siegler and Hellman 1964).

C. *Acute Salpingitis, Gonococcal*

Acute salpingitis in humans is most commonly (over 50%, Novak and Woodruff 1962) caused by the gonococcus and usually associated with other inflammatory processes as a part of "pelvic inflammatory disease" (PID), endometritis, urethritis, etc. Case report (Figs. 9 to 12): This 24-year-old woman had menorrhagia, increasing abdominal pain, a tender and cystic left abdominal mass, minimally elevated temperature, and WBC 15,000, raised sedimentation rate. Two years previously salpingitis, then a full-term pregnancy. At operation bilateral hydrosalpinx, partial pyosalpinx, retrouterine pelvic abscess, and massive adhesions. Gonococci were cultured. Histologically, this is an acute and chronic process and associated with typical chronic endometritis. The organisms were subsequently also isolated from the husband.

In the early phases of gonococcal salpingitis the oviduct is swollen, contains pus, and Neisseria may be cultured from the lumen. The plicae are edematous, and contain polyps, plasma cells, and lymphocytes. Fibrin exudation is common, epithelium degenerates and interplical adhesions occur; fimbriae adhere to one another and to the ovary. When peri-oophoritis and adhesions are established a tubo-ovarian abscess may be formed and pyosalpinx is associated. The serosal surface participates either by spillage in this pelvic peritonitis or by lymphatic spread. The exudate and fibrin are later replaced by granulation tissue and fibrous adhesions, at a time when plasma cells and lymphocytes predominate. Such events lead often to complete occlusion, the classical retort-shaped hydrosalpinx (Fig. 13) and follicular salpingitis (Fig. 36).

D. *Acute Salpingitis, Postpartum*

Next in numerical importance is puerperal or postabortal salpingitis. As antibiotics have become widely used, this complication has decreased in incidence, although many clinicians still regard it as a prevalent menace. Its importance in instrumental (criminal) abortion is considerable and it may even follow surgical procedures on the cervix, vagina, etc. (Novak and Woodruff 1962). Numerous organisms have been incriminated, especially pneumococci, streptococci, and *E. coli.* Nuckols and Hertig (1938) describe three pneumococcal cases (one nonpuerperal) and make a detailed survey of the literature of postpartum genital infections in general. Their account does not especially stress salpingitis as a sequel of postpartum metritis but this is well appreciated to occur. As the acute stage is overcome and a more chronic infection ensues the picture becomes very complex, particularly because organisms can neither be recognized histologically nor be cultured in many cases. The following illustrated case report may serve as an example of such a situation:

A 31-year-old white, married woman, one living child, suffered from prolonged menstrual periods over a year's time, for which a curettage was performed. Secondary sterility was diagnosed previously. An abundance of endometrium was curetted which was misinterpreted to show stromal sarcoma. A total hysterectomy was performed. Pelvic adhesions were found to enclose, in a saclike fashion, ovaries and oviducts. The latter were tortuous,

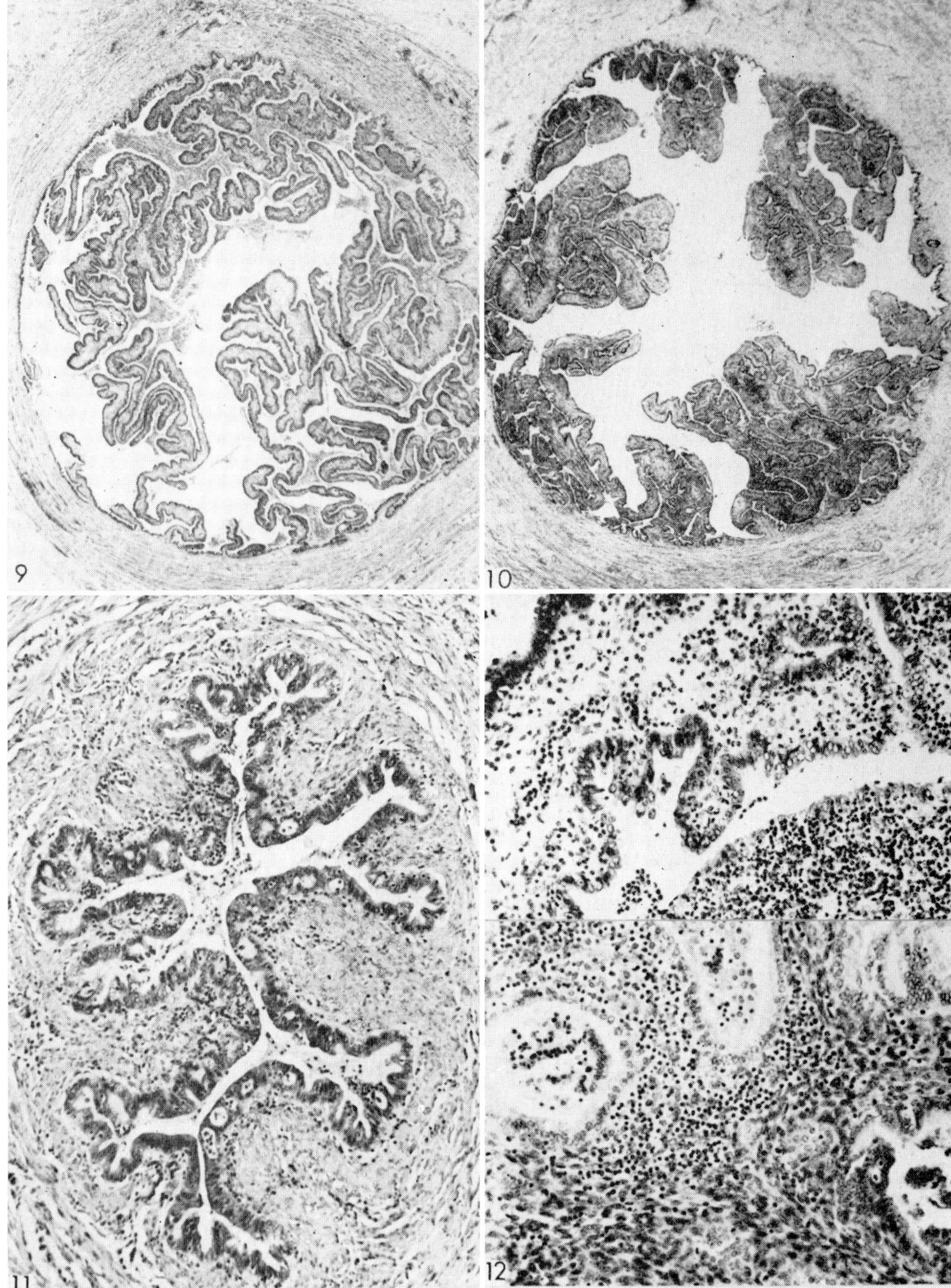
9
10
11
12

covered with papillary and stringy adhesions, and filled with enlarged, white plicae (Figs. 17 to 25). The photographs demonstrate severe chronic salpingitis, perisalpingitis, perioophoritis, salpingitis isthmica nodosa, and chronic endometritis. Plasma cells and lymphocytes are abundant and the peritoneal adhesions contained abundant psammoma bodies, a frequent degenerative event. Doubtless, sterility was due to an episode of unrecognized postpartal pelvic inflammatory disease. No history of gonorrhea, and stains failed to disclose organisms.

Lukasik (1963) reviews the literature of bacteriologic studies on inflamed oviducts and finds it confusing and in need of reevaluation, particularly in view of the therapeutic regimen to be employed. In 46 patients with adnexal inflammatory diseases no gonococci were found. Numerous cervical and oviductal cultures yielded primarily *E. coli* and *Streptococcus faecalis*, many resistant to 8 antibiotics; less often he found staphylococci and other streptococci. No positive cultures were obtained in control material and in patients with chronic adnexal inflammation but without a palpable tumor. Of these 46 cases with active disease 22 had identical pathogenic organisms obtained from uterus, cervix, and oviduct. The pathologic process was chronic in 38. Halbrecht (1965) states that latent nonspecific infections after normal delivery and abortion play an important role in causing sterility. In his study, 36% of 225 patients with secondary sterility had oviductal occlusion following abortion or delivery.

Robinson (1961) emphasizes the difficulties of therapy of these later stages of PID and describes 71 cases of pelvic abscess treated by aspiration and direct instillation of antibiotics. Streptococci were found most often (25 times) and in 29 cases no growth was reported. Symptomatology was cured by this treatment in 75%; 8 women had subsequent pregnancies. Khorshed and Ghosh (1966) endeavored to ascertain the bacteriologic content of oviducts in the puerperium. In 21 of 32 nonpuerperal oviducts, organisms were isolated (*Strept. faecalis* most often) and 7 showed inflammatory changes. This contrasts with 22 of 40 puerperal cases whose oviducts contained organisms. In 3 cases these were associated with subacute inflammation. *Staphylococcus aureus* and streptococci were found in the latter group. There was good correlation with uterine culture and histologic findings and, once more, the greater susceptibility of the postpartum genital tract to infection was noted.

E. *Sequelae of Salpingitis*

The acute and subacute phases of salpingitis may resolve without clinical symptoms having been recognized, to be followed by partial or complete obliteration of the oviductal lumen. Plicae fuse to one another and sequester bits of mucosa which lead to *follicular salpingitis* (Fig. 36). When the fimbriated end has become occluded or fused in a fibrous mass to the ovary, the oviduct may fill with a serous fluid and become greatly distended

---

FIGS. 9 and 10. Acute recurrent gonorrheal salpingitis. Different levels of ampullary portion of oviduct. The degree of lymphocyte and plasma cell infiltration of mucosa is much more extensive near fimbriae (*right*) and conglutination of plicae is apparent. Exudate is present in the mid-portion (*left*) but few inflammatory cells are present outside the mucosa (H and E ×14).

FIG. 11. Same case as previous, gonorrheal salpingitis, isthmic portion. The involvement of mucosa and lamina propria is apparent (H and E ×90).

FIG. 12. At top is an enlarged view of Figure 11 to show acute inflammatory type of exudate and destruction of epithelium. At bottom is a simultaneous curettage specimen with typical chronic endometritis, dominated by plasma cells (H and E ×144).

(hydrosalpinx, Fig. 16). In patients this is usually found only after the initial stage of pyosalpinx and extensive peritonitis (Fig. 15), however, experimentally mere ligation of the distal end of the oviduct has this result (Woskressensky 1891). The wall of the oviduct ultimately becomes paper-thin, and inflammatory changes are then difficult to recognize. Torsion occurs occasionally. The occlusion with moderate abdominal pain, a mass, or secondary infertility often leads to abdominal exploration and to attempts at oviductal reconstruction whose results are notoriously poor (Review by Resnick 1962). A refreshingly new look of this situation has been taken by McEwen (1966) who finds fault with many time-honored concepts and particularly with the despair in attempts at surgical repair. He also holds the view that *constrictive perisalpingitis* may lead to hydrosalpinx, a disease not frequently discussed in the American literature. The isthmic portions of the oviduct are often constricted in this manner and Resnick (1962) devises a method to deal with the problem surgically. The etiology is thought to be postabortal or postpartum pelvic peritonitis. Fibrosis and calcification ensuing are shown in Figure 21.

On rare occasions uterine or oviductal surgery is followed by a cutaneous fistula as reported and reviewed by Guixa and Cebollero (1963). Tubocutaneous fistulas occur also

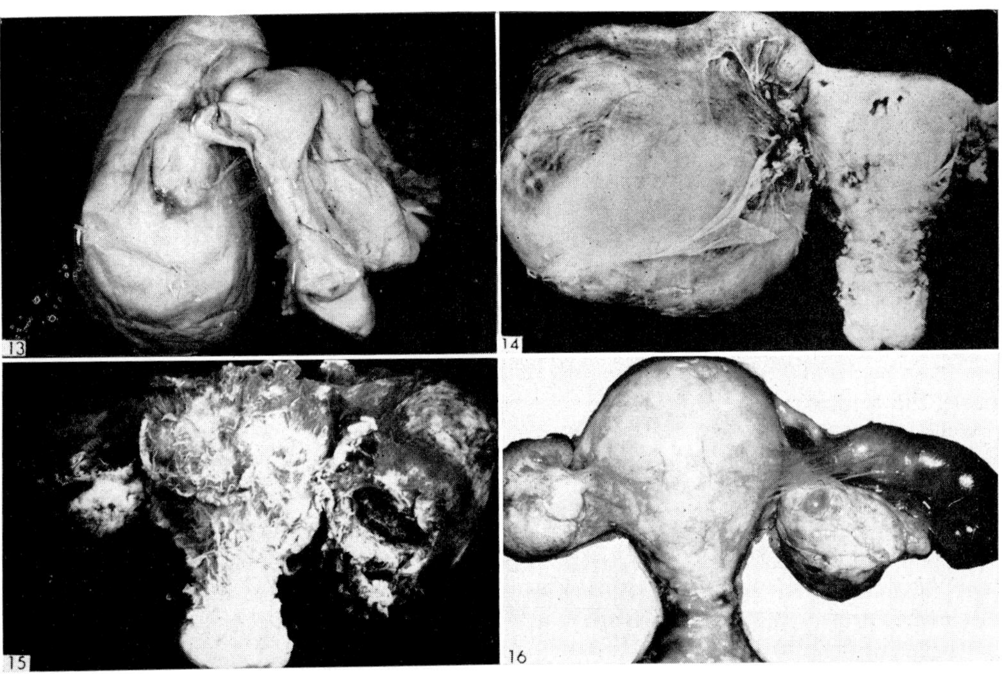

FIG. 13. Typical retort-shaped hydrosalpinx to left of opened human uterus. Note also the fibrous adhesions from surface of uterus to hydrosalpinx.

FIG. 14. Parovarian cyst in man, located in mesosalpinx and displacing and compressing oviduct. Uterus at right. Histologically, the epithelium of this solitary, serous fluid-filled cyst was flat but may be presumed to be of Wolffian duct origin.

FIG. 15. Acute pelvic inflammatory disease in man. Uterus and adnexa, posterior view. Note extensive fibrinous exudate covering all surfaces and pyosalpinx at right.

FIG. 16. Healed pelvic inflammatory disease in man. Extensive fibrous adhesions are present on both sides and early hydrosalpinx is present at right.

spontaneously from within due to infection by actinomyces (Frankl 1930). Finally, the ampulla of the oviduct occasionally becomes involved in cases of acute peritonitis. Serious involvement of the oviduct occurs at times with ruptured appendicitis (Peter and Veselý 1966), and Frankl (1930) points out that appendix-to-oviduct adhesions are common incidental findings at autopsy. Pyosalpinx is not uncommonly found when lower genital cancers obstruct the cervix or uterus (Henriksen 1956). So serious are salpingitis and its sequelae that Poluni and Saunders (1958) consider it to be the cause of gradual extinction of some North Bornean tribes. In this fascinating study the most likely causes were not venereal but involved contamination of the genital tract by the frequent vaginal examination during labor under primitive conditions.

F. *Salpingitis in Animals*

Among a variety of animals, salpingitis has been recognized as an important event. Puerperal sepsis in a chimpanzee, ascending salpingitis with peritonitis in a baboon, and peritonitis following abortion with retention of placenta in a *Cercopithecus* are all mentioned by Ruch (1959). Köhler (1958) has reviewed the relatively frequent occurrence of salpingitis in birds, associated often with foreign bodies and retention of eggs. Similarly, Nieberle and Cohrs (1962) find purulent salpingitis and peritonitis associated with trematodes. An astonishingly high frequency of oviductal abnormalities in "sterile" gilts and sows is reported by Day (1966, quoting Wilson 1949). Hydrosalpinx and pyosalpinx were found in 31% of sterile gilts (4% sows). Attempts to reproduce these diseases using the aspirate from affected animals were unsuccessful. Roberts (1956) makes scant reference to salpingitis in his discussion of postpartum metritis, however, it must occur in many animals, particularly since pyosalpinx and hydrosalpinx are well recognized (Smith and Jones 1966). The latter authors refer particularly to Gilman's older work in which cows with long-standing sterility were found to have salpingitis at autopsy with *Streptococcus viridans* and *Staphylococcus aureus*. Cases of acute salpingitis in cat and dog have been shown to this author by Jones from his surgical files. Roberts (1956) suggests that the reason why hydrosalpinx has not been seen in mares, despite the occurrence of salpingitis (22% of 83 mares with metritis), is the presence of a sphincter in the interstitial portions of the equine oviduct. The anatomic arrangement in the cow (blending of uterus into tube) may be an explanation why salpingitis is more common in cattle. In this species, hemorrhage, particularly after manipulation of the ovaries, is believed a common precursor of hydrosalpinx. Histologic descriptions have been furnished by Lombard, Morgan, and McNutt (1951), and Malik, Sengar, and Singh (1960) examined 1,000 Indian buffalos, slaughtered because they were no longer useful (age, sterility, etc.). In 14% salpingitis and pyosalpinx were identified which were difficult to diagnose by rectal examination and generally associated with pyometra, most often bilateral and more common in the summer. In 2.5% of cows hydrosalpinx was found, commonly bilateral, and in most the ampullae of both oviducts had fused with each other and peritoneal adhesions were present. Infection of the bovine genital tract with *Mycoplasma* organisms (PPLO) was first described by Hartmann *et al.* (1964). Among other lesions, acute salpingitis occurred. More recently, Hirth, Nielsen, and Plastridge (1966) describe the controlled infection of virgin heifers with *Mycoplasma*-infected semen. Various types of inflammatory lesions (acute, chronic, and follicular salpingitis, fibrous periovarian adhesions, oviductal occlusion, pyosalpinx) were observed and good correlation with sterility, prolonged estrus, and vaginal discharge was found. Organisms were cultured up to 8 months and these authors consider this organism as possibly important in the causation of lesions described by Lombard,

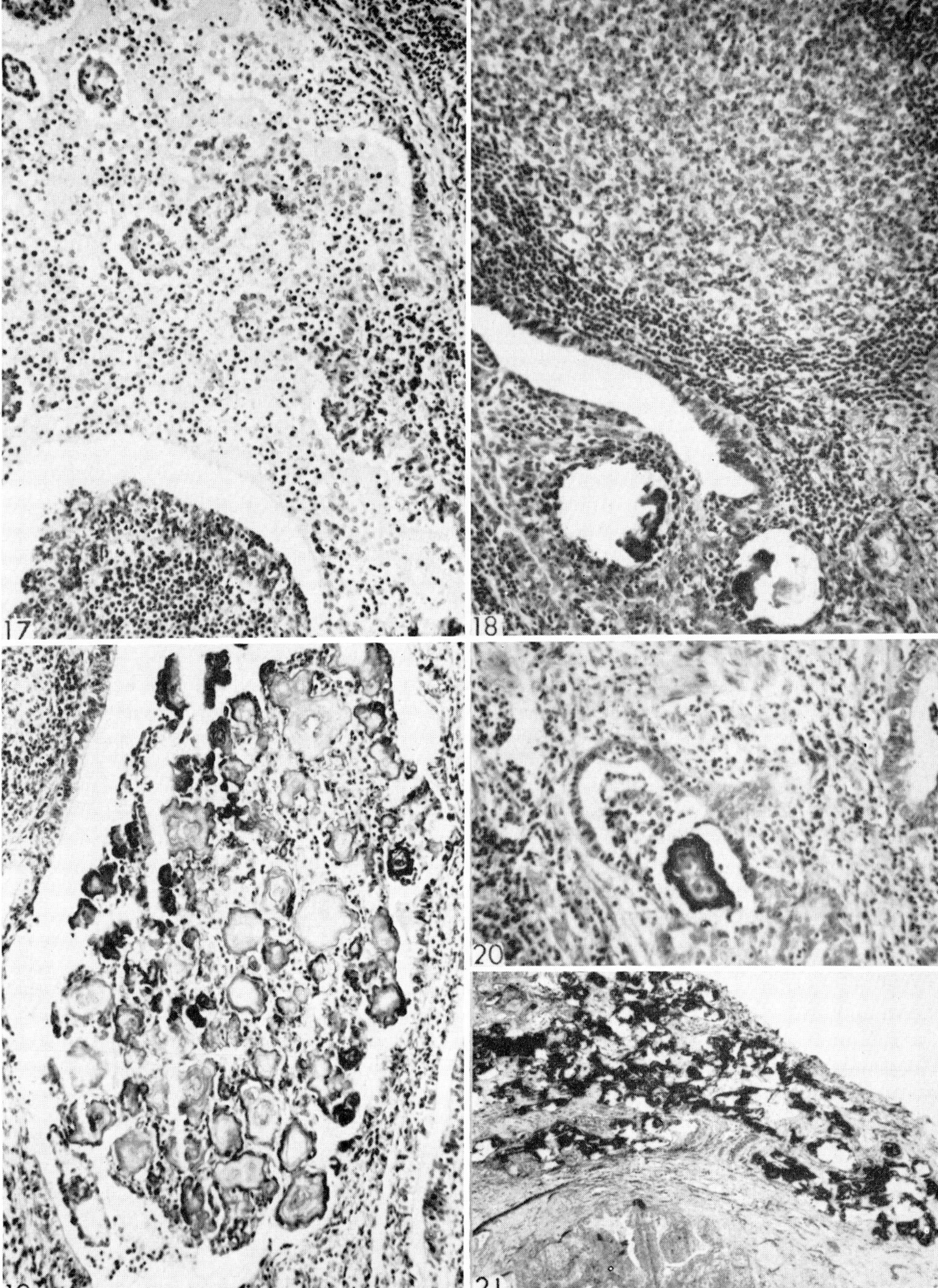
17
18
19
20
21

Morgan, and McNutt (1951). They also refer to reports of isolation of this organism by others in cattle and from the genital organs in man. For instance, Hoare and Haig (1964) report isolation from one or both oviducts in 20 of 33 cows slaughtered for infertility (22) and sundry reasons.

Acute salpingitis was induced experimentally by Woskressensky (1891) in rabbits and guinea pigs by staphylococci and croton oil. This author also found that hydrosalpinx was readily induced by ligation of the fimbriated end of the oviduct, whereas tying the uterine end was ineffective. Vetesi and Kemenes (1967) observed acute salpingitis in rabbits spontaneously and after experimental infection with *Listeria monocytogenes*.

Although a variety of bacterial infections interferes with pregnancy or causes infertility, and endometritis has been observed spontaneously or experimentally (e.g., Vibriosis toxins—Estes, Bryner, and O'Berry 1966), in few accounts is specific mention made of the state of the oviducts (Hoerlein 1967). *Brucella suis* is specifically singled out by this author as causing localized granulomatous lesions in endometrium and oviducts. The lesions of infectious bovine cervicovaginitis, a cause of epizootic infertility in cattle and probably a virus infection transmitted by copulation, are associated with chronic salpingitis and periovarian adhesions in approximately 25% of infected and untreated cows (Gibbons 1966).

Dawson (1964) investigated the suggested relationship between mucosal cysts in oviducts and inflammation in cows. Of 518 cattle slaughtered (460 because of failure to breed) 12 (2%) were found to have cysts in the oviducts, 3 bilaterally. They often involved the entire length of the duct and interfered with conception. This condition was associated with cystic endometrium and cystic ovaries, in other words with endocrine disorders, rather than with chronic inflammation, which was seen in 40% of this series.

Because of the difficulty of diagnosing oviductal patency in cattle, the intraperitoneal injection of starch, etc. (and cervical appearance in the case of patent oviducts) has been introduced, but Johari and Sharma (1964), who review this topic briefly, find the method useless.

G. *Salpingitis Isthmica Nodosa*

Salpingitis isthmica nodosa is a relatively common and classical condition in women, described often since its naming by Chiari in 1887. The reason for the frequent discussion of this peculiar lesion is its disputed etiology. Although Chiari has already suggested its true pathogenesis from chronic salpingitis, some authors are still dissatisfied with this explanation and wish to invoke endometriosis and adenomyosis as possible etiologies (Schenken and

---

FIG. 17. Chronic pelvic inflammatory disease with salpingitis isthmica nodosa, mistakenly diagnosed as endometrial sarcoma, described in text. Ampullary region with exudate in lumen, round cell infiltrate of plicae and destruction of epithelium (H and E ×157).

FIG. 18. Same case, ampullary portion with massive lymphoid hyperplasia ("germinal center" at top) and concrements in plical recesses (H and E ×157).

FIG. 19. Same case, massive concrements in ampulla; at top right it can be seen that some concrements form within plicae (H and E ×157).

FIG. 20. Same case, isthmic portion with gland spaces "herniated" in between muscle fibers ("diverticula") and surrounded by inflammatory cells. Mineralized exudate (H and E ×157).

FIG. 21. Same case, isthmic portion, lumen below. Note the extensive peritoneal adhesions with dense, calcific concretions, shown black (H and E ×16).

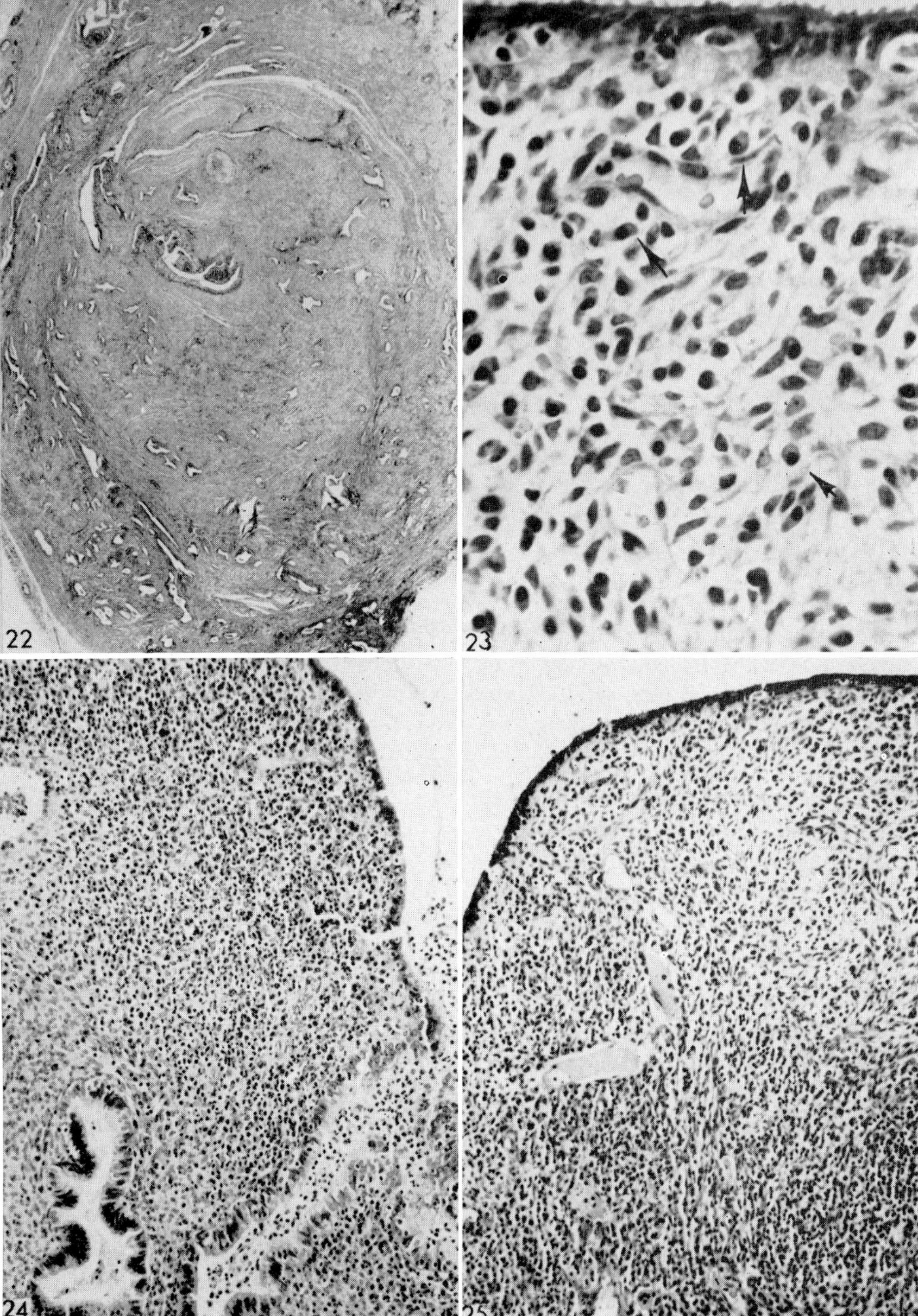

22
23
24
25

Burns 1943). The condition is bilateral, usually characterized by a fusiform, nodular swelling in the isthmic narrowing of the oviducts, but also, much more rarely, affecting the ampullary portion as well. The typical histologic findings are chronic salpingitis and cystic diverticula of mucosa within the muscular coat. Their connection with the lumen has been demonstrated by these authors on resected specimens (thorotrast, serial sectioning, reconstruction, X-ray) and can be seen in life during salpingography (Siegler 1967). A variety of associated findings such as abscess, foam cells, fibrosis, and cholesteatoma is discussed by Schenken and Burns (1943) (see also Fig. 22). In the large series of 208 patients presented by these authors a ratio of 8.9 to 1 colored to white patients is found which is reflected in other series. This is due to principally two factors, the greater frequency of gonorrhea, and myomata leading to hysterectomy in Negroes. The disease is identified most frequently in the 3d and 4th decades, a clear distinction from endometriosis is made, and particularly the nature of the diverticula is emphasized. As a pathogenetic mechanism they suggest what seems to us an important anatomic factor, namely, the absence of a muscularis mucosa in the oviduct. In striking parallel to this condition (Fig. 25), one finds in chronic cholecystitis similar lesions as the Aschoff-Rokitansky sinuses (Fig. 27) and, in ureters and bladder, the cystitis cystica (Fig. 26), both representing cystic spaces within the muscular coat and associated with chronic infection. It seems quite unlikely that either endometriosis (Fig. 29) or adenomyosis can account for these lesions, nor is it reasonable that reepithelialized abscess cavities give rise to this condition (see also Frankl 1930). For a different point of view see Bundey and Williams (1963). Since in their case and in others no chronic inflammation was present, and because a tumor-like swelling was found, they consider the lesion to be neoplastic. There was no episode of PID, as is also true in the patient here presented, although our patient still had the histologic changes of severe chronic inflammation (Fig. 22). It occurs in 85% bilaterally and may be found in 1–5% of the general population. In Siegler and Hellman's (1963) review of the results of 50 cases of tuboplasty, specimens were available in 30 cases. They find this disease in 9 instances; 12 additional cases had chronic salpingitis and 5 demonstrated merely fibrosis. These authors believe this to be an important disease preceding sterility.

H. *Tuberculosis*

The oviducts are the most common part of the female genital tract to become infected with tubercle bacilli. Next in frequency is the endometrium, and then the peritoneum. Sampling errors are formidable and the pathologic changes are quite variable (deBrux and Dupré-Froment 1966). Suspicion of tuberculosis must be kept in mind at all times when examining tissues from inflamed adnexa, for the classical picture of granulomatous endosalpingitis with Langhans giant cells is not always displayed. These authors describe and illustrate the pathologic changes so well that anything written would be repetitious. More-

---

FIG. 22. Same case as previous plate. Salpingitis isthmica nodosa, classical microscopic appearance. The oviductal lumen is severely reduced and throughout the thickened wall cystic, diverticulous spaces proliferate (H and E ×14).

FIG. 23. Same case, diagnostic curettage specimen. The presence of large numbers of plasma cells (arrows) should have led to a suspicion of an inflammatory condition (H and E ×576).

FIGS. 24 and 25. Same as previous, endometrial biopsy with extensive inflammatory infiltration and stromal proliferative response, mistaken as sarcoma (H and E ×144).

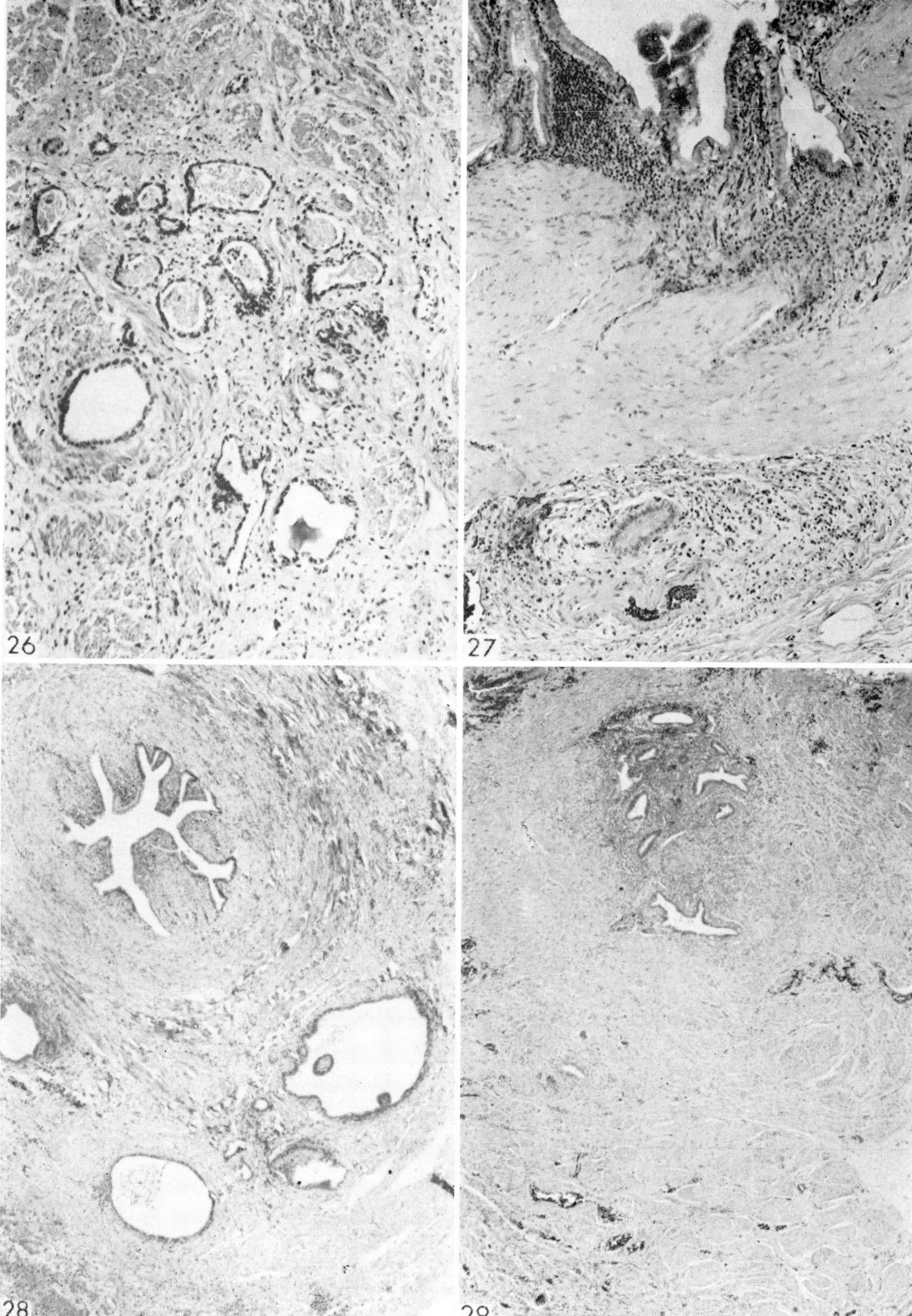
26
27
28
29

over, all aspects of this important genital infection are dealt with in this monograph, to which the reader is referred for details.

It is not well appreciated that pyosalpinx, caseous nodules, extensive fibrosis with very few granulomata, and salpingitis isthmica nodosa are all seen at times in pelvic tuberculosis. Its sequelae almost always lead to sterility, and one most interesting aspect in recent years has been to follow the literature on the effects of antibiotic treatment of this disease. In the past the disease was often fatal and surgical treatment often needed to be resorted to. In Sutherland's (1965) extensive experience this is now necessary in only about 20% of cases, usually after prolonged antibiotic therapy, and mortality is now quite low. The frequency of the disease differs greatly and is impossible to assess. Hertig and Gore (1966) state that it occurs in 5% of infertility cases (in the United States). Poland (1965) states that pelvic tuberculosis is uncommon in Ontario, however she presents an admirable study of 140 cases of pelvic tuberculosis, with special reference to infertility, which accrued during a period of approximately 10 years.

The disease has always been considered to be associated primarily with younger ages. Thus, Peter and Veselý (1966) still consider that in one-third of cases it is acquired around the time of menarche, perhaps being related to increased vascularity of genital organs or occurring for hormonal reasons. Rippman (1964) emphatically points out, however, that now more often older women are affected and the disease may even be mistaken for carcinoma. Despite a voluminous literature on female genital tuberculosis, the question of pathogenesis is uncertain. Most writers consider that hematogenous spread or lymphatic dissemination is likely while some find that tuberculous peritonitis is an almost constant intermediary step before endosalpingitis occurs. Certainly, ascending infection is extremely rare if it occurs, despite the fact that cervical tuberculosis is now more frequently recognized.

As mentioned, the disease almost always leads to sterility, or at least decreased fertility. DeBrux and Dupré-Froment (1966) find that in 300 ectopic pregnancies 5% had a tuberculous etiology, while of 282 patients with genital tuberculosis 16 ectopic pregnancies were found (5%). The association of tuberculous salpingitis with ectopic pregnancies has been studied in numerous case reports and many papers quoted here contain such material. Palmer and deBrux (1964) even have a patient with 3 oviductal ectopics and eventually a uterine pregnancy, probably a record. Other authors (Kistner, Hertig, and Rock 1951; Nokes *et al.* 1957) find this to be a gynecological rarity, each describing a case. The latter authors recognize only 68 reported cases, but we find many more discussed among recent papers describing pelvic tuberculosis. The case described by Nokes *et al.* (1957) is of additional interest because placental and fetal tuberculosis were diagnosed at autopsy as well. Tuberculous salpingitis and peritonitis are often found in cows. Here it resembles the human disease closely, and it is well described by Nieberle and Cohrs (1962).

FIG. 26. Cystitis cystica in a woman with chronic cystitis. Diverticula of bladder epithelium penetrate deeply into muscularis (H and E ×90).

FIG. 27. Chronic cholecystitis in man. Diverticula of gall bladder mucosa have penetrated muscularis (*below*) as so-called Aschoff-Rokitansky sinuses (H and E ×90).

FIG. 28. Salpingitis isthmica nodosa in man, nearly healed stage of chronic salpingitis, with deep recesses of mucosa into muscularis (H and E ×36). These three conditions (Figs. 26–28) are considered similar in pathogenesis and this similarity perhaps relates to the anatomic absence of a muscularis mucosae.

FIG. 29. Endometriosis of surface of oviduct in man, in contrast to previous figure, contains endometrial stroma and does not communicate with oviductal lumen (H and E ×36).

I. *Nontuberculous Granulomatous Salpingitis*

A few causes of granulomatous salpingitis other than those mentioned earlier are related to suture material and ova. Hofman, Shanberge, and Rubovits (1965) describe three cases of abscess formation 4 to 13 yr after oviduct ligation and due to silk sutures. They were acute processes, sterile upon culture, and had granulomatous foreign body reaction to silk. The second oviduct became involved later in one. Perhaps hypersensitivity is important, but the etiology of this rare process is not clear. Two types of schistosomiasis have been reported to affect the oviduct. Sedlis (1961) finds 5 previous cases and reports a woman who had 7 pregnancies after acquiring the infection. It was an incidental finding and the granulomata were within the plicae. In the case reported by Carpenter, Mozley, and Lewis (1964) the patient suffered from a tubo-ovarian mass, thought to be tuberculous. Ova were in the serosa and mucosa of the oviduct. Since Symmers' (1950) study it has also been established that the pinworm, *Enterobius vermicularis*, travels on occasion into the peritoneal cavity from the external genitalia via the uterus. As a result, ova and worms, occasionally with granulomata, have been observed in the endometrium, oviducts, and peritoneum. They have no clinical importance (Arthur and Tomlinson 1958) but occasionally they are associated with, and perhaps causes of, peritoneal abscesses (Campbell and Bowman 1961).

In routine sections of the genitalia of nine-banded armadillos (*Dasypus novemcinctus*) used experimentally, we have seen similar foreign body granulomata engulfing unidentified worms and ova (Fig. 31) in sporadic association with protein precipitate in the oviductal lumen (Fig. 30).

III. Ectopic Pregnancy of Oviducts

A. *Comparative Frequencies*

Extrauterine pregnancies are relatively common in man and most often they are found in the oviduct, much more rarely on the peritoneal surface, the ovary, and in the cervix. Usually, implantation occurs in the midportion of the oviduct but interstitial, as well as ampullary, pregnancies have been recorded frequently. Bilateral oviductal pregnancies, oviductal and intrauterine combined pregnancies, oviductal pregnancies with contralateral corpus luteum, those reaching term, twin to quintuplet ectopics, hydatidiform moles, and choriocarcinoma all have been described in a voluminous literature (Benirschke and Driscoll 1967). This is an important clinical problem in obstetrics and it is therefore very surprising

---

FIG. 30. Isthmic portion of oviduct of nonpregnant nine-banded armadillo (*Dasypus novemcinctus*) with protein exudate and minimal plasma cell infiltrate (H and E ×90).

FIG. 31. Same as previous, granuloma in mesosalpinx around presumed worm eggs. At lower left is higher magnification of plasma cells of oviductal mucosa. Since inflammation was observed only with such worm egg granulomas we presume a causal relationship (H and E ×144; ×576).

FIG. 32. Granulomatous, obstructive salpingitis with typical Langhans giant cells in man. Special stains, guinea pig inoculations, and clinical studies ruled out tuberculosis. Patient had one previous ectopic pregnancy for "tuberculous salpingitis," no organisms identified. Has healed subacute bacterial endocarditis. It is unknown whether she had lipiodol studies. Cause of granulomatous salpingitis unknown (H and E ×144).

FIG. 33. Granulomatous salpingitis in proved case of tuberculosis in woman. On first glance this appears less typical of tuberculosis than previous case, emphasizing the variability of this disease picture, discussed in text (H and E ×144).

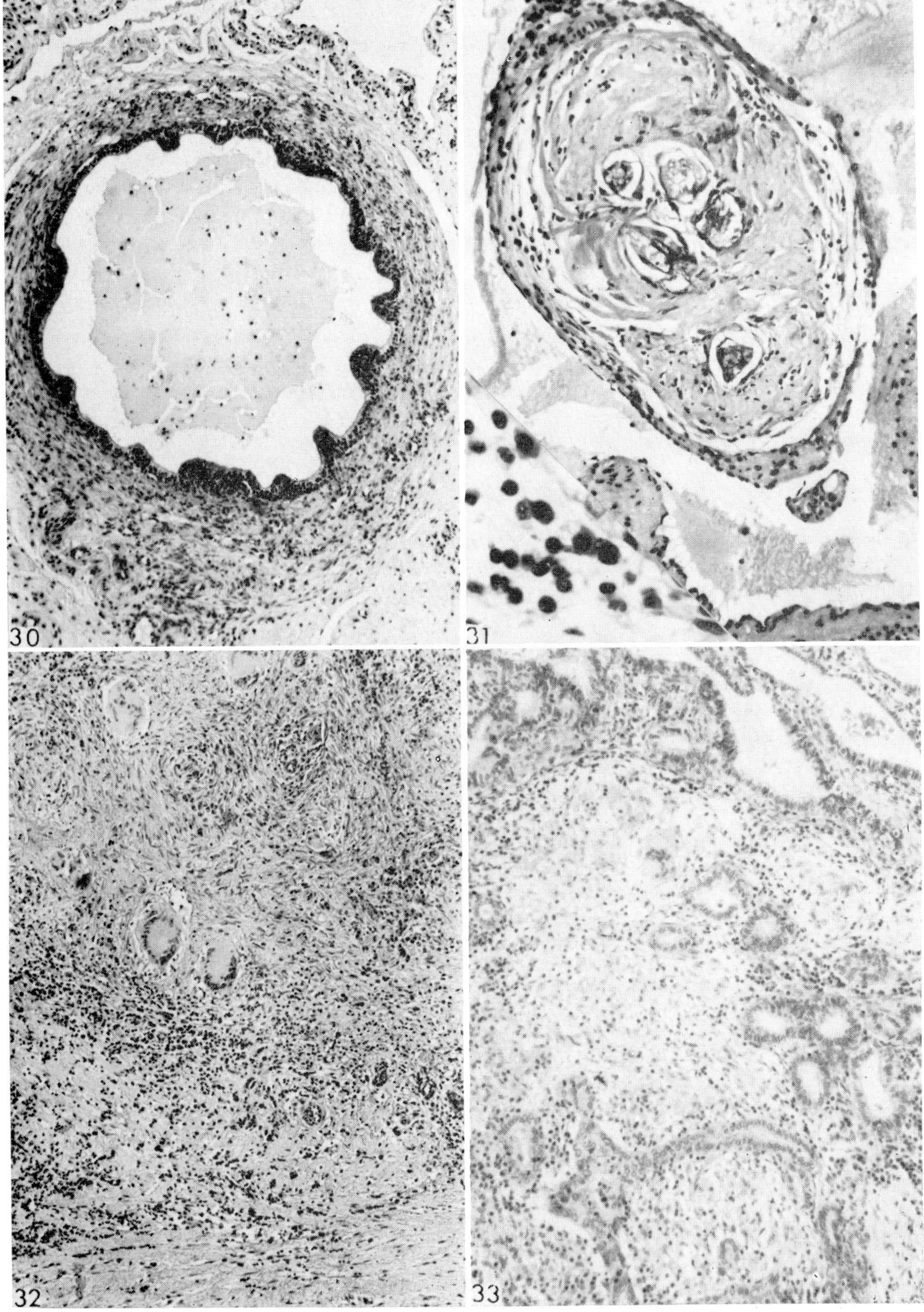

that both Roberts (1956) and Smith and Jones (1966) state that the condition does not exist in domestic animals. Similarly, in hundreds of pregnant nine-banded armadillos autopsied by us and in similar numbers of golden hamsters (*Mesocricetus auratus*) from Dr. V. H. Ferm's (personal communication) experience, no ectopic pregnancy has been observed. This striking contrast between man and other species is perplexing, particularly since in many polytocous animals like the pig spontaneous abortion or death of blastocysts is common, and since experimental overcrowding of uteri with multiple blastocysts is feasible in cattle (Hafez, Jainudeen, and Lindsey 1965). The most immediate thought is that placentation in some animals is possible only by specific interplay of placentomes between placenta (cotyledons) and uterine mucosa (caruncles), as in most ruminants, and that a delay of implantation favors eventual transport into the uterus in other species. This is not so in the mouse, however, and Kirby (1965), among others, can implant and grow for some time experimentally transplanted blastocysts in almost any part of the mouse, even testes. The discrepancy perhaps warrants some detailed study, e.g., the search for the existence of salpingitis in the mouse or its experimental production, and other experiments could be suggested which might help solve this riddle. Preissecker (1958) reviews the rare case of extrauterine (abdominal) pregnancies in lagomorphs and one case in the cat (Hunter 1931). It appears that these were secondary implantations after uterine rupture. Nicholas (1934) attempted to induce ectopic pregnancies in rats by severing and ligating oviducts containing segmenting eggs. Most blastocysts perished, but a few reached term as peritoneal implantations. Oviductal pregnancies are not mentioned. In hares and rabbits, and rarely in the ewe, abdominal pregnancies have been reported (Eales 1932) and this author describes a combined peritoneal, oviductal, and uterine pregnancy in a laboratory rabbit. The animal was healthy, albeit obese, and the oviduct was probe-patent.

In several monkeys oviductal pregnancies have been observed. Lapin and Yakovleva (1960) saw 3 ectopics in 1,892 pregnancies, once each in a rhesus monkey, a baboon, and a "green marmoset." They think the reason for the rarity of the condition may be the lower frequency of genital diseases in monkeys.

B. *Pathogenesis and Clinical Correlation*

Most oviductal pregnancies of women terminate in the first 2–3 months by hemorrhagic ablation of the placenta or rupture of the oviduct, or perhaps both. These events lead to

---

FIG. 34. A most instructive case of ruptured ectopic pregnancy: 27-year-old healthy gravida 1. Normal at supper (6 P.M.). Vomiting at 7 P.M. Physician administers i.v. fluids and Demerol at 8:30 P.M. Abdominal pain and vomiting do not cease, morphine injection at midnight, Death at 1 A.M., 6 hr after onset of complaints. At autopsy: 3,000 ml of blood in peritoneal cavity, 2 cm fresh embryo in cul-de-sac, 1 cm tear in bluish distended oviduct with cord protruding. Uterus and adnexa shown from behind; normal adnexa left. Right ovary (pale) with corpus luteum; bluish distension of right oviduct is obvious with small tear at apex. Grossly no evidence of salpingitis.

FIG. 35. Plasma cell (*arrow*) in another case of oviductal ectopic pregnancy, connoting chronic inflammation of oviduct (H and E ×608).

FIG. 36. Typical follicular salpingitis in another case of human ectopic pregnancy, probably the commonest associated condition. The cystic spaces are formed by fusion of plicae and continued secretion of oviduct mucosa (H and E ×6).

FIG. 37. Typical histologic appearance of oviductal pregnancy in woman. Muscular wall is at right and nearly penetrated by placenta near the top. The pale villi (*left*) are normal but are attached as placenta accreta, lacking proper decidua. Retroplacental hemorrhage, ablation of placenta has started at top and accounts for symptomatology (H and E ×15).

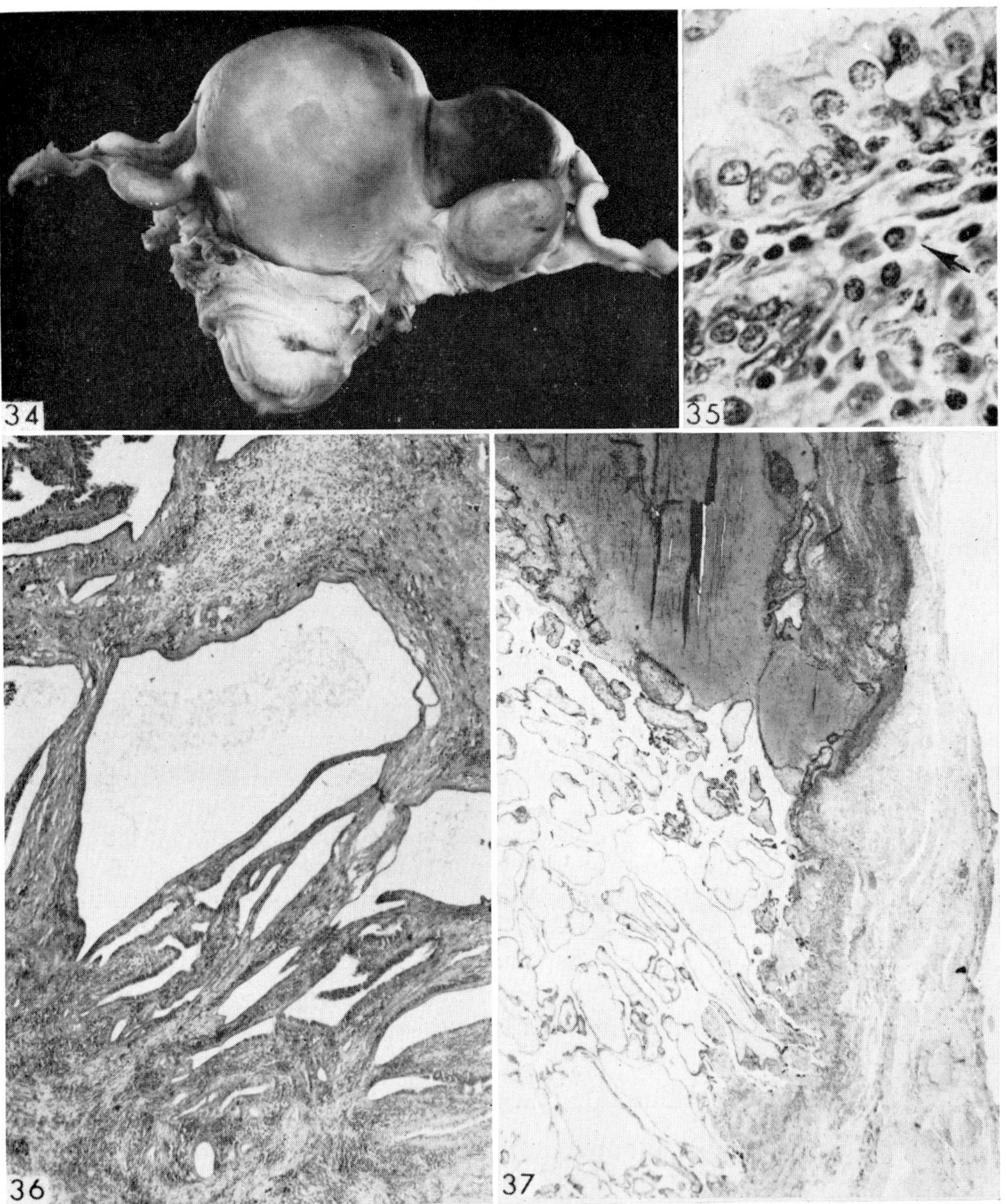
34
35
36
37

the predominant presenting symptom of abdominal pain, requiring skillful care by the obstetrician. The specimen received by the pathologist (Fig. 34) is usually a hemorrhagic, distended oviduct whose surface may be disrupted, containing a clot within which is frequently the degenerated ovisac. Also, the placenta often forms a placenta accreta or percreta (Fig. 38) and this, rather than the distension with blood, may be the cause of rupture (Fig. 37). Laufer, Sadovsky, and Sadovsky (1962), who reviewed the changes in the placental tissue, find them similar to those seen in spontaneous uterine abortions; avascular villi and hydatid swelling were frequent (12%), occurring perhaps twice as often as in the uterine specimens. These changes are probably secondary to embryonic death rather than connoting an intrinsic deficiency of placental tissue (Benirschke and Driscoll 1967). In view of the fact that a large number of uterine abortuses are aneuploid (Carr 1967) it is to be expected that the same may be true of oviductal specimens, but this has not been studied. If the proportions of aneuploids were less in oviductal abortions, greater weight could be laid on concomitant oviductal pathology as an etiological factor for the event. Care must be exercised not to interpret hydatid swelling of villi as hydatidiform moles, particularly in the presence of a large fetus (Westerhout 1964). Similarly, until additional cytogenetic findings are correlated with pathologic studies of both the placenta and oviducts in cases of ectopic pregnancy, the decision cannot be made whether primary oval or secondary degenerative events, caused by "nutritional failure" of placentation, are the etiologic factors involved (Krone and Jopp 1966). These authors find over 50% of oviductal conceptuses to be blighted and draw tenable hypothetical conclusions (ovum responsible; oviductal nutrition responsible).

Most authors agree that patients with one ectopic pregnancy often have a second. For instance, Blanchet, Sparling, and MacFarlane (1967), in reviewing 360 cases, find that 10% had a previous operation for this condition. Graber and O'Rourke (1965) report that only one-third of patients with an oviductal pregnancy may eventually have a child and they advise more conservative therapy (milking out rather than resecting the oviduct, etc.). Douglas (1963) reports that this disease is the commonest surgical emergency in women in Jamaica. He sees 1 ectopic per 28 liveborns, while in the United States the rate is purported to be 1:64. Oviductal ectopics account for 96%; 7% fimbriae, 42% ampulla, 28% isthmus, 13% interstitial, 2% stump, 8% unknown. The median age of 28 years was higher than that of controls (24) and the event was more frequent on the right (55%) than the left (41%). In 42% there was gross or microscopic evidence of previous infection; the recurrence rate was 10%. Kleiner and Roberts (1967) emphasize the need for a detailed study of the resected specimen and find 53% associated with follicular salpingitis (Fig. 36), the highest concurrence rate recorded yet. Other changes often held responsible, e.g., congenital anomalies

---

FIG. 38. Ectopic pregnancy of oviduct behaving as a placenta percreta which may well have been the cause of death in Figure 34. Into the mesosalpingeal fat permeate trophoblastic cells issuing from villi as seen in right lower quadrant. Oviduct wall has been destroyed completely (H and E ×128).

FIG. 39. Human oviductal plicae in term uterine pregnancy. The stromal cells have undergone an identical decidual reaction as the endometrium, and the epithelium, likewise, has become atrophic (H and E ×128).

FIG. 40. Papillary adenocarcinoma of human oviduct with superficial invasion of wall (*top left*). Death within 5 months (H and E ×80).

FIGS. 41 and 42. Adenocarcinoma of human oviduct with papillary and solid areas shown on top and anaplastic component shown below, from gross specimen in Figure 43 (H and E ×128).

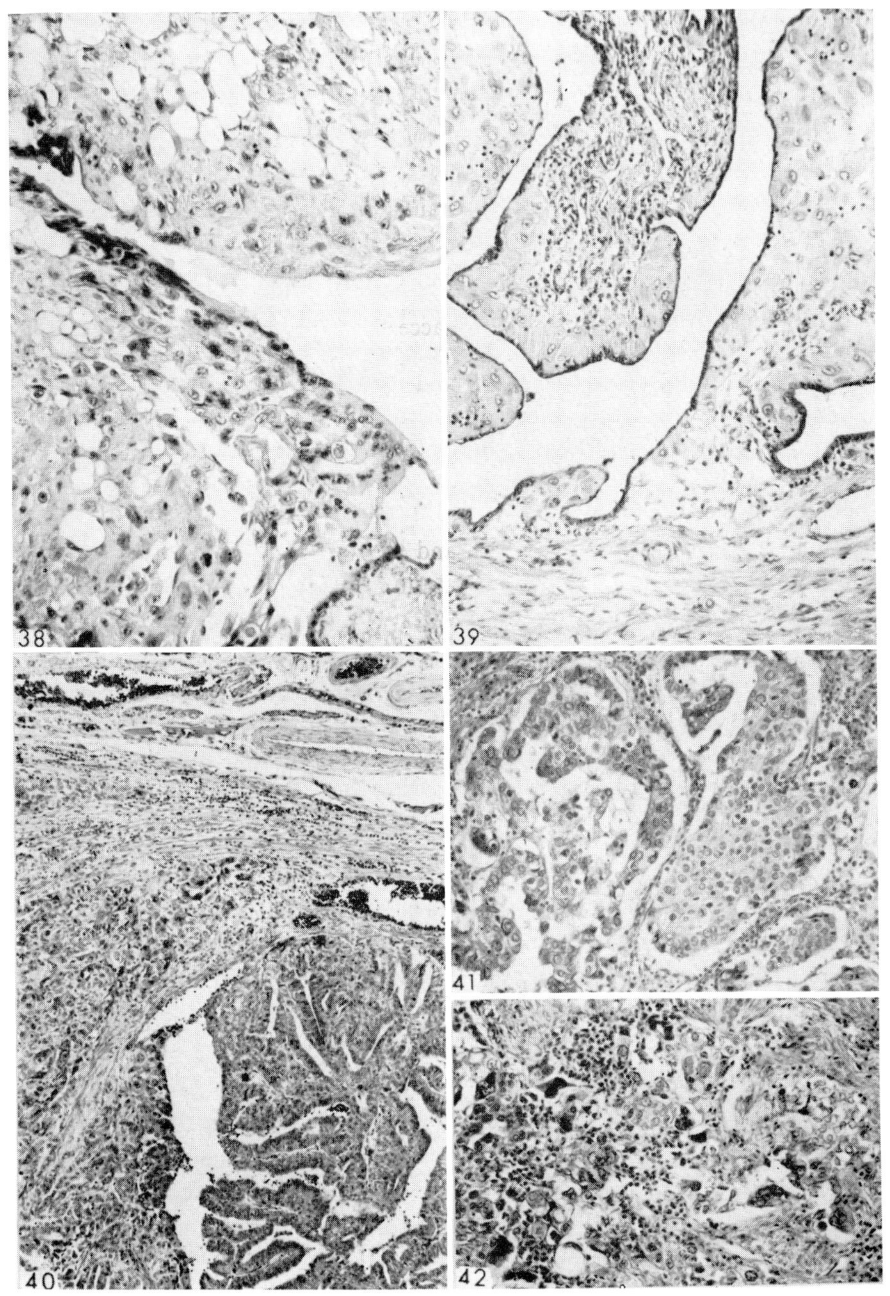
38
39
40
41
42

were conspicuously absent and only in a few cases could they elicit a history of PID. In a case with combined intra- and extrauterine pregnancy follicular salpingitis was found in *both* oviducts, and this, among other considerations, makes these authors cautious about accepting this finding as *the* important etiologic factor. Multiparous women over 35 years with significantly higher uterine abortion rate (47% vs. controls 18%) had the disease most commonly. The last point raises some important questions which can only be answered by detailed additional studies. Acute salpingitis is rarely associated with oviductal pregnancy as described by Scully (1967), and the question is unresolved which event came first. Occasionally an ectopic pregnancy is found in an accessory oviduct (Frankl 1930).

C. *Multiple Ectopic Pregnancy*

Multiple oviductal pregnancies have been of particular interest to this writer (Benirschke and Driscoll 1967). Well over 100 such cases have been reported (Hakim-Elahi 1965). Fox and Mevs (1963) believe that 1 in 113 ectopics occurs as a bilateral twin pregnancy. The interest in oviductal twin pregnancies and particularly the reported higher incidence of monozygous twins (compared with intrauterine twins) lies in the possibility that either twinning may be induced by the abnormal event of ectopic implantation or the monozygous twin blastocyst may have a greater chance to become implanted in the oviduct. The need for careful study of placental membrane relationships of these twins and their correlation with timing of the twinning event have been stressed.

D. *Ovarian Pregnancy*

While the exact relationship of oviductal disease, e.g., follicular salpingitis, to oviductal pregnancy is still debated, less controversy exists for *ovarian* pregnancies. Boronow *et al.* (1965) find only 8% to have had possible pelvic inflammatory disease as a preceding event. They invoke pathogenetic mechanisms for ovarian pregnancy other than salpingitis, as did other authors also. Some find normal oviducts with peritoneal pregnancies (Gustafson, Bowman, and Stout 1953), also suggesting mechanisms other than oviduct pathology as etiologic factors. Felbo and Fenger (1966) review the literature of combined intra- and extrauterine pregnancy and find 522 cases and add a new case. Both infants survived, a very rare event indeed, reported only 11 times before. Intramural or interstitial pregnancy is also uncommon; McGowan (1965) finds only 8 cases. Its etiology is even less certain, save for its more frequent occurrence after salpingectomy or other plastic oviduct surgery (Kalchman and Meltzer 1966, review; Komorski and Gerber 1965). The last report is particularly fascinating since it is followed by a paper on ovarian pregnancy after bilateral salpingectomy and unilateral oophorectomy! The occurrence of oviductal choriocarcinoma following ectopic pregnancy is described by Riggs *et al.* (1964).

E. *Diagnosis*

Inasmuch as the proper diagnosis of ectopic pregnancy has important implications for clinical management, many investigators have sought for endometrial changes, particularly decidua, which might be diagnostic of pregnancy. This has been found in a disappointingly low incidence (Moritz and Douglass 1928; Romney, Hertig, and Reid 1950; many others). A probable reason for this lack of correlation is the frequently degenerative state of the ectopic placenta (with reduced hormone output) at a time when symptoms appear (hemorrhage, rupture). The Arias-Stella phenomenon (1954) has become an additional tool for the pathologist. Atypical glands, characteristic infolding, large nuclei, etc. make a characteristic picture for the knowledgeable diagnostician. They are well shown and discussed also by Skulj *et al.* (1963) and are more frequently found (in 63% by Bernhardt, Bruns, and Drose 1966) in curettings than decidua. Presumably they are hormonally induced, as they are also seen with some types of hormone therapy, and it is of interest that Berge (1964) finds such glands as early as in a case of a 14-day pregnancy.

IV. Endometriosis; Decidua; Aging

Endometriosis may be defined as the presence of endometrium outside of the endometrial cavity. This endometrium must have, by definition, both epithelial and stromal components, and this ectopic endometrium may or may not partake in the menstrual cycle. Most commonly endometriosis occurs on the surface of ovary or pelvic peritoneum; however, it has been found on gut, ureters, umbilicus, lung, etc. The two theories of its pathogenesis, implantation of menstrual endometrial fragments, and coelomic epithelial metaplasia, have been reviewed by Madding and Kennedy (1963), who present a case which argues strongly in favor of the latter concept. A woman with rudimentary uteri, having never menstruated, developed an endometrioma 5 years after pelvic surgery. Metaplasia is much more common in adhesions of the pelvic peritoneum, and this is how we also envisage the pathogenesis of this frequent disease.

Peritoneal endometriosis is the most frequent form affecting the oviducts (Fig. 29), from without, by metaplastic events of the surface. This is not to deny that the disease may occur within the oviductal wall, but we have not seen it and believe it to be very rare. We can therefore not conceive that the lesions of salpingitis isthmica nodosa have this pathogenesis and cannot evaluate reviews such as Movers' (1963) which aim to ascertain the detrimental effect of oviductal endometriosis on fertility, since this author clearly describes, for the most part, salpingitis isthmica nodosa. If endometriosis of the oviduct has a primary effect on fertility, it is because of distortion of ovaries and oviducts or because of the frequent tubo-ovarian adhesions. The disease has been found in a variety of laboratory primates (Ruch 1959) and here, too, some fimbriae may be swollen but the lumen of the oviducts is not obstructed. Metaplastic changes sometimes follow estrogen stimulation, and Meissner, Sommers, and Sherman (1957) report endometrial as well as oviductal mucosal proliferation in rabbits after stilbestrol treatment.

The decidual reaction taking place within the plical stroma in some pregnancies (Hellman found it in 6–12% postpartum, 1949) should not be considered to represent endometriosis (Fig. 39). It is merely a reflection of the common reactivity of all subcoelomic mesenchyme derivatives and may be found in the endocervix as well.

The changes occurring with advancing age were studied by Pinero and Foraker (1963) with a variety of special stains and by planimetry. The latter technique disclosed a striking increase of total area occupied by plicae with older age, and a decrease of lumen size; also, plical connective tissue increased and collagen accumulated.

V. Tumors

A. *Papillary Carcinoma*

Neoplasms originating in the oviduct are relatively uncommon in women and even less common in animals (Green and Scully 1962; Hertig and Gore 1961). The commonest tumor is the adenocarcinoma which has usually, but not always, a papillary pattern (Figs. 40, 41, 42). Its usually being recognized only late in its development spells a poor prognosis, and it is for this reason that in recent years most emphasis has been laid on early diagnosis. Well over 800 cases have been described (Devambez, Benoit, and Cavrois 1964), and only a 5% survival rate is quoted by Schenck (1964). He finds a 5-year survival rate of 10% and quotes that 1.6% of all female genital cancers are adenocarcinomas of the oviduct. He reports diagnosis of bilateral cancer by uterine suction cytology in postmenopausal bleeding, with cure of the disease by early surgery. Many similar accounts are now being published and how important early diagnosis is may be appreciated from the study of Figure 43. Its frequent macroscopic resemblance to hydropyosalpinx is emphasized by Duckman, Cabaud, and Rosati (1963), who found 5 rapidly fatal cancers of the oviduct and suggest cytologic examination every 6 months after age 35. See also the good color photo of hydrosalpinx in the report of Anderson *et al.* (1954). Others, like Hu, Taymor, and Hertig (1950), reporting 12 cases (0.3% of female genital cancers), do not consider confusion with hydrosalpinx an

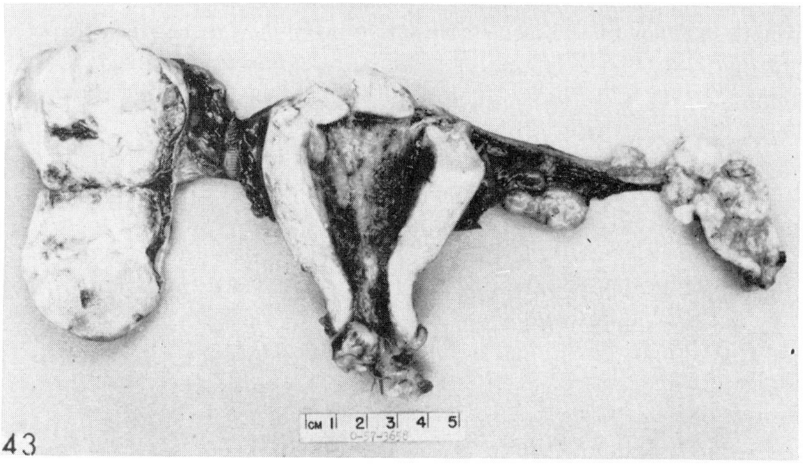

FIG. 43. Papillary and solid adenocarcinoma of left oviduct in 67-year-old woman operated upon 2 days after onset of postmenopausal bleeding, cobalt therapy later. Nine years later no evidence of recurrence. Same case as Figures 41 and 42.

important possibility. They find a fairly good correlation between histologic grading and prognosis. These authors also record coexistence of follicular salpingitis in the affected side of 9 cases, and Pauerstein and Woodruff (1966) study a variety of proliferative and anaplastic changes in various conditions of the oviduct. They believe they have identified two cases of carcinoma *in situ*. Goodlin (1962) is the first author to study the chromosome constitution of an oviductal cancer among 12 other female genital cancers. All metaphases were hyperdiploid-aneuploid and, while some unusual chromosomes are shown in the karyotype, no specific marker is recognized. Rarely, the cancer coexists with other neoplasms, although, of course, metastatic deposits are much more common. Thus Blaustein (1963) reports an oviductal cancer in a 17-year-old nullipara coexistent with a cystic teratoma, and another case with a mixed mesodermal tumor of the uterus at age 67. He thinks that cancer of the oviduct is only diagnosed by the pathologist, but cytology and aspiration cytology (Devambez, Benoit, and Cavrois 1964) as well as culdoscopy will change the picture in the future.

B. *Other Tumors; Tumors in Animals*

Other tumors of the oviducts are less common still. As the oviduct may become the seat of metastatic cancers (Hertig and Gore 1961, 1966) so it may be involved in lymphomatous spread and leukemia (Frankl 1930), and Smith and Jones (1966) remark that this is almost the only tumor seen in animals. Curiously, germ cell tumors, dysgerminoma, and cystic or solid teratomas (30 cases) have been observed more often than other tumors (Scully 1963). Indeed, at times they have solid thyroid or carcinoid components and in some "dermoids" contralateral ovarian dermoids were present (Grimes and Kornmesser 1960; Legerlotz 1964).

Fox (1912) and Ratcliffe (1933) reported cancers in birds, and Köhler (1958) refers to the report of an adenocarcinoma in a canary and fibrosarcoma in parrots, the only tumors reported in the zoological literature. Leiomyomas, sarcomas, adenomyomas are all either uncommon or disputed (Hertig and Gore 1961). Whether hilar cell tumors always originate in the ovary is undecided. Lewis (1964) presents a pertinent and fascinating case. A 58-year-old nullipara had extensive hilar cell hyperplasia (with typical Reinke crystalloids), and these cells were scattered throughout the fimbriae. These tumors must not be confused with misplaced adrenal glands, although they are similar to adrenal cortical tissue both structurally and functionally. While ectopic adrenal tissue is often found in the human mesosalpinx and nearly regularly in that of the nine-banded armadillo, it rarely gives rise to neoplasms. Also uncommonly found in the mesosalpinx are tumors of mesenchymal nature, mesenchymomas of a variety of histologic types and presumably derived from the plastic subcoelomic mesenchyme (Ober and Black 1955).

C. *Cysts; Adenomatoid Tumors*

A variety of cysts and tumors arises in the mesosalpinx which have their origin in structures related to the Wolffian duct remnants (Fig. 14). Such cysts are of clinical differential diagnostic importance and often give rise to torsion (MacDonald and Pratt 1967); rarely are these rests the presumed progenitors of mesonephromas or clear cell carcinomas (see Novak and Woodruff 1962). A peculiar tumor of much debated origin is the characteristic "adenomatoid tumor" which is depicted by Hertig and Gore (1961). This benign tumor has been mistaken for a malignant neoplasm and also occurs frequently along the tunica vaginalis in men (Evans 1943). These small nodules, composed of clefts lined by epithelium-like cells, often have an angiomatous appearance and are presumed to originate from mesothelium. Why they should be associated exclusively with tubes or the vas deferens is unclear, and it seems that the last word has not been said about their histogenesis. They are not uniform in appearance, indeed, cancers and pericytomas occasionally assume their histologic pattern (see Ross 1967).

VI. Conclusions

A survey of the literature and cases of diseases afflicting the oviducts discloses a great disparity between man and all domestic and wild animals. In particular, ectopic pregnancy and inflammatory conditions with various sequelae are very common in women, while they play a minor role in veterinary pathology. Indeed, ectopic pregnancies virtually never occur in the oviducts of animals, and salpingitis is rare or restricted to a few domestic species. Similarly, anomalies and tumors are well recognized in women; yet they rarely affect animals.

Several factors must be considered to explain these interesting discrepancies. One aspect relates undoubtedly to the greater effort spent on diagnosis in women, for instance in infertility problems. Furthermore, there are venereal diseases spreading at greater speed in man than would be feasible in either domestic or wild animals, although the risk of infected semen dissemination by artificial means may pose a potential threat. Of animals dying or slaughtered, only a fraction is thoroughly studied and few organs of the latter category are examined histologically. A more systematic inquiry may indeed disclose a greater frequency of salpingitis than is currently appreciated and the recent studies on *Mycoplasma* infection in cows point in this direction.

While inflammatory lesions may occur more often in animals than hitherto thought, neoplasms are presumably less common since most animals are not allowed to attain their full life span. A real disparity exists with respect to ectopic pregnancy. In women, this is one of the most important obstetric conditions, numerically and qualitatively. It is an emergency which, if not treated adequately, may be fatal (Fig. 34). Not so in animals; for all intents and purposes oviductal pregnancy does not occur. Surely this is not because one condition held to be a main precursor, chronic salpingitis with oviductal occlusion, does not exist. Smith and Jones (1966) suggest that the nonhuman blastocyst may be less aggressive, but this cannot hold exclusively, as rodent blastocysts become established readily when transplanted. More important may be the need for complex interactions between trophoblast and specialized areas of endometrium, absent from the oviduct, which prevent ectopic implantation in many animals. This is not true of all species, however, and a thorough study of these phenomena seems to be necessary. Perhaps through such investigations new facets of the poorly understood reasons for the frequent human oviductal ectopic pregnancies could be learned which would make this illness more amenable to prevention.

Acknowledgment

This work was supported by Grant GM 10210 from the National Institutes of Health.

References

1. Anderson, H. E.; Bantin, C. F.; Giffen, H. K.; Olson, L. J.; and Schack, C. B. 1954. Primary carcinoma of the Fallopian tube. *Obstet. Gynec.* 3:89.
2. Arey, L. B. 1943. The nature and significance of the grooved nuclei of Brenner tumors and Walthard cell islands. *Amer. J. Obstet. Gynec.* 45:614.
3. Arias-Stella, J. 1954. Atypical endometrial changes associated with the presence of chorionic tissue. *Arch. Path.* 58:112.
4. Arthur, H. R., and Tomlinson, B. E. 1958. Oxyuris granulomata of the Fallopian tube and peritoneal surface of an ovarian cyst. *J. Obstet. Gynec. Brit. Emp.* 65:996.
5. Benirschke, K. 1967. Cytogenetic abnormalities in reproduction. In *Progress in infertility*, ed. S. J. Behrman and R. W. Kistner. Boston: Little, Brown and Co.
6. Benirschke, K., and Driscoll, S. G. 1967. *The pathology of the human placenta.* New York: Springer-Verlag.
7. Berge, T. 1964. Arias-Stella's phenomenon. *Acta Path. Microbiol. Scand.* 61:152.
8. Bernhardt, R. N.; Bruns, P. D.; and Drose, V. E. 1966. Atypical endometrium associated with ectopic pregnancy: The Arias Stella phenomenon. *Obstet. Gynec.* 28:849.
9. Bertrand, M. 1965. Le freemartinisme. *Rev. Méd. Vét.* 66:575.

10. Biggers, J. D., and McFeely, R. A. 1966. Intersexuality in domestic mammals. In *Advances in reproductive physiology*, ed. A. McLaren. London: Logos Press.
11. Blanchet, J.; Sparling, D. W.; and MacFarlane, K. T. 1967. Ectopic pregnancy: A statistical review of 360 cases. *Canad. Med. Ass. J.* 96:71.
12. Blaustein, A. 1963. Tubal adenocarcinoma coexistent with other genital neoplasms. *Obstet. Gynec.* 21:62.
13. Bollack, C.; Kuhlmann, N.; and Papaevangelos, I. 1964. À propos d'un cas de dédoublement du vagin avec hématocolpos antérieur, utérus pseudodelphe symétrique et hématosalpinx gauche. *Strasbourg Méd.* 15:471.
14. Boronow, R. C.; McElin, T. W.; West, R. H.; and Buckingham, J. C. 1965. Ovarian pregnancy. *Amer. J. Obstet. Gynec.* 91:1095.
15. Braunstein, P. W.; Ryan, B. J.; and McCormick, R. C. 1961. Isolated torsion of the Fallopian tube. *N.Y. J. Med.* 61:1268.
16. Bundey, J. G., and Williams, J. D. 1963. Salpingitis isthmica nodosa with tumour formation resembling torsion of an ovarian cyst. *J. Obstet. Gynaec. Brit. Emp.* 70:519.
17. Campbell, C. G., and Bowman, J. 1961. *Enterobius vermicularis* granuloma of pelvis: A case report. *Amer. J. Obstet. Gynec.* 81:256.
18. Carpenter, C. G.; Mozley, P. D.; and Lewis, N. G. 1964. Schistosomiasis japonica involvement of the female genital tract. *J.A.M.A.* 188:647.
19. Carr, D. H. 1967. Cytogenetics of abortions. In *Comparative aspects of reproductive failure*, ed. K. Benirschke. New York: Springer-Verlag.
20. Chiari, H. 1887. Zur pathologischen Anatomie des Eileiterkatarrhs. *Prag. Z. Heilk.*, quoted by Frankl, 1930.
21. Clyman, M. J. 1966. Electron microscopy of the human Fallopian tube. *Fertil. Steril.* 17:281.
22. Dawson, F. L. M. 1964. Fertility relationships with the cystic condition in cow oviducts. *Fifth Int. Congr. Anim. Reprod. A. I. Trento.* Session IV, 20, p. 182.
23. Day, B. N. 1966. The reproduction of swine (Wilson *et al. J. Animal Sci.* 8:558, 1949). In *Reproduction in farm animals*, ed. E. S. E. Hafez. Philadelphia: Lea & Febiger.
24. DeBrux, J., and Dupré-Froment, J. 1966. An anatomo-pathological study of clinically "latent" female genital tuberculosis: Pathogenic and therapeutic conclusions. In *Latent female genital tract tuberculosis*, ed. E. T. Rippman and R. Wenner. Basel: S. Karger.
25. Deringer, M. K., and Heston, W. E. 1956. Abnormalities of urogenital system in strain A × C line 9935 rats. *Proc. Soc. Exp. Biol. Med.* 91:312.
26. Devambez, J.; Benoit, M.; and Cavrois, G. 1964. Cancer de la trompe: Guérison de cinq ans. *C.R. Soc. Franç. Gynéc.* 34:119.
27. Douglas, C. P. 1963. Tubal ectopic pregnancy. *Brit. Med. J.* 2:838.
28. Duckman, S.; Cabaud, P. G.; and Rosati, V. 1963. Primary carcinoma of the Fallopian tube. *Amer. J. Obstet. Gynec.* 86:401.
29. Eales, N. B. 1932. Abdominal pregnancy in animals, with an account of a case of multiple ectopic gestation in a rabbit. *J. Anat.* 67:109.
30. Estes, P. C.; Bryner, J. H.; and O'Berry, P. A. 1966. Histopathology of bovine vibriosis and the effects of *Vibrio fetus* extracts on the female genital tracts. *Cornell Vet.* 56:610.
31. Evans, N. 1943. Mesothelioma of the epididymis and tunica vaginalis. *J. Urol.* 50:249.

32. Felbo, M., and Fenger, H. J. 1966. Combined extra- and intrauterine pregnancy carried to term. *Acta Obstet. Gynec. Scand.* 45:140.
33. Felding, C. 1965. Obstetric studies in women with congenital solitary kidneys. *Acta Obstet. Gynec. Scand.* 44:555.
34. Fox, E. J., and Mevs, F. E. 1963. Simultaneous bilateral tubal pregnancies: Report of 2 cases. *Obstet. Gynec.* 21:499.
35. Fox, H. 1912. Observations upon neoplasms in wild animals in the Philadelphia Zoological Gardens. *J. Path.* 17:217.
36. Frankl, O. 1930. Tube. In *Handb. Spez. Path. Anat. Histol.*, ed. Henke-Lubarsch, 7:1. Berlin: Springer.
37. Gibbons, W. J. 1966. Viral, rickettsial and protozoan infections. In *Reproduction in farm animals*, ed. E. S. E. Hafez. Philadelphia: Lea & Febiger.
38. Goodlin, R. C. 1962. Karyotype analysis of gynecologic malignant tumors. *Amer. J. Obstet. Gynec.* 84:493.
39. Graber, E. A., and O'Rourke, J. J. 1965. Recent advances in gynecology. *New York J. Med.* 65:1110.
40. Green, T. H., and Scully, R. E. 1962. Tumors of the Fallopian tube. *Clin. Obstet. Gynec.* 5:886.
41. Greiss, F. C., and Mauzy, C. H. 1961. Genital anomalies in women: An evaluation of diagnosis, incidence, and obstetric performance. *Amer. J. Obstet. Gynec.* 82:330.
42. Grimes, H. G., and Kornmesser, J. G. 1960. Benign cystic teratoma of the oviduct: Report of a case and review of the literature. *Obstet. Gynec.* 16:85.
43. Grumbach, M. M.; Wyk, J. J. v.; and Wilkins, L. 1955. Chromosomal sex in gonadal dysgenesis (ovarian dysgenesis): Relationship to male pseudohermaphrodism and theories of human sex differentiation. *J. Clin. Endocr.* 15:1161.
44. Guixa, H. L., and Cebollero, L. M. 1963. Fistulas tubocutaneas: Consideraciones a proposito de una observacion. *Semana Med. (B. Air.)* 123:973.
45. Gustafson, G. W.; Bowman, H. E.; and Stout, F. E. 1953. Extrauterine pregnancy at term. *Obstet. Gynec.* 2:17.
46. Hafez, E. S. E. 1968. Female reproductive organs of farm animals. In *Reproduction in farm animals*, ed. E. S. E. Hafez, 2d ed. Philadelphia: Lea & Febiger.
47. Hafez, E. S. E., and Jainudeen, M. R. 1966. Intersexuality in farm animals. *Animal Breed. Abstr.* 34:1.
48. Hafez, E. S. E.; Jainudeen, M. R.; and Lindsey, D. R. 1965. Gonadotropin-induced twinning and related phenomena in beef cattle. *Acta Endocr.* 50: Suppl. 102.
49. Hakim-Elahi, E. 1965. Unruptured bilateral tubal pregnancy: Report of a case. *Obstet. Gynec.* 26:763.
50. Halbrecht, I. 1965. Endometrial and tubal sequelae of latent non-specific infections. *Int. J. Fertil.* 10:121.
51. Hartmann, H. A.; Tourtelotte, M. E.; Nielsen, S. W.; and Plastridge, W. N. 1964. Experimental bovine uterine mycoplasmosis. *Res. Vet. Sci.* 5:303.
52. Heinze, W. 1964. Uterus simplex mit Agenesia ovarii et tubae bei einem Kalb. *Anat. Anz.* 115:369.
53. Hellman, L. M. 1949. The morphology of the human Fallopian tube in the early puerperium. *Amer. J. Obstet. Gynec.* 57:154.
54. Henriksen, E. 1956. Pyometra associated with malignant lesions of the cervix and the uterus. *Amer. J. Obstet. Gynec.* 72:884.

55. Hertig, A. T., and Gore, H. 1961. *Tumors of the female sex organs. Part 3. Tumors of the ovary and Fallopian tube.* AFIP Atlas of Tumor Pathology IX/33. Washington, D.C.: Armed Forces Institute of Pathology.
56. ———. 1966. Female genitalia. In *Pathology*, vol. 2, 5th ed., ed. W. A. D. Anderson. St. Louis: C. V. Mosby Co.
57. Hirth, R. S.; Nielsen, S. W.; and Plastridge, W. N. 1966. Bovine salpingo-oophoritis produced with semen containing a *Mycoplasma. Path. Vet.* 3:616.
58. Hoare, M., and Haig, D. A. 1964. Isolation of *Mycoplasma* sp. from the oviducts of dairy cows. *Vet. Rec.* 76:956.
59. Hoerlein, A. B. 1967. Bacterial infertility in domestic animals. In *Comparative aspects of reproductive failure*, ed. K. Benirschke. New York: Springer-Verlag.
60. Hofman, W. I.; Shanberge, J. N.; and Rubovits, W. H. 1965. Salpingitis due to foreign-body reaction to silk sutures following tubal ligation. *Obstet. Gynec.* 25:112.
61. Hu, C. Y.; Taymor, M. L.; and Hertig, A. T. 1950. Primary carcinoma of the Fallopian tube. *Amer. J. Obstet. Gynec.* 59:58.
62. Hunter, R. H. 1931. Abdominal pregnancy in a cat. *J. Anat.* 66:261.
63. Hutt, F. B. 1967. Malformations and defects of genetic origin in domestic animals. In *Comparative aspects of reproductive failure*, ed. K. Benirschke. New York: Springer-Verlag.
64. Jainudeen, M. R., and Hafez, E. S. E. 1965. Attempts to induce bovine freemartinism experimentally. *J. Reprod. Fertil.* 10:281.
65. Johari, M. P., and Sharma, S. P. 1964. Fallopian tube lesions in farm animals: Diagnosis. Letter to the Editor. *Vet. Rec.* 76:293.
66. Jones, W. S. 1957. Obstetric significance of female genital anomalies. *Obstet. Gynec.* 10:113.
67. Jost, A. 1948. Le contrôle hormonal de la différenciation du sexe. *Biol. Rev.* 23:201.
68. Jost, A.; Chodkiewicz, M.; and Mauleon, P. 1963. Intersexualité du foetus de veau produite par des androgènes: Comparison entre l'hormone foetale responsable du free-martinisme et l'hormone testiculaire adulte. *C.R. Acad. Sci.* (*Par.*) 256:274.
69. Kalchman, G. G., and Meltzer, R. M. 1966. Interstitial pregnancy following homolateral salpingectomy. *Amer. J. Obstet. Gynec.* 96:1139.
70. Kamm, M. L., and Beernink, H. E. 1962. Uterine anomalies in habitual abortion and premature labor. *Obstet. Gynec.* 20:713.
71. Kent, S. W. 1963. Retroperitoneal tube and ovary associated with inguinal hernia: Report of a case. *Obstet. Gynec.* 21:234.
72. Khorshed, P., and Ghosh, B. N. 1966. Comparative study of the bacteriology and histology of normal and puerperal Fallopian tubes. *J. Obstet. Gynaec.* (*India*) 16:66.
73. Kirby, D. R. S. 1965. Endocrinological effects of experimentally induced extra-uterine pregnancies in virgin mice. *J. Reprod. Fertil.* 10:403.
74. Kistner, R. W. 1965. *Gynecology: Principles and practice.* Chicago: Year Book Medical Publ. Inc.
75. Kistner, R. W.; Hertig, A. T.; and Rock, J. 1951. Tubal pregnancy complicating tuberculous salpingitis. *Amer. J. Obstet. Gynec.* 62:1157.
76. Kleiner, G. J., and Roberts, T. W. 1967. Current factors in the causation of tubal pregnancy. A prospective clinico-pathologic study. *Amer. J. Obstet. Gynec.* 99:21.
77. Köhler, H. 1958. Krankheiten der Vögel. In *Pathologie der Laboratoriumstiere*, ed. H. Cohrs, R. Jaffé, and H. Meessen, vol. 2. Berlin: Springer-Verlag.

78. Komorski, J., and Gerber, J. 1965. Eine extrauterine Schwangerschaft an der Stelle der keilförmigen Tubenexzision bei Adnexexstirpation. *Zbl. Gynaek.* 87:358.
79. Krone, H. A., and Jopp, H. 1966. Über Vorkommen, Häufigkeit und kausale Pathogenese von Abortiveiern bei Tubargravidität. *Landarzt* 40:1553.
80. Lapin, B. A., and Yakovleva, L. A. 1960. *Comparative pathology in monkeys.* Springfield, Ill.: Charles C Thomas.
81. Lauchlan, S. C. 1966. Histogenesis and histogenetic relationships of Brenner tumors. *Cancer* 19:1628 (for ed. comments see: *Obstet. Gynec. Survey* 22:375, 1967.)
82. Laufer, A.; Sadovsky, A.; and Sadovsky, E. 1962. Histologic appearance of the placenta in ectopic pregnancy. *Obstet. Gynec.* 20:350.
83. Legerlotz, C. 1964. Teratomas of the Fallopian tube. *Zbl. Gynaek.* 86:137.
84. Lenz, W. 1964. Krankheiten des Urogenitalsystems. In *Humangenetik*, ed. P. E. Becker, 3/1. Stuttgart: G. Thieme.
85. Lewis, J. D. 1964. Hilus cell hyperplasia of ovaries and tubes: Report of a case. *Obstet. Gynec.* 24:728.
86. Lombard, L.; Morgan, B. B.; and McNutt, S. H. 1951. Some pathologic alterations of the bovine oviduct. *Amer. J. Vet. Res.* 12:43.
87. Lukasik, J. 1963. A comparative evaluation of the bacteriological flora of the uterine cervix and Fallopian tubes in cases of salpingitis. *Amer. J. Obstet. Gynec.* 87:1028.
88. MacDonald, C. J., and Pratt, J. H. 1967. Twisted parovarian cysts. *Obstet. Gynec.* 29:113.
89. McEwen, D. C. 1966. Reconstructive tubal surgery. *Fertil. Steril.* 17:39.
90. Madding, G. F., and Kennedy, P. A. 1963. Endometriosis: Case supporting coelomic metaplasia as possible cause. *J.A.M.A.* 183:686.
91. McGowan, L. 1965. Intramural pregnancy. *J.A.M.A.* 192:637.
92. Malik, P. S.; Sengar, O. P. S.; and Singh, S. N. 1960. Structure and abnormalities of the female genitalia in Indian buffalo *Bos* (*Bubalus*) *bubalis* L. *Agra Univ. J. Res.* (*Sci.*) 9:271.
93. Malouf, N.; Benirschke, K.; and Hoefnagel, D. 1967. XX/XY chimerism in a tricolored male cat. *Cytogenetics* 6:228.
94. Marin-Padilla, M.; Hoefnagel, D.; and Benirschke, K. 1964. Anatomic and histopathologic study of two cases of D_1 (13–15) trisomy. *Cytogenetics* 3:258.
95. Meissner, W. A.; Sommers, S. C.; and Sherman, G. 1957. Endometrial hyperplasia, endometrial carcinoma, and endometriosis produced experimentally by estrogen. *Cancer* 10:500.
96. Monroe, C. W., and Spector, B. 1962. The epithelium of the hydatid of Morgagni in the human adult female. *Anat. Rec.* 142:189.
97. Moritz, A. R., and Douglass, M. 1928. A study of uterine and tubal decidual reaction in tubal pregnancy: Based on the histological examination of the tubes and endometria of fifty-three cases of ectopic gestation. *Surg. Gynec. Obstet.* 47:785.
98. Movers, F. 1963. Die Tubenendometriosis in ihrer Auswirkung auf Fertilität und Sterilität. *Zbl. Gynaek.* 85:1409.
99. Murti, G. S.; Gilbert, D. L.; and Borgmann, A. R. 1966. Canine intersex states. *J. Amer. Vet. Med. Ass.* 149:1183.
100. Nachtsheim, H. 1958. Erbpathologie der Nagetiere. In *Pathologie der Laboratoriumstiere*, ed. H. Cohrs, R. Jaffé, and H. Meessen. Berlin: Springer-Verlag.

101. Nassberg, S.; McKay, D. G.; and Hertig, A. T. 1954. Physiologic salpingitis. *Amer. J. Obstet. Gynec.* 67:130.
102. Nes, N. 1966. Diploid-triploid chimerism in a true hermaphrodite mink (*Mustela vison*). *Hereditas* 56:159.
103. Nicholas, J. S. 1934. Experiments on developing rats. I. Limits of foetal regeneration; behavior of embryonic material in abnormal environments. *Anat. Rec.* 58:387.
104. Nieberle, K., and Cohrs, P. 1962. *Lehrbuch der speziellen pathologischen Anatomie der Haustiere*, 4th ed. Stuttgart: Gustav Fischer.
105. Nokes, J. M.; Claiborne, H. A.; Thornton, W. N.; and Yiu-Tang, H. 1957. Extra-uterine pregnancy associated with tuberculous salpingitis and congenital tuberculosis in the fetus. *Obstet. Gynec.* 9:206.
106. Novak, E. R., and Woodruff, J. D. 1962. *Gynecologic and obstetric pathology*. Philadelphia: Saunders Co.
107. Novak, J., and Rubin, I. C. 1952. *Anatomy and pathology of the Fallopian tubes.* Ciba Clinical Symposia 4, no. 6.
108. Nuckols, H. H., and Hertig, A. T. 1938. Pneumococcus infection of the genital tract in women, especially during pregnancy and the puerperium. *Amer. J. Obstet. Gynec.* 35:782.
109. Ober, W. B., and Black, M. B. 1955. Neoplasms of the subcoelomic mesenchyme. *Arch. Path.* 59:698.
110. Overzier, C. 1961. *Die Intersexualität*. Stuttgart: G. Thieme.
111. Palmer, R., and deBrux, J. 1964. À propos d'un cas de tuberculose tubaire démontrée par biopsie per-coelioscopique, avec trois grossesses tubaires et enfin une grossesse utérine. *C.R. Soc. Franç. Gynéc.* 34:207.
112. Pauerstein, C. J., and Woodruff, J. D. 1966. Cellular patterns in proliferative and anaplastic disease of the Fallopian tube. *Amer. J. Obstet. Gynec.* 96:486.
113. Peter, R., and Veselý, K. 1966. *Kindergynäkologie*. Stuttgart: G. Thieme.
114. Pinero, D. A., and Foraker, A. G. 1963. Aging in the Fallopian tube. *Amer. J. Obstet. Gynec.* 86:397.
115. Poland, B. J. 1965. Female pelvic tuberculosis with special reference to infertility: Study of 140 cases in Ontario. *Amer. J. Obstet. Gynec.* 91:350.
116. Poluni, I., and Saunders, M. 1958. Infertility and depopulation: A study of the Murut tribes of North Borneo. *Lancet* 2:1005.
117. Potter, E. L. 1961. *Pathology of the fetus and infant*, 2d ed. Chicago: Year Book Medical Publ. Inc.
118. Preissecker, E. 1958. Weibliche Genitalorgane. In *Pathologie der Laboratoriumstiere*, ed. P. Cohrs, R. Jaffé, and H. Meessen. Berlin: Springer-Verlag.
119. Proust, J.; deBrux, J.; and Palmer, R. 1964. Étude histologique des trompes après salpingographie aux organo-iodes avec poly-vinyl-pyrrolidone. *C.R. Soc. Franç. Gynéc.* 34:191.
120. Ratcliffe, H. L. 1933. Incidence and nature of tumors in captive wild mammals and birds. *Amer. J. Cancer* 17:116.
121. Resnick, L. 1962. Constrictive perisalpingitis: A preliminary report on another cause of non-patency of the proximal portion of the Fallopian tube with a short note on a more conservative approach to tubal plastic surgery. *S. Afr. Med. J.* 36:769.
122. Rewell, R. E. 1960. *Obstetrical and gynaecological pathology for postgraduate students.* Edinburgh: Livingstone Ltd.

123. Riggs, J. A.; Wainer, A. S.; Hahn, G. A.; and Farrell, D. M. 1964. Extrauterine tubal choriocarcinoma. *Amer. J. Obstet. Gynec.* 88:637.
124. Rippman, E. T. 1964. The clinical aspects of female genital tuberculosis today. *Gynaecologia* 157:77.
125. Roberts, S. J. 1956. *Veterinary obstetrics and genital diseases.* Ithaca, New York: S. J. Roberts.
126. Robinson, S. C. 1961. Pelvic abscess. *Amer. J. Obstet. Gynec.* 81:250.
127. Romney, S. L.; Hertig, A. T.; and Reid, D. E. 1950. The endometria associated with ectopic pregnancy: A study of 115 cases. *Surg. Gynec. Obstet.* 91:605.
128. Ross, W. M. 1967. Primary tumors of the Fallopian tube. *Canad. Med. Ass. J.* 96:328.
129. Ruch, T. C. 1959. *Diseases of laboratory primates.* Philadelphia: W. B. Saunders Co.
130. Schenck, S. B. 1964. Primary carcinoma of Fallopian tubes with positive smears: 5 year cure. *Amer. J. Obstet. Gynec.* 90:556.
131. Schenken, J. R., and Burns, E. L. 1943. A study and classification of nodular lesions of the Fallopian tubes: "Salpingitis isthmica nodosa." *Amer. J. Obstet. Gynec.* 45:624.
132. Schultz, L. R.; Newton, W. A.; and Clatworthy, H. W., Jr. 1963. Torsion of previously normal tube and ovary in children. *New Engl. J. Med.* 268:343.
133. Scully, R. E. 1963. Germ cell tumors of the ovary and Fallopian tube. In *Progress in gynecology*, vol. 4, ed. J. V. Meigs and S. H. Sturgis. New York: Grune & Stratton.
134. ———. 1967. Case 3-1967, MGH. *New Engl. J. Med.* 276:175.
135. Sedlis, A. 1961. Manson's schistosomiasis of the Fallopian tube. *Amer. J. Obstet. Gynec.* 81:254.
136. Siegler, A. M. 1967. Dangers of hysterosalpingography. *Obstet. Gynec. Survey* 22:284.
137. Siegler, A. M., and Hellman, L. M. 1963. Tubal plastic surgery: A retrospective study of 50 cases. *Amer. J. Obstet. Gynec.* 86:448.
138. ———. 1964. The effect of the intrauterine contraceptive coil on the oviduct. *Obstet. Gynec.* 23:173.
139. Skulj, V.; Bunarevic, A.; Drazancic, A.; and Stoiljkovic, C. 1963. The Arias-Stella phenomenon in the diagnosis of ectopic pregnancy. *Amer. J. Obstet. Gynec.* 87:499.
140. Smith, H. A., and Jones, T. C. 1966. *Veterinary pathology*, 3d ed. Philadelphia: Lea & Febiger.
141. Sutherland, A. M. 1965. The place of surgery in the treatment of genital tuberculosis in women. *Acta Obstet. Gynec. Scand.* 44:163.
142. Symmers, W. St. C. 1950. Pathology of oxyuriasis; with special reference to granulomas due to the presence of *Oxyuris vermicularis* (*Enterobius vermicularis*) and its ova in the tissues. *Arch. Path.* 50:475.
143. Teoh, T. B. 1951*a*. The structure and development of Walthard cell nests. *J. Path. Bact.* 66:433.
144. ———. 1951*b*. The histogenesis of Brenner tumors of the ovary. *J. Path. Bact.* 66:441.
145. Thuline, H. C., and Norby, D. E. Quoted by Hare, W. C. D.; Weber, W. T.; McFeely, R. A.; and Yang, T. J. 1966. Cytogenetics in the dog and cat. *J. Small Anim. Pract.* 7:575.
146. Vetesi, F., and Kemenes, F. 1967. Studies on listeriosis in pregnant rabbits. *Acta Vet. Acad. Sci. Hungar.* 17:27.
147. Westerhout, F. C. 1964. Ruptured tubal hydatidiform mole: Report of a case. *Obstet. Gynec.* 23:138.

148. Willis, R. A. 1958. *The borderland of embryology and pathology*. London: Butterworth Co.
149. Woskressensky, M. A. 1891. Experimentelle Untersuchungen über die Pyo- und Hydrosalpinxbildung bei den Tieren. *Zbl. Gynaek.* 15:849.
150. Youssef, A. F.; Fayad, M. M.; and Shafeek, M. A. 1962. Torsion of the Fallopian tube. *Acta Obstet. Gynec. Scand.* 41:292.
151. Zuckerman, S., and Krohn, P. L. 1938. The hydatids of Morgagni under normal and experimental conditions. *Phil. Trans. Roy. Soc. Lond. Ser. B* 228:147.

Part III Biochemistry

12

Histochemistry of the Oviduct

B. Fredricsson
Department of Obstetrics and Gynecology
Karolinska Sjukhuset
Stockholm, Sweden

I. Introduction

The aim of the present review is to throw some light on the milieu where fertilization and the first development of the egg takes place. We will emphasize particularly the occurrence and distribution of various substances and enzymes and point out their variations during the different phases of the female's reproductive life.

The low-power microscopic view of the human oviduct reveals an abundantly folded mucosa covered with an apparently uniform epithelium (Fig. 1*a*). Greater magnification discloses that there are at least two types of cells in this epithelium, ciliated and nonciliated (Fig. 1*b*). Other cell-types frequently referred to in the literature, such as "peg cells," "Stiftzellen" or intercalary cells, are considered to be secretory cells whose altered appearance may be due to depletion of secretory products (cf. Papanicolaou, Traut, and Marchetti 1948; Fredricsson 1959*c*).

The literature, however, is dominated by the apparently contradictory statement that the ciliated cells contribute to the secretion of the oviduct, either as such or by transformation to the ordinary type of secretory cell. The basic statement that two different functions, ciliary and secretory activity, are connected to two different types of cells is thus modified by a transformation theory. However, this theory is based on purely morphological studies, frequently using inadequate techniques when compared with the demand of today. In fact, ultrastructural, histochemical, and experimental studies give more support to the view of dualism in terms of morphology and cellular regeneration (cf. Fredricsson 1959*a*).

The contribution by Dr. Reinius in this volume (Chapter 3) presented us with much morphological data. The following conclusions are relevant in this connection: (1) the division of the oviduct into the infundibular, ampullar, and isthmic parts is not adequate in some species, e.g., the rat and mouse; (2) secretion granules and other cytological features of the nonciliated cell show greater species variation than the ciliated cells; (3) the terms

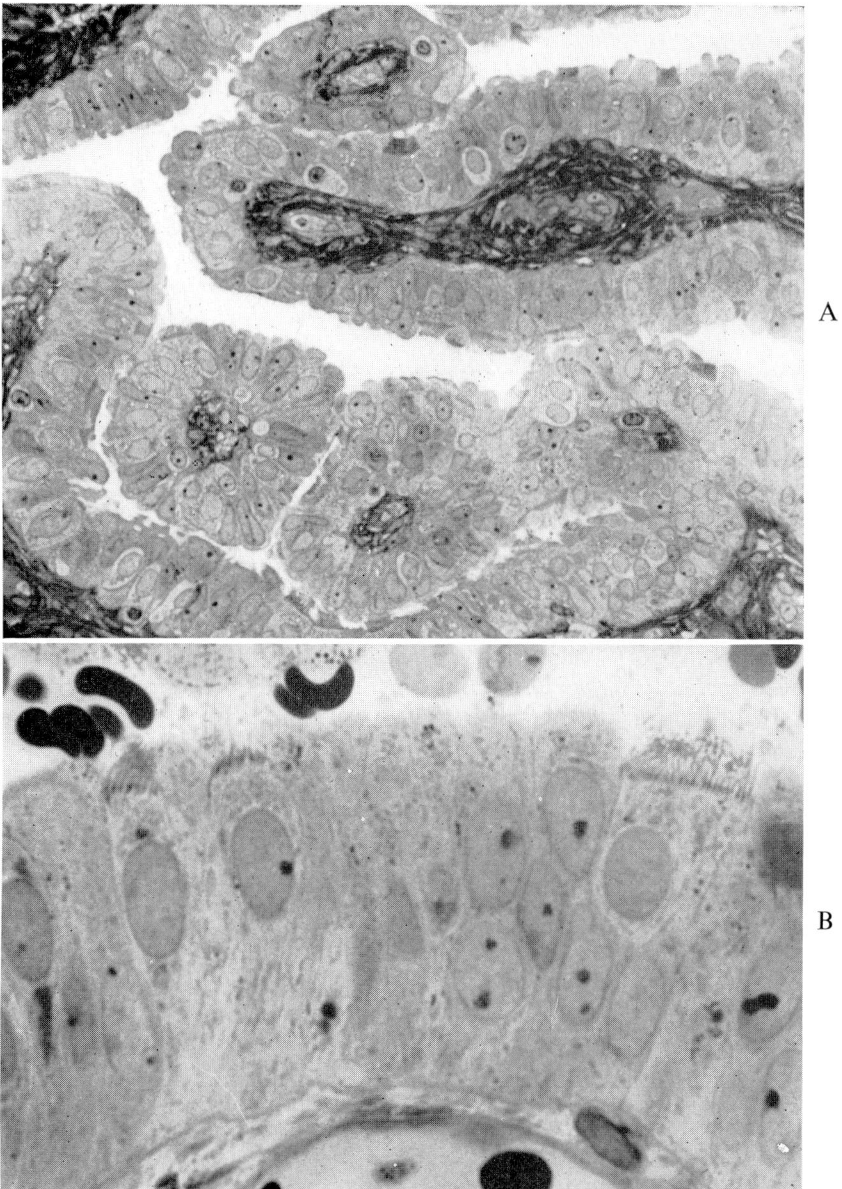

FIG. 1. Human oviduct, ampullary region. Toluidine blue stain: (*a*) 19th day of cycle. Glutaraldehyde-osmic acid fixation (×360). (Photomicrograph by Dr. S. Reinius, kindly placed at my disposal.)
(*b*) Ovulation time. Osmic acid fixation (×1350).

nonciliated and secretory are not always synonymous with regard to the cells of the oviduct epithelium.

Much information on the histochemistry of the oviductal epithelium originates from studies where ciliated and nonciliated cells were not distinguished. Many times the reason for this is that the methods of study could not be refined to such a degree of resolution as to permit this distinction. This justifies a section on the entire epithelium in this review although some of the material included actually may pertain to one or the other type of cell. The main references on the occurrence of glycogen, ribonucleic acid (RNA), lipids, acetal phosphatides, and some enzymes in the oviductal epithelium are presented in Tables 1 and 2.

TABLE 1

THE OCCURRENCE OF GLYCOGEN, RNA, LIPIDS, AND ACETAL PHOSPHATIDES IN THE OVIDUCTAL EPITHELIUM

Metabolite	Species	Location in Epithelium	Regional Difference	Hormonal Dependency	Main References
Glycogen	Man	Predominance in ciliated cells	Less in isthmus	Varies with cycle	Joel 1939*a*: Fredricsson 1959*b*
	Rabbit	Ciliated cells	Predominance in the first part(s)	Varies with cycle	Fredricsson 1959*a*; Fredricsson, this chapter; Graumann 1964
	Guinea pig	Not stated	"	"	
	Rat	Ciliated cells	"	"	
	Mouse	Not stated	"	"	
	Cattle	Not stated	Not stated	Varies with cycle	Weeth and Herman 1950
RNA	Man	Predominance in nonciliated cells	Not reported	Not reported	Balboni 1954
	Sheep	"	"	Varies with cycle	Hadek 1955
	Cattle	"	"	Not found	Björkman and Fredricsson 1961
Lipids	Man	All cells	Not reported	Suggested	Joel 1939*b*; Fredricsson 1959*b*
	Cattle	"	"	"	Weeth and Herman 1950
Acetal phosphatides	Man	Some ciliated and nonciliated cells	Suggested	Not found	Fredricsson 1959*b*
	Rat	Not reported	"	Suggested	Koch 1941

II. Methods

Histochemistry is a field of rapid development. This refers not only to the evolution of new methods but also to the identification of stained artifacts and the solution of other technical problems.

However, there is still a risk of unknown pitfalls in the performance and interpretation of the histochemical reaction. Furthermore, this often lacks in specificity, at its best enabling the demonstration of a reactive group and not of a chemical substance. Quantitation of the reaction product presents another difficulty and often is expressed subjectively as a number of plus signs. Unless the preparations are codified and thus unknown to the examiner, this is a poor method. Nevertheless, if most of the quantitative information concerning the oviductal histochemistry is based on such estimated data, the number of confirmatory reports may be of significance.

The present review is based on studies from various periods since 1939, using different methods for the demonstration of the same principle. Here we may expect to find inconsistent results and interpretations. In general there is good agreement, but we cannot know to what extent the investigator has been influenced by previous reports.

My own observations included here, and not published before, are based on frozen-dried or acetone-fixed oviducts of different mammals. These were mainly subjected to the

TABLE 2

THE OCCURRENCE OF SOME ENZYMES IN THE OVIDUCTAL EPITHELIUM

Enzyme	Species	Location in Epithelium	Regional Difference	Hormonal Dependency	Main References
Phospho-amidase	Man	Not reported	Not reported	Not found	Neumann, Oehlert, and Hansmann 1954
Alkaline phosphatase	Man	Nonciliated cell surface	Less activity in isthmus	Varies with cycle	Moschino 1954; Alamanni 1956; Augustin and Huwald 1956; Fredricsson 1959*b*
	Rat	Some ciliated and nonciliated cells	Present	Not reported	Augustin and Moser 1955; Bronzetti *et al.* 1963
	Mouse	"	"	"	Fredricsson, this chapter
	Hamster	Cilia	Not reported	"	Moog and Wenger 1952
	Cattle	"	"	Not found	Björkman and Fredricsson 1960
Acid phosphatase	Man	Near the free border	Not reported	Not reported	Goldberg and Jones 1953
Nonspecific esterase	Man	Nonciliated cells	Not found	Varies with cycle	Fredricsson 1959*b*
	Rat	Not reported	Not reported	Not reported	Nachlas and Seligman 1949
Lipase	Man	Not reported	Not reported	Varies with cycle	Sermann and Rigano 1960
ATP-ase	Man	Not reported	Not reported	Not reported	Jonek and Glenc 1963
γ-glutamylotranspeptidase	Guinea pig, hamster, mouse, rat, cat, rabbit	Not reported	Not reported	Not reported	Jonek *et al.* 1966
Succinodehydrogenase	Man	Not reported	Not reported	Not found	Foraker and Crespo 1962
	Monkey	"	"	Suggested	Velardo and Rosa 1963
	Rabbit	"	"	"	"
	Rat	"	Predominance in infundibulum	Not reported	Padykula 1952
Sterol-3β-ol dehydrogenase	Rat	Not reported	Not reported	Suggested	Levy, Deane, and Rubin 1959
DPN-diaphorase	Man	Not reported	Not reported	Not found	Foraker and Crespo 1962

periodic acid-Schiff (PAS) and alkaline phosphatase reactions. In these cases consistent results were obtained irrespective of the preparatory procedure used. Technical details and considerations have been published before (Fredricsson 1959*c*).

III. Observations on the Entire Epithelium

A. *Glycogen*

Glycogen is present in the oviductal epithelium of several species (cf. review by Graumann 1964). Frequently reference to the two cell types is lacking in the early reports. Later studies have revealed glycogen in both ciliated and nonciliated cells, where it appears differently during the sexual phases.

For this reason, a more extensive presentation of these results is given later for each type of cell.

B. *Lipids*

Lipids are present in the epithelium of the human oviduct in small amounts only (cf. review by Wolman 1964). They have about the same cytological distribution in both ciliated and nonciliated cells (Fredricsson 1959*b*, *c*), although the nonciliated cells seem to contain somewhat more (Tsuchiya 1956).

Sudanophilic particles in the form of mitochondria and lipid inclusions of various size occur regularly (Fawcett and Wislocki 1952; Deane 1952; Tsuchiya 1956; Fredricsson 1959*c;* Björkman and Fredricsson 1961). Vacuoles surrounded by lipids have been reported in the human oviduct only and are most frequently seen in the infundibular part (Fredricsson 1959*b*). The sudanophilic material is evident mostly in supranuclear position in man and in infranuclear location in cattle (Björkman and Fredricsson 1961). Otherwise species differences in the lipid distribution are not reported.

As the amount of lipid is small, cyclical changes are difficult to detect. In the woman lipids are more conspicuous at the time of ovulation (Sermann and Rigano 1960) or shortly thereafter (Joel 1939*b*, 1940; Fredricsson 1959*b*). They diminish after the menopause. In cattle the maximum appears postovulatory (Weeth and Herman 1950).

Attempts to define the lipid inclusions chemically indicate that they are mainly composed of neutral fats, since methods for unsaturated fatty acids and phospholipids fail to give a positive staining reaction (Fredricsson 1959*b*).

Acetal phosphatides are present in the apical parts of the human epithelium (Schäfer and Roloff 1950; Jonek 1957). There is some accentuation in the reaction in vacuole-containing cells in the infundibular part and in ciliated cells in the ampullar segment of the oviduct (Fredricsson 1959*b*, *c*). In the rat the phosphatides are present in the form of vesicles, mainly in the basal portions of the epithelium. The reaction is most evident during diestrus and proestrus, decreasing at estrus, particularly in the ampullar and isthmic regions. By metestrus the intensity of the reaction is restored. The reaction is moderate during pregnancy and negative during infancy (Koch 1941).

C. *Enzymes*

1. *Hydrolytic Enzymes.* There are a great number of reports on the general occurrence and cyclical changes of alkaline phosphatase in the human oviduct (Sani and Hanau 1952; Alamanni 1956; Augustin and Huwald 1956; Jonek and Glenc 1963), and in the rat (Augustin, Heidenreich, and Thilo 1954; Augustin and Moser 1955; Milio 1960; Bronzetti *et al.*

1963). As this enzyme is present distinctly in the nonciliated cells with few exceptions, it will be discussed further when the histochemical characteristics of the two cell types are presented separately.

Acid phosphatase is present mainly in the cytoplasm near the free border of the epithelium of the human oviduct (Goldberg and Jones 1953; Jonek and Glenc 1963). Phosphoamidase is also present, but cyclical changes like those in the uterine epithelium are not reported (Neumann, Oehlert, and Hansmann 1954). Adenosine triphosphatase was studied in the epithelium of the postmenopausal woman's oviduct only (Jonek and Glenc 1963).

Nonspecific esterase is present in the supranuclear region of the rat oviductal epithelium (Nachlas and Seligman 1949). It is predominant in the human female in the nonciliated cells and will be discussed in this connection. A lipase shows its maximum activity at the time of ovulation (Sermann and Rigano 1960).

A few biochemical studies on phosphatases and esterases of the oviduct should be mentioned. In homogenates of human oviducts, two pH-optima of phosphatases have been demonstrated, one in the acid and the other in the alkaline range, the hydrolyzing capacity being much higher in the acid range (Goldberg and Jones 1953). In homogenates of the rat oviduct, using zonal electrophoresis, 1 alkaline phosphatase, 4 types of acid phosphatases, and 7 different esterases have been defined (Robboy and Kahn 1964). The alkaline phosphatase showed maximal activity during proestrus and estrus. Only 2 of the acid phosphatases were present regularly, one showing dominance. Ovariectomy effected a decrease in its activity. Of the 7 esterases only 3 were present regularly. Great individual variation may have masked cyclical changes. Two of the esterases showed decreased activity after ovariectomy.

A γ-glutamylo-transpeptidase, a proteolytic enzyme, has been demonstrated in the guinea pig, hamster, mouse, rat, cat, and rabbit (Jonek *et al.* 1966).

2. *Oxidative and Glycogen-Synthesizing Enzymes.* As the oxidizing enzymes are mainly mitochondrial in origin, an equal intensity of the histochemical staining of the epithelial cells of the oviduct should be expected, which is in fact the case.

Succinodehydrogenase is present particularly in the infundibular region of the rat (Padykula 1952), while in the ampulla of the mouse most of the enzyme is localized basally in the epithelium (Rosa and Tsou 1962). The enzyme has been described also in man (Foraker, Denham, and Celi 1953). Neither the time of the cycle nor age alter the histochemical pattern of succinodehydrogenase and diphosphopyridine nucleotide-diaphorase in the woman (Foraker and Crespo 1962). The rhesus monkey also presents a low sensitivity to hormonal changes although some drop is noted after experimentally induced menstrual bleeding. In the rabbit, on the other hand, a decided drop in oxidative enzyme activity is produced by ovariectomy (Velardo and Rosa 1963).

A sterol-3β-ol dehydrogenase has been demonstrated in the epithelium of the oviduct of the rat. The intensity of the histochemical reaction is decreased by hypophysectomy (Levy, Deane, and Rubin 1959).

An attempt to demonstrate amylophosphorylase in the epithelium of the human oviduct failed, but, since the oviducts under study were devoid of glycogen, this finding does not exclude the presence of the enzyme, which is of importance in glycogen synthesis. It has thus far been demonstrated in other glycogen-containing epithelia of the woman's genital tract (Foraker and Crespo 1962).

In cases of oviductal pregnancy, acid and alkaline phosphatase, nonspecific esterase, β-glucuronidase, succinodehydrogenase, peroxidase, and cytochrome oxidase were studied histochemically in an attempt to analyze why decidual formation is inadequate in the mucosa of the oviduct (Kraus, Vacek, and Jirsová 1964).

D. *Uptake of Radioactive Compounds*

Studies on the uptake of radioactive isotopes have been restricted mainly to ^{35}S-containing compounds. When administered in the form of sulfate ^{35}S is rapidly incorporated by the oviductal epithelium of rat and rabbit (Boström and Odeblad 1952). The uptake is enhanced during estrus or by the administration of estrogen. After ovulation, the oviductal secretion of the rabbit contains radioactive material, and the mucin coat of the egg also becomes active (Moricard and Gothie 1956). Studied in this way, the secretion begins at the ovarian end of the oviduct and is last evident in the uterine end, thereby serving the egg with a maximum of secretion during its entire passage through the oviduct (Friz 1959; Koester 1964). The radioactive sulfate is incorporated much earlier in the oviductal epithelium than in the uterine epithelium of the rabbit (Zachariae 1958).

^{35}S-methionine and ^{35}S incorporated in yeast are taken up by the oviductal epithelium of the rat, mouse, and rabbit, but not to the same extent as by the uterine epithelium (Niklas and Oehlert 1956). In spite of its accumulation in the oviductal epithelium of the mouse, ^{35}S-methionine is not utilized by oviductal eggs after the 2-cell stage (Greenwald and Everett 1959).

IV. Observations on Ciliated Cells

A. *Cell Surface and Cilia*

In general the ciliated cells of the oviduct show few morphological characteristics and histochemical features differing from ciliated cells of other epithelia of the body. Species differences are not common either.

The cilia and their basal bodies are PAS-reactive in a number of species (Leblond 1950; Deane 1952; Fawcett and Wislocki 1952; Moog and Wenger 1952; Schultka 1963), and this stainability is resistant to the action of diastase. Using histochemical methods on the ultrastructural level glycogen has recently been demonstrated inside the basal bodies of the cilia of the bovine oviduct (Nicander *et al.* 1968). The situation is the same in man (Fredricsson and Björkman 1968). The cilia also take up stain by the Hale reaction for acid mucopolysaccharides; consequently they are also stained by Alcian blue. Sulfydryl and disulfide groups have been demonstrated. The periodic acid-Schiff reaction is positive in the basal bodies, indicating the presence of ethylene groups in unsaturated lipids (Schultka 1963; Schultka and Scharf 1963). Sudan black B stains the cilia grayish blue (Fawcett and Wislocki 1952; Fredricsson 1959*b*).

Some other staining reactions of the cilia are more difficult to evaluate, as they may be due to adhesion of secretory products and sometimes are interpreted as such; i.e., the case of toluidine blue metachromasia (Hølund 1946).

More of interest are the reports of alkaline phosphatase in the cilia (Figs. 3*c*, 4*c*) of the oviductal epithelium of the mouse, rat, hamster, and cow (Weeth and Herman 1950; Moog and Wenger 1952; Björkman and Fredricsson 1960; Clark 1961). In the mouse and cow the enzyme is not confined to the ciliary apparatus alone but is associated also with the rest of the cell surface including the microvilli (Björkman and Fredricsson 1960; Clark 1961). This finding is not peculiar to the ciliated cells of the oviduct, since a similar result is obtained in the tracheal epithelium of the rat. The intensity of the reaction is quite extraordinary in the bovine oviduct, where the enzyme is associated exclusively with the ciliated surface of the epithelium. Hence adhesion of secretion from other sources can be excluded as an explanation of this finding. This may be interpreted as an indication of secretory or perhaps absorptive capacity of this cell. More probably, however, the enzyme functions in the transduction of energy for ciliary movement in these species (Clark 1961).

Ciliary vesicles have attracted much attention in the oviductal epithelium (Mihálik 1935; Flerkó 1954). Microvillous vesicles occur also (Björkman and Fredricsson 1961), and therefore it may be difficult to determine the true nature of a vesicle in the histochemical preparation. PAS-reactive structures, which may represent ciliary vesicles in different stages of development, are described in the ampulla of the rabbit (Fredricsson 1959*a*) and can be seen in the ampulla of the mouse. In the human epithelium the vesicles are not PAS-reactive.

B. *Cytoplasmic Basophilia*

Cytoplasmic basophilia owing to RNA is less pronounced in ciliated than in nonciliated cells in man (Fawcett and Wislocki 1952; Balboni 1954), sheep (Hadek 1955), and cattle (Björkman and Fredricsson 1961). The RNA is distributed evenly within the cytoplasm. Cyclical changes are reported only in sheep, where the basophilia reaches a maximum during estrus and metestrus (Hadek 1955).

Metachromatic staining with toluidine blue, probably of the beta type, is seen frequently in the apical portion of the ciliated cells of the human oviduct, particularly during the luteal phase. Sometimes the basal portion of the cytoplasm exhibits this characteristic also (Hølund 1946).

C. *Glycogen*

There were several early attempts to demonstrate glycogen within the oviductal epithelium, but owing to inadequate techniques the results were not always consistent (cf. review by Graumann 1964). Iwata (1929) was the first to state that glycogen is present mainly in the ciliated cells of the human oviduct. This statement was confirmed by the use of the PAS-reaction and the diastase test (Tsuchiya 1956; Fredricsson 1959*b*, *c*). It has been recognized also in the electron microscopic picture (Björkman and Fredricsson 1962). The cyclical changes of glycogen in the human oviduct, first reported by Joel (1939*a*, 1940), have been confirmed in an extended study on frozen-dried pieces of a considerable number of human oviducts (Fredricsson 1959*b*, *c*). The changes are both morphological and quantitative. Before ovulation the glycogen is distributed equally around the nucleus; after ovulation the supranuclear deposit becomes depleted (Fig. 2*a*). In addition to this morphological change, there appears to be a maximum of glycogen present in the mid-luteal phase, which was also stated by Joel (1939*a*, 1940).

FIG. 2. (*a*) Human oviduct, infundibular region. Ovulation time. Frozen-dried preparation. Alkaline phosphatase reaction. Incubation period 15 min (×540).

(*b*) Human oviduct, ampullary region. Ovulation time. Frozen-dried preparation. Naphthyl acetate esterase reaction, incubation period 60 min (×540).

(*c*) Bovine oviduct, ampullary region. Luteal phase. Acetone fixation. Alkaline phosphatase reaction, incubation period 15 min (×540).

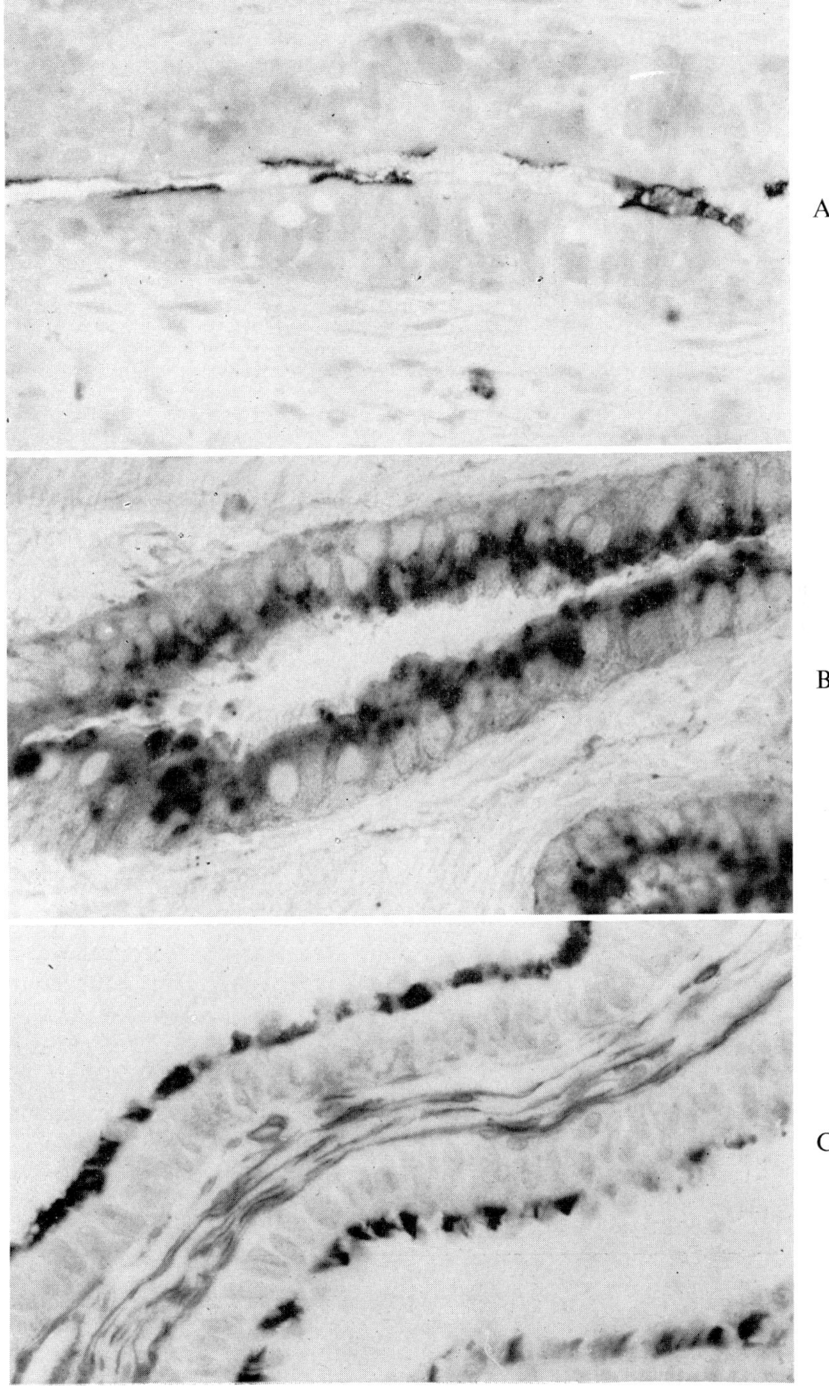
A
B
C

Glycogen is present also in the preovulatory phase in the infundibular epithelium of the rat (Fig. 4*a*) and in the epithelium of the ampulla of the rabbit under estrogen influence (Fredricsson 1959*a*). In the oviductal epithelium of the cow there is no glycogen at estrus. It appears later and reaches a maximum 8 days after estrus (Weeth and Herman 1950). Attempts to demonstrate glycogen in the mouse and sheep have failed.

The suggestion that glycogen is secreted from the ciliated cells into the oviductal lumen (Iwata 1929; Tsuchiya 1956; Schultka 1963; Schultka and Scharf 1963) has not been confirmed. The changes in the amount and distribution of glycogen within the ciliated cells in man may reflect hormone-dependent variations in ciliary activity (Fredricsson 1959*b*, *c*). The recent finding of glycogen inside the basal bodies of the cilia (Nicander *et al.* 1968) gives further support to this assumption.

V. Observations on Nonciliated Cells

A. *Cell Surface*

1. *PAS-Reactivity*. The surface of the nonciliated cell is more or less PAS-reactive. In some instances, particularly in the isthmus of the rat, secreted material seems to adhere to the cell surface, making the microvilli stick together into tuftlike formations (Fig. 4*b*).

2. *Alkaline Phosphatase*. There are several reports on the presence of alkaline phosphatase in the oviductal epithelium. As mentioned above, in some cases the enzyme is present in the ciliated cells. In general its activity is confined to the nonciliated cell, and by improving the histochemical technique, the enzyme was shown to be associated with the cell surface (Fig. 3*a*) (Fredricsson 1959*b*, *c*).

Some regional differences are reported. In man the phosphatase activity is lower in the isthmic region (Moschino 1954). In the rat the few nonciliated cells in the infundibulum show low activity, while in the ampulla the number and phosphatase activity of the nonciliated cells is increasing continually up to the transition to the isthmic part, where the activity suddenly ceases and only a few ciliated cells are active (Fig. 4*c*). In the distal isthmus again a high alkaline phosphatase activity is present (Augustin, Heidenreich, and Thilo 1954; Augustin and Moser 1955; Milio 1960; Bronzetti *et al.* 1963). In the mouse a number of both ciliated and nonciliated cells show alkaline phosphatase activity in the infundibulum. In the ampulla the frequency of active cells is less, increasing again in the isthmus, where there are few ciliated cells. In the distal isthmus all cells show phosphatase activity and none are ciliated (Fig. 5*b*, *c*).

The phosphatase activity is clearly dependent on the ovarian function. In the woman, there is a minimum at menstruation after which the activity increases reaching a maximum around the time of ovulation and then again decreases (Alamanni 1956; Augustin and Huwald 1956; Fredricsson 1959*b*). There are diverse opinions about the level of activity during pregnancy, but the activity is low during the puerperium and after menopause. In

---

FIG. 3. Rat oviduct. Estrus. Acetone fixation: (*a*) Infundibular part. Periodic acid-Schiff reaction. Diastase digestible inclusions in the infranuclear cytoplasm (×540).

(*b*) Isthmic part. Periodic acid-Schiff reaction. Secretory granules in the apical cytoplasm and secretion in the microvillous tufts and lumen (×540).

(*c*) Isthmic part. Alkaline phosphatase reaction, incubation period 15 min. Cilia in epithelial crypts are stained (×540).

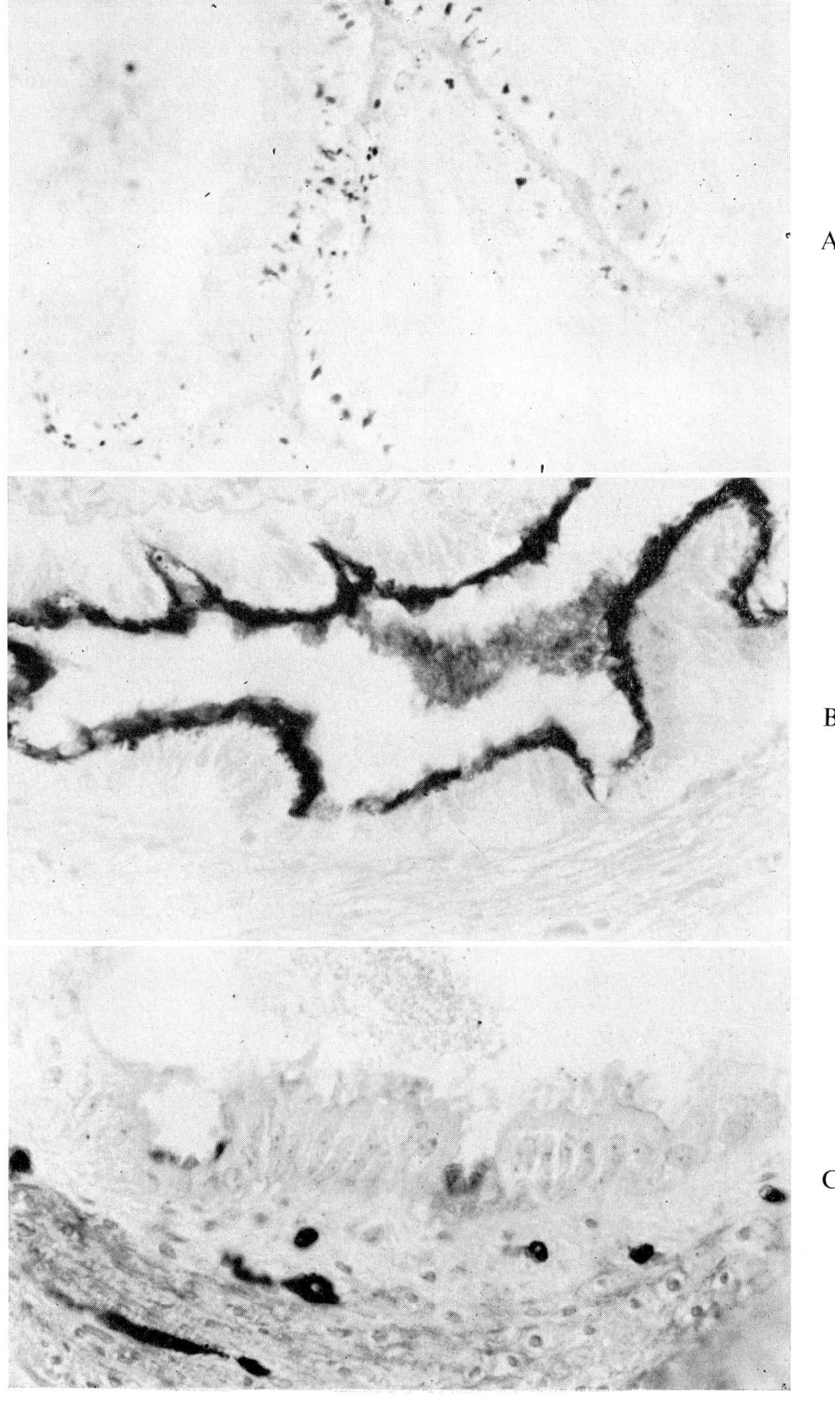
A
B
C

the rat a maximum activity may be present at estrus (Augustin and Huwald 1956), but cyclical changes have also been denied (Bronzetti *et al.* 1963).

Quite probably the activity of alkaline phosphatase is associated with the microvilli of the cell surface in man (Fredricsson 1959*b*). The luminal surface of the nonciliated cell, including the microvilli, is furnished with an extraneous coat, in contrast to the ciliated cell (Fredricsson and Björkman 1962). The microvilli and the extraneous coating are apparently under hormonal influence of the ovary. As the changes in amount of both more or less parallel the changes in the histochemical picture of alkaline phosphatase, the latter may reflect more changes in amount than in activity of the enzyme.

B. *Cytoplasmic Basophilia*

Cytoplasmic basophilia, due to RNA, is more pronounced in the nonciliated than in the ciliated cells in the human (Fawcett and Wislocki 1952; Balboni 1954), bovine (Björkman and Fredricsson 1961), and sheep oviducts (Hadek 1955). A maximum basophilia is seen in the proliferative phase in man and a reduction in some cells which are considered to have fulfilled their secretory activity (Balboni 1954); in the sheep the maximum basophilia is seen in estrus and metestrus. In the bovine oviduct basophilic cytoplasmic granules are most frequently seen at estrus and 5 days thereafter (Weeth and Herman 1950).

C. *Glycogen*

The presence of glycogen is characteristic for ciliated cell cytoplasm, but it is also present in smaller amounts in the nonciliated cells of the human oviduct (Fawcett and Wislocki 1952). It has been suggested that it is secreted into the oviductal lumen (Tsuchiya 1956). It is particularly evident in these cells during pregnancy (Fredricsson 1959*b*); otherwise cyclical changes in its presence have not been reported. Glycogen has been recognized also in the electron micrograph (Clyman 1966; Reinius, personal communication).

D. *Enzymes*

Esterases are the only cytoplasmic enzymes that have been characterized within the nonciliated cells of both human and cow oviducts (Fredricsson 1959*b;* Björkman and Fredricsson 1961).

There are no regional differences in the distribution of esterases in the species studied. Using α-naphthyl acetate as the substrate, the enzyme activity shows up in the cytoplasm as more or less distinct granules of variable size (Fig. 3*b*). Essentially the same picture is obtained if naphthol-AS acetate or 5-bromo-0-indoxyl acetate is used as the substrate. Diffusion is reduced and the reaction product appears more distinct.

The intracellular localization of the enzyme varies with ovarian function. In the follicular phase, the reaction product appears mainly in the apical portion of the cytoplasm; during the luteal phase, pregnancy, and the early puerperium, the activity is predominant in the basal part of the cell. The average enzyme activity apparently has two peaks, in the follicular and in the mid-luteal phases, separated by a distinct decrease in the early luteal

---

FIG. 4. Mouse oviduct. Diestrus. Acetone fixation: (*a*) Ampullary part. Periodic acid-Schiff reaction. A few secretory cells are seen (×540).

(*b*) Ampullary part. Alkaline phosphatase reaction, incubation period 15 min. Some of the ciliated cells are stained (×340).

(*c*) Isthmic part. Alkaline phosphatase reaction, incubation period 15 min. Staining of the entire epithelial surface (×340).

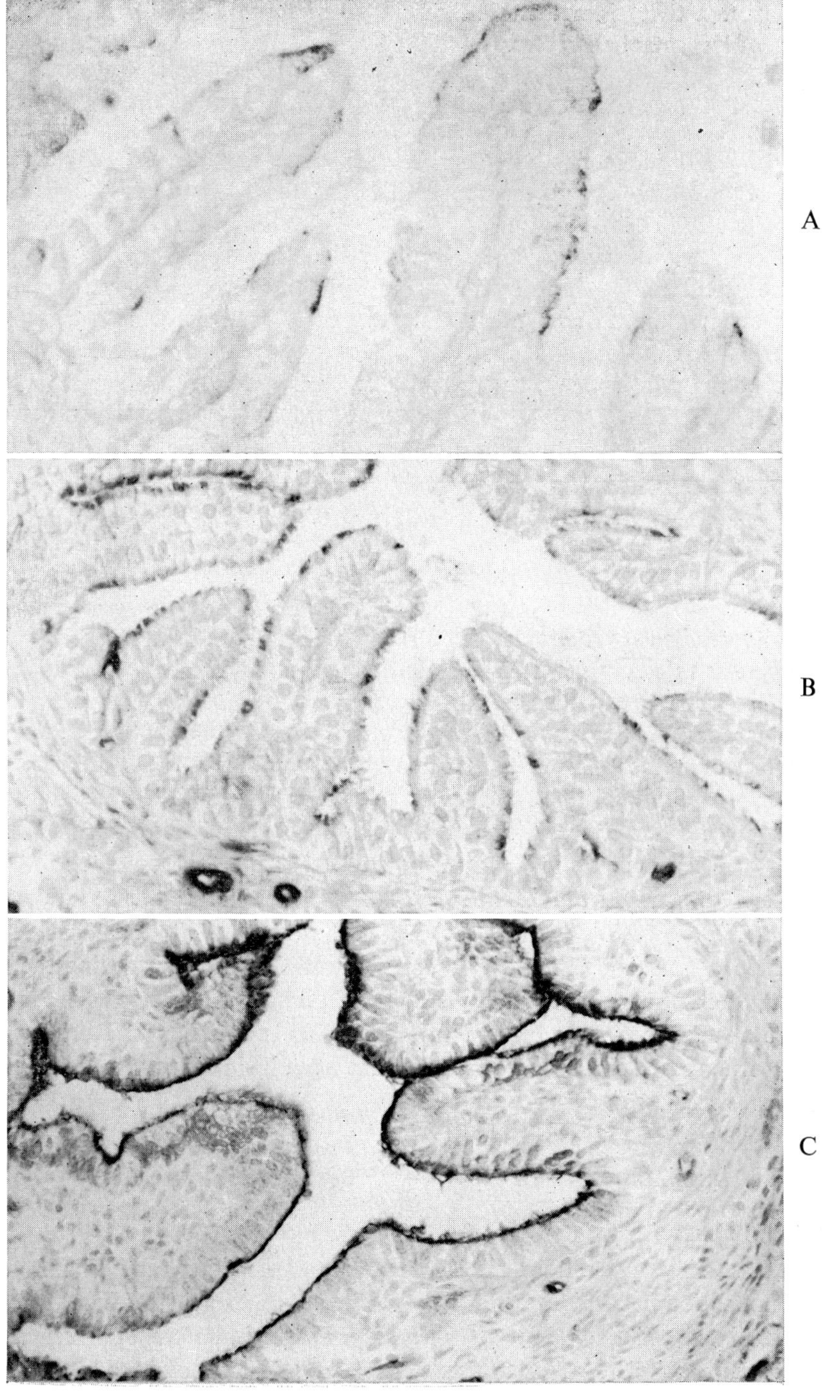
A
B
C

phase as evidenced by the staining intensity (Fredricsson 1959*b*). The morphological correlate of esterase activity is not established, but probably consists of multivesicular and dense bodies of lysosomal character.

E. *Secretory Granules*

The size and number of secretory granules vary widely in different species and during different phases of the reproductive cycle. As a rule, they increase in number toward the isthmic part of the oviduct, as do the nonciliated cells. Exceptions to this rule are found in the mouse, where the isthmic portion is devoid of granules (Reinius, this volume), and in the sheep where the nonciliated cells are scarce in the isthmus and have completely disappeared at the uterotubal junction (Hadek 1955). In the rabbit both size and number of granules are outstanding, in man they are small and few and have been overlooked frequently, as the size is very near the limit of resolution of the light microscope (Fig. 2*a*).

In spite of these differences, the granules of all species are PAS-reactive, viz., man (Balboni 1953, 1954; Tsuchiya 1956; Fredricsson 1957, 1959*b*), sheep (Hadek 1953, 1955), cattle (Björkman and Fredricsson 1961), rat (Deane 1952; Milio 1960; Bronzetti *et al.* 1963), guinea pig and rabbit (Lillie 1949; Varveri 1953), and mouse (Fig. 5*a*). Further, the granules are basophilic and sometimes stained slightly metachromatically (see Graumann 1964 for details and references). Generally they are resistant to the action of diastase but may be extracted nonspecifically from the section, which may lead to misinterpretation (Björkman and Fredricsson 1961).

The changes in density and structure of the granules as seen by the electron microscope may give a clue as to changes in stainability. Thus, the PAS-reaction increases in intensity as estrus and metestrus are approached in the sheep (Hadek 1955).

The appearance and amount of the secretory granules depend on the ovarian function. A great number of studies show this, particularly in the rabbit. Ovariectomy leads to their disappearance except in the isthmus, where some granules are retained long after the operation (Flerkó 1954). When estrogen is supplied they reappear. When estrogen is followed by progesterone or when ovulation is induced in the estrous rabbit, the granules are released into the oviduct (Greenwald 1958). The subject of the regulation of this secretion was dealt with in Chapter 7.

The degree of staining ascribed to secretory granules increases during the follicular phase in man, reaching a maximum in the early luteal phase, then decreasing in the midluteal phase. These changes are most evident in the ampulla and isthmus. During pregnancy and the puerperium small amounts are present in the ampulla in contrast to the high amounts in the isthmus (Fredricsson 1959*b*).

VI. Luminal Contents

A. *Secretion in General*

Since the studies of Westman (1930) and Westman, Jorpes, and Widström (1931), considerable attention has been paid to the secretory activity of the oviduct.

The nonciliated cells have been considered as the main source of secretion through the release of their granules, but other sources may contribute also to the luminal contents. The ciliated cells have been thought to contribute as well (Schultka 1963, and others), but this postulate is not settled in the absence of clear evidence with the electron microscope.

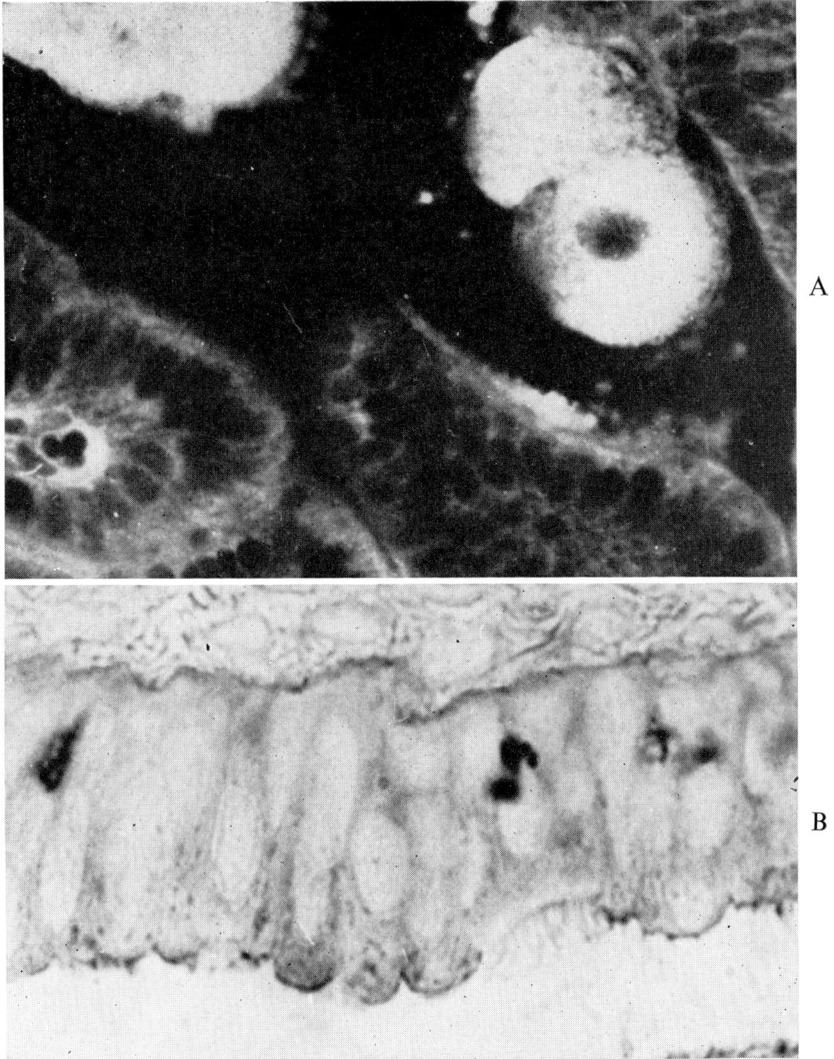

FIG. 5. (*a*) Human oviduct, ampullary part. Ovulation time. Preparation by freeze-drying. Periodic acid-Schiff reaction. No counterstain. Large inclusions (glycogen) in ciliated cells and small secretory granules in nonciliated cells are stained (×1,470).

(*b*) Mouse oviduct, isthmic portion with 2 early embryos. Fluorescence photomicrograph indicating the presence of antigens like those of maternal serum both in the embryos and in the nearby oviductal epithelium (from L. Glass 1963, by courtesy of the author and publisher).

The transfer of serum-like proteins to the egg engages the entire epithelium, making no difference between nonciliated or ciliated cells (Glass 1963).

The secreted material shows the same staining characteristics, PAS-reactivity, and resistance to diastase as the secretory granules although the intensity of staining is lower. In the human oviduct the secretion is resistant also to hyaluronidase (Hübner 1955), stains with Alcian blue at an acid pH, and contains some RNA. Metachromatic staining has not been reported. Positive reactions for tyrosine, tryptophane, disulfide and sulfydryl have been obtained (Schultka 1963). In accordance with biochemical data (Werner 1953), these staining characteristics suggest that the secreted material contains mucoproteins or mucopolysaccharides. The reports of glycogen in the secretion (Iwata 1929; Joel 1939*a;* Tsuchiya 1956; Schultka 1963) are difficult to confirm histochemically, as reduction in PAS-reactivity after the diastase test might well be due to a nonspecific extraction (Björkman and Fredricsson 1961; Józsa and Szederkényi 1967). Neither is there proof of lipids in the secretion.

Alkaline phosphatase is present in the luminal contents of both human and rat oviducts (Augustin and Moser 1955; Augustin and Huwald 1956; Fredricsson 1959*b*). The physiological implication of this observation is not known.

The amount of secretion is difficult to assess under physiological conditions. In the histological preparation, the size of the lumen is due mainly to shrinkage of the tissue components. In fact, opposing epithelial surfaces may touch one another *in vivo*, leaving only narrow clefts for the passage of the egg or sperm, particularly in the isthmus. In frozen-dried pieces of human ampulla any separation of the epithelial folds by fluid was evident only at the period of ovulation and around the 22d to 24th days of the cycle. At this time there is extensive uterine secretion also (Fredricsson 1959*b*, *c*), and the egg has left the oviduct.

When the oviductal fluids in rabbit and monkey are collected *in vivo* the production reaches rather high values, especially when under the influence of estrogen (Mastroianni *et al.* 1961; Mastroianni, Shah, and Abdul-Karim 1961). It is difficult to believe that this amount of fluid could be achieved by the release of secretory granules only; a good portion of the luminal fluid contents must be produced in other ways. Contribution from the peritoneal fluid (Gompper 1950) or some type of transudation through the epithelium (Glass 1963) would explain the close correlation between the protein components of oviductal fluid and serum (Marcus 1964; Marcus and Saravis 1965).

B. *Epithelial Activities Relative to the Ovum*

Except for the many studies on the rabbit oviduct and the essential mucin coating of its egg, little attention has been paid to the histochemical interrelation between the fertilized egg and the surrounding epithelium.

However, there are reports which indicate that the epithelium is more active in the vicinity of the egg. Thus, as the rabbit egg passes through the oviduct, there is an increased uptake of radioactive sulfate in the region containing the egg (Friz 1959; Koester 1964).

Using an immunofluorescence technique, Glass (1963) showed that the mouse egg, particularly in the 2- to 4-cell stages, takes up proteins similar to those in the mother's serum. Also, such proteins are accumulated in the epithelium, particularly in the region of the oviduct, containing the egg at the moment. The accumulation of these serum-like proteins in the epithelium is under hormonal control (Glass and McClure 1965). Since these proteins accumulate independently of cell type, the ordinary secretion granules cannot be their source. The transfer of maternal proteins to the egg and the incapacity of the egg to utilize ^{35}S-methionine for protein synthesis (Greenwald and Everett 1959), indicate that the oviductal egg is more dependent on external supply of protein than on internal synthesis.

We do not know how these epithelial activities are timed with the passage of the ovum. Toward the uterus there could be a gradually increasing delay in the response of the oviductal epithelium to the hormonal changes connected with ovulation. Also, the egg may release components inducing the epithelium in its vicinity to exert its specific activities. This would necessitate a resorptive function of the epithelium, which has been suggested a few times for other reasons. The only direct evidence in favor of such an assumption is the report that trypan blue particles are taken up by the nonciliated cells of the rabbit oviduct (Vokaer and Vanderbeken 1960).

VII. The Oviductal Wall

A. *Connective Tissue Elements*

The basement membrane and connective tissue fibers are PAS-reactive in the oviduct as in other places. However, visualized by the PAS-reaction, their appearance is affected by the cycle. In the human oviduct the fibers are thinner and more branched after ovulation than before, and the basement membrane becomes very thin, particularly beneath the nonciliated cells (Balboni 1955). Similar changes of the basement membrane have been demonstrated in the mouse uterus (Nilsson and Wirsén 1963).

Although connective tissue cells containing PAS-reactive granules are present in varying numbers, particularly during the luteal phase, their function is not established (Balboni 1955).

Alkaline phosphatase is quite active in the innermost layers of the isthmic muscle wall of the rat (Augustin, Heidenreich, and Thilo 1954; Augustin and Moser 1955). In cattle thin connective tissue fibers display phosphatase activity in the ampullary region (Fig. 3*c*) and even more in the isthmus.

Phosphatase activity of the capillary walls is reported in man, rat, and hamster. This activity appears decreased during the luteal phase and after menopause in the woman, but before ovulation and during pregnancy the activity is higher (Augustin and Huwald 1956).

B. *Musculature*

Glycogen is present in the smooth muscle wall of the human oviduct with more in the inner circular layer than in the outer longitudinal layer (Joel 1939*a*). The amount of glycogen present shows cyclical changes (Iwata 1929; Joel 1939*a*, 1940; Fredricsson 1959*b*), being low postmenstrually, rather high postovulatory, and particularly high during pregnancy and in the early puerperium. These variations are very similar to those in the human myometrium (Brody 1958).

A number of enzymes have been demonstrated in oviductal muscle but proof of a correlation to the sexual cycle is lacking. Activity of succinodehydrogenase and DPN-diaphorase has been demonstrated in man (Foraker and Crespo 1962). Amylophosphorylase, which participates in the synthesis of glycogen, occurs also in man (Takeuchi, Higashi, and Watanuki 1955; Foraker and Crespo 1962). Succinodehydrogenase activity in the rat

(Padykula 1952), lipase in man (Sermann and Rigano 1960) and an esterase in the dog (Nachlas and Seligman 1949) have been reported also.

The subject of the nervous regulation of the oviductal musculature is not discussed here, although important contributions to this field have been obtained by histochemical methods (cf. Brundin, this volume, Chapter 10).

Acknowledgment

This study has been made possible by a grant from the Åhlén Foundation, Stockholm, Sweden.

References

1. Alamanni, V. 1956. La fosfatasi alcalina nella salpinge di donna. *Riv. Ostet. Ginec.* 9:496.
2. Augustin, E.; Heidenreich, O.; and Thilo, A. 1954. Vorkommen und Aktivität der Phosphomonoesterasen im Genitaltrakt der weiblichen Ratte und im Blutserum und ihre Beeinflussung durch Ovarialhormone. *Arch. Gynaek.* 184:281.
3. Augustin, E., and Huwald, R. 1956. Vorkommen und Aktivität der alkalischen Phosphatase im Eileiter des Weibes. *Arch. Gynaek.* 187:406.
4. Augustin, E., and Moser, A. 1955. Vorkommen und Aktivität von alkalischer Phosphatase im Eileiter der Ratte und in unbefruchteten und befruchteten Eiern während der Tubenwanderung. *Arch. Gynaek.* 185:759.
5. Balboni, G. 1953. Ulteriori ricerche istochimiche sull' epitelio tubarico della donna. *Boll. Soc. Ital. Biol. Sper.* 29:1394.
6. ———. 1954. Ricerche istochimiche sull'epitelio tubarico della donna. *Riv. Ostet. Ginec.* 9:164.
7. ———. 1955. Lo stroma tubarico in rapporto al ciclo mestruale ed al momento istofunzionale delle cellule dell'epitelio. *Riv. Biol.* 47:153.
8. Björkman, N., and Fredricsson, B. 1960. The ultrastructural organization and the alkaline phosphatase activity of the epithelial surface of the bovine Fallopian tube. *Z. Zellforsch.* 51:589.
9. ———. 1961. The bovine oviduct epithelium and its secretory process as studied with the electron microscope and histochemical tests. *Z. Zellforsch.* 55:500.
10. ———. 1962. Ultrastructural features of the human oviduct epithelium. *Int. J. Fertil.* 7:259.
11. Boström, H., and Odeblad, E. 1952. Autoradiographic observations on the uptake of S^{35} in the genital organs of the female rat and rabbit after injection of labeled sodium sulphate. *Acta Endocr.* 10:89.
12. Brody, S. 1958. Hormonal influence on the glycogen content of the human myometrium. *Acta Endocr.* 27:377.
13. Bronzetti, P.; Mazza, E.; Milio, G.; and Motta, P. 1963. Su alcuni aspetti dell'attività fosfatasica alcalina e della P.A.S.-reattività dell'epitelio della tuba uterina di Mus Rattus albinus. *Biol. Lat.* 16:385.
14. Clark, S. L., Jr. 1961. The localization of alkaline phosphatase in tissues of mice, using the electron microscope. *Amer. J. Anat.* 109:57.

15. Clyman, M. J. 1966. Electron microscopy of the human Fallopian tube. *Fertil. Steril.* 17:281.
16. Deane, H. W. 1952. Histochemical observations on the ovary and oviduct of the albino rat during the estrous cycle. *Amer. J. Anat.* 91:363.
17. Fawcett, D. W., and Wislocki, G. B. 1952. Histochemical observations of the human Fallopian tube. *J. Nat. Cancer Inst.* 12:213.
18. Flerkó, B. 1954. Die Epithelien des Eileiters und ihre hormonalen Reaktionen. *Z. Mikroskopischanat. Forsch.* 61:99.
19. Foraker, A. G., and Crespo, J. Z. 1962. Oxidative and glycogen-synthesizing enzymes in the uterine tube. *Obstet. Gynec.* 19:64.
20. Foraker, A. G.; Denham, S. W.; and Celi, P. A. 1953. Dehydrogenase activity. II. In the Fallopian tube. *Obstet. Gynec.* 2:500.
21. Fredricsson, B. 1957. Histochemical studies on the epithelium in the human Fallopian tubes and comparison between different animal species. *Ark. Zool. Ser. 2.* 11:110.
22. ———. 1959*a*. Proliferation of rabbit oviduct epithelium after estrogenic stimulation with reference to the relationship between ciliated and secretory cells. *Acta Morph. Neerl. Scand.* 2:193.
23. ———. 1959*b*. Histochemical observations on the epithelium of human Fallopian tubes. *Acta Obstet. Gynec. Scand.* 38:109.
24. ———. 1959*c*. Studies on the morphology and histochemistry of the Fallopian tube epithelium. *Acta Anat. suppl.* 37, vol. 38.
25. Fredricsson, B., and Björkman, N. 1962. Studies on the ultrastructure of the human oviduct epithelium in different functional stages. *Z. Zellforsch.* 58:387.
26. Fredricsson, B., and Björkman, N. 1968. Distribution of glycogen within the mammalian oviduct epithelium. *Sixth Int. Cong. Anim. Reprod. A.I., Paris.*
27. Friz, M. 1959. Tierexperimentelle Untersuchungen zur Frage der Tubensekretion. *Z. Geburtsh. Gynaek.* 153:285.
28. Glass, L. E. 1963. Transfer of native and foreign serum antigens to oviducal mouse eggs. *Amer. Zool.* 3:135.
29. Glass, L. E., and McClure, T. R. 1965. Postnatal development of the mouse oviduct: Transfer of serum antigens to the tubal epithelium. In *Preimplantation stages of pregnancy*, ed. G. E. W. Wolstenholme and M. O'Connor, p. 294. London: J. & A. Churchill Ltd.
30. Goldberg, B., and Jones, H. W., Jr. 1953. Acid phosphatase in human female genital tract, a histochemical and biochemical study. *Proc. Soc. Exp. Biol. Med.* 83:45.
31. Gompper, H. J. 1950. Das Sekret des Eileiters. *Anat. Anz.* 97:391.
32. Graumann, W. 1964. Ergebnisse der Polysaccharidhistochemie. In *Handbuch der Histochemie*, ed. W. Graumann and K. Neumann, vol. 2, part 2, p. 403. Stuttgart: Gustav Fischer Verlag.
33. Greenwald, G. S. 1958. Endocrine regulation of the secretion of mucin in the tubal epithelium of the rabbit. *Anat. Rec.* 130:477.
34. Greenwald, G. S., and Everett, N. B. 1959. The incorporation of S^{35} methionine by the uterus and ova of the mouse. *Anat. Rec.* 134:171.
35. Hadek, R. 1953. Mucin secretion in the ewe's oviduct. *Nature* 171:750.
36. ———. 1955. The secretory process in the sheep's oviduct. *Anat. Rec.* 121:187.

37. Høllund, T. 1946. Tuba- og Uterinslimkindens Epithel. Thesis, Nyt Nordisk Forlag. Copenhagen: Arnold Busek.
38. Hübner, K. A. 1955. Histo-enzymatische Untersuchungen über Vorkommen und Bedeutung der Hyaluronsäure im weiblichen Genitaltrakt. *Z. Ges. Exp. Med.* 125:236.
39. Iwata, M. 1929. Beiträge zur Morphologie der menschlichen Tube. *Mschr. Geburtsh. Gynaek.* 81:283.
40. Joel, C. A. 1940. Zur Histologie und Histochemie der menschlichen Eileiter während Zyklus und Schwangerschaft. *Mschr. Geburtsh. Gynaek.* 110:252.
41. Joel, K. 1939*a*. The glycogen content of the Fallopian tubes during the menstrual cycle and during pregnancy. *J. Obstet. Gynaec. Brit. Emp.* 46:721.
42. ———. 1939*b*. The lipoid content of the Fallopian tubes during the menstrual cycle and during pregnancy. *J. Obstet. Gynaec. Brit. Emp.* 46:731.
43. Jonek, J. 1957. Histochemical investigations on the appearance of acetal phosphatides in the reproductive organs in women and its relation to the hormonal function of the ovaries. *Ginek. Pol.* 28:641.
44. Jonek, J., and Glenc, F. 1963. The reactions for alkaline phosphatase, acid phosphatase, and ATP-ase in the mucosa of the uterus in women during the climacterium. *Endokr. Pol.* 14:85.
45. Jonek, J.; Zieliński, Z.; Kochańska, D.; and Dzieciuchowicz, L. 1966. The localization of γ-glutamylo-transpeptidase in the mucosa of the uterus and oviducts in various laboratory animals. *Ginek. Pol.* 37:349.
46. Józsa, L., and Szederkényi, G. 1967. Über Verluste der Gewebsmukopolysaccharide während der Fixierung. *Acta Histochem.* 26:255.
47. Koch, W. 1941. Das Verhalten der Plasmalreaktion im Genitaltractus der weiblichen weissen Maus. *Z. Mikroskopischanat. Forsch.* 50:465.
48. Koester, H. 1964. Tierexperimentelle Untersuchungen zur Frage der Tubensekretion. *Beitr. Fertil. Steril. IV.* (Suppl. to *Z. Geburtsh. Gynaek.* vol. 162), p. 63.
49. Kraus, R.; Vacek, Z.; and Jirsová, Z. 1964. On the decidual transformation of the oviduct mucosa. *Cesk. Morf.* 12:74.
50. Leblond, C. P. 1950. Distribution of periodic acid-reactive carbohydrates in the adult rat. *Amer. J. Anat.* 86:1.
51. Levy, H.; Deane, H. W.; and Rubin, B. L. 1959. Visualization of steroid-3β-ol-dehydrogenase activity in tissues of intact and hypophysectomized rats. *Endocrinology* 65:932.
52. Lillie, R. D. 1949. Studies on the histochemistry of normal and pathological mucins in man and in laboratory animals. *Bull. Int. Ass. Med. Mus.* 29:1.
53. Marcus, S. L. 1964. Protein components of oviduct fluid in the primate. *Surg. Forum* 15:381.
54. Marcus, S. L., and Saravis, C. A. 1965. Oviduct fluid in the rhesus monkey: A study of its protein components and its origin. *Fertil. Steril.* 16:785.
55. Mastroianni, L., Jr.; Beer, F.; Shah, U.; and Clewe, T. H. 1961. Endocrine regulation of oviduct secretions in the rabbit. *Endocrinology* 68:92.
56. Mastroianni, L., Jr.; Shah, U.; and Abdul-Karim, R. 1961. Prolonged volumetric collection of oviduct fluid in the rhesus monkey. *Fertil. Steril.* 12:417.
57. Mihálik, P. von 1935. Über die Bildung des Flimmerapparates im Eileiterepithel. *Anat. Anz.* 79:259.

58. Milio, G. 1960. Attività fosfatasi alcalina e PAS-reattività dell'epitelio tubarico di Rattus albinus. *Boll. Soc. Ital. Biol. Sper.* 36:394.

59. Moog, F., and Wenger, E. L. 1952. The occurrence of a neutral mucopolysaccharide at sites of high alkaline phosphatase activity. *Amer. J. Anat.* 90:339.

60. Moricard, R., and Gothie, S. 1956. L'utilisation des traceurs (^{32}P et ^{35}S) dans l'étude de la maturation ovulaire, de la fécondation et de la segmentation de l'oeuf. *Ann. Ostet. Ginec.* 78:106.

61. Moschino, A. 1954. La fosfatasi alcalina nella tuba umana. Citotopografia e significato. *Riv. Ostet.* 36:21.

62. Nachlas, M. M., and Seligman, A. M. 1949. The comparative distribution of esterase in the tissues of five mammals by a histochemical technique. *Anat. Rec.* 105:677.

63. Neumann, K.; Oehlert, G.; and Hansmann, H. 1954. Histochemische Lokalisation des Enzyms Phosphoamidase im weiblichen Genitaltrakt und Vaginalschleim. *Z. Geburtsh. Gynaek.* 141:109.

64. Nicander, L.; Björkman, N.; Hellström, B.; and Selander, U. 1968. On the presence of glycogen inside ciliary basal bodies of bovine genital organs. (In preparation.)

65. Niklas, A., and Oehlert, W. 1956. Autoradiographische Untersuchung der Grösse des Eiweissstoffwechsels verschiedener Organe, Gewebe und Zellarten. *Beitr. Path. Anat.* 116:92.

66. Nilsson, O., and Wirsén, C. 1963. The effect of estrogen on the histology of the uterine epithelium of the mouse. II. Changes of PAS-reactive structures in the basement membrane and the glandular cell surface. *Exp. Cell. Res.* 29:144.

67. Padykula, H. A. 1952. The localization of succinic dehydrogenase in tissue sections of the rat. *Amer. J. Anat.* 91:107.

68. Papanicolaou, G. N.; Traut, H. F.; and Marchetti, A. A. 1948. *The epithelia of woman's reproductive organs: A correlative study of cyclic changes.* New York: Harvard University Press.

69. Robboy, S. J., and Kahn, R. H. 1964. Electrophoretic separation of hydrolytic enzymes of the female rat reproductive tract. *Endocrinology* 75:97.

70. Rosa, C. G., and Tsou, K. C. 1962. Cited by Velardo and Rosa 1963.

71. Sani, G., and Hanau, R. 1952. La fosfatasi alcalina nell'apparato genitale femminile del ratto in rapporto al ciclo sessuale ed alla somministrazione di ormoni. *Arch. Ital. Anat. Embriol.* 57:211.

72. Schäfer, G., and Roloff, H. E. 1950. Histochemische Untersuchungen über das Vorkommen von Plasmalogen im weiblichen Genitaltraktus. *Zbl. Gynaek.* 72:1583.

73. Schultka, R. 1963. Der Sekretionscyklus der Flimmerzellen der menschlichen Tuba uterina auf Grund cytologischer und cytotopochemischer Untersuchungen. *Acta Histochem.* 15:285.

74. Schultka, R., and Scharf, J. H. 1963. Sekretionszyklus der Tubenepithelzelle in Abhängigkeit vom ovariellen Zyklus. *Zbl. Gynaek.* 85:1601.

75. Sermann, R., and Rigano, A. 1960. Attività lipasica e metabolismo lipidico nella salpinge umana. *Arch. Ostet. Ginec.* 65:702.

76. Takeuchi, T.; Higashi, K.; and Watanuki, S. 1955. Distribution of amylophosphorylase in various tissues of human and mammalian organs. *J. Histochem. Cytochem.* 3:485.

77. Tsuchiya, K. 1956. Cytologische und histologische Untersuchungen über das Epithel des menschlichen Eileiters. *Arch. Histol. Jap.* 10:243.

78. Varveri, A. 1953. Il comportamento dell'epitelio della tuba di coniglia in estro di frone alla reazione di Hotchkiss McManus. *Boll. Soc. Ital. Biol. Sper.* 29:450.
79. Velardo, J. T., and Rosa, C. G. 1963. Female genital system. In *Handbuch der Histochemie*, ed. W. Graumann and K. Neumann, vol. 7, part 3, p. 45. Stuttgart: Gustav Fischer Verlag.
80. Vokaer, R., and Vanderbeken, Y. 1960. Étude histophysiologique des trompes de Fallope pendant la grossesse et le post-partum. *Bull. Soc. Roy. Belg. Gynec. Obstet.* 30:55.
81. Weeth, H. J., and Herman, H. A. 1950. A histological and histochemical study of the bovine oviducts, uterus and placenta. *Res. Bull. Missouri Agric. Exp. Station* 501:1.
82. Werner, I. 1953. Studies on glycoproteins from mucous epithelium and epithelial secretions. *Acta Soc. Med. Upsal.* 58:1.
83. Westman, A. 1930. Studies of the function of the mucous membrane of the uterine tube. *Acta Obstet. Gynec. Scand.* 10:288.
84. Westman, A.; Jorpes, E.; and Widström, G. 1931. Untersuchungen über den Schleimhautzyklus in der Tuba uterina, seine hormonale Regulierung und die Bedeutung des Tubensekrets für die Vitalität der befruchteten Eier. *Acta Obstet. Gynec. Scand.* 11:279.
85. Wolman, M. 1964. Lipids. II. Histochemistry of lipids in pathology. In *Handbuch der Histochemie*, ed. W. Graumann and K. Neumann, vol. 5, p. 459. Stuttgart: Gustav Fischer Verlag.
86. Zachariae, F. 1958. Autoradiographic (^{35}S) and histochemical studies of sulphomucopolysaccharides in the rabbit uterus, oviducts and vagina. *Acta Endocr.* 29:118.

13

Biochemistry of Oviductal Secretions

C. E. Hamner/S. B. Fox
Division of Reproductive Biology
Department of Obstetrics and Gynecology
School of Medicine
University of Virginia, Charlottesville

The oviduct holds a strategic position in the reproductive process. It serves as a transport organ between the ovary and the uterus and provides shelter and nourishment for the sperm and eggs before, during, and after fertilization. The oviduct is capable of capacitating sperm, and its secretions are necessary for normal segmentation and survival of the fertilized egg (Adams and Chang 1962; Hamner and Sojka 1967; Whitten 1930; Westman 1930).

Several reviews have indicated that the cyclic nature of the oviduct depends on the endocrine state of the animal (Schaffer 1908; Moreaux 1913; Novak and Everett 1928; Bourg 1931; Ober 1957; Olds and VanDemark 1957*c;* Lowi 1960; Mastroianni 1962; Lehto 1963). The aims in this review are to discuss the methods of collecting oviductal secretions; to bring together the many observations on the composition of this fluid; to indicate the influence of estrogen and progesterone on these secretions, and to comment on the effects of oviductal fluid on the gametes.

I. Collection

A. *Comparison of Methods*

The secretions of the oviduct have been investigated from many vantage points. Methods include the utilization of the light microscope with histochemical techniques and electron-microscopic studies of fresh specimens and of explants from tissue culture while the secretions are being formed in the oviductal epithelium during the estrous cycle.[1] The

[1] Frommel 1886; Schaffer 1908; Moreaux 1910; Jagerroos 1912; Moreaux 1913; Argand 1921; Allen 1922; Courrier and Gerlinger 1922; Sheyer 1926; Novak and Everett 1928; Guerriero 1929*a*, *b;* Westman 1930; Casida and McKenzie 1932; Caffier 1938; Joel 1941; Leblond 1950; Lombard, Morgan, and Menutt, 1950; Confalonieri 1951; Fawcett and Wislocki 1951; Deane 1952; Moog and Wenger 1952; Balboni 1953; Flerkó 1954; Hadek 1955; Borell *et al.* 1956; Hellström and Nilsson 1957; Nilsson and Hellström 1957; Greenwald 1958; Fredricsson 1959; Stegner 1961; Björkman and Fredricsson 1962; Hashimoto *et al.* 1962; Lehto 1963; Bousquet 1964; Hashimoto *et al.* 1964; Clyman 1966.

obvious difficulty with the histochemical approach is determining what materials are actually extruded into the lumen of the oviduct and what happens to the secretions once they are excreted. In order to observe the secretions in the lumen of the oviduct various investigators have tried utilizing radioisotopes (Zachariae 1958; Friz 1959), flushing the genital tract (Bond 1898; Heap 1962), stripping the oviduct after slaughter (Olds and VanDemark 1957*a*, *b;* White and Wallace 1961) and ligating the oviduct at the uterotubal junction and fimbriae (Woskressensky 1891; Bishop 1956*a;* Green 1957; Gregoire, Gongsakdi, and Rakoff 1961, 1962; Vishwakarma 1962; Gregoire and Rakoff 1963; Marcus 1964). Nonspecific labeling may occur in utilizing radioisotopes. Flushing the genital tract eliminates the possibility of measuring concentrations of components, and stripping the oviduct can cause damage to the epithelium and contaminate secretions with cytoplasm of other tissue. Double ligation is self-limited by the stretching and flattening of the cells and by the occlusion of the blood supply with fluid accumulation.

In 1956 Bishop cannulated the fimbriated end of the oviduct and connected the tubing to a manometer to measure the secretory pressure in the rabbit oviduct. This technique led to the development of various methods for the continuous collection of oviductal fluid which would be unadulterated by tissue reaction, cellular debris, or bacteria, and not inhibited by pressure. The cannulae from the oviduct are connected to an extra-abdominal or intra-abdominal collecting device (Clewe and Mastroianni 1960; Mastroianni, Beer, and Clewe 1960; Mastroianni and Wallach 1961; Mastroianni *et al.* 1961; Mastroianni, Shah, and Abdul-Karim 1961; Black, Duby, and Riesen 1963; Sugawara and Takeuchi 1964; Perkins *et al.* 1965; Edgerton *et al.* 1966; Perkins and Goode 1966; Restall 1966*a;* Hamner and Williams 1963, 1965). The method of continuous collection of oviductal fluid allows one to obtain large quantities for biochemical analysis, for studies in *in vitro* fertilization, embryo development, and metabolism of both sperm and eggs. Also the influence of various hormonal states on the female tract secretions can be investigated easily. The original extra-abdominal device (Clewe and Mastroianni 1960) has been modified (Holmdahl and Mastroianni 1965) to refrigerate the collecting device, enabling it to hold the accumulating fluid at +2° C to +4° C. This method prevents the enzymatic digestion or bacterial breakdown of the fluid which may occur in techniques where the oviductal fluid accumulates at body temperature.

An improved version of the original intra-abdominal device (Hamner and Williams 1963) has been developed which houses the collecting adapter of the flask in a hood so as to protect it from the cage environment (Hamner and Williams 1965). The hood greatly reduces the problem of keeping the flask sterile. The intra-abdominal device has the advantage of preventing the animal from biting or scratching off the oviductal cannulae. Another advantage is that it does not kink, collapse, or break.

Installation of the collecting device in the rabbit is accomplished as follows: A New Zealand White female rabbit, weighing 4 to 5½ kg, is anesthetized with sodium pentobarbital. The abdominal and flank areas of the rabbit are prepared for surgery. The animal is draped and the initial incision is made at the linea alba. The uterine horns of the rabbit are identified, and the uterotubal junction is located and doubly ligated; the area between the ligations is then transected. The fimbriated end of the oviduct is located and grasped gently between the thumb and forefinger. A curved P.E. 100 polyethylene tubing, which has been bulbed, is placed into the fimbriated end of the left oviduct and sutured in place. The suture used throughout this procedure is a synthetic protein known as Supramid, catgut size 0 (Jensen-Salsbery Laboratories). After the tubing is placed in the lumen of the oviduct, the ovary may be removed if the procedure or experimental design so dictates. A 20-gauge disposable needle is inserted into the polyethylene tube for attachment to the oviductal fluid collecting flask. A straight, bulbed P.E. 100 polyethylene tube is inserted into the fimbriated end of the right oviduct and sutured in place. After the infundibulum of the right oviduct has been cannulated, the ovary is removed and a 20-gauge needle is inserted into the cannula. The animal is then rotated from left to right in order to expose the left flank for the flank stab wound. The abdominal cavity is viewed through the initial incision out toward the flank to visualize the blood vessels that run across the flank area of the rabbit. Exercising care to avoid these vessels, one makes the stab wound about 10 cm below the backbone and about 2½ cm posterior to the last rib. The animal is rotated back to its original position and the disposable needle adapters are connected to the oviductal fluid collecting flask.

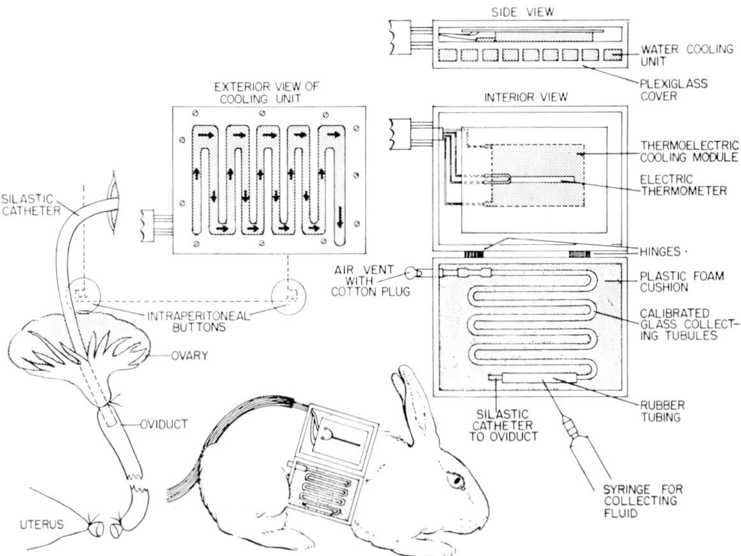

FIG. 1. Schematic representation of refrigerated extra-abdominal collecting system for oviductal fluid. (Adapted from Holmdahl and Mastroianni 1965.)

(At this point when the extra-abdominal device is used [Fig. 1], the cannulae are carried out through the stab wound and attached to the external collecting device, which is sutured in place to the external abdominal wall.) When the adapters are connected to the oviductal fluid collecting device, the device is then inserted into the abdominal cavity and the retrieving portion, or gooseneck section of the flask, is carried out through the stab wound in the flank area. The abdominal wall is closed by 2 layers of sutures, the first layer being simple interrupted sutures through the muscle and peritoneum and the second layer being through the skin. One should be sure to attach the skin to the underlying muscle area so as to avoid pockets which may become infected. The stab wound is closed in 2 layers around the neck of the collecting device in a manner similar to that for closing the abdominal cavity. Neomycin-sulpha powder should be placed around the stab wound area for at least 5 days following surgery or until healing is completed.

Collection of the oviductal fluid is accomplished by removing the front cotton ball from the flask globe and attaching a sterile syringe to the collecting adapter. After the fluid is withdrawn, a new sterile cotton ball is inserted into the flask globe (Fig. 2).

Oviductal fluid is a combination of a secretion and a transudate. Evidence is against the suggestion that oviductal fluid arises as an inflow of peritoneal fluid or simply as a blood filtrate. Fluid accumulates between the ligatures of the oviduct to rule out inflow of peritoneal or uterine fluids (Woskressensky 1891; Bishop 1957; Gregoire, Gongasakdi, and Rakoff 1961).

The rabbit oviduct will secrete against pressure up to 71 cm of water (Bishop 1956*a*). Both secretory rate and pressure are highest during estrus (average 0.79 ml/day/oviduct; 46 cm water pressure). Secretory activity is severely reduced after ovariectomy (0.14 ml/day/oviduct; 11.8 cm water pressure), but this decrease can be counteracted by administration of 5 μg/day of estradiol benzoate.

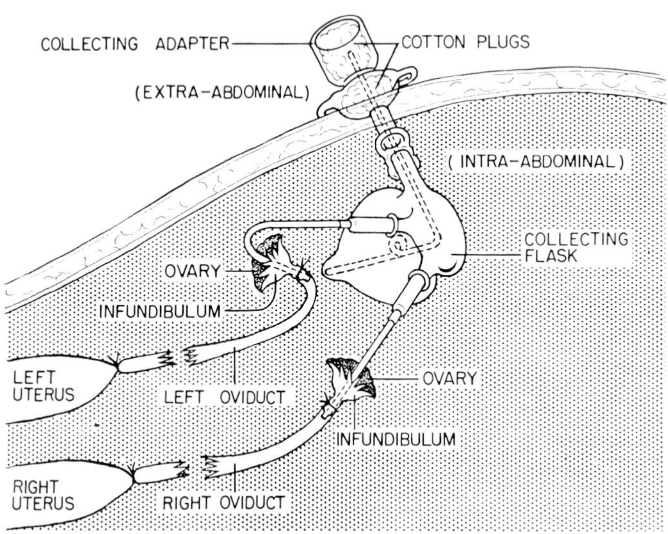

FIG. 2. Schematic representation of installed intra-abdominal oviductal fluid collecting flask

During estrus the secretory cells of the rabbit, rat, mouse, sheep, cow, and human oviduct epithelium produce granules which are extruded into the lumen at the time of ovulation (Novak and Everett 1928; Espinasse 1935; Kneer, Burger, and Simmer 1952; Borell *et al.* 1956; Fredricsson 1959; Clyman 1966). This secreted material is periodic acid-Schiff reactive and diastase resistant and appears to be a carbohydrate-protein complex (Leblond 1950; Braden 1952; Bacsich and Hamilton 1954; Greenwald 1958). Succinic dehydrogenase, an enzyme associated with tissues having high metabolic activity or engaging in absorptive or secretory activity (Padykula 1952), has been identified in the oviductal epithelium of man (Foraker, Denham, and Celi 1953).

On the other hand, oviductal fluid from rhesus monkeys contains albumin, alpha-1, alpha-2, beta, and gamma globulins. With the exception of gamma globulin these components are present in the same proportions of total protein as in serum (Marcus 1964). This information suggests that transudation of these proteins from the vascular system contributes to the formation of oviductal fluid.

B. *Hormonal Influence on Secretion Rates*

Oviductal secretion rate in the rabbit drops to 50% of the estrous level (1.6 ml/day/oviduct) by the 3d day of pregnancy and continues to fall until the 24th day of pregnancy to 0.3 ml/day/oviduct (Bishop 1956*a;* Mastroianni and Wallach 1961). Even though the rabbit oviductal secretion rate is very sensitive to estrogen, the estrogen surge reported in rats and mice at the time of implantation was not demonstrated in the rabbit by volume increase (Mastroianni and Wallach 1961). Ovariectomy of rabbits significantly reduces secretion rates from the estrous state and injection of 3 mg progesterone/day into castrate rabbits has no effect on the secretion rate. Injection of progesterone into estradiol-primed castrates causes a significant drop in the secretion rate by the 2d day. Injection of 5 μg estradiol benzoate into ovariectomized rabbits restores the secretion rate to precastration level within 48 hr (Mastroianni 1962; Mastroianni, *et al.* 1961). Oviductal fluid secretion rates in sheep range from 1.4 ml/day/oviduct at estrus to 0.4 ml/day/oviduct during the luteal phase (Black, Duby, and Riesen 1963; Perkins *et al.* 1965; Perkins and Goode 1966; Restall 1966*b*). The sow secretes at a maximum rate of 5 ml/day/oviduct at estrus (Edgerton *et al.* 1966). Rhesus monkeys have a very low rate of secretion at menses and up to 2 to 4 days prior to the peak of cornification, when secretion reaches 3.4 ml/day/oviduct (Mastroianni, Shah, and Abdul-Karim 1961).

Work in our laboratory with intact and ovariectomized rabbits indicates that intact does secrete an average volume of 1.2 ml/day/oviduct. The volume secreted ranges from 0.9 ml/day/oviduct to 1.4 ml/day/oviduct with the maximum rate being reached every 6½ days for the first 3 weeks. After the 3d week the periodicity of secretion becomes less distinct and tends to level out at a rate of 0.7 ml/day/oviduct (Fig. 3). This change in the secretion pattern may indicate a disturbance of the normal secretory physiology of the oviduct or a change in the rabbit's hormonal state. Subcutaneous injection of ovariectomized rabbits with 1 μg estradiol cypionate[2] (ECP)/kg/day in 1.0 ml cottonseed oil produces secretion rates equivalent to the intact estrous rabbit (average 1.2 ml/day/oviduct). Injection of 0.5 mg progesterone/kg/day in 1.0 ml cottonseed oil in conjunction with 0.5 μg ECP/kg/day to estrogen-primed rabbits reduces the secretion rate 50% (Fig. 3). Continuous injection of 1.0 mg progesterone/kg/day alone or 1.0 ml cottonseed oil/day reduces secretions to 1/6 (0.2 ml/day/oviduct) the estrogen-primed rate.

Ovariectomized rabbits were allowed to stabilize for 4 days before starting continuous subcutaneous injections of (1) 1 μg ECP/kg/day in 1.0 ml cottonseed oil or (2) 1.0 mg progesterone/kg/day in 1.0 ml cottonseed oil or (3) 1.0 ml cottonseed oil alone (Fig. 4). The ECP-injected rabbits increased their secretion 2½ times the first 48 hr after the initial injection and then dropped off slightly to an average rate of 1.2 ml/day/oviduct. Progesterone produced a slow decline in the secretion rate, reaching the cottonseed oil treatment level by the fourteenth day and remaining at that level. Cottonseed oil injection decreased the secretion rate to 0.25 ml/day/oviduct and the rate remained at this level indefinitely.

II. Composition

A. *Physical Properties*

The physical properties of oviductal fluid have not captured the interest of many investigators. However, a knowledge of such properties as viscosity, osmolarity, and specific gravity is essential for the development of a defined medium for *in vitro* fertilization and ova cleavage studies up to the blastocyst stage.

Oviductal fluid from women has been described as a colorless, watery fluid, neutral or slightly alkaline, of very low specific gravity, and containing albumin and mucin (Bond

[2] Estradiol cypionate is the product name used by the Upjohn Company.

1898). The oviductal fluid of the cow is 13.6% dry matter, and has a neutral to slightly acid pH and an osmolarity of 350 m osmols (Olds and VanDemark 1957*a*, *b*).

We determined that rabbit oviductal fluid is a clear, watery, alkaline fluid having a specific gravity of 0.9965, a viscosity of 0.0108 poise, an osmolarity of 375 m osmols, and a dry matter content of 13.1%. In ovariectomized rabbits injected subcutaneously with (1) 1 μg ECP/kg/day in 1.0 ml cottonseed oil, (2) 0.5 μg ECP/kg/day plus progesterone 0.5 mg/kg/day in 1.0 ml cottonseed oil, (3) progesterone 1 mg/kg/day in 1.0 ml cottonseed oil or (4) 1.0 ml cottonseed oil alone, there is no significant change in specific gravity or

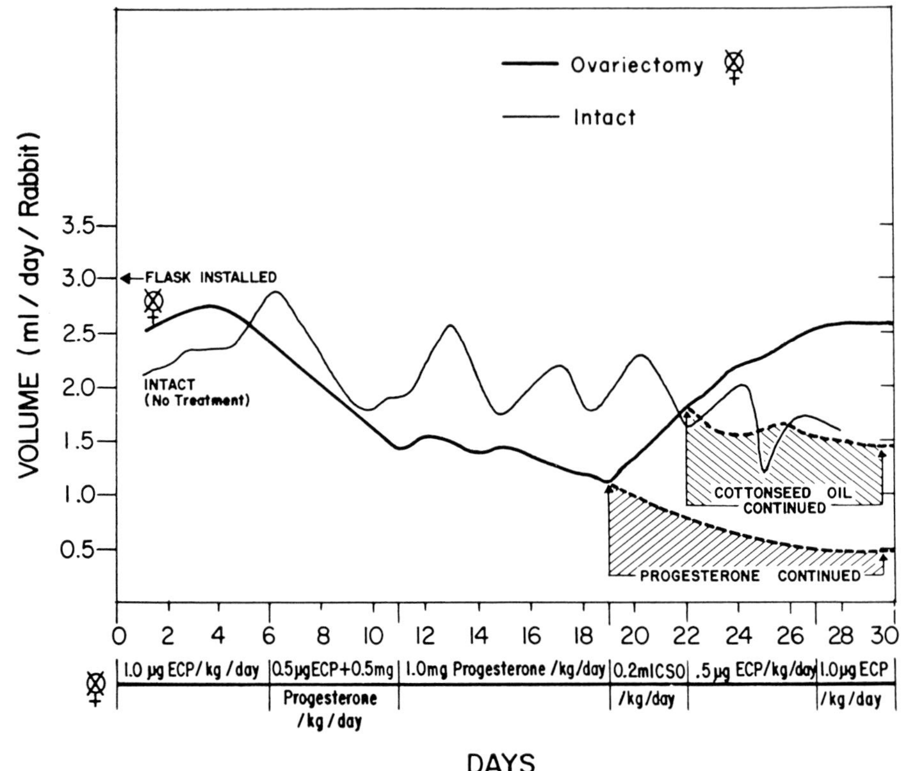

FIG. 3. Influence of sequential estrogen and progesterone on rabbit oviductal fluid secretion rate (5 rabbits/group).

viscosity of the oviductal fluid (Table 1). However, progesterone-injected rabbits have significantly more dry matter than ECP- or ECP plus progesterone-injected rabbits and the osmolarity of cottonseed oil-injected rabbits is significantly higher than that of ECP-treated rabbits.

B. *Constituents Identified*

Histochemical techniques have identified at least 4 types of cells in the oviductal epithelium: ciliated, nonciliated (secretory), reserve (round or club), and rod cells (peg) in the human, rabbit, rat, pig, cow, horse, sheep, dog, and cat (Courrier and Gerlinger 1922; Novak and Everett 1928; Lombard, Morgan, and Menutt 1950; Balboni 1953; Flerkó 1954; Hellström and Nilsson 1957; Nilsson and Hellström 1957). Three of the 4 types of

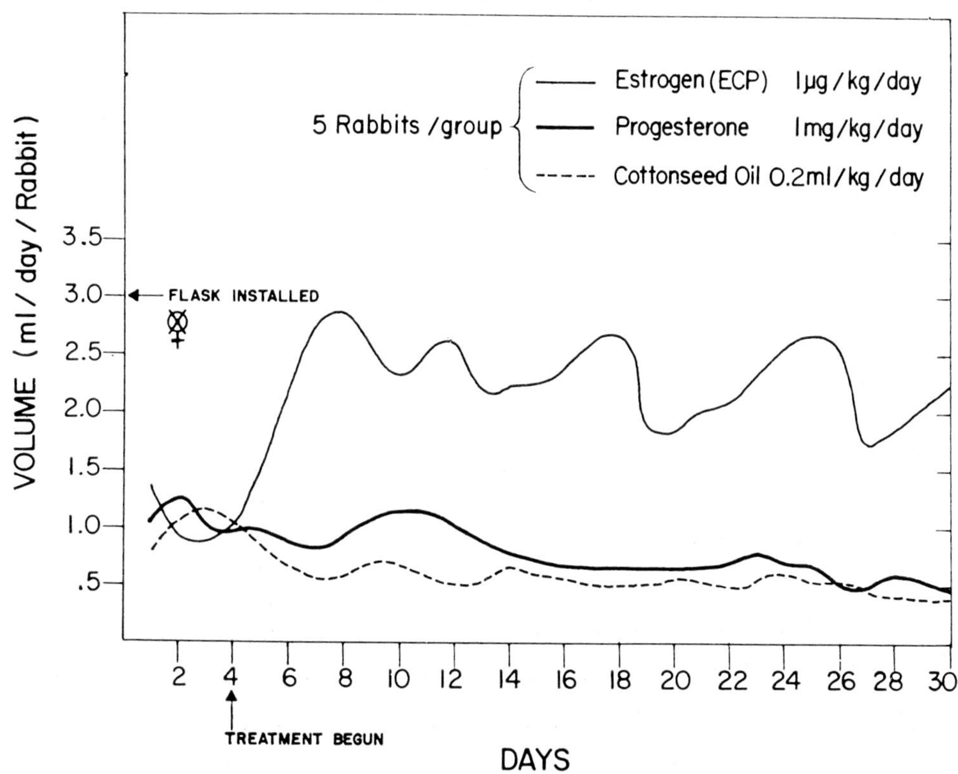

FIG. 4. Influence of continued estrogen and progesterone treatment on rabbit oviductal fluid secretion rate.

TABLE 1

PHYSICAL PROPERTIES OF OVIDUCTAL FLUID FROM OVARIECTOMIZED RABBITS

PHYSICAL PROPERTY	HORMONAL TREATMENT			
	Estrogen	Progesterone +Estrogen	Progesterone	Cottonseed Oil
Dry matter (mg/ml)......	11.9[a]	12.4	14.8[b]	13.4
Specific gravity..........	0.9962	0.9970	0.9959	0.9968
Viscosity (poise)..........	0.0112	0.0109	0.0111	0.0110
Osmolarity (m osmols)....	352	383	376	388[c]

[a] Values are means for 10 determinations for each hormonal treatment.

[b] Dry matter of progesterone-treated rabbits is significantly higher than in rabbits treated with ECP or ECP + progesterone. $P < 0.01$ by t-test (Snedecor 1961).

[c] Osmolarity of cottonseed oil-treated rabbits is significantly higher than ECP-treated rabbits by t-test: $P < 0.05$.

cells are associated with secretory processes in the epithelium. The round and rod cells are transitory forms of the secretory cells, which are holocrine in nature and very sensitive to estrogen (Argand 1921; Casida and McKenzie 1932; Clyman 1966). Synthesis of secretory materials can be detected histologically within 24 hr after an estradiol injection (Flerkó 1954). The same general pattern for secretion and extrusion in the oviductal epithelium appears in many mammals. In the human, monkey, rabbit, rat, pig, horse, cat, mouse, guinea pig, sheep, cow, dog, goat, and squirrel secretory granules (1 to 5 μ in diameter) are formed during proestrus under the influence of estrogen, reaching a maximum at estrus and crowding the secretory cells so that they bulge into the lumen of the oviduct.[3] After ovulation the secretory material is evacuated from the secretory cells under the influence of progesterone and the loss of estrogen dominance as the ovum passes down the tube. The secretory cells then pass through a resting stage until proestrus when a new secretory cycle begins. The hormonal effect on the height of the oviductal epithelium and production of PAS-positive material increases from the isthmus toward the infundibulum (Novak and Everett 1928; Caffier 1938).

Mitotic figures are almost never seen in the oviductal epithelium, and the normal tube does not contribute to the bleeding of menstruation (Novak and Everett 1928). In woman, sheep, goat, pig, rabbit, squirrel, guinea pig, and mouse, the epithelial cells of the ampulla of the oviduct increase in volume; while the nucleus and cytoplasm increase in a parallel manner. When the cell has attained a certain size, it presents a clear picture of amitosis: it produces a nuclear cleavage by formation of a narrow fissure. This increase in volume and number of epithelial cells causes a compression and an elimination of secretory material from the cells (Courrier 1921).

In the proliferation stage there is increased protein production, evidenced by RNA particles (ribosomes) and dilation of the endoplasmic reticulum with a fine amorphous substance (Björkman and Fredricsson 1962; Clyman 1966). There is also an increase in the concentration of the nonhistone protein fraction of the nucleus during the secretory phase (Lehto 1963). Ascorbic acid content of the oviduct is lower during the proliferation phase than during the secretory phase, but is always much higher in the oviduct than in blood serum (Joel 1941).

The presence (Joel 1941) and absence (Sheyer 1926; Hadek 1955) of lipids in the oviductal epithelium has been reported. Lipid droplets have been found in the lumen of the oviduct during the secretory phase (Tazawa 1958; Fredricsson 1959).

A periodic acid-Schiff, diastase-resistant material which is a carbohydrate-protein complex (Leblond 1950) is present in the secretory cells of the oviductal epithelium of women, rabbits, rats, hamsters, sheep, mice, and cows (Courrier 1921; Fawcett and Wislocki 1951; Balboni 1953; Assheton 1894; Greenwald 1957; Gregory 1930; Braden 1952; Friz 1959; Adams 1960; Leblond 1950; Moog and Wenger 1952; Hadek 1955; Fredricsson 1959). This material has been described as a neutral mucopolysaccharide (Moog and Wenger 1952) and an acid mucopolysaccharide (Hadek 1955). It contains sulfur (Zachariae 1958; Friz 1959) and is generally present at sites that are rich in alkaline phosphatase (Moog and

[3] Frommel 1886; Bouin and Limon 1900; Schaffer 1908; Moreaux 1910, 1913; Jagerroos 1912; Courrier 1921; Courrier and Gerlinger 1922; Novak and Everett 1928; Guerriero 1929*a*, *b*; Tietze 1929; Casida and McKenzie 1932; Espinasse 1935; McKenzie and Terrill 1937; Caffier 1938; Leblond 1950; Lombard, Morgan, and Menutt 1950; Confalonieri 1951; Hadek 1953*b*; Flerkó 1954; Borell *et al.* 1956; Fredricsson 1959; Stegner 1961; Fredricsson and Björkman 1962; Lehto 1963; Bousquet 1964; Hashimoto *et al.* 1964; Clyman 1966.

Wenger 1952). The PAS-positive, diastase-resistant material increases during the follicular phase and is secreted into the lumen of the oviduct after ovulation to coat the egg (Moreaux 1913; Courrier and Gerlinger 1922; Fredricsson 1959). The mucopolysaccharide envelopment of the egg is very pronounced in the rabbit and has been observed in several other mammals (Assheton 1894; Greenwald 1957; Adams 1960).

A thorough study of the luminal fluid of the oviduct has been confined to the rabbit and sheep (Haour, Conti, and Guyot 1956; Bishop 1957; Gregoire, Gongsakdi, and Rakoff 1961; Mastroianni and Wallach 1961; Gregoire and Rakoff 1963; Gregoire, Gongsakdi, and Rakoff 1962; Sugawara and Takeuchi 1964; Hamner and Williams 1965; Perkins and Goode 1966; Restall 1966*a*, *b*; Restall and Wales 1966). One laboratory has reported on the constituents of cow oviductal fluid (Olds and VanDemark 1957*a*, *b*, *d*), and some protein constituents of rhesus monkey oviductal fluid have been described (Marcus 1964; Marcus and Saravis 1965).

TABLE 2

COMPOSITION OF RABBIT OVIDUCTAL FLUID[a]

Component	Value (μg/ml)	Reference
Sodium	3,240 –3,320	2, 7, 8
Chloride	3,880 –4,070	2, 7, 8
Potassium	200 – 240	2, 7, 8
Magnesium	3.13– 3.76	8
Zinc	6.30– 6.66	8
Calcium	160 – 320	7, 8
Phosphate (PO_4^{-3})	3.7 – 6.1	8
pH	7.8 – 8.0	4, 8
Bicarbonate	1,662 –1,862	4, 6, 8
Carbohydrate	60 – 370	2, 8
Lactate	31 – 189	1, 7, 8
Pyruvate	14.5 – 16.7	7
Glucose	0 – 257	1, 7, 8
Inositol	26	3
Phospholipid	0 – 80	1
Urea	600	5
Total protein	2,130 –2,730	7, 8

References: 1. Bishop 1957; 2. Mastroianni and Wallach 1961; 3. Gregoire, Gongsakdi, and Rakoff 1962; 4. Vishwakarma 1962; 5. Gregoire and Rakoff 1963; 6. Hamner and Williams 1964*b*; 7. Holmdahl and Mastroianni 1965; 8. Hamner and Williams 1965.

[a] Intact does.

Constituents which have been identified in rabbit oviductal fluid are sodium, chloride, potassium, magnesium, zinc, calcium, phosphate, bicarbonate, lactate, pyruvate, glucose, inositol, phospholipid, urea, proteins (Table 2), alanine (3.2 mg%), glutamic acid (3.6 mg%), glycine (16.7 mg%), serine (1.1 mg%), and traces of thrionine, tryptophane, methionine, valine, taurine, glutamine (Gregoire, Gongsakdi, and Rakoff 1961), lysine, aspartic acid, and lucine (Hamner and Williams 1965). Other constituents identified are sialic acid (Yamazaki 1965), corona radiata dispersing factor (Shettles 1955; Mastroianni and Ehteshamzadeh 1964), diesterase (glycerylphosphorylcholine-splitting enzyme) (White and Wallace 1961; White *et al.* 1961; White, Wallace, and Stone 1963, 1964; Wallace and White 1965; Wallace, Stone, and White 1965), amylase (McGeachin *et al.* 1958), alkaline phosphatase (Augustin and Moser 1955), lysozyme (Dukelow *et al.* 1966), and oxygen (Bishop 1956*b*; Mastroianni and Jones 1965).

In sheep oviductal fluid the concentration of sodium, chloride, potassium, magnesium, calcium, phosphate, bicarbonate, carbohydrate, lactate, and protein has been determined (Table 3).

Components identified in cow oviductal fluid are sodium, chloride, potassium, calcium, phosphate, reducing sugars, fat, protein, and ash (Table 4).

Rabbit oviductal fluid, compared to sheep, appears to have the same concentrations of sodium, inorganic phosphate and lactate; there is less chloride, potassium, magnesium, carbohydrate, and protein, but more calcium and bicarbonate. A comparison of sheep, cow, and rabbit is not possible because the cow oviductal fluid was collected by stripping it from the tubes of sacrificed animals, whereas the sheep and rabbit oviductal fluids were collected by continuous methods.

In comparing rabbit blood serum and oviductal fluid one finds the same amount of chloride, potassium, and inorganic phosphate, and that it is more alkaline, has more bicarbonate, less sodium, calcium, and magnesium and much less protein. Sheep oviductal fluid contains less sodium, calcium, magnesium, bicarbonate, and phosphate but more potassium and chloride than sheep blood plasma (Restall and Wales 1966).

TABLE 3

COMPOSITION OF SHEEP OVIDUCTAL FLUID

Component[a]	Hormonal State			Reference
	Estrus	Metestrus	Diestrus	
Sodium	3,180	2,980	3,170	1
Chloride	4,580	4,430	4,760	1
Potassium	320	295	317	1
Magnesium	19.8	21.8	24.6	1
Calcium	122	114	121	1
Phosphorus (acid insoluble)	8.7	4.6	8.4	1
Phosphorus (total)	40.5	21.3	46.3	1
Bicarbonate	1,245	1,008	1,086	1
Carbohydrate	613	437	670	1
Lactate	150	114	382	1
	384	427	460	2
Protein	9,300	4,800	2,100	1
	28,600	27,900	32,400	2

References: 1. Restall and Wales 1966; 2. Perkins and Goode 1966.
[a] Mean value reported for each component (μg/ml).

TABLE 4

COMPOSITION OF COW OVIDUCTAL FLUID[a]

Component	Mean Value (μg/ml)
Sodium	2,080
Chloride	4,000
Potassium	2,230
Calcium	120
Phosphate (PO_4^{-3})	97
Reducing sugars	900
Fat	18,000
Protein (N $\times$ 6.25)	122,000
Dry matter	136,000
Ash	10,200

SOURCE: Olds and VanDemark 1957*a*.
[a] Intact cows.

C. *Hormonal Influence on Constituents*

The level of sodium in rabbit oviductal fluid is significantly lower in the first 3 days of pregnancy (142.6 meq/l) than in estrus (144.5 meq/l) but no change in potassium or chloride levels occurs between estrus and pseudopregnancy (Mastroianni and Wallach 1961).

The electrolyte composition of oviductal fluid from spayed ewes is similar to that from normal ewes except that the concentration of magnesium in normal ewes (0.77 meq/l) is significantly lower than in spayed animals (0.96 meq/l). In intact, normal ewes the levels of sodium, potassium, chloride, and bicarbonate are significantly lower at metestrus than at diestrus (Table 3). However, the magnesium content is lowest at estrus (Restall and Wales 1966).

At estrus cow oviductal fluid is lowest in sodium and potassium but highest in calcium content (Table 5).

The authors have recently investigated the effect of estrogen and progesterone on the concentration of sodium, chloride, potassium, magnesium, and calcium in rabbit oviductal fluid. Mature New Zealand White rabbits were ovariectomized and oviductal fluid collecting

TABLE 5

HORMONAL INFLUENCE ON COMPOSITION OF COW OVIDUCTAL FLUID

COMPONENT[a]	HORMONAL STATE		
	Estrus	Metestrus	Diestrus
Sodium	1,980	2,080	2,160
Potassium	1,670	2,570	2,250
Calcium	128	102	125

SOURCE: Olds and VanDemark 1957*a*.
[a] Mean values are reported in μg/ml.

flasks installed. The animals were injected with 1.0 μg/kg/day ECP for 7 days, 0.5 μg/kg/day ECP plus 0.5 mg/kg/day progesterone for the next 5 days, 1.0 mg/kg/day progesterone for the next 8 days, followed by 1.0 ml cottonseed oil alone for 4 days; they were then recycled. All hormones were suspended in 1.0 ml cottonseed oil and injected subcutaneously. The sequence of hormonal injections was selected because this level of hormonal treatment gave a histological response very similar to that seen in the course of pseudopregnancy in rabbits. On the first day (ovulation) of pseudopregnancy in the rabbit the epithelium is high, columnar, and rich in PAS-positive, diastase-resistant material found at the luminal edge of the epithelium. Seven days postovulation the epithelium is considerably lower and nearly all PAS-positive material is eliminated from the epithelium. The epithelium reaches its lowest point by the 11th day after ovulation and no trace of PAS-positive material is present. The epithelium recovers over the next 4 days and becomes noticeably columnar by the 16th day after ovulation, with faint traces of PAS-positive material present near the base of the epithelium. The epithelium returns to its original (day 1) height by the 19th day and is again rich in PAS-positive material. These data indicate that pseudopregnancy lasts approximately 19 days in the rabbit (Fig. 5). Cottonseed oil injected into the ovariectomized rabbit causes no detectable change in epithelium, whereas only 0.01 μg ECP for

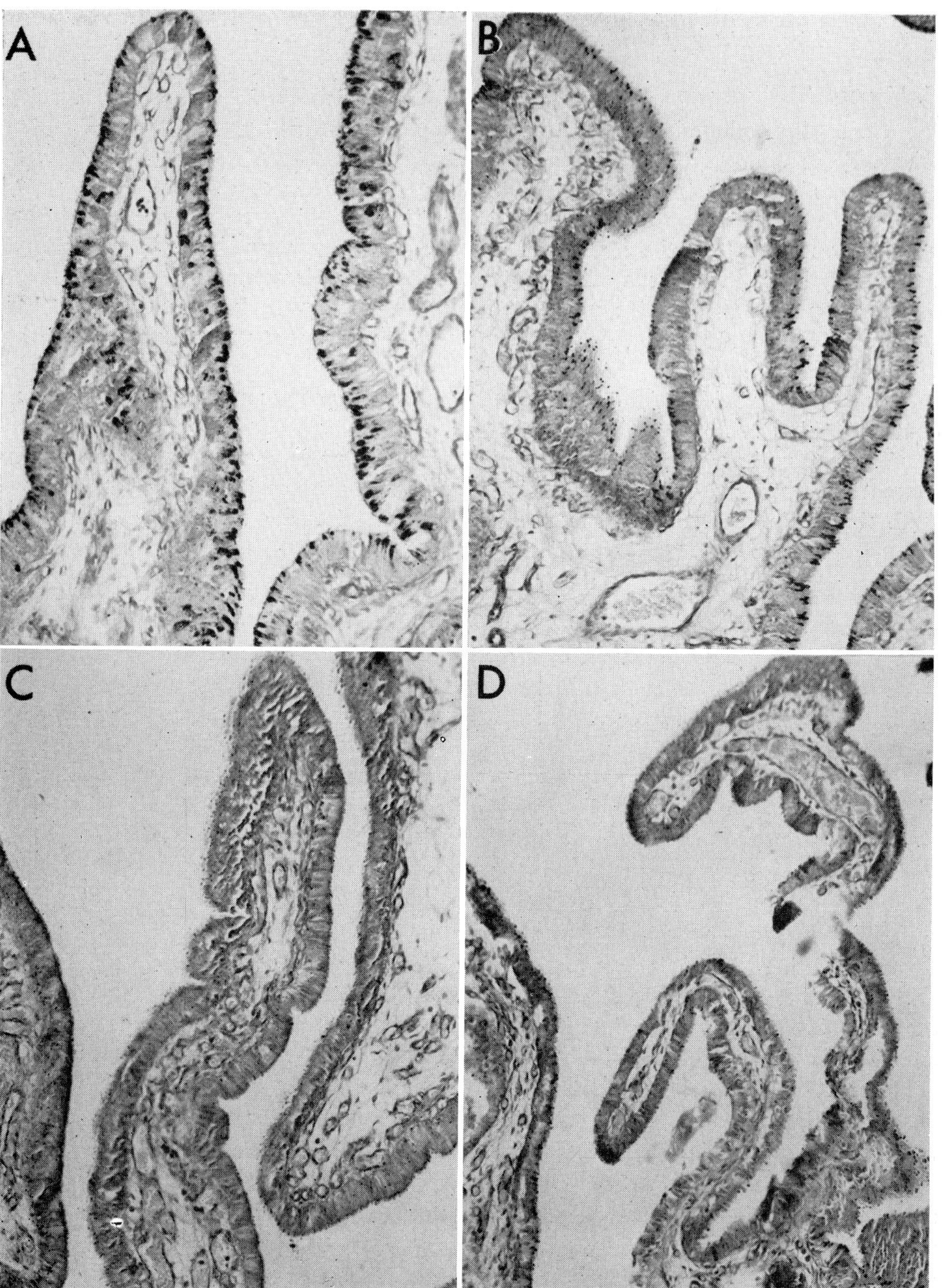

FIG. 5. Appearance of oviductal epithelium during pseudopregnancy of the rabbit: (*a*) Infundibulum of rabbit oviduct on first day (ovulation) of pseudopregnancy: note large accumulation of PAS-positive material at outer edge of epithelium (periodic acid-Schiff base with hematoxylin stain [PASH], ×122).

(*b*) Infundibulum of rabbit oviduct 5 days after ovulation (PASH, ×122).

(*c*) Infundibulum of rabbit oviduct 7 days after ovulation (PASH, ×122).

(*d*) Infundibulum of rabbit oviduct 11 days after ovulation: note degeneration of epithelium and absence of PAS-positive material (PASH, ×122).

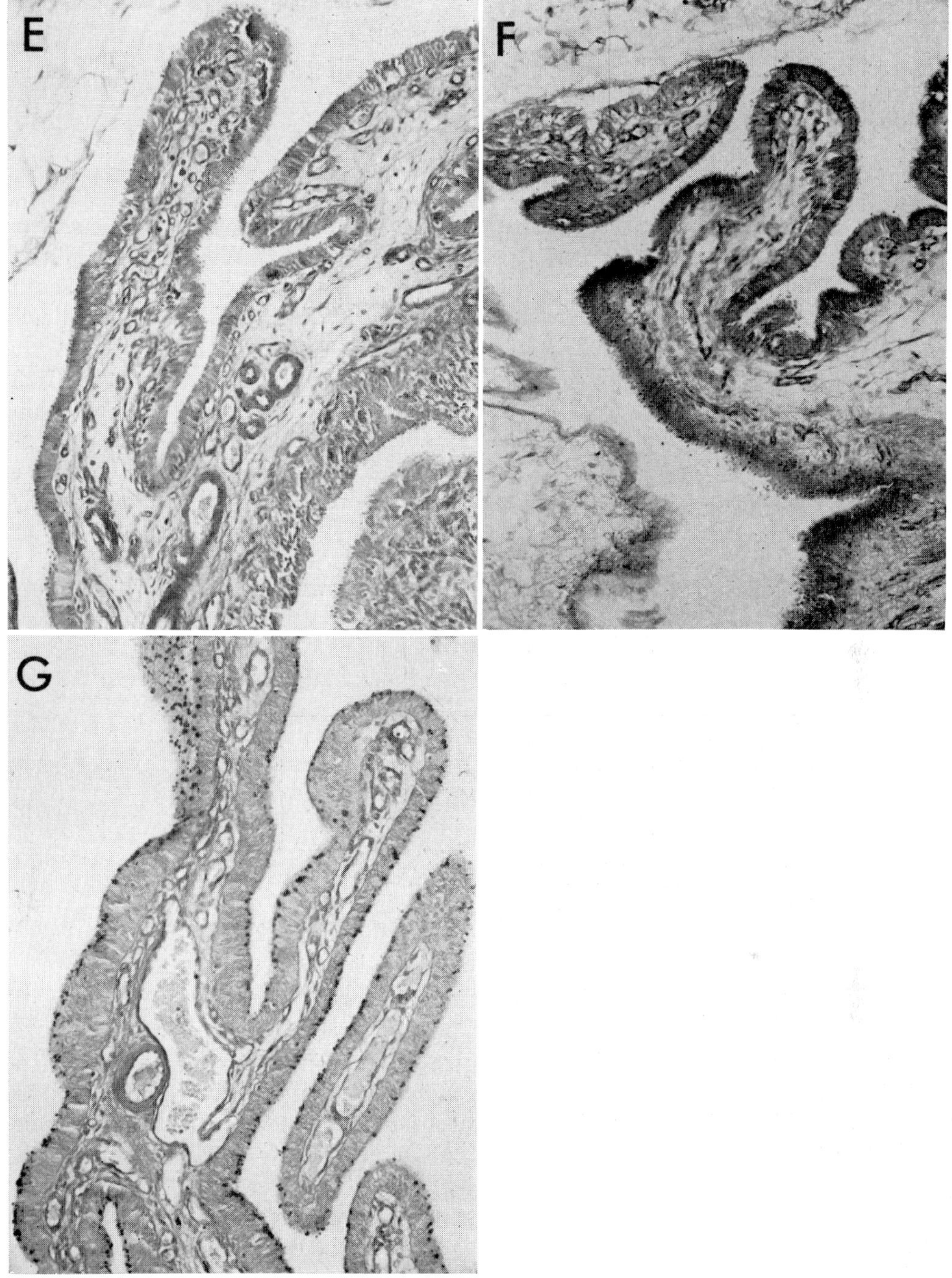

FIG. 5—*Continued*

(*e*) Infundibulum of rabbit oviduct 13 days after ovulation (PASH, ×122).

(*f*) Infundibulum of rabbit oviduct 16 days after ovulation: note regeneration of epithelium and PAS-positive material being formed near base of epithelium (PASH, ×122).

(*g*) Infundibulum of rabbit oviduct 19 days after ovulation: note tall, columnar epithelium and return of PAS-positive material to the outer edge of the epithelium (PASH, ×122).

3 days will stimulate oviductal epithelium and synthesis of PAS-positive, diastase-resistant material.

It was determined that following estrogen priming, progesterone treatment significantly lowers the chloride and raises the calcium content of rabbit oviductal fluid. Cottonseed oil treatment significantly lowers the potassium and raises the magnesium level (Table 6).

There is a doubling of the hydrogen ion concentration of rabbit oviductal fluid during pregnancy (pH 7.43) over the estrous state (pH 7.75) (Bishop 1957). Fluid from the dilated ampulla of the rat had a pH range of 7.3 to 8.5 (Blandau, Jensen, and Rumery 1958). Sheep oviductal fluid at estrus and metestrus has a pH of 6.8 to 7.0, diestrus pH 6.0 to 6.4 and proestrus pH 6.4 to 6.6 (Hadek 1953*a*).

The alkalinity of rabbit oviductal fluid is due to a high bicarbonate content (28.3×10^{-3} M) (Hamner and Williams 1964*a*, *b*). The oviductal fluid–plasma bicarbonate ratio is 2.2 and the epithelial cells of the oviduct are capable of concentrating and actively transporting bicarbonate (Vishwakarma 1962). Whether the source of bicarbonate is plasma or intracellular metabolism and in what form it is transported (HCO_3^- or hydrated CO_2) has not been determined. Since carbonic anhydrase is present in the oviduct (Lutwak-Mann 1955), hydration of CO_2 is probably the mechanism of transport. Carbonic anhydrase occurs in the oviduct of a variety of mammalian species. Progesterone stimulates carbonic anhydrase activity in the cow and rabbit but not in the ewe. No carbonic anhydrase is present in the nonpregnant oviduct of mice, cats, or dogs (Lutwak-Mann 1955).

TABLE 6

HORMONAL INFLUENCE ON ELECTROLYTES IN RABBIT OVIDUCTAL FLUID

Component	Estrogen	Progesterone +Estrogen	Progesterone	Cottonseed Oil
Sodium	2,600[a]	2,692	2,483	2,600
Chloride	2,840	2,610	2,340[b]	2,560
Potassium	201[c]	207[c]	187	185[c]
Calcium	116[d]	142[d]	138	131
Magnesium	4.0[e]	3.9[e]	4.1	4.6[e]

[a] Values are means (μg/ml) for 30 determinations for each hormonal treatment.

[b] Progesterone treatment significantly lowered chloride content after estrogen treatment, $P < 0.001$ by t-test (Snedecor 1961).

[c] Cottonseed oil treatment significantly lowered potassium content after estrogen treatment, $P < 0.05$.

[d] Progesterone + estrogen treatment significantly raised calcium content after estrogen treatment, $P < 0.05$.

[e] Cottonseed oil treatment significantly raised magnesium content after estrogen treatment, $P < 0.05$.

Peptidase activity in the rabbit oviduct is essentially constant regardless of the hormonal state, but there is a sharp increase in peptidase levels in the rabbit endometrium during pseudopregnancy (Albers, Bedford, and Chang 1961). This indicates that the biochemical response of the oviduct to hormonal influences is different from that of the uterus.

The oviductal epithelium produces an average of 290 μm lactate/gm dry tissue at estrus as compared to 430 μm 8 to 11 days into pseudopregnancy (Mastroianni and Wallach 1961; Mounib and Chang 1964, 1965). The doubling of lactate level from estrus to pregnancy would account for the increase in hydrogen ion concentration during pregnancy (Bishop 1956*a;* Bishop 1957; Mastroianni, Winternitz, and Lowi 1958; Mastroianni, Forrest, and Winternitz 1961).

Glucose, maltose, maltotriose, maltotetrose, and fructose are present in oviductal extracts (Gregoire and Gibbon 1965). Estrogen causes an increase in the oligosaccharide level, whereas progesterone and ovariectomy decrease the level. Glucose concentration (26 mg%) is significantly increased after ovulation (29 mg%), as are lactate (19 to 28 mg%) and pyruvate (1.6 to 2.2 mg%) in rabbit oviductal fluid (Holmdahl and Mastroianni 1965). The sialic acid content of rabbit oviductal fluid is highest during estrus, dropping after ovariectomy. Hexosamine content increases over estrus or pregnancy levels after castration (Yamazaki 1965). Inositol content (2.6 mg%) in intact female rabbit oviductal fluid, measured by observing the turbidity produced by growth of an inositol-dependent yeast, is not influenced by estrogen or ovariectomy but is doubled by progesterone injection (Gregoire, Gongsakdi, and Rakoff 1962).

The concentration of total phosphorus, acid insoluble phosphorus, carbohydrate, lactate, and protein in intact normal ewes does not change significantly between stages of the cycle (Table 3) but there is considerable variation among estimations within each stage.

The phospholipid content of rabbit oviductal fluid is dependent on estrus, when its level reaches 8 mg% (Bishop 1957). The oviduct of the human infant is free from lipids, which are confined mostly to the connective tissue cells in adult humans (Sheyer 1926). Oviductal lipids are due to fatty infiltration during the intramenstrual period, disappearing during pregnancy and menopause (Butomo 1927; Joel 1941). The amount of volatile fatty acids in rabbit oviductal fluid is similar to the value obtained for blood serum (2.6 to 5.2 mg/ml). Only acetic acid and an unidentified fatty acid are present (Sugawara and Takeuchi 1964).

The total nitrogen content of rabbit oviductal secretions remains constant at 6 to 7% dry matter, whether the rabbit is in estrus, pregnant, or ovariectomized. However, the protein content of the fluid is 25% of dry weight at estrus, 30% and 35% of dry weight during pregnancy and after ovariectomy, respectively (Yamazaki 1965). Urea content of oviductal fluid is 53.1 mg% in intact rabbits and is decreased to half that value by ovariectomy. Progesterone treatment has no effect on urea content (Gregoire and Rakoff 1963), but causes a doubling in glycine and serine content over estrogen treatment (Gregoire, Gongsakdi, and Rakoff 1961).

Amylase levels in the oviduct of woman, cow, rabbit, and sheep are greater than the corresponding serum levels, but the reverse is true for the rat, dog, hog, guinea pig, cat, and monkey (McGeachin *et al.* 1958). Very high levels of amylase are found in fluids from cysts associated with human oviducts (Green 1957). The enzyme splits alpha-1,4 glucosidic bonds in glycogen, and its presence in oviductal fluid could degrade glycogen to render glucose as an energy source for sperm and dividing eggs. Beta-amylase will partially capacitate rabbit sperm (Kirton and Hafs 1965). This partial capacitation is thought to be due to the removal of the decapacitation factor (Chang 1957) from the sperm head, since both alpha- and beta-amylase will destroy the decapacitation factor activity of seminal plasma (Dukelow, Chernoff, and Williams 1966).

Oviductal fluid from intact does contains 3.9 μg lysozyme/ml (Dukelow *et al.* 1966). Lysosomal sacs have been identified in the secretory cells of human oviducts, and have been observed to rupture late in the menstrual cycle (Clyman 1966). Alkaline phosphatase occurs in epithelium of the infundibulum and ampulla of the oviduct of several animals and is not influenced by cyclic changes (Bourne and Mackinnon 1943; Augustin and Moser 1955). Diesterase (glycerylphosphorylcholine splitting enzyme) has been isolated in rabbit, sow, rat, mouse, and cow oviductal fluid (White and Wallace 1961; White *et al.* 1961; White, Wallace, and Stone 1964; Wallace and White 1965; Wallace, Stone, and White

1965). Activity of the diesterase is highest at proestrus and estrus, least at metestrus and diestrus (White, Wallace, and Stone 1963). It was previously thought that a mechanical action of the oviduct was responsible for the removal of the corona radiata cells from the ova (Swyer 1947), but a corona cell-dispersing factor has now been described in oviductal fluids of woman and rabbit (Shettles 1953, 1955; Mastroianni and Ehteshamzadeh 1964).

III. Conclusion

There is little information concerning the mechanisms responsible for the secretions reaching the lumen of the oviduct. The role of active transfer of constituents from the oviductal epithelium to the oviductal lumen needs considerable clarification. Several facts lend support to the hypothesis that transudation contributes significantly to oviductal fluid at estrus: (1) the dry matter content and osmolarity of estrogen-treated rabbit oviductal fluid is low; (2) estrogen greatly enhances secretion rates; (3) the oviduct is edematous (Lombard, Morgan, and Menutt 1950) with the lymphatic system greatly expanded at estrus (Andersen 1927). A clear understanding of these mechanisms is needed for clinical treatment of sterility and for contraceptive methods because of the critical early phases of reproduction that take place in the oviduct and the influence that substances transported to the oviductal lumen could have on conception.

Since methods are now available for the continuous collection of oviductal fluid over prolonged periods, species comparisons of oviductal fluid composition and the utilization of oviductal secretions for *in vitro* fertilization, embryo development, and metabolism studies of both sperm and eggs are now possible.

The physiological role, if any, of several constituents of oviductal fluid would be of interest. The role of inositol may be important since female Syrian hamsters fed an inositol-deficient diet give birth to abnormal young and experience difficulty at parturition (Hamilton and Hogan 1944). Whether or not urea can be used by the blastocyst as a protein precursor (Villee 1960) would be valuable in understanding blastocyst development. One also needs to know the extent to which oviductal fluid is responsible for the growth in size of the blastocysts prior to implantation. Bicarbonate significantly stimulates respiration and glycolysis of rabbit, bull, rooster, and human sperm (Henle and Zittle 1942; Jones and Salisbury 1962; Lodge and Salisbury 1962; Hamner 1964; Hamner and Williams 1964*a*; Edgerton *et al.* 1966; Shelby and Foley 1966). Even though human sperm possess the components of a respiratory system (Mann 1951) and actively oxidize glucose (Terner 1960), their respiratory activity had been considered to be only 1 to 2 $\mu l/10^8$ sperm/hr (Ross, Miller, and Kurzvok 1941; MacLeod 1943). Washed human sperm respire at a rate of 13.7 $\mu l/10^8$ sperm/hr in the presence of 3.5×10^{-3} M HCO^{-3} and only at 1.4 $\mu l/10^8$ sperm/hr without HCO^{-3} (Hamner and Williams 1964*a*), indicating an important role for HCO^{-3} in human sperm metabolism.

Factors necessary for capacitation of rabbit sperm are present in the oviduct (Adams and Chang 1962) and are specific for the female reproductive tract (Hamner and Sojka 1967). *In vitro* capacitation of rabbit sperm may be prohibited in female tract secretions because the factors necessary for capacitation deteriorate rapidly or because the sperm require intimate contact with the epithelial cells of the female genital tract to achieve capacitation. The factors in oviductal fluid essential for *in vitro* fertilization may well undergo rapid degenerative changes when exposed to an external environment.

At present steroid hormones are being extensively utilized in fertility control without a basic knowledge of the mechanism by which these hormones affect the oviduct. Capacitation of rabbit sperm is best achieved in the intact estrous doe. However, estrogen alone may not be responsible for creating the conditions necessary for capacitation in the female reproductive tract. Both progesterone and the normal pseudopregnant state will inhibit the capacitation of sperm between the 4th and 12th days of pseudopregnancy in the rabbit (Chang 1958; Hafez and Ishibashi 1964; Hamner, Jones, and Sojka 1968).

The PAS-positive, diastase-resistant material in rabbits which coats the eggs and renders them unfertilizable within 6 hr after ovulation (Hammond 1934) may also coat the sperm deposited into pseudopregnant animals, causing the sperm to be incapable of fertilizing eggs in pseudopregnant animals (Hamner, Jones, and Sojka 1968). A thorough understanding of the proper hormonal balance needed for preparing the sperm for fertilization in the female reproductive tract would greatly enhance sterility treatment. Likewise the mechanism of inhibition of sperm fertilizing ability by progesterone would aid contraceptive approaches.

References

1. Adams, C. E. 1960. Development of the rabbit eggs with special reference to the mucin layer. *Acta Endocr., Adv. Abs.* 1:687.
2. Adams, C. E., and Chang, M. C. 1962. Capacitation of rabbit spermatozoa in the Fallopian tube and in the uterus. *J. Exp. Zool.* 151:159.
3. Albers, H. J.; Bedford, J. M.; and Chang, M. C. 1961. Uterine peptidase activity in the rat and rabbit during pseudopregnancy. *Amer. J. Physiol.* 201:554.
4. Allen, E. 1922. The estrous cycle in the mouse. *Amer. J. Anat.* 30:297.
5. Andersen, D. H. 1927. Lymphatics of the Fallopian tube of the sow. *Contr. Embryol. Carnegie Inst.* 19:135.
6. Argand, R. 1921. Sur le bourgeonnement nucléaire des épithéliums. *C.R. Soc. Biol.* 84:256.
7. Assheton, R. 1894. A re-investigation into the early stages of the development of the rabbit. *Quart. J. Micr. Sci.* 37:113.
8. Augustin, E., and Moser, A. 1955. Vorkommen und Aktivität von alkalischer Phosphatase im Eileiter der Ratte und in unbefruchteten und befruchteten Eiern während der Tubenwanderung. *Archiv Gynaek.* 185:759.
9. Bacsich, P., and Hamilton, W. J. 1954. Some observations on vitally stained rabbit ova with special reference to their albuminous coat. *J. Embryol. Exp. Morph.* 2:81.
10. Balboni, G. 1953. Uteriori ricerche istochimiche sull'epitelio tubarico della donna. *Boll. Soc. Ital. Biol.* 29:1394.
11. Bishop, D. W. 1956*a*. Active secretion in the rabbit oviduct. *Amer. J. Physiol.* 187:347.
12. ———. 1956*b*. Oxygen concentrations in the rabbit genital tract. *Int. Cong. Anim. Reprod.* 3:53.
13. ———. 1957. Metabolic conditions within the oviduct of the rabbit. *Int. J. Fertil.* 2:11.
14. Björkman, N., and Fredricsson, B. 1962. Ultrastructural features of the human oviduct epithelium. *Int. J. Fertil.* 7:259.
15. Black, D. L.; Duby, R. T.; and Riesen, J. 1963. Apparatus for the continuous collection of sheep oviduct fluid. *J. Reprod. Fertil.* 6:257.
16. Blandau, R.; Jensen, L.; and Rumery, R. 1958. Determination of the pH values of the reproduction-tract fluids of the rat during heat. *Fertil. Steril.* 9:207.
17. Bond, C. J. 1898. Preliminary note on certain undescribed features in the secretory

function of the uterus and Fallopian tubes in the human subject and in some of the mammalia. *J. Physiol.* 22:296.

18. Borell, U.; Nilsson, O.; Wersäll, J.; and Westman, A. 1956. Electronmicroscope studies of the epithelium of the rabbit Fallopian tube under different hormonal influences. *Acta Obstet. Gynec. Scand.* 35:35.
19. Bouin, P., and Limon, M. 1900. Fonction sécrétoire de l'épithélium tubaire chez le cobaye. *C.R. Soc. Biol.* 52:920.
20. Bourg, R. 1931. Recherches sur l'histophysiologie de l'ovaire, du testicule et des tractus génitaux du Rat et de la Souris. *Arch. Biol.* 41:245.
21. Bourne, G., and Mackinnon, M. 1943. The distribution of alkaline phosphatase in various tissues. *Quart. J. Exp. Physiol.* 32:1.
22. Bousquet, J. 1964. Culture organotypique de fragments d'oviducte de Ratte et de Lapine. Action des stéroïdes ovariens. *C.R. Soc. Biol.* 158:508.
23. Braden, A. W. H. 1952. Properties of the membranes of rat and rabbit eggs. *Aust. J. Sci. Res. Bull.* 5:460.
24. Butomo, W. 1927. Zur Frage von den zyklischen Veränderungen in den Tuben (über Tubenlipoide). *Archiv Gynaek.* 131:306.
25. Caffier, P. 1938. Über die hormonale Beeinflussung der menschlichen Tubenschleimhaut und ihre therapeutische Ausnutzung. *Z. Gynaek.* 64:1024.
26. Casida, L. E., and McKenzie, F. F. 1932. The oestrous cycle of the ewe; histology of the genital tract. *Mo. Agric. Exp. Sta. Res. Bull.* 170:4.
27. Chang, M. C. 1957. A detrimental effect of seminal plasma on the fertilizing capacity of sperm. *Nature* (*Lond.*) 179:258.
28. ———. 1958. Capacitation of rabbit spermatozoa in the uterus with special reference to the reproductive phases of the female. *Endocrinology* 63:619.
29. Clewe, T. H., and Mastroianni, L. Jr. 1960. A method for continuous volumetric collection of oviduct secretions. *J. Reprod. Fertil.* 1:146.
30. Clyman, M. J. 1966. Electronmicroscopy of the human Fallopian tube. *Fertil. Steril.* 17:281.
31. Confalonieri, C. 1951. Modificazioni cicliche e attività secretoria dell'epitelio tubarico nella donna. *Ann. Obstet. Ginec.* 73:627.
32. Courrier, R. 1921. Contribution à l'etude morphologique et fonctionelle de l'épithélium pavillon de l'oviducte chez mammifères, *C.R. Soc. Biol.* 84:571.
33. Courrier, R., and Gerlinger, H. 1922. Le cycle glandulaire de l'épithélium de l'oviducte chez la chienne. *C.R. Soc. Biol.* 87:1363.
34. Deane, H. W. 1952. Histochemical observations on the ovary and oviduct of the albino rat during the estrous cycle. *Amer. J. Anat.* 91:363.
35. Dukelow, W. R.; Chernoff, H. N.; Pinsker, M. C.; and Williams, W. L. 1966. Enzymatic activities at the time of sperm capacitation. *J. Dairy Sci.* 49:725.
36. Dukelow, W. R.; Chernoff, H. N.; and Williams, W. L. 1966. Stability of spermatozoan decapacitation factor. *Amer. J. Physiol.* 211:826.
37. Edgerton, L. A.; Martin, C. E.; Troutt, H. F.; and Foley, C. W. 1966. Collection of fluid from the uterus and oviducts. *J. Anim. Sci.* 25:1265.
38. Espinasse, P. G. 1935. The oviductal epithelium of the mouse. *J. Anat.* 69:363.
39. Fawcett, D. W., and Wislocki, G. B. 1951. Histochemical observations of the human Fallopian tube. *J. Nat. Cancer Inst.* 12:213.
40. Flerkó, B. 1954. Die Epithelien des Eileiters und ihre hormonalen Reaktionen. *Z. Microskopischanat. Forsch.* 61:99.

41. Foraker, A. G.; Denham, S. W.; and Celi, P. A. 1953. Dehydrogenase activity. II. In the Fallopian tube. *Obstet. Gynec.* 2:500.
42. Fredricsson, B. 1959. Histochemical observations on the epithelium of human Fallopian tubes. *Acta Obstet. Gynec. Scand.* 38:109.
43. Fredricsson, B., and Björkman, N. 1962. Studies on the ultrastructure of the human oviduct epithelium in different functional states. *Z. Zellforsch.* 58:387.
44. Friz, M. 1959. Tierexperimentelle Untersuchungen zur Frage der Tubensekretion. *Z. Geburtsh. Gynaek.* 153:285.
45. Friz, M., and Mey, R. 1959. Early embryonal death before implantation. *Int. J. Fertil.* 4:306.
46. Frommel, R. 1886. Beitrag zur Histologie der Eileiter. *Gesellsch. Gynaek.* 1:95.
47. Galstjan, S. 1935. Experimentell-histologische Untersuchungen über das Eileiterepithel. *Arch. Exp. Zellforsch.* 17:231.
48. Green, C. L. 1957. Identification of alpha-amylase as a secretion of the human Fallopian tube and "tubelike" epithelium of Müllerian and mesonephric duct origin. *Amer. J. Obstet. Gynec.* 73:402.
49. Greenwald, G. S. 1957. Interruption of pregnancy in the rabbit by the administration of estrogen. *J. Exp. Zool.* 135:461.
50. ———. 1958. Endocrine regulation of the secretion of mucin in the tubal epithelium of the rabbit. *Anat. Rec.* 130:477.
51. Gregoire, A. T., and Gibbon, R. 1965. Glucosyl oligosaccharides of the rabbit genital tract: Effects of ovarian hormone administration. *Int. J. Fertil.* 10:151.
52. Gregoire, A. T., and Rakoff, A. E. 1963. Urea content of the female rabbit genital tract fluids. *J. Reprod. Fertil.* 6:467.
53. Gregoire, A. T.; Gongsakdi, D.; and Rakoff, A. E. 1961. The free amino acid content of the female rabbit genital tract. *Fertil. Steril.* 12:322.
54. ———. 1962. The presence of inositol in genital tract secretions of the female rabbit. *Fertil. Steril.* 13:432.
55. Gregory, P. W. 1930. The early embryology of the rabbit. *Contr. Embryol. Carnegie Inst.* 21:141.
56. Guerriero, C. 1929*a*. Étude morphologique de l'épithélium de la trompe utérine au repos secuel chez la lapine adulte, impubère et castrée. *C.R. Soc. Biol.* 102:1072.
57. ———. 1929*b*. Sur la structure de l'épithélium de la trompe utérine pendant la période folliculaire et lutéinique de l'ovaire. *C.R. Soc. Biol.* 102:1074.
58. Hadek, R. 1953*a*. Alteration of pH in the sheep's oviduct. *Nature* (*Lond.*) 171:976.
59. ———. 1953*b*. Mucin secretion in the ewe's oviduct. *Nature* (*Lond.*) 111:750.
60. ———. 1955. The secretory process in the sheep's oviduct. *Anat. Rec.* 121:137.
61. Hafez, E. S. E., and Ishibashi, I. 1964. Block to fertilization in presence of corpora lutea in the rabbit. *Fifth Int. Cong. Anim. Reprod. A. I., Trento* 3:231.
62. Hamilton, W. J., and Hogan, A. G. 1944. Nutritional requirements of the Syrian hamster. *J. Nutr.* 27:213.
63. Hammond, J. 1934. The fertilization of rabbit ova in relation to time: A method of controlling the litter size, the duration of pregnancy and the weight of the young at birth. *J. Exp. Biol.* 11:140.
64. Hamner, C. E. 1964. Effect of the female reproductive tract on spermatozoa. Ph.D. diss., University of Georgia, Athens, Ga.
65. Hamner, C. E., and Sojka, N. J. 1967. Capacitation of rabbit spermatozoa: species specificity and organ specificity. *Proc. Soc. Exp. Biol. Med.* 124:689.

66. Hamner, C. E., and Williams, W. L. 1963. Effect of the female reproductive tract on sperm metabolism in the rabbit and fowl. *J. Reprod. Fertil.* 5:143.
67. ———. 1964*a*. Effect of bicarbonate on the respiration of spermatozoa. *Fed. Proc.* 23:430.
68. ———. 1964*b*. Identification of sperm stimulating factor of rabbit oviduct fluid. *Proc. Soc. Exp. Biol. Med.* 117:240.
69. ———. 1965. Composition of rabbit oviduct secretions. *Fertil. Steril.* 16:170.
70. Hamner, C. E.; Jones, J.; and Sojka, N. J. 1968. Influence of the hormonal state of the female on the fertilizing capacity of rabbit spermatozoa. *Fertil. Steril.* 19:137.
71. Haour, P.; Conti, C.; and Guyot, H. 1956. Étude chromatographique des acides amines dans les sécrétions génitales. *C.R. Soc. Biol.* 150:1963.
72. Hashimoto, M.; Shimoyama, I.; Kosaka, M.; Komori, A.; Hirasawa, T.; Yokoyama, Y.; and Akashi, K. 1962. Electronmicroscopic studies on the epithelial cells of the human Fallopian tube. *Int. J. Jap. Obstet. Gynec. Soc.* 9:200.
73. Hashimoto, M.; Shimoyama, T.; Kosaka, M.; Komori, A.; Hirasawa, T.; Yokoyama, Y.; Kawase, N.; and Nakamura, T. 1964. Electronmicroscopic studies on the epithelial cells of the human Fallopian tube. II. *Int. J. Jap. Obstet. Gynec. Soc.* 11:92.
74. Heap, R. B. 1962. Some chemical constituents of uterine washings: A method of analysis with results from various species. *J. Endocr.* 24:367.
75. Hellström, K. E., and Nilsson, O. 1957. *In vitro* investigation of the ciliated and secretory cells in the rabbit Fallopian tube. *Exp. Cell Res.* 12:180.
76. Henle, G., and Zittle, C. A. 1942. Studies of the metabolism of bovine epididymal spermatozoa. *Amer. J. Physiol.* 136:70.
77. Holmdahl, T. H., and Mastroianni, L., Jr. 1965. Continuous collection of rabbit oviduct secretions at low temperature. *Fertil. Steril.* 16:587.
78. Jagerroos, B. H. 1912. Zur Kenntnis der Veränderungen der Eileiterschwangerschaft während der Menstruation. *Z. Geburtsh. Gynaek.* 72:28.
79. Joel, C. A. 1941. Über den Ascorbinsäure-Gehalt der menschlichen Tube während Zyklus und Gravidität. *Schweiz. Med. Wschr.* 71:1286.
80. Jones, E. E., and Salisbury, G. W. 1962. The action of carbon dioxide as a reversible inhibitor of mammalian spermatozoan respiration. *Fed. Proc.* 21:86.
81. Kirton, K. T., and Hafs, H. D. 1965. Sperm capacitation by uterine fluid or beta-amylase *in vitro*. *Science* 150:618.
82. Kneer, M.; Burger, H.; and Simmer, H. 1952. Über die Atmung der Schleimhaut menschlicher Eileiter. *Archiv Gynaek.* 181:561.
83. Leblond, C. P. 1950. Distribution of periodic acid-reactive carbohydrates in the adult rat. *Amer. J. Anat.* 86:1.
84. Lehto, L. 1963. Cytology of the human Fallopian tube. *Acta Obstet. Gynec. Scand.* 42 (suppl. no. 4).
85. Lodge, J. R., and Salisbury, G. W. 1962. Initiation of anaerobic metabolism of mammalian spermatozoa by carbon dioxide. *Nature* (*Lond.*) 195:293.
86. Lombard, L.; Morgan, B. B.; and Menutt, S. H. 1950. The morphology of the oviduct of virgin heifers in relation to the estrous cycle. *J. Morph.* 86:1.
87. Lowi, R. N. P. 1960. Uterine tube physiology. *Obstet. Gynec.* 16:322.
88. Lutwak-Mann, C. 1955. Carbonic anhydrase in the female reproductive tract: Occurrence, distribution and hormonal dependence. *J. Endocr.* 13:26.

89. McGeachin, R. L.; Hargan, L. A.; Potter, B. A.; and Daus, A. T., Jr. 1958. Amylase in Fallopian tubes. *Proc. Soc. Exp. Biol. Med.* 99:130.
90. McKenzie, F. F., and Terrill, C. E. 1937. Estrus, ovulation, and related phenomena in the ewe. *Mo. Agric. Exp. Sta. Res. Bull.* 264:4.
91. MacLeod, J. 1943. The role of oxygen in the metabolism and motility of human spermatozoa. *Amer. J. Physiol.* 138:512.
92. Mann, T. 1951. Studies on the metabolism of semen. 7. Cytochrome in human spermatozoa. *Biochem. J.* 48:386.
93. Marcus, S. L. 1964. Protein components of oviduct fluid in the primate. *Surg. Forum* 15:381.
94. Marcus, S. L., and Saravis, C. A. 1965. Oviduct fluid in the rhesus monkey: A study of its protein components and its origin. *Fertil. Steril.* 16:785.
95. Mastroianni, L. Jr. 1962. The structure and function of the Fallopian tube. *Clin. Obstet. Gynec.* 5:781.
96. Mastroianni, L., Jr.; Beer, F. A.; and Clewe, T. H. 1960. Hormonal regulation of oviduct secretions. *First Int. Cong. Endocr. Copenhagen.* Session 7b 277.
97. Mastroianni, L., Jr.; Beer, F.; Shah, U.; and Clewe, T. H. 1961. Endocrine regulation of oviduct secretions in the rabbit. *Endocrinology* 68:92.
98. Mastroianni, L., Jr., and Ehteshamzadeh J., 1964. Corona cell dispersing properties of rabbit tubal fluid. *J. Reprod. Fertil.* 8:145.
99. Mastroianni, L., Jr.; Forrest, W.; and Winternitz, W. W. 1961. Some metabolic properties of the rabbit oviduct. *Proc. Soc. Exp. Biol. Med.* 107:86.
100. Mastroianni, L., Jr., and Jones, R. 1965. Oxygen tension within the rabbit Fallopian tube. *J. Reprod. Fertil.* 9:99.
101. Mastroianni, L., Jr.; Shah, U.; and Abdul-Karim, R. 1961. Prolonged volumetric collection of oviduct fluid in the rhesus monkey. *Fertil. Steril.* 12:417.
102. Mastroianni, L., Jr., and Wallach, R. C. 1961. Effect of ovulation and early gestation on oviduct secretions in the rabbit. *Amer. J. Physiol.* 200:815.
103. Mastroianni, L., Jr.; Winternitz, W. W.; and Lowi, N. P. 1958. The *in vitro* metabolism of the human endosalpinx. *Fertil. Steril.* 9:500.
104. Moog, F., and Wenger, E. L. 1952. The occurrence of a neutral mucopolysaccharide at sites of high alkaline phosphatase activity. *Amer. J. Anat.* 90:339.
105. Moreaux, R. 1910. Sur la structure et la fonction sécrétoire de l'épithélium de la trompe utérine chez les Mammifères. *C.R. Soc. Biol.* 68:142.
106. ———. 1913. Sur la morphologie et la fonction glandulaire. *Arch. Anat. Micr.* 14:515.
107. Mounib, M. S., and Chang, M. C. 1964. Metabolism of endometrium and Fallopian tube in the rabbit. *Feb. Proc.* 23:361.
108. ———. 1965. Metabolism of endometrium and Fallopian tube in the estrous and the pseudopregnant rabbit. *Endocrinology* 76:542.
109. Nilsson, O., and Hellström, K. E. 1957. Cell types identified in tissue cultures of epithelium of the rabbit Fallopian tube. *Acta Obstet. Gynec. Scand.* 36:340.
110. Novak, E., and Everett, H. S. 1928. Cyclical and other variations in the tubal epithelium. *Amer. J. Obstet. Gynec.* 16:499.
111. Ober, K. G. 1957. Ovary. In *Klinik der inneren Sekretion*, ed. A. Labhart. Berlin: Springer.
112. Olds, D., and VanDemark, N. L. 1957*a*. Composition of luminal fluids in bovine female genitalia. *Fertil. Steril.* 8:345.

113. Olds, D. and VanDemark, N. L. 1957*b*. Luminal fluids of bovine female genitalia. *J. Amer. Vet. Med. Ass.* 131:555.
114. ———. 1957*c*. Physiological aspects of fluids in female genitalia with special reference to cattle—a review. *Amer. J. Vet. Res.* 18:587.
115. ———. 1957*d*. The behavior of spermatozoa in luminal fluids of bovine female genitalia. *Amer. J. Vet. Res.* 18:603.
116. Padykula, H. A. 1952. The localization of succinic dehydrogenase in tissue sections of the rat. *Amer. J. Anat.* 91:107.
117. Perkins, J. L., and Goode, L. 1966. Effects of stage of the estrous cycle and exogenous hormones upon the volume and composition of oviduct fluid in ewes. *J. Anim. Sci.* 25:465.
118. Perkins, J. L.; Goode, L.; Wilder, W. A., Jr.; and Henson, D. B. 1965. Collection of secretions from the oviduct and uterus of the ewe. *J. Anim. Sci.* 24:383.
119. Restall, B. J. 1966*a*. The Fallopian tube of the sheep. I. Cannulation of the Fallopian tube. *Aust. J. Biol. Sci.* 19:181.
120. ———. 1966*b*. The Fallopian tube of the sheep. II. The influence of progesterone and oestrogen on the secretory activities of the Fallopian tube. *Aust. J. Biol. Sci.* 19:181.
121. Restall, B. J., and Wales, R. G. 1966. The Fallopian tube of the sheep. III. The chemical composition of the fluid from the Fallopian tube. *Aust. J. Biol. Sci.* 19:687.
122. Ross, V.; Miller, E. G.; Kurzvok, R. 1941. Metabolism of human sperm. *Endocrinology* 28:885.
123. Schaffer, J. 1908. Über Bau und Funktion des Eileiterepithels beim Menschen und bei Säugetieren. *Mschr. Geburtsh. Gynaek.* 28:526.
124. Shelby, D. R., and Foley, C. W. 1966. Influence of carbon dioxide absorbed on the consumption of oxygen by boar spermatozoa. *J. Anim. Sci.* 25:352.
125. Shettles, L. B. 1953. Observations on human follicular and tubal ova. *Amer. J. Obstet. Gynec.* 66:235.
126. ———. 1955. Further observations on living human oocytes and ova. *Amer. J. Obstet. Gynec.* 69:365.
127. Sheyer, H. E. 1926. Über die Lipoide der Tube. *Virchows Arch. Path. Anat.* 262:712.
128. Snedecor, G. W. 1961. *Statistical Methods*, pp. 37–50. 5th ed. Ames: Iowa State Press.
129. Stegner, H. E. 1961. Das Epithel der Tuba-Uterina des neugeborenen elektronenmikroskopische Befunde. *Z. Zellforsch.* 55:247.
130. Sugawara S., and Takeuchi, S. 1964. Identification of volatile fatty acid in genital tract of female rabbit. *Jap. J. Zool. Tech. Soc.* 35:283.
131. Swyer, G. I. M. 1947. A tubal factor concerned in the denudation of rabbit ova. *Nature* (*Lond.*) 159:873.
132. Tazawa, K. 1958. Studies on cyclic changes of the human Fallopian tube; mainly in reference to lipid fluctuation and fluorescent microscopical observations. *Acta Med. Biol.* 5:277.
133. Terner, C. 1960. Oxidation of exogenous substrates by isolated human spermatozoa. *Amer. J. Physiol.* 198:48.
134. Tietze, K. 1929. Zur Frage nach den zyklischen Veränderungen des menschlichen Tubenepithels. *Z. Gynaek.* 52:32.
135. Villee, L. A. 1960. *The placenta and fetal membranes.* Baltimore: Williams and Wilkins Co.

136. Vishwakarma, P. 1962. The pH and bicarbonate-ion content of the oviduct and uterine fluids. *Fertil. Steril.* 13:481.
137. Wallace, J. C., and White, I. G. 1965. Studies of glycerylphosphorylcholine diesterase in the female reproductive tract. *J. Reprod. Fertil.* 9:163.
138. Wallace, J. C.; Stone, G. M.; and White, I. G. 1965. The effect of the oestrous cycle and of estradiol on the breakdown of seminal glycerylphosphorylcholine by secretions of the rat uterus. *Aust. J. Biol. Sci.* 18:88.
139. Westman, A. 1930. Studies on the functions of the mucus membrane of the uterine tube. *Acta Obstet. Gynec. Scand.* 10:288.
140. White, I. G., and Wallace, J. C. 1961. Breakdown of seminal glycerylphosphorylcholine by secretions of the female reproductive tract. *Nature (Lond.)* 189:843.
141. White, I. G., Wallace, J. C.; and Stone, G. M. 1963. Studies of the glycerylphosphorylcholine diesterase activity of the female genital tract in the ewe, cow, sow and rat. *J. Reprod. Fertil.* 5:298.
142. ———. 1964. The metabolism of seminal glycerylphosphorylcholine by fluids of the female reproductive tract. *Fifth Int. Cong. Anim. Reprod. A.I., Trento* 4:526.
143. White, I. G.; Wallace, J. C.; Wales, R. G.; and Scott, T. W. 1961. The occurrence and metabolism of glycerylphosphorylcholine in semen and the genital tract. *Proc. Fourth Int. Cong. Anim. Reprod. A.I., The Hague* 2:266.
144. Whitten, W. K. 1930. Culture of tubal ova. *Nature (Lond.)* 179:1031.
145. Woskressensky, M. A. 1891. Experimentelle Untersuchungen über die Pyo- und Hydrosalpinxbildung bei den Tieren. *Zbl. Gynaek.* 15:849.
146. Yamazaki, K. 1965. Biochemical study of secretions from oviduct. *Jap. J. Zool. Tech. Soc.* 38:172.
147. Zachariae, F. 1958. Autoradiographic (^{35}S) and histochemical studies of sulphomucopolysaccharides in the rabbit uterus, oviducts and vagina. *Acta Endocrinol.* 29:118.

14

Immunology of Oviductal Secretions

S. J. Behrman

Department of Obstetrics and Gynecology
Medical Center
University of Michigan, Ann Arbor

I. Immunological Mechanisms

A. *Theoretical Aspects*

It is more than 60 years now since it was found that the sera of guinea pigs injected with macerated testicular materials agglutinated and immobilized sperm of various animals (Landsteiner 1899; Matchnikoff 1899). Subsequently numerous reports of studies on the antigenic properties of sperm, seminal fluid, testicular extracts, and ovary have been reported in a wide variety of animals (Tyler 1949, 1961; Weil and Finkler 1958, 1959; Weil 1960; Bishop and Tyler 1963; Katsh and Katsh 1965; Katsh 1967; Behrman 1965*a*, *b*). The results of immunization of animals by these antigenic substances and their effects on the inhibition of fertility varied considerably, depending upon the animal used, the preparation of the antigen, the method and route of immunization, type of antibody studied, and the interpretation of results.

Viewing the union of sperm and egg in fertilization as analogous to an antigen-antibody reaction, one appreciates the contribution Tyler (1961) made when he observed that sea urchin sperm were agglutinated when sea water from a suspension of unfertilized eggs was added to a suspension of homologous sperm. It was found that a gelatinous substance present on the surface of the sea urchin eggs slowly dissolved into the surrounding media, and a complementary receptor substance (antifertilizin) can be found on the surface of the sperm. Furthermore, when the sperm came into contact with the dissolved fertilizin from the sea urchin eggs, the sperm were agglutinated. These substances were believed to be vitally concerned in the attraction and attachment of the sperm to the eggs. Thus, removal of fertilizin from the surface of the egg lowers the possibility of fertilization. Treatment of sperm with an excess of this fertilizin results in their agglutination, followed by a spontaneous reversal of the agglutination. These sperm cells then, while motile and morphologically indistinguishable from untreated cells, are incapable of fertilization. Thus, it is apparent that there are antibodies that do not visually affect the morphologic nature of sperm but can impair their fertilizing capacity. It is possible, therefore, that antibodies to sperm may occur in both an agglutinating and nonagglutinating form, and that exposure of sperm to either form can render them incapable of fertilization. While this sequence of events seems quite clear in the invertebrate, it is the situation in the mammal that we are most interested in here.

The extensive literature on the antigenic properties of isologous male reproductive secretions upon the female of the same species is well summarized by Bishop and Tyler (1963). The animals used have been mouse (McLaren 1964), guinea pig (Katsh 1957, 1958), rabbit (Edwards 1960; Menge 1967*a*), monkey (Moyer 1965), sheep (Braden 1953), heifer (Menge 1967*b*), and man (Mancini *et al.* 1965). The antigen used has been whole semen, seminal plasma alone, washed sperm, and testicular extracts. The mode of immunization has varied between intramuscular, intradermal, intraperitoneal, and even transvaginal (Behrman and Otani 1963; Edwards 1960). The antibodies studied have been precipitating, hemagglutinating, complement-fixing, tissue-fixed, sperm-immobilizing, and agglutinating. It appears, therefore, that most mammals will show reduction in fertility when immunized with various fractions of the male reproductive secretions, by varying types and methods of immunization. There is, however, considerable question whether the titer of the antibody necessarily reflects the degree of fertility inhibition (Richter 1964; McLaren 1966). Consequently, at this stage it is necessary to ask, Is the reduction of fertility due to a generalized immunological reaction or a simple pharmacological one? In either event, it is still necessary to explain at what physiological level the fertility inhibition acts, and in this essay we are dealing specifically with the role of oviductal secretions.

B. *Immune Mechanisms*

Whether the animal is challenged by injection or by natural means, i.e., transvaginally by sperm or seminal plasma, the antigenic substance produces the immunologic response in the following classical way. At the site of immunization or invasion, the particular foreign protein is picked up by macrophages. It has been shown that sperm, after degeneration, are absorbed by macrophages transvaginally (Moyer 1965) or from the uterus (Austin 1957), from where these macrophages migrate along the lymphatics to the reticuloendothelial system (Fig. 1). Plasmoblasts form germinal centers in the spleen where large mature lymphocytes, containing a heavy 19S protein, are produced and released into the circulation. These wend their way back to the site of invasion and attack the foreign protein. Once the animal responds with this primary response, a second invasion by either the male reproductive secretions in the vagina following immunization or absorption of the foreign protein results in a secondary response. The large lymphocytes migrate to the site of antigenic or foreign protein invasion. Here again the macrophages invade the lymphatic system. At the time of their destruction enzymes and RNA messengers are released which are capable of stimulating the large mature lymphocytes to maximally proliferate into smaller lymphocytes containing a low molecular 7S-type antibody. This is the classic secondary response, as depicted in Figure 2. At this stage the proliferation of the smaller lymphocytes is such that there is a considerably larger defense mechanism than required. When the excess lymphocytes disintegrate there is a liberation of antibodies into the circulation. Detection of these antibodies in the circulation then indicates that the animal has been immunized. A third possibility has been suggested following prolonged stimulation by the antigen at a local level (Sawada and Behrman 1967; Paine and Behrman 1967; Behrman and Kistner

1968). This suggests that the circulating antibodies are either deposited in the tissue most exposed to the site of antigenic stimulation, having been brought there by lymphocytes, or produced locally by the genital structures (Fig. 3, 4*a*, *b*). This can be localized by direct or indirect immunofluorescence. It is also possible that the antibodies were produced locally by plasma cells under the basement membrane of the epithelium and migrated both into the epithelial surface and back into the general circulation (Tomasi 1967).

An immunological phenomenon pertinent to this discussion is the established fact

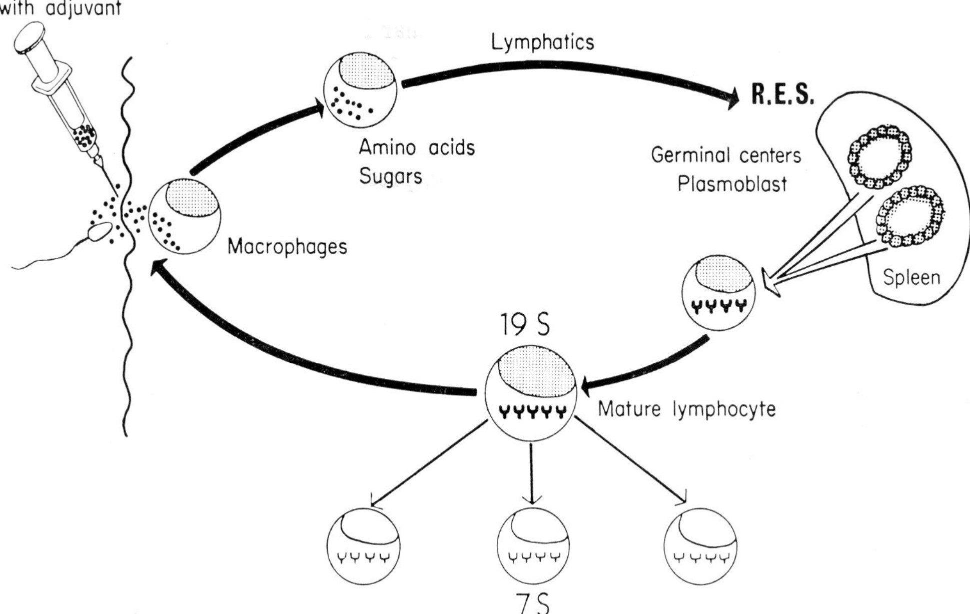

FIG. 1. Diagrammatic illustration of primary immunologic response to either injected or transvaginal application of antigen, in this case whole rabbit semen.

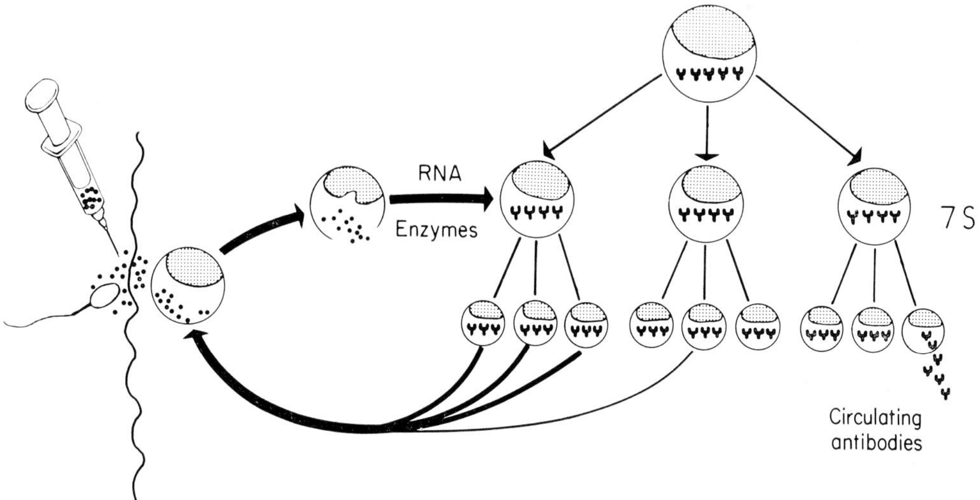

FIG. 2. Diagrammatic illustration of secondary immunologic response

(Shulman *et al.* 1964) that various immunization routes, either intravenous or intradermal, will result in different antibody responses. Thus, the heavy molecular (19S) antibody may fail to precipitate with antigen even when there is a high hemagglutinating titer; and the low molecular antibody (7S), in contrast, may give precipitation even when tested at much lower hemagglutinating titers. Similarly, intramuscular immunization of guinea pigs with homologous sperm shows high hemagglutinating titers in the serum, but very low titers in uterine or vaginal secretions (Otani and Behrman 1963). On the other hand, transvaginal immunization of the guinea pig with either testicular or epididymal sperm (Behrman and Otani 1963) shows much higher hemagglutinating, precipitating titers in the vaginal and uterine washings, as compared with the circulating serum antibodies. This factor will become important in our discussion.

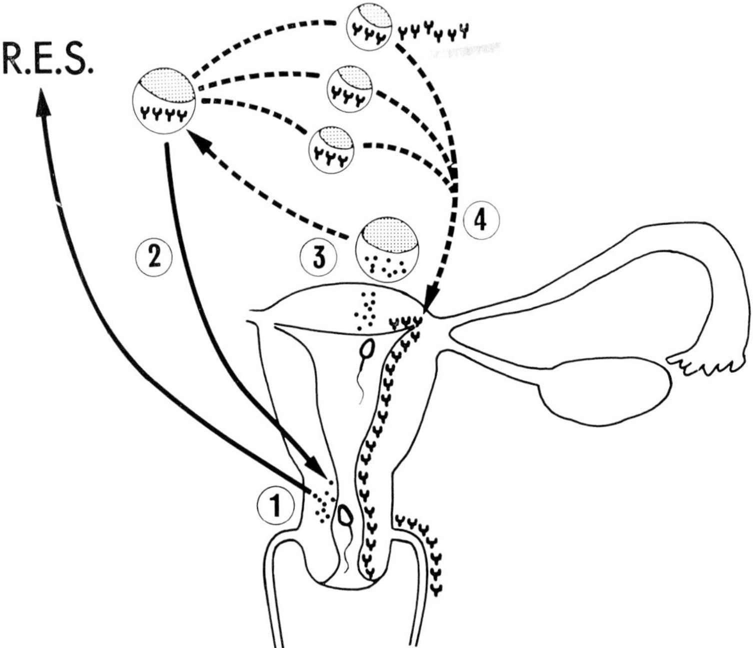

FIG. 3. Diagrammatic illustration of deposition of tissue-fixed antibodies. (1) Invading sperm and seminal plasma antigens (2) taken to reticuloendothelial system (R.E.S.); primary response of lymphocyte migrating to uterus followed by (3) a secondary invasion and response resulting eventually in (4) deposition of antibody in the tissues exposed to the antigen by migration of lymphocytes or due to local reaction.

C. *Physiological Mechanisms*

Immunization of the female with male reproductive secretions, whatever the precise antigen, results in various levels and types of antibodies depending on the route of immunization, but in all cases causes various degrees of fertility inhibition. It is now necessary to consider at what level of the reproductive cycle this mechanism operates. The question is schematically depicted in Figure 5. It will be clear from what follows that there is no unanimity of opinion, since these studies once again vary as to mode and route of immunization, animals used, and whether the antigen was epididymal sperm, washed sperm, whole semen, seminal plasma, or testicular extracts. Fundamentally, everyone seems to agree that ovulation is not inhibited. Some, however, feel that the uterus may be the main site of action

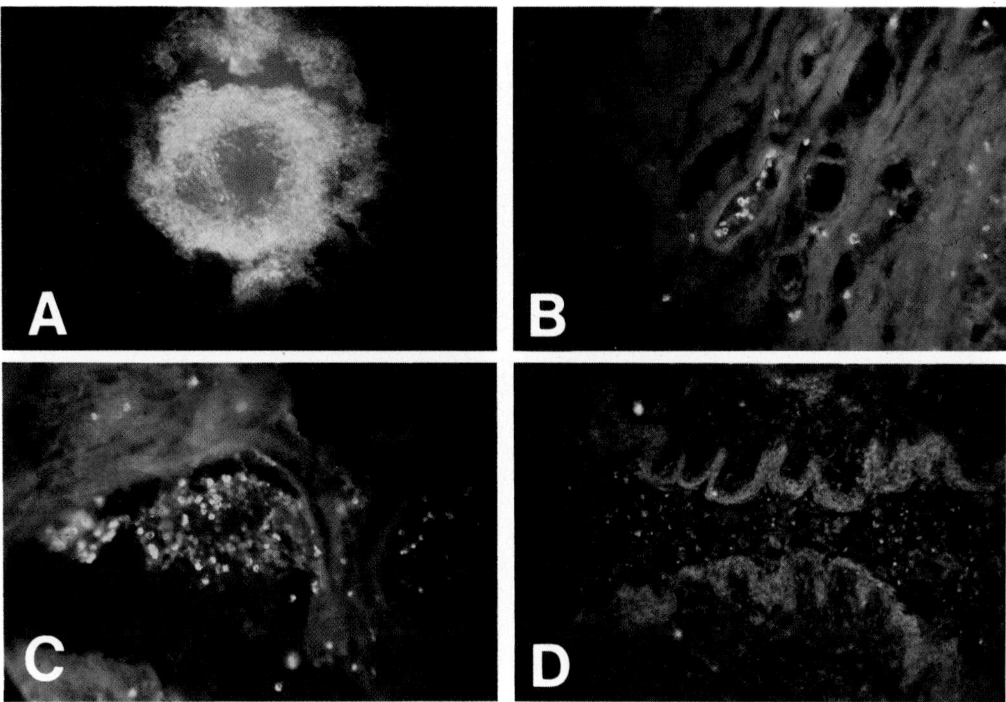

FIG. 4. Immunofluorescent localization of lymphocytes in guinea pig immunized with isologous whole semen. (*a*) Fluorescent localization in germinal centers. (*b*) Fluorescent staining of lymph vessel in uterus. Note lymphocytes within the vessel and scattered in tissue. (*c*) Fluorescent localization of lymphocytes in uterine glands and cavity. (*d*) Fluorescent localization of lymphocytes in vaginal wall and secretions of vagina. Note also the intense fluorescent staining of the vaginal epithelium.

(Katsh 1957, 1958; Kerr and Robertson 1947). The postulated theory is that the antigen reaches the uterus and reacts with the antibody present, releasing histamine, causing irregular uterine contractions, and so disturbs sperm migration. The uterus has been implicated as being unable to permit implantation because of the presence of tissue antibodies (Sawada and Behrman 1967; Behrman 1966; Schwimmer, Ustay, and Behrman 1967). Others have implied (Menge 1967*a*) that, depending on the dosage of antigen used for the immunization, one might get either failure of fertilization or early embryonic death. Failure of transport of sperm along the oviduct has been strongly championed (McLaren 1964), and it has been suggested also that the rate of sperm transport along the tube may be immunologically biased (Braden 1953; Mattner and Braden 1963).

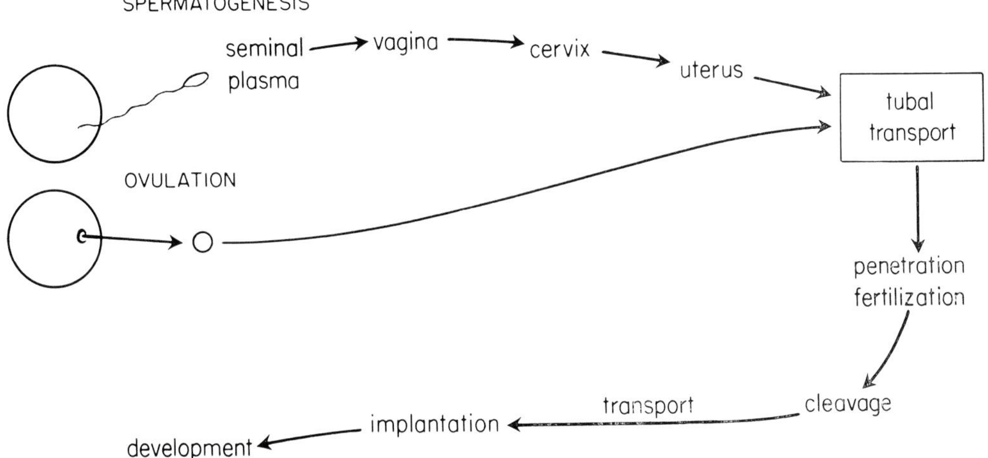

FIG. 5. Schematic illustration of physiological sites at which fertility could be reduced by immune mechanism.

If there is in fact failure of fertilization, due either to failure of sperm or egg transport, or to failure of their union, possibly on an immunologic basis due to antibodies in the oviductal fluid, then it becomes highly essential to investigate the immunological role of the oviductal secretions.

II. Egg Transport Studies

A. *Immunological Techniques*

Forty sexually matured female rabbits (3.5 kg to 4.5 kg) were used and separated into individual cages for 3 weeks to insure absence of pseudopregnancy and conception. Twenty rabbits were immunized intradermally 4 times at 2-week intervals with 2.4 ml of an equal volume of Freund's complete adjuvant and homologous whole semen (protein content, 10 mg per ml). The injections were made in 10 sites on the back of the animal and 2 on the hind pads. Another 20 rabbits were used as controls and were immunized with the same schedule, using an emulsion containing the same volume of saline plus adjuvant in place of the homologous semen. At the same time, 3×10^4 units of penicillin were given intramuscularly to prevent infection. Hemagglutination titers were performed weekly and, when the titer exceeded 320×, the experiments were instituted. Fresh semen was collected by means of an artificial vagina and inseminated into the vaginae of the experimental animals.

B. *Technique of Preparing the Oviduct*

Synchronous intravenous injections of 50 i.u. of HCG were given to each experimental rabbit in order to induce ovulation. After 30, 36, 48, 60, 72, and 96 hr, respectively, the animals were killed by pentobarbital injection and the number of ovulations was determined by counting the newly formed corpora lutea in each ovary. The oviducts were immediately removed and cleared. The original technique of Orsini (1962) was modified as follows: a solution composed of 30 ml 95% alcohol, 10 ml formalin, 10 ml glacial acetic acid added to 50 ml of water was used for fixation. For dehydration, the oviducts were successively put into 50, 70, and 80% alcohol solutions, containing 0.3 ml 30% hydrogen peroxide and absolute alcohol. For clearing, absolute alcohol diluted with the same volume of benzine followed by pure benzine was used. Finally, the specimens were placed in benzol-benzoate

TABLE 1

EGG TRANSPORT STUDIES IN 18 NONIMMUNIZED RABBITS SHOWING POSITION OF EGGS AT 30 TO 96 HR AFTER INSEMINATION; 90.3% OF EGGS WERE RECOVERED COMPARED TO OVULATION POINTS[a]

Egg Transport								Controls							
								Egg Position							
								←Ovary			Vagina→				
Rabbits	Hours after Insem.	Hemagglutinin Titer						Oviduct			Uterus		Eggs Recovered		
		Peak			Autopsy		Ovulation Points	A	B	C+D	E	F	No.	%	
3	30	0	0	0	0	0	0	35	8	12	11	0	0	31	88.6
3	36	80	0	0	0	0	0	34	2	18	12	0	0	32	94.1
3	48	0	20	0	0	0	0	23	4	5	14	0	0	23	100
3	60	0	0	0	0	0	0	37	0	5	21	8	1	35	94.6
3	72	40	0	0	0	0	0	25	0	0	2	14	5	21	84.0
3	96	0	0	0	0	0	0	31	0	0	0	18	7	25	80.6
18								185	Totals					167	90.3

[a] Positions A–F refer to the arbitrary divisions of the genital tract of Figure 6.

fluid. The fluid had to be changed carefully every 2 hr for the oviduct and every 4 hr for uterine specimens. The cleared specimens were viewed directly under a dissecting microscope where the position of the migrating eggs could be identified. The uterus was divided into 2 equal parts (E and F) and the oviducts into 3 approximately equal portions, i.e., A was the distal third, B was the middle third to the ampullary-isthmic junction, and C and D were the distance between the ampullary-isthmic junction and the uterotubal junction. The eggs in each section were counted and calculated in percentages of total eggs recovered, and were compared with the number of ovulation points in the ovary.

C. *Results*

Under direct vision it was possible to calculate the number, position, and speed of migration of the eggs. No attempt was made to distinguish between fertilized and unfertilized eggs. The results of the controls are seen in Table 1, and the immunized animals in Table 2. Notice that there was an equal number of ovulation points between the controls and immunized animals, again demonstrating that there was no influence on ovulation *per se*. The data obtained were calculated in terms of per cent at each position in order to express the rate of speed of egg migration. The stippled figures indicate the highest percentages of eggs in any particular position. It is apparent that in the controls 90.3% of the eggs were recovered, compared with 65.4% in the immunized animals. It appeared that the eggs in the control animals migrated down the oviduct more rapidly, but in neither the controls nor the immunized animals did the eggs enter the uterus before 60 hr after ovulation (cf. Greenwald 1961).

TABLE 2

EGG TRANSPORT STUDIES IN 18 RABBITS IMMUNIZED WITH WHOLE RABBIT SEMEN SHOWING POSITION OF EGGS AT 30 TO 96 HR AFTER INSEMINATION; ONLY 65.4% OF EGGS WERE RECOVERED[a]

EGG TRANSPORT					IMMUNIZED						
					Egg Position						
Rabbits	Hours after Insem.	Hemagglutinin Titer		Ovulation Points	←Ovary			Vagina→		Eggs Recovered	
					Oviduct			Uterus			
		Peak	Autopsy		A	B	C+D	E	F	No.	%
3	30	320 5,720 2,560	320 5,720 2,560	30	5	2	14	0	0	21	70
3	36	640 1,280 1,280	640 640 640	34	0	5	22	0	0	27	79.4
3	48	640 2,560 320	640 2,560 320	36	0	2	26	0	0	28	77.7
3	60	1,280 1,280 60	1,280 1,280 60	32	0	6	11	8	0	25	78.1
3	72	2,560 320 1,280	2,560 160 1,280	31	0	0	7	4	10	20	64.5
3	96	640 5,720 320	640 1,280 320	34	0	0	0	4	6	10	29.4
18				197	Totals					131	65.4

[a] Positions A–F refer to the arbitrary divisions of the genital tract of Figure 6.

III. Sperm Transport Studies

A. *Immunization Technique*

Twenty sexually mature female rabbits, 3.5 kg by weight, were separated into cages for over 3 weeks to insure the absence of pregnancy. Ten rabbits were immunized intradermally by injection 3 times at 2-week intervals with 2.4 ml of an equal volume of emulsion from Freund's adjuvant and homologous semen (protein content, 20 mg per ml) to 10 sites on the back and 1 pair of hind pads. Fresh semen from the same male rabbit was used. Ten control rabbits were injected as above, using the same volume of saline and Freund's adjuvant. The semen was collected by an artificial vagina from healthy male rabbits and diluted 10 times in saline.

B. *Technique*

The does were killed at 1, 3, 6, 9, and 12 hr after artificial insemination with diluted semen while they were concomitantly injected with 100 i.u. HCG intravenously to induce ovulation. After the animals were killed the entire genital tract was excised and divided in the same fashion as described above. The sections of each part of the genital tract were flushed with normal saline solution at least 3 times. The total number of sperm in each uterine portion was counted by means of a hemocytometer. Sections of the oviduct were centrifuged at 3,000 rpm for 10 min and the concentrated sperm suspension was then counted directly on a slide.

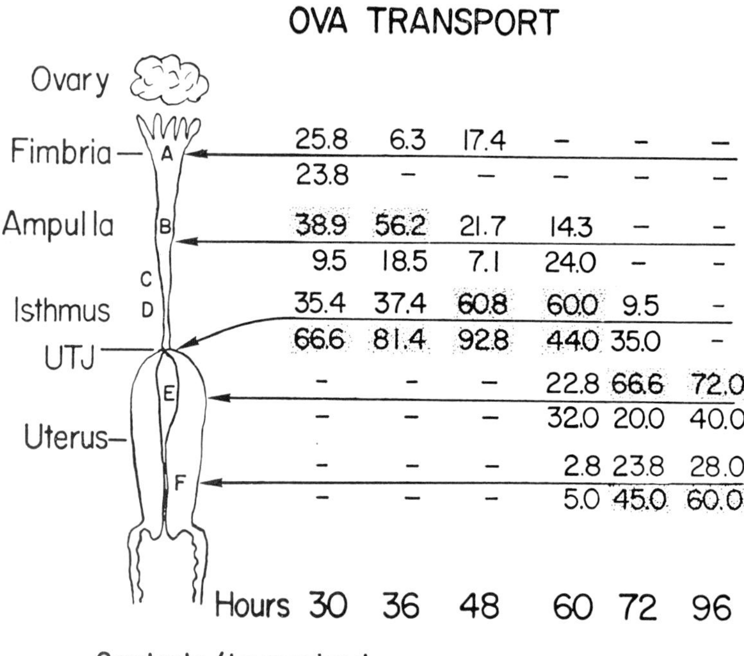

Fig. 6. Egg transport studies. Percentage of eggs in various sections of the oviduct and uterus at 30, 36, 48, 60, 72, and 96 hr. The controls are above the line, the immunized animals below. Stippled areas denote maximal percentage at any particular time.

C. *Results*

The results are interpreted in Figure 7. They show a slight reduction in the number of sperm that reach the fimbriated end of the oviduct, perhaps because of a slightly slower migration rate.

IV. Egg Transfer Studies

A. *Immunization Schedule*

The semen was collected with the artificial vagina, pooled, and stored in the freezer at −80° C. Whole semen was used, and the protein content in the pooled seminal plasma was measured. The pooled sample was diluted with normal saline in order to adjust the

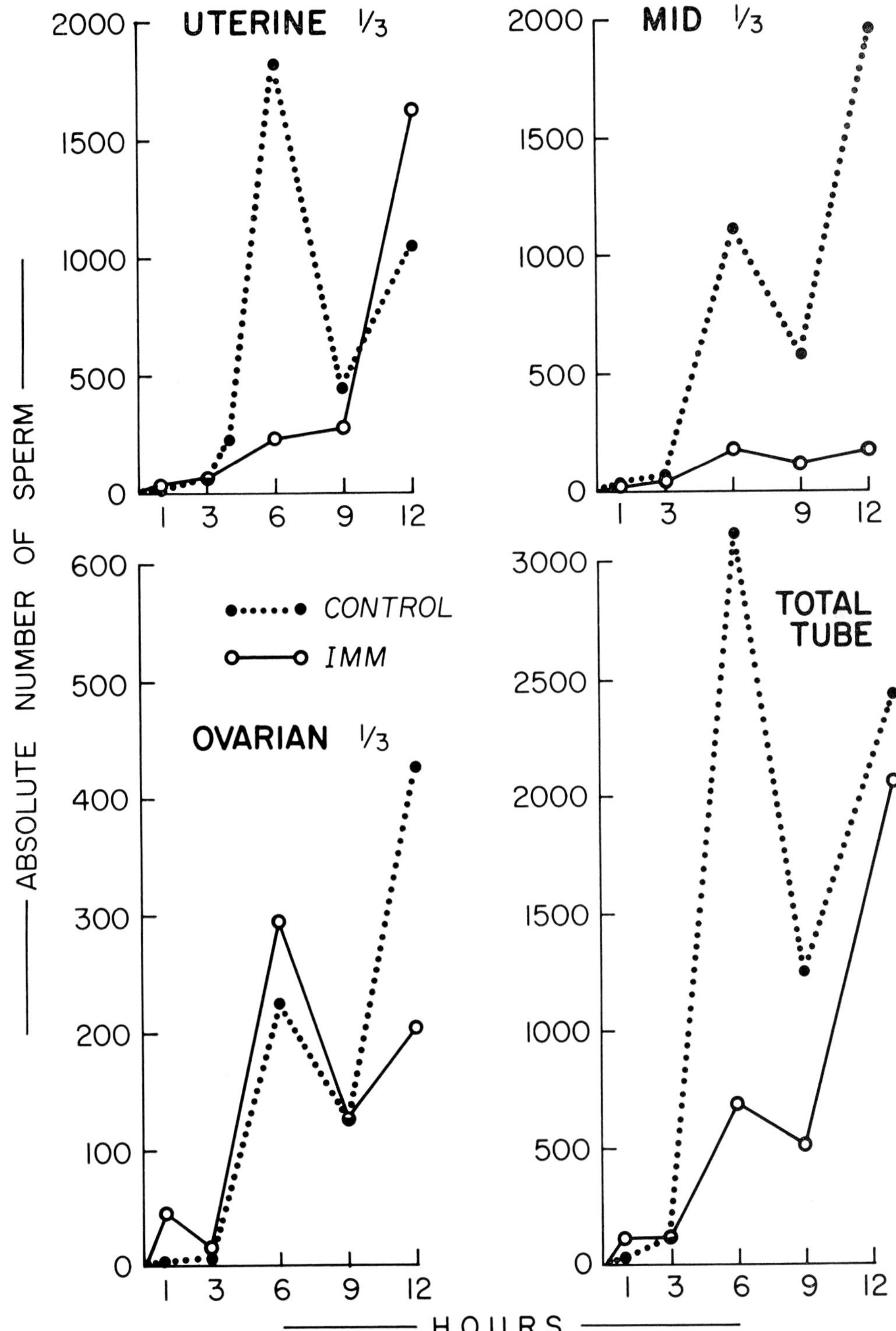

FIG. 7. Sperm migration studies showing absolute number of sperm in uterine, mid- and ovarian thirds of the rabbit oviduct in immunized (......) and control animals (———) at 1, 3, 6, 9, and 12 hr after mating. The lower right hand is a composite of number of sperm in the total tube.

protein content to 20 mg per ml. This suspension was mixed with complete Freund's adjuvant in equal volumes. 2.4 ml of this mixture, containing 24 mg of protein, were injected subcutaneously in 10 different sites over the back of the animal. In the beginning of the experiment the foot pads were used, but considering the multiple infections, this procedure was abandoned. Immunizations were repeated every 2 weeks until the antibody titer was at least 1/1,000. In most of the animals this was obtained after the first immunization.

Pregnant female New Zealand rabbits, 8 to 10 pounds, were purchased. These animals were kept in separate cages throughout the study and the number of offspring counted. In the controls, each delivered an average of 7, and in the immunized, the average number of young was 6.2. The animals were divided into two major groups. Group I was immunized with Freund's adjuvant plus rabbit whole semen, and the second group, the controls, received no treatment. Each major group had a subgroup, half used as recipients and half as donors.

The animals selected to be the recipients were made pseudopregnant by giving 200 i.u. of HCG intravenously, and 1 ml of frozen whole semen containing dead spermatozoa was introduced into the vagina in the same way as for the donors.

B. *Technique of Superovulation of the Donors*

Immunized and control animals were superovulated when the highest peak of the blood circulating antibody was determined. At that time 25 units of Pergonal®[1] were injected subcutaneously or intramuscularly for 3 days. Twenty-four hr later, 200 i.u. of HCG were injected intravenously and 36 to 40 hr after the injection, the eggs were recovered at the 4–8-cell stage. At the same time that the HCG was injected into these animals, 1 ml of fresh whole semen, with a high sperm count and good motility, was introduced into the vagina with a long tube and deposited near the cervix. Thus, we had fertilized eggs from both control and immunized donors ready to be inserted into control and immunized recipients.

C. *Technique of Ovum Transfer*

Thirty-six to 40 hr after induction of ovulation the recipient animals were taken to the operating room and egg transfer was done under sterile conditions. Pentobarbital, 50 mg per kilo, was given intravenously and anesthesia maintained with ether. A midline incision was made in 6 animals for each transfer; (1) an immunized donor, (2) a control donor, (3) 2 immunized recipients, and (4) 2 control recipients. When the donors were anesthetized, the animals were exsanguinated by heart puncture and killed immediately after the abdomen was opened. Both oviducts were dissected from the fat pad and flushed with 2 ml of normal saline containing 10% normal rabbit serum. Flushing was performed with a blunt 20 gauge needle inserted into the uterotubal junction. The fluid was flushed into a watch glass. The eggs were observed immediately under the dissecting microscope, counted, and examined for cleavage. The ovulation points on the ovary were counted also.

Five fertilized eggs from the immunized donors were introduced into the right oviduct of the recipient rabbits with a pipette syringe containing about 2 drops of fluid. Five fertilized eggs from a control donor were introduced into the left oviduct. The abdomen was closed immediately and penicillin and streptomycin were given for the subsequent 3–4 days.

On the 9th day the animals were killed by exsanguination and the genital tract was dissected in order to count the implantation sites and to study them histologically.

[1] Human menopausal gonadotrophin (Cutter Laboratories).

D. *Results*

Forty-nine eggs were transferred, of which 22, or 44.9%, implanted. Significant is the fact that there was no difference in the number of implantations coming from either the immunized donors or the control donors. Of the 40 eggs transferred into recipient immunized animals, only 4, or 10%, implanted. With the small numbers of implantations no comment can be made as to differences between immunized and control donors. Essentially, then, irrespective of whether the donor eggs came from control or immunized animals, if the recipient was immunized the implantation rate dropped significantly. A question that is not answered, however, is whether immunization caused a failure of implantation or a postzygotic death in the oviduct.

V. Hemagglutination and Immunodiffusion Tests of Serum and Tubal Fluid

A. *Technique*

The animals were immunized with whole rabbit semen as the antigen in the same way described above for 2 weeks, followed by bleeding for hemagglutination and immunodiffusion tests. All animals were operated on and a oviductal fluid collector inserted according to the method of Hamner, described elsewhere in this text.

1. *Hemagglutination Technique.* This method follows that of Stavitsky (1961) for microhemagglutination.

2. *Micro-Gel-Diffusion Test and Staining with Ponceau S.* A very hot solution of 2% Difco Bact agar (PBS,[2] pH 7.1) is pipetted rapidly into an immunodiffusion plate. After solidification of the agar, the puncher is used to stamp a central well with a ring of 6 wells around it. Well diameters are 3.0 mm. Clear distance between edges of the central well and peripheral wells is 5.0 mm. The antigen is placed in the central well. Rabbit sera and oviductal fluids are added to the peripheral wells. The plate is placed into a humidified 37° C incubator for development of precipitation. After 48 hr incubation, the immunodiffusion patterns may be viewed directly against a dark field or enhanced with a suitable histochemical dye.

After the pattern has developed the agar is removed from the slide and cut into a thin layer of 1 to 2 mm thickness. Unreacted protein is washed from the agar by placing the agar in a petri dish filled with PBS and distilled water. The fluid is changed about 3 times over a period of 1 to 2 days, after which the sections are dried on microslides. The dried agar is stained for 10 min in a jar filled with Ponceau S in 5% trichloracetic acid. The excess stain is removed by placing the slide into a jar filled with washing solution (5% acetic acid) changing the acetic acid 2 to 3 times over a period of 5 to 20 min.

B. *Results*

Figure 8 demonstrates the whole rabbit semen antibody titer in the serum as well as the oviductal fluid. It will be noted that the antibody titer rises steeply after the third immunization, and immunodiffusion becomes positive at titers above 1:2,000. At no stage was an immunodiffusion test positive in the oviductal fluid, nor did the oviductal fluid antibody titer rise above 1/240 under these experimental conditions.

[2] Phosphate Buffer Solution.

VI. Discussion

It has been claimed that immunizing an animal intramuscularly, intradermally, intravenously, and even transvaginally with various extracts of male reproductive tract secretions, i.e., testicular extract, whole semen, seminal plasma, or washed sperm, will result in varying degrees of fertility inhibition. Because of the eventual practical and clinical importance, it now becomes necessary to answer several questions: (1) Are the effects truly an immunological reaction? and, if so, (2) What is the specific antigen responsible? (3) What is the nature of the antibody? (4) What is the precise immune mechanism? (5) At what physiological level does the immune response operate, i.e., failure of ovulation, fertilization, implantation,

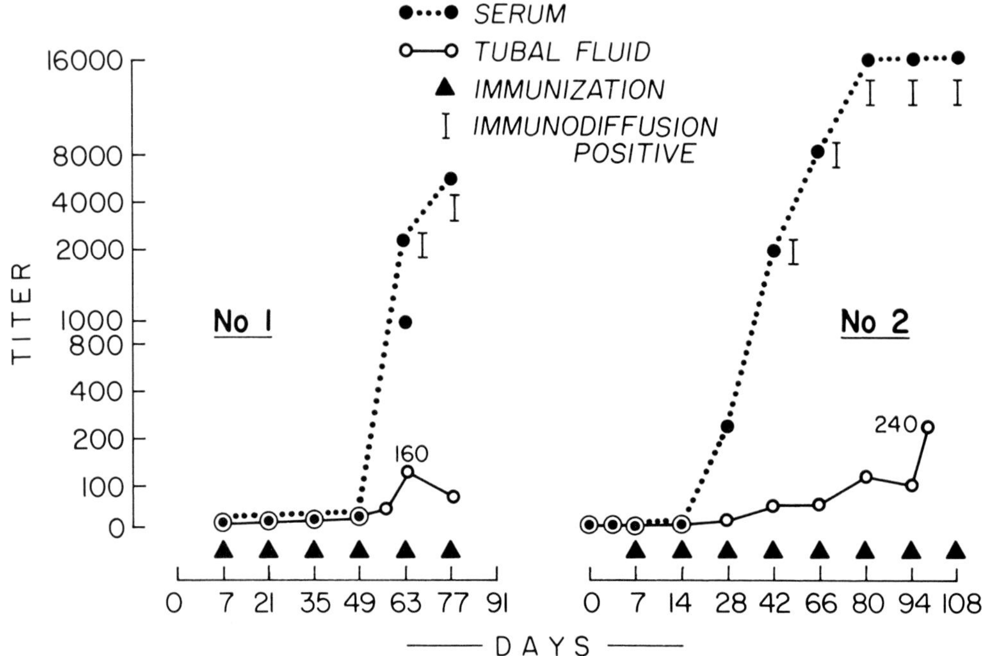

FIG. 8. Rabbits immunized with whole rabbit semen plus Freund's complete adjuvant in 2 representative animals. Note the rise in hemagglutination titer in the serum with positive immunodiffusion at titers above 1/2,000. In the oviductal fluid the titer becomes considerably positive later, does not exceed 1/240, and does not act positively on immunodiffusion.

sperm transport, etc.? (6) Does the oviductal fluid play a role at all, and if so, to what extent? (7) Is it a prezygotic or postzygotic effect? and, ultimately, (8) Are the results of lowered fertility perhaps not simply a generalized metabolic effect on the host animal due to repeated immunization?

There is little doubt that repeated experiments by us and others have illustrated that the classic primary and secondary effects occur and fertility is reduced. However, the precise nature of the antigen is still unknown though more and more evidence is appearing, suggesting that seminal plasma is not the major source of the antigen. We do not know whether the antibody responsible for the fertility reduction is an agglutinating, precipitating, immobilizing or tissue-fixed one. This is further complicated by the fact that different routes of immunization will give rise to different levels of antibody elevation. How, then, is one to assess the role of oviductal fluids when the ideal immunization schedule has not been determined, and the specific antigen has not been isolated? Which antibody or immunological response should be sought to study the role and effects of oviductal fluid?

As both sperm or seminal plasma are individually antigenic, we used whole isologous rabbit semen as the antigenic agent, being fully aware that the mixture of proteins does not make a particularly good antigen. The rationale was simply that if positive results were obtained then one could narrow down the specific antigen more logically. The method and route of immunization was admittedly excessive for the same reason, but special care was taken not to invalidate the results by creating a metabolically sick animal. Having immunized the animals, a series of experiments was designed to determine whether oviductal function, as opposed to oviductal fluid secretion, was affected by immunization. Thus, the first experiment was a study of egg transport, and the results show that 90.3% of eggs were picked up in the control animals, whereas only 65.4% of the eggs in the immunized group were transported. Migration down the tube was essentially the same in both groups. In the sperm transport studies there was very slight reduction in the actual number of sperm available for fertilization in the oviduct of the immunized group, probably because of a somewhat slower migration rate. However, in neither the egg nor the sperm transport studies was there a sufficient difference to implicate damage of oviductal function as a result of immunization with whole rabbit semen.

The egg transfer studies (Gonzales-Enders and Ackerman 1967) do show that, whether the blastocyst comes from immunized or control animals, there is no detectable difference in egg transport down the oviduct or implantation if the recipient animal has not been immunized. There is, however, significant difference when the recipient has been immunized, namely, 10% implantations in the immunized group against 44.9% implantations in the nonimmunized group. Was this due to tubal motility effects? In view of the previous experiments, this is unlikely. Was this perhaps an effect of antibodies in oviductal fluid upon the fertilized egg? To prove this, one would have to show the presence of antibodies in the oviductal fluid, and in these animals no antibodies could be found in the oviductal fluid. Was this, as previously postulated, perhaps a failure of implantation, or perhaps just a result of a metabolically sick animal? Certainly this can be answered only by demonstrating the presence of antibodies in oviductal fluid and noting their effects upon sperm, as well as upon both unfertilized and fertilized eggs.

Restall and Wales (1966), using sheep, showed that glucose utilization and lactate accumulation by sperm were stimulated by oviductal fluids. In the human, antibodies of the ABO system in anti-A or anti-B serum can affect the metabolism of sperm from men of A or AB secretor phenotypes (Ackerman 1967). In the sperm metabolism studies performed on immune rabbit sera, there was clear evidence of a depressing effect upon the glycolytic ability and function of washed rabbit sperm by rabbit whole semen antiserum. This inhibition, in turn, seems to be a function of the titer of the antisera. It might be inferred from this finding that wherever circulating antibodies are found in the female reproductive tract a corresponding effect of inhibition upon sperm function will be exerted. Consequently, it may be hypothesized also that immunological effects on fertility may be significant in both the pre- and postzygotic stages of reproductive physiology. However, within the framework of the last experiment (Fig. 8), no appreciable agglutinating or precipitating antibodies were found in the oviductal fluid, and therefore, no sperm metabolism studies have been done using immune oviductal fluid as yet. The failure to find antibodies in any quantity in oviductal fluid is a fascinating enigma. It is not strange that a transudate should contain only a fraction of the amount of globulin in the serum. Certainly the titer was sufficiently elevated that antibodies should have been present. It may be, however, that the quantities of antibody present were so small that the tests employed were not sufficiently sensitive to demonstrate them. Another possible explanation presents itself. These were virgin animals and, consequently, until the antigenic stimulus was presented to the reproductive tract, that is, whole semen, there would be no aggregation of antibody-bearing lymphocytes in the oviducts, thus bringing antibodies to the lumen. It is essential that this type of experiment be repeated after adding transvaginal immunization to the schedule or mating the animal prior to the final collection of oviductal fluid. If antibodies are found, then it will also be possible to do sperm metabolism studies on immune oviductal fluid. Such studies are now in progress in our laboratories. If, on the other hand, no significant amounts of antibody can be demonstrated in the oviductal fluid, the possibility of specific selectivity in the oviduct of the larger molecules and its significance must be entertained.

In final summation, therefore, it appears that immunization of rabbits with isologous whole rabbit semen produces circulating agglutinating and precipitating serum antibodies which have little, if any, effect on oviductal activity. There is, however, considerable reduction in implantations of transferred fertilized eggs to similarly immunized recipients, but it has not been definitely determined whether this is an effect of oviductal fluid on the zygote or failure of implantation. Sperm metabolism is greatly reduced when exposed to whole rabbit serum and whole rabbit semen antiserum, but as no antibodies were determined in oviductal fluid from such immunized animals, metabolism studies of sperm in oviductal fluid are still incomplete.

The use of whole semen as antigen probably results in antibodies to seminal plasma and very little to sperm. Thus, it is most important that, in the future, sperm or fractionated sperm be used as antigen.

References

1. Ackerman, D. R. 1967. Antibodies of the ABO system and the metabolism of human spermatozoa. *Nature* 213:253.
2. Austin, C. R. 1957. Fate of spermatozoa in the uterus of mouse and rat. *J. Endocr.* 14:335.
3. Behrman, S. J. 1965*a*. *Immunological aspects of fertility and infertility*. Biological Council Symposium on "Agents Affecting Fertility," p. 47. London: J. and A. Churchill, Ltd.
4. ———. 1965*b*. Agglutinins, antibodies, and immune reactions. *Clin. Obstet. Gynec.* 8:91.
5. ———. 1966. Immunologic phenomena in infertility: Some clinical applications. *Harper Hosp. Bull.* 24:147–55.
6. Behrman, S. J., and Kistner, R. W., eds. 1968. Immunologic mechanisms. In *Progress in infertility*. Boston: Little, Brown and Co.
7. Behrman, S. J., and Otani, Y. 1963. Transvaginal immunization of the guinea pig with homologous testis and epididymal sperm. *Int. J. Fertil.* 8:829–34.
8. Bishop, D. W., and Tyler, A. 1963. Mechanisms concerned with conception. In *Immunological phenomena*, ed. C. G. Hartman, chap. 8. New York: Macmillan Co.
9. Braden, A. W. H. 1953. Distribution of sperm in the genital tract of the female rabbits after coitus. *Aust. J. Biol. Sci.* 6:693.

10. Edwards, R. G. 1960. Antigenicity of rabbit semen, bull semen, and egg yolk after intravaginal or intramuscular injections into female rabbits. *J. Reprod. Fertil.* 1:385.
11. Gonzales-Enders, R., and Ackerman, D. R. 1967. Unpublished data.
12. Greenwald, G. S. 1961. A study of the transport of ova through the rabbit oviduct. *Fertil. Steril.* 12:80.
13. Katsh, S. 1957. *In vitro* demonstration of uterine anaphylaxis in guinea pigs sensitized with homologous testis or sperm. *Nature* (*Lond.*) 180:1047–48.
14. ———. 1958. Demonstration *in vitro* of an anaphylactoid response of the uterus and ileum of guinea pigs injected with testis or sperm. *J. Exp. Med.* 107:95–108.
15. ———. 1966. Immunological control of reproduction in experimental animals: implications regarding human fertility and infertility. *Ann. Allergy.* 24:615.
16. ———. 1967. Immunologic aspects of infertility and conception control. In *Advances in obstetrics and gynecology*, ed. S. L. Marcus and C. C. Marcus, chap. 34. Baltimore: Williams and Wilkins Co.
17. Katsh, S., and Katsh, G. F. 1965. Perspectives in immunological control of reproduction. *Pac. Med. Surg.* 73:28.
18. Kerr, W. R., and Robertson, M. 1947. A study of re-exposure to Tr. foetus in animals already exposed to the infection. *J. Comp. Path.* 57:301.
19. Landsteiner, K. 1899. Zur Kenntnis der spezifisch auf Blutkörperchen wirkenden Sera. *Zbl. Bakt. Paras.* 25:546.
20. McLaren, A. 1964. Immunological control of fertility in female mice. *Nature* 209:582.
21. ———. 1966. Studies on the iso-immunization of mice with spermatozoa. *Fertil. Steril.* 17:492–99.
22. Mancini, R. E.; Andrada, J. A.; Saraceni, D.; Bachmann, A. E.; Lavieri, J. C.; and Namirovsky, K. 1965. Immunological and testicular response in man sensitized with human testicular homogenate. *J. Clin. Endocr.* 25:859.
23. Mattner, P. E., and Braden, A. W. H. 1963. Spermatozoa in the genital tract of the ewe. I. Rapidity of transport. *Aust. J. Biol. Sci.* 16:473–81.
24. Menge, A. C. 1967*a*. Origin of the antigens in rabbit semen which induce antifertility antibodies. *J. Reprod. Fertil.* 13:31.
25. ———. 1967*b*. Induced infertility in cattle by iso-immunization with semen and testis. *J. Reprod. Fertil.* 13:445.
26. Metchnikoff, E. 1899. Études sur la résorption des cellules. *Ann. Inst. Pasteur* 12:737.
27. Moyer, D. L. 1965. Fate of spermatozoa in the female genital tract. Presented at the Pacific Coast Fertility Society, Oct. 1965.
28. Moyer, D. L.; Kunitake, G. M.; and Nakamura, R. M. 1965. Electron microscopic observations on phagocytosis of rabbit spermatozoa in the female genital tract. *Experientia* 21:1–4.
29. Orsini, M. W. 1962. Technique of preparation, study and photography of benzylbenzoate cleared material for embryological studies. *J. Reprod. Fertil.* 3:283–87.
30. Otani, Y., and Behrman, S. J. 1963. Immunization of the guinea pig with homologous testis and sperm. *Fertil. Steril.* 14:456–67.
31. Paine, P. and Behrman, S. J. 1967. Fluorescent antibody localization in the female reproductive tissues of the guinea pig. *Proc. Symp. Sperm Physiology, Bulgaria.* In press.
32. Restall, B. J., and Wales, R. G. 1966. The Fallopian tube of the sheep. IV. The metabolism of ram spermatozoa in the presence of fluid from the Fallopian tube. *Aust. J. Biol. Sci.* 19:883–93.

33. Richter, M. 1964. Lack of correlation of antibody titers as determined by the precipitin and hemagglutination technique. *Acta Allerg.* 19:1–10.
34. Sawada, Y., and Behrman, S. J. 1967. Reduction of fertility in rabbits by iso-immunization: Mechanism of action. *Proc. Fifth World Congr. Excerpta Medica Foundation,* 133:758.
35. Schwimmer, W. B.; Ustay, K. A.; and Behrman, S. J. 1967. Sperm-agglutinating antibodies and decreased fertility in prostitutes. *Obstet. Gynec.* 30:192.
36. Shulman, S.; Hubler, L.; and Witebsky, E. 1964. Antibody response to immunization by different routes. *Science* 145:815.
37. Stavitsky, A. B. 1961. *In vitro* studies of the antibody response. *Adv. Immun.* 1:211.
38. Tomasi, T. B. 1967. The gamma A globulins: First line of defense. *Hosp. Pract.* 2:26.
39. Tyler, A. 1949. Properties of fertilizin and related substances of eggs and sperm of marine animals. *Amer. Nat.* 83:195–219.
40. ———. 1961. Approaches to the control of fertility based on immunological phenomena. *J. Reprod. Fertil.* 2:473.
41. Weil, A. J. 1960. Immunological differentiation of epididymal and seminal spermatozoa of the rabbit. *Science* 121:1040–41.
42. Weil, A. J., and Finkler, A. E. 1958. Antigens of rabbit semen. *Proc. Soc. Exp. Biol. Med.* 98:794–97.
43. ———. 1959. Isoantigenicity of rabbit semen. *Proc. Soc. Exp. Biol. Med.* 102:624–26.

15

Biochemistry of Oviductal Eggs in Mammals

E. S. E. Hafez/S. Sugawara
Reproduction Laboratory, Department of Animal Sciences
Washington State University, Pullman
Faculty of Agriculture
Tohoku University
Sendai, Japan

Several major techniques have been used to study the biochemical characteristics of mammalian eggs, e.g., the use of histochemical or cytochemical techniques to observe the end products of reactions in intact cells, and the application of biochemical *in vitro* assays to extracts of disrupted cells. These techniques may involve autoradiography, the use of radioactive metabolites, isolation and homogenization procedures, and enzyme assays.

This chapter will deal briefly with the chemical composition, permeability, absorptive ability, biochemical characteristics, and nutritional requirements of oviductal eggs. (See chapters by Brinster and Glass in this volume.)

I. Chemical Composition of Eggs

Differences exist in the cytoplasmic structures of various species of mammalian eggs (Austin 1961; Boyd and Hamilton 1952). Some of the most remarkable aspects of comparative morphology of the egg are the apparent discrepancies observed at the light- and electron-microscopic levels (Hadek 1965).

Mammalian eggs can be classified into three general types from the distribution of histochemically detected lipid and glycogen. Eggs of cattle, sheep, goat, pig, ferret, dog, and horse (Type I) contain a large amount of lipid and either a large amount of glycogen or none at all. Eggs of the rat, mouse, and hamster (Type II) have little or no lipid, and either a large amount of glycogen or none at all. Eggs of the rabbit (Type III) contain small amounts of lipid and small amounts of glycogen (Ishida 1954, 1960).

The major chemical components of 1-cell ovulated eggs have been measured quantitatively in the mouse. The cytoplasm was 85% water and 15% solid matter; the latter was composed of 12.5% lipid, 71% protein, and 16% carbohydrate (Loewenstein and Cohen

1964). To date there is little information concerning the distribution of protein in mammalian eggs.

Between fertilization and the first cleavage, the egg undergoes deutoplasmolysis (Austin 1961). Other than the distributional changes of fine granular material in the fertilized egg, little is known concerning the biochemical relationship between the metabolic activities and deutoplasmolysis.

Cytoplasmic RNA and mucopolysaccharides remain relatively constant to the 8-cell stage and increase thereafter during development. Large amounts of glycogen are present during the cleavage of mouse eggs (2-cell to morula) but decrease during the expansion of the blastocyst (Thomson and Brinster 1966), suggesting that glycogen may serve as a source of energy during the 5th day of prenatal development (Fig. 1).

II. Permeability and Absorptive Ability of Eggs

The permeability and absorptive ability of oviductal eggs are determined by structural properties of the cell membrane. The major structural, physiochemical, and osmotic properties of the cell membrane have been described and reviewed (Dick 1959; Solomon 1960). Neither the ultrastructure nor the mechanisms controlling function is well understood.

The mammalian egg is covered by three concentric layers: the *primary* layer secreted by the oocyte itself; the *secondary* layer secreted by the follicle cells; and the *tertiary* layer secreted by the oviduct in certain species. The vitelline membrane is a cortical differentiation of the ovarian cytoplasm. It is similar to other animal cell membranes in functions of diffusion and active transport. The zona pellucida is a homogeneous, semipermeable membrane composed chiefly of neutral or weakly acidic mucoproteins (Braden 1952; Konecny 1959). The presence of polysaccharide in the zona pellucida is deduced from staining reactions: the action of proteases upon the membrane indicates the presence of proteins. The membrane can be removed by proteolytic enzymes—trypsin, chymotrypsin, and pronase—although the ease of digestion varies with the species (for eggs of mouse, see Smithberg 1953; Mintz 1962; and for eggs of rabbit, rat, and hamster, see Chang and Hunt 1956). Fine branching processes from the follicle cells pass through the zona pellucida, ending in contact with the oocyte plasma membrane. The surface membrane of the oocyte is studded with short microvilli which effectively increase its area. The physiochemical properties of the mammalian egg layers change at the time of ovulation, at fertilization, and during cleavage. For example, testicular hyaluronidase increases the permeability of a recently shed egg (Da Silva Sasso 1957).

A. *Permeability to Dyes*

A comparison of the permeability of fertilized and unfertilized eggs is lacking. In the rabbit, hamster, and rat the zona pellucida of the unfertilized egg is permeable to azo dyes

---

FIG. 1. Whole mounts of mouse embryos which developed *in vivo*, stained by the periodic acid-Schiff method (×115). Note the presence of glycogen in the cleavage stages, from the 2-cell to the morula, and the decrease in glycogen during the growth of blastocyst. (*a*) Fertilized egg with cumulus cells removed. (*b*) Two-cell embryo, approximately 36 hr postfertilization. (*c*) Eight-cell embryo, obtained 2½ days postfertilization. (*d*) Morula, obtained 3½ days postfertilization. (*e*) Early blastocyst showing the beginning of cavitation. (*f*) Blastocyst obtained approximately 4 days after mating. (*g*) Blastocyst obtained 4 days after fertilization, showing positive material confined to the inner cell mass. (*h*) Late blastocyst obtained on the 5th day after hatching. (From Thomson and Brinster 1966.)

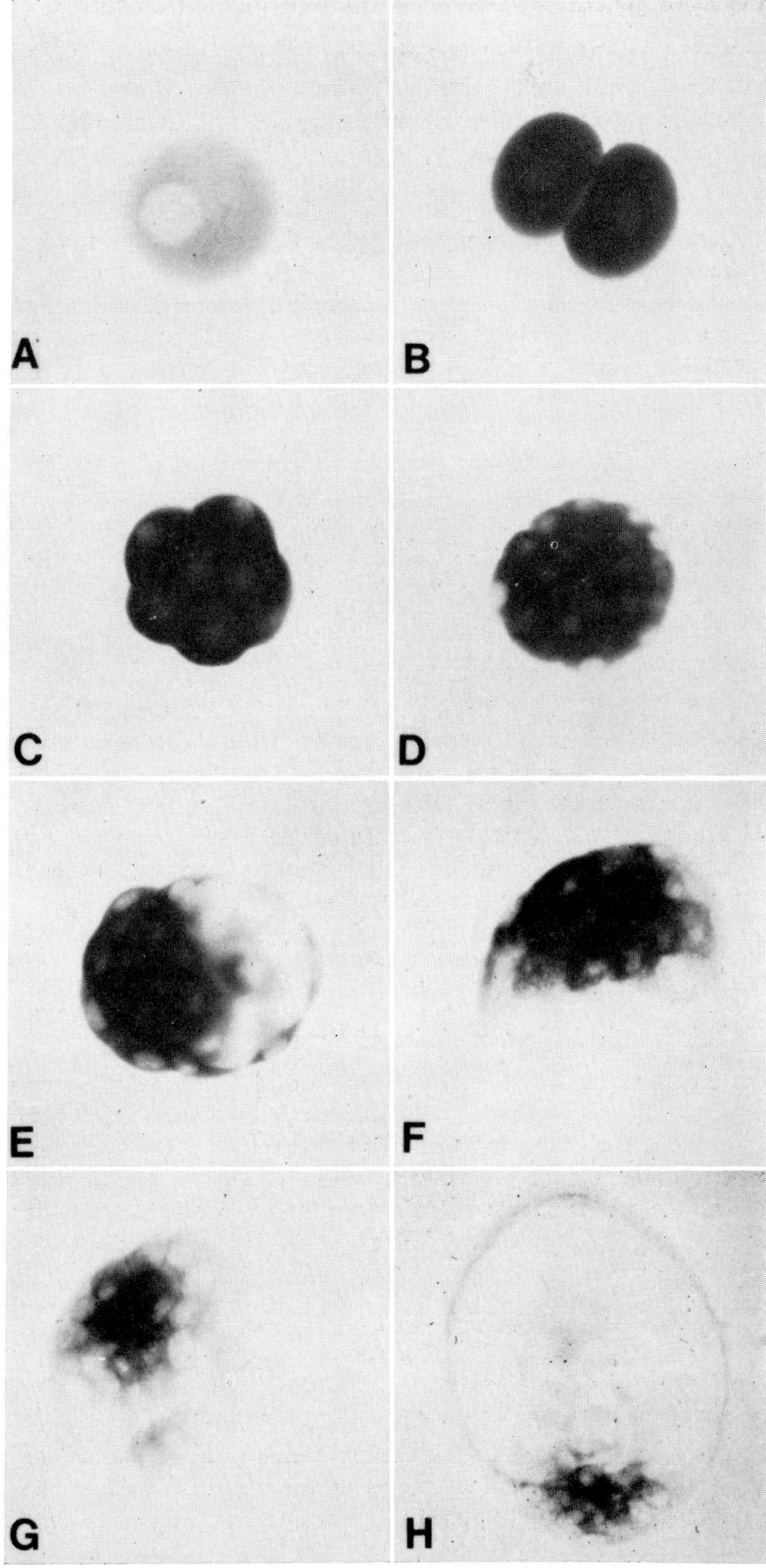
A
B
C
D
E
F
G
H

with a molecular weight of 1,200 or less, but not to substances with a molecular weight of 1,600 or more (Austin and Lovelock 1958). The permeability of the membrana granulosa is perhaps slightly less than that of the zona pellucida, but is still sufficient to allow passage of substances with molecular weights of 1,200.

B. *Transport of Inorganic Ions*

Extensive investigations have been conducted on the active transport of inorganic ions into frog and sea urchin eggs and many other animal cells and tissues (Keynes 1960). Active transport of inorganic ions and of substrates require energy from ATP (Past and Albright 1960). Enzymatic activity in the membrane itself may play an important role in transportation, as suggested by the very high phosphatase activity (Daniell 1952). The enzymatic basis for active transport of sodium and potassium ions (Na^+ and K^+) has been reviewed by Skou (1965).

The following experiments demonstrate that active transport of potassium ions seems to occur in mammalian eggs (Austin 1961). Two-cell eggs of rats maintained for 18 hr in isotonic solutions of different Na and K ratios show distinct volume differences. As the concentration of potassium ions is increased, the egg expands maximally within the zona pellucida. The pH in the egg also may play a role in absorption of ions, as suggested by studies in smooth muscle cells (Kostyuk and Sorokina 1960).

C. *Transport of Organic Substances*

The transport of organic substances, especially metabolic intermediates, is crucial for normal development. Membrane permeability is a fundamental change occurring in the egg as a result of sperm penetration. In sea urchin eggs the rate of transfer of labeled inorganic or organic molecules across the membrane increases considerably following fertilization (Monroy and Vittorelli 1962). To date there are no data concerning active transport of organic substances in mammalian eggs. The results discussed earlier on permeability of azo dyes suggest that metabolic intermediates with a maximum molecular weight of 1,200 can be absorbed. The role of glucose in metabolism of the developing blastocyst has been studied by Fridhandler, Wastila, and Palmer (1967).

III. Biochemical Characteristics of Eggs

Repressor substances manufactured by the egg in the later phases of maturation inhibit both the metabolic activities residing in the cytoplasm and the genetic activity of the nucleus. The key reaction at the time of sperm penetration is the removal of the repressor(s), releasing the cytoplasmic metabolic activities, and the activity of the nuclear genetic system (Monroy 1965). The block involves the energy-yielding systems; hence, all the reactions requiring a large amount of energy are inhibited before fertilization. This may be due to the accumulation of inhibitory substances during egg maturation. Little is known about the mechanisms which produce the inhibitors during the final phases of maturation, or how they are inactivated.

A. *Energy Requirements for Development*

Energy requirements of the embryo during prenatal development are related to size and shape of the individual cells, to the character, number, size, and shape of various organelles, and to the amount and nature of intercellular materials (Boell 1955). Biochemical parameters of differentiation may be detected by changes in: (1) the metabolic rate, (2) the input or output of specific substances, or (3) the appearance of, or change in, concentration of enzymes, metabolic intermediates, or products of specific metabolic pathways.

B. *Oxygen Consumption during Development*

Dragoiu, Benetato, and Oprean (1937) reported that oxygen uptake of the mature follicular eggs of cow and rat is 21–35 mμl/egg/hr, and 1.11 mμl/egg/hr, respectively. These values are of doubtful significance, since the follicular cells were still attached to the eggs at the time the measurements were made. For example, follicular pig eggs respire 5 times faster (4.31 mμl/egg/hr vs. 0.82 mμl/egg/hr) than pig eggs in which the follicular cells have been dispersed by hyaluronidase (Sugawara 1962). Even in this case the effect of hyaluronidase on the respiration is not clear.

The effect of sperm penetration on the respiratory activity of eggs varies with the species and the experimental conditions (Borei 1948; Brachet 1960). It is difficult to evaluate such effects in mammalian eggs. No differences were observed between fertilized and freshly ovulated unfertilized eggs of rabbits or rats (Fridhandler, Hafez, and Pincus 1956*a;* Sugawara and Umezu 1961), suggesting that the respiratory activity of the egg is not influenced by sperm penetration. In the rat, the respiratory activity increases gradually during cleavage from the 1-cell to the morula stage, but at the blastocyst stage there is a sudden two- to threefold increase (Sugawara and Umezu 1961).

C. *Respiratory Quotient*

The respiratory quotient during early development varies with the species. *In vitro* studies on chick embryos have shown that carbohydrate is the major energy source during early cleavage, whereas lipid is the predominant source from day 5 throughout the remainder of the incubation period (Boell 1955). The respiratory quotient of amphibian embryos increases from an initial value of 0.65 to approximately 1.0 at the gastrula stage, in close correlation with the beginning of carbohydrate utilization (Brachet 1957). In sea urchin eggs, the respiratory quotient after fertilization is 0.73; during subsequent development it increases as carbohydrate metabolism becomes increasingly predominant.

Little is reported about the respiratory quotient of mammalian eggs (Sugawara 1964). It increases from 1.2 to 1.4 in rat eggs when substrates, such as α-ketoglutaric acid, lactic acid, and glutamic acid, are added to the incubation medium. Such compounds are used both as oxidizable substrates for energy production and biosynthetic precursors of embryonic mass. The amount of carbon dioxide produced by rat eggs at all developmental stages, prior to the blastocyst, suggests that carbohydrate is the major energy source (Dalcq 1957; Ishida 1954).

D. *Metabolic Rate*

Respiratory activity based on size and weight is essential for the evaluation of metabolic activity of embryonic material during development. The rate of oxygen consumption in different species is proportional to the weight of the egg. The weight is calculated on the basis of a linear dimensional change in volume by the reconstruction method. The volume of rat eggs increases from 0.156 mμl at the zygote stage to 0.211 mμl at the 16-cell stage (Huber 1915). Detailed data on the eggs of other mammalian species are lacking. In general, the rate of oxygen consumed (mμl/mg/hr) by mammalian eggs increases during preimplantation and decreases considerably during the subsequent period of growth and differentiation (Boell and Nicholas 1948; Sugawara and Umezu 1961).

E. *Ion Concentration and Oxygen Consumption*

Osmotic requirements of rat eggs are adequately met by 0.15 M NaCl solution (Boell and Nicholas 1948). Brinster (1965*a*) has reported that the osmolarity of 2-cell mouse eggs is between 0.2002 and 0.3542 osmols and that the optimum osmolarity for the development of the 2-cell stage to the blastocyst is 0.2760. Clearly the oxygen consumption of eggs is affected by changes in the osmolarity of the medium.

The hydrogen ion concentration of the medium has a pronounced effect on the biophysical properties of eggs. The degree of sensitivity varies with the species, stage of development of the embryo, and conditions under which the eggs are examined. For example, the zona pellucida of rat eggs softens at pH more acid than 5 and dissolves at pH 4.5. The zona pellucida of mouse eggs dissolves in buffers of pH 3.58 to 4.50, whereas rabbit eggs require buffers of pH 3 or lower to accomplish the same effect (Hall 1935; Braden 1952; Gwatkin 1964). Little is known about the possible correlation between respiratory activity and hydrogen ion concentration of mammalian eggs.

F. *Energy Production and Synthetic Process*

During the course of development, metabolism is required for two major processes: energy production, and synthesis of macromolecules; the former being a prerequisite for the latter. The generation of energy-rich phosphate compounds is accomplished either through glycolysis or oxidative phosphorylation, the latter taking place in the mitochondria. In early development, energy is supplied primarily by glycolysis (see Brinster, this volume). In contrast, oviductal eggs derive most of their energy from oxidative phosphorylation, catalyzed by dehydrogenase enzymes (cf. Sugawara and Hafez 1968). The decrease in PAS-positive diastase-removable material during the blastocyst stages suggests that glycogen may serve as an energy source during the 5th day of development in the mouse (Fig. 1). Culture experiments and enzyme analyses have provided evidence that the cleaving embryo undergoes some basic changes in the pattern of energy metabolism (Thomson and Brinster 1966).

1. *DNA and RNA Synthesis.* Data on DNA and RNA synthesis in mammalian eggs have been obtained from cytochemical observation (Alfert 1950; Dalcq and Pasteels 1955; Szollosi 1964) and isotopic incorporation studies (Sirlin and Edwards 1959; Szollosi 1964; Mintz 1962, 1964, 1965; Oprescu and Thibault 1965). The first synthesis of DNA is initiated 1 to 2 hr after sperm penetration and persists for approximately 5 to 6 hr. The mean cycles of G_2 and G_1 stages are very short, and DNA synthesis begins immediately with the reconstitution of the nucleus. This time it lasts only for 1 to 2 hr (Szollosi 1964; Oprescu and Thibault 1965).

The discrepancies concerning synthesis of RNA during cleavage arise mostly from the methods used for analysis (Alfert 1950; Mintz 1964). Isotopic incorporation studies (^{3}H-uridine) demonstrate that synthesis of RNA is initiated at the time of fertilization.

2. *Protein Synthesis.* Monroy and Maggio (1964) reviewed the extensive studies on protein synthesis during early development in the sea urchin. The low level of amino acid incorporation suggests that the unfertilized eggs of these animals are in a state of rest with respect to protein synthesis (Monroy 1960). Immediately after fertilization protein synthesis begins and steadily increases, as judged by results of immunological techniques (Tyler 1963). The appearance of specific proteins in embryonic development may be the final product of complex biochemical synthesis by the embryo itself, or the result of the incorporation of pro-

teins of maternal origin into the embryo (Solomon 1965). Both mechanisms are presumed to be governed by complex genetic factors.

How and when protein synthesis occurs during early development of the mammalian embryo is not clear. ^{14}C-glycine accumulates in the nucleolus and vitellus of mouse eggs (Edwards and Sirlin 1958); ^{35}S-methionine is incorporated chiefly into the vitellus of ovarian and unfertilized eggs (Lin 1956). These findings suggest that protein synthesis occurs in very early stages of development. It has been implied that active protein synthesis occurs in follicular eggs and in mouse blastocysts (Greenwald and Everett 1959).

Protein synthesis can be monitored using ^{3}H-leucine. Synthesis occurs in the fertilized mouse eggs before cleavage and in greater amounts at later cleavage stages, e.g., the third

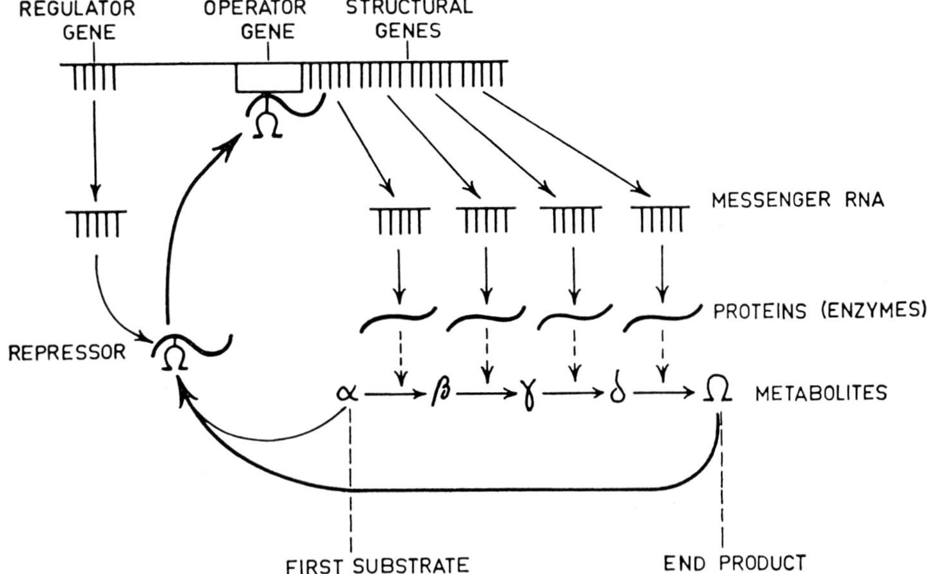

FIG. 2. Interpretation of the mechanism of regulation of the activity of a set of structural genes (an operon). An end product, Ω, in the metabolic sequence catalyzed by a series of enzymes reacts with a repressor gene which then becomes active and blocks the operator site of the operon, thereby preventing the release of messenger RNA from the structural genes. In the presence of an excess of the substrate, α, the activation of the repressor gene fails. (From Gustafson 1965; courtesy Academic Press, New York.)

cleavage and the blastocyst stage (Mintz 1964, 1965). Protein synthesis during these stages seems to be independent of RNA synthesis, as suggested by experiments in which actinomycin-D was employed. A decline eventually did occur but at a much slower rate than the decline in RNA synthesis.

G. *Enzyme Systems*

The transcription of the genetic information is controlled by metabolites in the cell (Fig. 2). The applicability of this concept to biochemical differentiation has been debated. However, evidence shows that substrate-induced formation of enzymes and product repression, as well as so-called repression of catabolites, occurs in embryos, adult organisms, and in explanted tissues and cells cultivated *in vitro*. It is not necessary to rely entirely on induction and repression of enzymes. Complex systems of enzymes have the ability to operate in alternative ways (Fig. 3). The critical factor that determines the choice between

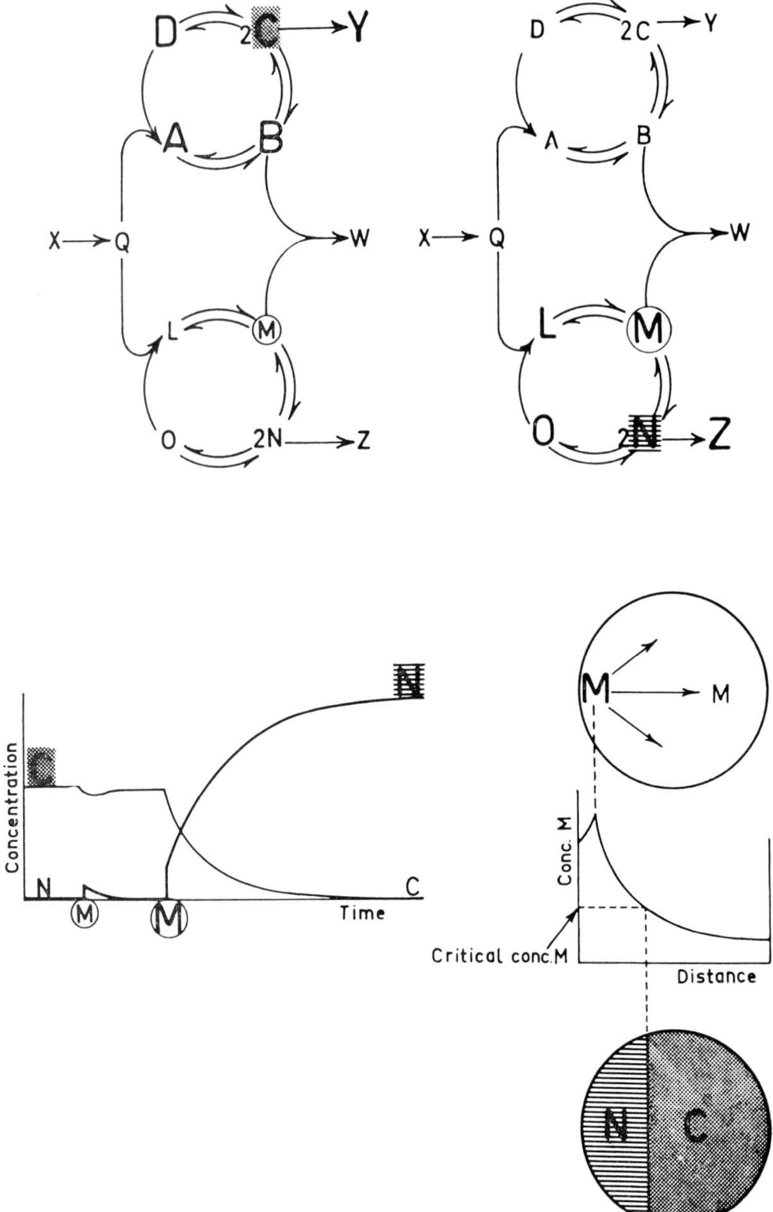

FIG. 3. Diagram of the mode of regulation of a complicated metabolic system where the metabolic flow can be funneled in two directions. The letters indicate various intermediary metabolites. The alternative metabolic patterns are characterized by a high concentration of *C* or *N*, respectively. The first pattern is realized when the concentration of the metabolite *M* is low and tends to stabilize itself, even if a small amount of *M* is added. If the concentration of *M* exceeds a critical level, the pattern changes into the *C* type which tends to be stable. In an egg or embryo where *M* is produced at one pole and a concentration gradient appears, the *N* pattern becomes established within one region sharply delineated from a region with a *C* pattern. (From Gustafson 1965; courtesy Academic Press, New York.)

the metabolic patterns may be the concentration of a single metabolite. A definite respiratory activity is present in embryos at all stages, indicating the existence of the enzymes responsible for respiration. For example, respiration is inhibited by 10^{-3} M cyanide, indicating the presence of the cytochrome oxidase enzyme system (Fridhandler, Hafez, and Pincus 1956*a*, *b;* Sugawara 1962).

The activity of alkaline phosphatase, acid phosphatase, and ATP-ase varies with the stage of development (Dalcq 1954, 1957; Mulnard 1955; Tondour 1959). These activities were determined using various substrates such as glycerophosphate, glucose-1-phosphate, glucose-6-phosphate, and fructose diphosphate.

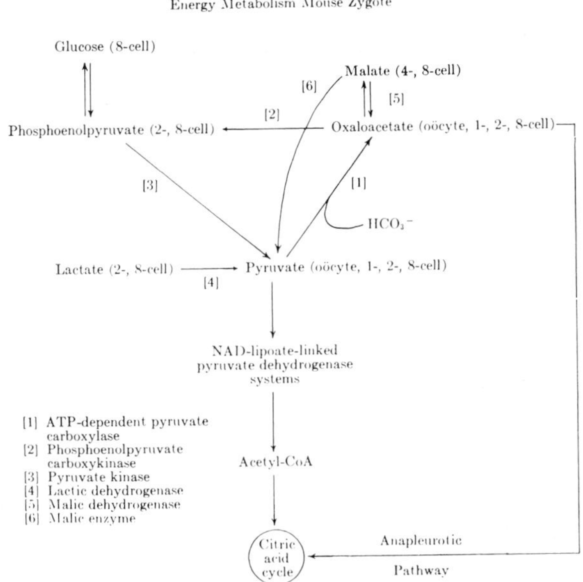

FIG. 4. The compounds able to act as sources of energy to mouse oocytes and early cleavage stages, and their metabolic relations. The stages supported by each compound are shown in parentheses. (From Biggers, Whittingham, and Donahue 1967.)

Succinic and glutamic acid dehydrogenase activities have been demonstrated histochemically in cleaving eggs of the rat, rabbit (Sugawara, 1962), mouse (Lin, Huang, and Chang 1964), and hamster (Ishida and Chang 1965). Other dehydrogenase activities (malate, β-glycerophosphate, glucose-6-phosphate, isocitrate, and lactate) have been observed in rabbit eggs (Sugawara and Hafez 1967). The degree of activity of these enzymes appears to change during cleavage. The activity of succinic and glutamic dehydrogenase in the zygote or the 8- to 16-cell eggs is higher than that of the other enzymes at these stages. The activity is very weak in unfertilized eggs.

A single lactate dehydrogenase enzyme has been observed in mouse eggs from the unfertilized to the blastocyst stage (Rapola and Koskimies 1967), as characterized by electrophoresis, substrate inhibition, and subunit dissociation-recombination experiments; it is identical to mouse lactate dehydrogenase. The UDPG-glycogen transferase and monoamine oxidase activities have been demonstrated in hamster eggs (Ishida and Chang, personal communication). The UDPG-glycogen transferase activity is present in the cytoplasm of the egg but not in the polar body (Fig. 4).

H. *Utilization of Substrates*

It is not known whether the energy required for development of the mammalian egg is supplied by the oxidation of endogenous substrates, the degradation of exogenous substrates, or both. The concept that mammalian embryos develop nutritionally in a closed system is supported by respiratory activity in the presence of several substrates. The addition of intermediates to the TCA cycle did not increase the respiratory activity of morula or earlier stages (Fridhandler, Hafez, and Pincus 1957; Sugawara 1962). However, the respiratory activity was increased at the early blastocyst stage by the addition of substrates; the rate of increase depending upon the specific substrates. Oviductal eggs reduce nitro-blue tetrazolium (Nitro-BT) when incubated in succinate or glutamate, suggesting the utilization of these two compounds.

In the rabbit, pyruvate, cystine, and glutathione stimulated blastocyst growth; whereas glucose and lactate did not (Pincus 1941). On the other hand, mouse eggs were successfully cultured *in vitro* from the 2-cell stage to the blastocyst stage when lactate was present in the medium (Whitten 1957; Mulnard 1964; Brinster 1965*a*, *b*). Amino acid requirement studies of mouse embryos in culture demonstrated that the omission of no single amino acid from the medium completely prevented the development of some 2-cell eggs into blastocysts, although the omission of cystine resulted in a significant decrease in development (Brinster 1965*b*). This finding suggests that sulfhydryl compounds, such as cystine, are essential to maintain maximum activity and synthesis of enzymes.

IV. Conclusions

The oocyte may be considered as an independent cell, but more often it is surrounded by follicular cells which are very active in supplying the oocyte with nutrients. The prominent endoplasmic reticulum of some follicular cells may be responsible for this activity. Little is known about the energy sources in mammalian eggs, which contain very small amounts of yolk material. Such sources could be identified by using a combination of biochemical and tissue culture methods, e.g., the type of substrates required for respiration, the rate of incorporation of labeled compounds, the respiratory quotient, and the pattern of growth and differentiation during culture in synthetic chemically defined media.

Comparative biochemical studies are needed in the following areas:

a) techniques for identification of metabolites and enzymes, together with electron microscopy, for the study of the physical and biochemical properties of the egg vitellus;

b) the mechanisms by which the blastomeres differentiate biochemically and morphologically; the role of biochemical inhibitors in the unfertilized egg and in eggs with retarded cleavage rate;

c) how the differentiating cells become properly adjusted to the *milieu intérieur;* how their spatial distribution is determined; and how they differentiate into tissues and organs characteristic of the species.

References

1. Alfert, A. A. 1950. A cytochemical study of oogenesis and cleavage in the mouse. *J. Cell. Comp. Physiol.* 36:381.
2. Austin, C. R. 1961. *The mammalian egg*. Springfield, Illinois: Charles C Thomas.
3. Austin, C. R., and Lovelock, J. E. 1958. Permeability of rabbit, rat and hamster egg membranes. *Exp. Cell Res.* 15:261.

4. Biggers, J. D.; Whittingham, D. G.; and Donahue, R. P. 1967. The pattern of energy metabolism in the mouse oocyte and zygote. *Proc. Nat. Acad. Sci.* 58:560–67.
5. Boell, E. J., and Nicholas, J. S. 1948. Respiratory metabolism of the mammalian egg. *J. Exp. Zool.* 109:267.
6. Boell, E. J. 1955. Energy exchange and enzyme development during embryogenesis. In *Analysis of development*, sect. 8, ed. B. J. Willier, P. A. Weiss, and V. Hamberger. Philadelphia: Saunders.
7. Borei, H. 1948. Respiration of oocytes, unfertilized eggs and fertilized eggs from Psammechinus and Asterias. *Biol. Bull.* 95:124.
8. Boyd, J. D., and Hamilton, W. J. 1952. Cleavage, early development and implantation of the egg. In *Marshall's physiology of reproduction*, ed. A. S. Parkes. London: Longmans.
9. Brachet, J. 1957. *Biochemical cytology*. New York: Academic Press.
10. ———. 1960. *The biochemistry of development*. London: Pergamon Press.
11. Braden, A. W. H. 1952. Properties of the membranes of rat and rabbit eggs. *Aust. J. Sci. Res., Ser. B* 5:460.
12. Brinster, R. L. 1965*a*. Studies on the development of mouse embryos *in vitro*. I. The effect of osmolarity and hydrogen ion concentration. *J. Exp. Zool.* 158:49.
13. ———. 1965*b*. Studies on the development of mouse embryos *in vitro*. III. The effect of fixed-nitrogen source. *J. Exp. Zool.* 158:69.
14. Chang, M. C., and Hunt, D. M. 1956. Effect of proteolytic enzymes on the zona pellucida of fertilized and unfertilized mammalian eggs. *Exp. Cell Res.* 11:497.
15. Dalcq, A. M. 1954. Fonctions cellulaires et cytochimie structurale dans l'oeuf de quelques Rongeures. *C.R. Biol. (Belg.)* 148:1332.
16. ———. 1957. *Introduction to general embryology*. Oxford: Oxford University Press.
17. Dalcq, A. M., and Pasteels, J. 1955. Détermination photométrique de la teneur relative en DNA des noyaux dans les oeufs en segmentation du rat et de la souris. *Exp. Cell. Res. Suppl.* 3:72.
18. Daniell, J. F. 1952. Structural factors in cell permeability and secretion. *Soc. Exp. Biol. Symposia* 6:1.
19. Da Silva Sasso, W. 1957. Permeability of the egg. *Ann. Fac. Farm Odent., Univ. S. Paulo* 14:111.
20. Dick, D. A. T. 1959. Osmotic properties of living cells. *Int. Rev. Cytol.* 8:387.
21. Dragoiu, I.; Benetato, G.; and Oprean, R. 1937. Recherches sur la respiration des ovocytes des Mammiferes. *C.R. Soc. Biol.* 126:1044.
22. Edwards, R. G., and Sirlin, J. L. 1958. Radioactive tracers and fertilization in mammals. *Endeavour* 17:42.
23. Fridhandler, L.; Hafez, E. S. E.; and Pincus, G. 1956*a*. Oxygen uptake of rabbit ova. *Proc. Third Int. Congr. Anim. Reprod., Cambridge*. Sec. 1:48.
24. ———. 1956*b*. Respiratory metabolism of mammalian eggs. *Proc. Soc. Exp. Biol. Med.* 92:127.
25. ———. 1957. Developmental changes in the respiratory activity of rabbit ova. *Exp. Cell Res.* 13:132.
26. Fridhandler, L.; Wastila, W. B.; and Palmer, W. M. 1967. The role of glucose in metabolism of the developing mammalian preimplantation conceptus. *Fertil. Steril.* 18:819.
27. Greenwald, G. S., and Everett, N. B. 1959. The incorporation of S^{35}-methionine by the uterus and ova of the mouse. *Anat. Rec.* 130:171.

28. Gustafson, T. 1965. Morphogenetic significance of biochemical patterns in sea urchin embryos. In *The biochemistry of animal development*, vol. 1, ed. Rudolf Weber, pp. 139–202. New York and London: Academic Press.
29. Gwatkin, R. B. L. 1964. Effect of enzymes and acidity on the zona pellucida of the mouse egg before and after fertilization. *J. Reprod. Fertil.* 7:99.
30. Hadek, R. 1965. The structure of the mammalian egg. *Int. Rev. Cytol.* 18:29.
31. Hall, B. V. 1935. The reactions of rat and mouse eggs to hydrogen ions. *Proc. Soc. Exp. Biol. Med.* 32:747.
32. Huber, G. C. 1915. The development of the albino rat, *Mus norvegicus albinus. J. Morph.* 26:247.
33. Ishida, K. 1954. Cytochemical studies of rat tubal ova. *Tohoku J. Agric. Res.* 5:1.
34. ———. 1960. Histochemical studies of lipids in the ovaries of domestic animals. *Arch. Histochem. Jap.* 19:547.
35. Ishida, K., and Chang, M. C. 1965. Histochemical demonstration of succinic dehydrogenase in hamster and rabbit eggs. *J. Histochem. Cytochem.* 13:470.
36. Keynes, R. D. 1960. The energy source for active transport in nerve and muscle. In *Membrane transport and metabolism*, ed. A. Kleinzeller and A. Kotyk. New York: Academic Press.
37. Konecny, M. 1959. Étude histochimique de la zone pellucide des ovules de chatte. *C.R. Soc. Biol.* 153:893.
38. Kostyuk, P. G., and Sorokina, Z. A. 1960. On the mechanism of hydrogen ion distribution between cell protoplasm and medium. In *Membrane transport and metabolism*, ed. A. Kleinzeller and A. Kotyk. New York: Academic Press.
39. Lin, C. H.; Huang, C. H.; and Chang, T. K. 1964. Histochemical changes in mouse eggs before and after fertilization during their early development. I. Dehydrogenases. *Acta Biol. Exp. Sinica* 9:281.
40. Lin, T. P. 1956. DL-Methionine (sulphur-35) for labelling unfertilized mouse eggs in transportation. *Nature* (*Lond.*) 178:1175.
41. Loewenstein, J. E., and Cohen, A. I. 1964. Dry mass, lipid content and protein content of the intact and zona-free mouse ovum. *J. Embryol. Exp. Morph.* 12:113.
42. Mintz, B. 1962. Experimental study of the developing mammalian egg: Removal of the zona pellucida. *Science* 138:574.
43. ———. 1964. Synthetic processes and early development in the mammalian egg. *J. Exp. Zool.* 157:85.
44. ———. 1965. Nucleic acid and protein synthesis in the developing mouse embryo. In *Preimplantation stages of pregnancy*, ed. G. E. W. Wolstenholme and M. O'Connor. Boston: Little, Brown and Co.
45. Monroy, A. 1960. Incorporation of S^{35}-methionine in the microsomes and soluble proteins during the early development of the sea urchin egg. *Experientia* 16:114.
46. ———. 1965. Biochemical aspects of fertilization. In *The biochemistry of animal development*, ed. Rudolf Weber, pp. 73–135. New York and London: Academic Press.
47. Monroy, A., and Maggio, R. 1964. Biochemical studies on the early development of the sea urchin. In *Advances in morphogenesis*, ed. M. Abercrombie and J. Brachet. New York: Academic Press.
48. Monroy, A., and Vittorelli, M. L. 1962. Utilization of C^{14}-glucose for amino acids and protein synthesis by the sea urchin embryo. *J. Cell. Comp. Physiol.* 60:285.

49. Mulnard, J. 1955. Contribution à la connaissance des enzymes dans l'ontogénise. Des phosphomonoestérase acide et alcaline dans le développement du rat et de la souris. *Arch. Biol.* 66:527.
50. ———. 1964. Obtention *in vitro* du développement continu de l'oeuf de souris du stade II au stade du blastocyste. *C.R. Acad. Sci.* 258:6228.
51. Operescu, St., and Thibault, C. 1965. Duplication de l'adn dans les oeufs de lapine après la fécondation. *Ann. Biol. Anim. Biochim. Biophys.* 5:151.
52. Past, R. L., and Albright, C. D. 1960. Membrane adenosine triphosphatase system as a part of a system for active sodium and potassium transport. In *Membrane transport and metabolism*, ed. A. Kleinzeller and A. Kotyk. New York: Academic Press.
53. Pincus, G. 1941. Factors controlling the growth of rabbit blastocyst. *Amer. J. Physiol.* 133:412.
54. Rapola, J., and Koskimies, O. 1967. Embryonic enzyme patterns: Characterization of the single lactate dehydrogenase isozyme in preimplanted mouse ova. *Science* 157:1311.
55. Sirlin, J. H., and Edwards, R. G. 1959. Timing of DNA synthesis in ovarian oocyte nuclei and pronuclei of mouse. *Exp. Cell Res.* 18:190.
56. Skou, J. C. 1965. Enzymatic basis for active transport of Na and K across cell membrane. *Physiol. Rev.* 45:596.
57. Smithberg, M. 1953. The effect of different proteolytic enzymes on the zona pellucida of mouse ova. *Anat. Rec.* 117:554.
58. Solomon, A. K. 1960. Measurement of the equivalent pore radius in cell membranes. In *Membrane transport and metabolism*, ed. A. Kleinzeller and A. Kotyk. New York: Academic Press.
59. Solomon, J. B. 1965. Development of nonenzymatic proteins in relation to functional differentiation. In *The biochemistry of animal development*, ed. Rudolf Weber, pp. 367–440. New York and London: Academic Press.
60. Sugawara, S. 1962. Metabolism in the mammalian ova. *Jap. J. Zootech. Sci.* 33:1.
61. ———. S. 1964. On the CO_2 evolution of the cleaved ova in rats. *Jap. J. Anim. Reprod.* 9:123.
62. Sugawara, S., and Hafez, E. S. E. 1967. Developmental changes in dehydrogenase activities in rabbit eggs. *Proc. Soc. Exp. Biol. Med.* N.Y. 126:849.
63. ———. 1968. Metabolism of mammalian eggs. *Int. Rev. Gen. Exp. Zool.*: 171.
64. Sugawara, S., and Umezu, M. 1961. Studies on the metabolism of the mammalian ova. II. Oxygen consumption of the cleaved ova of the rat. *Tohoku J. Agric. Res.* 12:17.
65. Szollosi, D. 1964. Time of DNA synthesis in mammalian egg after sperm penetration. *J. Cell Biol.* 23:92A.
66. Thomson, J. L., and Brinster, R. L. 1966. Glycogen content of preimplantation mouse embryos. *Anat. Rec.* 155:97.
67. Tondour, M. 1959. Effets de l'imprégnation osmique *in toto* des oeufs vierges, fécondes ou segmentes du rat et de la souris. *Meded. Klasse Wetenschappen* 45:487.
68. Tyler, A. 1963. Survey of some investigations of problems of reproduction by immunologic and protein-biosynthetic methods. *Excerpta Med. Intern. Cong. Series* no. 72, pp. 527–38.
69. Whitten, W. K. 1957. Culture of tubal ova. *Nature* (*Lond.*) 179:1081.

Part IV Experimental Manipulation of Oviductal Eggs and Oviduct

16

Egg Transfer in the Laboratory Animal

M. C. Chang/S. Pickworth
Worcester Foundation for Experimental Biology
Shrewsbury, Massachusetts

The first successful transfer of two 4-cell eggs from an Angora rabbit to the oviduct of a mated Belgian hare rabbit was made by Heape (1890) and demonstrated the possibility of development within a uterine foster mother. Two Dutch young were developed in other Belgian rabbits, and Heape (1897) concluded that the recipient female had no power of modifying the breed of her foster-children. Successful egg transfers were later carried out in the rat (Nicholas 1933) and in the mouse (Fekete and Little 1942), but intensive study and use of the technique did not start until the late 1940's. It was soon well established that egg transfer offered both a method of improvement of domestic livestock and a tool for the study of animal reproduction using laboratory species.

References to the literature up to 1961 have been published previously in tabulated form (Austin 1961; Beatty 1962). An attempt is made here to survey the achievements in laboratory animals up to the present time.

I. Techniques of Egg Transfer

A. *Donors*

Throughout this paper "day 1" and "1-day" refer to the 24 hr following mating or ovulation-inducing injection.

For convenience in experiments, eggs at particular stages of development should be available at prearranged times, and human chorionic gonadotrophin (HCG) is commonly used to induce ovulation. In order to obtain a large number of eggs, superovulation treatments of donors are also employed by many investigators. In general, these techniques consist of one or a series of injections of either pituitary extracts or pregnant mare's serum (PMS) to induce the growth of a large number of follicles, and then a final injection of HCG to bring about ovulation. In the rabbit, these treatments are usually subcutaneous and intravenous, respectively; both PMS and HCG are frequently given by the intraperitoneal route in the smaller species. Techniques for inducing superovulation in rabbits have been known for many years (Pincus 1940; Parkes 1943). Superovulation of mice (Runner and Palm 1953; Fowler and Edwards 1957), rats (Rowlands 1944; Austin 1950; Edwards and Austin 1959), and hamsters (Greenwald 1962*a;* Yanagimachi and Chang 1964) has also been described. Many slight modifications of the basic technique have been adopted by investigators. Success depends on the gonadotrophic preparation used, the dosage, the route, and the times of administration. Females may be mated or artificially inseminated to provide fertilized eggs.

B. *Recipients*

The recipient may be pregnant, pseudopregnant (induced by mating with a vasectomized male, by cervical stimulation or, in the rabbit, by injection of HCG), or may be under continuous hormone therapy. Pregnancy of the recipient means that the transferred and native eggs will be in competition, and genetic markers are needed for identification unless unilateral ovariectomy and transfer are feasible. Hormonal methods of inducing pseudopregnancy in the rat have been reported and discussed (Banik and Ketchel 1965; Mantalenakis and Ketchel 1966).

C. *Egg Recovery*

Eggs have been recovered from follicles for transfer in the rat (Noyes 1952) and rabbit (Chang 1955*a*, *b*) and for culture in other species (Edwards 1962). The follicles are broken under fluid in a small watch glass and the eggs picked up with a pipette. After ovulation in small rodents, eggs which are still in a bulky cumulus clot are most easily recovered from the oviduct by tearing the wall at the distended position. Later stages can be collected by flushing the oviduct; or it can be divided into sections and the contents of each pressed out under fluid. In the rabbit flushing is the most convenient method. Rabbit eggs can also be recovered under anesthesia, without sacrifice of the donor, by inserting a curved glass cannula into the ovarian end of the oviduct and injecting fluid at the uterine end (Avis and Sawin 1951; Hafez 1963). A method for collection of uterine blastocysts *in vivo* was described by Venge (1952).

Guinea pig eggs can be collected in the same way as those of the rabbit. Ferret eggs can also be flushed out, but dissection of the oviduct around the ovary requires special care.

Before flushing out eggs or blastocysts from uteri, it is best to remove connective tissue to allow straightening of the uterine horn; and blood should be blotted off to avoid contamination of the collected washings. It is desirable that only morphologically normal eggs be selected for transfer (Hafez 1962*a*). A slow rate of cleavage of eggs may be a symptom of reduced viability in rabbits (Hafez 1962*b*) and mice (Gates 1965).

D. *Transfer Media*

Various media have been used by different workers, but the most common are serum, physiological saline, various tissue culture media, and serum-saline combinations. Media used in experiments up to 1961 are listed by Austin (1961). Media used for hamster eggs since then include tissue culture medium TC 199 with glycine, or Tyrode's solution (Yanagimachi and Chang 1964), and Hank's solution (Blaha 1964). Rat eggs have been transferred in rat plasma (Duncan and Forbes 1965), Ringer's phosphate buffer (Mantalenakis and Ketchel 1966), and saline and rat plasma 2:1 (Dickmann and De Feo 1967). Conditioned Eagle's culture medium (Noyes, Doyle, Gates, and Bentley 1961) and horse serum with saline 1:1 (Talbert and Krohn 1966) have been used for mouse eggs.

Use of heterologous sera may be convenient, but some types are lethal to eggs and embryos of other species unless heat treated (Chang 1949; New 1966).

E. *Transfer Techniques*

Eggs can be transferred into the rabbit oviduct either through a midline incision or through two flank incisions. The latter method is preferred by the authors because the oviduct is easier to find and suffers less disturbance. By pulling out the fat tissue surrounding the ampulla, both the ovary and the infundibulum can be exposed outside the body cavity. The eggs are transferred in a pipette (2 mm diameter tip, fire-polished) which is inserted through the infundibulum into the ampulla where the eggs are deposited (Fig. 1*a*). For transfer into the uterus a similar pipette is used, but the tip is left unpolished so that it can puncture the wall of the uterus, and the eggs are then deposited in the lumen (Fig. 1*b*). A larger pipette (3 mm diameter tip) is used for transfer of large blastocysts, and a small incision in the uterine wall is necessary (Fig. 1*c*), which is subsequently closed with catgut.

The ovaries of the laboratory rodents are surrounded by a membranous capsule. A foramen is present in the rat and can be used as a passageway for the transferring pipette (Noyes 1952). Eggs in cumulus can be placed in the bursa in this way (Noyes and Dickmann 1961), or a fine pipette containing denuded eggs can be passed directly through the membrane or through the fat adjacent to the ovary. The latter method has been used for transfer of eggs to the bursa in mice (Runner and Palm 1953) and to the infundibulum in mice (Tarkowski 1959*a*) and rats (Noyes and Dickmann 1961). The same methods can be used for the hamster (Yang, unpublished) and the ferret (Chang, 1966*b*). Transfer of eggs to the uterus is performed through a small puncture in the uterine wall. Details of transfer procedures in the rat were well described and illustrated by Noyes and Dickmann (1960). In all transfers it is generally agreed that the volume of fluid should be minimal. Deposition of large air bubbles should also be avoided, although very small bubbles can be used to mark the position of eggs during transfer with a fine pipette. Nonsurgical transfers have been made in the mouse via the cervix—"inovulation"—(Beatty 1951; Tarkowski 1959*a*) and also in the rabbit (Hafez 1962*c;* Dauzier 1962, 1964). Only a small proportion of eggs injected directly into the peritoneal cavity of the rabbit were later recovered from the genital tract (Dauzier 1962).

II. Results Obtained by Egg Transfer

A. *Transfer of Ovarian Eggs*

Follicular rat eggs at various stages of maturation were transferred to the ovarian bursa of mated females by Noyes (1952). With one possible exception, development to term occurred only in eggs which had reached at least the tetrad formation stage of meiotic division in the ovary. In later experiments ovarian tissue was transplanted to the eye, and eggs were recovered while maturing there under hormone treatment. These gave similar results after transfer, although the percentage which developed was much lower than in the earlier experiments (Noyes, Yamate, and Clewe 1958).

The same problem was investigated in the rabbit by Chang (1955*a*). Follicular eggs were recovered at various times after an ovulation-inducing injection and were transferred to the oviducts of mated rabbits. Of the eggs which were recovered from the ovaries between

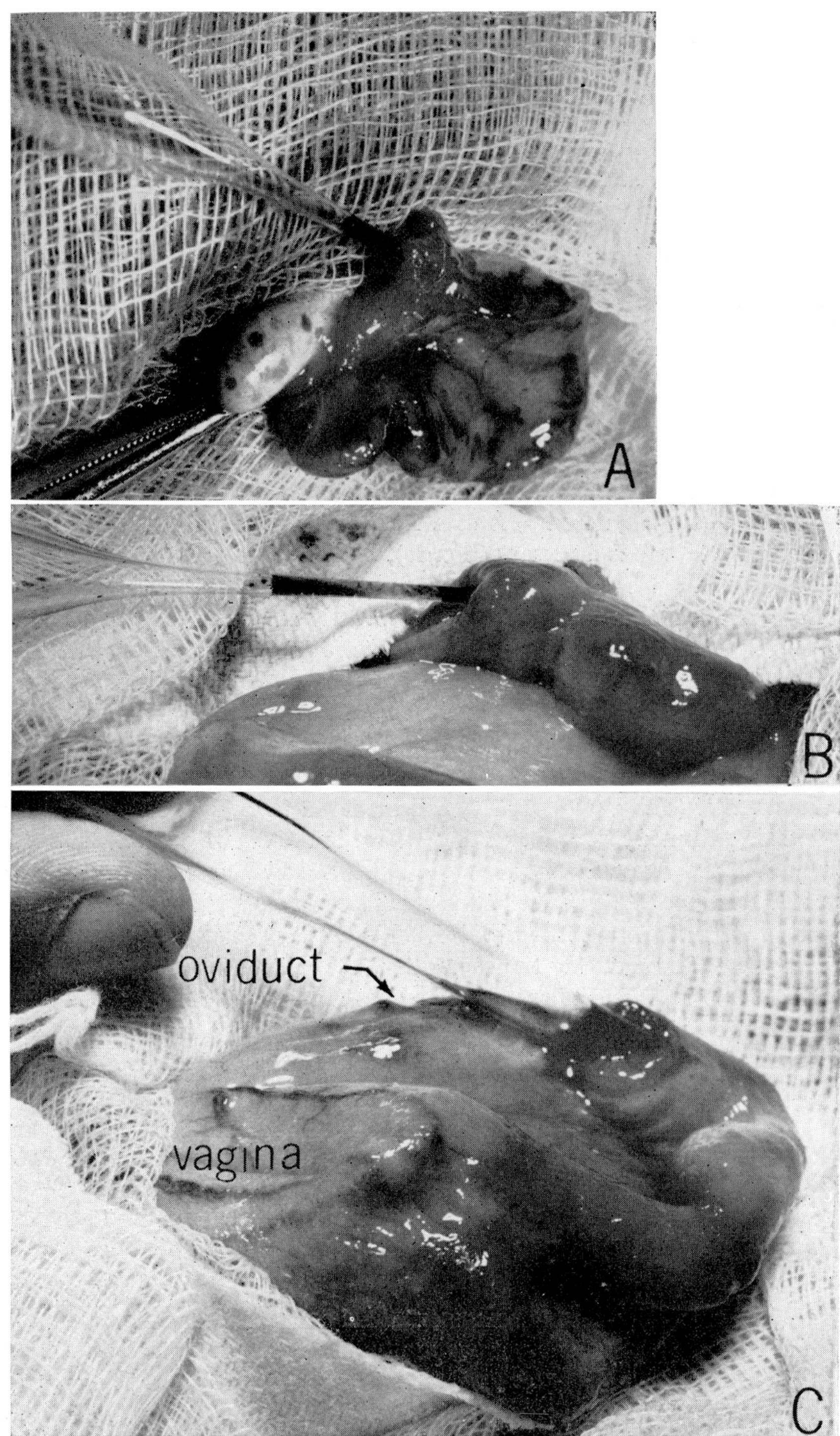

FIG. 1. Techniques of egg transfer in the rabbit: (*a*) Transfer of 1-day eggs into the oviduct by insertion of the pipette into the infundibulum (flank incision). (*b*) Transfer of early blastocysts into the uterus by puncture of the uterine wall close to the uterotubal junction (midline incision). (*c*) Transfer of 6-day blastocysts into the uterus through an incision in the uterine wall (midline incision).

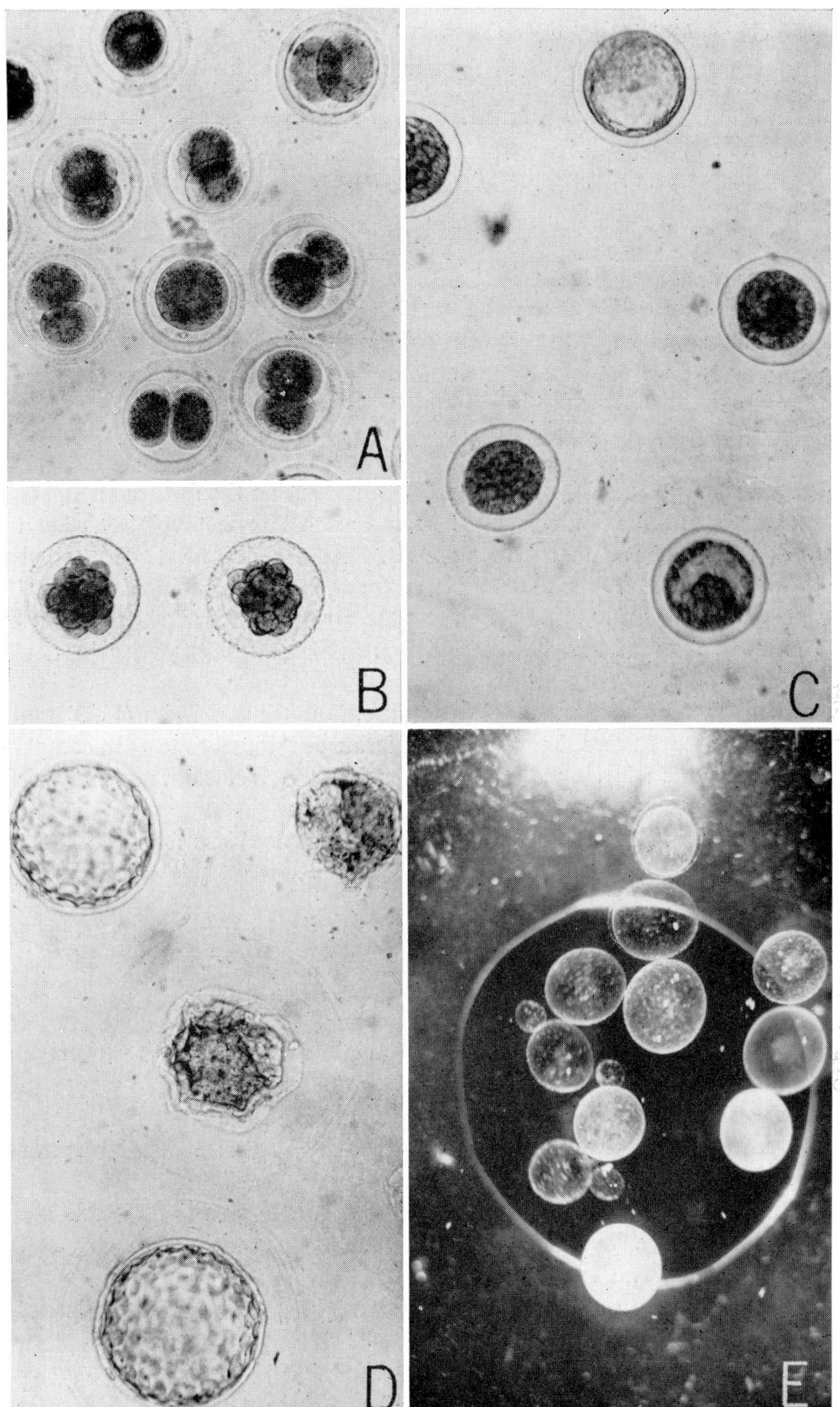

FIG. 2. Rabbit eggs recovered at different times after insemination: (*a*) One day, 1–4 cells, from the oviduct (×93). (*b*) Two days, 16–32 cells, from the oviduct (×93). (*c*) Three days, morulae and formation of the blastocoele, from oviduct and uterus (×93). (*d*) Four days, small blastocysts, from the uterus (×93). (*e*) Six days, large blastocysts, from the uterus (×3).

0 and 5 hr after injection, not more than 4% developed into normal fetuses. Recovery between 6 and 12 hr after injection raised the survival rate to 14–28%. Other experiments showed that the majority of oocytes from large follicles of unstimulated rabbits could mature (first polar body formation) in culture or after transfer to the oviduct (Chang 1955*b*). The fertilization rate of such oocytes transferred directly to the oviducts was 19–35%, but only 1 of 39 eggs developed to term. When oocytes cultured for 8–10 hr were transferred, not one developed (Chang 1955*a*, *b*).

These studies showed that the ability to become fertilized and to develop is acquired over a definite time period by the egg. Completion of the first meiotic division, as in culture or in the oviduct, does not represent the full physiological maturation of the egg in this sense.

B. *Transfer of Unfertilized Oviductal Eggs*

The egg transfer technique offers one way of measuring the fertilizable life of eggs after ovulation. In rabbits some eggs could be fertilized up to 8 hr postovulation when transferred to the oviducts of mated animals (Chang 1952). The fertilization rate dropped sharply 4 hr after ovulation. On the basis of the number of young born after delayed mating, Hammond (1934) concluded that 6 hr postovulation was the upper limit for fertilization in the rabbit; but it is not to be expected that all fertilized eggs will develop into young, particularly when aged eggs are fertilized.

Following a similar experimental procedure, Runner and Palm (1953) reported that in mice 14% of eggs transferred soon after ovulation developed to term in competition with native eggs. Few young developed after transfer 4 hr postovulation, and none was present at term when the interval was 8 hr.

Chang (1955*c*) showed that transferred rabbit eggs could be fertilized in the uterus of mated recipients.

C. *Effects of the Maternal Environment of the Recipient*

The effects of differing degrees of synchronization between donor and recipient were commented on by the earliest workers in rats (Nicholas 1933) and mice (Fekete and Little 1942). The first systematic study was made in the rabbit by Chang (1950). Eggs of different ages, from 1-day to 6-day, were transferred into oviducts and uteri of rabbits on various days before or after an ovulation-inducing injection. One-day eggs could not survive in the uterus before or after the pseudopregnancy commenced, although about 20% of 2-day eggs developed to term after transfer to the 2- or 3-day uterus. Synchronous transfer gave, in general, the best results, but uterine blastocysts could survive equally well in uteri 1 day "younger." Survival rates decreased sharply on either side of the favorable period. High survival rates have since been reported for a technically convenient timing of 60-hr eggs into a 72-hr uterus (Adams 1962).

The same problem was approached by Hafez (1962*c*) who made unilateral transfers of rabbit eggs into earlier and later stage oviducts and compared survival and growth with synchronized transfers into the opposite side of the tract. The implantation rate was reduced after transfer of early 1-day eggs into 2-day oviducts. Within the range examined, the transfer of eggs to "younger" oviducts neither reduced nor improved survival rates.

The factors affecting survival of native and transferred eggs in the mouse were studied by McLaren and Michie (1956). Of the possible egg-uterine combinations of 2½ and 3½ days *post coitum*, they found that transfer of 3½-day eggs to 2½-day uteri gave the best results. Asynchronous transfer to a recipient at an earlier stage of endometrial development was also recommended for mice by Tarkowski (1959*a*).

Further study of this subject has since been made by Noyes and co-workers in rats (Noyes and Dickmann 1960, 1961; Dickmann and Noyes 1960) and mice (Doyle, Gates, and Noyes 1963). Best survival rates were obtained from synchronous transfers or, for some egg stages, transfer to either synchronized or 1 day "younger" recipients. A 70% survival rate to term was achieved in the rat after transfer of 4- and 5-day eggs to the 4-day uterus. Transfer to "older" uteri gave poor results in both species. After transfer to "younger" uteri, eggs delayed their implantation until the normal time for the endometrial development, and a 24-hr delay resulted in fetuses approximately 25% heavier than synchronized littermates in both rats (Noyes, Doyle, and Bentley 1961) and mice (Noyes *et al.* 1961). This work was reviewed later by Noyes *et al.* (1963).

In the hamster, Dr. W. H. Yang in our laboratory has obtained 54% development of 4-day eggs into fetuses after transfer to a 4-day uterus and 35% development after transfer to a 3-day uterus. Transfer of 3-day eggs to 3- and 4-day uteri resulted in 49% and 14% development, respectively.

The general conclusions are similar for those species examined in detail—highest survival rates of eggs result from transfer to approximately synchronized or slightly earlier stage recipients. Asynchrony is tolerated much better when the eggs are "older" than the uterus than when the reverse is true. Eggs "younger" than their environment usually fail to implant in rats and mice. The endometrium of the rabbit seems to have less well-defined changes at the time of implantation and some younger eggs will survive if the timing difference is not great. Transfer of 60-hr morulae to the 8-day uterus results in their early degeneration (Adams 1965).

Growth and implantation of 1- and 6-day rabbit eggs occurred in intact nonovulated females treated with a large dose of progesterone (Chang 1951). Egg transfer has been used to demonstrate that progesterone treatment is adequate for implantation in the ovariectomized hamster (Orsini and Psychoyos 1965), but that in ovariectomized progesterone-treated rats the eggs delay their implantation until estrogen is given (Psychoyos 1961). The uteri of mice ovariectomized several weeks previously are still competent to support growth of normal embryos when appropriate progesterone and estrogen treatment is given (Humphrey 1967). The transfer technique has also been used to examine the effects of estrogens or progestational compounds on egg transport and development in the rabbit (Chang 1964; Chang 1966*a;* Chang and Harper 1966), effects of norethynodrel in mice (Davis 1963), and of various drugs, including estrogen antagonists (Duncan and Forbes 1965) and clomiphene (Staples 1966) on nidation in rats.

Egg transfer provides a useful means of studying female reproductive senescence. Reciprocal transfers of eggs between young and aged females have been made in the hamster (Blaha 1964), rabbit (Adams 1964), and mouse (Talbert and Krohn 1966). In each case the uterine environment of the aged female was inadequate for normal support of pregnancy. The viability of eggs from aged donors is not so well established, nor is it clear whether fertilization and the oviductal environment are normal or not. At the other end of the scale, inadequacy of the maternal environment of immature rabbits was demonstrated by Adams (1953) who made reciprocal transfers between immature and mature females. Eggs from immature donors were capable of development in mature recipients.

Transfer of eggs between different breeds of rabbits was made by Venge (1950) to investigate effects of genetically large or small recipients on fetal growth. Interbreed transfer has shown genetic differences in the ability to support large numbers of fertilized eggs through pregnancy (Hafez 1965*a*). The amount and distribution of prenatal mortality after transfer of varying numbers of eggs was studied directly in the rabbit by Adams (1960, 1962) and Hafez (1964) and in mice by McLaren and Michie (1959).

D. *Effects of* in Vitro *Manipulation of Eggs*

Chang (1952, 1953) transferred unfertilized rabbit eggs to the oviducts of mated females after storage and found that a period of 30–72 hr at 0°–10° C did not destroy their ability to be fertilized and to develop. More extreme freezing treatments of unfertilized mouse eggs were described by Sherman and Lin (1958, 1959). In early experiments on transfer of 2-cell rabbit eggs, Chang (1947, 1948*a*) reported that 28% of the eggs developed to term after storage at 10° C for 77–101 hr and culture for 1 day. In later work on storage without culture, 46% of 2-cell eggs reached term after storage for 3 days at 10° C and 6% of the eggs did so after 4 days at 0° C. Culture in rabbit serum for 2 days at 38° C without low temperature treatment resulted in 27% development to term (Chang 1948*b*).

Storage methods for rabbit eggs have since been investigated by Hafez (1961, 1965*b*), who found that addition of gelatin to the medium improved the subsequent survival of the eggs. After recovery at 40–48 hr *post coitum* and storage for 11 days at 10° C, 10 out of 25 eggs implanted and 7 were alive at autopsy 15 days *post coitum* (Hafez 1965*b*). Successful development after culture of 5-day blastocysts has been reported (Staples 1967). Seven young survived out of 25 blastocysts that had been cultured for 16 hr in modified culture medium F10 with 10% serum.

Great progress in culturing of mouse ova has been made over the past few years; but even in this species, development from 1-cell to blastocyst has not yet been achieved in a completely artificial medium, although successful transfer after culture between 2-cell and blastocyst stages was reported by Biggers, Moore, and Whittingham (1965). Some factor provided by the oviduct is required at a stage between the 1st and 2nd cleavage. Eggs cultured free in medium until after 1st cleavage and within the explanted oviduct up to blastocyst stage will develop into young after transfer (Whittingham and Biggers 1967).

Egg transfer has been important in supplying convincing proof of fertilization *in vitro* in the rabbit (Chang 1959; Thibault and Dauzier 1961; Bedford and Chang 1962; Brackett and Williams 1965). A variation on the method of Thibault and Dauzier has been used to produce heteroploid rabbit eggs which will develop to postimplantation stages (Bomsel-Helmreich and Thibault 1962; Bomsel-Helmreich 1965).

Birth of 7 young was reported by Pincus and Enzmann (1934) after 10 unfertilized rabbit eggs treated with sperm *in vitro* for 20 min were transferred to pseudopregnant females. When 615 eggs artificially activated *in vitro* by various treatments were transferred into the oviduct of 19 pseudopregnant rabbits, 4 of them produced young which were thought to have developed parthenogenetically (Pincus 1939). Chang (1954) later carried out transfer of unfertilized rabbit eggs after storage at 10° C. Although 18% of 145 eggs developed into parthenogenetic blastocysts, not one of 230 eggs implanted.

Included among other applications of the transfer technique are studies on effects of irradiation (Chang, Hunt, and Romanoff 1958; Chang and Hunt 1960; Lin and Glass 1962; Glass and Lin 1963; Glass and McClure 1964), high ambient temperature (Shah 1956; Alliston, Howarth, and Ulberg 1965; Alliston and Ulberg 1965), removal of the cumulus (Dickmann 1964) and the corona radiata (Chang and Bedford 1962), and reduction of the mucin coat of rabbit eggs (Greenwald 1962*b*).

Live young have been born after destruction of 1 blastomere of 2-cell rabbit eggs (Seidel 1952) and implantation has occurred when 3 cells of 4-cell eggs were destroyed (Seidel 1956). Fertile males and females were obtained from mouse eggs treated similarly (Tarkowski 1959*b*, *c*). Ingenious experiments have resulted in the development of chimeras of two mouse strains by fusion of cleaved eggs after removal of the zona (Tarkowski 1961; Mintz 1962, 1965*a*, *b*). The potential range of *in vitro* treatments has been extended by the success of a technique for injection of 1-cell eggs in mice (Lin 1966).

E. *Interspecific Transfer of Eggs*

The first attempts at interspecific transfers, between sheep and goats, were reported in 1934 by Warwick, Berry, and Horlacher. No full account dealing with laboratory species appeared until 20 years later when Briones and Beatty (1954) made transfers between mice, rats, rabbits, and guinea pigs. Eggs at various stages were placed in the reproductive tract or peritoneal cavity of the hosts and recovered 1 to 2 days later. Cleavage and blastocyst growth of rabbit and mouse eggs took place within the species tested but was slower than normal.

Reciprocal transfer between mice and rats was examined in further detail by Tarkowski (1962), who showed that development of both species could take place up to early implantation stages. A normal decidual reaction was provoked by foreign eggs, but erosion of the maternal epithelium was inadequate and all eggs had degenerated by the 7th or 8th day of pregnancy.

Transfers between rats and hamsters have also had only limited success. Blaha and De Feo (1964) placed 4- and 5-day eggs from rats into the uteri of 5 hamsters 1 day "younger," but no decidua were formed. The reverse transfer of sixty-eight 3-day hamster eggs into rats on day 4 of pseudopregnancy resulted in 26 implantation sites, but the eggs degenerated shortly afterward. Further work on these species has been carried out recently by Yang in our laboratory. He found that when newly fertilized and segmenting rat eggs were transferred into the ovarian capsule of pseudopregnant hamsters, they could develop into morulae or blastocysts in the hamster oviduct. The majority of rat morulae transferred into the uterus of pseudopregnant hamsters could develop into blastocysts, but very few (3%) could induce decidual tissue. When newly fertilized or segmenting hamster eggs were transferred into the ovarian capsule of pseudopregnant rats, cleavage of the eggs was observed within 2 days, but only degenerated eggs were found on day 3. The majority of hamster morulae degenerated when transferred into the uterus of pseudopregnant rats, but about 17% could induce decidua without fetal tissue.

The rabbit has received some attention as a temporary incubator for eggs of sheep (Averill, Adams, and Rowson 1955; Averill 1956) and cattle (Hafez and Sugie 1963). When eleven 2-cell hare eggs were transferred into the oviducts of 4 rabbits, development to the blastocyst stage was possible, but decidua formation and implantation failed (Chang 1965). Reciprocal transfer of eggs between rabbit and ferret was investigated by Chang (1966*b*). Ferret eggs could develop if introduced into the oviduct of the rabbit but not if introduced into the uterus, whereas rabbit eggs did not survive in either the oviduct or uterus of the ferret. Transfers of blastocysts between ferret and mink were made recently (Chang, unpublished). The majority of the mink blastocysts could develop and become implanted in the ferret uterus, but degeneration of embryos occurred soon after implantation. When ferret blastocysts were transferred into the mink uterus they could develop only for a few days. Besides the possible incompatibility between blastocysts and endometrium in this case, the development of corpora lutea in relation to the development of blastocysts and implantation may also be involved, because delayed implantation is a common feature in the mink, but not in the ferret.

Greater insight into the problem of interspecific research may come from the further use of "double transfer" procedures where eggs are put into the genital tract of other species for a short period and then replaced in their normal environment.

F. *Ectopic Transfer of Eggs*

Transfers of mouse eggs into the anterior chamber of the eye were made by Runner (1947) and Fawcett, Wislocki, and Waldo (1947). Much of the work on ectopic transfer during the past few years has also been carried out in mice, notably the experiments of Kirby (1960, 1963*a*, *b*) on growth of trophoblast and embryonic tissue after transfer of eggs to various abdominal sites in both sexes. Growth of embryonic shield derivatives under the kidney capsule occurred in some cases after transfer of blastocysts, but not oviductal eggs. Kirby (1962*a*) suggested that exposure to the uterine environment was essential for normal development. Other experiments with mice included transfer following immunization of the host animal (Simmons and Russell 1962, 1965; Kirby, Billington, and James 1966) and a study of some effects of the implants on the host (Kirby 1965, 1966).

Reciprocal transfers of rat and mouse blastocysts beneath the kidney capsule were also described by Kirby (1962*b*), who found that proportionately more mouse eggs than rat eggs were capable of implantation in this site. Contrary results have been obtained after transferring eggs into the testis (Mayer and Duluc 1966), where both normal and "delayed" rat blastocysts were able to develop in mice. No implantations of mouse blastocysts were found in the rat testis.

Transfers of eggs to the kidney and to abdominal muscle were carried out in guinea pigs, but only a small proportion of the eggs implanted (Bland and Donovan 1965). Fertilized rabbit eggs transferred to the anterior chamber of the eye developed normally up to morula formation, after which growth became retarded (Ahlgren and Bengtsson 1962).

III. Conclusions

In early experiments on egg transfer, the successful development of young was of interest in itself. More recently, intensive investigations in laboratory animals have led to improved methods, and egg transfer is now well established as a valuable experimental technique for the study of many problems in reproduction. Experiments in different species of laboratory animals provide a useful background for practical applications in the improvement of livestock and perhaps also in human gynecological practice.

The experimental techniques and results referred to in this paper demonstrate the wide scope of the problems which can be studied by means of egg transfer. We know now that follicular eggs can be transferred from one animal to another and that unfertilized eggs can be stored or subjected to various treatments. Fertilized eggs at different stages of development or after exposure to different experimental conditions can be transferred to recipients at a particular reproductive phase or under hormonal treatment. Fertilized eggs can also be transferred from one species to another or to an ectopic site in order to study their development, implantation, and differentiation. Some blastomeres in a fertilized egg can be destroyed, or two eggs can be fused together before transfer, thus introducing many interesting problems in genetics and developmental biology.

It is clear that the applications of egg transfer in research are not limited to studies on problems such as egg maturation, fertilization, and development. The range of the subject already includes wider aspects of general biological interest such as interactions between the embryo and uterus, immunological relationships between the conceptus and the mother, and studies on developmental genetics. Much progress has already been made and the potential for future development is great.

References

1. Adams, C. E. 1953. Some aspects of ovulation, recovery and transplantation of ova in the immature rabbit. In *Mammalian germ cells*, ed. G. E. W. Wolstenholme, pp. 198–212. London: J. & A. Churchill, Ltd.
2. ———. 1960. Early embryonic mortality in the rabbit. *J. Reprod. Fertil.* 1:315–16.
3. ———. 1962. Studies on prenatal mortality in the rabbit, *Oryctolagus cuniculus:* The effect of transferring varying numbers of eggs. *J. Endocr.* 24:471–90.
4. ———. 1964. The influence of advanced maternal age on embryo survival in the rabbit. *Fifth Int. Cong. Anim. Reprod. A.I., Trento* 1:305–8.
5. ———. 1965. The influence of maternal environment on preimplantation stages of pregnancy in the rabbit. In *Preimplantation stages of pregnancy*, ed. G. E. W. Wolstenholme and M. O'Connor, pp. 345–73. Boston: Little, Brown and Co.
6. Ahlgren, M., and Bengtsson, L. P. 1962. Transplantation of fertilized rabbit eggs to the anterior chamber of the eye. *J. Reprod. Fertil.* 3:89–92.
7. Alliston, C. W., and Ulberg, L. C. 1965. *In vitro* culture temperatures and subsequent viability of rabbit ova. *J. Anim. Sci.* 24:912 (abstr.).
8. Alliston, C. W.; Howarth, B.; and Ulberg, L. C. 1965. Embryonic mortality following culture *in vitro* of one- and two-cell rabbit eggs at elevated temperatures. *J. Reprod. Fertil.* 9:337–41.
9. Austin, C. R. 1950. The fecundity of the immature rat following induced superovulation. *J. Endocr.* 6:293–301.
10. ———. 1961. *The mammalian egg.* Springfield, Illinois: Charles C Thomas.
11. Averill, R. L. W. 1956. The transfer and storage of sheep ova. *Proc. Third Int. Congr. Anim. Reprod., Cambridge* Section 3, pp. 7–9.
12. Averill, R. L. W.; Adams, C. E.; and Rowson, L. E. A. 1955. Transfer of mammalian ova between species. *Nature* (*Lond.*) 176:167–68.
13. Avis, F. R., and Sawin, P. B. 1951. A surgical technique for the reciprocal transplantation of fertilized eggs in the rabbit. *J. Hered.* 42:259–60.
14. Banik, U. K., and Ketchel, M. M. 1965. Hormonal induction of pseudopregnancy in rats. *J. Reprod. Fertil.* 10:85–91.
15. Beatty, R. A. 1951. Transplantation of mouse eggs. *Nature* (*Lond.*) 168:995.
16. ———. 1962. Intra- and inter-specific zygote transfer. In *Growth* (Biological Handbooks), ed. P. L. Altman and D. S. Dittmer, pp. 177–79. Washington, D.C.: Fed. Am. Soc. for Exp. Biol.
17. Bedford, J. M., and Chang, M. C. 1962. Fertilization of rabbit ova *in vitro. Nature* (*Lond.*) 193:898–99.
18. Biggers, J. D.; Moore, B. D.; and Whittingham, D. G. 1965. Development of mouse embryos *in vivo* after cultivation from two-cell ova to blastocysts *in vitro. Nature* (*Lond.*) 206:734–35.

19. Blaha, G. C. 1964. Effect of age of the donor and recipient on the development of transferred golden hamster ova. *Anat. Rec.* 150:413–16.
20. Blaha, G. C., and De Feo, V. J. 1964. Interspecies ova transfer between hamsters and rats. *Anat. Rec.* 148:261–62 (abstr.).
21. Bland, K. P., and Donovan, B. T. 1965. Experimental ectopic implantation of eggs and early embryos in guinea-pigs. *J. Reprod. Fertil.* 10:189–96.
22. Bomsel-Helmreich, O. 1965. Heteroploidy and embryonic death. In *Preimplantation stages of pregnancy*, ed. G. E. W. Wolstenholme and M. O'Connor, pp. 246–67. Boston: Little, Brown and Co.
23. Bomsel-Helmreich, O., and Thibault, C. 1962. Développement d'oeufs triploïdes expérimentaux chez la lapine. *Ann. Biol. Anim. Biochim. Biophys.* 2:265–66.
24. Brackett, B. G., and Williams, W. L. 1965. *In vitro* fertilization of rabbit ova. *J. Exp. Zool.* 160:271–82.
25. Briones, H., and Beatty, R. A. 1954. Interspecific transfers of rodent eggs. *J. Exp. Zool.* 125:99–118.
26. Chang, M. C. 1947. Normal development of fertilized rabbit ova stored at low temperature for several days. *Nature* (*Lond.*) 159:602–3.
27. ———. 1948*a*. The effects of low temperature on fertilized rabbit ova *in vitro*, and the normal development of ova kept at low temperature for several days. *J. Gen. Physiol.* 31:385–410.
28. ———. 1948*b*. Transplantation of fertilized rabbit ova—the effect on viability of age, *in vitro* storage period, and storage temperature. *Nature* (*Lond.*) 161:978–79.
29. ———. 1949. Effect of heterologous sera on fertilized rabbit ova. *J. Gen. Physiol.* 32:291–300.
30. ———. 1950. Development and fate of transferred rabbit ova or blastocyst in relation to the ovulation time of recipients. *J. Exp. Zool.* 114:197–225.
31. ———. 1951. Maintenance of pregnancy in intact rabbits in the absence of corpora lutea. *Endocrinology* 48:17–24.
32. ———. 1952. Fertilizability of rabbit ova and the effects of temperature *in vitro* on their subsequent fertilization and activation *in vivo*. *J. Exp. Zool.* 121:351–82.
33. ———. 1953. Storage of unfertilized rabbit ova: Subsequent fertilization and the probability of normal development. *Nature* (*Lond.*) 172:353–54.
34. ———. 1954. Development of parthenogenetic rabbit blastocysts induced by low temperature storage of unfertilized ova. *J. Exp. Zool.* 125:127–50.
35. ———. 1955*a*. Fertilization and normal development of follicular oocytes in the rabbit. *Science* 121:867–69.
36. ———. 1955*b*. The maturation of rabbit oocytes in culture and their maturation, activation, fertilization and subsequent development in the Fallopian tubes. *J. Exp. Zool.* 128:379–406.
37. ———. 1955*c*. Développement de la capacité fertilisatrice des spermatozoïdes du lapin à l'intérieur du tractus génital femelle et fécondabilité des oeufs de lapine. In *La fonction tubaire et ses troubles*. Paris: Masson et Cie, pp. 40–52.
38. ———. 1959. Fertilization of rabbit ova *in vitro*. *Nature* (*Lond.*) 184:466–67.
39. ———. 1964. Effects of certain antifertility agents on the development of rabbit ova. *Fertil. Steril.* 15:97–106.
40. ———. 1965. Artificial insemination of snowshoe hares (*Lepus americanus*) and the

transfer of their fertilized eggs to the rabbit (*Oryctolagus cuniculus*). *J. Reprod. Fertil.* 10:447–49.

41. ———. 1966*a*. Transport of eggs from the Fallopian tube to the uterus as a function of oestrogen. *Nature* (*Lond.*) 212:1048–49.
42. ———. 1966*b*. Reciprocal transplantation of eggs between rabbit and ferret. *J. Exp. Zool.* 161:297–305.
43. Chang, M. C., and Bedford, J. M. 1962. Fertilizability of rabbit ova after removal of the corona radiata. *Fertil. Steril.* 13:421–25.
44. Chang, M. C., and Harper, M. J. K. 1966. Effects of ethinyl estradiol on egg transport and development in the rabbit. *Endocrinology* 78:860–72.
45. Chang, M. C., and Hunt, D. M. 1960. Effects of *in vitro* radiocobalt irradiation of rabbit ova on subsequent development *in vivo*, with special reference to the irradiation of maternal organism. *Anat. Rec.* 137:511–20.
46. Chang, M. C.; Hunt, D. M.; and Romanoff, E. B. 1958. Effects of radiocobalt irradiation on unfertilized or fertilized rabbit ova *in vitro* on subsequent fertilization and development *in vivo*. *Anat. Rec.* 132:161–80.
47. Dauzier, L. 1962. Nouvelles données sur la transplantation des oeufs, chez la lapine, par voie vaginale ou intrapéritonéale. *Ann. Biol. Anim. Biochim. Biophys.* 2:17–23.
48. ———. 1964. Resultats sur la transplantation des oeufs, chez la lapine, par voie vaginale ou intrapéritonéale. *Fifth Int. Congr. Anim. Reprod. A.I., Trento* 3:438–42.
49. Davis, B. K. 1963. Studies on the termination of pregnancy with norethynodrel. *J. Endocr.* 27:99–106.
50. Dickmann, Z. 1964. Fertilization and development of rabbit eggs following the removal of the cumulus oophorus. *J. Anat.* 98:397–402.
51. Dickmann, Z., and De Feo, V. J. 1967. The rat blastocyst during normal pregnancy and during delayed implantation, including an observation on the shedding of the zona pellucida. *J. Reprod. Fertil.* 13:3–9.
52. Dickmann, Z., and Noyes, R. W. 1960. The fate of ova transferred into the uterus of the rat. *J. Reprod. Fertil.* 1:197–212.
53. Doyle, L. L.; Gates, A. H.; and Noyes, R. W. 1963. Asynchronous transfer of mouse ova. *Fertil. Steril.* 14:215–25.
54. Duncan, G. W., and Forbes, A. D. 1965. Blastocyst survival and nidation in rats treated with oestrogen antagonists. *J. Reprod. Fertil.* 10:161–67.
55. Edwards, R. G. 1962. Meiosis in ovarian oocytes of adult mammals. *Nature* (*Lond.*) 196:446–50.
56. Edwards, R. G., and Austin, C. R. 1959. Induction of oestrus and ovulation in adult rats. *J. Endocr.* 18:vii–viii.
57. Fawcett, D. W.; Wislocki, G. B.; and Waldo, C. M. 1947. The development of mouse ova in the anterior chamber of the eye and in the abdominal cavity. *Amer. J. Anat.* 81:413–32.
58. Fekete, E., and Little, C. C. 1942. Observations on the mammary tumor incidence in mice born from transferred ova. *Cancer Res.* 2:525–30.
59. Fowler, R. E., and Edwards, R. G. 1957. Induction of superovulation and pregnancy in mature mice by gonadotrophins. *J. Endocr.* 15:374–84.
60. Gates, A. H. 1965. Rate of ovular development as a factor in embryonic survival. In *Preimplantation stages of pregnancy*, ed. G. E. W. Wolstenholme and M. O'Connor, pp. 270–88. Boston: Little, Brown and Co.

61. Glass, L. E., and Lin, T. P. 1963. Development of X-irradiated and non-irradiated mouse oocytes transplanted to X-irradiated and non-irradiated recipient females. *J. Cell. Comp. Physiol.* 61:53–60.
62. Glass, L. E., and McClure, T. R. 1964. Equivalence of X-irradiation *in vivo* or *in vitro* on mouse oocyte survival. *J. Cell. Comp. Physiol.* 64:347–54.
63. Greenwald, G. S. 1962*a*. Analysis of superovulation in the adult hamster. *Endocrinology* 71:378–89.
64. ———. 1962*b*. The role of the mucin layer in development of the rabbit blastocyst. *Anat. Rec.* 142:407–15.
65. Hafez, E. S. E. 1961. Storage of rabbit ova in gelled media at 10° C. *J. Reprod. Fertil.* 2:163–78.
66. ———. 1962*a*. *In vitro* and *in vivo* survival of morphologically atypical embryos in rabbits. *Nature* (*Lond.*) 196:1226–27.
67. ———. 1962*b*. "Differential cleavage rate" in 2-day litter-mate rabbit embryos. *Proc. Soc. Exp. Biol. Med.* 110:142–45.
68. ———. 1962*c*. Effect of progestational stage of the endometrium on implantation, fetal survival and fetal size in the rabbit, *Oryctolagus cuniculus*. *J. Exp. Zool.* 151:217–26.
69. ———. 1963. Physio-genetic interaction between mammalian blastocyst and endometrium. *J. Exp. Zool.* 154:163–68.
70. ———. 1964. Effects of overcrowding *in utero* on implantation and fetal development in the rabbit. *J. Exp. Zool.* 156:269–88.
71. ———. 1965*a*. Maternal effects on implantation and related phenomena in the rabbit. *Experientia* 21:234–37.
72. ———. 1965*b*. Storage media for rabbit ova. *J. Appl. Physiol.* 20:731–36.
73. Hafez, E. S. E., and Sugie, T. 1963. Reciprocal transfer of cattle and rabbit embryos. *J. Anim. Sci.* 22:30–35.
74. Hammond, J. 1934. The fertilization of rabbit ova in relation to time. A method of controlling the litter size, the duration of pregnancy and the weight of the young at birth. *J. Exp. Biol.* 11:140–61.
75. Heape, W. 1890. Preliminary note on the transplantation and growth of mammalian ova within a uterine foster-mother. *Proc. Roy. Soc.* 48:457–58.
76. ———. 1897. Further note on the transplantation and growth of mammalian ova within a uterine foster-mother. *Proc. Roy. Soc.* 62:178.
77. Humphrey, K. 1967. The development of viable embryos after ovum transfers to long-term ovariectomized mice. *Steroids* 9:53–56.
78. Kirby, D. R. S. 1960. Development of mouse eggs beneath the kidney capsule. *Nature* (*Lond.*) 187:707–8.
79. ———. 1962*a*. The influence of the uterine environment on the development of mouse eggs. *J. Embryol. Exp. Morph.* 10:496–506.
80. ———. 1962*b*. Reciprocal transplantation of blastocysts between rats and mice. *Nature* (*Lond.*) 194:785–86.
81. ———. 1963*a*. Development of the mouse blastocyst transplanted to the spleen. *J. Reprod. Fertil.* 5:1–12.
82. ———. 1963*b*. The development of mouse blastocysts transplanted to the scrotal or cryptorchid testis. *J. Anat.* (*Lond.*) 97:119–30.

83. ———. 1965. Endocrinological effects of experimentally induced extra-uterine pregnancies in virgin mice. *J. Reprod. Fertil.* 10:403–12.
84. ———. 1966. The difference in response by the mouse adrenal X-zone to trophoblast derived from transplanted tubal eggs and uterine blastocysts. *J. Endocr.* 36:85–92.
85. Kirby, D. R. S.; Billington, W. D.; and James, D. A. 1966. Transplantation of eggs to the kidney and uterus of immunised mice. *Transplantation* 4:713–18.
86. Lin, T. P. 1966. Microinjection of mouse eggs. *Science* 151:333–37.
87. Lin, T. P., and Glass, L. E. 1962. Effects of *in vitro* X-irradiation on the survival of mouse eggs. *Radiat. Res.* 16:736–45.
88. McLaren, Anne, and Michie, D. 1956. Studies on the transfer of fertilized mouse eggs to uterine foster mothers. I. Factors affecting the implantation and survival of native and transferred eggs. *J. Exp. Biol.* 33:394–416.
89. ———. 1959. Studies on the transfer of fertilized mouse eggs to uterine foster mothers. II. The effect of transferring large numbers of eggs. *J. Exp. Biol.* 36:40–50.
90. Mantalenakis, S. J., and Ketchel, M. M. 1966. Pseudopregnant recipients for blastocyst transfer in rats. *Int. J. Fertil.* 11:318–21.
91. Mayer, G., and Duluc, A.-J. 1966. Homotransplantation et hétérotransplantation intratesticulaires chez le rat et la souris de blastocystes en phase de préimplantation et de léthargie. *C.R. Acad. Sci.* (*Par.*) 263:1111–14.
92. Mintz, B. 1962. Formation of genotypically mosaic mouse embryos. *Amer. Zool.* 2:432.
93. ———. 1965*a*. Genetic mosaicism in adult mice of quadriparental lineage. *Science* 148:1232–33.
94. ———. 1965*b*. Experimental genetic mosaicism in the mouse. In *Preimplantation stages of pregnancy*, ed. G. E. W. Wolstenholme and M. O'Connor, pp. 194–207. Boston: Little, Brown and Co.
95. New, D. A. T. 1966. Development of rat embryos cultured in blood sera. *J. Reprod. Fertil.* 12:509–24.
96. Nicholas, J. S. 1933. Development of transplanted rat eggs. *Proc. Soc. Exp. Biol. Med.* 30:1111–13.
97. Noyes, R. W. 1952. Fertilization of follicular ova. *Fertil. Steril.* 3:1–12.
98. Noyes, R. W., and Dickmann, Z. 1960. Relationship of ovular age to endometrial development. *J. Reprod. Fertil.* 1:186–96.
99. ———. 1961. Survival of ova transferred into the oviduct of the rat. *Fertil. Steril.* 12:67–79.
100. Noyes, R. W.; Dickmann, Z.; Doyle, L. L.; and Gates, A. H. 1963. Ovum transfers, synchronous and asynchronous, in the study of implantation. In *Delayed implantation*, ed. A. C. Enders, pp. 197–209. Chicago: University of Chicago Press.
101. Noyes, R. W.; Doyle, L. L.; and Bentley, D. L. 1961. Effects of preimplantation development on foetal weight in the rat. *J. Reprod. Fertil.* 2:238–45.
102. Noyes, R. W.; Doyle, L. L.; Gates, A. H.; and Bentley, D. L. 1961. Ovular maturation and fetal development. *Fertil. Steril.* 12:405–16.
103. Noyes, R. W.; Yamate, A. M.; and Clewe, T. H. 1958. Ovarian transplants to the anterior chamber of the eye. *Fertil. Steril.* 9:99–113.
104. Orsini, M. W., and Psychoyos, A. 1965. Implantation of blastocysts transferred into progesterone-treated virgin hamsters previously ovariectomized. *J. Reprod. Fertil.* 10:300–301.

105. Parkes, A. S. 1943. Induction of superovulation and superfecundation in rabbits. *J. Endocr.* 3:268–79.
106. Pincus, G. 1939. The comparative behavior of mammalian eggs *in vivo* and *in vitro*. IV. The development of fertilized and artificially activated rabbit eggs. *J. Exp. Zool.* 82:85–130.
107. ———. 1940. Superovulation in rabbits. *Anat. Rec.* 77:1–8.
108. Pincus, G., and Enzmann, E. V. 1934. Can mammalian eggs undergo normal development *in vitro? Proc. Nat. Acad. Sci.* 20:121–22.
109. Psychoyos, A. 1961. Nouvelles recherches sur l'ovoimplantation. *C.R. Acad. Sci.* (*Par.*) 252:2306–7.
110. Rowlands, I. W. 1944. The production of ovulation in the immature rat. *J. Endocr.* 3:384–91.
111. Runner, M. N. 1947. Development of mouse eggs in the anterior chamber of the eye. *Anat. Rec.* 98:1–13.
112. Runner, M. N., and Palm, J. 1953. Transplantation and survival of unfertilized ova of the mouse in relation to postovulatory age. *J. Exp. Zool.* 124:303–16.
113. Seidel, F. 1952. Die Entwicklungspotenzen einer isolierten Blastomere des Zweizellenstadiums im Säugetierei. *Naturwissenschaften* 39:355–56.
114. ———. 1956. Nachweis eines Zentrums zur Bildung der Keimscheibe im Säugetierei. *Naturwissenschaften* 43:306–7.
115. Shah, M. K. 1956. Reciprocal egg transplantations to study the embryo-uterine relationship in heat-induced failure of pregnancy in rabbits. *Nature* (*Lond*) 177:1134–35.
116. Sherman, J. K., and Lin, T. P. 1958. Survival of unfertilized mouse eggs during freezing and thawing. *Proc. Soc. Exp. Biol. Med.* 98:902–5.
117. ———. 1959. Temperature shock and cold storage of unfertilized mouse eggs. *Fertil. Steril.* 10:384–96.
118. Simmons, R. L., and Russell, P. S. 1962. The antigenicity of mouse trophoblast. *Ann. N.Y. Acad. Sci.* 99:717–32.
119. ———. 1965. Histocompatibility antigens in transplanted mouse eggs. *Nature* (*Lond.*) 208:698–99.
120. Staples, R. E. 1966. The effect of clomiphene on blastocyst nidation in the rat. *Endocrinology* 78:82–86.
121. ———. 1967. Development of 5-day rabbit blastocysts after culture at 37° C. *J. Reprod. Fertil.* 13:369–72.
122. Talbert, G. B., and Krohn, P. L. 1966. Effect of maternal age on viability of ova and uterine support of pregnancy in mice. *J. Reprod. Fertil.* 11:399–406.
123. Tarkowski, A. K. 1959*a*. Experiments on the transplantation of ova in mice. *Acta Theriol.* 2:251–67.
124. ———. 1959*b*. Experiments on the development of isolated blastomeres of mouse eggs. *Nature* (*Lond.*) 184:1286–87.
125. ———. 1959*c*. Experimental studies on regulation in the development of isolated blastomeres of mouse eggs. *Acta Theriol.* 3:191–267.
126. ———. 1961. Mouse chimaeras developed from fused eggs. *Nature* (*Lond.*) 190:857–60.
127. ———. 1962. Inter-specific transfers of eggs between rat and mouse. *J. Embryol. Exp. Morph.* 10:476–95.
128. Thibault, C., and Dauzier, L. 1961. Analyse des conditions de la fécondation *in vitro* de l'oeuf de la lapine. *Ann. Biol. Anim. Biochim. Biophys.* 1:277–94.

129. Venge, O. 1950. Studies of the maternal influence on the birth weight in rabbits. *Acta Zool.* 31:1–148.
130. ———. 1952. A method for continuous chromosome control of growing rabbits. *Nature* (*Lond.*) 169:590–91.
131. Warwick, B. L.; Berry, R. O.; and Horlacher, W. R. 1934. Results of mating rams to angora female goats. *Proc. Amer. Soc. Anim. Prod.* pp. 225–27.
132. Whittingham, D. G., and Biggers, J. D. 1967. Fallopian tube and early cleavage in the mouse. *Nature* (*Lond.*) 213:942–43.
133. Yanagimachi, R., and Chang, M. C. 1964. *In vitro* fertilization of golden hamster ova. *J. Exp. Zool.* 156:361–75.

17

Egg Transfer in Cattle, Sheep, and Pigs

P. J. Dziuk
Department of Animal Science
University of Illinois, Urbana

Fertilized eggs can be transferred from the "genetic mother" to a recipient host who then provides the prenatal uterine environment. Egg transfer is a requisite for definitive research on maternal genetic and physiological influence in viviparous mammals. Egg transfer is also useful in studies on embryonal implantation and embryonal survival, and could be useful in animal breeding by increasing the number of potential offspring from certain females.

With the knowledge presently at hand, several rather rigid conditions must be met before transferred eggs survive and grow. Estrus of both donors and recipients of eggs must be very nearly synchronous, fertilizable eggs must be produced and recovered, and recovered eggs must be stored *extra utero* for only a few hours. We shall outline in this chapter some of the means by which egg transfers have been accomplished in cattle, sheep, and pigs.

I. Egg Production

A. *Timing of Ovulation*

1. *Cattle.* Ovulation time must be nearly synchronous between egg donor and recipient. Under usual conditions only 5% of cows will ovulate on any one date. Because of the vagaries of ovulation date, some form of control is imperative for synchrony except in experimental herds of 50 to 100 cows where, by chance, cows might be in synchronous estrus.

The classic means of control of estrus has been through the withdrawal of the inhibitory effect of progesterone by eliminating the corpus luteum or by stopping regular, usually daily, administrations of progesterone or a derivative about 3 to 6 days before the desired heat date. No one method is perfectly reliable but several are sufficiently efficacious to be useful.

Expression of the corpus luteum by manual pressure applied *per rectum* is effective, but only during the phase of the estrous cycle when the corpus luteum is discrete and functional (Avery, Cole, and Graham 1962). This in effect limits its usefulness to the time between the 6th and 16th days of the estrous cycle. Heat occurs 3 days later.

Daily injections of 50 mg of progesterone in oil beyond the end of the luteal phase delays estrus until about 5 days after the last injection (Donker *et al.* 1958; Dziuk *et al.* 1958; Avery, Cole, and Graham 1962). Injection of 150 mg of progesterone every third day is also effective, but the interval between the last injection and estrus is lengthened slightly to about 6 days.

The daily incorporation of 150 to 180 mg of 6α-methyl-17α-acetoxyprogesterone (MAP) into the diet will inhibit ovulation. Estrus will occur 2 to 3 days after the last of 20 days of treatment (Zimbelman 1963; Nestel *et al.* 1963; Fahning *et al.* 1966; Dhindsa, Hoversland, and Smith 1967). Daily feeding of 0.2 mg of 6-chloro-Δ^6-dehydro-17-acetoxyprogesterone (CAP) also inhibits estrus. Estrus occurs about 5 or 6 days after withdrawal (Van Blake, Brunner, and Hansel 1963). Injection of an estrogen with progesterone or a derivative to shorten the functional life of the corpus luteum effectively controls time of estrus (Wiltbank *et al.* 1965).

Since each of the methods just described is imperfect and response is variable, about 20 to 25% more cows should be started on treatment than are needed for egg transfer (Hansel 1961; Hansel, Malven, and Black 1961). This will provide enough cows for egg transfer even though some are not in estrus or do not ovulate.

2. *Sheep*. In flocks of sufficient size, donors and recipients with spontaneously synchronous estrus may be paired (Schmidt 1961; Hancock and Hovell 1961*a*, *b;* Niswender 1967). Daily injections of 5 to 10 mg of progesterone for 12 to 14 days will successfully control estrus (Shelton 1965; Shelton and Moore 1966). Estrus occurs 3 to 5 days after withdrawal of treatment (Braden, Lamond, and Radford 1960; Lamond and Bindon 1962; Findlay and Vaughan 1964; Moore and Shelton 1964*b;* Foote and Waite 1965). Injection of 20 mg of progesterone on alternate days is also effective. Incorporation of 60 mg of MAP into the daily diet for 14 days has been shown to effectively control estrus and ovulation (Hogue, Hansel, and Bratton 1962; Southcott, Braden, and Moule 1962; Hinds, Dziuk, and Lewis 1964; Dhindsa, Hoversland, and Smith 1966). Onset of heat occurs about 60 to 96 hr after withdrawal of MAP. The intramuscular injection of 250 or 500 i.u. of human chorionic gonadotrophin (HCG) 48 to 54 hr after MAP will cause ovulation in 25 to 26 hr (Dziuk, Polge, and Rowson 1964; Brunner, Hansel, and Hogue 1964; Dziuk 1965). Therefore, with judicious use of progestogens and ovulating hormones, one will achieve not only synchronous estrus but also synchronous ovulation.

Intravaginal tampons impregnated with progestogen serve as still another way of inhibiting heat. Heat occurs 36 to 54 hr after removal (Robinson 1964, 1965*a*, *b;* Curl *et al.* 1966; Roberts 1966; Roberts and Edgar 1966).

3. *Pigs*. Several alternatives are available for obtaining synchrony between donors and recipients. In a herd, 1 of 21 (5%) gilts should be in estrus on any one day if estrus is uniformly distributed. In a herd of 126 gilts, 6 would be in estrus on one day. This is a sufficient number for most purposes. Synchronous ovulation could be induced by injection of 500 i.u. of HCG intramuscularly during late proestrus (Dziuk, Polge, and Rowson 1964; Baker 1965; Hunter and Polge 1966).

Estrus will occur about 5 to 7 days after the last of 18 or 20 daily injections of 100 mg of progesterone (Gerrits *et al.* 1962). Oral administration of MAP or other progestogens will control time of ovulation, but for optimum fertility gilts should be used at the spontaneous heat occurring 21 days later (Dziuk and Baker 1962; Steinbach and Smidt 1964; Pond *et al.* 1965; Dziuk and Polge 1965; Ray and Seerley 1966). Injection of 500 i.u. of PMS followed in 96 hr by 500 i.u. of HCG will group heats about 21 days after HCG (Day *et al.* 1965).

A nonsteroidal antigonadotrophin, methallibure, has been found to be very effective in inhibiting and controlling estrus in gilts. Six days after the last of 18 to 20 daily oral doses of 100 to 150 mg of methallibure, about 80% of gilts will be in estrus (Polge 1964, 1965; Gerrits and Johnson 1964; Hafez *et al.* 1966; Groves 1967). Time of ovulation can be controlled precisely by injection of 500 i.u. of HCG on the 5th day after last treatment during proestrus (Baker 1965).

The methods outlined for each of the three species may not be always perfect or convenient, but they have been used successfully as part of egg transfer studies. The reader is directed to recent reviews for further details (Anderson, Schultz, and Melampy 1963; Dziuk 1966).

B. *Superovulation*

1. *Cattle.* An unmated cow produces 1 egg each 21 days throughout her reproductive life. Potentially, her ovaries contain sufficient oocytes that she might produce many times the normal number of eggs if properly stimulated (Hafez 1961; Sugie and Hafez 1963). Because follicle-stimulating hormone (FSH) is associated with egg maturation and follicular development, most procedures for superovulation depend on administration of exogenous FSH at levels assumed to be in excess of normal endogenous levels. The source of FSH is usually pregnant mare's serum gonadotrophin (PMS) or an extract of the anterior pituitary.

To be effective, the FSH must be administered at the critical times during the follicular phase of the estrous cycle. Administration of FSH in the face of a functional corpus luteum may stimulate follicular growth, but fertility is impaired. Hence, FSH is given immediately after manual expression of the corpus luteum, after the last of a series of injections or feedings of a progestogen or on day 15 or 16 of a normal estrous cycle (Jainudeen and Hafez 1966).

Purified swine pituitary extracts are effective superovulators in both cows and prepubertal calves (Avery, Fahning, and Graham 1962). Treatment is most effective following expression of the corpus luteum or a series of injections of progesterone rather than when begun on day 16 of the estrous cycle (Avery, Fahning, and Graham 1962). Injection of a single dose of 2,000 i.u. PMS stimulates follicular development, and the follicles ovulate from endogenous LH. Fertility is adequate in some cases (Gordon, Williams, and Edwards 1962) but very low in others (Hafez, Jainudeen, and Lindsay 1965; Jainudeen and Hafez 1966; Jainudeen, Hafez and Lineweaver 1966). PMS also has the disadvantage of inducing an antibody reaction (Dziuk *et al.* 1958), thereby reducing the stimulatory response of follicles to subsequent PMS injections (Hafez and Sugie 1961; Hafez, Sugie, and Gordon 1963; Hafez *et al.* 1964; Hafez, Jainudeen, and Lindsay 1965). This is apparently no problem with repeated injections of purified swine pituitary extracts. Ovaries respond repeatedly and anaphylaxis is absent (Nichols 1957). Superovulation reduces fertility and in some way interferes with the possibility of recovering more than 30% of eggs (Nichols 1957; Dziuk *et al.* 1958; Avery *et al.* 1962; Hafez *et al.* 1964). The results are uncertain when one tries to induce simultaneous ovulation of several follicles in the cow, which normally ovulates 1 or 2 follicles at the expense of others.

2. *Sheep.* PMS has been used extensively in producing superovulation in ewes (Averill 1958; Gordon 1958; Averill and Rowson 1958; Hancock and Hovell 1961*a, b;* Lamond 1962). As the dose level increases from 700 to 1,300 i.u., the number of eggs recovered increases from 3 to 9 (Averill 1958). Horse anterior pituitary extract is also a very effective superovulator, with the number of corpora lutea increasing from 4 to 9 with doses of 60 to 135 mg divided among 3 daily injections begun on day 12 after estrus (Moore and Shelton, 1964*a*). FSH will also stimulate superovulation when given at the time of last progestogen $\pm$(24 hr) following treatment to control time of estrus (Robinson 1961, 1965*b*).

Modification of the methods briefly described will work and can be used to produce eggs for transfer (Averill and Rowson 1958; Alliston and Ulberg 1961; McDonald and Rowson 1961; Niswender 1967). Lambs can also be induced to produce eggs, but no fertility has been recorded (Mansour 1959).

3. *Pigs.* Although the pig is normally polyovular, superovulation still provides the same advantages as for uniovular species. The pig is quite responsive to PMS. Injection of 1,500 i.u. on day 15 of the estrous cycle will produce 33 ovulations per gilt with estrus occurring 4 days after PMS (Hunter 1964, 1966). PMS is also effective when 1,250 i.u. are given in a single subcutaneous injection 24 hr after the last day of methallibure administration (Dziuk, unpublished). Prepuberal gilts can be a source of fertilized eggs when injected with 500 i.u. of PMS followed by 500 i.u. of HCG intramuscularly 96 hr later (Dziuk and Gehlbach 1966).

Potentially, there are many more eggs available in ovaries than are realized even by the best superovulation methods available. Much work needs to be done. Even with these limitations sufficient eggs can be produced to meet most needs for egg transfer.

II. Egg Recovery and Isolation

A. *Recovery Methods*

The small size of the egg relative to the size of the donor animal makes the search for eggs a little like the search for the proverbial needle in the haystack. Fortunately the search can be narrowed down to a rather restricted area of the oviduct or uterus, particularly when ovulation time is known. Egg recovery has always been based on a scheme of flushing the uterus or oviduct with a fluid medium, thereby carrying the eggs along. Because the specific gravity of an egg is greater than that of most media, the eggs settle to the bottom of the fluid, limiting the search to one focal plane. A stereoscopic, binocular microscope with a range of magnification between 8$\times$ and 30$\times$ is essential for identification and isolation of the egg. Flat-bottomed plastic petri dishes marked off with lines 10 mm apart permit a systematic search without changing the focal plane. They will hold as much as 40 ml of fluid in a relatively thin layer. This is especially important if the fluid contains blood or other debris. The search can best be done with light transmitted through frosted glass or translucent petri dishes.

Eggs for transfer can be picked up with a micropipette with a vernier control or by a Pasteur pipette attached to a 1 or 2 cc syringe. We will now deal separately with the peculiarities of each of the three species.

In the cow, excision and flushing of the genital tract obtained either at hysterectomy or slaughter is the most direct and effective method for isolating the eggs (Nichols 1957; Avery and Graham 1962; Rowson and Moor 1966*b*). In cows with single ovulations, 85% of eggs are recovered. In striving to preserve the integrity of the donor, nonsurgical egg recovery has been attempted (Dowling 1950; Dracy and Petersen 1950; Dziuk *et al.* 1958), but success has been rather limited (Sugie 1965).

Eggs can be removed from sheep at laparotomy by flushing the oviduct toward the ovarian end or into the uterus. When flushing toward the ovary a cannula is placed in the upper oviduct, directed into a dish which serves to collect fluid and eggs (Averill and Rowson 1958; Alliston and Ulberg 1961). If fluid is flushed through the oviduct into the uterus, a cannula can be inserted into the uterus to collect the eggs and fluid (Rowson and Moor 1966*a;* Niswender 1967). Approximately 80% of eggs are recovered by these methods.

In the pig, eggs can be flushed up the oviduct, through a cannula inserted into the infundibulum and then collected (Vincent, Robison, and Ulberg 1964; Smidt, Steinbach, and Scheven 1965). If the donor is expendable her oviduct can be cut and flushed directly into a test tube (Baker 1965). Four-cell eggs may be in the uterus or in the oviduct. The most suitable method to recover them requires flushing the fluid through the oviduct into the uterus and trapping it in a segment of the uterus. A cannula or large needle is then inserted through the uterine wall to recover fluid and eggs (Hancock and Hovell 1962; Dziuk, Polge, and Rowson 1964; Smidt, Steinbach, and Scheven 1965). About 80% of pig eggs can be recovered by each of the methods.

B. *Recovery Media*

Homologous blood serum alone or with equal portions of some physiological salt solution is the basis for most media for egg recovery and storage in sheep and cattle. The serum is usually heated to destroy possible ovicidal properties and filtered to remove bacteria and foreign matter (Moore and Rowson 1960; Averill and Rowson 1959; Alliston and Ulberg 1961). Antibiotics such as penicillin and streptomycin can be added at levels of 1,000 units/ml and 500 to 1,000 μgm/ml, respectively (Buttle and Hancock 1964; Avery and Graham 1962; Dziuk *et al.* 1958; Nichols 1957). Commercially available sterile lamb's serum is convenient and effective (Niswender 1967). The uterus of the rabbit provides an effective medium for storage of sheep eggs (Adams *et al.* 1961; Hunter *et al.* 1962). Goat-sheep interspecies transfers survive temporarily but have not gone to term (Bowerman and Hancock 1963; Hancock 1964).

While it seems advisable to maintain the recovered eggs at near body temperature if they are to be stored for only a short period, there is only nonconclusive evidence showing superiority of body temperature over room temperature for storage of eggs of cattle, sheep, or pigs (Averill and Rowson 1959; Hancock and Hovell 1961*b;* Buttle and Hancock 1964; Dziuk, Polge, and Rowson 1964). Storage of eggs up to 10 or 12 hr seems to have little adverse effect on their survival, but longer storage times are detrimental to successful development of the stored egg (Harper and Rowson 1963).

The recovery medium for pig eggs may be either Tyrode's solution with 0.1% bovine plasma albumin, fraction V, Hanks's solution with 10% horse serum, or Tyrode's solution with 10% pig serum (Hancock and Hovell 1962; Dziuk, Polge, and Rowson 1964; Vincent, Robison, and Ulberg 1964; Smidt, Steinbach, and Scheven 1965). Tissue culture medium 199 with bicarbonate works as well for pig eggs as those just mentioned, but without bicarbonate no eggs survive (Niswender 1967). This culture medium has the advantage of being readily available commercially in sterile form.

III. Transfer

A. *Nonsurgical*

To reach the uterus in a cow by surgery is difficult. The cervical passage is quite easily penetrated by a pipette and the uterus can be manipulated *per rectum*. For these reasons one is sorely tempted to transfer eggs in cattle by nonsurgical means. While many attempts have been made (Nichols 1957; Dziuk *et al.* 1958; Avery and Graham 1962), only a few successes have been reported (Sugie 1965; Rowson and Moor 1966*b*). The insufflation of the uterus with gaseous carbon dioxide is the apparent key to success. The egg can be transferred by either puncturing the vaginal wall and penetrating the uterus (Sugie 1965) or by going directly through the cervix (Rowson and Moor 1966*b*). Nonsurgical egg transfer in the pig, although successful in a few cases, has limited usefulness due to difficulty in penetrating the cervical opening without puncturing the cervical wall (Polge and Day, personal communication).

B. *Surgical*

The surgical approaches in the cow are by flank or midventral laparotomy (Willett, Buckner, and Larson 1953; Avery *et al.* 1962). In the sheep and pig midventral laparotomies are routine and require a minimum of surgical intervention (Averill and Rowson 1958; Hancock and Hovell 1961*a*, *b;* Dziuk, Polge, and Rowson 1964; Woody and Ulberg 1964; Smidt, Steinbach, and Scheven 1965). The isolated egg(s) is drawn up into a fine pipette with a small amount of fluid. The pipette is directed into the upper oviduct or a very small puncture wound in the uterus. The egg and fluid are expelled, care being taken that the pipette is actually in the uterine lumen and not still in the uterine wall. The site of deposition of the egg should correspond to the site of recovery.

Egg transfer requires the successful integration of many separate steps, each of which must be successful by itself. No one link in the chain of events can fail or all is lost. I will draw conclusions for each species separately, selecting the method that appears at this time to be most suitable for each of the steps.

1. *Cattle.* Oral administration of 180 mg of MAP for 18 days (Zimbelman 1963) will control time of estrus in donors and recipients. Four daily doses of purified swine pituitary extract will superovulate the donor (Avery, Fahning, and Graham 1962). Eggs are recovered 96 hr after ovulation by flushing the uterus of the killed donor with autologous serum. Eggs are transferred via the cervix to the uterus, which is insufflated with gaseous carbon dioxide (Rowson and Moor 1966*b*).

2. *Sheep.* Control of ovulation is effected by 14 daily feedings of 60 mg of MAP (Dziuk *et al.* 1964). Ewes are superovulated by an injection of 750 i.u. of PMS 24 hr after last MAP. Seventy-two hr after ovulation the uterus and oviduct of the donors are flushed with lamb's serum during surgery to recover eggs (Niswender 1967). The eggs are placed in the upper uterus through a small puncture wound during laparotomy.

3. *Pigs.* Daily oral administration of 125 mg of methallibure for 18 days will control heat (Groves 1967). Injection of 1500 i.u. of PMS (Hunter 1966) 24 hr after last day of treated diet followed in 100 hr with 500 i.u. of HCG (Dziuk and Baker 1962) will control ovulation time and numbers of ovulations. The uterus and oviducts can be flushed with tissue culture medium 199 with bicarbonate to obtain eggs 48 hr after ovulation (Niswender 1967).

Transfer of eggs is effected during laparatomy by placing eggs in the uterus with a narrow-bore pipette via a small puncture wound (Dziuk, Polge, and Rowson 1964).

Eggs have been transferred successfully in pigs, sheep, and cattle by the present methods. These methods will undoubtedly be refined in the future to increase their usefulness.

References

1. Adams, C. E.; Rowson, L. E. A.; Hunter, G. L.; and Bishop, G. P. 1961. Long distance transport of sheep ova. *Proc. Fourth Int. Congr. Anim. Reprod., The Hague.* 2:381.
2. Alliston, C. W., and Ulberg, L. C. 1961. Early pregnancy loss in sheep at ambient temperatures of 70° and 90° F as determined by embryo transfer. *J. Anim. Sci.* 20:608.
3. Anderson, L. L.; Schultz, J. R.; and Melampy, R. M. 1963. Pharmacological control of ovarian function and estrus in domestic animals. In *Gonadotrophins: Their chemical and biological properties and secretory control*, ed. H. H. Cole, chap. 5. San Francisco: W. H. Freeman.
4. Averill, R. L. W. 1958. The production of living sheep eggs. *J. Agric. Sci.* 50:17.
5. Averill, R. L. W., and Rowson, L. E. A. 1958. Ovum transfer in the sheep. *J. Endocr.* 16:326.
6. ———. 1959. Attempts at storage of sheep ova at low temperatures. *J. Agric. Sci.* 52:393.
7. Avery, T. L.; Cole, C. L.; and Graham, E. F. 1962. Investigations associated with the transplantation of bovine ova. I. Synchronization of oestrus. *J. Reprod. Fertil.* 3:206.
8. Avery, T. L.; Fahning, M. L.; and Graham, E. F. 1962. Investigations associated with the transplantation of bovine ova. II. Superovulation. *J. Reprod. Fertil.* 3:212.
9. Avery, T. L.; Fahning, M. L.; Pursel, V. G.; and Graham, E. F. 1962. Investigations associated with the transplantation of bovine ova. IV. Transplantation of ova. *J. Reprod. Fertil.* 3:229.
10. Avery, T. L., and Graham, E. F. 1962. Investigations associated with the transplantation of bovine ova. III. Recovery and fertilization. *J. Reprod. Fertil.* 3:218.
11. Baker, R. D. 1965. Factors influencing the transport of spermatozoa in artificially inseminated gilts. Ph.D. diss., University of Illinois.
12. Bowerman, H. R. L., and Hancock, J. L. 1963. Sheep-goat hybrids. *J. Reprod. Fertil.* 6:326.
13. Braden, A. W. H.; Lamond, D. R.; and Radford, H. M. 1960. The control of the time of ovulation in sheep. *Aust. J. Agric. Res.* 11:389.
14. Brunner, M. A.; Hansel, W.; and Hogue, D. E. 1964. Use of 6-methyl-17-acetoxyprogesterone and pregnant mare serum to induce and synchronize estrus in ewes. *J. Anim. Sci.* 23:32.
15. Buttle, H. R. L., and Hancock, J. L. 1964. Birth of lambs after storage of sheep eggs *in vitro*. *J. Reprod. Fertil.* 7:417.
16. Curl, S. E.; Cockrell, T.; Bogard, G.; and Hudson, F. 1966. Use of intravaginal progestin to synchronize estrus in sheep. *J. Anim. Sci.* 25:921.
17. Day, B. N.; Neill, J. D.; Oxenreider, S. L.; Waite, A. B.; and Lasley, J. F. 1965. Use of gonadotrophins to synchronize estrous cycles in swine. *J. Anim. Sci.* 24:1075.
18. Dhindsa, D. S.; Hoversland, A. S.; and Smith, E. P. 1966. Estrous synchronization and lambing rate in ewes treated with MAP. *Vet. Med./Small Anim. Clin.* 11:1094.

19. Dhindsa, D. S.; Hoversland, A. S.; and Smith, E. P. 1967. Estrous control and calving performance in beef cattle fed 6-methyl-17-acetoxy-progesterone under ranch conditions. *J. Anim. Sci.* 26:167.
20. Donker, J. D.; Nichols, J. R.; Graham, E. F.; and Petersen, W. E. 1958. Controlled estrus in cattle. In *Reproduction and infertility*, ed. F. X. Gassner, chap. 4. New York: Pergamon Press.
21. Dowling, D. F. 1950. Problems of the transplantation of fertilized ova. *J. Agric. Sci.* 39:374.
22. Dracy, A. E., and Petersen, W. E. 1950. Isolation of ova from the living bovine. *J. Dairy Sci.* 33:797.
23. Dziuk, P. J. 1965. Timing of maturation and fertilization of the sheep egg. *Anat. Rec.* 153:211.
24. ———. 1966. Hormones in reproduction. In *Hormonal relationships and applications in the production of meats, milk, and eggs.* Washington, D.C.: National Academy of Sciences.
25. Dziuk, P. J., and Baker, R. D. 1962. Induction and control of ovulation in swine. *J. Anim. Sci.* 21:697.
26. Dziuk, P. J.; Donker, J. D.; Nichols, J. R.; and Petersen, W. E. 1958. Problems associated with the transfer of ova between cattle. *University of Minnesota Agricultural Experiment Station, Tech. Bull. 222.* April.
27. Dziuk, P. J., and Gehlbach, G. D. 1966. Induction of ovulation and fertilization in the immature gilt. *J. Anim. Sci.* 25:410.
28. Dziuk, P. J.; Hinds, F. C.; Mansfield, M. E.; and Baker, R. D. 1964. Follicle growth and control of ovulation in the ewe following treatment with 6-methyl-17-acetoxyprogesterone. *J. Anim. Sci.* 23:787.
29. Dziuk, P. J., and Polge, C. 1965. Fertility in gilts following induced ovulation. *Vet. Rec.* 77:236.
30. Dziuk, P. J.; Polge, C.; and Rowson, L. E. 1964. Intra-uterine migration and mixing of embryos in swine following egg transfer. *J. Anim. Sci.* 23:37.
31. Fahning, M. L.; Schultz, R. H.; Graham, E. F.; Donker, J. D.; and Mohrenweiser, H. W. 1966. Synchronization of oestrus in dairy heifers with 6α-methyl-17α-acetoxyprogesterone and its effect on conception rate. *J. Reprod. Fertil.* 12:569.
32. Findlay, A. L. R., and Vaughan, S. M. 1964. The practicability of methods of controlling the oestrous cycle and breeding season in the sheep: A review. *Vet. Rec.* 76:880.
33. Foote, W. C., and Waite, A. B. 1965. Some effects of progesterone on estrous behavior and fertility in the ewe. *J. Anim. Sci.* 24:151.
34. Gerrits, R. J.; Fahning, M. L.; Meade, R. J.; and Graham, E. F. 1962. Effect of synchronization of estrus on fertility in gilts. *J. Anim. Sci.* 21:1022.
35. Gerrits, R. J., and Johnson, L. A. 1964. The effect of an orally administered non-steroid on estrus, ovulation and fertility in gilts. *Proc. Fifth Int. Congr. Anim. Reprod. A.I., Trento* 3:455.
36. Gordon, I. 1958. The use of progesterone and serum gonadotrophin (P.M.S.) in the control of fertility in sheep. *J. Agric. Sci.* 50:123.
37. Gordon, I.; Williams, G.; and Edwards, J. 1962. The use of serum gonadotrophin (P.M.S.) in the induction of twin-pregnancy in the cow. *J. Agric. Sci.* 59:143.
38. Groves, T. W. 1967. Methallibure in the synchronisation of oestrus in gilts. *Vet. Rec.* 80:470.

39. Hafez, E. S. E. 1961. Procedures and problems of manipulation, selection, storage, and transfer of mammalian ova. *Cornell Vet.* 51:299.
40. Hafez, E. S. E.; Jainudeen, M. R.; Kroening, G. H.; and El-Banna, A. A. 1966. Use of progesterone and thiocarbamoyl-hydrazine compounds for estrous synchronization in gilts. *J.A.V.M.A.* 149:35.
41. Hafez, E. S. E.; Jainudeen, M. R.; and Lindsay, D. R. 1965. Gonadotropin-induced twinning and related phenomena in beef cattle. *Acta Endocr.* (suppl. no. 102):5.
42. Hafez, E. S. E.; Rajakoski, E.; Anderson, P. B.; Frost, O. L.; and Smith, G. 1964. Problems of gonadotropin-induced multiple pregnancy in beef cattle. *Amer. J. Vet. Res.* 25:107.
43. Hafez, E. S. E., and Sugie, T. 1961. Superovulatory responses in beef cattle and an experimental approach for non-surgical ova transfer. *Proc. Fourth Int. Congr. Anim. Reprod., The Hague* 2:387.
44. Hafez, E. S. E.; Sugie, T.; and Gordon, I. 1963. Superovulation and related phenomena in the beef cow. I. Superovulatory responses following PMS and HCG injections. *J. Reprod. Fertil.* 5:359.
45. Hancock, J. L. 1963. Survival *in vitro* of sheep eggs. *Anim. Prod.* 5:237.
46. ———. 1964. Attempted hybridization of sheep and goats. *Proc. Fifth Int. Congr. Anim. Reprod. A.I., Trento* 3:445.
47. Hancock, J. L., and Hovell, G. J. R. 1961*a*. Transfer of sheep ova. *J. Reprod. Fertil.* 2:295.
48. ———. 1961*b*. Transfer of sheep ova. *J. Reprod. Fertil.* 2:520.
49. ———. 1962. Egg transfer in the sow. *J. Reprod. Fertil.* 4:195.
50. Hansel, W. 1961. Estrous cycle and ovulation control in cattle. *J. Dairy Sci.* 44:2307.
51. Hansel, W.; Malven, P. V.; and Black, D. L. 1961. Estrous cycle regulation in the bovine. *J. Anim. Sci.* 20:621.
52. Harper, M. J. K., and Rowson, L. E. A. 1963. Attempted storage of sheep ova at 7° centigrade. *J. Reprod. Fertil.* 6:183.
53. Hinds, F. C.; Dziuk, P. J.; and Lewis, J. M. 1964. Control of estrus and lambing performance in cycling ewes fed 6-methyl-17-acetoxyprogesterone. *J. Anim. Sci.* 23:782.
54. Hogue, D. E.; Hansel, W.; and Bratton, R. W. 1962. Fertility of ewes bred naturally and artificially after estrous cycle synchronization with an oral progestational agent. *J. Anim. Sci.* 21:625.
55. Hunter, G. L.; Bishop, G. P.; Adams, C. E.; and Rowson, L. E. 1962. Successful long-distance aerial transport of fertilized sheep ova. *J. Reprod. Fertil.* 3:33.
56. Hunter, R. H. F. 1964. Superovulation and fertility in the pig. *Anim. Prod.* 6:189.
57. ———. 1966. The effect of superovulation on fertilisation and embryonic survival in the pig. *Anim. Prod.* 8:457.
58. Hunter, R. H. F., and Polge, C. 1966. Maturation of follicular oocytes in the pig after injection of human chorionic gonadotrophin. *J. Reprod. Fertil.* 12:525.
59. Jainudeen, M. R., and Hafez, E. S. E. 1966. Control of estrus and ovulation in cattle with orally active progestin and gonadotropins. *Int. J. Fertil.* 11:47.
60. Jainudeen, M. R.; Hafez, E. S. E.; and Lineweaver, J. A. 1966. Superovulation in the calf. *J. Reprod. Fertil.* 12:149.
61. Lamond, D. R. 1962. Oestrus and ovulation following administration of placental gonadotrophins to Merino ewes. *Aust. J. Agric. Res.* 13:707.

62. Lamond, D. R., and Bindon, B. M. 1962. Oestrus, ovulation and fertility following suppression of ovarian cycles in Merino ewes by progesterone. *J. Reprod. Fertil.* 4:57.
63. McDonald, M. F., and Rowson, L. E. A. 1961. Ovum transfer to lactating ewes. *Proc. Fourth Int. Congr. Anim. Reprod., The Hague* 2:392.
64. Mansour, A. M. 1959. The hormonal control of ovulation in the immature lamb. *J. Agric. Sci.* 52:88.
65. Moore, N. W., and Rowson, L. E. A. 1960. Egg transfer in sheep: Factors affecting the survival and development of transferred eggs. *J. Reprod. Fertil.* 1:332.
66. Moore, N. W., and Shelton, J. N. 1964*a*. Response of the ewe to a horse anterior pituitary extract. *J. Reprod. Fertil.* 7:79.
67. ———. 1964*b*. Egg transfer in sheep: Effect of degree of synchronization between donor and recipient, age of egg, and site of transfer on the survival of transferred eggs. *J. Reprod. Fertil.* 7:145.
68. Nestel, B. L.; Creek, M. J.; Wiggan, L. G. S.; and Murtagh, J. E. 1963. Oestrus synchronization in hybrid beef heifers following the oral use of 6-methyl-17-acetoxy-progesterone. *Brit. Vet. J.* 119:23.
69. Nichols, J. R. 1957. Superovulation and ova transplantation in the bovine. Ph.D. diss., University of Minnesota.
70. Niswender, G. D. 1967. Direct effects of embryos and the uterus on the function of corpora lutea in sheep and swine. Ph.D. diss., University of Illinois.
71. Polge, C. 1964. Synchronisation of oestrus in pigs by oral administration of ICI compound 33828. *Proc. Fifth Int. Congr. Anim. Reprod. A.I., Trento* 2:388.
72. ———. 1965. Experiments on egg transplantation in the pig. *Proc. Roy. Soc. Med.* 58:907.
73. Pond, W. G.; Hansel, W.; Dunn, J. A.; Bratton, R. W.; and Foote, R. H. 1965. Estrous cycle synchronization and fertility of gilts fed progestational and estrogenic compounds. *J. Anim. Sci.* 24:536.
74. Ray, D. E., and Seerley, R. W. 1966. Oestrus and ovarian morphology in gilts following treatment with orally effective steroids. *Nature* 211:1102.
75. Roberts, E. M. 1966. The use of intra-vaginal sponges impregnated with 6-methyl-17-acetoxyprogesterone (MAP) to synchronize ovarian activity in cyclic Merino ewes. *Proc. Aust. Soc. Anim. Prod.* 6:32.
76. Roberts, E. M., and Edgar, D. G. 1966. The stimulation of fertile oestrus in anoestrous Romney ewes. II. *J. Reprod. Fertil.* 12:565.
77. Robinson, T. J. 1961. The time of ovulation and efficiency of fertilization following progesterone and pregnant mare serum treatment in the cyclic ewe. *J. Agric. Sci.* 57:129.
78. ———. 1964. Synchronization of oestrus in sheep by intravaginal and subcutaneous application of progestin impregnated sponges. *Proc. Aust. Soc. Anim. Prod.* 5:47.
79. ———. 1965*a*. Use of progestagen-impregnated sponges inserted intravaginally or subcutaneously for the control of the oestrous cycle in the sheep. *Nature (Lond.)* 206:39.
80. ———. 1965*b*. The practical control of ovulation and oestrus in domestic animals using new synthetic progestogens. In *Recent advances in ovarian and synthetic steroids and the control of ovarian function*, ed. R. P. Shearman. Searle, High Wycombe, England.
81. Rowson, L. E. A., and Moor, R. M. 1966*a*. Embryo transfer in the sheep: The sig-

nificance of synchronizing oestrus in the donor and recipient animal. *J. Reprod. Fertil.* 11:207.

82. ———. 1966*b*. Non-surgical transfer of cow eggs. *J. Reprod. Fertil.* 11:311.
83. Schmidt, K. 1961. Eitransplantation beim Schaf nach Progesteron-PMS-Synchronisation. *Proc. Fourth Int. Congr. Anim. Reprod., The Hague* 2:398.
84. Shelton, J. N. 1965. Control of oestrus in sheep. *Aust. Vet. J.* 41:112.
85. Shelton, J. N., and Moore, N. W. 1966. Survival of fertilized eggs transferred to ewes after progesterone treatment. *J. Reprod. Fertil.* 11:149.
86. Smidt, D.; Steinbach, J.; and Scheven, B. 1965. Modified method for the *in vivo* recovery of fertilized ova in swine. *J. Reprod. Fertil.* 10:153.
87. Southcott, W. H.; Braden, A. W. H.; and Moule, G. R. 1962. Synchronization of oestrus in sheep by an orally active progesterone derivative. *Aust. J. Agric. Res.* 13:901.
88. Steinbach, J., and Smidt, D. 1964. Untersuchungen zur hormonalen Beeinflussung von Brunst und Ovulation beim Schwein. *Proc. Fifth Int. Congr. Anim. Reprod. A.I., Trento* 7:482.
89. Sugie, T. 1965. Successful transfer of a fertilized bovine egg by non-surgical techniques. *J. Reprod. Fertil.* 10:197.
90. Sugie, T., and Hafez, E. S. E. 1963. Induction of superovulation in beef cattle by means of PMS and HCG application. Invitational paper to Third Symposium of Physiology of Reproduction, University of Tokyo, February 26, 1963.
91. Van Blake, H.; Brunner, M. A.; and Hansel, W. 1963. Use of 6-chloro-Δ^6-dehydro-17-acetoxyprogesterone (CAP) in estrous cycle synchronization of dairy cattle. *J. Dairy Sci.* 46:459.
92. Vincent, C. K.; Robison, O. W.; and Ulberg, L. C. 1964. A technique for reciprocal embryo transfer in swine. *J. Anim. Sci.* 23:1084.
93. Willett, E. L.; Buckner, P. J.; and Larson, G. L. 1953. Three successful transplantations of fertilized bovine eggs. *J. Dairy Sci.* 36:520.
94. Wiltbank, J. N.; Zimmerman, D. R.; Ingalls, J. E.; and Rowden, W. W. 1965. Use of progestational compounds alone or in combination with estrogen for synchronization of estrus. *J. Anim. Sci.* 24:990.
95. Woody, C. O., and Ulberg, L. C. 1964. Viability of one-cell sheep ova as affected by high environmental temperature. *J. Reprod. Fertil.* 7:275.
96. Zimbelman, R. G. 1963. Determination of the minimal effective dose of 6α-methyl-17α-acetoxyprogesterone for control of the estrual cycle of cattle. *J. Anim. Sci.* 22:1051.

18

Mammalian Embryo Culture

R. L. Brinster

King Ranch Laboratory of Reproductive Physiology
Department of Animal Biology
School of Veterinary Medicine
University of Pennsylvania

Eighty-seven years ago a Viennese embryologist (Schenk 1880) published the first account of the cultivation of a mammalian embryo *in vitro*. He reported that he had inseminated ovarian ova of the rabbit and guinea pig, and observed the first cleavage division. According to a posthumous report in 1893, Onanoff was able to effect *in vitro* fertilization and subsequent cleavage of uterine ova of the rabbit and the guinea pig. He was able to culture these ova to the 8-cell stage *in vitro* and claimed that they developed to the primitive streak stage when placed in the abdominal cavity of females or males of the same or other species. These examples serve to establish the long history of attempts to study embryos outside the body.

Most of the work in the field of *in vitro* cultivation of mammalian ova has been done during the past 35 years. The increasing interest in the cultivation of fertilized mammalian embryos has paralleled the rise of tissue culture techniques. The embryos of many mammalian species have been subjected to various culture techniques with unequal success; among the eutherian mammals that have been studied are the rabbit, mouse, rat, guinea pig, sheep, goat, cow, monkey, and man. On the basis of the success of the attempts, these species may be divided into two groups: the first group, which has experienced a fair amount of success, includes the rabbit and the mouse and the second group, which has experienced, at best, very limited success, includes all the other species mentioned. It seems best to consider first the techniques used to handle and grow the embryos, and to consider second, in detail, the various media which have been used to culture these embryos.

I. Techniques

A. *Embryo Manipulation*

Embryos are obtained from the reproductive tract by one of two methods: (1) by incising the wall of the oviduct or shredding the oviduct, or (2) by flushing an appropriate solution through the reproductive tract. Method 2 offers several advantages: (1) less damage is likely to be done to the recovered embryos by the pressure of the fluid; (2) it is quicker; and (3) there is less debris mixed with the embryos.

The entire reproductive tract need not be subjected to the recovery methods, since one quickly determines where the embryos of the desired stage are located and confines his activity to this area. The volume of fluid in which the embryos are flushed depends on the size of the tubes from which they must be recovered, but it is essential to keep this volume to a minimum in order that the embryos may be found rapidly in the flushing fluid. There is always a certain amount of cellular debris and blood fluid obtained in the flushing of the reproductive tract, and since this is not a beneficial environment for the embryos, rapid recovery is desirable. Figure 5*a* shows the effect of the adverse conditions which may be encountered in the flushing fluid. The 2-cell bovine embryo pictured had remained in the flushing fluid for approximately 45 min before it was found. This embryo subsequently recovered and went on to divide, but the abnormal morphology of the blastomeres suggests a deleterious effect of the flushing fluid. The length of time the embryos can remain in the recovery fluid is open to considerable speculation, since it is impossible to determine the exact nature of the fluids in every case. Consequently, one is well advised to keep this to a minimum.

In general, glass pipettes drawn to the proper diameter have been found to be best for handling the embryos. These pipettes are drawn from standard soft-glass tubing of the appropriate diameter. In selecting the size of the glass tubing, one should keep in mind that the ratio of the glass wall thickness to the bore diameter does not change during the process of drawing. Pasteur pipettes are a readily available source of short glass tubing which can be easily drawn to various bore diameters. The tip bore diameter should be 25–50% larger than the diameter of the embryo, and the tip should be broken off smoothly and at right angles to the long axis of the tubing. If this minimum bore diameter is continued along the tube for a considerable distance (1.5 to 2.5 in), it is possible to pick up large numbers of embryos in very small volumes. It is possible to transfer several hundred embryos in a volume of 2 μl with considerable speed.

If the embryos are to be handled in a protein-free solution, with or without a synthetic polymer such as Ficoll, then the pipettes should be coated with silicone to prevent adherence of the embryos to the glass. This may be done before the pipettes are drawn to the proper diameter, but the coating is considerably better if it is done after the drawing of the glass. It is a general rule that, when embryos are to be handled in solutions which do not contain protein, all glassware should be siliconized. This will immensely facilitate handling the embryos and has been shown to have no deleterious effect on the development of mouse embryos (Brinster, unpublished). More detailed information on handling embryos in protein-free solution can be found in Brinster and Thomson (1966) and Brinster (1967*a*).

Capillary action will draw fluid containing embryos into the first part of the pipette, but additional suction must be applied to draw fluid and embryos further into the pipette. This suction can be applied either mechanically or by mouth. For embryos measuring less than 1 mm in diameter, the suction applied by mouth is easier to control and faster for the manipulations. On the other hand, for embryos over 1 mm in diameter, such as 5-day rabbit blastocysts, mechanical means of inducing suction are superior. The mechanical means which may be used are a rubber bulb, a glass syringe, or a micrometer syringe. The micrometer syringe is the easiest to use. Figure 1 shows the pipettes and apparatus used in our laboratory.

If it is desired to study the blastomeres of the embryo devoid of external layers, these may be removed by exposure to the appropriate enzyme solutions or by mechanical manipulation. To remove the cumulus cells which surround 1-cell embryos, hyaluronidase at a concentration of 300 units per ml in a phosphate-buffered salt solution can be used (Brinster 1965*e*). In addition to the cumulus cells, in the mouse and the rat, the solution will remove the corona radiata. However, in the rabbit and human, hyaluronidase is not effective in removing the corona cells.

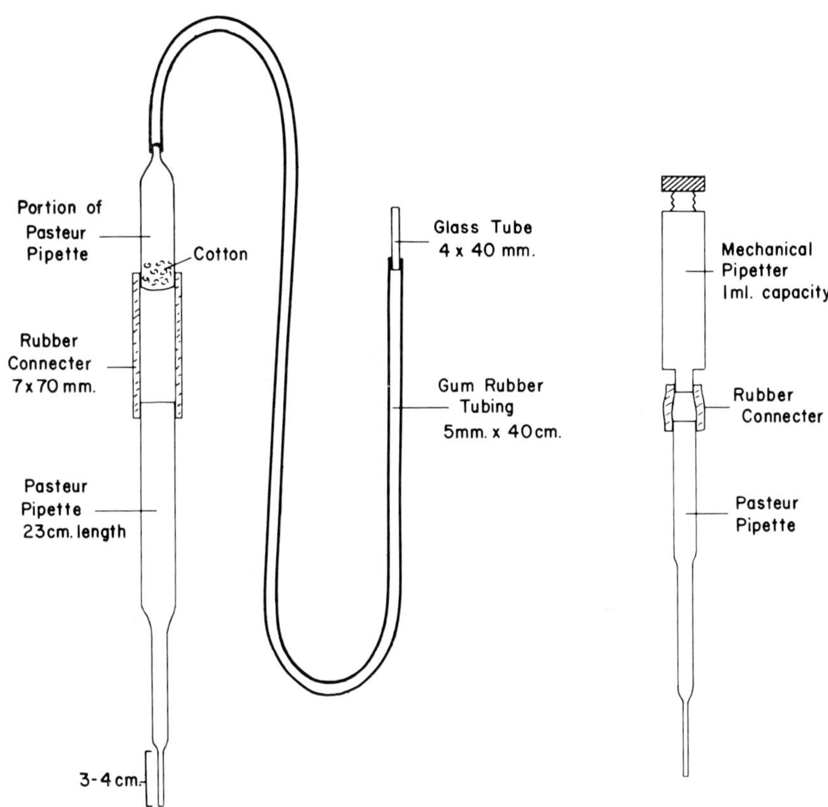

FIG. 1. Pipettes and apparatus used to handle embryos. See text for further details

To remove the corona radiata cells one can draw the embryos back and forth in a narrow pipette with a bore diameter the same size as the external diameter of the zona pellucida (Brinster 1968). Bedford and Chang (1962) have removed the corona cells from the rabbit embryo by shaking the embryos in 1 ml of culture medium in a serological test tube, and Mastroianni and Ehteshamzadeh (1964) have extracted from the rabbit oviduct a factor which will remove corona radiata cells.

To remove the zona pellucida of the mammalian embryo or the mucinous coat of the rabbit embryo, pronase at a concentration of 2.5 mg/ml in phosphate-buffered salt solution can be used (Mintz 1962; Brinster 1965*d*). It is desirable to watch the embryos during the action of hyaluronidase and pronase, and to remove the embryos from the enzyme solutions as soon as the desired effect has been achieved. This will prevent excessive damage to the embryos by the enzymes, particularly by the proteolytic enzyme pronase. Mouse embryos develop normally after exposure to hyaluronidase at the above concentration for 30 min (Brinster, unpublished), but information is not available concerning the length of time embryos may remain in pronase unharmed.

B. *Culture Methods*

A large number of vessel types and different methods have been employed to grow mammalian embryos *in vitro*. Some of these are shown in Figure 2. Method A has been developed by Brinster (1963) to cultivate the mouse embryo during the preimplantation period (Fig. 3). In this method 10 ml of sterile paraffin oil is placed in a 15 × 60 mm disposable plastic tissue culture dish (Falcon Plastics, Inc.), and microdrops (50 to 100 μl) of medium are placed under the oil. The oil prevents evaporation and aids in maintaining

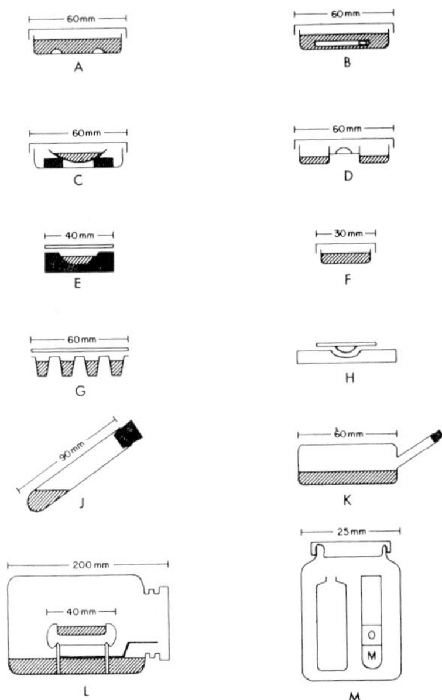

FIG. 2. Culture methods that have been used for *in vitro* studies on mammalian embryos. See text for details.

sterility of the culture medium. The dish is then placed in a chamber with a controlled temperature and atmosphere. In method B the petri dish again contains 10 ml of sterile paraffin oil, and on the bottom of the dish are small capillary tubes with an internal diameter of 1 to 2 mm. The embryos are placed in culture medium at the bottom of the tube, and a small air bubble is left at the top of the tube (Mulnard 1965). In method C, a 15 × 60 mm glass petri dish serves as a container to hold a sponge or glass wool on the bottom; on top of this is placed the watch glass to hold the culture medium (Chen 1954). The glass wool or sponge is moistened and serves to saturate the atmosphere inside the dish with water vapor. Method D is similar to method C, but the entire dish, with a sponge in the surrounding well, is available as a disposable plastic dish from Falcon Plastics. The culture medium can be as small as 50 μl or as large as 1 ml. Method F is an embryological watch glass with a culture medium volume of 1 to 4 ml. The dish may be left uncovered, covered with a glass slide, or the medium may be covered with paraffin oil (Brinster, unpublished). Method F is a small plastic disposable culture dish, approximately 25 to 30 mm in diameter, containing 1 ml of culture medium. The culture medium may be covered with sterile paraffin oil or may be left exposed to the atmosphere. In method G, a portion of a disposable agglutination titer-board is used for a culture dish (Brinster, unpublished). The total volume of the individual wells is approximately 0.5 ml, and 100 to 300 μl of medium are placed in each well. This is covered either with paraffin oil or with a glass slide. Method H is the standard hanging-drop method for culturing small numbers of cells in a small volume. Method J is a serological or small tube containing the culture medium and has been used by Hammond (1949) and Whitten (1956). The tube may be left open and placed in a controlled atmosphere, or gassed and stoppered. Method K is a standard Carrel flask, as used by Pincus (1941*a*, *b*),

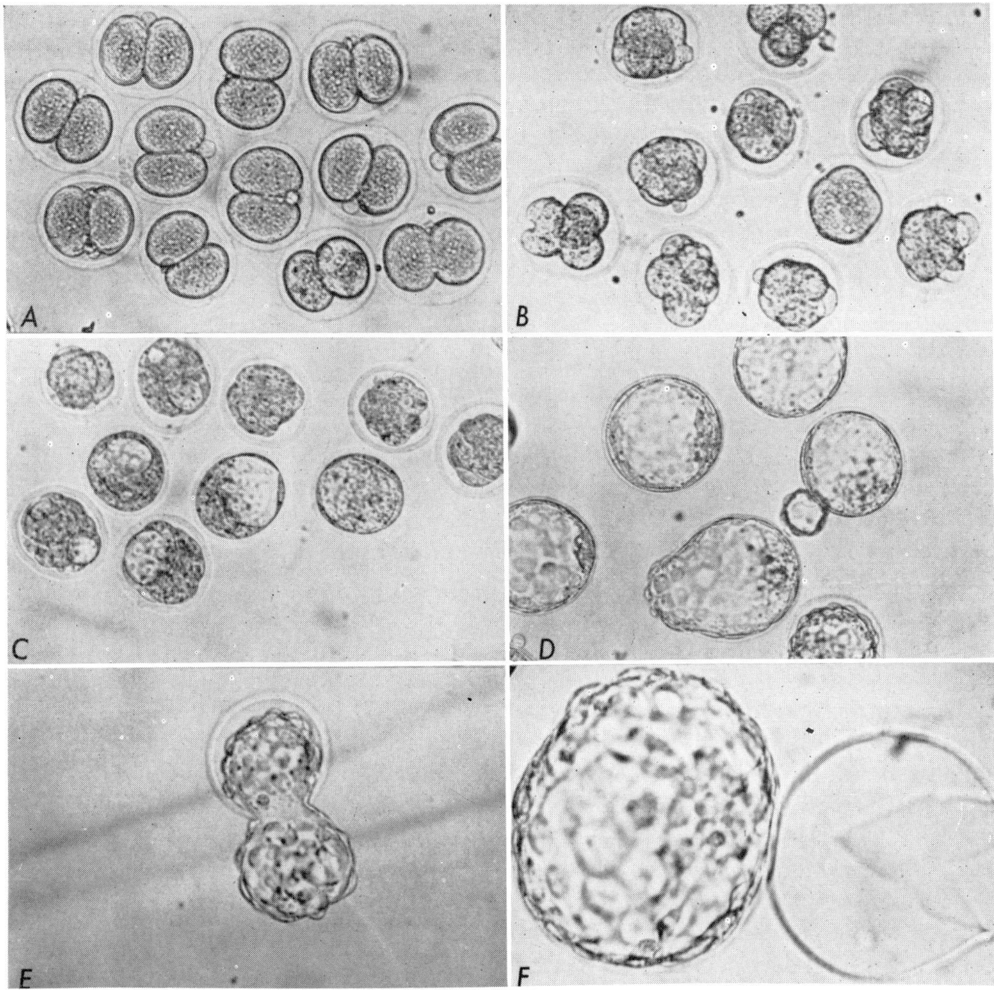

FIG. 3. Mouse embryos grown *in vitro*. (*a*) 2-cell embryos at beginning of culture (×36). (*b*) 8-cell stage. Developed from (*a*) after 24 hr in culture (×36). (*c*) Morulae and early blastocysts. Developed from (*a*) after 48 hr in culture (×36). (*d*) Blastocyst developed from (*a*) after approximately 72 hr in culture. Beginning to escape from zona pellucida (×72). (*e*) Late blastocyst escaping from zona pellucida after 72 hr in culture (×72). (*f*) Late blastocyst completely free of zona. Slightly later than (*e*) (×72).

containing 2 to 10 ml of culture medium, and it can be gassed as in method J. The advantage is that the embryos can be more easily seen on a flat surface. Method L has been used by Mintz (1964) for cultivation of mouse embryos and consists of a large bottle which contains a bicarbonate-buffered balanced salt solution on the bottom to maintain humidity. The embryos are then grown in a small dish on a platform over the salt solution; the atmosphere in the bottle is controlled by gassing with the appropriate gas mixture. Method M has been developed by Brinster (1967*b*) to measure carbon dioxide production from radioactively labeled substrates. It consists of a scintillation vial containing a 2 ml ampule with the top broken off and a serological test tube 7 mm by 45 mm. Fifty to 100 μl of culture medium are placed in the serological tube under 3 drops of sterile paraffin oil. The ampule is used later in the experiment to contain Hyamine. The top of the scintillation vial is closed with a rubber serum bottle stopper which has a plug diameter of 14 mm, and the vial may be gassed by inserting two 20-gauge needles through the rubber stopper. One 20-gauge needle is connected to a millipore filter which is in turn connected to a tank containing the appropriate gas mixture, and the other needle acts as an exhaust vent.

Several investigators have employed culture systems in which the fluid moves. Pincus and Werthessen (1938) employed a perfusion chamber, as they felt that rabbit embryos developed better in those culture vessels where the medium was in motion. Purshottam and Pincus (1961) employed Carrel flasks fixed to a rocking platform. Chang (1959) has also used this system to achieve fertilization in rabbit eggs. Brinster (unpublished) employed roller tubes, but was unable to show a beneficial effect of motion.

Mammalian embryos have also been grown in vessels designed to measure respiratory activity. Dragoiu, Benetato, and Oprean (1937) employed the Warburg apparatus to culture ovarian ova of a cow. The Cartesian diver has been used to grow embryos and measure respiration of embryos from the rat (Boell and Nicholas 1948), the rabbit (Fridhandler, Hafez, and Pincus 1957; Fridhandler 1961), and the mouse (Mills and Brinster 1967). In these studies, from 1 embryo (in the case of large rabbit blastocysts) to 200 embryos (in the case of unfertilized mouse ova) are cultivated in 2 or 3 μl of culture medium in a Cartesian diver. The total volume of the diver is 7 to 15 μl. In the studies of Mills and Brinster (1967) the embryos were recovered from the diver after the respiratory measurements were made, and the subsequent development was studied. They found that there was little difference between the development of the embryos which had been in the diver for 4 hr and control embryos which had not been placed in the diver.

Several of the methods shown in Figure 2 (*A*, *D*, *E*, *G*, *J*) have been examined for their relative ability to support development of the 2-cell mouse embryo into a blastocyst. The data are shown in Table 1. Method A produced the greatest number of blastocysts, and E the smallest number of blastocysts. Further studies indicated that the paraffin oil exercised a beneficial effect on the culture system. For instance, in an embryological watch glass, such as E, development is very poor if the medium is exposed, even if the atmosphere is completely humidified and controlled. This is true even if the watch glass is covered with a glass slide, or with a slide and a petroleum jelly seal. However, development with method E is not different from development with method A if paraffin oil is placed over the medium in E. In fact, most of the vessels shown in Figure 2 allow good development of the embryos, provided the surface of the medium is covered with paraffin oil.

If it is undesirable to cover the surface of the medium with paraffin oil, as in the case of studies on fat-soluble compounds, then a method such as J, K, or L should be used. In these cases, it is desirable to carefully control the atmosphere and to keep the ratio of the atmosphere to culture medium as small as possible. Even so, the time necessary for manipulation of the embryos and for observation of the embryos is considerably more than in a system such as A, E, or G. Although the volume of the culture medium in the experiment shown in Table 1 varied considerably, it has been shown by Brinster (1964) that changes in the volume of the culture medium between 10 μl and 500 μl have no effect on the development of 2-cell mouse embryos into blastocysts. Although there certainly is a minimum volume in which the embryo can exist, it seems possible that there may be a wide range of volumes in which development is possible.

C. *Incubation*

One of the most difficult and yet most important aspects of cultivating mammalian embryos is the control of the atmosphere and environment to which the culture vessels are exposed. The system used in our laboratory to control the environment of the cultures is shown in Figure 4. The characteristics of the environment controlled are: (1) temperature

TABLE 1

COMPARISON OF DIFFERENT CULTURE METHODS IN FIGURE 2 FOR *In Vitro* DEVELOPMENT OF MOUSE EMBRYOS

Treatment Method	Response[a] Replicate Number				Mean[b] Angular Response	Per Cent Response
	1	2	3	4		
1. Petri dish—A	7	6	9	10	55	66
2. Tubes—J	8	5	7	6	47	54
3. Watch glass—E	8	3	1	0	27	25
4. Agglutination board—G	10	7	2	2	40	43
5. Center-well dish—D	5	4	6	4	38	39

[a] Response is the number of normal blastocysts from twelve 2-cell embryos after 3 days cultivation *in vitro* in BMOC-2. Medium volume was 50 μl in 1, 2 ml in 2, 2 ml in 3, 250 μl in 4, and 1 ml in 5.

[b] The mean angular response is the statistical transformation used in order to compare treatment effects. The standard error of the mean in this example is ±6.14 angular degrees.

of the gas and the environment; (2) humidity of the gas phase; (3) flow of gas; and (4) actual gas composition.

Cultivation temperature has not been studied to a great extent, but Alliston (1965) has shown that rabbit embryos cultivated at 40° C for 6 hr do not develop as well as controls cultivated at 37° C, when transferred into foster mothers. Two-cell mouse embryos begin to show irreversible morphological changes when the temperature reaches 42° to 45° C (Brinster, unpublished). Several studies (Chang 1948; Hafez 1963) have found that rabbit embryos can be stored at low temperatures, but development appears to be arrested during storage. Subsequent development of the stored embryos occurs in foster mothers, but is not as good as in controls. In the absence of contradictory evidence, it is generally considered that a temperature of 37° to 37.5° C is the best temperature in which to maintain the cultures. It also seems desirable to maintain the embryos as close as possible to 37° C during the flushing and recovery period.

The humidity of the culture chambers is extremely important when cultures are maintained unsealed in the chamber, such as in methods C and D. If humidity is not maintained at 100%, evaporation will occur when the culture medium is exposed to the atmosphere, and changes in osmotic pressure and concentrations of essential nutrients will result. Under these conditions the embryos will die rapidly. However, if the culture dish is sealed, as in methods J and K, or is covered with paraffin oil, as in methods A and M, humidity control is not essential, since evaporation is eliminated by the seal or the paraffin oil cover.

To determine a desirable flow rate of the gas through the culture chamber, two things must be considered. First, the size of the chamber and, second, the time which one can wait for equilibration. The formula for the change in gas composition within the culture chamber is a differential equation, and can be solved for any of the desired values. For instance, the time necessary for the gas in a chamber to equilibrate can be obtained from the following formula:

$$V\frac{dC}{dt} = FC_x - FC\,,$$

where

$$C = \text{the concentration in the chamber}\,,$$

$$C_x = \text{the concentration of the entering gas}\,,$$

$$V = \text{the volume of the culture chamber}\,,$$

$$F = \text{the flow rate in liters per minute}\,,$$

$$\frac{dC}{dt} = \text{the change in concentration with time}\,.$$

If the initial concentration in the chamber is 0, the above formula reduces to

$$C = [1 - e^{-(F/V)T}]C_x\,;$$

For C to be 99% of C_x, $e^{-(F/V)T}$ must be

$$0.01\,, \quad \text{or} \quad \frac{F}{V}T = 4.60\,.$$

If F = 1 liter/min and V = 25 liter, F/V = 0.04/min; therefore,

$$T = \frac{4.60}{0.04} = 115 \text{ min for } 99\% \text{ equilibration}\,.$$

The time necessary for partial equilibration can be calculated for various chamber sizes and flow rates.

This is an important formula, since the composition of the gas within the culture medium is dependent upon the composition of the gas in the gas phase of the culture chamber. Equilibration of the culture medium with the atmosphere can take place no faster than the equilibration of the culture chamber volume with the composition of the inflow of gas. In fact, equilibration of the culture medium will show a definite time lag behind the equilibration of the gas phase in the culture chamber. This time lag is dependent on the physical barriers to diffusion of the gas into the medium. Diffusion of the gas into the medium, or from the medium, is slowed by the layer of paraffin oil as in method A. This may account for part of the beneficial effects of the paraffin oil. This interference is due primarily to the interphase between the medium and the oil and between the oil and the atmosphere, and not primarily to a decrease in mobility of the gas molecules within the oil. This increase in diffusion time of the gas in those culture vessels where paraffin oil covers the culture medium is beneficial when the vessels are to be used for short periods of time outside the incubator. It takes several times as long for precipitation of the medium to occur in a vessel where paraffin oil covers the surface as it does in a vessel where the medium is exposed directly to the atmosphere.

The composition of the gas to which the culture medium and embryos should be exposed has received only cursory attention. There are three major components of the gas: nitrogen, oxygen, and carbon dioxide. Nitrogen is considered inert and is varied depending on the amount of the other components desired. The oxygen tension of the culture medium has received some attention. Whitten (1956) showed that the 8-cell mouse embryo would develop in an atmosphere of 5% CO_2 in air or nitrogen, but not in 5% CO_2 in oxygen. In our laboratory, using a continuous flow system, we obtained slightly different results. Two-cell mouse embryos, when grown in 5% CO_2 in oxygen, degenerate within 24 hr. However, we found that the 2-cell mouse embryo had a definite and measurable requirement for atmospheric oxygen during the culture period. The embryos invariably degenerated

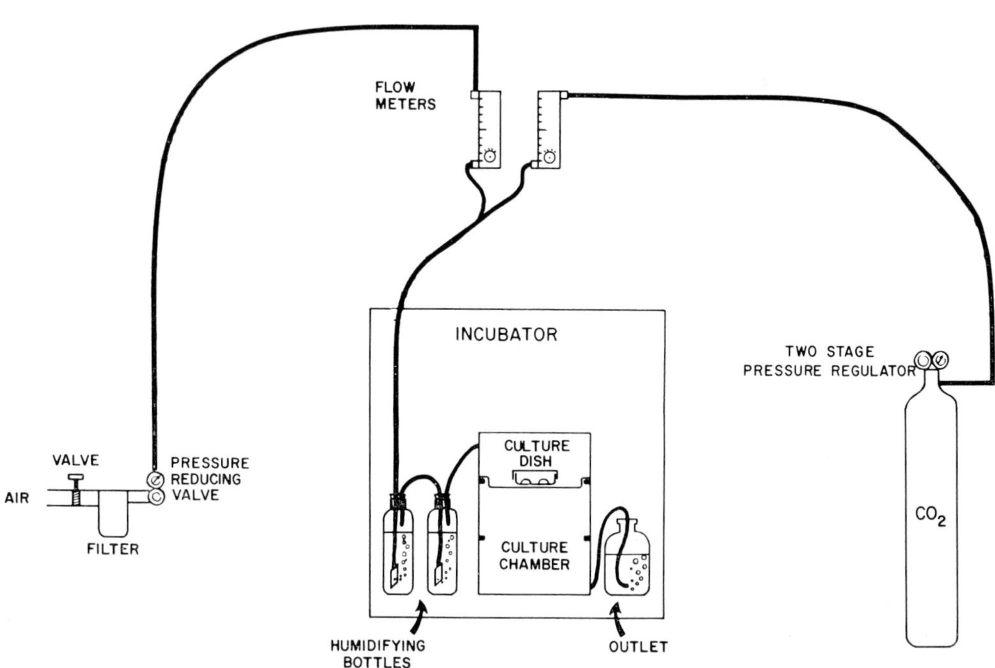

FIG. 4. System used to control the environment of the embryo cultures (not drawn to scale). The air is supplied under pressure from a central source or laboratory compressor. A valve is placed in the line to turn the flow on or off, and a filter is used to remove dirt and oil residues from the air. The filtration necessary depends on the quality of air supplied. A single-stage reducing valve regulates the pressure. Five to 7.5 pounds per square inch is desirable for flexible rubber hose lines. The capacity of the air flow meter is determined by the culture chamber size (see text). Carbon dioxide (100%) is delivered from a commercial tank by means of a 2-stage regulator at a pressure of 5 to 7.5 pounds per square inch. The capacity of the CO_2 flow meter is determined by the capacity of the air flow meter and the required composition of the final gas mixture. For 5% CO_2 in air, the CO_2 meter is one-twentieth the capacity of the air meter. Mixing of the gas occurs in the delivery lines. Gas washing tubes in 1-liter bottles are used to humidify and warm the gas mixture. The culture chamber must be airtight, and the size is determined by experimental requirements. The outlet tube is submerged in water and serves to indicate continuous flow. Biggers (1965) described the use of a similar system for organ culture.

in pure nitrogen, without undergoing even one cleavage. However, at a concentration of 1.15% oxygen, 2-cell mouse embryos developed normally to blastocysts. At 0.56% oxygen, there is a markedly reduced response, but a few embryos will develop to the blastocyst stage (Auerbach and Brinster, unpublished).

The amount of carbon dioxide which should be contained in the atmosphere is determined by whether or not the cells require CO_2 in the medium and by the buffer system employed. Thomson (personal communication) studied a large number of noncarboxylic buffering compounds, the most important of which were Tris and phosphate, to determine if they could be substituted for bicarbonate in Brinster's medium for ovum culture (Table 2). When Tris or phosphate replaced bicarbonate in the medium, 2-cell mouse embryos underwent 1, 2, or 3 cleavages and stopped dividing. However, the control embryos in BMOC-2 with bicarbonate developed into blastocysts. The Tris and phosphate media were kept in air and the bicarbonate medium in 5% CO_2. This information suggests that carbon

TABLE 2

BRINSTER'S MEDIUM FOR OVUM CULTURE (BMOC-2)

Component	mM	gm/l	ml of 0.154 M Stock in 13 ml
NaCl	94.88	5.546	5.90
NaLactate	25.00	[a]2.253	2.10
NaPyruvate	0.25	0.028	[b]2.10
KCl	4.78	0.356	0.40
$CaCl_2$	1.71	0.189	[c]0.20
KH_2PO_4	1.19	0.162	0.10
$MgSO_4$–$7H_2O$	1.19	0.294	0.10
$NaHCO_3$	25.00	2.106	2.10
Pen Strep	100 U/ml of Pen, 50 μg/ml of Strep		
Bovine serum albumin	1 mg/ml=1 gm/l		

SOURCE: Brinster 1965*d*.

[a] NaLac added as liquid prepared as follows: Add 1.82 ml of concentrated lactic acid (85–90%) to 200 ml of double distilled H_2O. Neutralize to pH 7.4 (about 15–20 ml of 1 N NaOH). This is enough lactic acid to make 1 liter of medium.

[b] NaPyr stock is 0.00154 M pyruvate in 0.154 M NaCl (17 mg/100 ml).

[c] $CaCl_2$ stock is 0.11 M.

dioxide is beneficial to the development of the embryo, but more extensive studies are needed to answer this fundamental question. Phosphate buffer seems to be an acceptable substitute for bicarbonate when the medium is to be employed for short-term experiments, or during the manipulation of the embryos. Mills and Brinster (1967) found very little effect on the subsequent development of embryos kept in this medium for periods ranging up to 4 hr.

If a bicarbonate buffer system is employed, then close attention must be paid to both bicarbonate concentration and the gas composition of the atmosphere over the culture medium, since both of these affect pH. For example, with a bicarbonate concentration of 25 mM, the pH of the medium in Table 2 is 7.4 in 5% CO_2. Lowering or raising the bicarbonate concentration tenfold lowers or raises the pH 1 unit respectively. If bicarbonate is held constant at 25 mM, and CO_2 concentration in the atmosphere is raised from 5% to 50%, then pH falls approximately 1 unit from 7.4 to 6.4. On the other hand, if the CO_2 concentration in the atmosphere falls from 5% to 0.5%, the pH of the culture medium rises approximately 1 pH unit, from 7.4 to 8.4.

A recently examined characteristic of the embryo's environment is light intensity. Daniel (1964) demonstrated that rabbit embryos subjected to light during culture showed a marked decrease in their development. This work suggests that the intensity of illumination during the manipulation and cultivation of rabbit embryos may affect their subsequent ability to cleave and develop normally.

It is important to be able to assess the effects of cultivation and manipulation on the development of the embryos. The more accurately this assessment can be made, the more valuable will be the information gained. One of the most desirable morphologic criteria of successful development is the attainment of a specific stage—the formation of a blastocyst or cleavage in experiments on fertilization. However, such obvious end-points are not always available, and one must be satisfied with other methods. The most obvious of these is whether the cell lives or dies during a specified period of time. This is easily discerned, but unfortunately measures only the severe effects, such as whether a compound is essential for development or whether a situation is intolerable for the embryo. However, it is incapable of demonstrating the graded response which is so often necessary to determine which of a series of treatments is the best. A frequently used parameter of development is whether cleavage occurs, and if cleavage occurs, how many cells are formed before development stops. In the case of blastocyst development, the increase in diameter of the blastocyst is often used as a measure of development.

The most desirable yardsticks of development are cytochemical parameters. These include such things as increases in DNA, RNA, and protein content, or increases in enzyme activity, or increases in the incorporation of radioactive isotopes. These parameters are direct and accurate indicators of development; they are quantitative, and can be treated statistically. Unfortunately, quite frequently the number of embryos available is so small that cytochemical studies cannot readily be made. Microtechniques are rapidly improving and, in some cases, may be applied to the quantities available for studies on mammalian preimplantation embryos. The assessment of development in a quantitative manner is desirable in order to permit use of the information in statistical comparisons of different treatments of the embryo. Recently, the application of the angular transformation and the probit transformation to the analysis of experimental results from experiments on preimplantation mouse embryos has been discussed (Biggers and Brinster 1965).

II. Culture Medium

A. *Introduction*

Much of the early work which was done on preimplantation stages of mammalian embryos was directed toward observing development outside the body. More recently, considerable emphasis has been placed on specific environmental factors and requirements necessary for development. The earlier work will be discussed by species and the later work will be discussed in terms of specific environmental factors.

Classical studies on the early cleavage stages of rabbit embryos were made in 1929 by Lewis and Gregory. They obtained embryos either at the 2-cell stage or the morula stage, and the embryos were cultivated in autologous or homologous plasma, with or without embryo juice, in warm chambers. They found that the 1- or 2-cell stage would develop into a morula, and that embryos taken at the morula stage would develop into normal-appearing blastocysts and expand. However, embryos taken at the 1-cell stage did not develop into normal-appearing blastocysts. Since that time a number of investigators have examined cleavage of the rabbit embryo *in vitro*. Most of the culture media employed contained 50% or more serum, and this has made it difficult to determine the specific requirements for development of the rabbit embryo.

Mark and Long (1912) first reported studying the mouse embryo *in vitro*. However, they observed the ova for only 12 hr and did not see any cleavage divisions. Lewis and Wright (1935) employed drops of plasma and embryo extract to observe and photograph early stages of the mouse embryo, but again, they did not observe any cleavage divisions. Hammond (1949) was the first investigator to successfully cultivate mouse embryos through several stages *in vitro*. As a medium he employed a salt solution containing sodium chloride, potassium chloride, and magnesium chloride with a glucose concentration of 1 mg per ml; to this he added about 5% egg white. Hammond put 1 to 6 embryos in small vessels in 2 to 3 ml of medium under 1 to 2 ml of air. The vessels were sealed and incubated at 37° C. The embryos were examined 48 hr after the beginning of cultivation. No 2-cell embryos developed beyond the 4-cell stage; but most 8-cell embryos and some late 4-cell embryos developed into blastocysts. He also observed the escape of the blastocyst from the zona pellucida.

During the last few years the preimplantation mouse embryo has been studied extensively. Considerable information has become available concerning its *in vitro* requirements. Brinster (1965*c*) has developed a completely defined medium for the development of the 2-cell mouse embryo into a blastocyst, and a semidefined medium (1965*d*) which allows maximum development (Table 2). These studies will be discussed under the specific environmental factors.

Defrise (1933) attempted to cultivate tubal cleavage stages of the rat in a variety of media. One of the media he used contained sodium chloride, potassium chloride, calcium chloride, magnesium chloride, sodium bicarbonate (pH 7.2), and 40% blood serum. In this medium he observed 3 cleavages in rat embryos. Washburn (1951) attempted to cultivate cleavage stages of the rat embryo from the 1st to the 5th day. He used Ringer's, Tyrode's, and Gey's physiological salines alone, after putting them in the peritoneal cavity for several hours, or in combination with various tissue components. For his cultures he used a hanging-drop method (Figure 3*h*) or small cups made in agar. Although a few of the embryos that he used underwent 1 cleavage division, he concluded that the cleavage stages of the rat were very difficult to culture. Wrba (1956) used a clot of chicken embryo extract to cultivate 2- and 4-cell rat embryos. Of 48 embryos cultured, less than half cleaved, and the cleavages that took place occurred within a few hours after being placed in culture.

Studies in our laboratory also have indicated that rat embryos are difficult to cultivate (Brinster, unpublished). If the embryos are recovered on day 4 (day 1 being the day sperm are found in the vagina) and placed in culture, some will continue developing for 2 days. When the embryos are recovered, they are generally in the 8- to 12-cell stage, and after 2 days *in vitro* under optimum conditions, blastocysts develop. Occasionally a blastocyst will hatch from the zona pellucida and attach to the surface of the petri dish. The cells of the blastocyst spread on the dish and form a monolayer. Our studies indicate that development from the 8-cell stage to the blastocyst stage will occur in a variety of media containing a protein source, with or without an energy source. However, if no protein source is available in a culture medium, and the medium contains only an energy source, the 8- to 12-cell embryo invariably degenerates in the first 24 hr. The best medium was BMOC-2 plus 10%

calf serum or Eagle's plus 10% calf serum. When placed in culture in BMOC-2 fertilized 1-cell rat ova will develop only to the 2- or 4-cell stage and occasionally to the 8-cell stage.

Of sixty-two 2-cell guinea pig embryos cultivated *in vitro* by Squier (1932), only four 2-cell ova divided and these progressed only to the 4-cell stage. These were grown in heparinized homologous plasma; several salt solutions supplemented with plasma and embryo extract were also tried, but without success.

Yanagimachi and Chang (1964) attempted to culture hamster ova which had been inseminated *in vitro*. As a culture medium they used the following combination: 10 parts of tissue culture medium 199, 10 parts of Dulbecco's solution, and 1 part of bovine serum. The ova were grown in small Carrel flasks (Fig. 2*k*) containing 1 ml of fresh culture medium. The mouth of the flask was tightly closed with a rubber stopper and kept at 37° C. When the embryos were examined after 24 hr in culture, 12 of the 27 were at the 2-cell stage. Most of the ova, however, showed signs of deterioration when they were carefully examined. So far, all attempts to culture the embryos beyond the 2-cell stage have failed.

Attempts to cultivate early cleavage stages of the large domestic animals have been reported for the sheep, goat, and cow. Wintenberger, Dauzier, and Thibault (1953) employed homologous blood serum as a medium and were able to obtain the cleavage of fertilized 1-cell sheep ova *in vitro*, but these did not progress beyond the 9-cell stage. When stages of 8 to 12 cells were placed *in vitro*, all but 2 of 27 degenerated; 1 of these 2 developed to 26 cells, and the other to 42 cells. However, embryos with a minimum of 15 to 20 blastomeres usually progress to the blastocyst stage. The experiments of Wintenberger, Dauzier, and Thibault with the cleavage stages of the goat were performed on a very limited number of ova, but the results appeared to be similar to those obtained with the sheep. Hancock (1963) used Krebs-Ringer bicarbonate plus 1 mg per ml glucose and 1 mg per ml of bovine serum albumin to culture sheep ova. A total of 46 ova from 12 ewes were cultured for up to 48 hr, and 21 were judged on examination to have developed in culture. The number of blastomeres at the end of culture in apparently normal embryos was never more than twice the number at the beginning of culture. When these embryos were transferred to foster mothers after 48 hr of culture, further development did not occur in those that had undergone cleavage during cultivation. Three eggs which were transferred after 24 hr of culture, however, continued to develop when transferred. These 3 were recovered at the 8-cell stage and were at the same stage when transferred.

Pincus (1951) reported that when cow embryos were placed *in vitro* some of them continued to cleave for 48 hr. Brock and Rowson (1952) used follicular fluid or bovine serum to culture 1- to 8-cell embryos *in vitro*. No divisions were obtained in follicular fluid, but in serum three out of four 4-cell stages went to 8 cells. The division of no other stage was obtained. Hafez, Sugie, and Gordon (1963) cultured 97 fertilized cow embryos of various stages between 1 and 32 cells. They used 5 ml Carrel flasks and autologous blood serum diluted with different amounts of physiological saline. In 4 embryos, 1 cleavage took place, but in these cases the blastomeres were not morphologically normal. In our laboratory, BMOC-2 was utilized to culture two 2-cell cow embryos. One of these went to a 4-cell stage and the other developed to a 9- to 12-cell embryo (Fig. 5). Thibault (1966) cultivated 46 early cleavage stages of cow embryos for periods up to 48 hr in the liquid of large Graafian follicles. In most cases the number of cells doubled, but none developed beyond the 24-cell stage.

Two fertilized ova of the monkey, *Macacus rhesus*, have been cultivated *in vitro* by

Lewis and Hartman (1933). One ovum was at the 1-cell stage at the beginning of the culture period and progressed to the 8-cell stage; the second developed from the 4-cell stage to the 8-cell stage *in vitro*. Homologous plasma was used as the culture medium.

A number of attempts to culture cleavage stages of the human ova *in vitro* have been made. Pincus (1939) incubated ovarian ova in human serum and was able to induce formation of the polar body, but he observed no cleavage. Rock and Menken (1944) and Menken and Rock (1948) incubated human ovarian ova and sperm in autologous serum for 1 hr, and then cultured the ova for 2 days in human plasma. From a large number of ova treated only two 2-cell stages and one 3-cell stage developed. Shettles (1955 *a*, *b*), in a series of

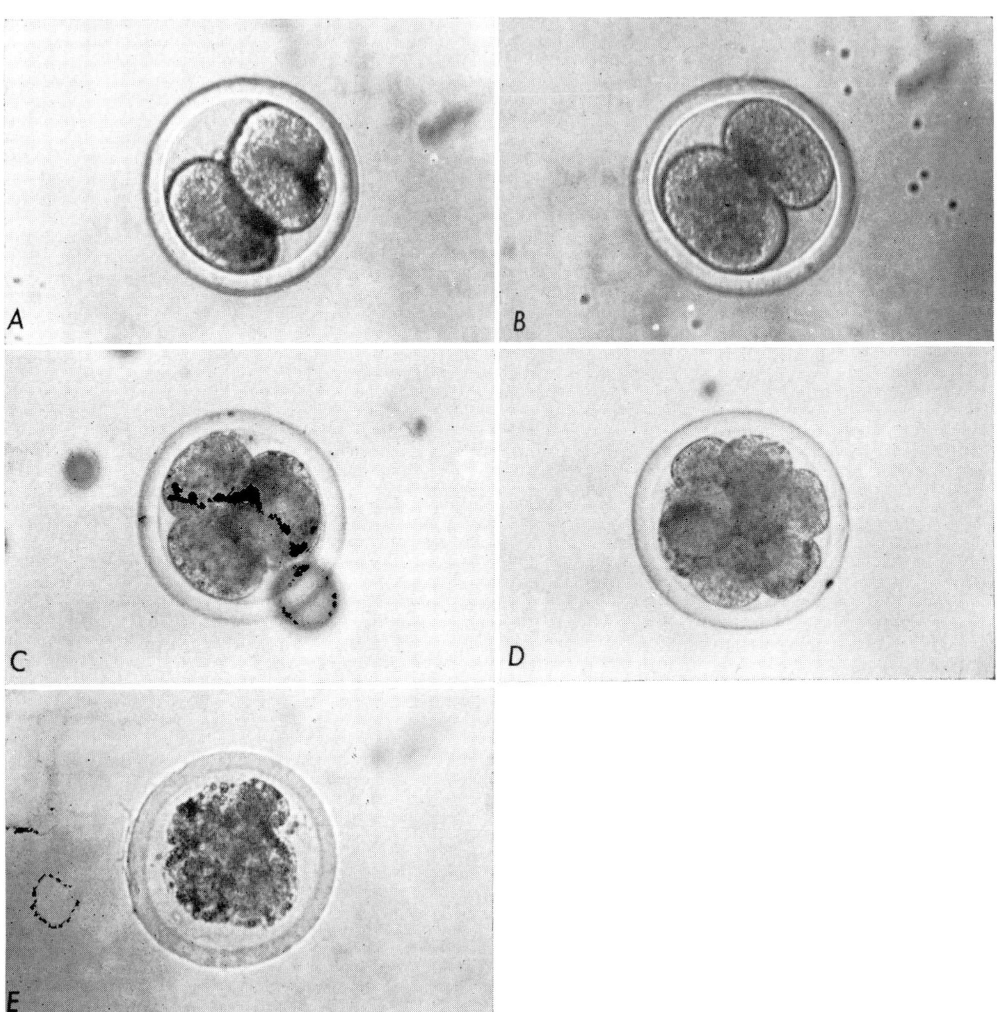

FIG. 5. *In vitro* cleavage of bovine embryo. (*a*) 2-cell embryo immediately after recovery. The misshapen blastomeres probably resulted from unfavorable environmental conditions during recovery. (*b*) The same embryo 2 hr after being placed in culture medium BMOC-2 plus glucose 1 mg/ml. (*c*) 4-cell embryo developed from (*b*) after 24 hr in culture medium. (*d*) 12-cell embryo developed from (*b*) after 48 hr in culture. (*e*) The same embryo 48 hr after (*d*). Original magnification ×36 for all pictures.

investigations, attempted to fertilize and cultivate human ovarian ova *in vitro*. He employed oviductal scrapings in human sera to culture the oocytes, and obtained the cleavage of several ova with these techniques, one as far as the 32-cell stage. However, in photographs of the cleaved ova the morphology of the blastomeres appears somewhat abnormal. More recently, Edwards (1965, 1966) has cultivated ovarian ova of the human and a number of other species *in vitro* to induce maturation of the oocytes. He has used medium 199, supplemented with 15% inactivated fetal calf serum, to which is added 100 units of penicillin and 100 micrograms of streptomycin per ml. The medium is buffered with sodium bicarbonate, and the cultures are incubated in 5% CO_2 in air, giving a pH of 7.1. Human oocytes have undergone maturation in this culture system, but attempts to fertilize the oocytes have not been successful (Edwards 1966). He has found that alkaline medium destroys the ability of oocytes to mature in culture (Edwards 1965).

B. *Ionic Composition*

The salts of all culture media closely resemble the ions contained in blood. Very little work has been done to determine what ions are essential or beneficial to the embryo, and most of the work which has been done has been on mouse embryos. Whitten (1956), using 8-cell mouse embryos, found that omission of calcium, magnesium, or potassium from the medium prevented growth, and that development was delayed without phosphate. He also found that growth continued even when the osmolarity was reduced to 0.09. Brinster (1965*a*) showed that the maximum response, as measured by the development of blastocysts from 2-cell mouse embryos, occurred at an osmolarity of 0.276, although development occurred between 0.200 and 0.354 osmols. The normal osmolarity of tissue culture media and blood serum is approximately 0.308. This optimum at low osmolarities *in vitro* has been found for other body cells. Eagle (1956) found an optimum osmolarity of 0.277 for Hela cells and mouse fibroblasts, and Trowell (1963) found that the optimum osmolarity for lymphocyte survival *in vitro* was approximately 0.274 osmols.

The importance of calcium and magnesium in the culture medium has been examined by Brinster (1964), and it was shown that a reduction in calcium concentration resulted in a significant decrease in the number of blastocysts which develop from 2-cell mouse embryos *in vitro*. However, the reduction of magnesium to one-fourth the level contained in the culture medium (BMOC-2) did not have a significant effect on development. Furthermore, there was no interaction between the effects of calcium and magnesium in these studies.

The reason for the difference in optimum ionic environment *in vitro* and *in vivo* is not known, but it has been suggested that *in vivo* the balance between extracellular fluid and intracellular ionic environment may be mediated by hormonal influences on ionic transport systems. Since the *in vitro* culture system cannot be expected to duplicate the *in vivo* hormonal environment, variations in the ionic character of the extracellular fluid may have marked effects on cellular metabolism (Stubblefield and Mueller 1960). Therefore, mammalian cells *in vitro* may have different requirements from those *in vivo*. It seems probable that the effect of osmolarity on development in these experiments may be a result of *in vitro* conditions rather than a difference between the osmolarities of normal oviductal fluid and blood serum. Most analyses of oviductal secretions have shown that the salt concentrations are very close to those of blood serum.

No experimental data are available concerning the effect of variations in the ionic composition of the culture media used in growing rabbit embryos. However, work on oviductal secretions in the rabbit is much more extensive than in any other species. These studies have shown that the salt composition of oviductal fluid is very similar to that found in blood plasma and in Krebs-Ringer bicarbonate. Until we have further information on this subject, it is probably safe to assume that salt concentrations of this nature provide an adequate environment for the rabbit embryo as well as for other embryos.

The effect of hydrogen ion concentration on the development of mammalian embryos has been studied primarily in the mouse. Whitten (1956) found that the development of 8-cell mouse embryos occurred between pH 6.9 and 7.7. Brinster (1965*a*) found that the development of 2-cell mouse embryos into blastocysts occurred between pH 5.87 and 7.78 in an atmosphere of 5% CO_2 in air. The equivalent bicarbonate concentrations were between 1 mM and 63 mM. Brinster found that there was a wide range over which optimum development appeared to occur. However, further studies (Brinster 1965*b*) demonstrated that the optimum pH for development of the 2-cell mouse embryo was dependent on the concentration of pyruvate or lactate in the medium. The higher the pH the higher the concentration of substrate necessary to obtain optimum development, suggesting that uptake of these compounds is related to the amount of the compound in the acid form. These results further suggest that the membrane of the developing embryo is able to show selective permeability to the substrates necessary for its development and that this selective permeability is dependent on hydrogen ion concentration in the case of these compounds. An outcome of this relationship is that the effect of hydrogen ion concentration on the developing embryo cannot be easily separated from the effect of the hydrogen ion concentration on the availability of the metabolite. Therefore, up to the present time it has not been possible to determine pH effect alone.

No pH measurements are available for mouse oviductal fluid. However, Blandau, Jensen, and Rumery (1958) found that the pH of reproductive tract fluids of the rat was significantly higher than the pH of the peritoneal fluid. They reported a mean pH of 7.74 for fluids accumulated in the ligated uterus, 8.04 for fluids from the dilated ampulla, 8.05 for the fluids from the periovarian sac, and 7.47 for the fluid from the peritoneal cavity. It should be emphasized that the concentration of bicarbonate, as well as the concentration of CO_2 in the atmosphere, determines the final pH. In the case of pH measurements of oviductal fluids, great care must be taken to obtain the samples in a manner such that CO_2 is not lost from the fluid. Otherwise, the pH indicated by the measurements will be higher than that which actually exists in the oviduct.

There is no information concerning the effect of hydrogen ion concentration on the development of rabbit embryos *in vitro*. However, there is some information available on oviductal secretions of the rabbit, which may provide an idea of normal hydrogen ion environment of the embryo. Vishwakarma (1962) found that the mean pH of the fluids from the ligated rabbit uterus was 7.86 and that of fluids from the ligated oviduct was 7.91. Hamner and Williams (1965) found that the bicarbonate concentration of rabbit oviductal fluid was 1.76 mg per ml, which is close to the concentration found in blood serum. Similar values have been found by Restall (1966) for sheep oviductal fluid. If CO_2 tension in the oviductal fluid is close to that of blood, then the pH would be approximately 7.4 for these fluids.

C. *Amino Nitrogen Sources*

In 1956, Whitten showed that 8-cell mouse embryos would develop into blastocysts when cultivated in Krebs-Ringer bicarbonate containing 1 mg per ml of glucose and crystalline bovine serum albumin at a concentration between 0.03 and 6%. He also demonstrated

(1957) that development would occur when the bovine serum albumin was replaced with glycine or other amino acids and simple peptides.

Brinster (1965*c*) has studied the effects of various exogenous amino nitrogen sources on the development of the 2-cell mouse embryo into the blastocyst. He found that a concentration between 1 mg per ml and 10 mg per ml of bovine serum albumin allowed maximum development. This requirement for protein can be supplied by mouse serum or bovine serum (Brinster, unpublished), as well as by the constituent amino acids of bovine serum albumin. However, there is no essential amino acid for development of the 2-cell mouse embryo. Brinster (1965*c*) demonstrated that the removal of any single amino acid from the culture medium does not prevent the development of the 2-cell mouse embryo into a blastocyst. Cystine is the only amino acid whose removal from the culture medium results in a significantly decreased number of blastocysts forming from 2-cell embryos. Recent studies indicate that either oxidized or reduced glutathione allows development of 2-cell mouse embryos into blastocysts without any other exogenous amino nitrogen source (Brinster, unpublished).

It appears that the embryo can rely to a substantial degree on endogenous amino acids, but requires exogenous amino nitrogen to supplement endogenous stores. Furthermore, the mouse embryo normally may use endogenous nitrogen material for *in vivo* development. Brinster (1967*a*) has shown that the protein content of the developing mouse embryo decreases by 25% from the 1-cell stage to the morula. Brinster and Thomson (1966) have shown that many single amino acids will allow the development of some 8-cell embryos into blastocysts. This is further evidence that the early mouse embryo can rely to a substantial degree on endogenous protein sources.

Gwatkin (1966) has studied the amino acid requirement for attachment and outgrowth of mouse blastocysts *in vitro*. He used blastocysts which had been cultured from the 2-cell stage by the method of Brinster (1963, 1965*d*). These blastocysts were then grown in a modified Eagle's medium in the culture vessel shown in Figure 3*d*. He found that arginine, cystine, histidine, leucine, and threonine were needed for outgrowth. Omission of lysine, methionine, phenylalanine, tryptophan, and tyrosine from the culture medium reduced the outgrowth, but did not inhibit it completely, while omission of isoleucine and valine reduced the extent of outgrowth only slightly.

The amount of work that has been done on the specific amino nitrogen requirements of the rabbit embryo is not as extensive as for the mouse embryo. Most workers have employed complex protein components in the culture medium for the rabbit embryo, and the embryos seem to do best when serum forms a substantial part of the culture medium. Chang (1949) studied the effect of heterologous serum on the development of rabbit embryos. He found that sera from fowl, goat, cattle, sheep, and man contained an ovicidal factor, whereas the sera of rabbit, horse, dog, guinea pig, and rat were free of this factor. The factor was undialyzable and heat labile and was lost after storage of the sera for 17 days at 3° C.

Adams (1956) employed a simple culture medium of Krebs-Ringer bicarbonate and 0.2% bovine plasma albumin, fraction 5. He observed that cleavage of the rabbit embryo would occur from the 2- to the 16-cell stage when it was incubated at 37° C for 1 to 2 days in this medium. More recently, attempts to culture rabbit embryos have been made by Purshottam and Pincus (1961) using Carrel flasks which were fixed to a rocking platform and incubated at 37° C. They cultured a total of 407 embryos at 4 different stages in 7 different media. In all, there were 22 treatments, and the 407 eggs were divided among these treatments. Results are given in the form of the percentage of embryos showing growth and development over a period of 4 days. Although the number of embryos was small, it appeared that 100% rabbit serum allowed the best growth and development.

Daniel (1965) studied the effect of single amino acids on the growth of 5-day rabbit blastocysts by adding single amino acids to Ham's medium F10 plus 15% normal rabbit serum. The criterion of development was the change in volume determined at 4 hr and 24 hr after the beginning of culture. Daniel selected for study those amino acids found in the uterus in high concentration by Gregoire, Gongsakdi, and Rakoff (1961). In some cases during culture 4-hr development was faster under *in vitro* conditions than would be expected *in vivo*. However, after 24 hr there was no case where blastocyst development *in vitro* was as good as would be expected *in vivo*. The concentrations of amino acids which Daniel recommends as optimum are shown in Table 3. Staples (1967) has employed Daniel's medium plus 5 to 10% rabbit serum to grow 5-day blastocysts for 8-, 16- and 24-hr periods, after which he transferred the blastocysts into foster mothers. Implantation of cultured blastocysts was significantly lower than controls for the 24-hr group but not the 8- or 16-hr group.

TABLE 3

OPTIMUM CONCENTRATION OF 5 AMINO ACIDS FOR 5-DAY RABBIT BLASTOCYSTS

Amino Acid	Concentration in Uterine Fluid mg/ml	Concentration in F10 as Modified by Daniel mg/ml
Alanine	0.064	0.1
Glycine	0.268	0.2
Glutamic acid	0.119	0.2
Threonine	0.040	0.1
Serine	0.104	0.1

SOURCE: Gregoire, Gongsakdi, and Rakoff 1961; Daniel 1965.

In our laboratory it has been possible to cultivate 1-cell rabbit ova to the morula, using the standard culture medium (BMOC-2) containing 1 mg per ml of bovine serum albumin, 2.5×10^{-4} M pyruvate, and 2.5×10^{-2} M lactate. Thus it appears that an amino nitrogen source is essential for development of the early rabbit embryo, and that development is improved by the addition of whole serum to the culture medium.

D. *Energy Sources*

Hammond (1949) showed that 8-cell and some late 4-cell embryos could develop into blastocysts when the medium contained glucose as an energy source, but no 2-cell embryos developed into blastocysts. Whitten (1956, 1957) confirmed Hammond's findings and demonstrated that a number of other compounds such as mannose, lactate, pyruvate, and malate could provide energy for the 8-cell embryo. However, compounds such as fructose, lactose, maltose, galactase, acetate, propionate, citrate, glycerol and glycine could not provide the energy for development. He further demonstrated that 2-cell mouse embryos developed into blastocysts in the presence of calcium lactate or isotonic lactic acid in the culture medium. Brinster (1965*b*, *d*) extended Whitten's studies on energy sources for developing mouse embryos and determined the optimum for those compounds which will allow development (Table 4). He found that the compounds which will allow development of the 2-cell mouse embryo form a group with pyruvate as the central compound.

Cole and Paul (1965) used Waymouth's medium, supplemented with nucleosides, ATP, and serum, to culture 1-cell, 2-cell, and 4-cell mouse embryos. This medium worked best when employed with a feeder layer of HeLa cells, and the authors considered that one of the contributions of the feeder layer was to supply lactate and pyruvate. Whittingham has shown that pyruvate and oxaloacetate, but not lactate or phosphoenolpyruvate, will allow cleavage of the 1-cell embryo into the 2-cell embryo (personal communication).

Development of the 8-cell embryo into a blastocyst occurs when the culture medium contains singly a number of energy sources (Table 4) which do not allow development of 2-cell embryos (Brinster and Thomson 1966). This suggests that there are marked changes in the metabolic capability of the mouse embryo with successive developmental stages. Of

TABLE 4

POSSIBLE ENERGY SOURCES FOR THE DEVELOPING 2-CELL MOUSE EMBRYO

I. Compounds which will not support development

Malate*	Glucose*
Fumarate	Fructose*
Succinate	Ribose
Iso-citrate	D-glyceraldehyde
Citrate*	Glucose-6-phosphate
Acetate*	Fructose-1-6-diphosphate
Cis-aconitate	Bovine serum albumin*
α-ketoglutarate	

(All compounds tested at 10^{-2}, 10^{-3}, 10^{-4}, 10^{-5} molar except glucose and fructose, which were tested at 2.78×10^{-2}, 5.56×10^{-3}, and 2.78×10^{-3} molar.)

II. Compounds which will support development and their optimum concentration

Pyruvate	(5×10^{-4} M)	Oxaloacetate	(5×10^{-4} M)
Lactate DL	(5×10^{-2} M)	Phosphoenolpyruvate	(1×10^{-2} M)

SOURCE: Brinster 1965*b;* Brinster and Thomson 1966.

NOTE: Compounds which will support development of the 8-cell but not the 2-cell embryo are indicated with an asterisk. Fumarate, Iso-citrate, and Cis-aconitate were not tried with 8-cell embryos.

considerable interest in this respect are glucose and malate, which allow development of 8-cell embryos, but not 2-cell embryos. Studies in our laboratory have shown that the difference, using glucose as an energy source, is not due to changes in permeability of the embryo to glucose. Wales and Brinster (1968) have shown that the permeability of the 8-cell embryo to glucose is only slightly greater than that of the 2-cell embryo. Therefore, it appears that the ability of the 8-cell embryo to survive on glucose depends on the development of the necessary enzyme systems, as has been suggested by Brinster (1965*a*, *d*, *e*).

Other studies in our laboratory by Wales and Biggers (1968) have shown that, in the case of malate, there is a marked difference in permeability between the 2-cell embryo and the 8-cell embryo. The 8-cell embryo is approximately twice as permeable to malate as the 2-cell embryo. This may account for the ability of the 8-cell embryo to develop on malate. The reason that oxaloacetate, although similar in structure to malate, allows development of the 2-cell embryo is that the oxaloacetate is decarboxylated to pyruvate in the culture medium. This occurs under normal conditions when oxaloacetate is in solution. However, the rate of decarboxylation appears to be increased in the presence of the embryos, thus suggesting that the embryo contains the enzyme to facilitate this conversion (Brinster, unpublished).

Recently, a simple system has been developed by Brinster (1967*b*) for determining carbon dioxide production from labeled substrates in the culture medium. Figure 2*m* shows the culture vessel used for this procedure. Carbon dioxide production from glucose, pyruvate, and lactate are shown in Table 5. Pyruvate appears to be the best energy source for the early stages of development in the mouse (Brinster 1965*b*, *d*, 1967*c*). Later stages oxidize pyruvate, glucose, and lactate equally well.

The metabolic rate of the mouse embryo has been determined for all stages of development from ovulation to implantation. The rate is based on oxygen consumption (Mills and Brinster 1967) and protein content of the embryo (Brinster 1967*a*). Oxygen consumption and the Q_{O_2} of the mouse embryo at various stages of development are shown in Table 6. The oxygen consumption of the embryo does not change with fertilization, and remains level until the 8-cell stage, when it rises sharply and continues to rise until implantation. During the first 2 days of development, the mouse embryo has a relatively low Q_{O_2}, which

TABLE 5

CARBON DIOXIDE PRODUCTION BY MOUSE EMBRYOS FROM GLUCOSE, LACTATE, AND PYRUVATE AT OPTIMUM CONCENTRATION

Stage of Development	Hours after Ovulation	CO_2 from Glucose μmoles $\times 10^6$/embr/hr	CO_2 from Lactate μmoles $\times 10^6$/embr/hr	CO_2 from Pyruvate μmoles $\times 10^6$/embr/hr
Unfertilized	12	0.13	3.09	7.24
Fertilized	12	0.68	3.31	6.95
2-cell	36	1.19	2.77	6.03
8-cell	60	2.16	4.54	7.25
Morula	84	6.73	9.31	10.00
Blastocyst	84	10.94	13.08	13.92
Late blastocyst	108	14.69	15.06	15.73

SOURCE: Brinster 1967*c*.

NOTE: The values for glucose are based on 6 determinations, for lactate on 4 determinations, and for pyruvate on 4 determinations. Glucose concentration was 5.56×10^{-3} M, lactate concentration was 5.00×10^{-2} M, and pyruvate concentration was 5.00×10^{-4} M.

TABLE 6

RESPIRATORY ACTIVITY OF PREIMPLANTATION MOUSE EMBRYOS

Stage of Development	O_2 Uptake $\mu l \times 10^6$	Q_{O_2} μl/mg Dry Wt.
Unfertilized	155	3.73
Fertilized	156	3.77
2-cell	150	3.84
8-cell	191	5.44
Morula	351	11.39
Blastocyst	460	12.88
Late blastocyst	534	16.32

SOURCE: Mills and Brinster 1967; Brinster 1967*a*.

NOTE: All volumes adjusted to standard temperature and pressure. Dry weight of the stages was calculated from values for the protein content of each stage assuming protein = $0.66 \times$ dry weight.

is comparable to values obtained on tissues such as skin. However, the blastocyst has a Q_{O_2} value that is high and is comparable to such tissues as brain.

The carbon dioxide formed from various substrates by mouse embryos cultivated *in vitro* compared to total oxygen consumption of the embryo is shown in Table 7. Only 2% of the total oxygen consumption can be accounted for by CO_2 formed from glucose by the unfertilized ovum. On the other hand, 65% of the total oxygen consumption at the blastocyst stage can be accounted for by exogenous glucose. One hundred percent of oxygen consumption of the unfertilized ovum can be accounted for by CO_2 from pyruvate. However, this value drops to approximately 65% at the blastocyst stage. The reason for this apparently constant 65% of total oxygen consumption being accounted for by exogenous substrate at the blastocyst stage probably is utilization of endogenous glycogen by the blastocyst.

Thomson and Brinster (1966) have shown that the mouse embryo contains a large store of glycogen in the early stages of preimplantation development. This store of glycogen is not markedly reduced until blastocyst formation begins, and then the glycogen is rapidly

TABLE 7

COMPARISON OF CARBON DIOXIDE FROM DIFFERENT SUBSTRATES TO OXYGEN CONSUMPTION

Stage of Development	CO_2 from Pyruvate / O_2 Consumption	CO_2 from Lactate / O_2 Consumption	CO_2 from Glucose / O_2 Consumption
Unfertilized	1.04	0.45	0.02
Fertilized	0.99	0.47	0.10
2-cell	0.90	0.41	0.18
8-cell	0.85	0.53	0.25
Morula	0.64	0.59	0.42
Blastocyst	0.68	0.64	0.53
Late blastocyst	0.66	0.63	0.62

NOTE: Values are based on data from previous tables.

utilized during the last day and a half of preimplantation development. It appears that the 35% of oxygen consumption which cannot be accounted for by exogenous substrate in the late blastocyst could be accounted for by oxidation of the endogenous glycogen. It has further been demonstrated by Thomson and Brinster (1966) that this endogenous store of glycogen is unavailable even to embryos which are maintained in culture media with suboptimal energy source concentrations. This suggests that the necessary enzymes are not available for the utilization of the stored glycogen during the early preimplantation period.

Culture media energy sources have not been studied in as much detail for the rabbit embryo as they have for the mouse embryo. Pincus (1941*a*, *b*) studied the development of the rabbit morula into the blastocyst, as he considered this to be a critical phase in the development of the preimplantation rabbit embryo. The conclusions he drew from his work were: (1) that the energy for growth is derived (a) at least in part by oxidative processes, (b) by glycolytic degradation of carbohydrates, and (c) chiefly by the action of phosphorylating enzyme systems; and (2) that the limiting process appears to involve the metabolism of pyruvic acid since (a) fluoride, which inhibits the production of pyruvate, can be counteracted by the addition of pyruvate, (b) vitamin B_1 probably functions as part of the pyruvate-splitting system, and (c) the sulfhydral compounds act as cocarboxylases to a system effecting the decarboxylation of pyruvate.

Fridhandler, Hafez, and Pincus (1957) and Fridhandler (1961) studied the energy metabolism of the rabbit embryo in detail using Cartesian diver techniques. They could find no evidence for glycolytic activity in the 1- to 16-cell rabbit embryo, but the late morula and the blastocyst showed glycolytic activity. Since cyanide blocked respiration, and exogenous substrate had no effect on respiration, they postulated an endogenous energy store for the early cleavage stages of the rabbit embryos. At the time of blastocyst formation, glycolytic activity and Krebs cycle activity is markedly increased. In addition, Fridhandler found that the oxygen consumption of the rabbit embryo was approximately 0.61 mμl per embryo per hr for the 1-cell stage. This figure is roughly 3 to 4 times what was found for the mouse embryo by Mills and Brinster (1967). This is not surprising since the 1-cell rabbit embryo is approximately 3.5 times as large as the mouse embryo. After blastocyst formation in the rabbit, there is a marked expansion, and the oxygen consumption increases rapidly.

Fridhandler (1961) found that the C_1 to C_6 ratio was high in the rabbit embryo before blastocyst formation, indicating an active pentose shunt. After blastocyst formation the C_1 to C_6 ratio was approximately 1, indicating the pentose shunt had become less important in relation to the Krebs cycle. In mouse embryos (Brinster 1967*b*) the C_1 to C_6 ratio remains at about 1.6 throughout the preimplantation period, indicating that the pentose shunt is not very important relative to the Krebs cycle in this species.

Daniel (1965) examined the effect of pyruvate, lactate, and glucose on the development of the 5-day rabbit blastocyst. The methods used are described in the section on amino nitrogen sources. All three compounds improved development for a 4-hr period, but the *in vitro* blastocysts at 24 hr were smaller than comparable *in vivo* blastocysts. On the basis of his work, Daniel suggests that the concentration of pyruvate be raised to 1 mg per ml and the concentration of glucose be reduced from 1 to 0.2 mg per ml in the F10 medium, in order to obtain optimum growth of the 5-day rabbit blastocyst. He also suggested that 0.8 mg per ml of sodium lactate and 0.1 mg per ml of glycogen be added to this medium.

Mounib and Chang (1965) cultured 6-day rabbit blastocysts in Warburg vessels in Krebs-Ringer phosphate (pH 7.4) in the presence of labeled glucose, fructose, and pyruvate. They found that glucose was oxidized to a greater degree than fructose at low concentrations, but not at high concentrations. Pyruvate C_1 oxidation was greater than pyruvate C_2 oxidation, perhaps indicating that acetate was used for synthetic processes.

In respect to availability of energy sources to the rabbit embryo *in vivo*, it has been shown (Bishop 1956; Mastroianni and Wallach 1961; and Holmdahl and Mastroianni 1965) that the glucose concentration is very low in the rabbit oviduct, whereas the lactate and pyruvate concentrations are high. Lutwak-Mann (1962) has shown similar conditions to exist in the uterus of the rabbit.

III. Conclusions

The information on nutrient requirements of preimplantation stages of mammalian embryos is still incomplete. Considerable work must yet be done before we have a comprehensive knowledge of what these embryos need to develop normally *in vitro*. In the coming years emphasis should be placed on well-designed studies which will yield quantitative information. It seems logical to make the largest number and most detailed studies on the embryos of laboratory animals, and then to attempt to extrapolate this information to the embryos of the larger species. Certainly, if a good foundation of knowledge is available about the embryos of laboratory animals, it will be possible to design more meaningful experiments in which the eggs of larger animals are used. In this way, the most efficient use can be made of embryos from the costly large domestic animals and primates.

Until we have more information, a basic medium for the cultivation of mammalian embryos might consist of the following: (1) the salts contained in blood plasma, or Krebs-Ringer bicarbonate; (2) a sodium bicarbonate buffer with a concentration of approximately 25 m moles; (3) an atmosphere containing 5% CO_2 in the gas phase; (4) a pH of approximately 7.4 which would be the result of items 2 and 3; (5) a protein concentration of 1 to 10 mg per ml; (6) a glucose concentration of 1 mg per ml; (7) a pyruvate concentration of about 5×10^{-4} M. Supplementary compounds such as serum, embryo extract, lactate, amino acids, vitamins, and cofactors could be added to such a preparation. However, it should be emphasized that the use of complex natural or undefined substances in the medium makes it very difficult to determine the exact effect of other omissions or additions to the medium.

Acknowledgment

The author would like to thank Dr. Joan Thomson for reading and commenting on the manuscript and Mrs. Peggy Earnest for the typing. Financial support for some of the research reported in this chapter has come from the National Science Foundation (GB 4465), the Population Council, the Pennsylvania Department of Agriculture, and the National Institute of Child Health and Human Development (HD03071).

References

1. Adams, C. E. 1956. Egg transfer and fertility in the rabbit. *Proc. Third Int. Cong. Anim. Reprod. (Cambridge)* Section 3:5.
2. Alliston, C. W. 1965. Embryonic mortality following culture *in vitro* of one- and two-cell rabbit eggs at elevated temperatures. *J. Reprod. Fertil.* 9:337.
3. Bedford, J. M., and Chang, M. C. 1962. Fertilization of rabbit ova *in vitro. Nature* 193:898.
4. Biggers, J. D. 1965. Cartilage and bone. In *Cells and tissues in culture*, vol. 2, ed. E. N. Willmer. London: Academic Press.
5. Biggers, J. D., and Brinster, R. L. 1965. Biometrical problems in the study of early mammalian embryos *in vitro. J. Exp. Zool.* 158:39.
6. Bishop, D. 1956. Metabolic conditions within the oviduct of the rabbit. *Int. J. Fertil.* 2:11.
7. Blandau, R.; Jensen, L.; and Rumery, R. 1958. Determination of the pH values of the reproductive tract fluids of the rat during heat. *Fertil. Steril.* 9:207.
8. Boell, E. J., and Nicholas, J. S. 1948. Respiratory metabolism of the mammalian egg. *J. Exp. Zool.* 109:267.
9. Brinster, R. L. 1963. A method for *in vitro* cultivation of mouse ova from two-cell to blastocyst. *Exp. Cell Res.* 32:205.
10. ———. 1964. Studies on the development of mouse embryos *in vitro*. Ph.D. diss., University of Pennsylvania.
11. ———. 1965*a*. Studies on the development of mouse embryos *in vitro*. I. The effect of osmolarity and hydrogen ion concentration. *J. Exp. Zool.* 158:49.
12. ———. 1965*b*. Studies on the development of mouse embryos *in vitro*. II. The effect of energy source. *J. Exp. Zool.* 158:59.

13. Brinster, R. L. 1965*c*. Studies on the development of mouse embryos *in vitro*. III. The effect of fixed-nitrogen source. *J. Exp. Zool.* 158:69.
14. ———. 1965*d*. Studies on the development of mouse embryos *in vitro*. IV. Interaction of energy sources. *J. Reprod. Fertil.* 10:227.
15. ———. 1965*e*. Lactic dehydrogenase activity in the preimplanted mouse embryo. *Biochim. Biophys. Acta* 10:439.
16. ———. 1967*a*. Protein content of the mouse embryo during the first five days of development. *J. Reprod. Fertil.* 13:413.
17. ———. 1967*b*. Carbon dioxide production from glucose by the preimplantation mouse embryo. *Exp. Cell Res.* 47:271.
18. ———. 1967*c*. Carbon dioxide production from lactate and pyruvate by the preimplantation mouse embryo. *Exp. Cell Res.* 47:634.
19. ———. 1968. Lactate dehydrogenase activity in the oocytes of mammals. *J. Reprod. Fertil.* In press.
20. Brinster, R. L., and Thomson, J. L. 1966. *In vitro* culture requirements of the eight-cell mouse ovum. *Exp. Cell Res.* 42:308.
21. Brock, H., and Rowson, L. E. 1952. The production of viable bovine ova. *J. Agric. Sci.*, 42:479.
22. Chang, M. C. 1948. The effects of low temperature on fertilized rabbit ova *in vitro* and the normal development of ova kept at low temperature for several days. *J. Gen. Physiol.* 31:385.
23. ———. 1949. Effects of heterologous sera on fertilized rabbit ova. *J. Gen. Physiol.* 32:291.
24. ———. 1959. Fertilization of rabbit ova *in vitro*. *Nature* (*Lond.*) 184:466.
25. Chen, J. M. 1954. The cultivation in fluid medium of organized liver, pancreas, and other tissues of foetal rats. *Exp. Cell Res.* 7:518.
26. Cole, R. J., and Paul, J. 1965. Properties of cultured preimplantation mouse and rabbit embryos, and cell strains derived from them. In *Preimplantation stages of pregnancy*, ed. G. E. W. Wolstenholme and M. O'Connor. London: J. & A. Churchill Ltd.
27. Daniel, J. 1964. Cleavage of mammalian ova inhibited by visible light, *Nature* 201:316.
28. ———. 1965. Studies on the growth of 5-day-old rabbit blastocysts *in vitro*. *J. Embryol. Morph.* 13:83.
29. Defrise, A. 1933. Some observations on living eggs and blastulae of the albino rat. *Anat. Rec.* 57:239.
30. Dragoiu, I.; Benetato, G.; and Oprean, R. 1937. Research on the respiration of mammalian oocytes. *C. R. Soc. Biol.* 126:1044.
31. Eagle, H. 1956. The salt requirements of mammalian cells in tissue culture. *Arch. Biochem. Biophys.* 61:356.
32. Edwards, R. G. 1965. Maturation *in vitro* of human ovarian oocytes. *Lancet* 2:926.
33. ———. 1966. Preliminary attempts to fertilize human oocytes matured *in vitro*. *Amer. J. Obstet. Gynec.* 96:192.
34. Fridhandler, L. 1961. Pathways of glucose metabolism in fertilized rabbit ova at various preimplantation stages. *Exp. Cell Res.* 22:303.
35. Fridhandler, L.; Hafez, E. S. E.; and Pincus, G. 1957. Developmental changes in the respiratory activity of rabbit ova. *Exp. Cell Res.* 13:132.
36. Gregoire, A. T.; Gongsakdi, D.; and Rakoff, A. 1961. The free amino acid content of the female rabbit genital tract. *Fertil. Steril.* 12:322.

37. Gwatkin, R. B. L. 1966. Amino acid requirement for attachment and outgrowth of mouse blastocyst *in vitro*. *J. Cell. Physiol.* 68:335.
38. Hafez, E. 1963. Storage of fertilized ova. *Int. J. Fertil.* 8:459.
39. Hafez, E. S. E.; Sugie, T.; and Gordon, I. 1963. Superovulation and related phenomena in the beef cow. I. Superovulatory responses following PMS and HCG injections. *J. Reprod. Fertil.* 5:359.
40. Hammond, J., Jr. 1949. Recovery and culture of tubal mouse ova. *Nature* 163:28.
41. Hamner, C. E., and Williams, W. L. 1965. Composition of rabbit oviduct secretions. *Fertil. Steril.* 16:170.
42. Hancock, J. 1963. Survival *in vitro* of sheep eggs. *Anim. Prod.* 5:237.
43. Holmdahl, T. H. S., and Mastroianni, L., Jr. 1965. Continuous collection of rabbit oviduct secretions at low temperature. *Fertil. Steril.* 16:587.
44. Lewis, W. H., and Gregory, P. W. 1929. Cinematographs of living developing rabbit eggs. *Science* 69:226.
45. Lewis, W. H., and Hartman, C. G. 1933. Early cleavage stages of the eggs of the monkey (*Macacus rhesus*). *Contr. Embryol. Carnegie Inst.* 24:187.
46. Lewis, W. H., and Wright, E. S. 1935. On the development of the mouse. *Contr. Embryol. Carnegie Inst.* 25:113.
47. Lutwak-Mann, C. 1962. Glucose, lactic acid, and bicarbonate in rabbit blastocyst fluid. *Nature* (*Lond.*) 193:653.
48. Mark, F. L., and Long, J. A. 1912. Studies on early stages of development in rats and mice. *Univ. Calif. Pub. Zool.* 9:105.
49. Mastroianni, L. Jr., and Ehteshamzadeh, J. 1964. Corona cell dispersing properties of rabbit tubal fluid. *J. Reprod. Fertil.* 8:145.
50. Mastroianni, L. Jr., and Wallach, R. C. 1961. Effect of ovulation and early gestation on oviduct secretions in the rabbit. *Amer. J. Physiol.* 200:815.
51. Menkin, M. F., and Rock, J. 1948. *In vitro* fertilization and cleavage of human ovarian eggs. *Amer. J. Obstet. Gynec.* 55:440.
52. Mills, R. M. Jr., and Brinster, R. L. 1967. Oxygen consumption of preimplantation mouse embryos. *Exp. Cell Res.* 47:337.
53. Mintz, B. 1962. Experimental study of developing mammalian eggs. Removal of the zona pellucida (mice). *Science* 138:594.
54. ———. 1964. Formation of genetically mosaic mouse embryos, and early development of lethal (t^{12}/t^{12}): Normal mosaics. *J. Exp. Zool.* 157:273.
55. Mounib, M. and Chang, M. C. 1965. Metabolism of glucose, fructose and pyruvate in the 6-day rabbit blastocyst. *Exp. Cell Res.* 38:201.
56. Mulnard, J. G. 1965. Studies of regulation of mouse ova *in vitro*. In *Preimplantation stages of pregnancy*, ed. G. E. W. Wolstenholme and M. O'Connor. London: J. & A. Churchill Ltd.
57. Onanoff, J. 1893. Recherches sur la fécondation et la gestation des mammifères. *C. R. Soc. Biol.* (*Par.*) 45:719.
58. Pincus, G. 1939. The maturation of explanted human ovarian ova. *Amer. J. Physiol.* 126:600.
59. ———. 1941*a*. The control of ovum growth. *Science* 93:438.
60. ———. 1941*b*. Factors controlling the growth of rabbit blastocysts. *Amer. J. Physiol.* 133:412.

61. Pincus, G. 1951. Observations on the development of cow ova, *in vivo* and *in vitro*. *Proc. First Nat. Egg-transfer Breed Conf.* P. 18. San Antonio, Texas.
62. Pincus, G., and Werthessen, N. T. 1938. The comparative behaviour of mammalian eggs *in vivo* and *in vitro*. III. Factors controlling the growth of the rabbit blastocyst. *J. Exp. Zool.* 78:1.
63. Purshottam, N., and Pincus, G. 1961. *In vitro* cultivation of mammalian eggs. *Anat. Rec.* 140:51.
64. Restall, B. J. 1966. The Fallopian tube of the sheep. III. The chemical composition of the fluid from the Fallopian tube. *Aust. J. Biol. Sci.* 19:687.
65. Rock, J., and Menkin, M. F. 1944. *In vitro* fertilization and cleavage of human ovarian eggs. *Science* 100:105.
66. Schenk, S. L. 1880. Das Säugerthierei künstlich befruchtet ausserhalb des Mutterthieres. *Mitt. Embr. Inst. K. K. Univ. Wien* 1:107.
67. Shettles, L. B., 1955*a*. A morula stage of human ova developed *in vitro*. *Fertil. Steril.* 6:287.
68. ———. 1955*b*. Further observations on living human oocytes and ova. *Amer. J. Obstet. Gynec.* 69:365.
69. Squier, R. R. 1932. The living egg and early stages of its development in the guinea pig. *Contr. Embryol. Carnegie Inst.* 23:223.
70. Staples, R. E. 1967. Development of 5-day rabbit blastocysts after culture at 37° C. *J. Reprod. Fertil.* 13:369.
71. Stubblefield, E., and Mueller, G. C. 1960. Effects of sodium chloride concentration on growth, biochemical composition, and metabolism of HeLa cells. *Cancer Res.* 20:1646–55.
72. Thibault, C. 1966. La culture *in vitro* de l'oeuf de vache. *Ann. Biol. Anim. Biochim. Biophys.* 6:159.
73. Thomson, J. L., and Brinster, R. L. 1966. Glycogen content of preimplantation mouse embryos. *Anat. Rec.* 155:97.
74. Trowell, O. A. 1963. The optimum concentration of sodium chloride for the survival of lymphocytes *in vitro*. *Exp. Cell Res.* 29:220.
75. Vishwakarma, P. 1962. The pH and bicarbonate-ion content of the oviduct and uterine fluids. *Fertil. Steril.* 13:481.
76. Wales, R. G., and Biggers, J. D. 1968. The permeability of two- and eight-cell mouse embryos to L-malic acid. *J. Reprod. Fertil.* 15:103.
77. Wales, R. G., and Brinster, R. L. 1968. The uptake of hexoses by mouse embryos. *J. Reprod. Fertil.* 15:415.
78. Washburn, W. W., Jr. 1951. A study of the modifications in rat eggs observed *in vitro* and following tubal retention. *Arch. Biol.* (*Par.*) 62:439.
79. Whitten, W. K. 1956. Culture of tubal mouse ova. *Nature* (*Lond.*) 176:96.
80. ———. 1957. Culture of tubal ova. *Nature* (*Lond.*) 179:1081.
81. Wintenberger, S.; Dauzier, L.; and Thibault, C. 1953. Le développement *in vitro* de l'oeuf de la brébis et de celui de la chèvre. *C. R. Soc. Biol.* (*Par.*) 147:1971.
82. Wrba, H. 1956. Zum Verhalten des befruchteten Ratteneies *in vitro*. *Naturwissenschaften* 43:334.
83. Yanagimachi, R., and Chang, M. C. 1964. *In vitro* fertilization of golden hamster ova. *J. Exp. Zool.* 156:361.

19

The Fetal Mouse Oviduct in Organ and Tissue Culture

R. E. Rumery
Department of Biological Structure
School of Medicine
University of Washington, Seattle

I. Historical Background

A survey of the literature reveals that there are only a limited number of investigations concerned with mammalian oviducts in culture. The development of Müllerian ducts of mammals has been studied in organ culture in order to evaluate the effects of fetal gonads on sexual differentiation (Jost and Bergerard 1949; Jost and Bozic 1951; Price and Pannabecker 1958, 1959; Price and Ortiz 1965). Mature oviducts have been studied in both organ and tissue cultures, although only a few reports have been published.

Using whole organ cultures, Gwatkin and Biggers (1963) cultivated mouse oviducts containing zygotes. After 4 days in culture, the zygotes had developed into blastocysts, the ciliated epithelia of the oviducts had been maintained and there was very little necrosis in the muscle layers. In another study, oviducts from mature rabbits and rats were grown in organ cultures in order to evaluate the action of ovarian steroids on the epithelia (Bousquet 1964). The epithelial cells and cilia in the oviducts taken from untreated, control animals were well preserved even after 6 days in culture. Explanted oviducts from castrated female rats which had been injected with estrogen followed by progesterone showed both a reduction in secretion and a loss of cilia. Fragments of oviducts from 2- to 5-month-old rabbits were maintained in culture for 1 month by Galstjan (1935). He described groups of epithelial cells growing out from the explants and forming a unicellular membrane. Many of these cells had beating cilia. Galstjan concluded that the epithelial cells were already highly differentiated at the time of explantation, and at no time did they dedifferentiate into more primitive cells.

The growth characteristics of cultured tissue from mature rabbit oviducts were described also by Hellström and Nilsson (1957). They observed both ciliated and secretory cells in the epithelial membranes growing out from the tissue fragments. Transformation of ciliated cells into secretory cells did not occur. In a further study of cultured rabbit oviducts, these authors found three types of cells lining the oviductal lumina: ciliated cells, those containing granules which stained specifically with PAS, and undifferentiated cells (Nilsson and Hellström 1957).

A problem in evaluating cultured oviductal organs is the great variety and pleomorphism of cells in the outgrowing connective tissues, muscle, and epithelium. Ordinarily it is most difficult to obtain pure cultures of oviductal epithelium. Valenti (1963) published a technic for trypsinizing hollow reproductive organs of rabbits, dogs, and man. By controlled trypsination, epithelia of the oviducts were harvested and cultured in Rose chambers for 30 days. Cilia present on the epithelial cells at the time of culturing were beating for at least 5 days.

From these reports it is apparent that the oviduct of mature animals can be maintained *in vitro* under controlled conditions, and will even continue to function to some extent. The prenatal oviducts, on the other hand, have not been cultured.

Therefore, the investigation to be reported here has been designed (1) to study the growth characteristics and differentiation of 16-day fetal mouse oviducts in tissue and organ cultures; (2) to observe the development of cilia *in vitro* and to characterize their initial beat; (3) to evaluate the development of the musculature in culture and to determine the time of onset of contractions; (4) to study the differentiation of oviducts when gonadal tissue is added to the culture; and (5) to evaluate the influence of various hormones on oviductal differentiation.

Oviducts from 16-day-old fetal mice were selected for this study because at this stage of development they are simple tubular organs and their cells are not yet fully differentiated.

II. Material and Methods

A. *The Culture Medium*

The culture fluid used throughout this investigation was a modification of Eagle's medium. The following substances were added to 1000 ml of water triple distilled in glass: NaCl, 8.0 gm; KCl, 0.38 gm; $Na_2HPO_4 \cdot 7H_2O$, 0.30 gm; KH_2PO_4, 0.025 gm; glucose, 4.0 gm; $MgCl_2 \cdot 6H_2O$, 0.21 gm; $CaCl_2 \cdot 2H_2O$, 0.13 mg; and $NaHCO_3$, 0.25 gm. To this basic solution were added 10.0 ml MEM[1] essential amino acids (50×), 10.0 ml MEM vitamin mixture (100×); 10.0 ml MEM nonessential amino acids (100×); 10.0 ml sodium pyruvate 100 mM (100×); 20.0 ml L-glutamine[1] 200 mM (100×); and 1.25 ml of a 1% solution of phenol red,[2] as well as penicillin (200 U/ml) and mycifradin sulfate[3] (100 μg/ml). Finally heat-inactivated horse serum (56° C for 30 min) was combined with the medium in a total concentration of 10%. Just prior to use, the hydrogen ion concentration of the medium was always adjusted to pH 7.0–7.2. The medium in all cultures was replaced every 2 or 3 days.

B. *Preparation of Tissue Culture*

Mature female mice of the Swiss Webster strain were placed with males and examined for vaginal plugs at 8:00 A.M. each day. When a plug was found, the female was considered to be in day zero of gestation. On day 16 the females were killed by decapitation and their uteri removed aseptically to sterile petri dishes. As soon as the fetuses were freed from the

[1] Microbiological Associates, Albany, California.

[2] Difco, Detroit, Michigan.

[3] Upjohn Co., Kalamazoo, Michigan.

cornua, their ovaries and attached oviducts were removed with iridectomy scissors and transferred to a 16 mm petri dish containing 3 to 4 ml of the modified Eagle's medium. When the ovaries had been cut away, the ampullar and isthmic portions of the oviducts were separated, placed in separate petri dishes containing fresh medium, and minced into approximately 1 mm pieces.

Eight to 10 of these fragments were transferred to the center of a 43 × 50 mm coverslip[4] with a fine pipette. Excess medium was withdrawn, leaving just sufficient fluid to prevent the tissues from drying. A strip of dialysis membrane[5] was placed over the tissues and drawn quite taut in order to hold them in place. The Rose chamber was assembled, filled with 1.8 ml of the modified Eagle's medium (leaving a small air bubble), and incubated at 37° C.

The cultures were examined daily with the phase microscope and detailed observations were made of their appearance and differentiation. Various phases in the growth and differentiation of the tissues were recorded by means of cinematography.

For histologic study, some of the cultures were fixed at specific time intervals after explantation. The Rose chambers were dismantled and the coverslips on which the explants were growing were removed, immersed in physiological saline for 10 min and then placed in chilled Hsu's fixative.[6] The fixation procedure was continued for 45 min in the refrigerator and then the tissues, as whole mounts, were stained with PAS and hematoxylin.

C. *Preparation of Membranes*

Dialysis membranes were cut into strips 51 mm long and 13 mm wide from dialysis tubing (24 Å, average pore size). These strips were sterilized by soaking in 70% ethanol, rinsed twice in triple glass-distilled water, and washed several times in Hanks's balanced salt solution. Before use in the Rose chambers, the membranes were stored overnight in the modified Eagle's medium.

D. *Preparation of Organ Cultures*

Oviducts which were to be cultivated as organ cultures were divided into their ampullar and isthmic segments and placed in separate petri dishes. Six or 8 of each of these segments were transferred with a fine pipette to a strip of agar which was prepared in the following manner: Bactoagar[7] was dissolved at a concentration of 2% in 0.7% NaCl and poured into 120 mm petri dishes to a depth of about 4 mm. After the agar had cooled and solidified, modified Eagle's medium was poured over its surface and the petri dishes were placed in the refrigerator overnight to condition the agar with the medium. The following morning the medium was withdrawn and the agar cut into 26 × 7 mm strips. These were transferred to 16 mm petri dishes, 1 or 2 strips to each dish. Modified Eagle's medium was added slowly into the bottom of the dish until the level of the fluid had almost reached the upper surface of the agar strip. The proper relationship of fluid to the surface of the agar is important since too much medium could wash the explants off the agar. The cultures were incubated at 37° C in an oven having an atmosphere of 5% CO_2 and 70% humidity.

The dishes containing the whole organ cultures were examined every other day with a dissecting microscope maintaining sterile conditions. On occasion, squash preparations

[4] Gold Seal coverglasses, Clay Adams, New York.

[5] Union Carbide Corporation, Chicago.

[6] Hsu's fixative: 75 cc distilled water; 15 cc 40% formalin; 10 cc glacial acetic acid; 1 gm picric acid; 1 gm chromic acid; 1 gm urea.

[7] Difco, Detroit, Michigan.

were made and examined with the phase microscope. Representative cultures were fixed in Hsu's solution and sections were stained with PAS and hematoxylin.

E. *Addition of Fetal Gonads*

In order to observe the effects of fetal gonads on the oviducts growing *in vitro*, ovaries and testes from 16-day-old fetal mice were cultured with the oviducts. Minced gonads were explanted in the tissue cultures, while the whole gonads were added to the organ cultures.

F. *Hormones Added to Cultures*

The influence of estradiol benzoate and progesterone[8] was tested in two separate groups of cultures. The crystalline hormones were dissolved in chemically pure ethanol and then added to the culture medium to obtain concentrations of 2.5 μg per ml.

III. Observations

A. *Appearance of Oviducts in the 16-Day-Old Fetal Mouse*

At the 16th day of gestation the fetal mouse oviducts are simple tubular structures measuring 2.0 to 2.5 mm in length (Fig. 1). At this stage of development they are suspended by a delicate mesosalpinx. The lumina are completely open. The fimbriated end of each oviduct is divided into two lips by a median cleft. Morphologically, there is no difference in appearance between the ampullar and isthmic regions, but there is a noticeable difference in size between the isthmus and the cornua of the uterus so that the uterotubal junction can be discerned easily. Histological sections of the 16-day-old oviduct reveal a single layer of nonciliated epithelial cells surrounded by a simple connective tissue stroma, and there is no evidence of a distinct muscle layer.

B. *Oviducts in Tissue Culture*

The minced tissue, when placed below the dialysis membrane in the Rose chambers, became flattened against the surface of the coverglass. Fibroblasts grew out rapidly from the explants, forming cellular sheets which served to anchor the explants firmly to the glass. By the 2d day in culture, the luminal areas were lined with large round epithelial cells with prominent central nuclei. These areas varied in size and shape: some long and narrow, others quite irregular in appearance (Fig. 2*a*). Even at the light microscopic level, cilia could be seen appearing on these epithelial cells as short, thread-like structures which were clearly visible at least 24 hr before they began to beat.

In the majority of cultures the cilia began to beat between the 5th and 8th day after explantation. At first, their movements were slow and sporadic, but within a few hours their beat became more regular. Not all ciliated cells became active at the same time; often only a few were beating in any one field. Frequently small groups of ciliated cells were observed to stop beating for a few minutes, while those on adjoining cells remained active. Once the cilia had matured, their beat was very rapid and consistently in one direction. In older cultures the lumina appeared to be wider and filled with accumulations of debris. The cilia lining the lumina could be seen moving this material to and fro (Fig. 2*b*). Sometimes when the cultures needed to be fed with fresh medium, the cilia stopped beating temporarily, and then began again after the medium had been renewed. As the explants aged in culture and began to degenerate, the cilia ceased beating. Even though they were visible for several days, they finally disappeared completely.

[8] Nutritional Biochemical Corporation, Cleveland, Ohio.

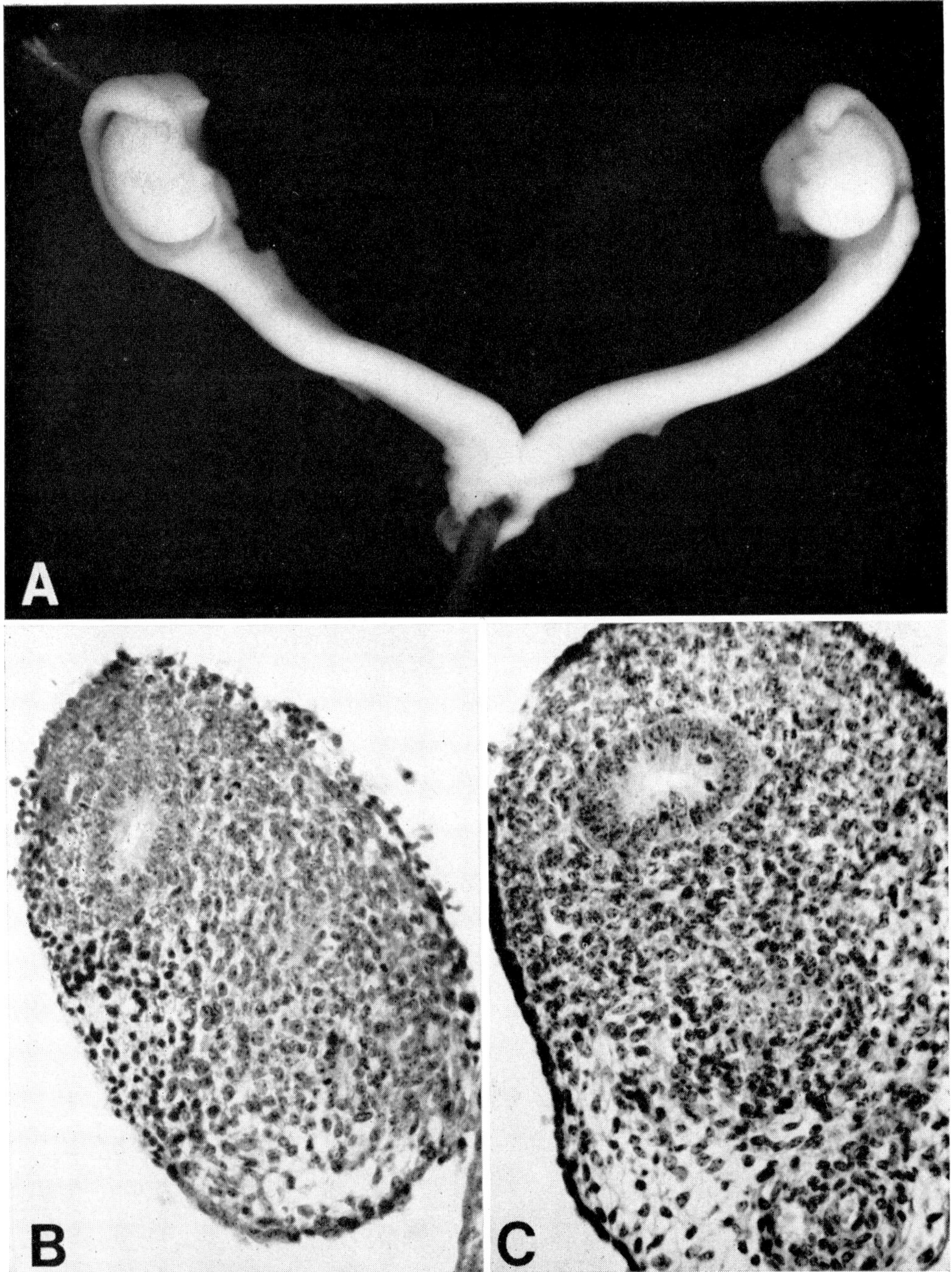

FIG. 1. (*a*) Intact reproductive tract of the 16-day fetal mouse. The oviducts, at this stage of development, are simple, uncoiled tubes. (Fixed, unstained specimen ×30.)

(*b*) and (*c*) Cross sections of the 16-day fetal oviduct. The section at the left was located near the fimbriated end of the ampullary segment, while the section on the right was from the isthmic segment. The lumina in both sections are lined by a single layer of epithelial cells which have not yet become ciliated. (Light micrographs, hematoxylin and eosin ×210.)

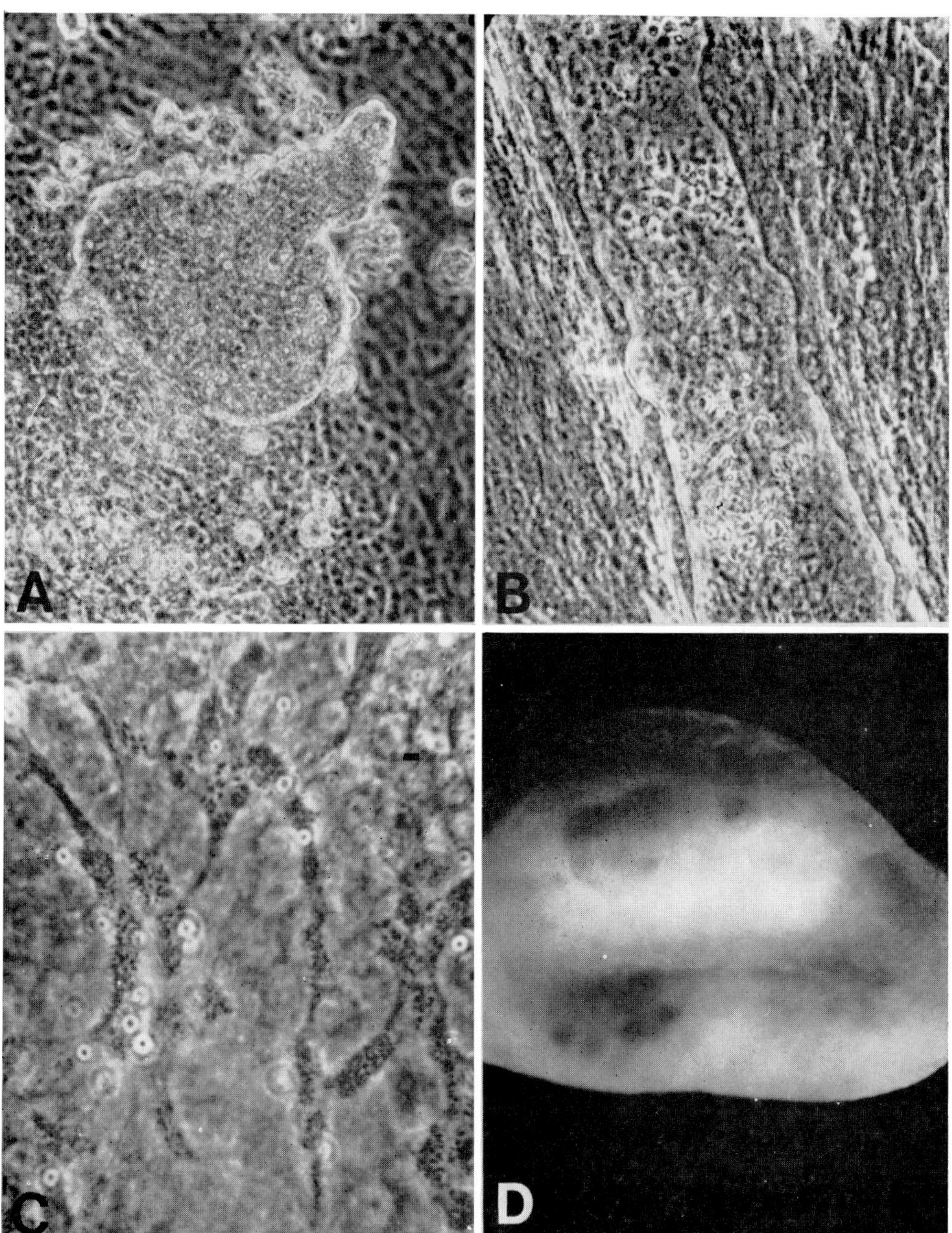

FIG. 2. (*a*) Ampullar tissue from the fetal oviduct in culture for 14 days. A piece of fimbriae is well delineated in the center of the photograph. (Phase micrograph ×412.)

(*b*) Surface view of cells lining the lumen of the ampullar region of a fetal oviduct which has been in tissue culture for 8 days. Note the irregular pattern formed by the ciliated cells. (Phase micrograph, unstained ×412.)

(*c*) Luminal area from the ampullar region of the fetal oviduct after 16 days in culture. Debris within the lumen was being propelled by ciliary action when photographed. (Phase micrograph, unstained ×412.)

(*d*) A single large vesicle which has formed within an ampullar segment of a fetal oviduct after 12 days in organ culture. Beating cilia were visible within the vesicle and the segment was contracting when photographed (×54).

In explants of the ampulla, cilia continued to be active for 2 to 3 weeks, whereas in the isthmic explants the cilia stopped beating after 8 to 10 days (Table 1). In a few isthmic explants, ciliated cells were never actually observed, but when they did occur they were found in small isolated clusters in the lining epithelium. In contrast, the ciliated cells in ampullar explants were much more numerous and appeared to be arranged in anastomosing cords (Fig. 2*c*).

Muscular contractions were seen in almost half the cultures of oviductal fragments, but the time of onset of the contractions was variable. The origin of the explant, whether from the ampullar or the isthmic segment, had no significant effect on the ability to contract, but there was a tendency for ampullar explants to begin to contract earlier than the isthmic explants (Table 1).

TABLE 1

SIXTEEN-DAY FETAL MOUSE OVIDUCTS IN TISSUE CULTURE

EXPERIMENTAL CONDITIONS	REGION OF OVIDUCT	CILIARY ACTIVITY		MUSCULAR CONTRACTIONS		LIFE SPAN OF CULTURES (WEEKS)
		Onset (Days)	Duration (Weeks)	Onset (Days)	Duration (Weeks)	
Controls (oviductal tissue alone)	Ampulla	5–8	2–3	7–16	2–3	2–4
	Isthmus	6–8	< 2	13–16	3–4	2–4
Ovarian tissue added	Ampulla	6	2–3	11–13	3–4	3–4
	Isthmus	....	....	12	< 1	2–3
Testicular tissue added	Ampulla	5–10	< 1	....	....	2–3
	Isthmus	8	< 1	....	....	2
Estradiol benzoate (2.5 μgm/ml medium)	Ampulla	8–10	< 1	15	< 1	2–3
	Isthmus	....	....	....	....	2–3
Progesterone (2.5 μgm/ml medium)	Ampulla	5–10	3–4	....	....	3–4
	Isthmus	6–8	3–4	11	< 1	3–4

Initially the contractions were so weak and intermittent that they could be seen only under the higher powers of the microscope, but after 24 hr they became more regular and vigorous. Often there were prolonged intervals between contractions, but once initiated they were slow and prolonged. At the peak of contractile activity they occurred every 90 sec, each lasting 4 to 6 sec. In the older cultures the interval between contractions lengthened to 2 to 3 min. Within a single explant two foci of contractions were not uncommon, one located along the luminal border while the other lay farther away in the stroma. Each focus contracted independently. In our experience contractions were similar in the ampullary and isthmic explants. In a number of explants the muscle cells continued to contract even after the connective tissue in the immediate vicinity appeared necrotic. The muscle in most cultures continued to contract as long as the explants were viable.

C. *Addition of Gonadal Tissue to Tissue Cultures*

When minced ovarian tissues were added to the explants of fetal oviducts, they appeared to have a beneficial effect on the growth and differentiation of the ampulla. The same procedure appeared to be detrimental to the isthmic cultures. Ciliary activity in the ampullar fragments was similar to that of the control cultures, whereas there was no ciliary action in the isthmic cultures despite the fact that cilia were visible on some of the luminal epithelial cells.

Adding ovarian tissue to the cultures enhanced muscular contractions. In the cultures of ampullae, nearly all the explants developed contracting musculature. Of significance is the fact that contractions were maintained for 3 to 4 weeks, which was at least a week longer than in the controls. Visible contractions occurred in only a few of the isthmic fragments but, when present, the contractions continued for less than a week (Table 1).

When testicular fragments were added to the oviductal cultures, the effect was quite dramatic. Although cilia developed and began to beat in cultures of each of the oviductal regions, they soon stopped. Muscular contractions were never observed (Table 1). The oviductal tissues in these cultures remained in reasonably good condition for at least 2 weeks. The fragments of testes, however, became necrotic within 6 or 7 days after explantation.

D. *Addition of Hormones to the Tissue Cultures*

When estradiol benzoate was added to the culture medium at a concentration of 2.5 μg per ml, it had an adverse effect on all cultures. Ciliary activity of the ampullar explants decreased and cilia within the isthmic fragments never became active. Muscular contractions were seen only rarely in the ampullar explants and not at all in cultures of isthmic tissues. This concentration of estradiol benzoate did not appear to injure the cultured tissues, for they all remained viable for 2 to 3 weeks (Table 1).

The addition of progesterone to the medium appeared to stimulate ciliary activity; they beat so rapidly and constantly that the whole culture seemed to be flickering. Cilia in all of the cultures continued to beat longer than those in the controls. No muscular contractions were seen in the cultures of ampullar tissues and, although they were noted in a few of the isthmic cultures, they lasted for only a few days (Table 1).

E. *Oviducts in Organ Culture*

When the 16-day-old fetal oviducts were dissected from the ovaries, they were simple, straight tubes. After they had been divided and each segment explanted intact in organ cultures, their appearance changed quite rapidly. The most noticeable changes occurred in the ampullar segments. In each case the lumen began to enlarge and by the 3d day in culture 2 ridges appeared along the walls of the lumen which extended the full length of the segment. These ridges, which at first were quite straight, then became folded at intervals along their length and gave the impression of finger-like projections into the lumen. At the same time the whole segment began curving into a crescent, which eventually formed a complete circle. The lumen became distended with fluid, probably secreted by the mucosa. As the pressure of the fluid increased, several small vesicles appeared in the tissue. These vesicles soon increased in size and often coalesced, in many cases forming a single large vesicle whose walls were very thin and transparent (Fig. 2*d*). The rest of the explant now appeared to serve as a base for the vesicle.

Ciliary activity was observed first in ampullar segments on the 6th day after cultivation, and continued usually for as long as the cultures remained viable (Table 2). The ciliated cells were distributed in the epithelium lining the lumen and they were most conspicuous along the folds.

The muscle cells within the ampullar explants also began contracting on the 6th day and continued for about 3 weeks. These contractions were vigorous, rhythmic, and peristaltic in character. When the explant had formed a circle, as described, occasionally a piece of the fimbriae was left free and this was seen to contract from side to side.

The isthmic explants developed in organ culture in a very similar manner, but the changes were not as pronounced as in the ampullar explants. In each case the lumen enlarged, and the explant began to expand into a crescent, soon forming a complete circle. The lumen became distended with fluid and vesicles formed. Many of these vesicles were smaller than those seen in ampullar explants and very few of the extremely large vesicles occurred. For the most part, the walls appeared thicker and more opaque than those of ampullary vesicles.

TABLE 2

SIXTEEN-DAY FETAL MOUSE OVIDUCTS IN ORGAN CULTURE

EXPERIMENTAL CONDITIONS	REGION OF OVIDUCT	CILIARY ACTIVITY		MUSCULAR CONTRACTIONS		LIFE SPAN OF CULTURES (WEEKS)
		Onset (Days)	Duration (Weeks)	Onset (Days)	Duration (Weeks)	
Controls (oviductal segments alone)	Ampulla	6	2–3	6–8	2	3
	Isthmus	6	1–3	7–10	2–3	3–4
Fetal ovaries added	Ampulla	6	2–4	6–9	2–3	3–4
	Isthmus	6	2–3	7–14	2	3
Fetal testes added	Ampulla	7	2	7–16	2	2–3
	Isthmus	7–13	1–2	....	....	2–3
Estradiol benzoate (2.5 μgm/ml medium)	Ampulla	5–7	3	9–11	< 1	3
	Isthmus	6–8	2–3	11	< 1	3
Progesterone (2.5 μgm/ml medium)	Ampulla	5–8	1	....	....	2–3
	Isthmus	....	....	....	....	1–2

Cilia within the isthmic region began to beat at the same time as those observed in the ampullae, but the ciliated cells were fewer in number and were seen less frequently (Table 2). In some explants they appeared to be arranged in a linear fashion for short distances along the walls of the lumen. In other segments they were localized in small groups and were not distributed throughout the entire segment. Some explants never showed ciliary activity.

Muscle tissue began to contract in isthmic explants about the same time after cultivation as that in the ampullar segments (Table 2). The contractions were strong peristaltic waves which occurred at regular intervals. As the lumen became greatly distended with fluid, the walls in some explants became so stretched that contractions became weaker and finally stopped, while in others which were less distended, contractions continued as long as the tissues remained alive.

Contractions within all the explants were very sensitive to temperatures lower than

37° C. When cultures were left at room temperature for more than 10 min, the muscle no longer contracted.

F. *Addition of Gonadal Tissue to Organ Cultures*

Whole ovaries from 16-day-old fetal mice were placed adjacent to the oviductal segments on the agar strips. As the segments started to form crescents, the ovary was often completely surrounded by oviductal tissue. If the ovary was not encircled, it tended to become attached to the oviduct so that the tissues of the two organs merged. Frequently the ovaries became necrotic before the oviductal tissue showed any gross degenerative changes.

Cilia in both ampullar and isthmic segments of the oviduct began to beat on the 6th day, but in each segment the cilia continued to beat for a longer period of time than in controls where ovaries were absent (Table 2).

Muscular contractions were seen in all explants of both ampullar and isthmic segments. The contractile characteristics resembled those seen in control cultures, but contractile activity was sometimes maintained for a longer period of time than in the controls.

Sixteen-day fetal testes cultured with segments of oviduct became moderately distended with fluid after they had been in culture for about a week, although no vesicles were formed. They became fused with the oviductal segments. Testes cultured with ampullar segments seemed to have little effect on the activity of either the cilia or the muscle elements. Cilia within the isthmic segments did not continue beating generally for as long a period as those of the controls, and no muscular contractions were seen in any of the explants of the isthmic region (Table 2).

G. *Addition of Hormones to the Organ Cultures*

When estradiol benzoate was added to the culture medium in a concentration of 2.5 μg per ml, the action of cilia in both the ampullar and isthmic segments was unaffected. Muscular contractions began in most cultures, but they stopped within 6 days. The explants, however, remained in good condition for 3 weeks (Table 2).

With the addition of progesterone at a concentration of 2.5 μg per ml of medium, beating cilia were observed briefly in ampullar segments, but they stopped at the end of 5 days. No ciliary activity was noted in explants of the isthmic region. Muscular contractions were inhibited also in all cultures. These cultures did not survive for as long a period of time as either the controls or other experimental cultures (Table 2).

IV. Discussion

From the observations reported here, some insight has been gained into the functional differentiation of 16-day-old fetal mouse oviducts maintained *in vitro*. Two aspects of development were noted, namely, the appearance of cilia and the onset of their beat, and the time of appearance of muscular contractions. Cilia appeared first on the epithelial cells lining the oviductal lumina and began to beat on the 6th day after explantation. Fresh squash preparations of oviducts of mice taken at daily intervals after birth revealed that cilia, though present on the 2d day, did not begin to beat until the 3d day postpartum. Since mice in our strain are born on the 19th day of gestation, the day when ciliary activity began in the cultures corresponds to the 3d day of postnatal life. Thus the onset of ciliary activity in culture compares very well with that seen *in vivo*.

Muscular contractions were observed in less than 50% of the tissue culture preparations. The onset and rate of contractions varied considerably in the various preparations. When present they were regular, vigorous, and peristaltic in nature. In the organ cultures muscular contractions began on the 6th or 7th day after explantation. The development of contracting, smooth muscle has not been reported previously in cultures of oviductal tissues. Although illustrations of oviductal cultures in Galstjan's (1935) publication show layers of smooth muscle cells beneath the epithelia he does not comment on their contractility. In all of the organ cultures muscle contraction began within 1 week after explantation. At the time of onset of contractions the oviductal segments were filled with fluid. Evidence for contraction is more readily visualized under these circumstances. In addition a distended lumen may enhance contractility.

There was no effect on ciliary beat or muscle contraction when fetal ovaries were added to the oviductal tissues in organ cultures. These results agree with the reports of Price and co-workers who noted that the differentiation of Müllerian ducts in organ culture proceeded quite normally with or without the presence of ovaries (Price and Pannabecker 1959; Price and Ortiz 1965; Price, Ortiz, and Zaaijer 1967).

The presence of fetal testicular tissue inhibited muscular contractions in all the tissue culture preparations as well as in the isthmic segments growing in organ culture. Whether or not this lack of contractility is related to a failure of the muscle cells to undergo sufficient differentiation or, having differentiated, not being able to contract, is under investigation. Ortiz, Price, and Zaaijer (1966) have demonstrated that the fetal testes secrete androgens in organ cultures. Even though Jost (1950) showed that if a fetal testis is grafted to the mesosalpinx adjacent to the ovary there is a suppression of development of the Müllerian duct, we cannot be certain that the inhibitory results described in our cultures are related to the action of fetal androgens or to some other substance. When estradiol benzoate was added to the culture fluid, both ciliary activity and muscular contractions appeared to be inhibited. Organ cultures, on the other hand, were not similarly affected. The differences may be related to toxic or pharmacological effects of estrogens when they come into direct contact with cells growing in monolayers. It should be recalled, however, that the cells appeared to be alive for approximately the same period as the controls. Hamilton (1961), who cultured Müllerian ducts from chick embryos, found that large doses of estradiol benzoate both inhibited and suppressed the growth of the tissues.

Progesterone inhibited muscular contractions in both organ and tissue cultures. Even though Harper (1966) reported that muscular contractions in the mature rabbit oviducts were depressed after injections of progesterone and Lehto (1963) reported similar results in human oviducts, the exact mode of action of this hormone remains unknown (Boling 1968). Further research is needed to define more accurately the effects of various hormones on the differentiation of fetal oviducts.

V. Conclusions

The functional differentiation of the 16-day-old mouse oviducts was observed in tissue and organ cultures. In all cultures cilia began to beat by the 6th day after explantation. Though muscular contractions were first observed on the 6th and 7th days in organ cultures, their onset in the tissue cultures varied considerably. Muscle contractions were seen in less than 50% of the tissue cultures. Fetal ovaries cultured in conjunction with fetal oviducts had no effect on the organ cultures, but muscular contractions were inhibited somewhat in the tissue cultures. Fetal testicular tissue added to the cultures reduced ciliary activity in the tissue cultures and inhibited muscular contractions. Estradiol benzoate added to the culture medium caused a suppression of muscular contractions in all cultures and lessened ciliary activity in the tissue culture preparations. When progesterone was added to the culture fluid it inhibited muscle contractions in all cultures. The cilia, however, were stimulated in the tissue cultures and they beat for a longer period than did the controls. These observations have shown that 16-day-old fetal oviducts in culture may develop cilia which beat and muscle cells which contract.

Acknowledgment

Unpublished observations referred to in this chapter were supported by grants from the National Institutes of Health, United States Public Health Service. The author wishes to acknowledge the expert assistance of Mrs. Lynn Goldner, technician, and Mr. Roy Hayashi, photographer.

References

1. Boling, J. L. 1968. Endocrinology of oviductal musculature. In *The mammalian oviduct*, ed. E. S. E. Hafez and R. J. Blandau, chap. 6. Chicago: University of Chicago Press.
2. Bousquet, J. 1964. Culture organotypique de fragments d'oviducte de ratte et de lapine. Action des stéroïdes ovariens. *C.R. Soc. Biol. (Par.)* 158:508.
3. Galstjan, S. 1935. Experimentell-histologische Untersuchungen über das Eileiterepithel. *Arch. Exp. Zellforsch.* 17 (Nr. 3):231.
4. Gwatkin, R. B. L., and Biggers, J. D. 1963. Histology of mouse Fallopian tubes maintained as organ cultures on a chemically defined medium. *Int. J. Fertil.* 8:453.
5. Hamilton, T. H. 1961. Studies on the physiology of urogenital differentiation in the chick embryo. I. Hormonal control of sex differentiation of Müllerian ducts. *J. Exp. Zool.* 146:265.
6. Harper, M. J. K. 1966. Hormonal control of transport of eggs in cumulus through the ampulla of the rabbit oviduct. *Endocrinology* 78:568.
7. Hellström, K. E., and Nilsson, O. 1957. *In vitro* investigation of the ciliated and secretory cells in the rabbit Fallopian tube. *Exp. Cell Res.* 12:180.
8. Jost, A. 1950. Sur le contrôle hormonal de la différenciation sexuelle du lapin. *Arch. Anat. Micr. Morph. Exp.* 39:577.
9. Jost, A., and Bergerard, Y. 1949. Culture *in vitro* d'ébauches du tractus génital du foetus de rat. *C.R. Soc. Biol. (Par.)* 143:608.
10. Jost, A., and Bozic, B. 1951. Données sur la différenciation des conduits génitaux du foetus de rat, étudiée *in vitro*. *C.R. Soc. Biol. (Par.)* 145:647.
11. Lehto, L. 1963. Cytology of the human Fallopian tube. *Acta Obstet. Gynec. Scand.* 42 (suppl. no. 4):3.
12. Nilsson, O., and Hellström, K. E. 1957. Cell types identified in tissue cultures of epithelium of the rabbit Fallopian tube. *Acta Obstet. Gynec. Scand.* 36:340.
13. Ortiz, E.; Price, D.; and Zaaijer, J. J. P. 1966. Organ culture studies of hormone secretion in endocrine glands of fetal guinea pigs. II. Secretion of androgenic hormone in

adrenals and testes during early stages of development. *Proc. Kon. Med. Akad. Wetensch.* 3:400.
14. Price, D., and Ortiz, E. 1965. The role of fetal androgen in sex differentiation in mammals. In *Organogenesis*, ed. R. L. DeHaan and H. Ursprung, chap. 25. New York: Holt, Rinehart & Winston.
15. Price, D.; Ortiz, E.; and Zaaijer, J. J. P. 1967. Organ culture studies of hormone secretion in endocrine glands of fetal guinea pigs. III. The relation of testicular hormone to sex differentiation of the reproductive ducts. *Anat. Rec.* 157:27.
16. Price, D., and Pannabecker, R. 1958. A study of sex differentiation in the foetal rat. In *A symposium on the chemical basis of development*, ed. W. D. McElroy and B. Glass. Baltimore: Johns Hopkins Press.
17. ———. 1959. Comparative responsiveness of homologous sex ducts and accessory glands of foetal rats in culture. *Arch. Anat. Micr. Morph. Exp.* 48:223.
18. Valenti, C. 1963. Harvesting and culture of epithelial cells from hollow organs of the female reproductive system. *Z. Zellforsch.* 60:850.

20

Immunocytological Studies of the Mouse Oviduct

L. E. Glass

Department of Anatomy

University of California Medical School, San Francisco

Some years ago A. M. Schechtman (1955) wrote "... new molecules like new morphological structures make their appearance during the course of development ... (and) ... the *de novo* appearance of antigens or other molecules tends to create a false impression of the synthesizing capacities of the embryonic organism. The embryo is definitely deficient in some types of molecular synthesis and is dependent upon presynthesized molecules supplied normally by the maternal body .... The embryo admits complex macromolecules and such admission occurs prior to and during the time when the basic differentiations of the vertebrate body are established."

Originally heretical, the idea that complex macromolecules, presynthesized by the maternal organism, are transferred into ovarian eggs and oviductal embryos is becoming commonplace. In insects, a sex-linked protein is present in the blood of females but absent from the blood of males; during vitellogenesis, the protein passes between the follicle cells, enters the egg and apparently is deposited in the carbohydrate-protein complex which makes up the yolk platelet (Telfer 1954; Telfer 1961; Telfer and Melius 1963; Roth and Porter 1964). In the chicken, 3 sex-associated proteins are present in the blood of laying hens; the proteins are absent from the blood of nonlaying hens and cockerels but can be stimulated to appear by estrogen treatment (reviewed Schjeide *et al.*, 1963). Synthesized in the liver, these proteins are transported via the serum, through (or between) the ovarian follicle cells and are deposited in the fluid and granular yolk of the hen's egg (Schjeide *et al.* 1963). Other blood molecules, albumin, and globulins, which are not sex-limited, also are deposited in the hen's yolk (Martin, Vandegaar, and Cook 1957; Schechtman 1955); all told, at least 8 blood components are transferred into the hen's egg during yolk formation. Serum molecules are transferred into the frog egg, too, and initially are associated with the sites of yolk formation and finally with the full-grown yolk platelets (Flickinger and Rounds 1956; Glass 1959). Less adequate data indicate that transfer of presynthesized macromolecules from the blood to the yolk probably occurs in reptiles and fish as well (Schjeide *et al.* 1963; Maung 1963). In the mammal, several kinds of blood molecules pass from the vessels into the ovarian oocytes. The transfer is selective as to molecular type, and serum albumin, at least, passes into the nucleus of oocytes at particular stages to become associated with the chromatin and nucleolus. Moreover, in the mammalian ovary, transfer is selective as to the follicle stage at which it occurs (Glass 1960, 1961, 1966*a*, *b*).

Such macromolecules, presynthesized by the mother and transferred intact (or nearly so) into the egg or embryo are called "heterosynthetic" (Schechtman 1955). In contradistinction, macromolecules synthesized *de novo* within the egg or embryo are called "autosynthetic."

The existence of heterosynthetic transfer in such a breadth of phyla suggests that the presence of presynthesized maternal molecules may have conveyed a significant selective advantage on the eggs and embryos in which it occurred. At the very least, since it affects the egg, heterosynthetic transfer cannot be harmful to the developing organism. In lower phyla, oviparous and with yolky eggs, the presynthesized maternal macromolecules generally are associated with the yolk. Clearly, they will not be used until yolk degradation begins much later in development. In contrast, however, the entire development of the mammalian embryo occurs within the mother's body and mammalian eggs are relatively nonyolky. If heterosynthetic molecules are important for normal mammalian development, transfer of the complex, presynthesized heterosynthetic molecules must occur not only in the ovary but, presumably, in the oviduct and uterus as well.

Uterine transfer has been demonstrated in the mammal. In the rabbit, for example, maternal serum molecules pass into the uterine lumen during and for a short while after implantation and are transferred via the yolk sac splanchnopleure into the embryo (e.g., Brambell, Hemmings, and Henderson 1951). Near the end of gestation in many mammalian species, maternal antibody γ-globulins are transferred through the placenta to the fetus (Brambell and Hemmings 1960).

Oviductal transfer also occurs. This phenomenon is less well known and the data presently available are described below.

I. Basic Methodology

Most of the information about heterosynthetic transfer in the oviduct has been obtained by use of fluorescent antibody methods. For necessary background, these procedures will be described briefly.

As is well known, the injection of proteins or other large molecules into a rabbit or other animals causes the production of antibodies; the molecular configuration of the antibody thus formed is such that it recognizes and adheres to molecules like those against which it was formed. In many circumstances, the specific affinity between an antigen and its antibody results in their union and the formation of an antigen-antibody precipitate either in solution or in an agar gel. This characteristic led to the recognition that antigen in fixed position in a tissue or cell still might be able to react with antibody and that, in this case, the antigen-antibody precipitate would be formed *in situ* on the tissue section or cell. Such reactions do occur and when the antibody is labeled with fluorescein and the reaction subsequently is visualized by a fluorescence microscope, it gives a very precise marker for the location of a specific antigen in the cells (Coons and Kaplan 1950; Mellors 1959).

In the studies to be reported, rabbit antisera were prepared against whole mouse serum or against the foreign plasma protein, bovine plasma albumin (BPA). The antisera were characterized by interface precipitin, agar diffusion, and microimmunoelectrophoresis procedures using homologous and heterologous test antigens. Absorptions to delimit the antiserum specificities were carried out.

The specific antisera thus obtained were used for indirect fluorescent antibody tests (Mellors 1959) on serial sections of mouse oviduct obtained from uninjected mice or from mice injected intravenously with 10 mg of BPA. Hydrated tissue sections were reacted with the unlabeled specific antiserum, and unreacted antibody was washed thoroughly from the slide, leaving unlabeled tissue antigen-antibody complexes affixed to the section. Then the sections were reacted with fluorescein-labeled sheep antiserum directed against rabbit antibody globulin. After thorough rinsing, coverslips were mounted in 50% glycerine and the section was viewed with a Zeiss fluorescence microscope.

Controls for the immunological specificity of the fluorescence used various combinations of absorbed and unabsorbed antisera on adjacent serial sections from the same animal or on sections from similarly and differently treated animals. In order to define regions of immunologically nonspecific tissue affinities for rabbit serum and fluorescein, normal rabbit serum controls were used routinely. Tissue autofluorescence was determined by observation of unreacted hydrated sections. In combination these tests, plus the fact that antisera of different specificities were localized at consistently different tissue sites, provided adequate controls for the specificity of the fluorescent antibody localizations.

Tissues were obtained from ether-anesthetized mice, fixed in Carnoy II, embedded in low temperature paraffin (M.P. 50°–53° C), and sectioned serially at 5–7 micra. This treatment did not cause degradation nor translocation of the serum antigens studied.

II. Transfer of Serum Antigens to Oviductal Eggs

Initially, it was demonstrated that heterosynthetic transfer occurs in the mouse ovary (Glass 1960, 1961). Native mouse serum antigens were detected in the cytoplasm of ovarian eggs at all stages of follicle development. When a foreign protein, bovine plasma albumin, was injected systemically, BPA activity appeared in the ooplasm too. Since mouse eggs do not synthesize bovine specific protein, the latter study demonstrated that the blood proteins were transferred to the egg rather than synthesized *de novo* within the egg.

At ovulation, the ooplasm of the egg remained brightly fluorescent, indicating that the heterosynthetic proteins transferred into the oocyte during its ovarian existence were still present (Glass 1963). An oocyte just after ovulation is pictured in Figure 1. Whether fertilization occurred or not, the serum-like antigens in the oviductal egg could not be detected by 2 or 3 hr after ovulation (Fig. 2) (Glass 1963). Serum-like antigens reappeared in the cytoplasm of the pronuclear egg after follicle cell dispersal, which occurred 9 or 10 hr after ovulation (Fig. 3) and were present in embryonic blastomeres at all subsequent preimplantation stages (Figs. 4–9) (Glass 1963).

When bovine albumin was injected intravenously into the female mouse before ovulation, BPA antigen was present in the ooplasm of the ovulating egg, had disappeared by 2 or 3 hr later, and had reappeared after follicle cell dispersal (Glass 1963). If albumin was injected into the female after ovulation but before the follicle cells dispersed, BPA antigens did not appear in the ooplasm until after follicle cell dispersal from around the egg (Figs. 4–9).

Clearly, the presynthesized maternal molecules are transferred to the nude pronuclear ootid in the ampulla. Equally clear is the fact that serum protein, though available, is not

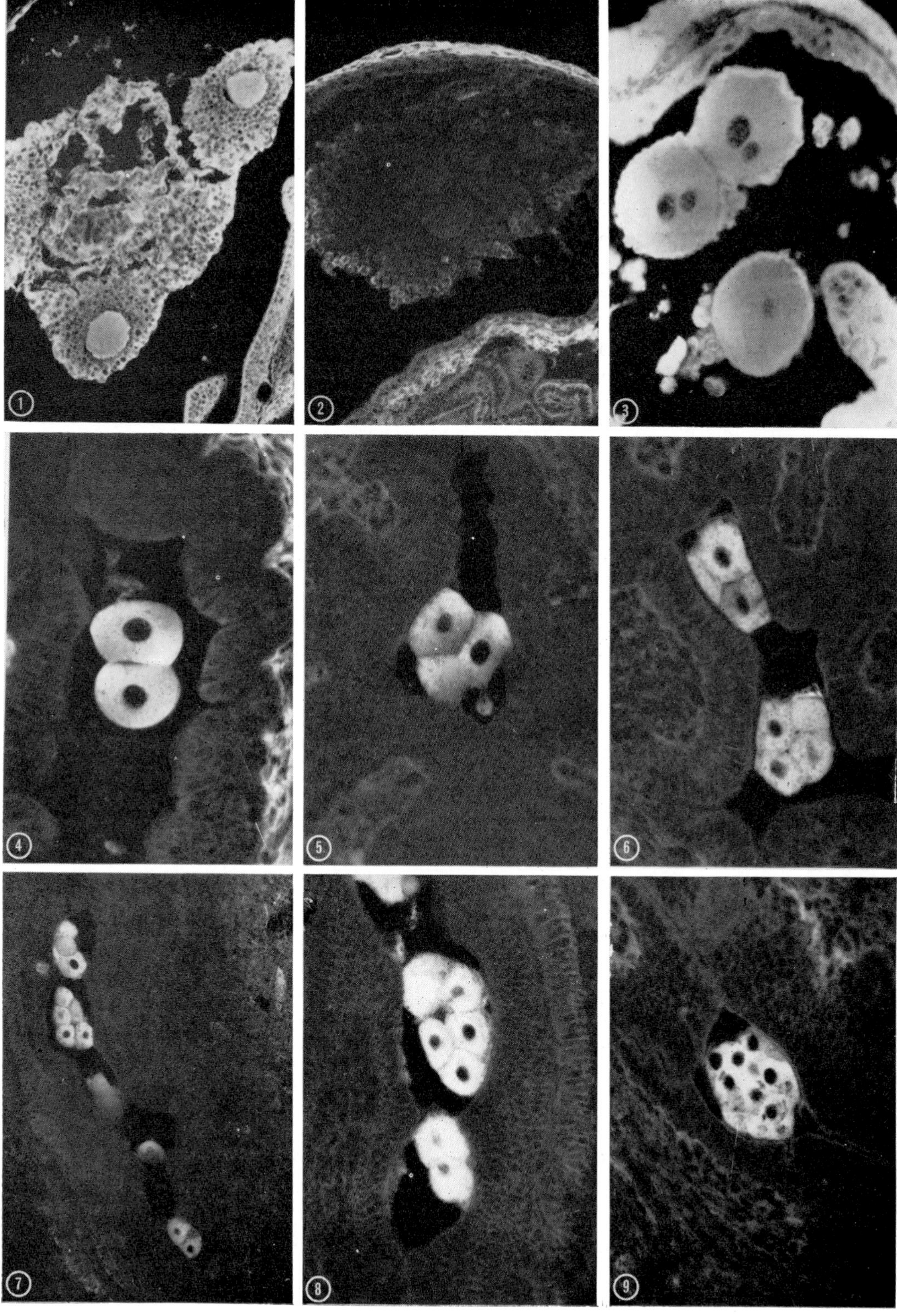
1
2
3
4
5
6
7
8
9

transferred until the follicle cells have dispersed. Electron micrographs (Odor 1960) indicate that follicle cell processes and oocyte microvilli retract from the zona pellucida at the stage when the oviductal oocyte and fertilized ovum becomes nonfluorescent. That association of follicle cell and ovum membranes aids macromolecular transfer is suggested not only by their morphology but also by recent experiments in which the relative amounts of human serum albumin-^{131}I (HSA) in the ooplasm were compared for 9 ovarian follicle stages (Glass and Cons 1968). At the stage before follicle cells surrounded the oocyte and at stages after follicle cell processes and oocyte microvilli interdigitated, relatively large amounts of HSA per unit area were present in the ooplasm. However, when the oocyte was completely surrounded by follicle cells but before the surface adaptations between follicle cells and oocyte had differentiated, the ooplasm contained little or no HSA. These data indicate that the complex membrane interdigitations facilitate heterosynthetic transfer from follicle cells to oocyte in the ovary; the data support the suggestion that their loss after ovulation makes heterosynthetic transfer difficult or impossible until the follicle cells have dispersed from around the oviductal ootid.

III. Serum Antigens in the Oviductal Epithelium

Fluorescence, indicating the presence of antigens similar to or identical with those in female serum, was very intense in the ampullary epithelium before and after ovulation. Dur-

---

FIGS. 1–3. Mouse oviductal ampullae treated with rabbit antiserum specific for mouse serum antigens. Fluorescence indicates the presence of molecules similar to or identical with those of mouse blood. (Figs. 1 and 2 from Glass 1963; by permission of publishers.)

(1) Oviductal eggs in Ampulla III shortly after ovulation. Serum antigens are present in the ooplasm, in the cytoplasm of follicle cells surrounding the oocyte, and in the oviductal epithelium (×49).

(2) Fertilized oviductal eggs in Ampulla III 6½ hr after mating. Serum antigens are no longer detectable in the ooplasm or in the follicle cells adjacent the eggs. Fluorescence of the oviductal epithelium is less intense than in Figures 1 or 3 (×49).

(3) Fluorescent ootids in Ampulla III 10 hr after ovulation and after follicle cells had dispersed from the egg. Oviductal epithelium is brightly fluorescent (×98).

FIGS. 4–9. Isthmus from pregnant mice which had received an intravenous injection of bovine plasma albumin (BPA) before ovulation or shortly thereafter. The absorbed antisera used in the immunocytological tests were specific for bovine plasma albumin and did not cross-react with mouse serum antigens. Fluorescence indicates the presence of antigens similar to or identical with the systemically injected BPA. (Fig. 9 from Glass 1963; by permission of publisher.)

(4) Two-cell embryo in Isthmus I of mouse injected with BPA before ovulation. Albumin antigen in the blastomeres is slightly less homogeneous in distribution than in ootids and just ovulated eggs from albumin-injected mice. Apical fluorescence of low intensity is seen in the epithelium adjacent to the embryo (×98).

(5, 6) Three- and 4-cell embryos from Isthmus II of different animals 53 hr after mating. Cytoplasmic distribution of albumin antigen is uneven in the blastomeres. Albumin antigen is essentially undetectable in the epithelium (×98).

(7) Photomicrograph showing embryos at several stages in Isthmus II. Albumin antigen is present in the blastomeres but absent from the epithelium (×49).

(8) Higher-powered view of different embryos from the same oviduct pictured in Figure 7. Photograph was printed so that epithelial details would show more clearly. Fluorescent, granular debris is at the epithelial surface, but there is no evidence of apical or generalized fluorescence in the epithelial cytoplasm (×98).

(9) Morula at junction of oviduct and uterus in albumin-injected animal. There is some albumin-specific fluorescence in the connective tissue underlying the epithelial cells but there is no evidence of albumin within them (×98).

ing the stage when serum molecules could not be detected in the egg, the epithelium was a little less fluorescent. After follicle cell dispersal, when serum molecules were passing again into the egg, the ampullary epithelium fluoresced brightly. Most important, the apical end of the cells fluoresced more intensely than did the basal pole. After the egg had passed into the isthmus from the ampulla, epithelial fluorescence in the ampulla decreased and was absent by 72 hr after mating, when the embryo was in the uterus and beginning to implant (Glass 1963 and unpublished).

Fluorescence was always of low intensity in the isthmus, although in regions containing embryos the cell apices often fluoresced slightly. The uterine epithelium contained serum-like molecules at all times, but they were especially apparent at the time of implantation (Glass 1963 and unpublished).

The data just cited support the idea that transfer occurs mainly in the oviductal ampulla and that little or no transfer occurs in the isthmus. This interpretation is consistent with observations on the distribution of serum molecules in embryonic blastomeres during cleavage and transit through the oviduct. Serum molecules were distributed evenly in the cytoplasm of the ovarian oocyte, in the just-ovulated egg and in the nude ootid in the ampulla (Figs. 1–3). As cleavage proceeded, the embryos passed out of the ampulla and through the isthmus and serum molecules in the blastomeres of these older embryos were distributed less evenly (Figs. 4–9). It can be suggested that some of the presynthesized "maternal" molecules were used up during cleavage but that new molecules were not (could not be?) transferred to the embryo while it was in the isthmus.

There was a slight apical fluorescence in the isthmus epithelium in the region of the embryo, and serum molecules were distributed irregularly in the blastomere cytoplasm of embryos in the isthmus. Both these observations could be accounted for by the hypothesis that some of the maternal molecules pass in a reverse direction, from the egg to the epithelium, during embryo passage through the isthmus. However, experiments in which BPA was administered by a single injection just before ovulation do not support this suggestion. The ovulated eggs contained BPA and the cycle of serum antigen disappearance from the egg and reappearance in the nude ootid occurred with the foreign BPA just as with the native serum molecules. Bovine protein was present in the embryonic blastomeres throughout cleavage (Figs. 4–9). However, with the exception of the 2-cell stage, albumin was not detected in the apices of oviductal cells adjacent to the embryo as would be expected if reverse transfer from the embryo to the epithelium were occurring (Figs. 4–9). These data indicate that the concentration of the systemically injected bovine antigen decreased with time and by the 2d and 3d days of pregnancy was not available for transfer to the oviduct.

Evidence that reverse transfer from the lumen to the oviductal epithelium does not occur even in the ampulla was obtained from experiments in which ovarian eggs and their surrounding follicle cells were radio-labeled *in vivo* by methionine-^{35}S (Lin and Glass, unpublished). After ovulation, the unfertilized, radio-labeled eggs were removed from the oviductal ampulla and transferred to mated recipient females for fertilization and development. Figure 9*a* pictures the ampulla of a recipient oviduct containing ^{35}S-labeled donor embryos and unlabeled recipient embryos. There was no evidence that radio-labeled protein had passed from the donor oocyte, embryo, or follicle cells into the ampullary epithelium.

Since serum molecules are present in the cytoplasm of the epithelial cells and especially at the end toward the oviductal lumen, it is probable that serum antigens were transferred through the cells. Transfer between the epithelial cells may also occur, but electron micrographic methods, which might demonstrate such transfer, have not been applied.

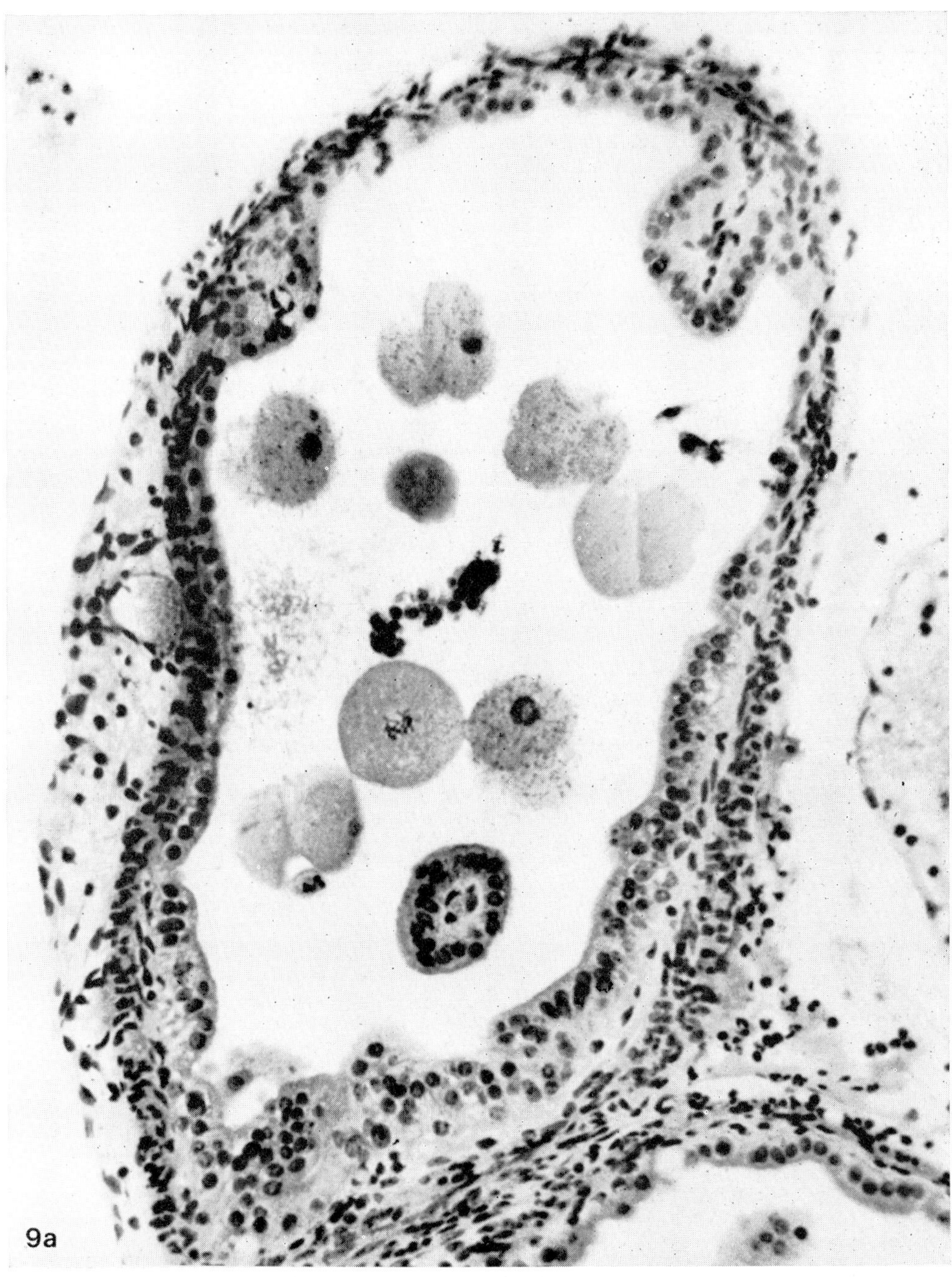

FIG. 9*a*. Autoradiograph of Ampulla III from the oviduct of a recipient female mouse containing radiolabeled donor embryos and unlabeled native embryos. Epithelium of the oviduct has not taken up radiolabeled material from its lumen (×68). (Courtesy of Teh Ping Lin and Laurel E. Glass, unpublished.)

IV. Regional and Age Differences in Transfer during Postnatal Development of the Mouse Oviduct

Studies on the transfer of blood antigens to the oviductal epithelium during postnatal development gave particular evidence of both regional and age-dependent differences in transfer activity. Oviducts from newborn, 1-, 2-, 4-, 6-, 8-, and 12–16-week-old mice were examined by fluorescent antibody techniques for the presence and absence of blood antigens (Glass and McClure 1965).

These data on the transfer of heterosynthetic molecules and those in the following section on DNA synthesis indicate that there is a complex regional differentiation in the mouse oviduct, at least functionally. The morphological characteristics of these oviductal regions are given in Table 1; they conform in general to previous descriptions (Agduhr 1927).

Rather surprisingly, all regions of the newborn oviduct contained relatively large amounts of serum molecules as indicated by bright fluorescence in the epithelium (Figs. 10, 11). By 1 week of age, however, the serum-like molecules could not be detected in the ovi-

TABLE 1

OVIDUCTAL REGIONS IN THE MOUSE[a]

Region		Description
FIMBRIAE		
	F	Complex folds of epithelium, projecting into ovarian bursa at end of oviduct, surround the ostium and first portion of oviductal ampulla. Folds are blunt or club-shaped and may branch; have a connective tissue "core." Epithelium of ciliated columnar and cuboidal cells, the former with basal nuclei which appear pseudostratified. Many nonciliated "peg cells" with nuclei bulging above the surface of the epithelium.
AMPULLA		
	Amp I	First portion of oviduct internal to ostium; surrounded by fimbrial folds. Unbranched, regular folds of epithelium; with connective tissue core; folds project into lumen and nearly fill it. Low columnar cells with basal nuclei cover folds; cells may be ciliated or nonciliated ("secretory"); ciliated cuboidal cells in grooves between folds. Fewer peg cells than in fimbrial region; tend to be near tips of folds.
	Amp II	Longer, more finger-like folds than in Amp I, with a connective tissue core; folds fill lumen completely; unbranched except at bends in oviduct. Columnar cells are lower and more regular in shape than in Amp I; most cells are ciliated with few nonciliated; nuclei are basal or central. Ciliated, cuboidal cells in grooves between folds. Many nonciliated peg cells, especially at tips of folds.
	Amp III	Folds lower and less regular in shape than in Amp II or Amp I; seem not to fill lumen completely; most with connective tissue core. In contrast to finger-like shape of Amp II folds, Amp III folds often are bulb-like in shape, being wider in the middle than at their base or tip. Low columnar cells of uneven height cover folds; nuclei usually are basal but may be apical, especially at tips of folds. Few ciliated cells on folds. Most of low cuboidal cells between the folds are ciliated. Between longer folds, some shorter folds present, most of which do not have a connective tissue core. Few peg cells present.
(TRANSITIONAL)		
	Tr	Folds lower and thicker than in ampulla; fewer and fewer folds with connective tissue core; very few cilia on folds; peg cells absent.
ISTHMUS		
	Isth I	Folds irregular in shape and size and do not fill lumen; may have square or blunt tips; most without connective tissue core. Epithelium is pseudostratified, columnar; nuclei apical with light staining cytoplasm extending to basement membrane. Low cuboidal cells in grooves between folds, occurring less frequently passing from ovarian to uterine end. Lumen irregular shape, due primarily to uneven height of epithelium; many crypts, usually lined by ciliated cells.
	Isth II	Heavy, regular columnar epithelium; folds with dense connective tissue core; nuclei regularly aligned, in apical position at tip of fold and in middle of cells on side of fold. Epithelium very ordered in appearance. Isth II includes intrauterine oviduct though lumen of latter much smaller.

[a] These regions are equivalent to the following Agduhr (1927) regions: Amp I and Amp II = Agduhr *g;* Amp III = *f;* Tr = *e* (perhaps *d*)*;* Isth I = *c*, *b;* Isth II = *a*.

ductal epithelium either in the ampulla (Fig. 12) or in the isthmus (Fig. 13). The only cells where serum molecules were present were in the fimbriated, ovarian end of oviduct.

Serum molecules were absent from Amp I and from Isth I in animals 2 weeks of age. Cells in Amp II were moderately fluorescent and there was low intensity fluorescence in Amp III in some animals and none in others. The epithelium of Isth II fluoresced at moderate intensity in most mice. A few follicles with antra were present in the ovaries of the 2-week-old animals.

By 4 weeks of age, variable amounts of serum molecules were present in the epithelium of the ampulla, for ampullary fluorescence varied from animal to animal and even within an animal (Fig. 14), probably due to the changing hormonal status of the growing animals. In some mice, the isthmus epithelium fluoresced at low intensity, but in others no portion of the isthmus fluoresced (Fig. 15). The fimbrial epithelium fluoresced brightly in all the animals.

Fertile matings occur in this mouse strain by 5 weeks of age and, in the older animals (6 weeks, 8 weeks, and 12–16 weeks of age), fluorescence of the oviductal epithelium apparently varied with the reproductive state of the animals. Large tertiary follicles were present in the ovaries of all these animals and ovulation had occurred in several. The epithelium of the ampulla fluoresced brightly in those animals in which ovulation had occurred but was nonfluorescent in animals containing embryos at the uterine end of the oviduct or uterus. Serum molecules either were absent from the isthmus or were present in very small amounts, particularly in the apical region of the cells.

The fimbrial epithelium fluoresced brightly at every age examined (e.g., Figs. 10, 12).

That transfer of blood macromolecules into the oviductal epithelium might be correlated with the hormonal status of the animal, perhaps with the estrogen levels, was suggested from these studies. Maternal estrogen levels are high at parturition in mice. Relatively, many serum molecules are present in the oviductal epithelium of the neonate, which has just emerged from the influence of the high maternal estrogen levels. At birth, the pup leaves the hormonal environment of the mother. Relatively little estrogen or progesterone is produced yet by its own ovaries; few or no serum molecules are present in the oviductal epithelium of 1- and 2-week-old animals. When follicular antra develop and ovarian hormone production stabilizes, serum molecules are detected again in the oviductal epithelium; however, they reappear in quantity only in the ampulla and are in less abundance in the isthmus. The interpretation that estrogen is associated with transfer is consistent with the observation that ampullary fluorescence is high at ovulation and during early cleavage but becomes less and finally is absent when the embryo enters the uterus. It is consistent also with the fact that the epithelium of the ampulla is higher and shows more histological evidence of secretory activity during estrus, metestrus, or early pregnancy than during diestrus or in castrate animals, e.g., mouse (Allen 1922; Espinasse 1935).

Hormonal control may be exerted at any of several points in the transfer sequence. An indirect effect, modifying the rate of synthesis of serum molecules in the liver, could be exerted. Such an estrogen effect on liver synthesis of yolk protein has been demonstrated conclusively in fowl (e.g., Schjeide *et al.* 1963). In the mammal, estrogen does modify the rate of synthesis of serum proteins (Williams-Ashman 1965), but studies have only begun of estrogen effects on heterosynthetic transfer to the oviduct in castrate mice. Alternatively, the hormone might stimulate the transfer process directly; there is some evidence that estrogen modifies membrane permeability in target organs (Williams-Ashman 1965). Another possi-

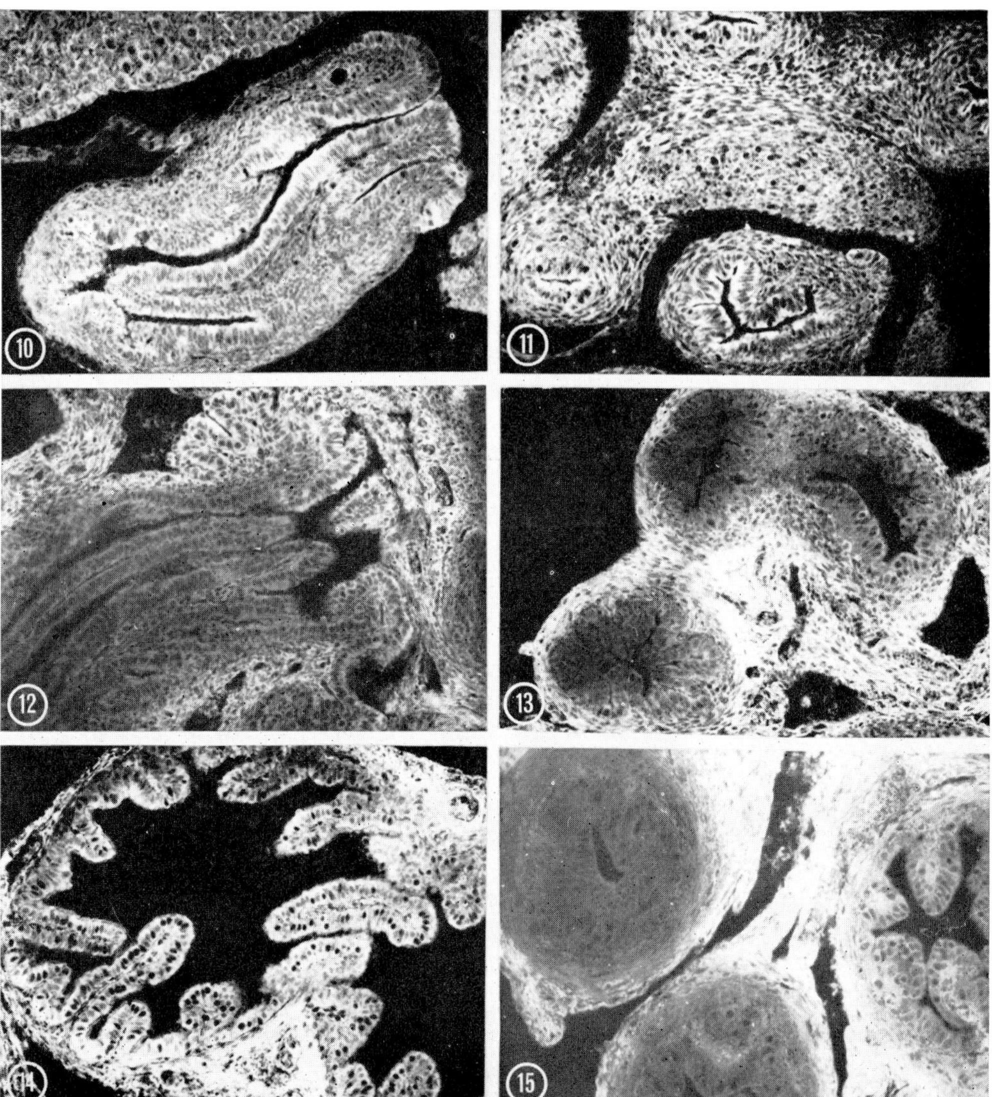

FIGS. 10–15. All oviducts pictured were treated with rabbit antiserum specific for mouse serum antigens. Fluorescence indicates the presence of molecules similar to or identical with those of the blood.

(10, 11) Ampulla (Fig. 10) and isthmus (Fig. 11) from a newborn mouse. There is generalized epithelial fluorescence in the ampulla, the isthmus, and the fimbria (×67).

(12, 13) Ampulla I and fimbriae (Fig. 12) and isthmus (Fig. 13) from the oviduct of a 1-week-old mouse. Although the fimbrial epithelium is brightly fluorescent, neither the ampullary nor the isthmus epithelia contain detectable serum molecules (×67). (Fig. 12 from Glass and McClure 1965; by permission of publishers.)

(14) Ampulla II or the first part of Ampulla III from a 4-week-old mouse. The variability in distribution of serum antigens within the epithelium pictured here is characteristic for this age group (×67).

(15) Oviduct from 4-week-old mouse. Cross sections of 2 nonfluorescent loops of Isthmus I (left) and a portion of Ampulla III (right), the last with a moderately fluorescent epithelium. Characteristically, serum antigens are present in Ampulla II and III but generally are absent in the isthmus epithelium (×67).

bility is that hormone modifies the rate at which the transferred molecules are utilized within the epithelium. For example, progesterone, higher during the 2d and 3d days of pregnancy, may stimulate the rapid utilization of the transferred serum molecules by the isthmus epithelium; this fits with the observation that very few serum molecules are present in the oviductal epithelium at stages when blood progesterone levels are elevated. No experimental data yet allow a clear choice between these (or other) alternatives.

Although there are regional differences and age-dependent differences in macromolecular transfer to the oviductal epithelium, cellular selectivity does not occur. If transfer was observed in the ampulla; secretory, peg, and ciliated cells all contained the heterosynthetic molecules; if there was no transfer in a region, none of the cell types contained serum antigens in its cytoplasm (Glass and McClure 1965). Apparently whatever controls selectivity is generalized to a region rather than to specializations in the cellular epithelium. Likely candidates for mediation of the transfer process are the epithelial basement membrane, the connective tissue of the lamina propria, or characteristics of the capillary bed.

TABLE 2

PERCENTAGE OF CELLS INCORPORATING THYMIDINE^{3H} BY AGE AND OVIDUCTAL REGION[a]

AGE	OVIDUCTAL REGION		
	Amp II (Per Cent Cells Labeled)	Isth I (Per Cent Cells Labeled)	Isth II (Per Cent Cells Labeled)
Newborn	10–15%	. . . . 4–10%	
1 week	3–6	8–12%	no data
2 weeks	2–5	5–10	8%[b]
4 weeks	<0.5	<0.25	0.3 –2
6 weeks	<0.5	<0.1	0.25–1
18 weeks[c]	(1) <0.5	<0.1	0.1 –0.3
	(2) 0.1–2	0.5–1	1 –3

[a] Percentages are based on counts of labeled cells in 2 animals at each age.
[b] Only 1 animal.
[c] (1) Ovulation had not occurred recently.
(2) Recently ovulated eggs in Amp III.

V. Regional Differences in Synthetic Activity in the Oviduct—DNA Synthesis

Newborn, 1-, 2-, 4-, 6- and 18-week-old mice were injected intraperitoneally with tritiated thymidine in amounts of 1 μg per gram of body weight. One and a half or 24 hr after isotope injection, the mice were killed and their oviducts were fixed in Carnoy II and paraffin embedded. After sectioning at 5 micra and mounting, autoradiographs were prepared using Kodak NTB3 liquid emulsion. Thymidine is incorporated specifically into DNA providing that the tritiation is recent as in this case; therefore, the sites where the radio label was incorporated into nuclei were interpreted as sites of DNA synthesis.

Some data from these studies are recorded in Table 2 and pictured in Figures 16–23. Both regional and age differences in DNA synthesis occur. DNA synthesis was very active in both the ampulla (Fig. 16) and the isthmus (Fig. 17) of the newborn although the rate was lower in the isthmus. By 1 week of age, the rate of synthesis by ampullary cells had decreased half or more (Fig. 18) and by 2 weeks of age had decreased slightly further (Fig. 20). In the isthmus, synthesis was more active at 1 week of age (Fig. 19) than at birth (Fig. 17), but was again about at birth level in 2-week-old animals (Fig. 21). At 4 weeks of age fewer than 1% of the cells were synthesizing DNA in either the ampulla (Fig. 22) or the isthmus (Fig. 23). Both Amp II and Isth II were more active than Isth I in 4-, 6-, and 18-week-old animals (Table 2). In the 18-week-old animal which contained recently ovulated eggs in Amp III, the number of cells incorporating thymidine was elevated over that in the non-ovulated animal of the same age in both Amp II and Isth II, though the response was more marked in Isth II. Clearly, there are both age and regional differences in DNA synthesis by the mouse oviductal epithelium.

VI. Hormonal Effects on DNA Synthesis by the Oviductal Ephithelium

Thymidine incorporation by the oviductal epithelium in the noncastrate adult (12–16 weeks old) mouse was examined at several stages during the hormonal induction of ovulation. Two to 3 i.u. of pregnant mare's serum (PMS) were administered intraperitoneally 72 hr before the desired time of ovulation. Four to 6 i.u. of human chorionic gonadotrophin were administered 60 hr later, that is, about 12 hr before the desired time of ovulation. A sample, consisting of two mice, was killed at each of several intervals after PMS injection, PMS, and HCG injections, and after mating. Autoradiographs were prepared of their oviducts. Table 3 records these data. There is no evidence that thymidine incorporation by the oviduct of the intact mouse was stimulated by such hormonal treatment.

Thymidine incorporation in the mouse oviduct follows different patterns during postnatal development than does the transfer of heterosynthetic macromolecules from the blood to the oviductal epithelium. It is clear that these processes have different control mechanisms.

VII. Characteristics and Functions of the Heterosynthetic Molecules

No direct evidence is available on the nature or function of heterosynthetic molecules in the mouse oviduct. Immunoelectrophoretic studies of rabbit oviductal fluids demonstrated the presence of significant amounts of serum albumin and very small amounts of α-globulin; β- and γ-globulins were never detected (Chapter 14, this volume). Apparently, therefore, oviductal transfer is selective, as is heterosynthetic transfer in the ovary (Glass 1961, 1966*a*) and uterus (Brambell and Hemmings 1960). Bovine albumin is transferred to the mouse ovarian oocytes although bovine γ-globulin and rabbit γ-globulin are not (Glass 1961, 1966*a*), and mouse plasma albumin is transferred into ovarian oocytes more readily than are mouse plasma γ-globulins (Glass 1969). Since albumin has a strong affinity for ions, it

---

FIGS. 16–19. All photographs are of mouse oviducts from animals injected intraperitoneally with tritiated thymidine. Sites of incorporation are interpreted as sites of DNA synthesis. All magnifications ×284.

(16) Autoradiograph of ampulla from a 1-day-old mouse. Thymidine was administered 24 hr before sacrifice.

(17) Autoradiograph of the isthmus from a 1-day-old mouse. Thymidine was administered 24 hr before sacrifice.

(18) Ampulla from 1-week-old mouse. Note that peg cells are the predominant cell type labeled. Killed 1½ hr after thymidine administration.

(19) Isthmus from 1-week-old mouse. Killed 1½ hr after thymidine administration.

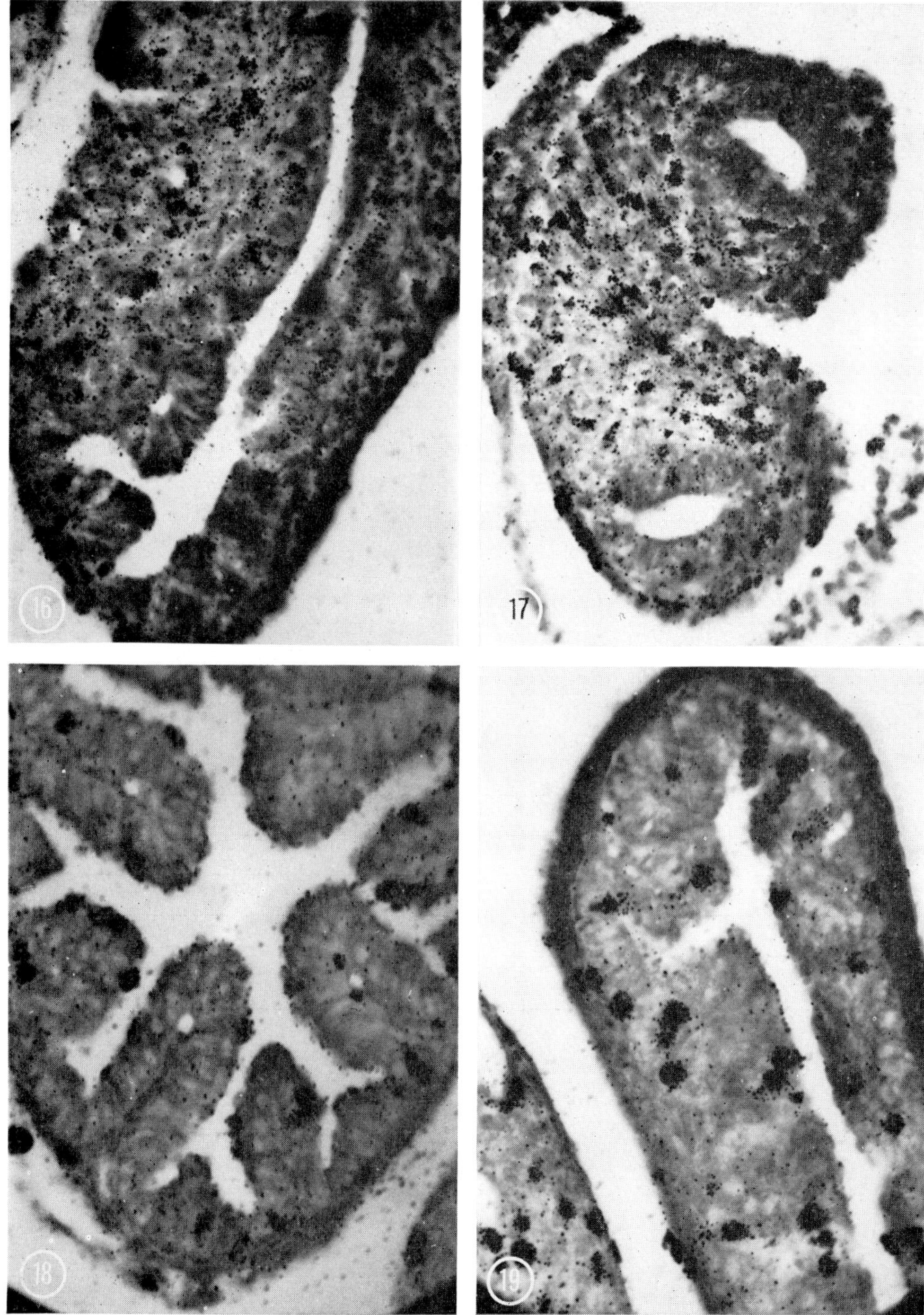
16
17
18
19

functions rather ubiquitously in plasma as a carrier molecule for a variety of substances (Foster 1960). Probably, therefore, in the blood-to-oviduct system, the transferred albumin may be acting as a carrier for other, as yet undetected, molecules, e.g., nucleic acids, hormones, or even other proteins.

Except for Behrman's data, little more is known now about the nature of macromolecules in oviductal secretions than 2 years ago (reviewed, Glass and McClure 1965). Electrophoretic studies of proteins and enzymes in the serum of mice at various stages of the estrous cycle and pregnancy showed that activity of one nonspecific esterase component in the β-globulin region decreased during the first 4 days of pregnancy (Cons and Glass 1963). (Day of mating was defined as day 0.) Immunoelectrophoretic studies demonstrated a component in the fast α-1-globulin region which was present in the serum of nonpregnant mice and in 3 of 6 mice examined during the 1st day of pregnancy; the component was absent from the serum of 21 mice examined at days 2, 3, and 4 of pregnancy (Hanson and Glass, unpublished). This constituent stained with Sudan Black and carried nonspecific esterase, cholinesterase, alkaline phosphatase and aminopeptidase activities; it was inactive for acid phosphatase and cytochrome oxidase. It is tempting to speculate that these components are among those transferred to the nude oviductal ootid during the 1st day of development.

All present data indicate that the transferred serum molecules are macromolecules. In fluorescent antibody tests, hapten-labeled protein was not able to localize antiserum directed

TABLE 3

THYMIDINE INCORPORATION BY MOUSE OVIDUCTAL EPITHELIUM DURING HORMONE INDUCTION OF OVULATION[a]

HORMONE TREATMENT	OVIDUCTAL REGION		
	Amp II	Isth I	Isth II
60 hr post PMS	0.5%	0	0
60 hr post PMS; 6 hr post HCG	0.2	0	0–3%
60 hr post PMS; 11 hr post HCG[b]	0–1	0.05%	0–0.2
10–12 hr after mating[c] (pronuclear stage)	0.2	0[d]	0[d]
48 hr after mating[c] (4-cell stage)	0.4	0[d]	0[d]

[a] Two animals at each stage.
[b] Recently ovulated eggs surrounded by follicle cells were present in Amp III.
[c] Stromal incorporation was increased in all regions of the oviduct.
[d] A few basal cells were labeled, on the order of 0.01—0.1%.

FIGS. 20–23. All photographs are of mouse oviducts from animals injected intraperitoneally with tritiated thymidine. Sites of incorporation are interpreted as sites of DNA synthesis. All magnifications ×287.

(20) Oviductal ampulla from 2-week-old mouse. Peg cells comprise about half of the epithelial cells labeled. Killed 1½ hr after thymidine administration.

(21) Isthmus I from oviduct of 2-week-old mouse. DNA synthesis is not limited to the basal cells. Killed 1½ hr after thymidine administration.

(22) Oviductal ampulla from 4-week-old mouse. Two labeled cells are visible in the epithelium; both are peg cells. Killed 24 hr after thymidine administration.

(23) Isthmus I from 4-week-old mouse. Killed 24 hr after thymidine administration.

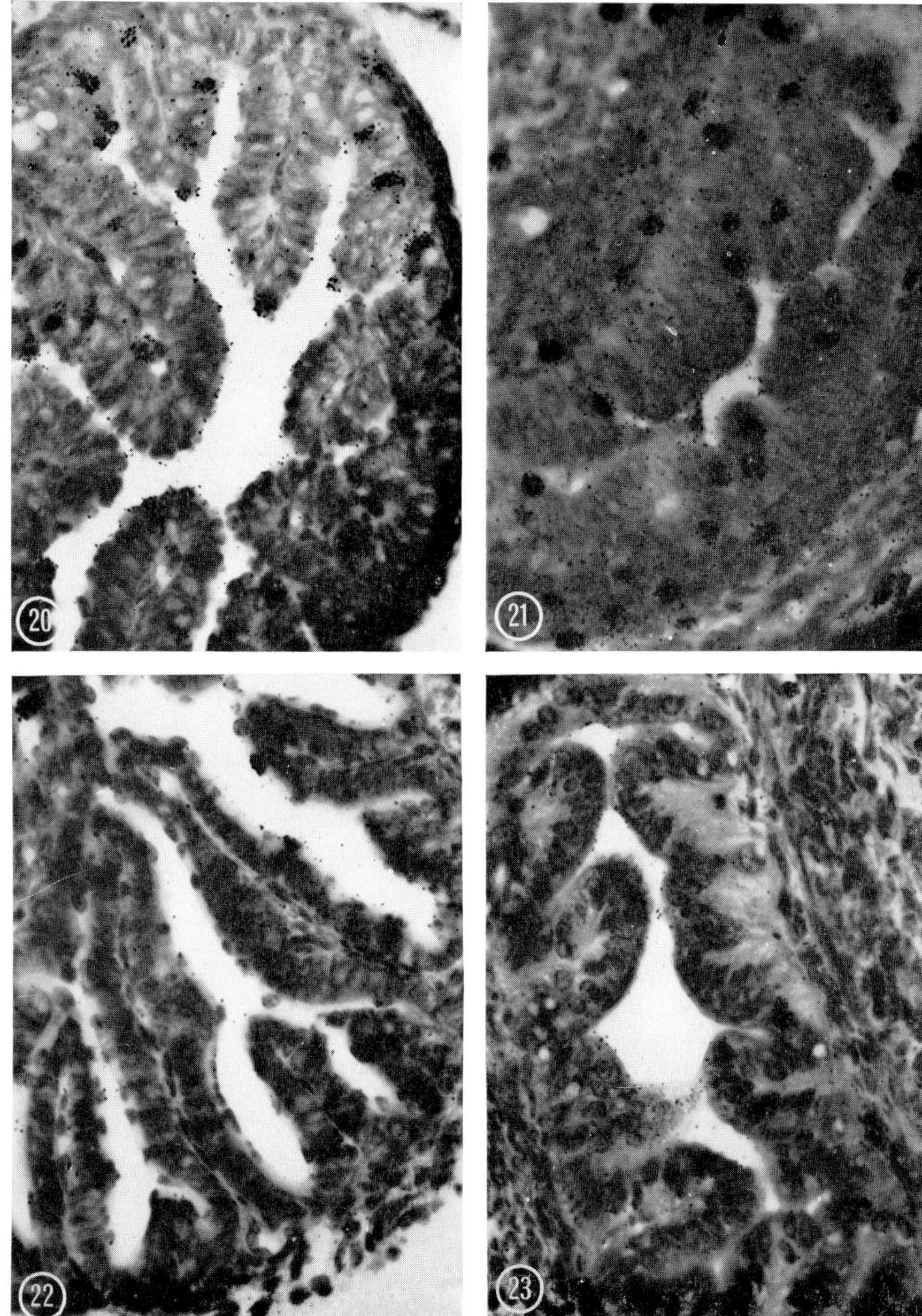
20
21
22
23

against a different, hapten-labeled protein, although it reacted with antiserum directed against the homologous unlabeled protein (Table 4; Glass and Cons 1965). In other experiments, human serum albumin was injected intravenously into female mice. Autoradiographs detected ^{131}I at similar or identical sites in the mouse ovary as fluorescent antibody tests localized the antigenic "label" (Glass and Cons 1968). The simultaneous transfer of both radiolabel and antigenic label markedly increases the probability that no significant degradation of the protein occurred during transfer. Data from the hen (Martin, Vandegaar, and Cook 1957; Schjeide *et al.* 1963) indicate that the heterosynthetic molecules in hen's egg yolk are nearly identical to (not identical to) those in the blood. This may be true also in the mammal, although the methods available are not refined enough to be certain. Data in all species, including mammals, strongly support the working hypothesis that normal heterosynthetic transfer involves macromolecules which are transferred without major degradation and, therefore, which retain their capacity for biological function virtually intact.

TABLE 4

FLUORESCENT ANTIBODY STUDIES WITH PROTEIN-BOUND HAPTEN IN TISSUE VS. ANTIHAPTEN SERA[a, b]

ANTI-SERUM	TISSUE ANTIGEN			
	A	Azo A	G	Azo G
Anti-A......	+	+	−	−
Anti-Azo A..	+	+	−	−
Anti-G......	−	−	+	+
Anti-Azo G..	−	−	+	+

Glass and Cons, unpublished.

[a] A: Bovine plasma albumin; Azo A: Bovine plasma albumin-sulfanilic acid conjugate; G: Bovine gamma globulin; Azo G: Bovine gamma globulin-sulfanilic acid conjugate.

[b] + means fluorescent antibody was localized by antigen; − means antigen did not localize antibody.

Recent fluorescent antibody studies suggest that (some of) the transferred heterosynthetic molecules may play a role in the control of genetic activity during mouse oogenesis. Most strikingly, components migrating in the albumin region in immunoelectrophoresis were demonstrated by fluorescent antibody techniques to be associated with the nucleolus but not with the chromatin at some ovarian follicle stages, with both chromatin and nucleolus at other stages, and with neither chromatin nor nucleolus at still other stages (Glass 1966*b*, 1969). Serum molecules were associated with the pronuclei of oviductal ootids and with delicate chromatin strands in some blastomeres of cleaving embryos (Glass 1963). Fluorescence was particularly bright in the nuclei of some cells of the blastocyst. Clearly, the role of the heterosynthetic molecules observed in the nucleus of oviductal embryos should be investigated more extensively.

Whatever its function, heterosynthetic transfer occurs in many phyla. The phylogenetic persistence of the phenomenon plus the association of the presynthesized maternal macromolecules with the cytoplasm and nucleus of ovarian oocytes and oviductal embryos lend weight to the assumption that heterosynthetic transfer is important for normal development. In the mouse, molecules are transferred from the blood to the oviductal epithelium, oviductal lumen, and oviductal embryo. Transfer is selective as to molecular species, oviductal region, and the age and hormonal status of the animals in which it occurs. It is hoped that future experiments will clarify the functional significance of this fascinating process.

Acknowledgment

New research was supported by United States Atomic Energy Commission Contract AT (11-1)-34 Project Agreement 53.

References

1. Agduhr, E. 1927. Studies on the structure and development of the bursa ovarica and the tuba uterina in the mouse. *Acta Zool.* 8:1.
2. Allen, E. 1922. The oestrous cycle in the mouse. *Am. J. Anat.* 30:297.
3. Brambell, F. W. R., and Hemmings, W. A. 1960. The transmission of antibodies from mother to fetus. In *The placenta and fetal membranes*, ed. C. A. Villee, pp. 71–84, Baltimore: Williams and Wilkins Co.
4. Brambell, F. W. R.; Hemmings, W. A.; and Henderson, M. 1951. *Antibodies and embryos.* London: The Athlone Press.
5. Cons, J. M., and Glass, L. E. 1963. Electrophoresis of serum proteins and selected enzymes in males, non-pregnant, pregnant, and lactating female mice. *Proc. Soc. Exp. Biol.* 113:893.
6. Coons, A. H., and Kaplan, M. H. 1950. Localization of antigen in tissue cells. II. Improvements in a method for the detection of antigen by means of fluorescent antibody. *J. Exp. Med.* 91:1.
7. Espinasse, P. G. 1935. The oviductal epithelium of the mouse. *J. Anat.* 69:363.
8. Flickinger, R., and Rounds, D. E. 1956. The maternal synthesis of egg yolk proteins as demonstrated by isotopic and serological means. *Biochem. Biophys. Acta* 22:38.
9. Foster, J. F. 1960. Plasma albumin. In *The plasma proteins*, ed. Frank W. Putnam, vol. 1, pp. 179–239. New York and London: Academic Press.
10. Glass, L. E. 1959. Immuno-histological localization of serum-like molecules in frog oocytes. *J. Exp. Zool.* 141:257.
11. ———. 1960. Immuno-histological localization of systemically injected foreign protein in the ovary of the mouse. In *Symposium on the germ cells and earliest stages of development.* Pallanza, Italy: Int. Inst. Embryol. and Fondazione A. Baselli.
12. ———. 1961. Localization of autologous and heterologous serum antigens in the mouse ovary. *Develop. Biol.* 3:787.
13. ———. 1963. Transfer of native and foreign serum antigens to oviductal mouse eggs. *Amer. Zool.* 3:135.
14. ———. 1966*a*. Serum antigen transfer in the mouse ovary: Dissimilar localization of bovine albumin and globulin. *Fertil. Steril.* 17:226.
15. ———. 1966*b*. Nucleolar localization of serum-like antigens in mouse ovarian oocytes. *J. Cell Biol.* 31:39A.
16. ———. 1969. Differential nuclear and cytoplasmic localization of serum-like components in mouse ovarian oocytes. *Devel. Biol.* (submitted).
17. Glass, L. E., and Cons, J. M. 1965. Failure of immunohistological techniques to demonstrate hapten in the mouse ovary. *Anat. Rec.* 151:452 (Abstr.).

18. Glass, L. E., and Cons, J. M. 1968. Stage dependent transfer of systemically injected foreign protein antigen and radiolabel into mouse ovarian oocytes. *Anat. Rec.* 162:1–17.
19. Glass, L. E., and McClure, T. R. 1965. Postnatal development of the mouse oviduct: Transfer of serum antigens to the tubal epithelium. In *Preimplantation stages of pregnancy*, ed. G. E. W. Wolstenholme and M. O'Connor. London: J. & A. Churchill Ltd.
20. Martin, W. G.; Vandegaar, J. E.; and Cook, W. H. 1957. Fractionation of livetin and the molecular weights of the α and β components. *Can. J. Biochem. Physiol.* 35:241.
21. Maung, R. T. 1963. Immunity in the tortoise *Testudo ibera*. *J. Path. Bact.* 85:51.
22. Mellors, R. C. 1959. Fluorescent-antibody method. In *Analytical cytology*, ed. R. C. Mellors, 2d ed., pp. 1–67. New York: McGraw-Hill.
23. Odor, D. L. 1960. Electron microscopic studies on ovarian oocytes and unfertilized tubal ova in the rat. *J. Biophys. Biochem. Cytol.* 7:567.
24. Roth, T. F., and Porter, K. R. 1964. Yolk protein uptake in the oocyte of the mosquito *Aedes aegypti* L. *J. Cell Biol.* 20:313.
25. Schechtman, A. M. 1955. Ontogeny of the blood and related antigens and their significance for the theory of differentiation. In *Biological specificity and growth*, ed. E. G. Butler, pp. 3–31. Princeton, N.J.: Princeton University Press.
26. Schjeide, O. A.; Wilkens, M.; McCandless, R. G.; Munn, R.; Peterson, M.; and Carlsen, E. 1963. Liver synthesis, plasma transport, and structural alterations accompanying passage of yolk proteins. *Amer. Zool.* 3:167.
27. Telfer, W. H. 1954. Immunological studies of insect metamorphosis. II. The role of a sex-linked protein in egg formation by the *Cecropia* silkworm. *J. Gen. Physiol.* 37:539.
28. ———. 1961. The route of entry and localization of blood protein in the oocytes of saturniid moths. *J. Biophys. Biochem. Cytol.* 9:747.
29. Telfer, W. H., and Melius, M. E., Jr. 1963. The mechanism of blood protein uptake by insect oocytes. *Amer. Zool.* 3:185.
30. Williams-Ashman, H. G. 1965. New facets of the biochemistry of steroid hormone action. *Cancer Res.* 25:1096.

21

Mechanisms Affecting Embryo Development

A. McLaren

Institute of Animal Genetics
University of Edinburgh, Scotland

The present chapter aims to draw attention to various phenomena concerned with the life of the oviductal embryo which have not yet been sufficiently emphasized in this volume. By "oviductal embryos" is meant not only embryos within the oviduct, but also embryos which would have been within the oviduct had they not been removed.

I. Positive Environmental Factors

The first question to be considered is whether the embryo gains anything, in a positive sense, from the oviduct.

A. *Rate of Cleavage*

Cleavage is perhaps the most characteristic activity of embryos in the oviduct, and the rate of cleavage is one of the few properties that can easily be measured. As an example, Figure 1 shows the rate of cleavage of a randomly bred strain of mice *in vivo* and *in vitro* (Bowman and McLaren, unpublished). *In vivo* the doubling time remains steady at about 11 hr from the first cleavage until shortly before implantation. *In vitro*, the rate becomes progressively slower, with plateaus corresponding to 8- and 16-cells. For the early developmental stages, cell number was ascertained on living material, but after about the 8-cell stage, embryos were squashed and stained with acetocarmine. Cell counts were carried out by the squash technique on samples of embryos from four inbred strains at 3½ days *post coitum*. The strains were found to differ strikingly, both in the mean number of cells at this time and in the proportion of embryos which had reached the blastocyst stage (Fig. 2).

What happens if a fast- and a slow-developing strain are crossed? Whitten and Dagg (1962) reported that the cleavage rate of F_1 hybrid embryos from crosses between strains 129 and BALB/C was significantly different from that of embryos of the maternal strain, suggesting that the paternal genes, brought in by the sperm, were exerting an effect even at this very early stage. In Table 1, part of their data is extracted to show that in BALB/C females, both at $57\frac{1}{2}$ and at 77 hr, the F_1 embryos were consistently further advanced in development than the inbreds. In the reciprocal cross no consistent difference was observed. These experiments need to be confirmed and extended, since they are the sole evidence that genes are active so early in embryonic development.

Reciprocal crosses between the C57BL and C3H inbred strains show a different picture (McLaren and Michie, unpublished data) (Table 2). A slight but significant paternal effect is

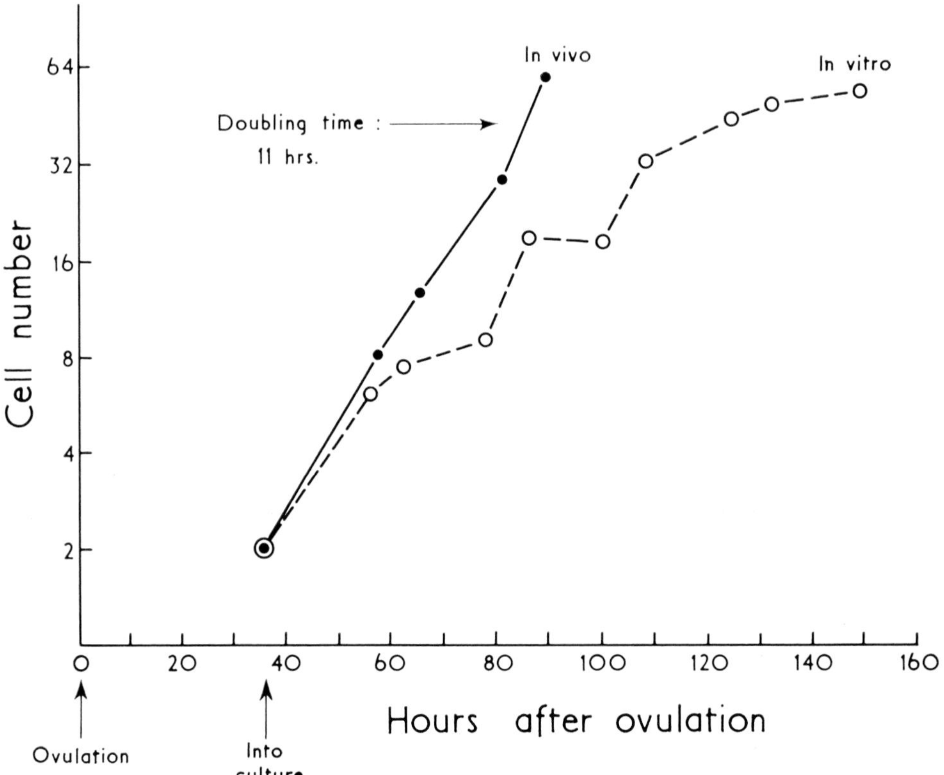

FIG. 1. The rate of cleavage of embryos belonging to the randomly bred *Q* strain, *in vivo* and during *in vitro* culture from the 2-cell stage onward by the technique of Brinster (1963). (Bowman and McLaren, unpublished data.)

again evident in the offspring of C3H females, but the striking feature of the data is a strong maternal effect: inbred and F_1 embryos both develop fast in C57BL and slowly in C3H mothers. This could indicate either an effect of the cytoplasm, or an effect of the oviductal environment. These two possibilities could be distinguished by growing the four classes of embryo in culture, from the 2-cell stage onward. If the reciprocal F_1 hybrids still differed from one another, it would suggest a cytoplasmic effect; but if the two F_1s resembled one another in their rate of development *in vitro*, it would indicate that the oviductal environment is affecting rate of development directly, a conclusion which would be of much greater relevance to the present volume.

TABLE 1

DATA ON RATE OF DEVELOPMENT OF MOUSE EMBRYOS BELONGING TO STRAINS 129 AND BALB/C AND THEIR RECIPROCAL F_1 HYBRIDS

♀ ♂	At $57\frac{1}{2}$ hr			At 77 hr		
	Number of ♀♀	Number of Ova	Proportion of 8-Cell	Number of ♀♀	Number of Ova	Proportion of Blastocysts
129 129.......	12	73	44%	11	76	26%
129 BALB.......	7	64	50%	5	32	16%
BALB BALB.......	8	90	21%	4	36	11%
BALB 129.......	8	94	57%	5	57	47%

SOURCE: Whitten and Dagg 1962.

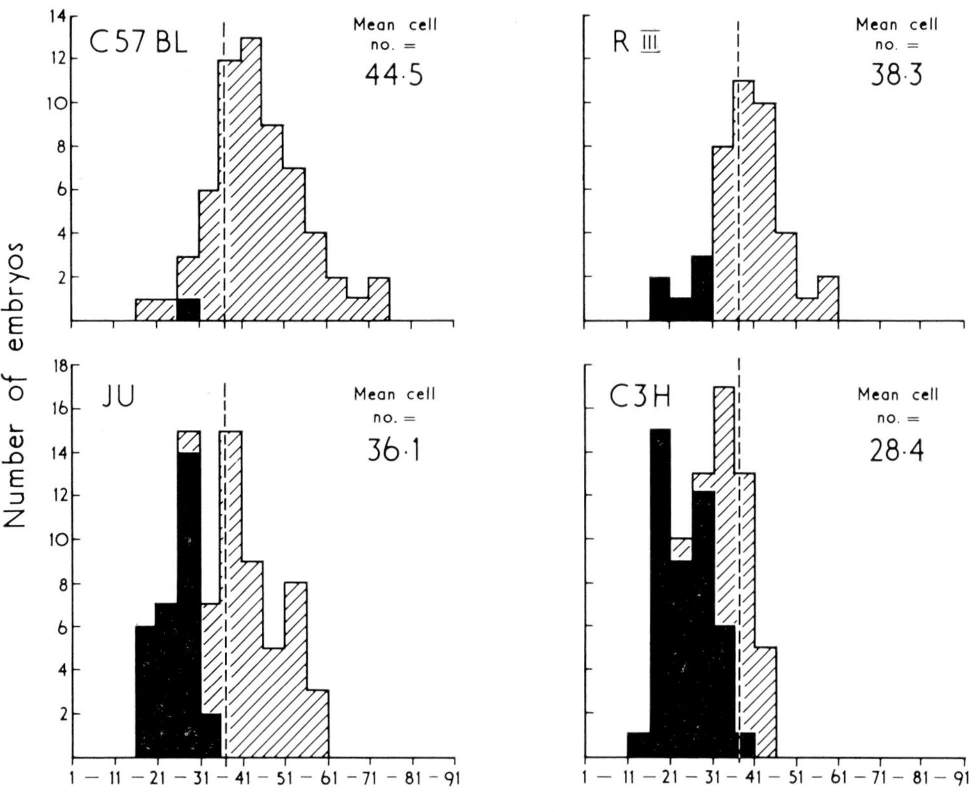

FIG. 2. The distribution of cell number 3½ days *post coitum* in mice of 4 inbred strains. Counts were made from acetocarmine-stained squashes. *Black areas*, morulae; *hatched areas*, blastocysts. The broken lines indicate the overall mean for 4 strains. (Bowman and McLaren, unpublished data.)

B. *Transfer of Macromolecules to Oviductal Embryos*

The likelihood of embryonic development being affected directly by the oviductal environment is increased now that it is known that the zona pellucida can be penetrated by macromolecules, and even by viruses, as well as by lower molecular weight substances. Table 3 shows that 2-cell embryos or morulae of mice cultured in the presence of *Mengovirus* became infected with virus whether the zona was left on or artificially removed (Gwatkin 1963, 1967). Once infected, the embryos were able to support virus multiplication (Gwatkin and Auerbach 1966). This suggests the possibility that viruses and pathogens play a role in the oviduct under natural conditions and may indeed be responsible for some of the naturally occurring mortality found among preimplantation embryos. At an earlier stage of development, Glass (1963) has shown by fluorescence studies that both native and foreign serum antigens are taken up by 1-cell eggs, and can be demonstrated in all subsequent preimplantation stages. These studies are illustrated and more fully discussed by Glass in Chapter 20 of the present volume.

TABLE 2

DATA ON RATE OF DEVELOPMENT AND TIME OF ENTRY INTO THE UTERUS OF MOUSE EMBRYOS BELONGING TO THE C57BL AND C3H STRAINS AND THEIR RECIPROCAL F_1 HYBRIDS $3\frac{1}{2}$ DAYS P.C.

♀	♂	Number of		Blastocysts	In Oviduct
		♀ ♀	Ova		
C57	C57	13	61	84%	5%
C57	C3H	33	205	80%	0%
C3H	C3H	22	147	16%	31%
C3H	C57	31	102	31%	36%

SOURCE: McLaren and Michie, unpublished data.

II. Negative Environmental Factors

Three examples will now be considered of negative environmental factors associated with the oviduct; that is, factors which the oviduct lacks in comparison with the uterus.

Sheep eggs retained in the oviduct by means of a ligature continued to develop at the normal rate for the first 7 days of pregnancy (Wintenberger-Torres 1956). From the 7th to the 9th day, their development slowed down, but as soon as they were transferred to the uterus they resumed their normal rate of development, as judged by autopsy at 11 days. After the 9th day they had been damaged irreversibly and would not develop further even if transferred to the uterus. It seems clear that from the 7th day onward, something is required for the development of the sheep embryo which is lacking in the oviductal environment.

The second example concerns Kirby's experiments on the extrauterine transfer of embryos in mice (Kirby 1962, 1965). Table 4 gives data on the transfer of embryos beneath the kidney capsule; similar results have been obtained with transfers to the testis. When oviductal stages were used, morulae or earlier, abundant outgrowth of trophoblast and extra-embryonic membranes occurred, but no embryonic tissue ever developed. When uterine blastocysts were used, embryonic tissue as well as trophoblast developed in a high proportion of cases. But when the blastocysts were retained in the oviduct by a ligature, they behaved like oviductal stages, not like uterine blastocysts, and again no embryonic development took place. These experiments led Kirby to postulate the existence of a uterine factor, present in the uterus between 3¼ and 4¼ days *post coitum* but absent from the oviduct, and required for normal development of the embryo.[1]

Kirby (1965) has also suggested that there may be hormonal differences between the actual trophoblast that grows out from transplanted oviductal and uterine stages. Blastocysts transplanted to the kidney of ovariectomized recipients caused the X zone of the adrenals to degenerate even if only trophoblast developed from the transplants. Morulae from the oviduct, on the other hand, had no such effect: trophoblast outgrowth occurred, but the X zone was apparently unaffected.

TABLE 3

THE EFFECT OF CULTURING 2-CELL MOUSE EMBRYOS OR MORULAE WITH *Mengovirus*. REMOVAL OF THE ZONA PELLUCIDA WITH PRONASE DID NOT SIGNIFICANTLY INCREASE THE PERCENTAGE OF EMBRYOS INFECTED

	Embryos Infected	
	Zona On	Zona Off
2-cell......	97%	97%
Morula.....	73%	90%

SOURCE: Gwatkin 1963, 1967.

TABLE 4

THE RESULTS OF TRANSPLANTING BENEATH THE KIDNEY CAPSULE MOUSE EMBRYOS AT VARIOUS STAGES OF DEVELOPMENT

Stage	Number of Transfers	Number of Implants	With Embryos	Trophoblast Only
1–4-cell...........	26	20	0	20
Morulae...........	38	26	0	26
Blastocysts........	166	121	38	83
Tube-locked blastocysts............	32	20	0	20

SOURCE: Kirby 1962, 1965.

The last example of the inadequacies of the oviductal environment concerns the loss of the zona pellucida in mice. Studies on the time of loss of the zona pellucida from blastocysts in normal pregnancies and under various experimental conditions (Orsini and McLaren 1967; McLaren 1967) showed that blastocysts retained in the oviduct by a ligature kept the

[1] This interpretation is not supported, however, by recent experiments in which oviductal mouse embryos (2- or 8-cell) were grown to blastocyst stage *in vitro*. After transfer to the kidneys of host mice, the percentage which gave rise to embryonic structures did not differ significantly from that observed using uterine blastocysts (McLaren, unpublished).

zona for at least 12 hr longer than they would have done in the uterus (Table 5). In a normal pregnancy, in which the blastocysts are in a receptive, estrogen-sensitized uterus, the majority have lost the zona by the beginning of the 5th day; but in the oviduct, the majority of the blastocysts are still in intact zonae 12 hr later, just as they are in the non-estrogen-sensitized uterus of ovariectomized or lactating females. Moreover, in a normal pregnancy the zonae disappear within 1 or 2 hr of being shed, but under these various abnormal conditions the zonae persist. This suggests the intervention of a *lytic* factor in normal pregnancy, acting on the zona to weaken it and hence to expedite the hatching of the blastocyst, and leading subsequently to the total dissolution of the zona. This lytic factor is produced by the uterus and not by the blastocyst itself, but it is not yet known whether an actual lytic enzyme is involved, or some other factor such as a drastic localized decline in uterine pH.

TABLE 5

THE EFFECT OF LIGATION OF THE OVIDUCT, OVARIECTOMY ON THE 2ND OR 3RD DAY OF PREGNANCY, OR CONCURRENT LACTATION ON THE TIME OF LOSS OF THE ZONA PELLUCIDA FROM MOUSE BLASTOCYSTS

GROUP	TIME OF AUTOPSY	NUMBER OF ♀♀	BLASTOCYSTS	
			In Zona	Free
Control..........	3d, 22h 4d, 2h	15	13	34
Tube-locked......	4d, 8h 4d, 17h	9	22	16
Ovariectomized...	"	7	39	2
Suckling..........	"	13	86	67

SOURCE: Orsini and McLaren 1967; McLaren 1967.

The conclusion from all three sets of experiments summarized in this section is that the oviduct is a splendid place for the embryo to be, provided it does not stay there too long. It is not yet known whether the mouse uterine lysin, Kirby's "uterine factor," and Wintenberger-Torres' effect in sheep involve three separate factors, or whether possibly a single shift in environmental conditions between the oviduct and the uterus might underlie all three.

III. Experimental Interference

A. *Irradiation*

Very great variations in radiosensitivity exist during the early cleavage stages in the mouse (Russell 1965). Damaged embryos do not die immediately, but continue to develop until implantation, or shortly after implantation. Survivors show few abnormalities, suggesting that the early stages possess considerable powers of regulation, and only succumb when more than a threshold proportion of cells have been destroyed.

B. *Injection of Eggs*

Injection of substances into the eggs of mice has been achieved by Lin (1966). He injected bovine gamma globulin into 1-cell eggs with a micropipette (Fig. 3) and showed that the embryos continued to develop normally. Within the next few years, some exciting experiments are likely to be performed using this technique.

C. *Reduction of Blastomere Number*

If one blastomere of a 2-cell embryo is destroyed, the other may continue developing to produce a normal individual. Seidel (1952) succeeded in bringing to term two rabbits derived from single blastomeres. In mice, Tarkowski (1959*a*, *b;* 1965) has carried out extensive experiments along these lines. Single blastomeres from the 2-cell stage form normal-sized blastocysts, but the inner cell mass tends to be only about half the normal size. Embryonic size continues to be approximately halved until about the 10th day of gestation, but by full-term, regulation is complete and birth weight is normal. Sometimes the inner cell mass becomes very small indeed, or fails completely to develop, so that the remaining blastomere

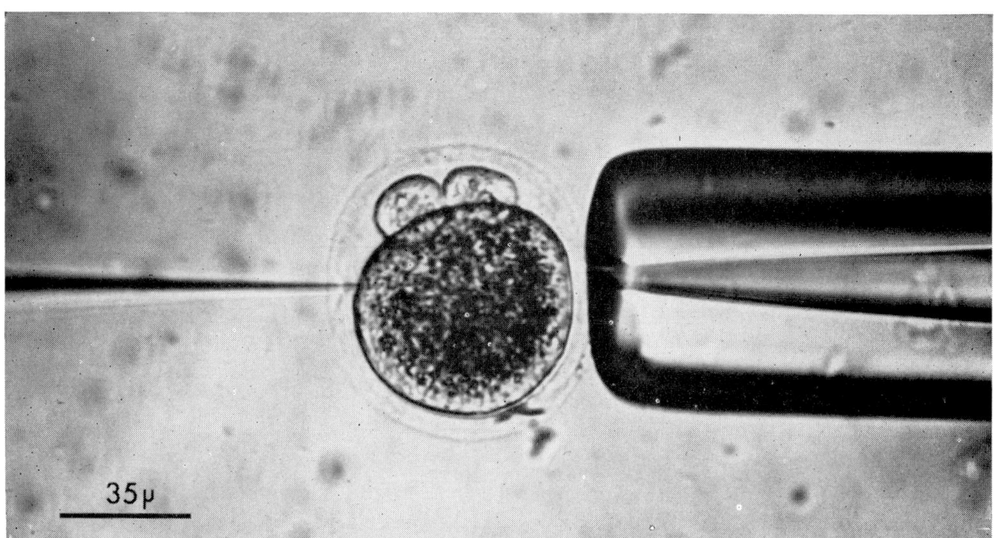

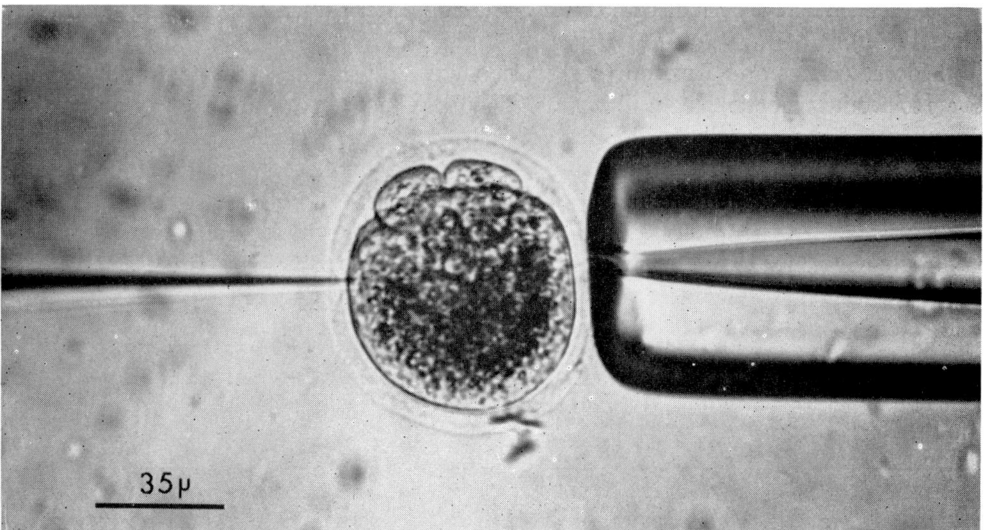

FIG. 3. Microinjection of bovine gamma globulin into mouse eggs. *Top*, the micropipette enters the egg, which is steadied by suction. *Bottom*, the vitellus is distended by over-injection. (From Lin 1966.)

forms a so-called trophoblastic vesicle, devoid of inner cell mass and incapable of further development.

In order to see whether there was segregation of potential trophoblastic and inner cell mass material between the first two blastomeres, Mulnard (1965) separated the blastomeres at the 2-cell stage in mice, and obtained 16 cases in which both partners continued to develop in culture. If segregation had already occurred, the two partners should have developed in a complementary fashion, so that if inner cell mass material was strongly represented in one, it should have been weak or absent in the other. But no complementarity could be detected: in some pairs both members showed some inner cell mass material, while in others both developed into trophoblastic vesicles (Table 6). In fact the data (4 pairs with both embryos showing inner cell mass material, 4 pairs with neither showing such material, and 8 pairs consisting of one of each type) are exactly as would be expected if the probability of developing inner cell mass material was equal for each blastomere. The embryological implications of these experiments have been discussed (see discussions after Mulnard 1965; Tarkowski 1965; Mintz 1965*b*).

TABLE 6

PRESENCE (+) OR ABSENCE (−) OF "POSITIVE" CELLS (I.E., THOSE SHOWING THE PRESENCE OF INNER CELL MASS MATERIAL) IN PAIRED EMBRYOS DEVELOPING FROM SEPARATED BLASTOMERES

Embryo 1	Embryo 2	Number of Pairs
+	+	4
+	−	8
−	−	4

SOURCE: Mulnard 1965.

Moore, Adams and Rowson (personal communication) have been trying by similar means to produce identical twins in rabbits. They have had rabbits born from a single blastomere surviving at the 2-cell, 4-cell, or even 8-cell stage (Fig. 4; Table 7), so it seems that the embryo of the rabbit can regulate up to an even later stage of development than that of the mouse. Unfortunately their attempts to induce development of both blastomeres, separated at the 2-cell stage, were unsuccessful because the rabbit embryo, again unlike the mouse, is apparently unable to develop without a zona pellucida. The somewhat heroic procedure of transferring one blastomere to a previously evacuated zona was also unsuccessful.

D. *Augmentation of Blastomere Number*

The final section of this chapter considers augmentation of blastomere number by fusing two embryos to form a single chimeric individual. The first report of mice born as a result of

---

FIG. 4. (*a*) On the left, a normal 4-cell embryo of rabbit: on the right, similar embryos with 3 out of 4 blastomeres destroyed by piercing with a pipette. Much of the blastomeric debris was extruded from the embryo when the pipette was withdrawn.

(*b*) Young rabbits born from single blastomeres remaining at the 8-cell stage, after the other 7 cells had been destroyed. (*a* and *b* from Moore, Adams, and Rowson, unpublished.)

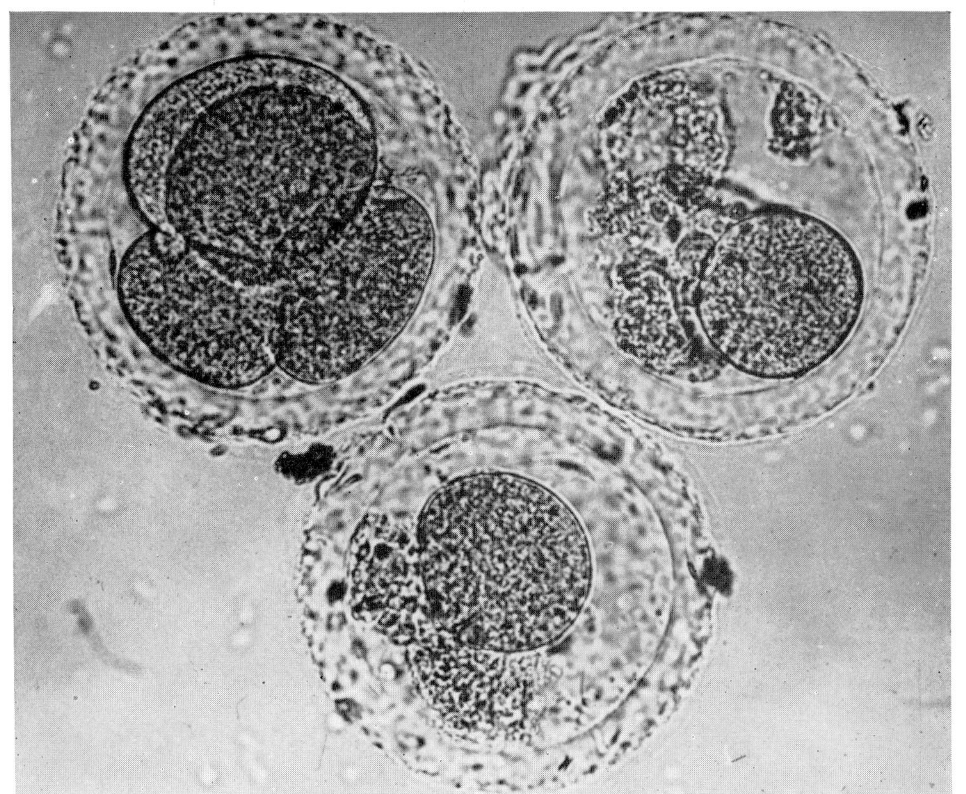

A

B

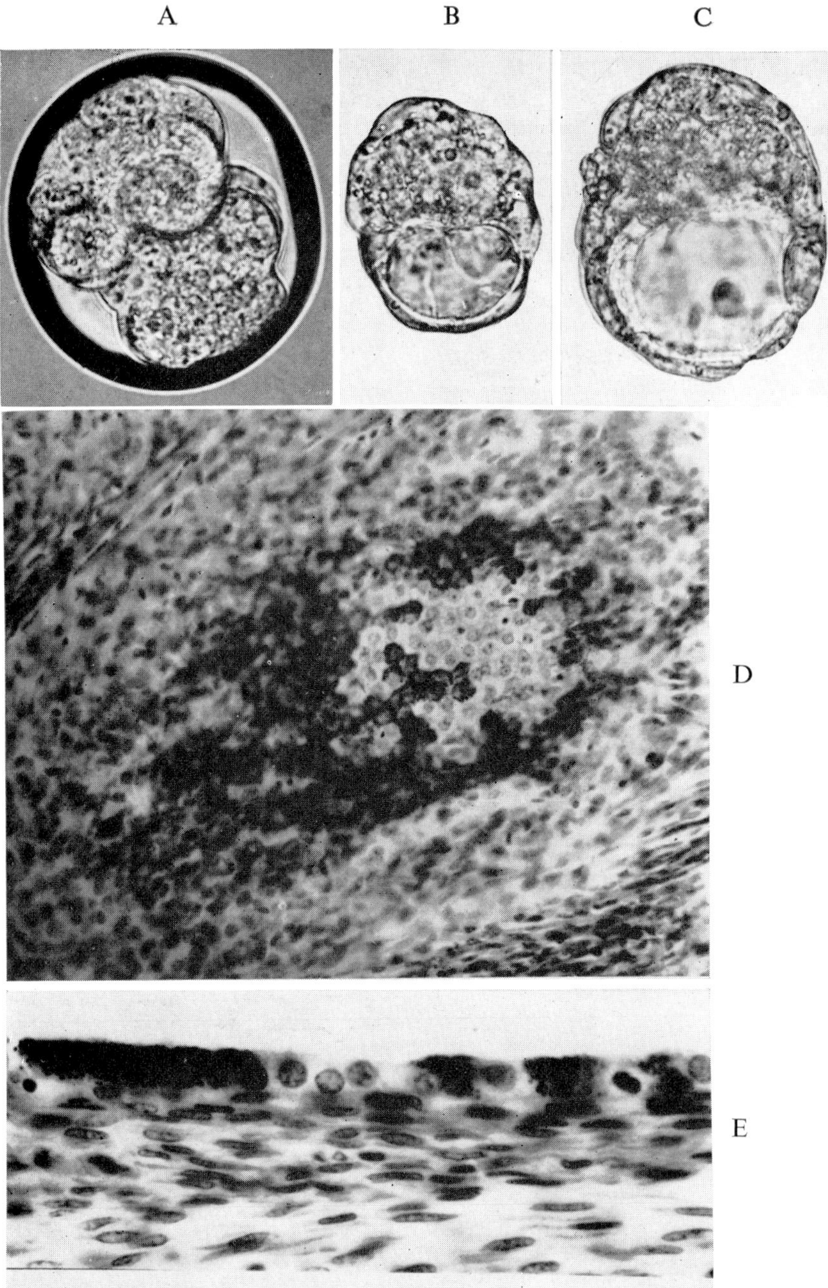

FIG. 5. (*a*) Two 8-cell mouse embryos compressed in a small drop of medium suspended in liquid paraffin. (*b*) Single naked mouse blastocyst developed from 8-cell stage cultured for 24 hr. (*c*) Double blastocyst developed from two fused 8-cell mouse embryos after 24 hr in culture. (*a*, *b*, and *c* from Tarkowski 1961.) (*d*) Tangential section of the outer layer of retina of a newborn chimeric mouse, developed from LAB Grey and (LAB Grey × $A_2G)F_1$ embryos fused. Pigmented and nonpigmented areas are intermingled. (From Tarkowski 1963.) (*e*) Perpendicular section through the mosaic outer layer of retina of a chimeric mouse, as (*d*). While the cells of the hybrid genotype are already overloaded with pigment granules, no pigment is detectable in the other cells. (From Tarkowski 1964*a*.)

this ingenious and unlikely technique was from Tarkowski (1961) (Fig. 5). The following year Mintz published an abstract (Mintz 1962), showing that she had independently been working along very similar lines, and in 1964 she described results of various studies in which chimeras were allowed to develop to the blastocyst stage, fusing lethal with normal and radioactively labeled with unlabeled embryos (Mintz 1964, 1965*a*). Her technique is more convenient than Tarkowski's in at least two respects, namely, enzymatic rather than mechanical removal of the zona, and aggregation of blastomeres at 37° C rather than at room temperature. By now she and her colleagues must have raised several hundred adult mice from fused embryos.

The technique of embryo fusion has been slow to spread to species other than the mouse, but recently J. L. Hancock and Eleanor Pighills (personal communication) have attempted to fuse sheep embryos. Because the zona was resistant to pronase treatment, it had

TABLE 7

SURVIVAL OF SINGLE BLASTOMERES OF RABBIT EMBRYOS, AFTER DESTRUCTION OF ALL OTHER BLASTOMERES IN ZONA AT VARIOUS CELL STAGES

Cell Stage	Number of Embryos		Per Cent Survival
	Transferred	Survived to 10 Days	
2	70	26	37
4	72	19	26
8	72	11	15

SOURCE: Moore, Adams, and Rowson, unpublished.

to be removed mechanically, and total fusion between pairs of embryos proved hard to obtain. After transfer to pseudopregnant ewes, several embryos in the experimental series developed to term, but only in a single case, where the lamb died late in gestation, were the transferrin types of the supposed chimera and its four parents such as to suggest that chimerism had in fact been obtained (Table 8).

The two major fields of study which have been opened up by Tarkowski's and Mintz's experiments are, first, the occurrence of hermaphrodites and the disturbed sex ratios among the chimeras and, second, the distribution of tissue derived from the two chimera components.

Tarkowski's original series contained 3 hermaphrodites out of a total of 14 presumed chimeras (Tarkowski 1961, 1964*b*). The remaining mice comprised 9 males and only 2 females. Mintz (1965*b*) confirmed the rare occurrence of hermaphrodites, and also the preponderance of males. On a priori grounds, one might expect that 50% of all embryo fusions would be between male and female embryos, and would therefore show some evidence of hermaphroditism in their later development. That the incidence of hermaphrodites is so much below 50% could be explained in at least four ways: (1) All the cells derived from one component might have been concentrated in the trophoblast, so that the embryo proper would be nonchimeric. In this case not only would no hermaphroditism develop, but genetic and chromosomal analysis would yield no evidence of chimerism. (2) Canalization of sexual development might be so extreme that a very small overall majority of male or female cells in the body might suffice to switch sexual development into the male or female channel respectively. (3) The sex of an individual might be determined by the sex of a relatively small number of cells (e.g. in the primitive gonad) at a particular stage of embryonic development, small enough to have a good chance of all stemming from either one component or the other. (4) The majority of potentially hermaphroditic chimeras might develop in a phenotypically male direction. This, which was Tarkowski's original hypothesis, would account not only for the scarcity of hermaphrodites, but also for the predominantly male sex ratio.

On the spatial distribution of the genetically differing cell types within the chimeric individual, we so far have two pieces of information only. First, Tarkowski (1961, 1963, 1964*a*) has looked at the retinas of chimeras between agouti and pink-eyed strains. The chimerism is astonishingly fine-grained, with patches of pigmented and nonpigmented cells containing not

TABLE 8

PARENTAGE OF 3 LAMBS BORN TO EWES RECEIVING FUSED EGGS

Recipient Ewe	Eggs Transferred after Fusion	Lambs Born
Blackface (BD)	Blackface (AC×AB) Blackface (AC×CE)	Blackface ♂ (AC)
Blackface (CD)	Blackface (AC×CE) Blackface (AC×AB)	Blackface ♂ (BCE)
Blackface (CD)	Blackface (AC×DD) Welsh (BC×BB or BC)	Blackface ♀ (AD)

SOURCE: Hancock and Pighills, unpublished.

NOTE: Transferrin types are given in parenthesis; for the transferred eggs, the mother's and father's types are given for both of the fused partners. Normally, an individual is characterized by 2 transferrin types only; the supposed chimeric lamb showed 3.

more than 100 or so cells in each (Fig. 5*d*,*e*). This must imply an enormous amount of cell mixing in the course of ontogeny. A very similar mosaic appearance of the retina is seen in the mottled mice which arises when an autosomal segment of chromosome heterozygous for a color marker is translocated onto the X chromosome, and hence, according to the Lyon hypothesis, becomes involved in X-chromosome inactivation (Cattanach, personal communication).

Second, Mintz (1965*a*, *b*) reported that chimeric individuals derived from components differing for coat color markers often showed a variegated coat color, with characteristic striped markings. How does this pattern come about? One possibility is that it reflects major embryological cell movements, but this seems unlikely in view of the fine-grained nature of the retinal chimerism. The problem is not unlike that which Grüneberg (1966) has discussed in relation to the ordered pattern shown by heterozygotes for coat color markers located on the X chromosome. Grüneberg is puzzled by how *random* inactivation of one X chromosome per cell (the Lyon or "inactive-X" hypothesis) could bring about an *ordered* coat color pattern, and suggests that the transverse striping may have its developmental basis in the transverse skin wrinkling shown by mouse fetuses, or the transverse skin folds which arise postnatally. Grüneberg goes further, and argues that the inactive-X hypothesis is untenable. It seems likely that, both in experimental chimeras and in X-chromosome heterozygotes, what one is dealing with is an underlying fine-grained mosaicism of the type demonstrated in the retina of both types of mice; and that this enables a genetically determined "prepattern" to be manifested through minor, secondary shifts in cell distribution or gene expression. In other words, the existence of more than one genetically distinct cell population, whether due to embryo fusion or to X-chromosome inactivation, is a *necessary* condition for the manifestation of the pattern, but it may not, by itself, be a sufficient condition.

IV. Conclusions

In the course of looking at several thousand oviductal embryos of the mouse, I have not seen one which had spontaneously lost the zona pellucida. It therefore seems unlikely that, in this species at least, embryonic fusion could take place *in vivo*. On the other hand hormonal influences and transfer of macromolecules almost certainly play a part in normal development, and infection by pathogens at this early stage may be more common than is realized. As can be judged by the very incomplete and sketchy nature of the observations described in the present chapter, this field is still largely unexplored.

Acknowledgments

I am grateful to the Wellcome Foundation for a travel grant enabling me to attend this symposium; to Drs. N. W. Moore, C. E. Adams, and L. E. A. Rowson of the Agricultural Research Council Unit of Reproductive Physiology and Biochemistry, Cambridge, for allowing me to describe their unpublished work and to reproduce Table 7 and Figure 4; to Dr. J. L. Hancock and Miss Eleanor Pighills for allowing me to describe their unpublished work and to reproduce Table 8; to Dr. A. K. Tarkowski for allowing me to reproduce Figure 5; to Dr. T. P. Lin for Figure 3; to the Lalor Foundation for financial support of the work on rate of development being carried out by Dr. Patricia Bowman and myself; and to the Ford Foundation for financial support.

References

1. Brinster, R. L. 1963. A method for *in vitro* cultivation of mouse ova from two-cell to blastocyst. *Exp. Cell Res.* 32:205.
2. Glass, L. E. 1963. Transfer of native and foreign serum antigens to oviducal mouse eggs. *Amer. Zool.* 3:135.
3. Grüneberg, H. 1966. More about the tabby mouse and about the Lyon hypothesis. *J. Embryol. Exp. Morph.* 16:569.
4. Gwatkin, R. B. L. 1963. Effect of viruses on early mammalian development. I. Action of Mengo encephalitis virus on mouse ova *in vitro*. *Proc. Nat. Acad. Sci. U.S.A.* 50:576.
5. ———. 1967. Passage of Mengovirus through the zona pellucida of the mouse morula. *J. Reprod. Fertil.* 13:577.
6. Gwatkin, R. B. L., and Auerbach, S. 1966. Synthesis of a ribonucleic acid virus by the mammalian ovum. *Nature* (*Lond.*) 209:993.

7. Kirby, D. R. S. 1962. The influence of the uterine environment on the development of mouse eggs. *J. Embryol. Exp. Morph.* 10:496.
8. ———. 1965. The role of the uterus in the early stages of mouse development. In *Preimplantation stages of pregnancy*, ed. G. E. W. Wolstenholme and M. O'Connor. London: J. & A. Churchill.
9. Lin, T. P. 1966. Microinjection of mouse eggs. *Science* 151:333.
10. McLaren, A. 1967. Delayed loss of the zona pellucida from blastocysts of suckling mice. *J. Reprod. Fertil.* 14 159.
11. McLaren, A. 1969. Recent studies on developmental regulation in Vertebrates. In *Handbook of molecular cytology*, ed. A. Lima-de-Faria. North-Holland Publishing Co.
12. Mintz, B. 1962. Formation of genotypically mosaic mouse embryos. *Amer. Zool.* 2, Abstr. 310.
13. ———. 1964. Formation of genetically mosaic mouse embryos, and early development of "lethal (t^{12}/t^{12})–normal" mosaics. *J. Exp. Zool.* 157:273.
14. ———. 1965*a*. Genetic mosaicism in adult mice of quadriparental lineage. *Science* 148: 1232.
15. ———. 1965*b*. Experimental genetic mosaicism in the mouse. In *Preimplantation stages of pregnancy*, ed. G. E. W. Wolstenholme and M. O'Connor. London: J. & A. Churchill.
16. Mulnard, J. G. 1965. Studies of regulation of mouse ova *in vitro*. In *Preimplantation stages of pregnancy*, ed. G. E. W. Wolstenholme and M. O'Connor. London: J. & A. Churchill.
17. Orsini, M. W., and McLaren, A. 1967. Loss of the zona pellucida in mice, and the effect of tubal ligation and ovariectomy. *J. Reprod. Fertil.* 13:485.
18. Russell, L. B. 1965. Death and chromosome damage from irradiation of preimplantation stage. In *Preimplantation stages of pregnancy*, ed. G. E. W. Wolstenholme and M. O'Connor. London: J. & A. Churchill.
19. Seidel, F. 1952. Die Entwicklungspotenzen einer isolierten Blastomere des Zweizellenstadiums im Säugetierei. *Naturwissenschaften* 39:355.
20. Tarkowski, A. K. 1959*a*. Experiments on the development of isolated blastomeres of mouse eggs. *Nature* (*Lond.*) 184:1286.
21. ———. 1959*b*. Experimental studies on regulation in the development of isolated blastomeres of mouse eggs. *Acta Theriol.* 3:191.
22. ———. 1961. Mouse chimaeras developed from fused eggs. *Nature* (*Lond.*) 190:857.
23. ———. 1963. Studies on mouse chimeras developed from eggs fused *in vitro*. *Nat. Cancer Inst. Monogr.* No. 11:51.
24. ———. 1964*a*. Patterns of pigmentation in experimentally produced mouse chimaerae. *J. Embryol. Exp. Morph.* 12:575.
25. ———. 1964*b*. True hermaphroditism in chimaeric mice. *J. Embryol. Exp. Morph.* 12: 735.
26. ———. 1965. Embryonic and postnatal development of mouse chimeras. In *Preimplantation stages of pregnancy*, ed. G. E. W. Wolstenholme and M. O'Connor. London: J. & A. Churchill.
27. Whitten, W. K., and Dagg, C. P. 1962. Influence of spermatozoa on the cleavage rate of mouse eggs. *J. Exp. Zool.* 148:173.
28. Wintenberger-Torres, S. 1956. Les rapports entre l'oeuf en segmentation et le tractus maternel chez la brebis. *Proc. Third Int. Congr. Anim. Repr., Cambridge* 1:62.

Part V Clinical Aspects

22

The Clinical Evaluation of Oviductal Function

W. L. Herrmann/L. R. Spadoni/D. C. Smith
Department of Obstetrics and Gynecology
School of Medicine
University of Washington, Seattle

The clinical concept of oviductal function is limited. Evaluation is directed toward patency and gross structural integrity of the oviduct and its immediate surroundings. While there is, in general, awareness of the fact that microanatomy, motility, and biochemistry must play a role of considerable importance, a search for tools to be applied clinically must await a better understanding of the basic physiology of the Fallopian tube. In the absence of knowledge of the normal, diagnosis of the abnormal is, at best, a thing of the future.

Indeed, one must find it difficult to remain scientific or even overly serious when discussing the history and current status of the procedures directed toward evaluating the function of the human Fallopian tube. In fact, there are a great number of clinical observations which can be explained only if speculative imagination is allowed to run free. The following are examples. Buxton and Southam reported 151 patients with evidence for closed oviducts (oviductal insufflation or salpingography); 28 of these conceived without further therapy, a figure not too dissimilar from the one found when evaluating the results of reconstructive surgery of the oviducts (Buxton and Southam 1958). Pregnancy is also expected to occur in 0.5 to 0.7% of all patients subjected to oviductal ligation for sterilizing purposes after crushing, cutting, and tying of the delicate tissue (Garb 1957; Thomas 1953). The proposed explanation is "recanalization." And, finally, pregnancy has been reported by a number of authors after direct implantation of ovarian tissue into the uterine cavity, in the absence of oviductal epithelium, except for remnants of the interstitial portion (Estes and Heitmeyer 1934; Preston 1953; Sturma 1959; Ghosal 1966). While the incidence of pregnancies is low—2.5%, Reiprich (Reiprich 1933); 8%, Estes and Heitmeyer; 14%, Preston—and the percentage of abortions quite high following the procedure, there are numerous records of normal, full-term deliveries resulting. In the face of such observations, one must question the necessity of an active biological or biochemical contribution by the oviduct to the process of fertilization other than that of gamete transport.

Regardless of the above, testing oviductal patency (and, in a limited sense, the function) is a required and recommended step to be undertaken as part of an infertility workup.

I. Indication for Testing for Oviductal Patency

Any medical intervention requiring the manipulation of the patient must be looked at in terms of the involved risks and expected usefulness. This seems particularly true when the intervention has become part of a routine, as is the case for an oviductal evaluation in an infertile patient.

Usefulness must be defined in consideration of the percentage of positive findings, and not merely yield reassurance of the patent oviducts, as part of a time factor, which might contribute to the occurrence of a pregnancy for whatever reasons. Risks and disadvantages (other than cost and discomfort) include the following.

A. *Erroneous Diagnosis of Nonpatent Oviducts*

Both of the more common procedures, oviductal insufflation and salpingography, may yield misleading evidence as the result of faulty technique or perhaps uncontrollable spasm at the uterotubal junction.

B. *Erroneous Diagnosis of Adequate Oviductal Function*

This actually represents the key problem. Patency does not necessarily reflect function. For instance, in the presence of a patent oviduct, peritubal pathology (adhesions, endometriosis), pathological changes of the oviductal epithelium, etc., are easily overlooked.

C. *Inflammatory Reactions Directly Attributable to the Procedure*

These may occur either through flareup of a chronic process, spread of cervical flora, or as a reaction to the agent used (contrast material employed during hysterosalpingography).

D. *Embolization of the Injected Medium*

Immediate and delayed reactions to this complication of variable severity (including death) have been observed and reported with both gaseous (Buxton 1957) and liquid media (aqueous and oil suspensions). Levinson recently reviewed 25 reported nonfatal cases and 1 fatal case of pulmonary embolism resulting from injection of an oily medium, and stressed the ever-present danger of this entity (Levinson 1963).

E. *Vagal Reaction (Cardiovascular Collapse) to Cervical Manipulation*

Calculation or estimation of the true incidence of these and other undesirable results remains a futile undertaking. They are not always recognized. Clinical manifestations associated with one or the other may be ill defined, therefore making a correct diagnosis difficult. However, the fact that at least the first two are of common occurrence should caution the clinician not only at the time of interpretation of the tests, but perhaps even when planning his workup.

In our opinion, a logical approach constitutes limitation of these procedures to patients with positive history (infection, surgical trauma, endometriosis, etc., or as a final step of the investigation after other abnormalities not requiring traumatic intervention have been corrected or ruled out.

A final comment and analogy may be in order. When discussing the problems surrounding clinical and laboratory procedures held valuable in establishing the fact that ovulation has taken place, it is often felt that pregnancy provides the only undisputable proof. At the present moment, this also holds true for oviductal function.

II. Timing of Oviductal Patency Tests

All tests for oviductal patency should be performed in the time period between the cessation of the menstrual flow and ovulation. This will prevent interference with a chance pregnancy; decrease the possibility of embolism due to extravasation of injected media; and, finally, diminish the possibility of misinterpretation of nonpatency as the result of the thick secretory endometrium which surrounds the uterine opening of the Fallopian tubes.

III. Diagnostic Procedures

A. *Oviductal Insufflation*

Of the commonly used tests for evaluating oviductal function, oviductal insufflation (Rubin's test) is the easiest to perform, is most subject to technical errors, and is most difficult to interpret. Most physicians today have not attained Rubin's degree of accuracy and sophistication in interpreting the data obtained by oviductal insufflation. Therefore the usefulness of this test is limited, and other techniques such as hysterosalpingography and direct visualization are being employed for definitive diagnosis (Weir 1962).

The basic principle is relatively naïve. Air or carbon dioxide, when delivered under pressure through the cervical canal, will normally escape through the fimbriated end of the oviduct, thus proving patency—at least for a gaseous medium. A variety of gadgets have been designed to facilitate this procedure, and perhaps lend it greater accuracy. They involve such devices as balloons, CO_2-discharging Alka-Seltzer tablets, safety valve equipped pressure cartridges, and sophisticated recorders equipped with flowmeters, lights, etc. The concept of what constitutes proof of normality varies with the imagination of different authors from the pitch of the sound of gas passing through or leaving the fimbriated end (as determined by specially designed stethoscopes), to palpation of crepitating masses (aerosalpinx) following the procedure, to the demonstration of gas collection under the diaphragm by fluoroscopy, to shoulder pain, or finally to a pressure tracing. Under optimal circumstances the latter will show an initial rise in intrauterine tension, followed by a drop-off to a maintenance pressure level which, in view of the continuing gas flow, indicates that the system has a leak—hopefully at the fimbriated end. Fluctuations in the recorded pressure of the system have been interpreted as representing oviductal peristalsis. This, however, appears to represent a misconception, as uterine irritability and contractility offer a more likely explanation.

As indicated before, misinterpretation is common, with spasm in the area of the isthmic portion of the oviduct probably representing the major factor. The use of various sedatives, tranquilizers, amylnitrate, local nerve blocks, etc., does not seem to improve this outlook significantly.

Obviously, bilateral assessment of oviductal patency is not possible. No information is gained concerning oviductal motility or the possibility of egg transport, and finally, the peritubal space cannot be assessed adequately.

With these considerations, we doubt that this procedure constitutes even an adequate screening maneuver. As an example of what can be expected, in 1965 Sweeney and Gepfert reported on a series of 453 patients who, having undergone oviductal insufflation at the

Women's Clinic of the New York Hospital, were followed for at least 2 years after insufflation (Sweeney and Gepfert 1965). They found that, in 84 patients who had no demonstrable oviductal patency on the basis of insufflation, 42 or 50% subsequently became pregnant. A similar incidence of false positive results conceivably might occur also.

1. *Technique.* The technique of oviductal insufflation is simple and requires a minimum of equipment. This includes a source of carbon dioxide, a flowmeter, a cannula, and a recording device for making a permanent record of the results.

Carbon dioxide is preferred over oxygen or air because the latter have been associated with fatal embolism. In addition to its safety, the possible discomfort, due to the accumulation of gas in the peritoneal cavity following the procedure, is of much shorter duration, as carbon dioxide is more rapidly absorbed than air or oxygen.

First, the cervix is grasped with a tenaculum and cleansed with an antiseptic solution. Then carbon dioxide is allowed to flow through the system, including the tubing and cannula, to displace air, in order to diminish the danger of introducing air into the uterine cavity. The cannula tip is placed carefully into the external cervical os, and, after a tight fit is accomplished, carbon dioxide is allowed to flow through the system at a rate of 30 to 60 cc per min. As the gas flows through the system, the pressure of the carbon dioxide in the intrauterine cavity rises until it overcomes the resistance in the Fallopian tubes, and the gas will then pass through the oviducts and out the fimbriated ends. The pressure will drop at the instant the carbon dioxide flows through the oviducts. If the oviducts are obstructed, the pressure will continue to rise, and, in most instances, the procedure should be discontinued when the pressure reaches a level of 200 mm Hg.

Although the above described pressure changes are indicative of patency or lack of patency of the Fallopian tubes, other evidence should be sought. Auscultation of the lower abdomen to determine whether the carbon dioxide can be heard passing through the fimbriated end often can be of help. More definitive evidence, however, would be the shoulder pain experienced by the patient on sitting up following the procedure, resulting from accumulation of carbon dioxide in the subdiaphragmatic area. Additional information regarding oviductal patency can be obtained by detecting gas beneath the diaphragm on X-ray examination or fluoroscopy.

B. *Hysterosalpingography*

The injection of radiopaque medium dissolved in an oily or aqueous vehicle, combined with a variety of radiological techniques ranging from fluoroscopy and multiple spot films to cinefluoroscopy, utilizing image-intensifying equipment (and thus reducing radiation hazards), constitutes an alternate and, at least in our opinion, preferable approach to the problem of testing for oviductal function. The advantages are obviously in the pictorial representation of the endometrial and salpingeal lumen, as well as in the additional information gained through radiographs following the distribution of dye in the peritubal space, which thus provides at least some evidence of the presence or absence of adhesions in this critical area. The excellent book by Rozin is exhaustive in describing normal and pathological findings (Rozin 1965).

The amount of gamma radiation delivered to the gonads varies from 200 to 500 milliroentgens, according to required fluoroscopy time, number of spot films, etc. (For reasons of comparison, 575 milliroentgens are delivered to the gonads during pelvimetry and 425 milliroentgens during intravenous pyelogram, using similar apparatus.) This obviously constitutes a disadvantage.

In competent hands, erroneous diagnosis is less likely to occur than with oviductal insufflation, but inflammatory reactions, allergic reactions, and embolization of dye all have been reported. No true figure of incidence seems available; however, it appears that these complications are sufficiently common to have occurred in the history of every major infertility clinic.

1. *Technique.* Pelvic examination should be performed prior to hysterosalpingography in order to assess the position of the uterus and to detect any pathology which would contraindicate the procedure at a given time. The contrast medium, previously warmed to decrease the incidence of uterotubal spasm, is placed into a sterile syringe attached to the uterine cannula, and flushed through the entire cannula system to displace the air. The cervix is exposed, and, after cleansing with an antiseptic solution, is grasped with a tenaculum, and the cannula tip inserted through the external cervical os. Countertraction is applied between the tenaculum and cannula to insure against leakage of the contrast medium from the

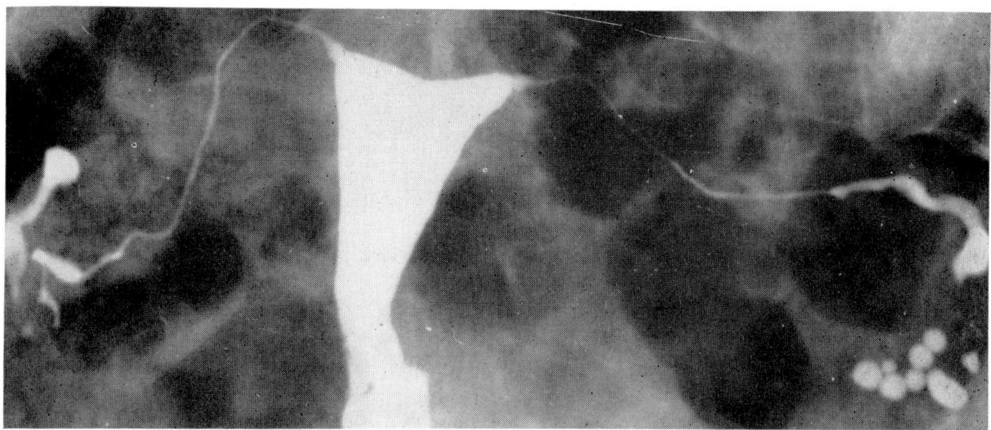

FIG. 1. Normal hysterosalpingogram: The uterine cavity is outlined as a regular, triangular structure with the Fallopian tubes visualized bilaterally as fine, radiopaque lines in the intramural-isthmic portions, gradually dilating as the ampullary portion is reached. Note the coalescence of the oily medium as it spills from the right oviduct into the peritoneal cavity.

external os. Finally, the vaginal speculum is removed, and the patient is allowed to resume a recumbent position.

If circumstances permit, fluoroscopy should be used during injection of the material to obtain optimal results. The contrast medium is injected slowly in increments of 1 to 2 cc. This will diminish the incidence of uterine cramps and oviductal spasm. Traction on the cervix with the tenaculum will vary the position of the uterus to obtain optimum visualization. Periodic spot films are taken for review and permanent record of the procedure. For an overall view of the uterus, oviducts, and peritubal area, oblique films of the pelvis are usually taken, especially in those situations where there may be some question as to the patency of the oviducts or to the presence of loculations of dye due to adhesions. A 24-hr follow-up film is obtained, which is important in detecting loculations or accumulations of dye which would be indicative of peritubal or perifimbrial adhesions.

Figures 1 through 9 show examples of a normal hysterosalpingogram and various physiologic and pathologic variations which may be encountered with this technique.

Summarizing this brief critique, a hysterosalpingogram, while it is not a test for oviductal function in the proper sense, undoubtedly constitutes a useful diagnostic procedure, the advantages of which far outweigh its potential hazards and complications. It must be done, however, by the specialist; and the technique and interpretation require skills which are all too often underestimated. Common mistakes, including omission of spot films during the filling phase, delayed films, improper timing, application of inappropriate pressure during dye injection, have led to errors in diagnosis which are then compounded by a sequence of further mistakes, an unnecessary laparotomy being only one example.

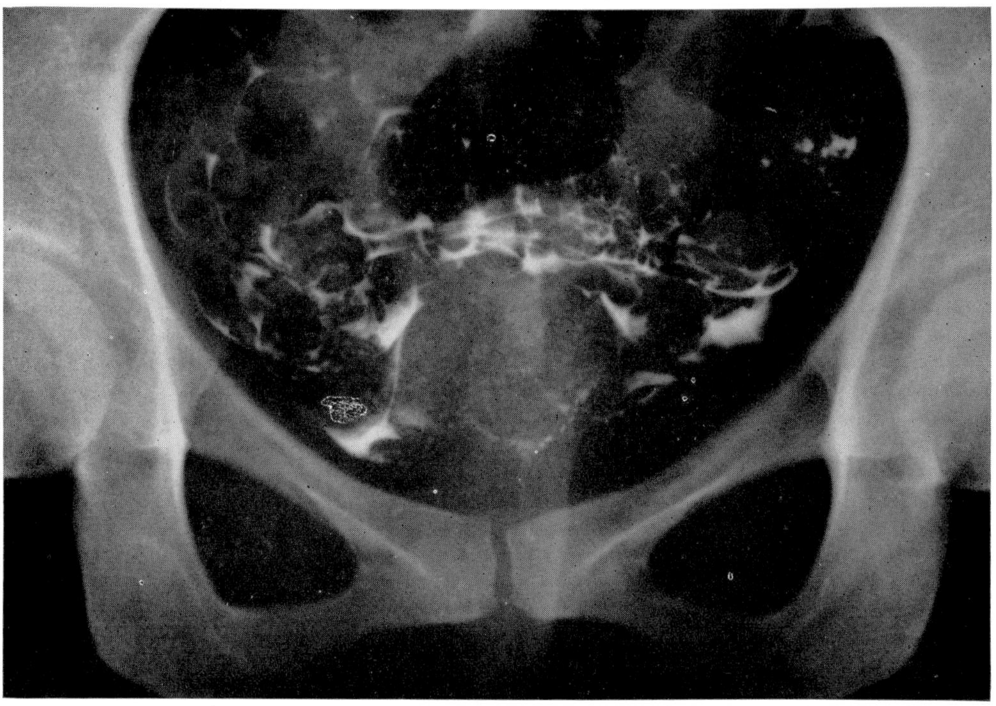

FIG. 2. Normal 24-hr follow-up film: The oily medium is disseminated widely throughout the pelvis and between loops of intestine in a rather delicate, flocculated pattern. Please contrast this radiograph to Figure 7, where heavy concentrations of injected material remain on both sides, indicating peritubal adhesions.

C. *Speck Test*

Brief mention should be made of the less well-known Speck test, which consists of the introduction of solution of phenolsulfonphthalein (PSP) into the uterine cavity, followed by catheterization of the bladder after approximately 30 min (Speck 1948). In the presence of patent oviducts, the PSP solution will reach the peritoneal cavity and, after being absorbed, will find its way via blood stream and kidneys to the bladder. In the presence of PSP in urine, alkalinization will yield a red or pink color. The principle is one of selective absorption through the peritoneal surface, no absorption taking place through either vaginal mucosa, endometrium, or endosalpinx.

As far as one is able to establish from the literature or through direct inquiry with any number of infertility clinics, the Speck test has never enjoyed great popularity. The reasons for this are not entirely clear. In any case, however, no information concerning peritubal

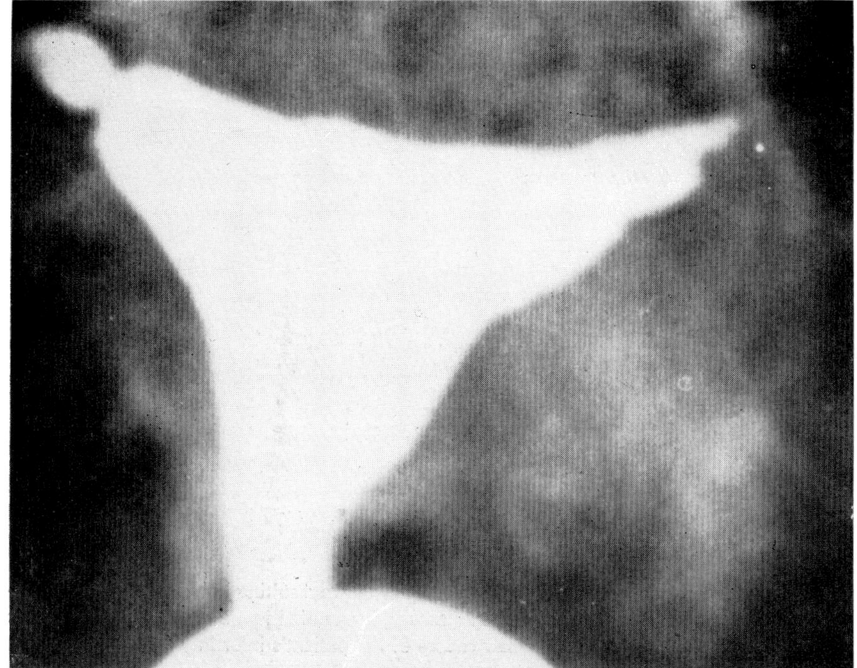

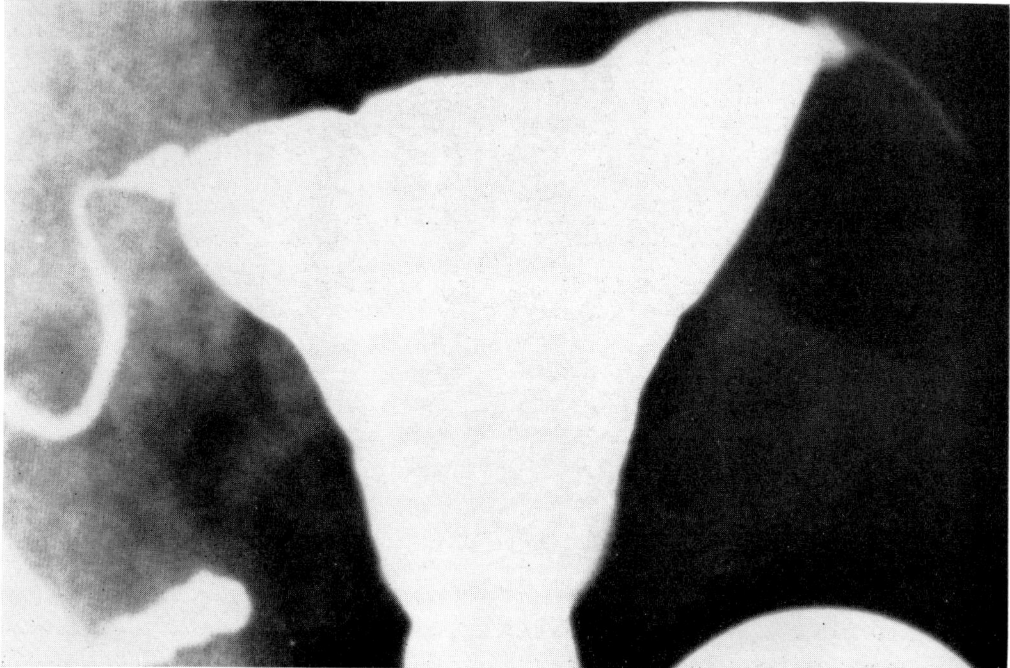

FIG. 3 (*top*). Cornual spasm: A diagnosis of unilateral or bilateral oviductal occlusion may be made erroneously on the basis of spasm in or near the uterotubal junction. Partial or complete "constriction rings" may be visualized on one or both sides, as shown above, inhibiting the passage of dye beyond this point. Subsequent examinations often show normal oviductal patency. See below.

FIG. 4 (*bottom*). Repeat hysterosalpingogram performed on the patient shown in Figure 3: During this examination, contrast medium was noted to pass freely beyond the uterotubal junctions bilaterally, although partial shadow-free areas still remain in these regions.

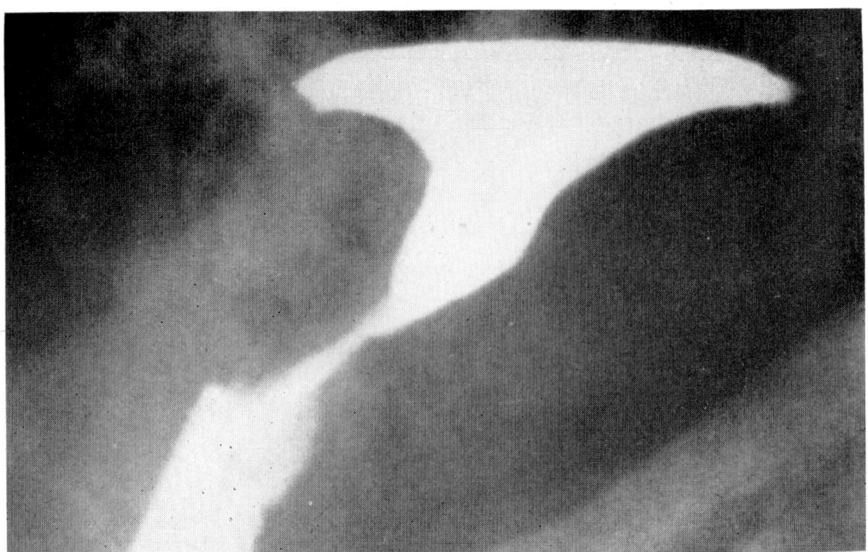

Fig. 5. Cornual spasm: Another example of complete cornual spasm resulting in the erroneous diagnosis of bilateral oviductal occlusion. During a subsequent laparoscopic examination, a normal passage of dye was visualized through both Fallopian tubes, and other causes of the patient's infertility were sought.

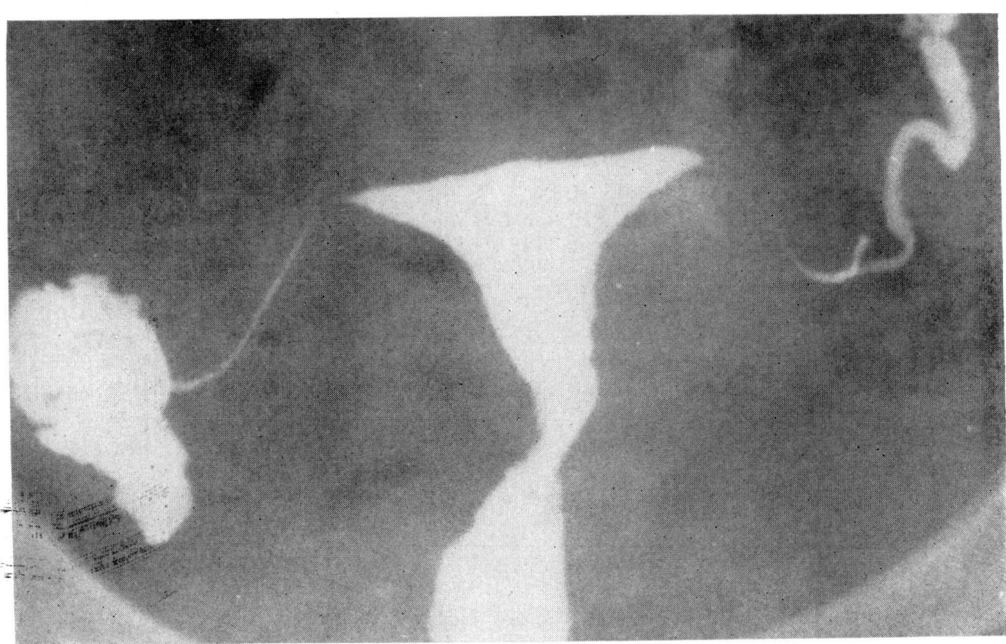

Fig. 6. Peritubal adhesions: A variety of pelvic and other intraperitoneal inflammatory processes may result in peritubal scarring and adhesion formation while preserving oviductal patency. As shown in this example, the radiopaque solution passes through the oviductal lumina, but pools in dense collections near the fimbriated ends. A 24-hr follow-up film is important in defining the extent of such pockets of adhesions causing the trapping of dye.

pathology can be expected. On the other hand, the small but definitely present hazards associated with this type of injection, as well as catheterization of the bladder, must be taken into account.

D. *Endoscopic Procedures*

Direct visualization of the uterus, oviducts, and ovaries, as well as their immediate surroundings, is easily achieved through an endoscope introduced either through the vaginal cul-de-sac (culdoscopy), or through the abdominal wall (laparoscopy) (Green 1962; Steptoe 1967). While these procedures can be utilized widely for a variety of gynecological conditions, they are particularly valuable in the evaluation of the infertile patient, as the Fallopian tubes,

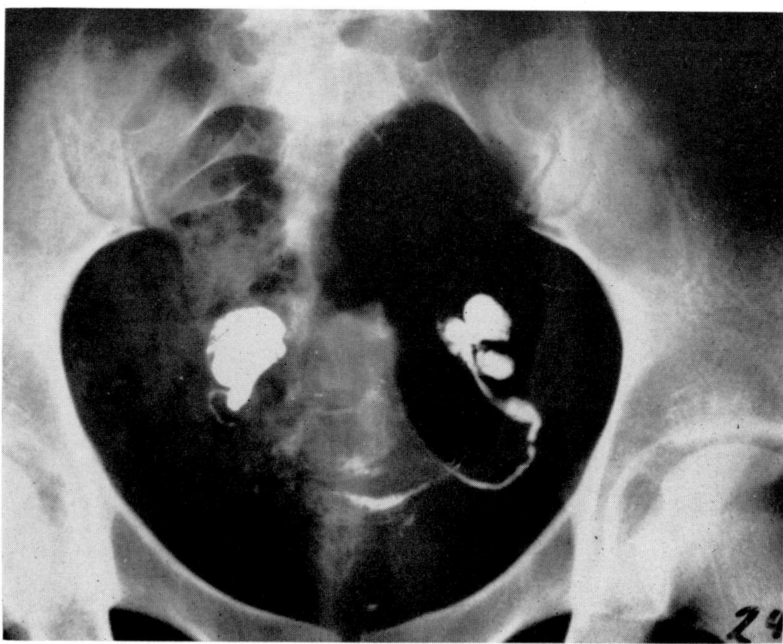

Fig. 7. Peritubal adhesions—24-hr follow-up film: Irregular collections of medium are readily visualized bilaterally in marked contrast to the regular disseminated pattern in the normal. In spite of oviductal patency, ovarian-tubal function may be impaired by scarring and adhesions, which may cause oviductal kinking or blockage of ovum transport into the oviductal ostia.

unless marked pathologic, reveal little information of their structural and functional relationships to other pelvic organs by more indirect methods. In this regard, endoscopic evaluations may uncover subtleties of oviductal, peritubal, or adnexal pathology not previously documented without performing more extensive surgery. Additional information concerning oviductal patency can also be obtained by injecting a dye through a cervical cannula and directly following its mode of passage through the oviduct and its subsequent appearance at the fimbriated end.

Much of the success of this procedure depends on the operative skill of the examiner, adequate anesthesia, and the types of endoscopic instruments available. Recent advances in instrumentation and the development of fiber-optic illumination systems (Balin, Wan, and Israel 1966) have made endoscopic examination, manipulation, and photography a practical adjunct to the gynecologist's armamentarium (Fig. 10). In our hands, laparoscopy rather than culdoscopy is the preferred approach. Ease of patient positioning (supine vs. knee-chest) and easier manipulation of pelvic organs assure a more adequate examination. Also, additional instruments may be inserted through the laparoscope shaft or at a distant site, giving extra mobility in performing the examination, biopsy, or other minor surgical procedures deemed necessary at the time. General anesthesia is recommended for the laparoscopic (abdominal) approach.

The patient is placed in the dorsal supine or a modified dorsolithotomy position, and a stab wound made in the infraumbilical skin fold to the depth of the fascial layer. A pneumoperitoneum needle is inserted through this small incision, and the peritoneal cavity distended with 2 to 4 liters of nitrous oxide or carbon dioxide. Steep Trendelenburg position assures clearance of the pelvis of all nonadhesed intestine. A large bore (11 mm) trocar and cannula

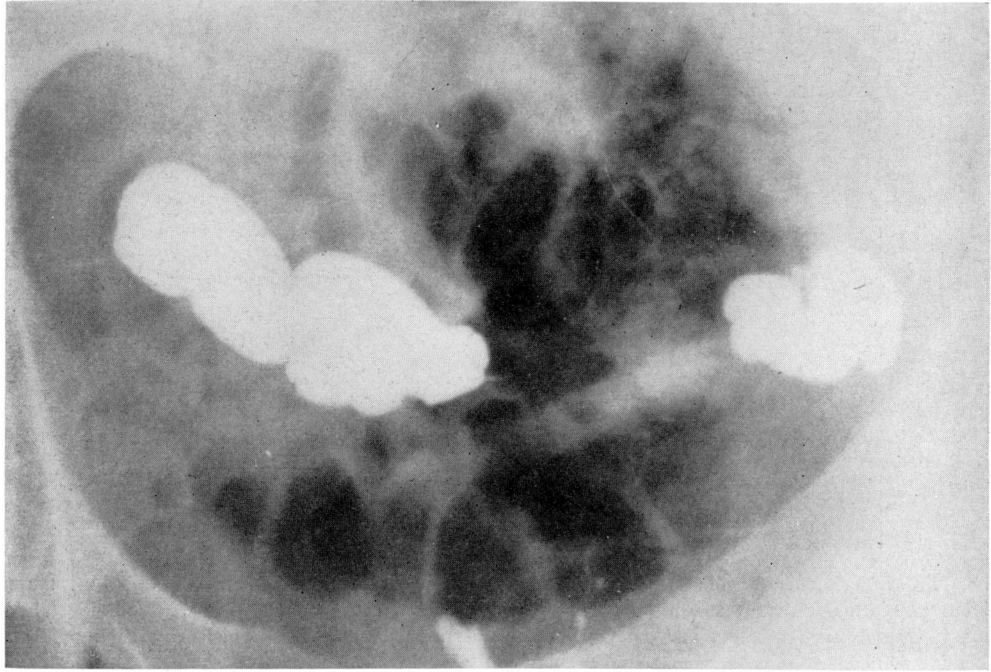

FIG. 8. Bilateral hydrosalpinx: Marked oviductal distortion, occlusion, and sac-like dilatation of the ampullary portion may be encountered following various inflammatory reactions affecting the salpinges. In this example, there are large bilateral sacculations, which fill with contrast medium as it is injected under slight pressure. Interestingly, in this case, only the dilated oviducts are visualized as all of the radiopaque solution passes immediately into these structures, preventing visualization of the uterine cavity.

FIG. 9 (*top*). Bilateral hydrosalpinx—24-hr follow-up film: Note the nearly identical pattern of dye collection as seen in Figure 8. Oviductal occlusion is evidenced by the fact that no injected medium has reached the peritoneal cavity on this delayed film. The contour and pattern of dye dispersion (or failure thereof) on the 24-hr radiogram is an important differential diagnostic aid in oviductal and peritubal pathology.

FIG. 10 (*bottom*). Fiber-optic illuminated endoscope, fiber-optic light cable, and light source used in laparoscopic examination: The separate light source provides intense, cool illumination, enabling excellent visualization and photography with less danger to the patient than older distal bulb instruments.

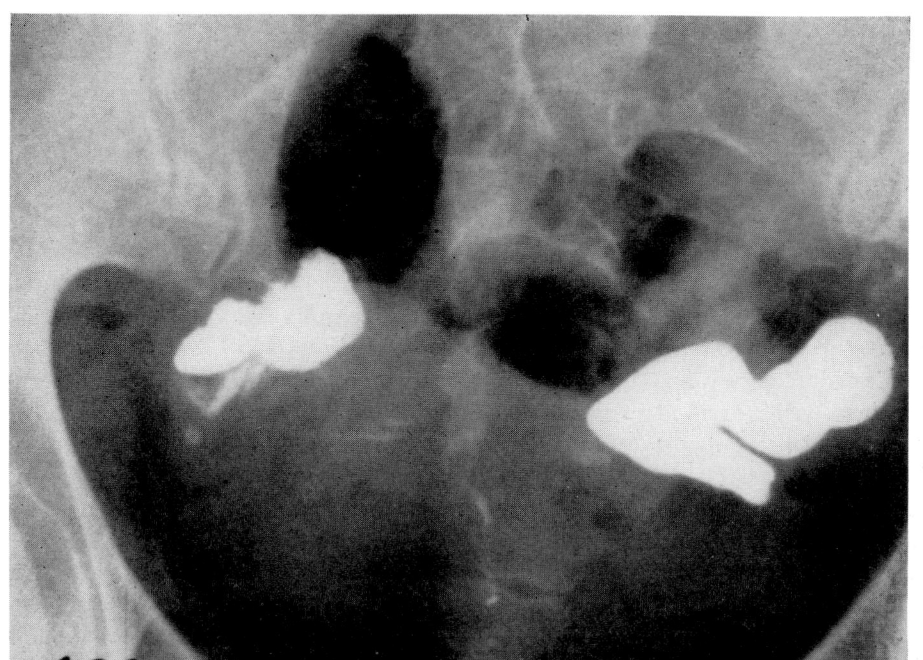

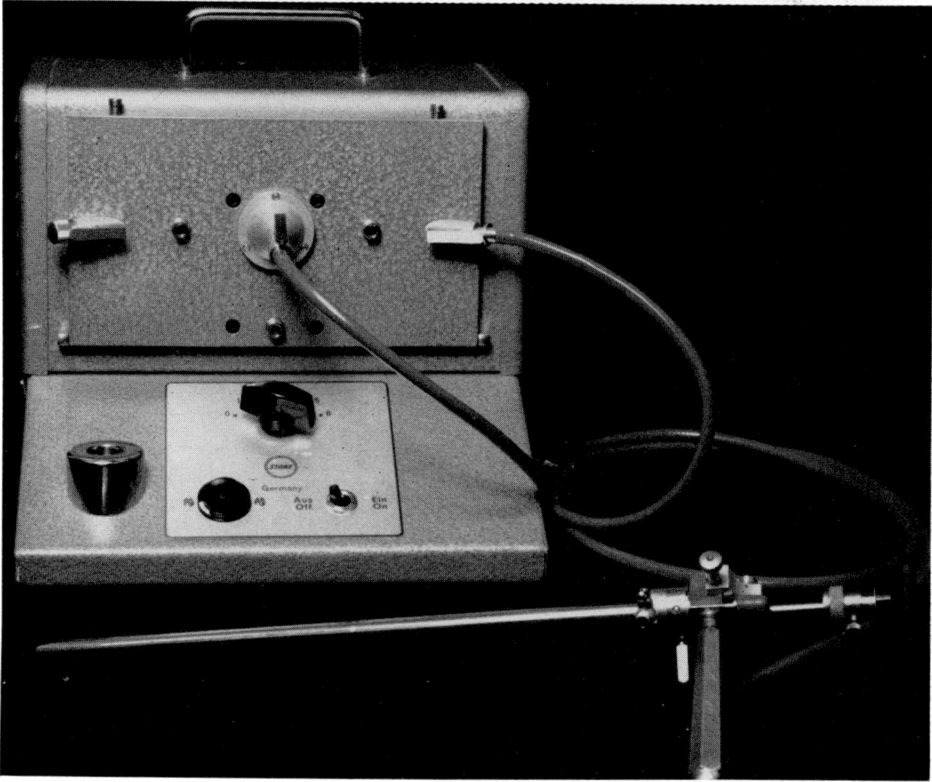
Germany
Aus
Off
Ein
On

are then placed through the stab wound and, after entry into the abdominal cavity, the trocar is removed and the laparoscope is inserted through the cannula shaft. Excellent opportunity to view the pelvic viscera is usually afforded by this technique. Access to the Fallopian tubes and adnexal structures is facilitated further with the aid of manipulating instruments, which can be inserted as described above, or by uterine manipulation with a previously placed cervical sound or cannula. The preoperative placement of a hysterogram cannula and cervical tenaculum affords an excellent "handle" for uterine manipulation, as well as permitting instillation of dye for evaluating oviductal patency. As the procedure is terminated, excess gas is expressed from the peritoneal cavity, and the skin incision closed with a single suture or skin clip.

While endoscopic evaluations undoubtedly yield the most complete information, these must be considered as surgical procedures which involve all the risks of anesthesia in addition to the inconvenience and cost of hospitalization (1 to 2 days). As an immediate solution to this, local anesthesia has been recommended, and the use of outpatient facilities instead of an operating room is certainly feasible, although probably less than optimal. The major argument against the use of local anesthesia is the fact that this can be applied only to the site of entrance of the scope, while the highly sensitive peritoneal surface remains unanesthetized. The administration of narcotics under such circumstances is of some help, but it seems fair to say that some discomfort to the patient is unavoidable, the direct implication being that the examination is usually abbreviated and the inspection less than optimal.

Bowel perforations have occurred with both procedures; most of them, however, have healed without major sequelae.

The one major shortcoming of intraperitoneal inspection is the fact that no details concerning the uterine or salpingeal lumen are revealed by this route.

In summarizing these concepts, it can be said that direct inspection probably offers the greatest amount of valuable information concerning oviductal integrity and patency, including information dealing with the anatomical possibilities of ovum transport. While evaluation of oviductal motility remains a matter of speculation (according to the presence or absence of adhesions, scarring, etc.), one should also consider that this approach holds some promise for future investigation, inasmuch as modern instrumentation allows the application of microelectrodes, aspiration of cysts, and other interventions which might yield useful information with a minimum of trauma.

IV. Conclusions

Clinical evaluation of oviductal function has as yet not reached beyond the state of morphological observation. Even at this level, it remains in many ways inadequate. The literature on the subject is voluminous, but repetitive, in its contents, with little attention given to the shortcomings of all procedures. According to our interpretation of available data, serious consideration should be given to the question of whether oviductal insufflation should not be abandoned altogether. Other procedures require a modicum of indication beyond the fact that the patient has been unable to conceive during the last 1 or 2 years.

References

1. Balin, H.; Wan, L. S.; and Israel, S. L. 1966. Recent advances in pelvic endoscopy. *Obstet. Gynec.* 27:30.

2. Buxton, C. L. 1957. Death from air embolus: Report of a case following tubal insufflation. *Amer. J. Obstet. Gynec.* 74:430.
3. Buxton, C. L., and Southam, A. L. 1958. *Human infertility*. New York: Paul B. Hoeber.
4. Estes, W. L., and Heitmeyer, P. L. 1934. Pregnancy following ovarian implantation. *Amer. J. Surg.* 24:563.
5. Garb, A. E. 1957. A review of tubal sterilization failures. *Obstet. Gynec. Survey* 12:291.
6. Ghosal, K. K. 1966. Ovarian function after Estes' operation. *J. Obstet. Gynec. India* 16:540.
7. Green, T. H., Jr. 1962. The role of culdoscopy in tubal disorders. *Clin. Obstet. Gynec.* 5:799.
8. Levinson, J. M. 1963. Pulmonary oil embolism following hysterosalpingography. *Fertil. Steril.* 14:21.
9. Preston, P. G. 1953. Transplantation of the ovary into the uterine cavity for the treatment of sterility in women. *J. Obstet. Gynaec. Brit. Emp.* 60:862.
10. Reiprich, W. 1933. Die operative Behandlung der Tubensterilität und experimentelle Studien über die Erfolgsaussichten der freien Eileitverpflanzung. *Z. Geburtsh. Gynaek.* 104:1.
11. Rozin, S. 1965. *Uterosalpingography in gynecology*. Springfield, Illinois: Charles C Thomas.
12. Speck, G. 1948. Phenolsulfonphthalein as a test for the determination of tubal patency. *Amer. J. Obstet. Gynec.* 55:1048.
13. Steptoe, P. C. 1967. *Laparoscopy in gynaecology*. Baltimore: Williams and Wilkins Co.
14. Sturma, J. 1959. Gravidität und Geburt nach der Implantation des Ovar in die Gebärmütter. *Zbl. Gynaek.* 81:1260.
15. Sweeney, W. J., III, and Gepfert, R. 1965. The Fallopian tube. *Clin. Obstet. Gynec.* 8:32.
16. Thomas, W. L. 1953. Prevenception insurance: Pan-hysterectomy versus tubectomy. *Southern Med. J.* 46:787.
17. Weir, W. C. 1962. Rubin's test and hysterosalpingography. *Clin. Obstet. Gynec.* 5:260.

Epilogue

L. Mastroianni, Jr.
Department of Obstetrics and Gynecology
School of Medicine
University of Pennsylvania, Philadelphia

In this volume the structure and function of a vital organ, the oviduct, have been probed in depth. That the oviduct is a worthy subject for such concentrated attention is evident when one considers the various biological processes which occur at the oviductal level and are clearly prerequisites to a normal pregnancy. These include successful transfer of the egg from the site of ovulation into the oviductal lumen, conditioning of spermatozoa precedent to fertilization, the fertilization process, cleavage through many cell divisions, and, finally, delivery of the developing zygote into the uterus after a discreetly timed interval. The function of the oviduct may be conveniently divided into two categories: oviductal transport of gametes, and provision of a suitable environment for the gametes and for early development of the fertilized egg. With reference to these activities, the oviduct functions periodically and only during a relatively brief period of time which, in most mammalian species, spans the first 3 postovulatory days. The structural characteristics of the oviduct which may relate to function have been emphasized. In addition, certain cyclic modifications in structure have received attention. The latter may or may not relate to function but, in some cases, hold promise of providing insight into some basic biological processes of broader interest.

The first paper of the symposium, by Price, Ortiz, and Zaaijer, provided a fitting introduction, since it reviewed the prenatal development of the oviduct. It included comments on the rather striking similarity in oviductal organogenesis among such species as the hamster, mouse, rat, guinea pig, and man, suggesting that, at least in the beginning, there are common denominators in structural development. Evidence was presented from early ontogeny in the female guinea pig that the oviduct is not simply an extension of the uterus, although admittedly both are derived from the Müllerian duct. The prospective oviductal region, with the development of coils of a smaller size, early becomes clearly distinguishable from that which will become the uterus. Later, discrete segments of oviduct are identifiable as infundibulum, ampulla, and isthmus. The timing of these events is variable among species. In the hamster, for example, the oviduct is not distinguishable from the uterus until after birth. The structural differences among species begin to become apparent. In *in vitro* studies, the fetal oviduct of the guinea pig fails to respond to testicular and adreno- and reno-genic hormones, and in this species early morphogenesis does not depend upon specific hormonal stimulation. The status of the oviducts of guinea pig and man at birth does, however, suggest a prenatal hormonal influence toward the end of gestation, with the appearance of secretions and cilia. This is of special interest in light of the work presented by Brenner, from which it is clear that in the adult subhuman primate ciliogenesis in the oviduct is estrogen-dependent. Rumery has provided additional information on organogenesis of the mouse oviduct. Cilia are also present in most species, and their activity in the fetal oviduct may be modified by exogenous hormones.

In the discussion which follows, an attempt will be made to relate structure to function with reference to (1) oviductal transport of gametes, (2) creation of the oviductal environment, and (3) relationship between the oviductal environment and the egg.

I. Oviductal Transport of Eggs

The mechanism by which eggs are transferred into the oviductal ostium varies considerably among different species. This subject has been explored in depth in only a few species, with the rabbit receiving rather concentrated attention. The rabbit oviduct is particularly suited to experimentation because of its size and, in certain aspects, because of the resemblance of its fimbriated extremity to that of man. The rabbit oviduct is endowed with luxuriant fimbriae covered with cilia. The relationship between the fimbriae and the ovary during ovulation was first observed by Axel Westman, a man whose name is as intimately connected with the Fallopian tube as that of Fallopius himself. He observed the action of the fimbriae in the rabbit through an abdominal window, noting that at ovulation the fimbriae came into contact with the ovary. They move back and forth over the ovulating follicles, which practically assures egg pickup. This mechanism has now been magnificently documented with cinematography by Blandau. Eggs, as they are released with their accompanying granulosa cells, remain adherent to the ruptured follicle unless they are swept free by the fimbriae. Once in contact with the delicate fimbrial cilia, they pass rapidly into the oviductal ostium. The cumulus oophorus plays an important role in this mechanism since the cilia apparently direct the cumulus mass toward the ostium. In contrast, fimbriae are apparently unable to direct spheres made of inert material of the same size as rabbit eggs. These show a rotary motion and hence remain essentially stationary. If the fimbriated end of the oviduct is amputated or is separated from the ovary, eggs remain attached to the follicle for several hours following ovulation. That a suction mechanism is not essential to egg pickup in the rabbit has been demonstrated by the ability of the oviducts, ligated just beyond the fimbriae, to pick up eggs. Clewe has likened the action of the rabbit fimbriae to a carpet sweeper, rather than to a vacuum cleaner which would suck eggs into the oviductal lumen or to a hand which would reach out and grasp eggs, inasmuch as the fimbriae move over the rupturing follicles, directing the eggs enclosed in cumulus toward the ostium.

In the rat and mouse the anatomical arrangement in the fimbrial-ovarian area is quite different, since the infundibulum projects into a periovarial sac. Eggs are not adherent to the follicles following ovulation but are released into the fluid contained in the sac. By virtue of fluid currents they are directed to the fimbriae where, once in contact, they pass into the ostium.

Blandau has emphasized the importance of the ligaments of the rabbit oviduct. The activity of the mesovarium and mesosalpinx is responsible for the movement of the fimbriae over the surface of the ovary. The muscular elements contained in these structures contract vigorously in a rhythmic fashion. The sequence of events which brings about this movement is unknown and is among the important unanswered questions to which attention should be directed in the future. Are these ligaments under hormonal control and, if so, can their action at the time of ovulation be reproduced by administration of exogenous hormones? Are they influenced by some nonhormonal product of the ovary, perhaps a substance contained in the follicular fluid itself? Can hormones be found to modify these mechanisms and prevent egg pickup?

The action of the fimbriae has not been observed satisfactorily in the human female. The anatomical arrangement, however, is such that a realignment of fimbriae over the ovulating follicle is certainly within the realm of possibility, and a few observations support this. The observations of Strange indicate that in man there are muscular elements which could play a role in the mechanism of egg pickup. There is a separate muscle bundle, the musculus tubae attrahens, extending from the fimbriae along the fimbriae ovariae to one pole of the ovary, as well as some muscular elements in the paraovarium. The latter might, on contraction, lift the ovary in the direction of the oviductal ostium, and the former would direct ovary and fimbriae toward one another. Preliminary observations on the monkey by Blandau suggest a mechanism somewhat resembling that of the rabbit.

In all species studied, the movement of eggs along the first few millimeters of the ampulla occurs as the result of ciliary activity. The cilia of the monkey fimbriae have been a subject of special attention. Interesting data presented by Brenner indicate that fimbrial cilia disappear in the postovulatory phase of the cycle and ciliogenesis renews them under the influence of estrogen in the following cycle. This interesting biological process has provided a model for the study of ciliogenesis. The rabbit oviduct, in contrast to that of the monkey, retains its cilia throughout the cycle.

The fimbriated extremity of mouse, rat, guinea pig, rabbit, sheep, cow, and man, has been studied with light and electron microscopy by Nilsson and Reinius. In these species cilia were found and, except in the sheep, the largest concentration of cilia was noted at the fimbriae.

Once the egg is safely within the oviductal ostium, in all species studied, it passes rapidly to a point well within the ampullary portion. Several methods have been used to time this event. The most dramatic of these involves the use of a vital dye which permits the progress of the eggs to be observed through the oviductal wall. This approach, first used by Harper, has been documented with cinematography by Blandau. The film shows that eggs reach the ampullary-isthmic junction in 4 to 12 min when observations are carried out *in vitro* under special experimental conditions designed to protect the oviduct from external influences. During the course of transport, muscular contractions are observed.

The influence of exogenous hormones on the oviductal transport of eggs was described by Boling. He presented evidence which would support the hypothesis that increased contractility of the rabbit oviduct is brought about by estrogen withdrawal rather than by estrogen, as previously supposed. Oviductal contractions can be augmented by administration of progesterone. In studies of the fetal mouse oviduct, Rumery has noted suppression of muscular activity in organ culture by estradiol. The role of the cilia in the transport of eggs along the oviduct is more difficult to evaluate, but the appearance of vigorous muscular contraction leaves little doubt that muscular activity is an important aspect of oviductal transport.

In most mammalian species fertilization occurs well within the ampullary portion of the oviduct, and the egg is retained within the oviduct for an interval of approximately 3 days. There are exceptions to this rule, as, for example, in marsupials and montremes, in which fertilization occurs at the ovarian level, and in some species, such as swine, in which eggs are transferred more rapidly from the oviduct. Interest in the mechanism for retention of eggs has focused attention on the isthmus and the uterotubal junction. Available information on this latter area in various species was reviewed by Hafez and Black. In several species there is a physiological closure in that area of the oviduct which has been termed the ampullary-isthmic junction. The rabbit provides a notable example of this type of closure, which is sufficiently strong to retain fluid under pressure. Such physiological closure is probably responsible for the retention of eggs within the oviduct during early development. The retention of eggs for 3 days does have importance in the reproductive economy of the rabbit, as was so clearly demonstrated by Chang. He has pointed out that when eggs of rabbit are transferred prematurely from the oviduct into the uterus they fail to implant.

Brundin evaluated the isthmus of the rabbit oviduct by use of kymographic techniques. The isthmus is endowed with a rich supply of noradrenergic nerve endings which decrease in number following section of the hypogastric nerve. Norepinephrine produces an increased muscular tone in the region. The approach used by Brundin has allowed evaluation of the effect of various pharmacological agents on the isthmus. The exact mechanism for the temporary closure of this area is a matter for further exploration. Although there is evidence that it comes under endocrine control, what is the relationship between neural influences and the hormonal status of the ovary? What is it which influences the accurately timed process by which eggs are finally transferred into the uterus? In considering the uterotubal junction, Hafez and Black indicated also the importance of evaluating the relationship between the myometrium and the intramural portion of the oviduct. Indeed the uterine musculature may very well influence the events at this level. The anatomical characteristics of the oviduct at the isthmus and the uterotubal junction vary among species and the mechanism for egg retention may very well differ among the species.

II. The Oviductal Environment

The oviduct has long been recognized as a potential secretory organ. It may exert a strong influence on sperm, eggs, and the developing embryo through the fluid which is present in its lumen. It is of some importance to evaluate the milieu provided by the oviduct from the hours preceding fertilization through the time when the fertilized egg is retained within the oviductal lumen. In mammalian species, the oviductal epithelium contains secretory cells whose characteristics change in the estrous or menstrual cycle. Nilsson and Reinius have studied the ultrastructure of the oviductal epithelium in several species, introducing the fixative by perfusion technique. They have documented the interesting morphologic events associated with secretory activity and have delineated the structural differences in various segments of the oviduct. Common to all species studied is evidence of intracellular secretory activity, but there is also a possibility that a transudate may contribute to the oviductal fluid. Changes in the oviductal epithelium were further evaluated by Fredricsson with histochemical techniques, which showed that the secretory granules and the intraluminal secretions were PAS-positive and diastase-resistant. This and various other staining characteristics suggest that the material contains mucoproteins or mucopolysaccharides. The oviductal egg of the rabbit is in fact gradually encased in a mucopolysaccharide coat, presumably the product of the oviductal epithelium. Greenwald reviewed his experiments in which estrogen administration resulted in synthesis and retention of mucin in the oviductal epithelium and progesterone caused its release. The importance, if any, of the mucin coat which surrounds the rabbit ovum during its residence in the oviduct remains to be explored.

In recent years, methods have been devised to collect oviductal fluid. Hamner has described his ingenious approach in which an intra-abdominal glass chamber is employed. Another method involves the use of an external collecting chamber, a recent modification of which allows collection of fluid under refrigeration. In the ovariectomized rabbit, fluid accumulation is increased following estrogen administration and diminished in the estrogen-primed castrate following progesterone treatment. A 50% reduction in daily fluid accumulation is observed during the first 3 days of pregnancy. Certain variations in electrolyte content of oviductal fluid have been noted. Perhaps of greater interest are the relatively high concentrations of lactate, pyruvate, and bicarbonate in oviductal fluid of the rabbit.

III. Influence of Oviductal Environment on the Egg

The possibility that substances may be transferred to the oviductal egg via the oviductal fluid was reviewed by Glass. Using fluorescent techniques, she has demonstrated that serum antigens, both native and foreign, may be taken up by the egg. This and other work raises the possibility of influencing the course of events in the development of the egg at the oviductal level.

The environmental conditions which favor normal development have been explored *in vitro*. Some of the outstanding contributions to this field have been made by McLaren and Brinster. The metabolic substrates which seem to be important in the development of the early mouse ovum, pyruvate and lactate, have in fact been found in oviductal fluid. One of the principal functions of the oviduct, then, is to provide these and other substances in a ratio which would most favor early developmental stages. McLaren pointed out that the oviduct does not continue to constitute the most suitable environment for eggs. At a given time in development transfer to the uterus is important if development is to continue. Such observations serve to highlight the importance of timing in the endogenous transfer of eggs. The egg transfer techniques which have been worked out in various species, as reviewed by Chang, and in domestic animals by Dziuk, provide useful tools for the continued investigation of the relationship between early development and the oviductal environment in various species.

The relationship between sperm and the oviductal environment has received but scant attention. Behrman presented data which suggest that immunologically induced infertility in a rabbit does not have its origin at the oviductal level. Oviductal influence on capacitation of sperm and the effect of oviductal fluid on fertilization are areas in which sufficient information is not available to permit an extensive review.

During the course of the symposium, the oviduct has been referred to repeatedly as a neglected organ, but many of the observations and interpretations presented are of relatively recent vintage. Techniques have been developed which allow more extensive exploration of reproductive processes at the oviductal level. Little imagination is required to appreciate the potential importance of such research in fertility and fertility control, and in our understanding of certain basic biological processes whose scope extends beyond reproduction *per se*.

Contributors

S. J. BEHRMAN
Department of Obstetrics and Gynecology
Medical Center, University of Michigan
Ann Arbor, Michigan 48104

K. BENIRSCHKE
Department of Pathology
Dartmouth Medical School
Hanover, New Hampshire 03755

D. W. BISHOP
Department of Physiology
Medical College of Ohio at Toledo
Toledo, Ohio 43614

D. L. BLACK
Department of Veterinary and Animal Science
University of Massachusetts
Amherst, Massachusetts 01002

R. J. BLANDAU
Department of Biological Structure
School of Medicine
University of Washington
Seattle, Washington 98105

C. W. BODEMER
Department of Biomedical History
School of Medicine
University of Washington
Seattle, Washington 98105

J. L. BOLING
Department of Biology
Linfield College
McMinnville, Oregon 97128

R. M. BRENNER
Department of Electron Microscopy
Oregon Regional Primate Research Center
Beaverton, Oregon 97006

R. L. BRINSTER
King Ranch Laboratory of Reproductive Physiology
Department of Animal Biology
School of Veterinary Medicine
University of Pennsylvania
Philadelphia, Pennsylvania 19104

J. BRUNDIN
Department of Physiology
Karolinska Institutet
Stockholm 60, Sweden

M. C. CHANG
Worcester Foundation for Experimental Biology
Shrewsbury, Massachusetts 01545

P. J. DZIUK
Department of Animal Science
University of Illinois
Urbana, Illinois 61801

S. B. FOX
Department of Surgery
School of Medicine
University of Virginia
Charlottesville, Virginia 22901

B. FREDRICSSON
Department of Obstetrics and Gynecology
Sjukhuset
Stockholm 60, Sweden

L. E. GLASS
Department of Anatomy
University of California Medical School
San Francisco, California 94122

G. S. GREENWALD
Departments of Gynecology and Obstetrics and Anatomy
University of Kansas Medical Center
Kansas City, Kansas 66103

E. S. E. HAFEZ
Reproduction Laboratory
Department of Animal Sciences
Washington State University
Pullman, Washington 99163

C. E. HAMNER
Division of Reproductive Biology
Department of Obstetrics and Gynecology
School of Medicine
University of Virginia
Charlottesville, Virginia 22901

W. L. HERRMANN
Department of Obstetrics and Gynecology
School of Medicine
University of Washington
Seattle, Washington 98105

A. McLAREN
Institute of Animal Genetics
University of Edinburgh
Edinburgh 9, Scotland

L. MASTROIANNI, JR.
Department of Obstetrics and Gynecology
School of Medicine
University of Pennsylvania
Philadelphia, Pennsylvania 19104

A. V. NALBANDOV
Department of Animal Science
University of Illinois
Urbana, Illinois 61801

O. NILSSON
Department of Human Anatomy
University of Uppsala
Uppsala, Sweden

E. ORTIZ
Biology Department
University of Puerto Rico
Rio Piedras, Puerto Rico 00931

S. PICKWORTH
Worcester Foundation for Experimental Biology
Shrewsbury, Massachusetts 01545

D. PRICE
Department of Zoölogy
University of Chicago
Chicago, Illinois 60637

S. REINIUS
Department of Human Anatomy
University of Uppsala
Uppsala, Sweden

R. E. RUMERY
Department of Biological Structure
School of Medicine
University of Washington
Seattle, Washington 98105

D. C. SMITH
Department of Obstetrics and Gynecology
School of Medicine
University of Washington
Seattle, Washington 98105

L. R. SPADONI
Department of Obstetrics and Gynecology
School of Medicine
University of Washington
Seattle, Washington 98105

S. SUGAWARA
Faculty of Agriculture
Tohoku University
Sendai, Japan

J. J. P. ZAAIJER
Laboratory for Cell Biology and Histology
University of Leiden
Leiden, The Netherlands

Author Index

Subject Index